RENEWABLE ENERGY SYSTEMS

Advanced Conversion Technologies and Applications

INDUSTRIAL ELECTRONICS SERIES

Series Editors:
Bogdan M. Wilamowski
J. David Irwin

PUBLISHED TITLES

Renewable Energy Systems: Advanced Conversion Technologies and Applications,
Fang Lin Luo and Hong Ye

Multiobjective Optimization Methodology: A Jumping Gene Approach,
K.S. Tang, T.M. Chan, R.J. Yin, and K.F. Man

The Industrial Information Technology Handbook, *Richard Zurawski,*

The Power Electronics Handbook, *Timothy L. Skvarenina*

Supervised and Unsupervised Pattern Recognition: Feature Extraction and
Computational Intelligence, *Evangelia Micheli-Tzanakou*

Switched Reluctance Motor Drives: Modeling, Simulation, Analysis, Design,
and Applications, *R. Krishnan*

FORTHCOMIING TITLES

Extreme Environment Electronics, *John D. Cressler and H. Alan Mantooth*

Power Electronics and Control Techniques for Maximum Energy Harvesting in
Photovoltaic Systems, *Giovanni Spagnuolo, Nicola Femia, Giovanni Petrone,
and Massimo Vitelli*

Industrial Wireless Sensor Networks: Applications, Protocols, Standards, and Products,
Vehbi Cagri Gungor and Gerhard P. Hancke

Multilevel Converters for Industrial Applications, *Sergio Alberto Gonzale,
Santiago Andres Verne, and Maria Ines Valla*

Smart Grid Technologies: Applications, Architectures, Protocols, and Standards,
*Vehbi Cagri Gungor, Carlo Cecati, Gerhard P. Hancke, Concettina Buccella,
and Pierluigi Siano*

Sensorless Control Systems for AC Machines: A Multiscalar Model-Based Approach,
Zbigniew Krzeminski

RENEWABLE ENERGY SYSTEMS

Advanced Conversion Technologies and Applications

FANG LIN LUO
HONG YE

CRC Press
Taylor & Francis Group
Boca Raton London New York

CRC Press is an imprint of the
Taylor & Francis Group, an **informa** business

CRC Press
Taylor & Francis Group
6000 Broken Sound Parkway NW, Suite 300
Boca Raton, FL 33487-2742

First issued in paperback 2017

© 2013 by Taylor & Francis Group, LLC
CRC Press is an imprint of Taylor & Francis Group, an Informa business

No claim to original U.S. Government works

Version Date: 20120615

ISBN 13: 978-1-138-07758-4 (pbk)
ISBN 13: 978-1-4398-9109-4 (hbk)

Library of Congress Cataloging-in-Publication Data

Luo, Fang Lin.
    Renewable energy systems : advanced conversion technologies and applications / Fang Lin Luo, Hong Ye.
        p. cm. -- (Industrial electronics)
    Includes bibliographical references and index.
    ISBN 978-1-4398-9109-4 (hardback)
    1. Electric current converters. 2. Renewable energy sources. 3. Electric power systems--Energy conservation. I. Ye, Hong. II. Title.

TK2796.L86 2012
621.31'3--dc23

2012021351

Visit the Taylor & Francis Web site at
http://www.taylorandfrancis.com

and the CRC Press Web site at
http://www.crcpress.com

Contents

Preface

The aim of this book is to describe advanced conversion technologies that are useful for engineering students and practicing professionals. It consists of approximately 800 pages, 550 diagrams, and 50 algorithms and introduces more than 200 topologies of advanced converters originally developed by us, including 150 updated circuits on modern conversion technologies. Some new topologies published recently have also been discussed. All prototypes are novel approaches and great contributions to modern power electronics. Conversion technologies have evolved rapidly in recent years. Many new converter circuits, which were not well analyzed earlier, have been covered in detail. The book also describes appropriate methods to determine accurate solutions and provides examples to design converters for industrial applications. Some historic problems have been resolved, for example, the accurate determination of the conduction angle of single-phase rectifiers and power factor correction.

Due to the shortage of energy worldwide, we have to search for new sources of energy. To study energy, we must explore the universe. In this book, we first review some of the basics of astronomy and earth physics, study certain pioneering methods, and finally return to our main topic—the development of renewable sources of energy, energy-saving methods, and improvement of power supply quality.

Power electronics is the technology that converts electrical energy from a source to users. It is of vital importance for all branches of industry and to every family. Energy conversion technique is the principal topic of power electronics. The corresponding equipment can be divided into four groups:

- AC/DC rectifiers
- DC/DC converters
- DC/AC inverters
- AC/AC transformers/converters

AC/DC rectifiers were the earliest developed converters. Most traditional circuits are discussed. Unfortunately, some circuits were not analyzed accurately. A typical example is the single-phase diode rectifiers with R-C load. The novel approach on the AC/DC rectifiers is the new topologies with power factor correction (PFC) and unity power factor (UPF). Several methods have been introduced in this book to implement the techniques of PFC and UPF.

DC/DC conversion technology has been rapidly developed in recent decades. According to latest figures, there exist more than 600 topologies of DC/DC converters. Many new topologies are created every year. The large number of DC/DC converters has primarily been divided into six generations since 2000. This systematic organization is very helpful to trace the evolution and development of DC/DC converters. The outstanding contributions are voltage-lift and super-lift techniques. Furthermore, we have created novel approaches—split-capacitor and split-inductor—to enhance super-lift techniques. This book introduces more than 100 topologies of voltage-lift and super-lift converters. DC/DC conversion technologies have been so successfully developed that their fundamental techniques have been used in other kinds of converters as well. Boost methods for AC/DC converters and DC-modulated AC/AC converters have been discussed. These technical methodologies can implement the UPF.

DC/AC inverters can be divided into two groups: pulse-width-modulation (PWM) inverters and multilevel inverters. People are familiar with PWM inverters such as the voltage source inverter (VSI) and the current source inverter (CSI). Impedance source inverter (ZSI) was introduced in 2003. It attracted the attention of most power electronics experts, leading to hundreds of papers devoted to its research and industrial applications in recent years. Multilevel inverters were invented in the early 1980s, and they have since developed rapidly. Many new topologies of multilevel inverters have been designed and applied in industrial applications, especially in renewable energy systems. The typical circuits are the diode-clamped inverters, capacitor-clamped inverters, and hybrid H-bridge multilevel inverters. We have also created a new series—laddered multilevel inverters—that uses fewer devices to produce more levels. Multilevel inverters overcome the drawbacks of the PWM inverter and provide great scope for industrial applications.

Traditional AC/AC converters can be divided into three groups: voltage-modulation AC/AC converters, cycloconverters, and matrix converters. All traditional AC/AC converters convert high voltage to lower voltage with adjustable voltage and frequency. Their drawbacks are the lower output voltage and poor total harmonic distortion (THD). Matrix converters were introduced in 1980. Unfortunately, the high THD restricts its applications. New types of AC/AC converters, such as sub-envelope modulation AC/AC converters and DC-modulated AC/AC converters, have thus been created. These techniques have successfully overcome the disadvantage of high THD. DC-modulated AC/AC converters have other advantages such as high output voltages and multiphase outputs.

Renewable energy source systems require a large number of converters. Many new types of converters have been created in recent decades, including Vienna rectifiers and z-source inverters. New AC/DC/AC converters are required in wind-turbine power systems. DC/AC/DC converters are also required in solar panel power systems. Therefore, these converters are investigated in the last two chapters of this book.

This book is organized in 18 chapters. The basic concepts of energy sources are introduced in Chapter 1. New energy sources are described in Chapter 2. 3G and renewable energy sources are covered in Chapter 3. Chapter 4 deals with power electronics. Traditional AC/DC diode rectifiers and controlled AC/DC rectifiers are discussed in Chapters 5 and 6, respectively. Chapter 7 describes power factor correction and unity power factor techniques, while Chapter 8 deals with DC/DC converters. Voltage lift and super-lift techniques are introduced in Chapters 9 and 10, respectively. The novel approaches of split-capacitor and split-inductor methods are introduced in Chapter 11. Pulse-width modulation DC/AC inverters and multilevel DC/AC inverters are covered in Chapters 12 and 13, respectively. Advanced multilevel DC/AC inverters are introduced in Chapter 14. Traditional AC/AC converters are discussed in Chapter 15, while Chapter 16 describes improved AC/AC converters. AC/DC/AC converters and DC/AC/DC converters used in renewable energy source systems are presented in Chapter 17. Finally, designs of solar-panel and wind-turbine energy systems are introduced in Chapter 18.

We are the pioneers in advanced conversion technology and have devoted many years of our life conducting research in this area. We have created a large number of converters; the series of DC/DC converters named Luo converters, which cover all six converter generations, are recognized worldwide. Super-lift converters are our best-known achievement in the 25 years that we have been conducting research.

Dr. Fang Lin Luo
Nanyang Technological University
Singapore, Singapore

Dr. Hong Ye
Nanyang Technological University
Singapore, Singapore

MATLAB® is a registered trademark of The MathWorks, Inc. For product information, please contact:

The MathWorks, Inc.
3 Apple Hill Drive
Natick, MA, 01760-2098 USA
Tel: 508-647-7000
Fax: 508-647-7001
E-mail: info@mathworks.com
Web: www.mathworks.com

Author

Dr. Fang Lin Luo is an associate professor at the School of Electrical and Electronic Engineering, Nanyang Technological University (NTU), Singapore. He received his BSc, first class with honors (magna cum laude), in radio-electronic physics from Sichuan University, Chengdu, China, and his PhD in electrical engineering and computer science from Cambridge University, England, United Kingdom, in 1986.

After graduating from Sichuan University, Dr. Luo joined the Chinese Automation Research Institute of Metallurgy (CARIM), Beijing, China, as a senior engineer. He then joined Entreprises Saunier Duval, Paris, France, as a project engineer in 1981–1982. He also worked with Hocking NDT Ltd., Allen-Bradley IAP Ltd., and Simplatroll Ltd. in England as a senior engineer after receiving his PhD from Cambridge University. Dr. Luo is a fellow of Cambridge Philosophical Society and a senior member of IEEE. He has published 12 textbooks and 308 technical papers in IEE/IET Proceedings and IEEE Transactions as well as in various international conferences. His research interests include power electronics and DC and AC motor drives with computerized artificial intelligent (AIC) control and digital signal processing (DSP) as well as AC/DC, DC/DC, and AC/AC converters and DC/AC inverters, renewable energy systems, and electrical vehicles.

Dr. Luo is currently the associate editor of *IEEE Transactions on Power Electronics* and *IEEE Transactions on Industrial Electronics*. He is also the editor of the international journal *Advanced Technology of Electrical Engineering and Energy*. Dr. Luo was the chief editor of international journal *Power Supply Technologies and Applications* in 1998–2003. He is the general chairman of the *First IEEE Conference on Industrial Electronics and Applications (ICIEA 2006)* and the *Third IEEE Conference on Industrial Electronics and Applications (ICIEA 2008)*.

Dr. Hong Ye (S'00-M'03) received her BSc, first class, from Xi'an University of Technology, China, in 1995, MEng from Xi'an Jiaotong University, China, in 1999, and PhD from Nanyang Technological University (NTU), Singapore, in 2005.

She was with the R&D Institute, XIYI Company, Ltd., China, as a research engineer from 1995 to 1997. She then moved on to NTU as a research associate in 2003–2004 and has been a research fellow since 2005.

Dr. Ye is a member of the IEEE and has coauthored 12 books. She has published more than 80 technical papers in IEEE Transactions, IEE Proceedings, and other international journals, as well as in various international conferences. Her research interests include power electronics and conversion technologies, signal processing, operations research, and structural biology.

1 Introduction

"Renewable energy sources" is a topic of much discussion in the present day. We cannot help but ask the following questions:

- What is the source of renewable energy?
- Where does the new energy come from?
- What are energy sources?

Such questions do not have straightforward answers. By investigating the energy sources in use, we can summarize as follows:

1. Fossil energy sources, such as coal, oil, and gas
2. Bioenergy sources, including corn oil and sugar alcohol
3. Renewable energy sources, for example, solar panel, wind turbine, and hydraulic power
4. Others such as tidal-wave energy, geothermic heat, and so on
5. Nuclear energy sources

We review these energy sources and find that fossil energy is the energy that is stored in the Sun for thousands of years; biologic and renewable energies are forms of energy transferred from the Sun; nuclear energy comes from nuclear power processes; and other types of energies are derived from gravitation. The application of energy sources goes through the cycle from "energy source" to "energy applications," via "energy storage and transmission," and to "remaining treatment" (ash or pollution treatment), as shown in Figure 1.1. Renewable energies usually have no remaining material and therefore the last step does not apply.

We can now answer the first question at the beginning of this chapter. Renewable energy comes from the Sun. The energy of the Sun is nuclear energy generated by the fusion process. The Sun is one of the billions of stars in the universe. Thus, there are enormous sources of energy in outer space. The rest of the questions are answered in the following chapters.

All energies in the universe come from gravitation and nuclear energy (by fusion and fission processes). Therefore, we need to review some concepts in astronomy and geography. We introduce some fundamental concepts before that:

- *Astronomical unit (AU)*: the average distance between the Sun and the Earth, which is a unit of length equal to about 150,000,000 km (i.e., 1.5×10^8 km)
- *Light year (LY)*: the distance that light travels in a year, around 9.4608×10^{12} km (or 63,072 AU $\approx 6.3 \times 10^4$ AU)

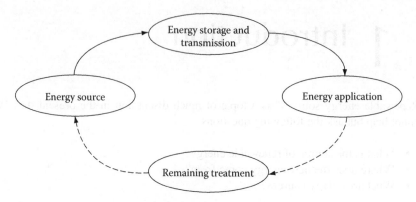

FIGURE 1.1 Energy sources' application train.

1.1   STARS IN THE UNIVERSE

We look at the sky at the night and see stars, as shown in Figure 1.2. All stars are huge balls of glowing gas with diameters ranging from hundreds of thousands (10^5) to millions (10^9) of kilometers. They are powerful and radiate large amounts of energies into space.

The Sun is a medium-sized star with a diameter of about 1,400,000 km. Some stars are smaller than the Sun, whereas some stars are much larger.

The Alexandrian scientist Ptolemy or Claudius Ptolemaeus (c. AD 90–168) observed and named more than 6000 stars. He sorted them into 48 constellations for convenience (at present there are 88 constellations). The Sun is more than one million times larger than the Earth, which is a sphere with a diameter of about 12,732 km, as shown in the Figure 1.3a [1]. If we show the Earth and the Sun together (Figure 1.3), we can see that the Sun is much larger than the Earth.

However the size of the Sun is insignificant when compared with other larger stars, for example, Eta Carinae with a diameter of about 239,400,000 km (its radius is 0.8 AU), which is the largest star in our Mercury galaxy [1] and is five million times larger than the Sun (Figure 1.4). If it were located in the solar system, it would occupy the size of Venus's orbit.

FIGURE 1.2 Stars in the universe. (Courtesy of European Space Agency, Hubble, Munchen, Germany.)

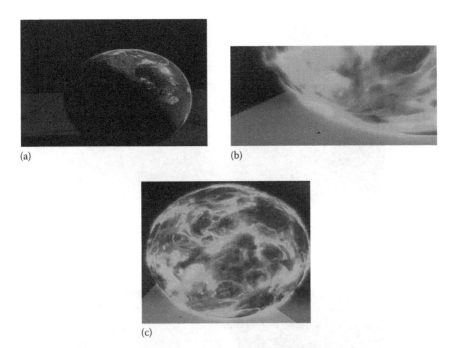

FIGURE 1.3 Size comparison of the Earth and the Sun. (a) The Earth. (b) Comparing the size of the Earth with that of the Sun. (c) Comparing the size of the Earth with that of the Sun: zoom out.

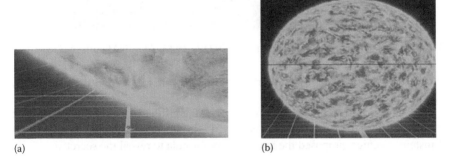

FIGURE 1.4 Size comparison of the Sun and Eta Carinae. (a) The Sun and Eta Carinae. (b) Comparing the size of the Sun to that of Eta Carinae: zoom out.

α Orionis with a diameter of about 1,602,600,000 km (its radius is 5.4 AU), which is the largest star in Constellation Orion [2]. It is 300 times larger than Eta Carinae (Figure 1.5). If it were located in the solar system, it would occupy the size of Jupiter's orbit.

Moreover, VY Canis Majoris in Constellation Canis Majoris (VY star) is the largest star we have ever known. Its diameter is 3,000,000,000 km (its radius is 10 AU). It is 6.6 times larger than α Orionis, and it is nearly 10^{10} times larger than the Sun.

FIGURE 1.5   Comparison of size between Eta Carinae and α Orionis.

FIGURE 1.6   Comparison of Eta Carinae, α Orionis, and VY Canis Majoris in size.

Their size comparison is shown in Figure 1.6. If it were located in the solar system, it would occupy the size of Saturn's orbit.

All stars contain a large amount of energy. What is the source of this energy that can last billions of years? This had confused astronomers for centuries until Albert Einstein, a genius, proposed the mass-energy formula to reveal the secret:

$$E = mc^2 \qquad (1.1)$$

where
   E is the energy
   m is the mass
   c is the light speed equal to 300,000 km/s

The energy of a star is mostly generated by the fusion process and is clean energy. In fusion reactions, two light atomic nuclei fuse together to form a heavier nucleus.

In the process, a little mass is lost and is transformed into a large energy (we discuss this process in Chapter 2). According to Einstein's theory, the nuclear energy can also be generated by fission power. Marie Skłodowska Curie discovered that a heavy nucleus can be split into two (or more) lighter atomic nuclei and releases a large amount of energy. Using the fission process, Julius Robert Oppenheimer made the first atomic bomb in 1945.

1.2   OUR MERCURY GALAXY, NEBULAE, AND BLACK HOLE

Stars were gestated in nebulae. Nebulae are the cradle of stars. There are billions of nebulae in the universe. Stars have gathered together to form galaxies. Each galaxy contains more than millions of stars. There is a black hole in the centre of each nebula and galaxy. A black hole is formed by the supernova collapse via strong gravitation. It absorbs stars in the galaxy. All stars surround the black hole, which governs the life of the nebula and galaxy. We can see a few nebulae in Figure 1.7.

The Mercury galaxy is one of the millions of galaxies. It contains billions of stars in the size of 100,000 LY. It looks like an iron cake as shown in Figure 1.8.

We have observed that the black hole locates at the center of our galaxy [1] (Figure 1.9).

The Earth is safe because it is 25,000 LY away from the black hole (Figure 1.10) and will not be absorbed in billions of years.

1.3   REDSHIFT AND BIG BANG

After centuries of debate between geocentrism and heliocentrism, heliocentrism has been finally recognized, because it is comparably correct. Johannes Kepler discovered three laws of planetary motion called Kepler's laws. People think that the universe works calmly in order and all stars run in their stable orbits. Until early twentieth century, in 1929, Edwin Hubble observed the phenomena of Redshift [3] and discovered that all stars run away from each other at a high speed of about 30 km/s. The redshift is shown in Figure 1.11a. Since then people know that the universe changes continuously. Edwin Hubble established the theory of redshift and proposed the Hubble's law:

$$v = H_0 D \tag{1.2}$$

where
   v is the recessional velocity of the galaxy or other distant object
   D is the comoving distance to the object
   H_0 is Hubble's constant, measured to be $70.4^{+1.3}_{-1.4}$ km/s/pc
   1 pc $= 3.26$ light years $= 206,265$ AU $= 30.857 \times 1012$ km
   The redshift formula derives the location of stars in a reverse fashion.

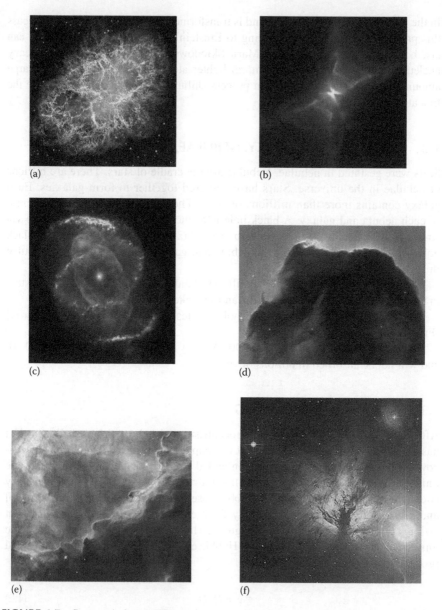

FIGURE 1.7 Some nebulae. (a) The crab nebula, (b) the red rectangle nebula, (c) the cat's eye nebula, (d) the horsehead nebula, (e) the omega nebula, (f) NGC 2024, the flame nebula,

Our universe was born 13.7 billion years ago as a result of a Big Bang [4]. The Big Bang is the origin of the time and the space of the universe. It is shown in Figure 1.11b.

Although the Big Bang theory is widely accepted, other contrary ideas exist. They state that the Big Bang is an absurdity, because nobody can answer a very simple question: what happened before the Big Bang?

FIGURE 1.7 (continued) (g) the eagle nebula, (h) the "Pillars of Creation" from the eagle nebula, (i) the triangulum emission garren nebula, (j) the strawhat nebula, and (k) the Eddie nebula. (Courtesy of NASA, Washington, DC.)

1.4 SOLAR SYSTEM

The Sun, planets, and other objects together are called the solar system. The Sun and the whole solar system were formed about nine billion years after the Big Bang. It is supposed that the Sun's age is 4.67 billion years. If we observe our solar system from 20 LY away, we can only see a small vague point (brightness of a third magnitude star) because the Sun is surrounded by the Oort cloud.

Further, the solar system consists of an average-size star, the Sun; the inner planets (Rocky Planets or terrestrial planets) Mercury, Venus, Earth, and Mars; the

FIGURE 1.8   Our Mercury galaxy. (Courtesy of NASA, Washington, DC.)

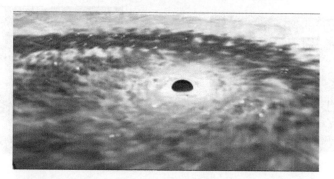

FIGURE 1.9   A large black hole in Mercury galaxy.

FIGURE 1.10   The location of the Earth in Mercury galaxy. (Courtesy of NASA, Washington, DC.)

outer planets (the gas giant) Jupiter, Saturn, Uranus, and Neptune; and dwarf planets including Ceres, Pluto, Haumea, Makemake, and Eris in the Kuiper belt. The Oort cloud is extended to 1 LY away from the Sun. The solar system also includes the satellites of the planets, numerous comets, asteroids, and meteoroids, and the interplanetary medium. The Sun is the richest source of electromagnetic energy

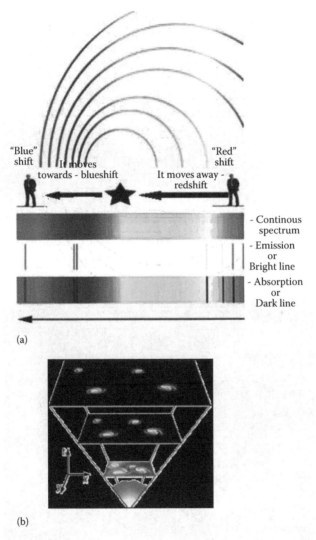

- Continous spectrum
- Emission or Bright line
- Absorption or Dark line

(a)

(b)

FIGURE 1.11 (a) Redshift and (b) Big Bang.

(mostly in the form of heat and light) in the solar system. The Sun's nearest known stellar neighbor is a red dwarf star called Proxima Centauri, at a distance of 4.3 LYs. The whole solar system, together with the local stars visible in a clear night, orbits the center of our home galaxy, as a spiral disk of 200 billion stars, which we call the Milky Way. The Milky Way has two small galaxies orbiting nearby, which are visible from the southern hemisphere. They are called the Large Magellanic Cloud and the Small Magellanic Cloud. The nearest large galaxy is the Andromeda galaxy. It is a spiral galaxy like the Milky Way but is four times more massive and is two million LYs away. Our galaxy, one of billions of galaxies known, is traveling

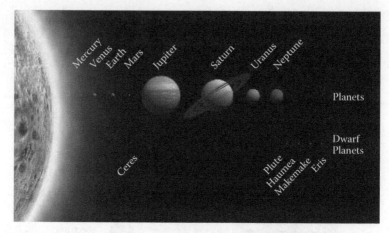

FIGURE 1.12 Planets and dwarf planets of the solar system (sizes are to scale but the relative distances from the Sun). (Courtesy of NASA, Washington, DC.)

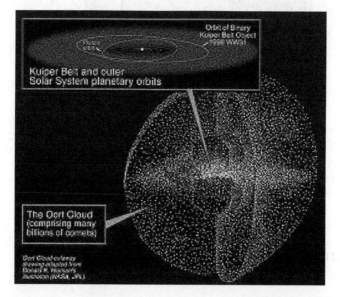

FIGURE 1.13 The Oort cloud. (Courtesy of NASA, Washington, DC.)

through intergalactic space. The solar system and the Oort cloud are shown in Figures 1.12 and 1.13, respectively.

Table 1.1 lists statistical information for the Sun and planets.

The Sun contains 99.85% of all the matter in the solar system. The planets contain only 0.135% of the solar system mass. Jupiter contains more than twice the matter of all the other planets combined. Satellites of the planets, comets, asteroids,

TABLE 1.1

Statistical Information for the Sun and Planets

	Distance (AU)	Radius (Earth's)	Mass (Earth's)	Rotation (Earth's)	#Moons	Orbital Inclination	Orbital Eccentricity	Obliquity	Density (g/cm³)
Sun	0	109	332,800	25–36[a]	9	—	—	—	1.410
Mercury	0.39	0.38	0.05	58.8	0	7	0.2056	0.1°	5.43
Venus	0.72	0.95	0.89	244	0	3.394	0.0068	177.4°	5.25
Earth	1.0	1.00	1.00	1.00	1	0.000	0.0167	23.45°	5.52
Mars	1.5	0.53	0.11	1.029	2	1.850	0.0934	25.19°	3.95
Minor planets	1.6–5.0	—	0.06665	—	—	—	—	—	2.64
Jupiter	5.2	11	318	0.411	16	1.308	0.0483	3.12°	1.33
Saturn	9.5	9	95	0.428	18	2.488	0.0560	26.73°	0.69
Uranus	19.2	4	17	0.748	15	0.774	0.0461	97.86°	1.29
Neptune	30.1	4	17	0.802	8	1.774	0.0097	29.56°	1.64
Pluto	39.5	0.18	0.002	0.267	1	17.15	0.2482	119.6°	2.03
Kuiper belts	30–50×60	—	—	—	—	—	—	—	—
Oort cloud	300–60.000	—	—	—	—	—	—	—	—

[a] The Sun's period of rotation at the surface varies from approximately 25 days at the equator to 36 days at the poles. Deep down, below the convective zone, everything appears to rotate with a period of 27 days.

TABLE 1.2
Mass Distribution within Our
Solar System

Sun	99.85%
Planets	0.135%
Comets	0.01%
Satellites	0.00005%
Minor planets	0.0000002%
Meteoroids	0.0000001%
Interplanetary medium	0.0000001%

meteoroids, and the interplanetary medium constitute the remaining 0.015%. Table 1.2 lists the mass distribution within our solar system.

There is no scale model or picture of the solar system that is correct in both planetary diameter and distance. Let us image the solar system with the scale of 1: 1,000,000,000. If we put a medicine ball with a diameter of 1.4 m at a corner of a football field to represent the Sun, our Earth should be a bean with a diameter 1.28 cm almost at the other end of the diagonal in the football field (about 150 m). Mercury should be a small pea with a diameter of 0.486 cm at a distance of 58.5 m, likely the length of the diagonal of a basketball field, away from the medicine ball (The Sun). Venus should be a red bean with a diameter of 1.2 cm at a distance of 108 m away from the medicine ball. Mars should be a yellow bean with a diameter of 0.68 cm at a position 225 m away from the medicine ball (likely the length of one and a half the diagonal of a football field). Jupiter should be a Pomelo with a diameter of 14.08 cm and 780 m away from the Sun (likely six times the length of the diagonal of a football field). Saturn should be a big orange with a diameter of 11.52 cm at a position 1425 m away from the Sun (likely 11 times the length of the diagonal of a football field). Uranus and Neptune should be duck eggs with a diameter of 5.12 cm at the positions of 2880 and 4515 m away from the Sun, respectively. Pluto should be a small sand in the Kuiper belt with a diameter of 0.18 cm and 5925 m away from the Sun. The Kuiper belts (including Dwarf planets Pluto, Haumea, Makemake, Eris, and so on) should be at a distance between 5,250 and 45,000 m away from the Sun. The Oort cloud should at a distance between 45,000 and 9,000,000 m away from the Sun. The model is shown in Table 1.3.

The Sun is still in its young age. It will spend a total of approximately 10 billion years as a main sequence star. The Sun does not have enough mass to explode as a supernova. Instead, in about 5.5 billion years, it will become a red giant. The Sun's life cycle is shown in Figure 1.14.

1.5 THE EARTH

Our Earth revolves around the Sun at an average speed of about 30 km/s, which is derived from the following calculation:

TABLE 1.3
Model of the Solar System

Celestial Body	Sun	Mercury	Venus	Earth	Mars	Jupiter	Saturn	Uranus	Neptune	Pluto	Kuiper Belts	Oort Cloud
Diameter	1.4 m	0.486 cm	1.2 cm	1.28 cm	0.68 cm	14.08 cm	11.52 cm	5.12 cm	5.12 cm	0.18 cm	—	—
Likeness	Medicine ball	Small pea	Red bean	Bean	Yellow bean	Pomelo	Big orange	Duck egg	Duck egg	Small sand	—	—
Distance	0	58.5 m	108 m	150 m	225 m	780 m	1425 m	2880 m	4515 m	5925 m	>5250 m	>45 km

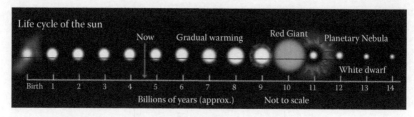

FIGURE 1.14 Life cycle of the Sun. (From Wikimedia Commons.)

$$\text{Speed} = \frac{2\pi R}{t} = \frac{2\pi \times 150,000,000\,\text{km}}{365 \times 24 \times 3600\,\text{s}} = \frac{94.24778 \times 10^7\,\text{km}}{3.1536 \times 10^7\,\text{s}} \approx 30\,\text{km/s} \quad (1.3)$$

where

R is the distance between the Earth and the Sun, which is 150,000,000 km, or 1 AU

t is the time of 1 year

1.5.1 THE EARTH IS ROUND

Our Earth was formed about 4.54 billion years ago. For a long time, people believed in Ptolemy's theory: the Earth is a plane and the center of the universe (geocentric theory). Nicolaus Copernicus established the heliocentric theory in 1543. Christopher Columbus discovered Newfoundland in 1495 and Fernando de Magallanes circumnavigated the Earth in 1520. Now we know the Earth is really round and also a planet in the solar system.

1.5.2 REVOLUTION AND ROTATION

The Earth revolves around the Sun in an elliptical orbit at a moving speed of about 30 km/s, making one revolution every 365.25 days. The eccentricity of the ellipse is small and the orbit is, in fact, quite nearly circular. Perihelion, the point at which the Earth is nearest to the sun, occurs on January 4 on average. At perihelion, the Earth is a little over 147 million kilometers away from the Sun. At the other extreme, the aphelion, which occurs on July 3, the Earth is about 152 million kilometers away from the sun. This variation in distance has a certain relationship.

Each day, the Earth rotates on its own axis with the maximum line velocity of about 0.463 km/s (on the equator), which just like what Mr. Mao said "Motionless, by earth I travel eighty thousand miles a day." If the Earth spins only 360° a day, after 6 months our clocks would be off by 12 h. To keep synchronized, the Earth needs to rotate one extra turn each year. It means that in a 24 h day the Earth actually rotates 360.9863° (i.e., 360° + 360°/365).

As shown in Figure 1.15, the plane swept out by the Earth in its orbit is called the ecliptic plane [5]. The Earth's spin axis is currently tilted 23.45° with respect to the ecliptic plane, and that tilt is what causes our seasons. On March 21 and

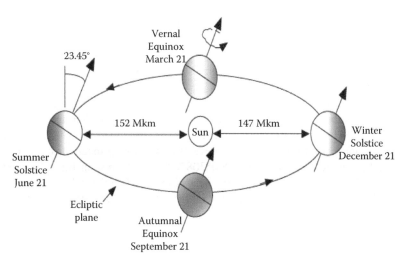

FIGURE 1.15 The Sun and the Earth. (From Masters, G.M., *Renewable and Efficient Electric Power Systems*, John Wiley & Sons, Inc., Hoboken, NJ, 2004. With permission.)

September 21, a line from the center of the sun to the center of the Earth passes through the equator and everywhere on Earth we have 12 h of daytime and 12 h of night, hence the term *equinox* (equal day and night). On December 21, the winter *solstice* in the Northern Hemisphere, the inclination of the North Pole reaches its highest angle away from the sun (23.45°), while on June 21 the opposite occurs. For convenience, we use the 21st day of the month for the solstices and equinoxes even though the actual days vary slightly from year to year. For solar energy applications, the characteristics of the Earth's orbit are considered to be unchanging, but over longer periods of time, measured in thousands of years, orbital variations are extremely important as they significantly affect climate. The shape of the orbit oscillates from elliptical to more nearly circular with a period of 100,000 years (*eccentricity*). The Earth's tilt angle with respect to the ecliptic plane fluctuates from 21.5° to 24.5° with a period of 41,000 years (*obliquity*). Finally, there is a 23,000 year period associated with the *precession* of the Earth's spin axis. This precession determines, for example, where in the Earth's orbit a given hemisphere's summer occurs. Changes in the orbit affect the amount of sunlight striking the Earth as well as the distribution of sunlight both geographically and seasonally. Those variations are thought to be influential in the timing of the coming and going of ice ages and interglacial periods. In fact, careful analysis of the historical record of global temperatures does show a primary cycle between glacial episodes of about 100,000 years, mixed with secondary oscillations with periods of 23,000 years and 41,000 years that match these orbital changes. This connection between orbital variations and climate was first proposed in the 1930s by an astronomer named Milutin Milankovitch, and the orbital cycles are now referred to as *Milankovitch oscillations*. Sorting out the impact of human activities on climate from those caused by natural variations such as the Milankovitch oscillations is a critical part of the current climate change discussion.

The Earth's rotation period relative to the Sun—its mean solar day—is 86,400 s of mean solar time (86,400.0025 s). As the Earth's solar day is now slightly longer than it was during the nineteenth century because of tidal acceleration, each day varies between 0 and 2 ms longer. The Earth rotates 361° a day (considering the revolution surrounding the Sun).

The Earth's rotation period relative to the fixed stars, called its stellar day by the International Earth Rotation and Reference Systems Service (IERS), is 86164.098903691 s of mean solar time (UT1) or 23 h 56 m 4.098903691 s. The Earth's rotation period relative to the precessing or moving mean vernal equinox, misnamed its sidereal day, is 86,164.09053083288 s of mean solar time (UT1) (23 h 56 m 4.09053083288 s). Thus, the sidereal day is shorter than the stellar day by about 8.4 ms. The length of the mean solar day in seconds is available from the IERS for the periods 1623–2005 and 1962–2005.

Apart from meteors within the atmosphere and low-orbiting satellites, the main apparent motion of celestial bodies in the Earth's sky is to the west at a rate of $15°/h = 15'/min$. For bodies near the celestial equator, this is equivalent to an apparent diameter of the Sun or the Moon every 2 min; from the planet's surface, the apparent sizes of the Sun and the Moon are approximately the same.

The Earth revolves (orbits) the Sun at an average distance of about 150 million kilometers every 365.2564 mean solar days, or one sidereal year. From the Earth, this gives an apparent movement of the Sun eastward with respect to the stars at a rate of about 1°/day (360°/365 days), or a Sun or Moon diameter, every 12 h. Because of this motion, on average it takes 24 h—a solar day—for the Earth to complete a full rotation about its axis so that the Sun returns to the meridian. The orbital speed of the Earth averages about 29.8 km/s (107,000 km/h), which is fast enough to cover the planet's diameter (about 12,600 km) in 7 min, and the distance to the Moon (384,000 km) in 4 h.

1.5.3   THE EARTH IS A PLANET IN THE SOLAR SYSTEM

The Earth is the third planet from the Sun with the mass M_e (M_e—Earth mass is the unit of mass equal to that of the Earth: 5.9722×10^{24} kg), and it is the densest and fifth largest of the eight planets in the solar system. It is also the largest of the solar system's four terrestrial planets. It is sometimes referred to as the World, the Blue Planet, or by its Latin name, Terra.

Home to millions of species, including humans, the Earth is the only place in the universe where biosphere (the area life) exists. The planet formed 4.54 billion years ago, and life appeared on its surface in less than one billion years. The Earth's biosphere has significantly altered the atmosphere and other abiotic conditions on the planet, enabling the proliferation of aerobic organisms as well as the formation of the ozone layer which, together with the Earth's magnetic field, blocks harmful solar radiation, permitting life on land. The physical properties of the Earth, as well as its geological history and orbit, have allowed life to persist during this period. The planet is expected to continue supporting life for at least another 500 million years.

1.5.4 Layers of the Earth

The Earth's outer surface is divided into several rigid segments, or tectonic plates, that migrate across the surface over periods of many millions of years. About 71% of the surface is covered by salt water oceans, with the remainder consisting of continents and islands, which together have many lakes and other sources of water that contribute to the hydrosphere. Liquid water, necessary for all known life, is not known to exist in equilibrium on any other planet's surface. The Earth's poles are mostly covered with solid ice (Antarctic ice sheet) or sea ice (Arctic ice cap). The planet's interior remains active, with a thick layer of relatively solid mantle, a liquid outer core that generates a magnetic field, and a solid iron inner core.

The interior of the Earth, like that of the other terrestrial planets, is divided into layers by their chemical or physical (rheological) properties, but unlike the other terrestrial planets, it has a distinct outer and inner core. The outer layer of the Earth is a chemically distinct silicate solid crust, which is underlain by a highly viscous solid mantle. The crust is separated from the mantle by the Mohorovičić discontinuity, and the thickness of the crust varies: averaging 6 km under the oceans and 30–50 km on the continents. The crust and the cold, rigid, top of the upper mantle are collectively known as the lithosphere, and it is of the lithosphere that the tectonic plates are comprised. Beneath the lithosphere is the asthenosphere, a relatively low-viscosity layer on which the lithosphere rides. Important changes in crystal structure within the mantle occur at 410 and 660 km below the surface, spanning a transition zone that separates the upper and lower mantle. Beneath the mantle, an extremely low-viscosity liquid outer core lies above a solid inner core. The inner core may rotate at a slightly higher angular velocity than the remainder of the planet, advancing by 0.1°–0.5° per year.

The Earth's layers are shown in Figure 1.16 and Table 1.4.

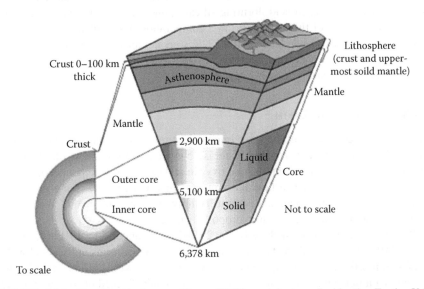

FIGURE 1.16 The Earth's layers. (From USGS contributors, Inside the Earth, *U.S. Geological Survey*, http://pubs.usgs.gov/gip/dynamic/inside.html, accessed May 7, 2012.)

TABLE 1.4

Geologic Layers of the Earth

Depth (km)	Component Layer	Density (g/cm³)
0–60	Lithosphere	—
0–35	Crust	2.2–2.9
35–60	Upper mantle	3.4–4.4
35–2890	Mantle	3.4–5.6
100–700	Asthenosphere	—
2890–5100	Outer core	9.9–12.2
5100–6378	Inner core	12.8–13.1

1.5.5 CHEMICAL COMPOSITION OF THE EARTH'S CRUST

The mass of the Earth is approximately 5.98×1024 kg. It is composed mostly of iron (32.1%), oxygen (30.1%), silicon (15.1%), magnesium (13.9%), sulfur (2.9%), nickel (1.8%), calcium (1.5%), and aluminum (1.4%), with the remaining 1.2% consisting of trace amounts of other elements. Due to mass segregation, the core region is believed to be primarily composed of iron (88.8%), with smaller amounts of nickel (5.8%), sulfur (4.5%), and less than 1% trace elements.

The geochemist F. W. Clarke calculated that a little more than 47% of the Earth's crust consists of oxygen. The more common rock constituents of the Earth's crust are nearly all oxides; chlorine, sulfur, and fluorine are the only important exceptions to this, and their total amount in any rock is usually much less than 1%. The principal oxides are silica, alumina, iron oxides, lime, magnesia, potash, and soda. The silica functions principally as an acid, forming silicates, and all the commonest minerals of igneous rocks are of this nature. From a computation based on 1672 analyses of all kinds of rocks, Clarke deduced that 99.22% were composed of 11 oxides (see the table at right). All the other constituents occur only in small quantities. The chemical composition of the Earth's crust is listed in Table 1.5.

1.5.6 WATER ON THE EARTH

Water is a source of life. It is a very impotent compound for the biosphere in the Earth. All kinds of life rely on water. Water is a chemical substance with the chemical formula H_2O. Its molecule contains one oxygen and two hydrogen atoms connected by covalent bonds. Water is a liquid at ambient conditions, but it often coexists on Earth with its solid state, ice, and gaseous state (water vapor or steam). Water also exists in a liquid crystal state near hydrophilic surfaces.

The area of our Earth's surface is 509,295,820 km² (about 5.1×10^8 km²). Water covers 70.9% of the Earth's surface and is vital for all known forms of life. On Earth, it is found mostly in oceans and other large water bodies, with 1.6% of water below ground in aquifers and 0.001% in the air as vapor, clouds (formed of solid and liquid water particles suspended in air), and precipitation. Oceans hold 97% of surface water; glaciers and polar ice caps 2.4%; and other land surface water such as rivers,

TABLE 1.5

Chemical Composition of the Crust

Compound	Formula	Composition (%)	
		Continental	Oceanic
Silica	SiO_2	60.2	48.6
Alumina	Al_2O_3	15.2	16.5
Lime	CaO	5.5	12.3
Magnesia	MgO	3.1	6.8
Iron(II) oxide	FeO	3.8	6.2
Sodium oxide	Na_2O	3.0	2.6
Potassium oxide	K_2O	2.8	0.4
Iron(III) oxide	Fe_2O_3	2.5	2.3
Water	H_2O	1.4	1.1
Carbon dioxide	CO_2	1.2	1.4
Titanium dioxide	TiO_2	0.7	1.4
Phosphorus pentoxide	P_2O_5	0.2	0.3
Total		99.6	99.9

lakes, and ponds 0.6%. A very small amount of the Earth's water is contained within biological bodies and manufactured products.

Water on Earth moves continually through a cycle of evaporation or transpiration (evapotranspiration) into the sky above the sea and oceans → via wind fly to continents and falls down as rain → precipitation and runoff in the rivers → usually reaching the sea. Over land, evaporation and transpiration contribute to the precipitation over land.

Clean drinking water is essential to humans and other life forms. Access to safe drinking water has improved steadily and substantially over the last decades in almost every part of the world. There is a clear correlation between access to safe water and GDP per capita. However, some observers have estimated that by 2025 more than half of the world population will be facing water-based vulnerability. A recent report (November 2009) suggests that by 2030, in some developing regions of the world, water demand will exceed supply by 50%. Water plays an important role in the world economy, as it functions as a solvent for a wide variety of chemical substances and facilitates industrial cooling and transportation. Approximately 70% of the fresh water that is actively handled by humans is consumed by agriculture.

Water is the carrier of hydraulic energy as well. All hydraulic power stations use fresh water potential from dams to the exhauster level at a height of about 100 m, but the fresh water drop from sky to the ground (the dam) wastes the potential energy at a height of about 10,000 m. It means that 99% of the water potential energy is wasted.

1.5.7 PLATES

The mechanically rigid outer layer of the Earth, the lithosphere, is broken into pieces called tectonic plates. There are seven primary plates; these seven plates comprise the bulk of the seven continents and the Pacific Ocean:

- African plate
- Antarctic plate
- Eurasian plate
- Indo–Australian plate
- North American plate
- Pacific plate
- South American plate

The Earth is the only known planet with life in the universe until now. Its location is perfectly right, not very far from and not very close to the Sun. The temperature and climate are suitable for life: not very cold and not very hot. The water on the Earth is just right, not too much and not too less. The Earth's magnetic field protects the atmosphere and water and avoids the Sun wind blowing them away.

1.5.8   THE EARTH IS VERY FRAGILE

In reverse thinking, our Earth is like an egg that is very fragile. The crust is only 35 km, that is, less than 0.275% of the Earth's diameter (12,732 km = 40,000 km/π). We normally think of eggshells as being very fragile, which is about 2% of the egg's diameter (the thickness of the eggshell is about 1 mm and the egg diameter is about 50 mm). The eggshell is seven times thicker than the crust of our Earth. Furthermore, egg has a close-grained shell; the Earth's crust consists of plates with gaps. For example, there was a small asteroid with a size of only 7 miles that ran into Earth about 6.5 million years ago. It ran with a speed of 72,000 km/h (20 km/s), fell in the sea near Mexico, and made the Mexico bay. About 70% of the life on the Earth was deracinated, for example, all dinosaurs were deracinated and the Jurassic period ended. Fortunately, we have Jupiter located outside the Earth. Its enormous gravitation largely protected the Earth from the impacts by small asteroids, for example, in 1994 a comet ran into Jupiter.

1.5.9   THE EARTH'S GEOLOGICAL AGE

There are two contradictory accounts of the age of Earth. From uranology, the Earth is about 4.5 billion years old; but by geology, we can get evidences only from the Cambrian period, about 500 million years ago. The Earth's geological age in the recent 500,000,000 years is listed in Table 1.6.

We do not know what happened before the Cambrian period. We call it the ice age. From 4.5 billion years ago till 500 million years ago, there is a big gap of about 4 billion years. It is possible that there were a few cycles in the Earth's past life.

1.5.10   PROTECTION OF THE EARTH

There is a popular slogan of "Protection of the Earth." Actually, the Earth has its own life cycles and would not need any protection from humans. A correct parlance should be "Protection of human beings' environment." As we know that our Earth

TABLE 1.6
The Earth's Age

Time (Million Years)	Epoch
500–450	Cambrian
450–400	Ordovician
400–370	Silurian period
370–320	Devonian
320–250	Carboniferous
250–210	Permian
210–150	Trias
150–65	Jurassic
65–35	Cretaceous
35–now	Anno mundi

is a unique planet with biosphere in the universe, it is necessary to develop renewable energy sources to protect human beings' environment and solve the problems of energy crisis, pollution, and greenhouse effect.

REFERENCES

1. Tudou contributors. How the universe works. *Tudou.* http://www.tudou.com/playlist/p/112632908.html (accessed on May 7, 2012).
2. Dibon-Smith, R. α Orionis. *The Constellations.* http://www.dibonsmith.com/ori_a.htm (accessed on May 7, 2012).
3. Cyberphysics contributors. Doppler effect and red shift. *My Cyber Notebook for Physics.* http://www.cyberphysics.co.uk/topics/space/redshift.htm
4. Universe Today contributors. Big Bang. *Universe Today.* http://www.universetoday.com/50782/big-bang/ (accessed on May 7, 2012)
5. Masters, G. M. 2004. *Renewable and Efficient Electric Power Systems.* USA: John Wiley & Sons, Inc.
6. USGS contributors. Inside the Earth. *U.S. Geological Survey.* http://pubs.usgs.gov/gip/dynamic/inside.html (accessed on May 7, 2012).

2 New Energy Sources

There is enormous energy in the universe. The energy is caused by the thermonuclear reaction and gravitation. Most young stars contain large amounts of energy that is yielded by the fusion process. In fusion process reaction, two light atomic nuclei fuse together to form a heavier nucleus (in contrast with fission process). In the process, a little mass is lost and a large amount of energy is generated. The dying stars also contain a large amount of energy that is yielded by the fission process. In fission process reaction, a heavier nucleus splits into two lighter atomic nuclei. Similarly, a little mass is lost and transformed into a large amount of energy.

At present, thermonuclear power occupies about 25% of the energy market of the whole world. Nearly all nuclear power generations take the nuclear fission processes. People live under the shadow of nuclear radiation. It is necessary to use new energy sources.

2.1 NUCLEAR FISSION

Many scientists devoted themselves to the fission process. **Julius Robert Oppenheimer** and **Marie Skłodowska Curie** are outstanding contributors in this area. They successfully implemented the fission process to obtain enormous energy. Their research opened a way to use atomic energy peacefully. After the World War II, hundreds of nuclear power generation stations have been built to supply about 25% of the world's energy. All nuclear power generation stations implement the fission process. Since nuclear fission process produces materials with strong radiation, it is possible to pollute people's living environment. Therefore, the fission power is sometimes called unclean energy or dangerous energy.

2.1.1 Fission Process

Nuclear fission is a nuclear reaction in which the nucleus of an atom splits into smaller parts (lighter nuclei). It often produces free neutrons and photons (in the form of gamma rays) and releases a tremendous amount of energy. The two nuclei produced are most often of comparable size, typically with a mass ratio around 3:2 for common fissile isotopes [1]. Most fissions are binary fissions, but occasionally (2–4 times per 1000 events) there can be three positively charged fragments produced in a ternary fission.

Fission is usually an energetic nuclear reaction induced by a neutron, although it is occasionally seen as a form of spontaneous radioactive decay, especially in very high-mass-number isotopes. The unpredictable composition of the products (which vary in a broad probabilistic and somewhat chaotic manner) distinguishes fission from purely quantum-tunneling processes such as proton emission, alpha decay, and cluster decay, which give the same products every time.

Fission of heavy nuclei is an exothermic reaction, which can release large amounts of energy both as electromagnetic radiation and as kinetic energy of the fragments (heating the bulk material where fission takes place). In order for fission to produce energy, the total binding energy of the resulting elements must be less than that of the starting element. Fission is a form of nuclear transmutation because the resulting fragments are not the same element as the original atom.

Nuclear fission produces energy for nuclear power and to drive the explosion of nuclear weapons [1]. Both uses are possible because certain substances called nuclear fuels undergo fission when struck by fission neutrons and in turn emit neutrons when they break apart. This makes possible a self-sustaining chain reaction that releases energy at a controlled rate in a nuclear reactor or at a very rapid uncontrolled rate in a nuclear weapon.

The amount of free energy contained in nuclear fuel is millions of times the amount of free energy contained in a similar mass of chemical fuel such as gasoline, making nuclear fission a very tempting source of energy. The products of nuclear fission, however, are on average far more radioactive than the heavy elements that are normally fissioned as fuel and remain so for a significant period of time, giving rise to a nuclear waste problem. Concerns over nuclear waste accumulation and over the destructive potential of nuclear weapons may counterbalance the desirable qualities of fission as an energy source and give rise to ongoing political debate over nuclear power.

The fission process is an induced fission reaction. A slow moving neutron is absorbed by a Uranium-235 nucleus turning it briefly into a Uranium-236 nucleus; this in turn splits into fast-moving lighter elements (fission products) and releases three free neutrons. The stages of binary fission are shown in a liquid drop model. Energy input deforms the nucleus into a fat "cigar" shape and then a "peanut" shape, followed by binary fission as the two lobes exceed the short-range strong force attraction distance, and then is pushed apart and away by their electrical charge. Note that in this model, the two fission fragments are the same size (Figure 2.1).

2.1.2 CHAIN REACTIONS

Several heavy elements, such as uranium, thorium, and plutonium, undergo both spontaneous fission, a form of radioactive decay, and induced fission, a form of nuclear reaction [1]. Elemental isotopes that undergo induced fission when struck by a free neutron are called fissionable; isotopes that undergo fission when struck by a thermal, slow moving neutron are also called fissile. A few particularly fissile and readily obtainable isotopes (notably ^{235}U and ^{239}Pu) are called nuclear fuels because they can sustain a chain reaction and can be obtained in large enough quantities to be useful.

All fissionable and fissile isotopes undergo a small amount of spontaneous fission, which releases a few free neutrons into any sample of nuclear fuel. Such neutrons would escape rapidly from the fuel and become a free neutron, with a mean lifetime of about 15 min before decaying to protons and beta particles. However, neutrons

FIGURE 2.1 An induced fission reaction. (From Wikimedia Commons.)

almost invariably impact and are absorbed by other nuclei in the vicinity long before this happens (newly created fission neutrons move at about 7% of the speed of light, and even moderated neutrons move at about eight times the speed of sound). Some neutrons will impact fuel nuclei and induce further fissions, releasing yet more neutrons. If enough nuclear fuel is assembled in one place, or if the escaping neutrons are sufficiently contained, then these freshly emitted neutrons outnumber the neutrons that escape from the assembly, and a sustained nuclear chain reaction will take place.

An assembly that supports a sustained nuclear chain reaction is called a critical assembly or, if the assembly is almost entirely made of a nuclear fuel, a critical mass. The word "critical" refers to a cusp in the behavior of the differential equation that governs the number of free neutrons present in the fuel: if less than a critical mass is present, then the amount of neutrons is determined by radioactive decay, but if a critical mass or more is present, then the amount of neutrons is controlled instead by the physics of the chain reaction. The actual mass of a critical mass of nuclear fuel depends strongly on the geometry and surrounding materials.

Not all fissionable isotopes can sustain a chain reaction. For example, ^{238}U, the most abundant form of uranium, is fissionable but not fissile: it undergoes induced fission when impacted by an energetic neutron with over 1 MeV of kinetic energy. However, A very few neutrons produced by ^{238}U fission are energetic enough to induce further fissions in ^{238}U, so no chain reaction is possible with this isotope. Instead, bombarding ^{238}U with slow neutrons causes it to absorb them (becoming ^{239}U) and decay by beta emission to ^{239}Np, which then decays again by the same process to ^{239}Pu; this process is used to manufacture ^{239}Pu in breeder reactors. In situ plutonium production also contributes to the neutron chain reaction in other types of reactors after sufficient plutonium-239 has been produced, since plutonium-239 is also a fissile element that serves as fuel. It is estimated that up to half of the power produced by a standard "nonbreeder" reactor is produced by the fission of plutonium-239 produced in place, over the total life cycle of a fuel load.

Fissionable, nonfissile isotopes can be used as fission energy source even without a chain reaction. Bombarding ^{238}U with fast neutrons induces fissions, releasing energy as long as the external neutron source is present. This is an important effect in all reactors where fast neutrons from the fissile isotope can cause the fission of nearby ^{238}U nuclei, which means that some small part of the ^{238}U is "burned up" in all nuclear fuels, especially in fast breeder reactors that operate with higher-energy neutrons. That same fast-fission effect is used to augment the energy released by modern thermonuclear weapons, by jacketing the weapon with ^{238}U to react with neutrons released by nuclear fusion at the center of the device.

2.2 NUCLEAR FUSION

Nuclear fusion process produces clean energy. *Nuclear fusion* is the process by which two or more atomic nuclei join together, or "fuse," to form a single heavier nucleus. This is usually accompanied by the release or absorption of large quantities of energy. Fusion is the process that powers active stars, the hydrogen bomb, and experimental devices examining fusion power for electrical generation [2].

The fusion of two nuclei with lower masses than iron (which, along with nickel, has the largest binding energy per nucleon) generally releases energy, while the fusion of nuclei heavier than iron *absorbs* energy. The opposite is true for the reverse process, nuclear fission. This means that fusion generally occurs for lighter elements only, and likewise, that fission normally occurs only for heavier elements. There are extreme astrophysical events that can lead to short periods of fusion with heavier nuclei. This is the process that gives rise to nucleosynthesis, the creation of heavy elements during events like supernovas.

Creating the required conditions for fusion on Earth is very difficult, to the point that it has not been accomplished at any scale for protium, the common light isotope of hydrogen that undergoes natural fusion in stars. In nuclear weapons, some of the energy released by an atomic bomb is used to compress and heat a fusion fuel containing heavier isotopes of hydrogen and also sometimes lithium, to the point of "ignition." At this point, the energy released in the fusion reactions is enough to briefly maintain the reaction. Fusion-based nuclear power experiments attempt to create similar conditions using less dramatic means, although to date these experiments have failed to maintain conditions needed for ignition long enough for fusion to be a viable commercial power source [2].

Building upon the nuclear transmutation experiments by Ernest Rutherford, carried out several years earlier, the laboratory fusion of heavy hydrogen isotopes was first accomplished by Mark Oliphant in 1932. During the remainder of that decade, the steps of the main cycle of nuclear fusion in stars were worked out by Hans Bethe. Research into fusion for military purposes began in the early 1940s as part of the Manhattan Project, but this was not accomplished until 1951 (see the Greenhouse Item nuclear test), and nuclear fusion on a large scale in an explosion was first carried out on November 1, 1952, in the Ivy Mike hydrogen bomb test.

Research into developing controlled thermonuclear fusion for civil purposes also began in earnest in the 1950s, and it continues to this day. Two projects, the National

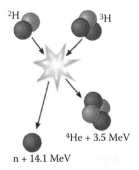

²H   ³H

⁴He + 3.5 MeV

n + 14.1 MeV

FIGURE 2.2 Fusion of deuterium with tritium creating helium-4. (From en.wikipedia.org/wiki/Nuclear_fusion)

Ignition Facility and ITER, are in the process of reaching breakeven after 60 years of design improvements developed from previous experiments (Figure 2.2).

2.2.1 Fusion Process

Nuclear fusion is the process by which two or more atomic nuclei join together, or "fuse," to form a single heavier nucleus. This is usually accompanied by the release or absorption of large quantities of energy. Fusion is the process that powers active stars, the hydrogen bomb, and experimental devices examining fusion power for electrical generation.

The fusion of two nuclei with lower masses than iron (which, along with nickel, has the largest binding energy per nucleon) generally releases energy, while the fusion of nuclei heavier than iron absorbs energy. The opposite is true for the reverse process, nuclear fission. This means that fusion generally occurs for lighter elements only, and likewise, that fission normally occurs only for heavier elements. There are extreme astrophysical events that can lead to short periods of fusion with heavier nuclei. This process gives rise to nucleosynthesis, the creation of the heavy elements during events like supernovas.

Creating the required conditions for fusion on Earth is very difficult. For nuclear weapons, some of the energy released by an atomic bomb is used to compress and heat a fusion fuel to the point of "ignition." At this point, the energy released in the fusion reactions is enough to maintain the reaction. Fusion-based nuclear power experiments attempt to create similar conditions using less dramatic means, although to date these experiments have failed to maintain conditions needed for ignition long enough for fusion to be a viable commercial power source.

Building upon the nuclear transmutation experiments by Ernest Rutherford, carried out several years earlier, the laboratory fusion of hydrogen isotopes was first accomplished by Mark Oliphant in 1932. Then, the steps of the main cycle of nuclear fusion in stars were first worked out by Hans Bethe throughout the remainder of that decade. Research into fusion for military purposes began in the early 1940s as part of the Manhattan Project, but this was not accomplished until 1951 (see the Greenhouse Item nuclear test), and nuclear fusion on a large scale in an explosion was first carried out on November 1, 1952, in the Ivy Mike hydrogen bomb test. Research into developing controlled thermonuclear fusion for civil purposes also began in earnest in the 1950s, and it continues to this day. Two projects, the National Ignition Facility and ITER, are in the process of reaching breakeven after 60 years of design improvements developed from previous experiments.

2.2.2 HYDROGEN

Hydrogen H has three stable isotopes ^{1}H (or $^1_1$H is called protium), ^{2}H (or $^2_1$D is called deuterium), and ^{3}H (or $^3_1$T is called tritium), and other four unstable isotopes ^{4}H to ^{7}H [3].

^{1}H is the most common hydrogen isotope with an abundance of more than 99.98%. Because the nucleus of this isotope consists of only a single proton, it is given the descriptive but rarely used formal name protium.

^{2}H, the other stable hydrogen isotope, is known as deuterium and contains one proton and one neutron in its nucleus. Essentially all deuterium in the universe is thought to have been produced at the time of the Big Bang and has endured since that time. Deuterium is not radioactive and does not represent a significant toxicity hazard. Water enriched in molecules that include deuterium instead of normal hydrogen is called heavy water. Deuterium and its compounds are used as a nonradioactive label in chemical experiments and in solvents for ^{1}H-NMR spectroscopy. Heavy water is used as a neutron moderator and coolant for nuclear reactors. Deuterium is also a potential fuel for commercial nuclear fusion.

^{3}H is known as tritium and contains one proton and two neutrons in its nucleus. It is radioactive, decaying into helium-3 through beta decay with a half-life of 12.32 years. It is so radioactive that it can be used in luminous paint, making it useful in such things as watches. The glass prevents the small amount of radiation from getting out. Small amounts of tritium occur naturally because of the interaction of cosmic rays with atmospheric gases; tritium has also been released during nuclear weapons tests. It is used in nuclear fusion reactions, as a tracer in isotope geochemistry, and specialized in self-powered lighting devices. Tritium has also been used in chemical and biological labeling experiments as a radiolabel. Some of the most stable isotopes are listed in Table 2.1.

The deuterium contained in 1 L water can release the energy of 4 L gasoline.

TABLE 2.1
Most Stable Isotopes

Isotopes	Percentage	Half-Life	DM	DE (MeV)	DP
^{1}H	99.985	^{1}H is stable with 0 neutrons			
^{2}H	0.015	^{2}H is stable with 1 neutron			
^{3}H	Trace	12.32 Years	β-	0.01861	^{3}He

2.2.3 FUSION REACTIONS

The following are the fusion reactions (fuel cycles) with a large cross section [2]:

$$_1^2D + _1^3T \rightarrow _2^4He\,(3.5\,\text{MeV}) + n^0\,(14.1\,\text{MeV}) \tag{2.1}$$

$$_1^2D + _1^2D \rightarrow _1^3T\,(1.01\,\text{MeV}) + p^+(3.02\,\text{MeV}) \quad 50\% \tag{2.2a}$$

$$_1^2D + _1^2D \rightarrow _2^3He\,(0.82\,\text{MeV}) + n^0(2.45\,\text{MeV}) \quad 50\% \tag{2.2b}$$

$$_1^2D + _2^3He \rightarrow _2^4He\,(3.6\,\text{MeV}) + p^+(14.7\,\text{MeV}) \tag{2.3}$$

$$_1^3T + _1^3T \rightarrow _2^4He + 2 \times n^0 + 11.3\,\text{MeV} \tag{2.4}$$

$$_2^3He + _2^3He \rightarrow _2^4He + 2 \times p^+ + 12.9\,\text{MeV} \tag{2.5}$$

$$_2^3He + _1^3T \rightarrow _2^4He + p^+ + n^0 + 12.1\,\text{MeV} \quad 51\% \tag{2.6a}$$

$$_2^3He + _1^3T \rightarrow _2^4He\,(4.8\,\text{MeV}) + _1^2D\,(9.5\,\text{MeV}) \quad 43\% \tag{2.6b}$$

$$_2^3He + _1^3T \rightarrow _2^4He\,(0.5\,\text{MeV}) + n^0\,(1.9\,\text{MeV}) + p^+(11.9\,\text{MeV}) \quad 6\% \tag{2.6c}$$

$$_1^2D + _3^6Li \rightarrow 2 \times _2^4He + 22.4\,\text{MeV} \tag{2.7a}$$

$$_1^2D + _3^6Li \rightarrow _2^3He + _2^4He + n^0 + 2.56\,\text{MeV} \tag{2.7b}$$

$$_1^2D + _3^6Li \rightarrow _3^7Li + p^+ + 5.0\,\text{MeV} \tag{2.7c}$$

$$_1^2D + _3^6Li \rightarrow _4^7Be + n^0 + 3.4\,\text{MeV} \tag{2.7d}$$

$$p^+ + _3^6Li \rightarrow _2^4He(1.7\,\text{MeV}) + _2^3He(2.3\,\text{MeV}) \tag{2.8}$$

$$_2^3He + _3^6Li \rightarrow 2 \times _2^4He + p^+ + 16.9\,\text{MeV} \tag{2.9}$$

$$p^{+} + {}^{11}_{5}B \rightarrow 3 \times {}^{4}_{2}He + 8.7\,\text{MeV} \qquad (2.10)$$

where

$\quad {}^{2}_{1}D$ is the deuterium

$\quad {}^{3}_{1}T$ is the tritium

$\quad {}^{3}_{2}He$ and ${}^{4}_{2}He$ are the helium

$\quad {}^{6}_{3}Li$ is the lithium

$\quad {}^{7}_{4}Be$ is the beryllium

$\quad {}^{11}_{5}B$ is the boron

$\quad n^{0}$ is the neutron

$\quad p^{+}$ is the proton

$\quad$ MeV is a million eV (electron-Volt—energy unit)

2.2.4 Hot Fusion

In hot fusion, the fuel reaches tremendous temperature and pressure inside a fusion reactor or star. The methods in the second group are examples of nonequilibrium systems, in which very high temperatures and pressures are produced in a relatively small region adjacent to material of much lower temperature. In his doctoral thesis for MIT, Todd Rider did a theoretical study of all quasineutral, isotropic, nonequilibrium fusion systems. He demonstrated that all such systems will leak energy at a rapid rate due to bremsstrahlung produced when electrons in the plasma hit other electrons or ions at a cooler temperature and suddenly decelerate. The problem is not as pronounced in hot plasma because the range of temperatures, and thus the magnitude of the deceleration, is much lower. Note that Rider's work does not apply to non-neutral or anisotropic nonequilibrium plasmas [4].

The most important fusion process in nature is the one that powers stars. The net result is the fusion of four protons into one alpha particle, with the release of two positrons, two neutrinos (which changes two of the protons into neutrons), and energy, but several individual reactions are involved, depending on the mass of the star. For stars the size of the Sun or smaller, the proton–proton chain dominates. In heavier stars, the CNO cycle is more important. Both types of processes are responsible for the creation of new elements as part of stellar nucleosynthesis (Figure 2.3).

At the temperatures and densities in stellar cores, the rates of fusion reactions are notoriously slow. For example, at solar core temperature ($T \approx 15\,\text{MK}$) and density ($160\,\text{g/cm}^3$), the energy release rate is only $276\,\mu\text{W/cm}^3$—about a quarter of the volumetric rate at which a resting human body generates heat. Thus, reproduction of stellar core conditions in a lab for nuclear fusion power production is completely impractical. Because nuclear reaction rates strongly depend on temperature ($\exp(-E/kT)$), achieving reasonable energy production rates in terrestrial fusion reactors requires 10–100 times higher temperatures (compared to stellar interiors): $T \approx 0.1$–$1.0\,\text{GK}$.

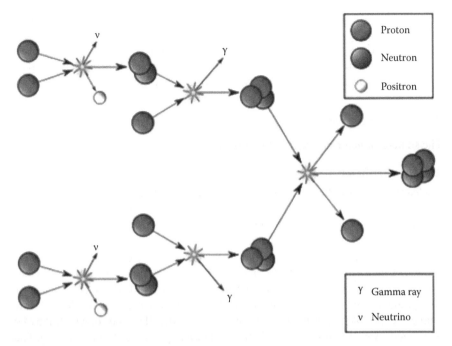

FIGURE 2.3 The proton–proton chain dominates in stars the size of the Sun or smaller. (From Canada Wide Virtual Science Fair 2009, Nuclear fusion, *Nuclear Power*. http://www. virtualsciencefair.org/2009/xing9d2/fusion.htm)

2.3 CAPTURE OF NEUTRINO

2.3.1 NEUTRINO

A neutrino, meaning "small neutral one," is an elementary particle that usually travels close to the speed of light, is electrically neutral, and is able to pass through ordinary matter almost unaffected, "like a bullet passing through a bank of fog." This makes neutrinos extremely difficult to detect. Neutrinos have a very small but nonzero mass. They are denoted by the Greek letter ν (nu). There are three types of neutrinos: Electron-neutrino νe, muon-neutrino $\nu\mu$, and tau-neutrino $\nu\tau$ [6].

Neutrinos are similar to the more familiar electron, with one crucial difference: neutrinos do not carry electric charge. Because neutrinos are electrically neutral, they are not affected by the electromagnetic forces that act on electrons. Neutrinos are affected only by a "weak" subatomic force of much shorter range than electromagnetism and are therefore able to pass through great distances within matter without being affected by it. Neutrinos also interact gravitationally with other particles.

Neutrinos are created as a result of certain types of radioactive decay or nuclear reactions such as those that take place in the Sun, in nuclear reactors, or when cosmic rays hit atoms. There are three types, or "flavors," of neutrinos: electron neutrinos, muon neutrinos, and tau neutrinos. Each type also has a corresponding antiparticle, called an antineutrino. Electron neutrinos (or antineutrinos) are generated whenever protons change into neutrons, or vice versa—the two forms of beta decay. Interactions involving neutrinos are mediated by the weak interaction.

Most neutrinos passing through the Earth emanate from the Sun. Every second, in the region of the Earth, about 65 billion (6.5×1010) solar neutrinos pass through every square centimeter perpendicular to the direction of the Sun.

2.3.2 Neutrino Sources

The sources of neutrino are listed in this section [6].

2.3.2.1 Artificial

Nuclear reactors are the major source of human-generated neutrinos. Antineutrinos are made in the beta decay of neutron-rich daughter fragments in the fission process. Generally, the four main isotopes contributing to the antineutrino flux are ^{235}U, ^{238}U, ^{239}Pu, and ^{241}Pu (i.e., the antineutrinos emitted during beta-minus decay of their respective fission fragments). The average nuclear fission releases about 200 MeV of energy, of which roughly 4.5% (or about 9 MeV) is radiated away as antineutrinos. For a typical nuclear reactor with a thermal power of 4000 MW, meaning that the core produces this much heat, and an electrical power generation of 1300 MW, the total power production from fissioning atoms is actually 4185 MW, of which 185 MW is radiated away as antineutrino radiation and never appears in the engineering. That is, 185 MW of fission energy is lost from this reactor and does not appear as heat available to run turbines, since the antineutrinos penetrate all building materials essentially tracelessly and disappear.

The antineutrino energy spectrum depends on the degree to which the fuel is burned (plutonium-239 fission antineutrinos on average have slightly more energy than those from uranium-235 fission), but in general, the detectable antineutrinos from fission have a peak energy between about 3.5 and 4 MeV, with a maximal energy of about 10 MeV. There is no established experimental method to measure the flux of low-energy antineutrinos. Only antineutrinos with energy above the threshold of 1.8 MeV can be uniquely identified (see neutrino detection in the following). An estimated 3% of all antineutrinos from a nuclear reactor carry energy above this threshold. An average nuclear power plant may generate over 1020 antineutrinos per second above this threshold, and a much larger number that cannot be seen with present detector technology.

Some particle accelerators have been used to make neutrino beams. The technique is to smash protons into a fixed target, producing charged pions or kaons. These unstable particles are then magnetically focused into a long tunnel where they decay while in flight. Because of the relativistic boost of the decaying particle, the neutrinos are produced as a beam rather than isotropically. Efforts to construct an accelerator facility where neutrinos are produced through muon decays are ongoing. Such a setup is generally known as a neutrino factory.

2.3.2.2 Geological

Neutrinos are part of the natural background radiation. In particular, the decay chains of ^{238}U and ^{232}Th isotopes, as well as 40 K, include beta decays that emit antineutrinos. These so-called geoneutrinos can provide valuable information on the Earth's interior. A first indication for geoneutrinos was found by the KamLAND experiment in 2005. KamLAND's main background in the geoneutrino measurement is the antineutrinos coming from reactors. Several future experiments aim at improving the geoneutrino measurement, and these will necessarily have to be far away from reactors.

2.3.2.3 Atmospheric

Atmospheric neutrinos result from the interaction of cosmic rays with atomic nuclei in the Earth's atmosphere, creating showers of particles, many of which are unstable and produce neutrinos when they decay. A collaboration of particle physicists from Tata Institute of Fundamental Research, India, Osaka City University, Japan, and Durham University, United Kingdom, recorded the first cosmic ray neutrino interaction in an underground laboratory in Kolar Gold Fields in India in 1965.

2.3.2.4 Solar

Solar neutrinos originate from the nuclear fusion powering the Sun and other stars. The details of the operation of the Sun are explained by the Standard Solar Model. In short, when four protons fuse to become one helium nucleus, two of them have to convert into neutrons, and each such conversion releases one electron neutrino.

The Sun sends enormous numbers of neutrinos in all directions. Every second, about 65 billion (6.5×10^{10}) solar neutrinos pass through every square centimeter on the part of the Earth that faces the Sun. Since neutrinos are insignificantly absorbed by the mass of the Earth, the surface area on the side of the Earth opposite the Sun receives about the same number of neutrinos as the side facing the Sun.

2.3.2.5 By Supernovae

Neutrinos are an important product of Types Ib, Ic, and II (core-collapse) supernovae. In such events, the density at the core becomes so high ($10^{17}\,\mathrm{kg/m^3}$) that the degeneracy of electrons is not enough to prevent protons and electrons from combining to form a neutron and an electron neutrino. A second and more important neutrino source is the thermal energy (100 billion kelvins) of the newly formed neutron core, which is dissipated via the formation of neutrino–antineutrino pairs of all flavors. Most of the energy produced in supernovas is thus radiated away in the form of an immense burst of neutrinos. The first experimental evidence of this phenomenon came in 1987, when neutrinos from supernova 1987A were detected. The water-based detectors Kamiokande II and IMB detected 11 and 8 antineutrinos of thermal origin, respectively, while the scintillator-based Baksan detector found 5 neutrinos (lepton number $= 1$) of either thermal or electron-capture origin, in a burst lasting less than 13 s. It is thought that neutrinos would also be produced from other events such as the collision of neutron stars. The neutrino signal from the supernova arrived at Earth several hours before the arrival of the first electromagnetic radiation, as expected from the evident fact that the latter emerges along with the shock wave. The exceptionally feeble interaction with normal matter allowed the neutrinos to pass through the churning mass of the exploding star, while the electromagnetic photons were slowed (Figure 2.4) [7].

FIGURE 2.4   Supernova 1987A. (Courtesy of NASA, Washington, DC.)

Because neutrinos interact so little with matter, it is thought that a supernova's neutrino emissions carry information about the innermost regions of the explosion. Much of the visible light comes from the decay of radioactive elements produced by the supernova shock wave, and even light from the explosion itself is scattered by dense and turbulent gases. Neutrinos, on the other hand, pass through these gases, providing information about the supernova core (where the densities were large enough to influence the neutrino signal). Furthermore, the neutrino burst is expected to reach Earth before any electromagnetic waves, including visible light, gamma rays, or radio waves. The exact time delay depends on the velocity of the shock wave and on the thickness of the outer layer of the star. For a Type II supernova, astronomers expect the neutrino flood to be released seconds after the stellar core collapse, while the first electromagnetic signal may emerge hours later. The SNEWS project uses a network of neutrino detectors to monitor the sky for candidate supernova events; the neutrino signal will provide a useful advance warning of a star exploding in the Milky Way.

2.3.2.6 By Supernova Remnants

The energy of supernova neutrinos ranges from a few to several tens of MeV. However, the sites where cosmic rays are accelerated are expected to produce neutrinos that are at least one million times more energetic, produced from turbulent gaseous environments left over by supernova explosions: the supernova remnants. The origin of the cosmic rays was attributed to supernovas by Walter Baade and Fritz Zwicky; this hypothesis was refined by Vitaly L. Ginzburg and Sergei I. Syrovatsky who attributed the origin to supernova remnants and supported their claim by the crucial remark that the cosmic ray losses of the Milky Way are compensated, if the efficiency of acceleration in supernova remnants is about 10%. Ginzburg and Syrovatskii's hypothesis is supported by the specific mechanism of "shock wave acceleration" happening in supernova remnants, which is consistent with the original theoretical picture drawn by Enrico Fermi, and is receiving support from observational data. Very high energy neutrinos are still to be seen, but this branch of neutrino astronomy is just in its infancy. The main existing or forthcoming experiments that aim at observing very high energy neutrinos from our galaxy are Baikal, AMANDA, IceCube, Antares, NEMO, and Nestor. Related information is provided by very high energy gamma ray observatories, such as VERITAS, HESS, and MAGIC. Indeed, the collisions of cosmic rays are supposed to produce charged pions, whose decay gives the neutrinos, and also neutral pions, whose decay gives gamma rays: the environment of a supernova remnant is transparent to both types of radiation.

Still higher energy neutrinos, resulting from the interactions of extragalactic cosmic rays, could be observed with the Pierre Auger Observatory or with the dedicated experiment named ANITA.

2.3.2.7 By the Big Bang

It is thought that, just as the cosmic microwave background radiation left over from the Big Bang, there is a background of low-energy neutrinos in our Universe. In the 1980s, it was proposed that these may be the explanation for the dark matter thought to exist in the universe. Neutrinos have one important advantage over most other dark matter candidates: We know they exist. However, they also have serious problems.

From particle experiments, it is known that neutrinos are very light. This means that they move at speeds close to the speed of light. Thus, dark matter made from neutrinos is termed "hot dark matter." The problem is that being fast moving the neutrinos would tend to have spread out evenly in the universe before cosmological expansion made them cold enough to congregate in clumps. This would cause the part of dark matter made of neutrinos to be smeared out and unable to cause the large galactic structures that we see.

Further, these same galaxies and groups of galaxies appear to be surrounded by dark matter that is not fast enough to escape from those galaxies. Presumably, this matter provided the gravitational nucleus for formation. This implies that neutrinos make up only a small part of the total amount of dark matter.

From cosmological arguments, relic background neutrinos are estimated to have a density of 56 of each type per cubic centimeter and a temperature of 1.9 K ($1.7 \times 10-4$ eV) if they are massless and much colder if their mass exceeds 0.001 eV. Although their density is quite high, due to extremely low neutrino cross-sections at sub-eV energies, the relic neutrino background has not yet been observed in the laboratory. In contrast, boron-8 solar neutrinos—which are emitted with a higher energy—have been detected definitively despite having a space density that is lower than that of relic neutrinos by some 6 orders of magnitude.

2.3.3 Neutrino Detection

Because neutrinos are very weakly interacting, neutrino detectors must be very large in order to detect a significant number of neutrinos. Neutrino detectors are often built underground in order to isolate the detector from cosmic rays and other background radiation. Antineutrinos were first detected in the 1950s near a nuclear reactor. Reines and Cowan used two targets containing a solution of cadmium chloride in water. Two scintillation detectors were placed next to the cadmium targets. Antineutrinos with energy above the threshold of 1.8 MeV caused charged current interactions with the

protons in the water, producing positrons and neutrons. The resulting positron anni-
hilations with electrons created photons with an energy of about 0.5 MeV. Pairs of
photons in coincidence could be detected by the two scintillation detectors above and
below the target. The neutrons were captured by cadmium nuclei resulting in gamma
rays of about 8 MeV that were detected a few microseconds after the photons from a
positron annihilation event. Since then, various detection methods have been used.
Super Kamiokande is a large volume of water surrounded by photomultiplier tubes
that watch for the Cherenkov radiation emitted when an incoming neutrino creates an
electron or muon in the water. The Sudbury Neutrino Observatory is similar, but uses
heavy water as the detecting medium, which uses the same effects but also allows
the additional reaction any-flavor neutrino photodissociation of deuterium, resulting
in a free neutron that is then detected from gamma radiation after chlorine capture.
Other detectors have consisted of large volumes of chlorine or gallium, which are
periodically checked for excesses of argon or germanium, respectively, which are
created by electron–neutrinos interacting with the original substance. MINOS uses a
solid plastic scintillator coupled to photomultiplier tubes, while Borexino uses a liquid pseudocumene scintillator also watched by photomultiplier tubes. The proposed
NOvA detector will use liquid scintillator watched by avalanche photodiodes. The
IceCube Neutrino Observatory uses 1 km³ of the Antarctic ice sheet near the South
Pole with photomultiplier tubes distributed throughout the volume.

2.4   CONCLUSION

At present, hydraulic power occupies about 35% of the world's energy market; ther-
moelectricity power occupies 30%; nuclear power (fission power) occupies about
25%; others including renewable energy source occupy 10%. The nuclear power
occupies a large percentage; we cannot stop using it immediately. Unfortunately,
nuclear and radiation accidents of nuclear power stations happen frequently worldwide, including the nuclear power plants Three Mile Island, the United States, in
1979, Chernobyl, Russia, in 1986, and Fukushima, Japan, in 2011. Heavy nuclear
radiation pollution strongly damaged the environment of the biosphere. We should
avoid constructing more nuclear power stations implementing the fission process.
We believe that the radical solution of the world energy crisis is to use fusion power
nuclear energy. The comparison of the three energy sources is shown in Table 2.2.
There may be other energy sources we can develop in the future.

TABLE 2.2
Comparison of the Three New Sources

Sources	Feasibility	Cost	Operation/Control	Pollution
Fusing power	Difficult	High	Complicated	Clean
Fission power	Less difficult	Lower	Easy	Heavy
Neutrino	Very difficult	Very high	In research	NA

REFERENCES

1. Attribute to Wikipedia contributors. Nuclear fission. Available under the Creative Commons Attribution-ShareAlike License
2. Attribute to Wikipedia contributors. Nuclear fusion. Available under the Creative Commons Attribution-ShareAlike License
3. Attribute to Wikipedia contributors. Hydrogen. Available under the Creative Commons Attribution-ShareAlike License
4. Attribute to Wikipedia contributors. Proton–proton chain reaction. Available under the Creative Commons Attribution-ShareAlike License
5. Canada Wide Virtual Science Fair 2009. Nuclear fusion. *Nuclear Power*. http://www. virtualsciencefair.org/2009/xing9d2/fusion.htm
6. Attribute to Wikipedia contributors. Neutrino. Available under the Creative Commons Attribution-ShareAlike License
7. Attribute to Wikipedia contributors. SN 1987A. Available under the Creative Commons Attribution-ShareAlike License

3 3G and Renewable Energies

3G means distributed generation (DG), microgrid, and smart grid. In recent years, 3G has become a popular topic and is developing rapidly due to energy shortage.

Renewable energies are derived from natural processes that are replenished constantly and have various forms. Included in the definition is electricity and heat produced from solar, wind, ocean, hydropower, biomass, geothermal resources, and biofuels, and hydrogen derived from renewable resources.

3.1 DISTRIBUTED GENERATION

Distributed generation, also called on-site generation, dispersed generation, embedded generation, decentralized generation, decentralized energy, or distributed energy, generates electricity from many small energy sources. A DG is the use of small-scale power generation technologies located close to the load being served. DG stakeholders include energy companies, equipment suppliers, regulators, energy users, and financial and supporting companies. For some customers DG can lower costs, enhance efficiency, improve reliability, reduce emissions, or expand their energy options. DG may add redundancy that increases grid security even while powering emergency lighting or other critical systems [1].

Currently, industrial countries generate most of their electricity in large centralized facilities, such as fossil fuel (coal, gas powered), nuclear, large solar power plants, or hydropower plants. These plants have excellent economies of scale, but they usually transmit electricity over long distances and negatively affect the environment.

3.1.1 ECONOMIES OF SCALE

Most plants are built this way due to a number of economic, health and safety, logistical, environmental, geographical, and geological factors. For example, coal power plants are built away from cities to prevent their heavy air pollution from affecting the populace. In addition, such plants are often built near collieries to minimize the cost of transporting coal. Hydroelectric plants are by their nature limited to operating at sites with sufficient water flow. Most power plants are often considered to be too far away for their waste heat to be used for heating buildings.

Low pollution is a crucial advantage of combined cycle plants that burn natural gas. The low pollution permits the plants to be near enough to a city to be used for district heating and cooling.

3.1.2 LOCALIZED GENERATION

Distributed generation is another approach. It reduces the amount of energy lost in
transmitting electricity because the electricity is generated around the place where it
is used, perhaps even in the same building. This also reduces the size and the number
of power lines that must be constructed.

Typical distributed power sources in a feed-in tariff (FIT) scheme have low maintenance, low pollution, and high efficiencies. In the past, these traits required dedicated operating engineers and large complex plants to reduce pollution. However,
modern embedded systems can provide these traits with automated operation and
renewables, such as sunlight, wind, and geothermal. This reduces the size of the
power plant that can show a profit.

3.1.3 DISTRIBUTED ENERGY RESOURCES

Distributed energy resource (DER) systems are small-scale power generation technologies (typically in the range of 3–10,000 kW) used to provide an alternative to or
an enhancement of the traditional electric power system. The usual problem with
distributed generators is their high costs.

One popular source is solar panels on the roofs of buildings. The production cost
is $0.99–$2.00/W (2007) plus installation and supporting equipment unless the
installation is Do it yourself (DIY) bringing the cost to $5.25–$7.50/W (2010) [1].
This is comparable to coal power plant costs of $0.582–$0.906/W (1979), adjusting
for inflation. Nuclear power is higher at $2.2–$6.00/W (2007) [1]. Some solar cells
("thin-film" type) also have waste disposal issues, since thin-film type solar cells
often contain heavy-metal electronic wastes, such as Cadmium telluride (CdTe) and
Copper indium gallium selenide (CuInGaSe), and need to be recycled, as opposed
to silicon semiconductor type solar cells, which are made from quartz. The positive
side is that unlike coal and nuclear, there are no fuel costs, pollution, and mining
or operating safety issues. Solar also has a low duty cycle, producing peak power at
local noon each day. Average duty cycle is typically 20%.

Another source is small wind turbines. These have low maintenance, and low
pollution. Construction costs are higher ($0.80/W, 2007) per watt than large power
plants, except in very windy areas. Wind towers and generators have substantial
insurable liabilities caused by high winds but good operating safety. In some areas
of the United States, there may also be property tax costs involved with wind turbines that are not offset by incentives or accelerated depreciation. Wind also tends to
be complementary to solar; on days there is no sun there tends to be wind and vice
versa. Many distributed generation sites combine wind power and solar power such
as Slippery Rock University.

In addition, solid oxide fuel cells using natural gas, such as the Bloom Energy
Server, have recently become a distributed energy resource.

Distributed cogeneration sources use natural gas-fired microturbines or recipro-
cating engines to turn generators. The hot exhaust is then used for space or water
heating or to drive an absorptive chiller for air conditioning. The clean fuel has only

low pollution. Designs currently have uneven reliability, with some makes having excellent maintenance costs and others being unacceptable.

3.1.4 Cost Factors

Cogenerators are also more expensive per watt than central generators. They find favor because most buildings already burn fuels, and the cogeneration can extract more value from the fuel.

Some larger installations utilize combined cycle generation. Usually this consists of a gas turbine whose exhaust boils water for a steam turbine in a Rankine cycle. The condenser of the steam cycle provides the heat for space heating or an absorptive chiller. Combined cycle plants with cogeneration have the highest known thermal efficiencies, often exceeding 85%.

In countries with high-pressure gas distribution, small turbines can be used to bring the gas pressure to domestic levels while extracting useful energy. If the United Kingdom were to implement this countrywide an additional 2–4 GWe would become available. (Note that the energy is already being generated elsewhere to provide the high initial gas pressure—this method simply distributes the energy via a different route.)

Future generations of electric vehicles will have the ability to deliver power from the battery into the grid when needed. An electric vehicle network could also be an important distributed generation resource.

3.2 MICROGRID

Microgrid is usually a reliable power in a small package. Microgrid mostly gathers distributed generation sources and forms a small power group to supply power in a small area. We know that the main grid supplies electrical energy to a large area covering millions of families. Once there is a fault, it always results in thousands of families losing the power supply. For example, in 1996, a sagging power line in Oregon brushed against a tree and within minutes 12 million customers in 8 states lost power. Such is the vulnerability of today's main power grid [2].

To address this weakness, Berkeley Lab scientists are helping to develop a new approach to power generation in which a cluster of small, on-site generators serves office buildings, industrial parks, and homes. Called a microgrid, the system could help shoulder the nation's growing thirst for electricity—estimated to jump by almost 400 GW by 2025—without overburdening aging transmission lines or building the 1000 new power plants required to meet this demand. In addition, it may make statewide blackouts a thing of the past or at least ensure that service to critical equipment is maintained.

Microgrids boast other advantages, but it is no coincidence that reliability is high on the list. The concept is being pioneered by the Consortium for Electric Reliability Technology Solutions, a national lab, university, and industry group convened by the Department of Energy in 1999 to explore ways to improve power reliability. The consortium, which is supported by the California Energy Commission and

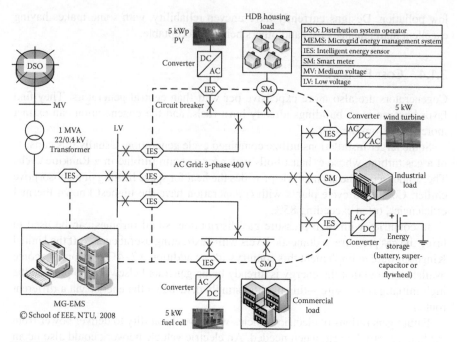

| DSO: Distribution system operator |
| MEMS: Microgrid energy management system |
| IES: Intelligent energy sensor |
| SM: Smart meter |
| MV: Medium voltage |
| LV: Low voltage |

FIGURE 3.1   MG-EMS.

centered at Berkeley Lab, is developing several innovative strategies in addition to microgrids, including managing power grids in real time and determining how an emerging open electricity market affects reliability. The group conducted the first microgrid bench test in early 2004, in which three microturbines and several end loads were linked together at a utility-grade testing facility. This was followed in late 2004 by the first microgrid field test, with Berkeley Lab researchers playing a key role in selecting the best site. The microgrid shown in Figure 3.1 is the "Microgrid Energy Management System (MG-EMS)" conducted by research labs in Nanyang Technological University (NTU).

3.3   SMART GRID

Smart grid is a type of electrical grid that attempts to predict and intelligently respond to the behavior and actions of all electric power users connected to it—suppliers, consumers, and those that do both—in order to efficiently deliver reliable, economic, and sustainable electricity services [3].

In Europe, the smart grid is conceived of as employing innovative products and services together with intelligent monitoring, control, communication, and self-healing technologies in order to

- Better facilitate the connection and operation of generators of all sizes and technologies
- Allow consumers to play a part in optimizing the operation of the system

- Provide consumers with greater information and options for choice of supply
- Significantly reduce the environmental impact of the whole electricity supply system
- Maintain or even improve the existing high levels of system reliability, quality, and security of supply
- Maintain and improve the existing services efficiently

In the United States, the smart grid concept is defined as the modernization of the nation's electricity transmission and distribution system to maintain a reliable and secure electricity infrastructure that can meet future demand growth and to achieve each of the following, which together characterize a smart grid [3]:

1. Increased use of digital information and control technology to improve reliability, security, and efficiency of the electric grid
2. Dynamic optimization of grid operations and resources, with full cyber-security
3. Deployment and integration of distributed resources and generation, including renewable resources
4. Development and incorporation of demand response, demand-side resources, and energy-efficiency resources
5. Deployment of "smart" technologies (real-time, automated, interactive technologies that optimize the physical operation of appliances and consumer devices) for metering, communications concerning grid operations and status, and distribution automation
6. Integration of "smart" appliances and consumer devices
7. Deployment and integration of advanced electricity storage and peak-shaving technologies, including plug-in electric and hybrid electric vehicles, and thermal storage air conditioning
8. Provision to consumers of timely information and control options
9. Development of standards for communication and interoperability of appliances and equipment connected to the electric grid, including the infrastructure serving the grid
10. Identification and lowering of unreasonable or unnecessary barriers to adoption of smart grid technologies, practices, and services

The particular system line graph of the NTU smart grid is shown in Figure 3.2. It is the MG-EMS microgrid in Figure 3.1 equipped with intelligent control equipment.

3.4   SOLAR ENERGY

Solar energy is a clean energy since it fuses hydrogen atoms into helium to radiate light and heat. It has been harnessed by human beings since ancient time using a range of ever-evolving technologies. Solar radiation, along with secondary solar-powered resources such as wind and wave power, hydroelectricity and biomass, is accounted for most of the available renewable energy on the Earth. Only a minuscule fraction of the available solar energy is used.

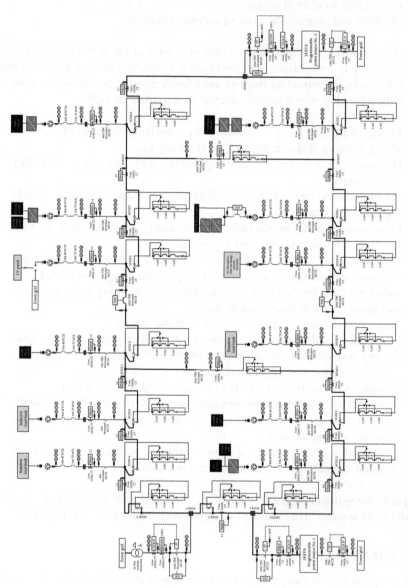

FIGURE 3.2 MG_EMS system line graph.

Solar-powered electrical generation relies on heat engines and photovoltaics. Solar energy's usages are limited only by human ingenuity. A partial list of solar applications includes space heating and cooling through solar architecture, potable water via distillation and disinfection, day lighting, solar hot water, solar cooking, and high temperature process heat for industrial purposes. To harvest the solar energy, the most common way is to use solar panels.

Solar technologies are broadly characterized as either passive solar or active solar depending on the way they capture, convert, and distribute solar energy. Active solar techniques include the use of photovoltaic panels and solar thermal collectors to harness the energy. Passive solar techniques include orienting a building to the Sun, selecting materials with favorable thermal mass or light dispersing properties, and designing spaces that naturally circulate air.

The Sun is a thermonuclear furnace fusing hydrogen atoms into helium. The resulting loss of mass is converted into about 3.8×10^{20} MW (P_S) of electromagnetic energy that radiates outward from the surface into space. Our earth (with the radius about 6366 km) is located at the surface of a big ball with the Radius 1 AU. Therefore, the power ($P_{E\text{-Sun}}$) our Earth receives from the Sun is

$$P_{E\text{-Sun}} = P_S \frac{\text{Earth section area}}{\text{Bigarea ball area}} = P_S \frac{\pi \times (6,366\,\text{km})^2}{4\pi \times (150,000,000\,\text{km})^2}$$

$$= 3.8 \times 10^{20}\,\text{MW} \frac{40,528,473}{9 \times 10^{16}} \approx 174 \times 10^9\,\text{MW or } 174\,\text{PW}. \tag{3.1}$$

The Earth receives 174 petawatts (PW) of incoming solar radiation (insolation) at the upper atmosphere. Approximately 30% is reflected back to space while the rest is absorbed by clouds, oceans, and landmasses. The spectrum of solar light at the Earth's surface is mostly spread across the visible and near-infrared ranges with a small part in the near ultraviolet. Figure 3.3 shows that the incoming solar energy reaches the Earth's surface [4].

The Earth's land surface, oceans, and atmosphere absorb solar radiation, and this raises their temperature. Warm air containing evaporated water from the oceans rises, causing atmospheric circulation or convection. When the air reaches a high altitude, where the temperature is low, water vapor condenses into clouds, which rain onto the Earth's surface, completing the water cycle. The latent heat of water condensation amplifies convection, producing atmospheric phenomena such as wind, cyclones, and anticyclones. Sunlight absorbed by the oceans and landmasses keeps the surface at an average temperature of 14°C. By photosynthesis, green plants convert solar energy into chemical energy, which produces food, wood, and the biomass from which fossil fuels are derived.

The total solar energy absorbed by the Earth's atmosphere, oceans, and land masses is approximately 3,850,000 Exa-Joules (EJ—10^{18} J) per year. In 2002, this was more energy in 1 h than the world used in 1 year. Photosynthesis captures approximately 3000 EJ per year in biomass. The amount of solar energy reaching the surface of the planet is so vast that in 1 year it is about twice as much as will

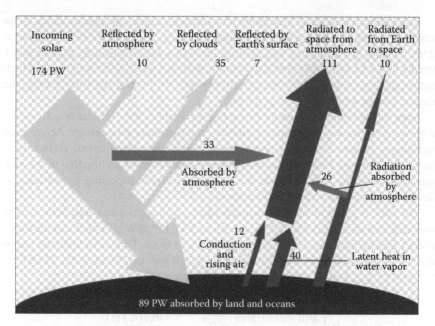

FIGURE 3.3   About half the incoming solar energy reaches the Earth's surface. (From Frank van Mierlo, available under the Creative Commons Attribution-Share Alike 3.0 Unported license.)

TABLE 3.1
Yearly Solar Fluxes and Human
Energy Consumption in 2005

Solar	3,850,000 EJ
Wind	2,250 EJ
Biomass	3,000 EJ
Primary energy use (2005)	487 EJ
Electricity (2005)	56.7 EJ

---

ever be obtained from all of the Earth's nonrenewable resources of coal, oil, natural gas, and mined uranium combined. Solar energy can be harnessed in different levels around the world. Depending on a geographical location, the closer to the equator the more "potential" the solar energy that is available. We can show some data for readers' reference. The yearly solar fluxes and human energy consumption in 2005 are listed in Table 3.1.

3.5   RENEWABLE ENERGY

Renewable energy can be biofuel, biomass, geothermal, hydroelectricity, solar energy, tidal power, wave power, and wind power [5]. Renewable energy is energy that comes from natural resources such as sunlight, wind, rain, tides, and geothermal heat, which are renewable (naturally replenished). About 16% of global final energy consumption comes from renewables, with 10% coming from traditional biomass, which is mainly used for heating, and 3.4% from hydroelectricity. New renewables (small hydro, modern biomass, wind, solar, geothermal, and biofuels) account for another 2.8% and are growing very rapidly. The share of renewables in electricity generation is around 19%, with 16% of global electricity coming from hydroelectricity and 3% from new renewables.

Wind power is growing at the rate of 30% annually, with a worldwide installed capacity of 198 GW in 2010, and is widely used in Europe, Asia, and the United States. At the end of 2010, cumulative global photovoltaic (PV) installations surpassed 40 W, and PV power stations are popular in Germany and Spain. Solar thermal power stations operate in the United States and Spain, and the largest of these is the 354 MW SEGS power plant in the Mojave Desert. The world's largest geothermal power installation is The Geysers in California, with a rated capacity of 750 MW. Brazil has one of the largest renewable energy programs in the world, involving production of ethanol fuel from sugar cane, and ethanol now provides 18% of the country's automotive fuel. Ethanol fuel is also widely available in the United States. Figure 3.4 illustrates the statistics of the world renewable energy in 2008.

While many renewable energy projects are large-scale, renewable technologies are also suited to rural and remote areas, where energy is often crucial in human development. As of 2011, small solar PV systems provide electricity to a few million households, and micro-hydro configured into minigrids serves many more. Over 44 million households use biogas made in household-scale digesters for lighting and

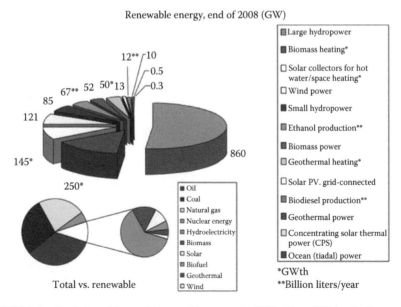

FIGURE 3.4   Statistics of the world renewable energy in 2008. (From Wikipedia Commons.)

cooking, and more than 166 million households rely on a new generation of more-efficient biomass cook stoves.

Climate change concerns, coupled with high oil prices, peak oil, and increasing government support, are driving increasing renewable energy legislation, incentives, and commercialization. New government spending, regulation, and policies help the industry weather the global financial crisis better than many other sectors.

REFERENCES

1. Attribute to Wikipedia contributors. Distributed generation. Available under the Creative Commons Attribution-ShareAlike License
2. D. Krotz. Microgrids: reliable power in a small package. *Berkely Lab Science Beat.*, http://www.lbl.gov/Science-Articles/Archive/EETD-microgrids.html (accessed on May 7, 2012).
3. Attribute to Wikipedia contributors. Smart grid. Available under the Creative Commons Attribution-ShareAlike License
4. Attribute to Wikipedia contributors. Solar energy. Available under the Creative Commons Attribution-ShareAlike License
5. Attribute to Wikipedia contributors. Renewable energy. Available under the Creative Commons Attribution-ShareAlike License

4 Power Electronics

Power electronics is the knowledge to process and control the flow of electric energy by supplying voltages and currents in a form that is optimally suited for user loads [1]. A typical block diagram is shown in Figure 4.1 [2]. The input power can be AC and DC sources. A general example is that the AC input power is from the electric utility. The output power to load can be AC and DC voltages. The power processor in the block diagram is usually called a converter. Conversion technologies are the knowledge to construct converters. Therefore, there are four categories of converters [3]:

1. AC/DC converters/rectifiers
2. DC/DC converters
3. DC/AC inverters/converters
4. AC/AC converters

We will use *converter* as a generic term to refer to a single power conversion stage that may perform any of the functions listed earlier. To be more specific, in AC to DC and DC to AC conversion, *rectifies* refers to a converter when the average power flow is from the AC to the DC side. *Inverter* refers to the converter when the average power flow is from the DC to the AC side. In fact, the power flow through the converter may be reversible. In that case, as shown in Figure 4.2 [2], we refer to that converter in terms of its rectifier and inverter modes of operation.

4.1 SYMBOLS AND FACTORS USED IN THIS BOOK

We list the factors and symbols used in this book here. If there are no specific descriptions, the parameters follow the meaning stated here.

4.1.1 SYMBOLS USED IN POWER SYSTEMS

For instantaneous values of variables such as voltage, current, and power that are functions of time, the symbols used are lowercase letters v, i, and p, respectively. They are time's functions performing in the time domain. We may or may not explicitly show that they are functions of time, for example, using v rather than $v(t)$. The uppercase symbols V and I refer to their computed values from their instantaneous waveforms. They generally refer to an average value in DC quantities and a root-mean-square (rms) value in AC quantities. If there is a possibility of confusion, the subscript avg or rms is added explicitly. The average power is always indicated by P.

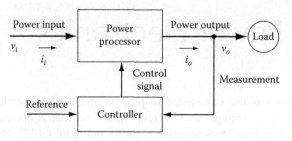

FIGURE 4.1 The block diagram of a power electronics system.

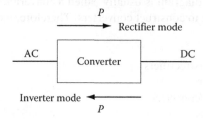

FIGURE 4.2 AC to DC converters.

Usually, the input voltage and current are represented by v_{in} and i_{in} (or v_1 and i_1), and the output voltage and current are represented by v_O and i_O (or v_2 and i_2). The input and output powers are represented by P_{in} and P_O. The power transfer efficiency (η) is defined as $\eta = P_O/P_{in}$.

Passive loads such as resistor R, inductor L, and capacitor C are generally used in circuits. We use R, L, and C to indicate their symbols and values as well. All these three parameters and their combination Z are linear load since the performance of the circuit constructed by these components is described by a linear deferential equation. Z is used as impedance of a linear load. If the circuit consists of a resistor R, an inductor L, and a capacitor C in series connected, the impedance Z is represented by

$$Z = R + j\omega L - j\frac{1}{\omega C} = |Z| \angle \phi \qquad (4.1)$$

where
  R is the resistance measured by Ω
  L is the inductance measured by H
  C is the capacitance measured by F
  ω is the AC supply angular frequency measured by rad/s
  $\omega = 2\pi f$, f is the AC supply frequency measured by Hz

For the calculation of Z, if there is no capacitor in a circuit, the item $j(1/\omega C)$ is omitted (do not take $c=0$ and $j(1/\omega C)=>\infty$). The absolute impedance $|Z|$ and the phase angle ϕ are determined by

$$|Z| = \sqrt{R^2 + \left(\omega L - \frac{1}{\omega C}\right)^2}$$
(4.2)

$$\phi = \tan^{-1}\frac{\omega L - (1/\omega C)}{R}$$

If a circuit consists of a resistor R and an inductor L only in series connected, the corresponding impedance Z is represented by

$$Z = R + j\omega L = |Z| \angle \phi$$
(4.3)

The absolute impedance $|Z|$ and phase angle ϕ are determined by

$$|Z| = \sqrt{R^2 + (\omega L)^2}$$
(4.4)

$$\phi = \tan^{-1}\frac{\omega L}{R}$$

We define the circuit time constant τ as

$$\tau = \frac{L}{R}$$
(4.5)

If a circuit consists of a resistor R and a capacitor C only in series connected, the impedance Z is represented by

$$Z = R - j\frac{1}{\omega C} = |Z| \angle \phi$$
(4.6)

The absolute impedance $|Z|$ and phase angle ϕ are determined by

$$|Z| = \sqrt{R^2 + \left(\frac{1}{\omega C}\right)^2}$$
(4.7)

$$\phi = -\tan^{-1}\frac{1}{\omega CR}$$

We define the circuit time constant τ as

$$\tau = RC$$
(4.8)

4.1.1.1 Summary of the Symbols

Symbol   Explanation (Measuring Unit)
C Capacitance (F)
F Frequency (Hz)

i, I	Instantaneous current, average/rms current (A)
L	Inductance (H)
R	Resistance (Ω)
p, P	Instantaneous power, rated/real power (W)
q, Q	Instantaneous reactive power, rated reactive power (VAR)
s, S	Instantaneous apparent power, rated apparent power (VA)
v, V	Instantaneous voltage, average/rms voltage (V)
Z	Impedance (Ω)
ϕ	Phase angle (degree or radian)
η	Efficiency (%)
τ	Time constant (s)
ω	Angular frequency (radian/s), $\omega = 2\pi f$

4.1.2 FACTORS AND SYMBOLS USED IN AC POWER SYSTEMS

The input AC voltage can be single-phase or three-phase voltages. They are usually pure sinusoidal wave function. For a single-phase input voltage $v(t)$, it can be expressed as [4]

$$v(t) = \sqrt{2}V \sin \omega t = V_m \sin \omega t \tag{4.9}$$

where
   v is the instantaneous input voltage
   V is its rms value
   V_m is its amplitude
   ω is the angular frequency, $\omega = 2\pi f$
   f is the supply frequency

Usually, the input current may not be a pure sinusoidal wave depending on the load. If the input voltage supplies a linear load (resistive, inductive, capacitive loads or their combination), the input current $i(t)$ is not distorted but may be delayed in a phase angle ϕ. In this case, it can be expressed as

$$i(t) = \sqrt{2}I \sin(\omega t - \phi) = I_m \sin(\omega t - \phi) \tag{4.10}$$

where
   i is the instantaneous input current
   I is its rms value
   I_m is its amplitude
   ϕ is the phase-delay angle

We define the power factor (PF) as

$$PF = \cos \phi \tag{4.11}$$

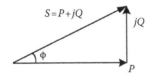

FIGURE 4.3   Power vector diagram.

PF is the ratio of the real power (P) over the apparent power (S). We have the relation $S = P + jQ$, where Q is the reactive power. The power vector diagram is shown in Figure 4.3. We have the relations between the powers:

$$S = VI^* = \frac{V^2}{Z^*} = P + jQ = |S| \angle \phi \qquad (4.12)$$

$$|S| = \sqrt{P^2 + Q^2} \qquad (4.13)$$

$$\phi = \tan^{-1} \frac{Q}{P} \qquad (4.14)$$

$$P = S \cos \phi \qquad (4.15)$$

$$Q = S \sin \phi \qquad (4.16)$$

If the input current is distorted, it consists of harmonics. Its fundamental harmonic can be expressed as

$$i_1 = \sqrt{2} I_1 \sin(\omega t - \phi_1) = I_{m1} \sin(\omega t - \phi_1) \qquad (4.17)$$

where
   i_1 is the fundamental harmonic instantaneous value
   I_1 is its rms value
   I_{m1} is its amplitude
   ϕ_1 is its phase-angle

In this case, the displacement power factor (DPF) is defined

$$DPF = \cos \phi_1 \qquad (4.18)$$

Correspondingly, the PF is defined as

$$PF = \frac{DPF}{\sqrt{1 + THD^2}} \qquad (4.19)$$

where THD is the total harmonic distortion. It can be used to measure both voltage and current waveforms. It is defined as

$$\text{THD} = \frac{\sqrt{\sum_{n=2}^{\infty} I_n^2}}{I_1} \quad \text{or} \quad \text{THD} = \frac{\sqrt{\sum_{n=2}^{\infty} V_n^2}}{V_1} \tag{4.20}$$

where I_n or V_n is the amplitude of the nth-order harmonic.

Harmonic factor (HF) is a variable to describe the weighted percent of the nth order harmonic referring to the amplitude of the fundamental harmonic V_1. It is defined as

$$\text{HF}_n = \frac{I_n}{I_1} \quad \text{or} \quad \text{HF}_n = \frac{V_n}{V_1} \tag{4.21}$$

$n=1$ corresponds to the fundamental harmonic. Therefore, $\text{HF}_1 = 1$. Total harmonic distortion (THD) can be written as

$$\text{THD} = \sqrt{\sum_{n=2}^{\infty} \text{HF}_n^2} \tag{4.22}$$

A pure sinusoidal waveform has THD=0.

Weighted total harmonic distortion (WTHD) is a variable to describe the waveform distortion. It is defined as

$$\text{WTHD} = \frac{\sqrt{\sum_{n=2}^{\infty} V_n^2/n}}{V_1} \tag{4.23}$$

Note that THD gives an immediate measure of the inverter output voltage waveform distortion. WTHD is often interpreted as the normalized current ripple expected into an inductive load when fed from the inverter output voltage.

4.1.2.1 Summary of the Symbols

Symbol   Explanation (Measuring Unit)
DPF   Displacement power factor (%)
HF_n nth-order harmonic factor
i_1, I_1 Instantaneous fundamental current, average/rms fundamental current (A)
i_n, I_n Instantaneous nth-order harmonic current, average/rms nth-order harmonic current (A)
I_m Current amplitude (A)
PF   Power factor (leading/lagging, %)
q, Q Instantaneous reactive power, rated reactive power (VAR)

s, S Instantaneous apparent power, rated apparent power (VA)
t Time (s)
THD   Total harmonic distortion (%)
v_1, V_1 Instantaneous fundamental voltage, average/rms
              fundamental voltage (V)
v_n, V_n Instantaneous nth-order harmonic voltage, average/rms nth-order
              harmonic voltage (V)
WTHD   Weighted total harmonic distortion (%)
ϕ_1 Phase angle of the fundamental harmonic (degree, or radian)

4.1.3 FACTORS AND SYMBOLS USED IN DC POWER SYSTEMS

We define the output DC voltage instantaneous value to be v_d and average value to be V_d (or V_{d0}) [5]. A pure DC voltage has no ripple; it then is called ripple-free DC voltage. Otherwise, a DC voltage is distorted, and it consists of DC component and AC harmonics. Its rms value is $V_{d\text{-}rms}$. For a distorted DC voltage, its rms value $V_{d\text{-}rms}$ is constantly higher than its average value V_d. The ripple factor (RF) is defined as

$$RF = \frac{\sqrt{\sum_{n=1}^{\infty} V_n^2}}{V_d} \tag{4.24}$$

where V_n is the nth-order harmonic. The form factor (FF) is defined as

$$FF = \frac{V_{d\text{-}rms}}{V_d} = \frac{\sqrt{\sum_{n=0}^{\infty} V_n^2}}{V_d} \tag{4.25}$$

where V_0 is the 0th-order harmonic, that is, the average component V_d. Therefore, we obtain FF > 1, and the relation

$$RF = \sqrt{FF^2 - 1} \tag{4.26}$$

FF and RF are used to describe the quality of a DC waveform (voltage and current parameters). For a pure DC voltage, FF = 1 and RF = 0.

4.1.3.1 Summary of the Symbols

Symbol   Explanation (Measuring Unit)
FF   Form factor (%)
RF   Ripple factor (%)
v_d, V_d Instantaneous DC voltage, average DC voltage (V)
$V_{d\text{-}rms}$ rms DC voltage (V)
v_n, V_n Instantaneous nth-order harmonic voltage, average/rms nth-order
              harmonic voltage (V)

4.1.4 FACTORS AND SYMBOLS USED IN SWITCHING POWER SYSTEMS

Switching power systems such as power DC/DC converters, power PWM DC/AC inverters, soft-switching converters, and resonant converters are widely used in power transfer equipment. In general, a switching power system has a pumping circuit and several energy-storage elements. It is likely an energy container to store certain energy during performance. The input energy is not smoothly flowing through the switching power system from input source to the load. The energy is quantified by the switching circuit and then pumped though the switching power system from input source to the load [6–8].

We assume the switching frequency is f and the corresponding period is $T = 1/f$. The pumping energy (PE) is used to count the input energy in a switching period T. Its calculation formula is

$$\text{PE} = \int_0^T P_{in}(t)dt = \int_0^T V_{in}i_{in}(t)dt = V_{in}I_{in}T \qquad (4.27)$$

where

$$I_{in} = \int_0^T i_{in}(t)dt \qquad (4.28)$$

is the average value of the input current if the input voltage V_1 is constant. Usually the input average current I_1 depends on the conduction duty cycle.

Energy storage in switching power systems has been paid attention long time ago. Unfortunately, there is no clear concept to describe the phenomena and reveal the relationship between the stored energy (SE) and the characteristics.
The SE in an inductor is

$$W_L = \frac{1}{2}LI_L^2 \qquad (4.29)$$

The SE across a capacitor is

$$W_C = \frac{1}{2}CV_C^2 \qquad (4.30)$$

Therefore, if there are n_L inductors and n_C capacitors the total SE in a DC/DC converter is

$$\text{SE} = \sum_{j=1}^{n_L} W_{Lj} + \sum_{j=1}^{n_C} W_{Cj} \qquad (4.31)$$

Usually, the SE is independent from the switching frequency f (as well as the switching period T). Since the inductor currents and the capacitor voltages rely on the conduction duty cycle k, the SE also relies on the conduction duty cycle k. We use the SE as a new parameter in further description.

Most switching power systems consist of inductors and capacitors. Therefore, we can define the capacitor–inductor stored energy ratio (CIR).

$$CIR = \frac{\sum_{j=1}^{n_C} W_{Cj}}{\sum_{j=1}^{n_L} W_{Lj}} \tag{4.32}$$

As described in previous parts, the input energy in a period T is $PE = P_{in} \times T = V_{in} I_{in} \times T$. We now define the energy factor (EF), which is the ratio of the SE over the PE:

$$EF = \frac{SE}{PE} = \frac{SE}{V_{in} I_{in} T} = \frac{\sum_{j=1}^{m} W_{Lj} + \sum_{j=1}^{n} W_{Cj}}{V_{in} I_{in} T} \tag{4.33}$$

EF is a very important factor of a switching power system. It is usually independent from the conduction duty cycle and inversely proportional to the switching frequency f since the PE is proportional to the switching period T.

The *time constant* τ of a switching power system is a new concept to describe the transient process. If there are no power losses in the system, it is defined as

$$\tau = \frac{2T \times EF}{1 + CIR} \tag{4.34}$$

This time constant τ is independent from the switching frequency f (or period $T = 1/f$). It is available to estimate the system responses for a unit-step function and impulse interference.

If there are power losses and $\eta < 1$, it is defined as

$$\tau = \frac{2T \times EF}{1 + CIR}\left(1 + CIR\frac{1-\eta}{\eta}\right) \tag{4.35}$$

If there is no power loss, $\eta = 1$, Equation 4.35 becomes (4.34). Usually, if the power losses (the lower efficiency η) are higher, the time constant τ is larger since $CIR > 1$.

The *damping time constant* τ_d of a switching power system is a new concept to describe the transient process. If there are no power losses, it is defined as

$$\tau_d = \frac{2T \times EF}{1 + CIR} CIR \tag{4.36}$$

This damping time constant τ_d is independent from switching frequency f (or period $T = 1/f$). It is available to estimate the oscillation responses for a unit-step function and impulse interference.

If there are power losses and $\eta < 1$, it is defined as

$$\tau_d = \frac{2T \times EF}{1 + CIR} \frac{CIR}{\eta + CIR(1 - \eta)} \tag{4.37}$$

If there is no power loss, $\eta = 1$, Equation 4.37 becomes (4.36). Usually, if the power losses (the lower efficiency η) are higher, the damping time constant τ_d is smaller since $CIR > 1$.

The *time constants ratio* ξ of a switching power system is a new concept to describe the transient process. If there are no power losses, it is defined as

$$\xi = \frac{\tau_d}{\tau} = CIR \tag{4.38}$$

This time constant ratio is independent from the switching frequency f (or period $T = 1/f$). It is available to estimate the oscillation responses for a unit-step function and impulse interference.

If there are power losses and $\eta < 1$, it is defined as

$$\xi = \frac{\tau_d}{\tau} = \frac{CIR}{\eta(1 + CIR(1 - \eta/\eta))^2} \tag{4.39}$$

If there is no power loss, $\eta = 1$, Equation 4.39 becomes (4.38). Usually, if the power losses (the lower efficiency η) are higher, the time constant ratio ξ is smaller since $CIR > 1$. From this analysis, most switching power systems with lower power losses possess larger output voltage oscillation when the converter operation state changes. On the contrary, switching power systems with high power losses will possess smoothening output voltage when the converter operation state changes.

By cybernetic theory, we can estimate the unit-step function response using the ratio ξ. If the ratio ξ is equal to or smaller than 0.25, the corresponding unit-step function response has no oscillation and overshot. Alternatively, if the ratio ξ is greater than 0.25 the corresponding unit-step function response has oscillation and overshot. The high the value of ratio ξ, the heavier the oscillation with higher overshot.

4.1.4.1 Summary of the Symbols

Symbol	Explanation (Measuring Unit)
CIR	Capacitor–inductor stored energy ratio
EF	Energy factor
f	Switching frequency (Hz)
k	Conduction duty cycle
PE	Pumping energy (J)
SE	Total stored energy (J)

W_L, W_C	Stored energy in an inductor/capacitor (J)
T	Switching period (s)
τ	Time constant (s)
τ_d	Damping time constant (s)
ξ	Time constants ratio

4.1.5 OTHER FACTORS AND SYMBOLS

Transfer function is the mathematical modeling of a circuit and system. It describes the dynamic characteristics of the circuit and system. Using transfer function, we can easily obtain the system step and impulse responses applying an input signal. A typical second-order transfer function is [6–8]

$$G(s) = \frac{M}{1 + s\tau + s^2 \tau\tau_d} = \frac{M}{1 + s\tau + s^2 \xi\tau^2} \tag{4.40}$$

where
  M is the voltage transfer gain: $M = V_O/V_{in}$
  τ is the time constant in (4.35)
  τ_d is the damping time constant in (4.37), $\tau_d = \xi\tau$ in (4.39)
  s is the Laplace operator in the s-domain

Using this mathematical model of a switching power system, it is significantly easy to describe its characteristics. In order to understand more characteristics of the transfer function, a few situations are analyzed in the following.

4.1.5.1 Very Small Damping Time Constant

If the damping time constant is very small (i.e., $\tau_d \ll \tau$, $\xi \ll 1$) and it can be ignored, the value of the damping time constant τ_d is omitted (i.e., $\tau_d = 0$, $\xi = 0$). The transfer function (4.40) is downgraded to the first order as

$$G(s) = \frac{M}{1 + s\tau} \tag{4.41}$$

The unit-step function response in the time domain is

$$g(t) = M(1 - e^{-t/\tau}) \tag{4.42}$$

The transient process (settling time) is nearly three times that of the time constant, 3τ, to produce $g(t) = g(3\tau) = 0.95\ M$. The response in time domain is shown in Figure 4.4 with $\tau_d = 0$.
  The impulse interference response is

$$\Delta g(t) = U \cdot e^{-t/\tau} \tag{4.43}$$

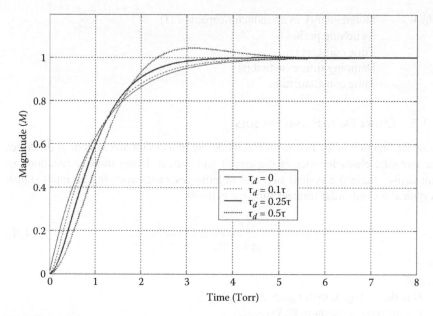

FIGURE 4.4   Unit-step function responses ($\tau_d=0$, 0.1τ, 0.25τ, and 0.5τ).

where U is the interference signal. The interference recovering progress is nearly three times that of the time constant, 3τ, and shown in Figure 4.5 with $\tau_d=0$.

4.1.5.2   Small Damping Time Constant

If the damping time constant is small (i.e., $\tau_d < \tau/4$, $\xi < 0.25$) and it cannot be ignored, the value of the damping time constant τ_d is not omitted. The transfer function (4.40) is retained in the second-order function with two real poles $-\sigma_1$ and $-\sigma_2$ as

$$G(s) = \frac{M}{1 + s\tau + s^2\tau\tau_d} = \frac{M/\tau\tau_d}{(s+\sigma_1)(s+\sigma_2)} \qquad (4.44)$$

where

$$\sigma_1 = \frac{\tau + \sqrt{\tau^2 - 4\tau\tau_d}}{2\tau\tau_d}$$

$$\sigma_2 = \frac{\tau - \sqrt{\tau^2 - 4\tau\tau_d}}{2\tau\tau_d}$$

There are two real poles in the transfer function, and we assume $\sigma_1 > \sigma_2$. The unit-step function response in the time domain is

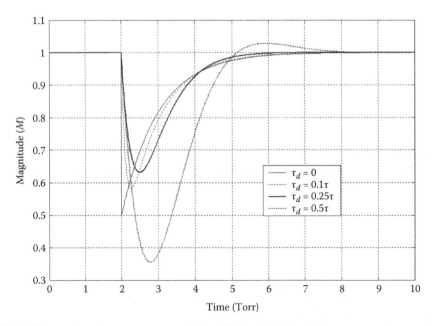

FIGURE 4.5 Impulse responses (τ_d=0, 0.1τ, 0.25τ, and 0.5τ).

$$g(t) = M\left(1 + K_1 e^{-\sigma_1 t} + K_2 e^{-\sigma_2 t}\right) \tag{4.45}$$

where

$$K_1 = -\frac{1}{2} + \frac{\tau}{2\sqrt{\tau^2 - 4\tau\tau_d}}$$

$$K_2 = -\frac{1}{2} - \frac{\tau}{2\sqrt{\tau^2 - 4\tau\tau_d}}$$

The transient process is nearly three times that of the time value $1/\sigma_1$, $3/\sigma_1 < 3\tau$. The response process is quick without oscillation. The corresponding waveform in time domain is shown in Figure 4.4 with τ_d=0.1τ.

The impulse interference response is

$$\Delta g(t) = \frac{U}{\sqrt{1 - 4\tau_d/\tau}}\left(e^{-\sigma_2 t} - e^{-\sigma_1 t}\right) \tag{4.46}$$

where U is the interference signal. The transient process is nearly three times that of the time value $1/\sigma_1$, $3/\sigma_1 < 3\tau$. The response waveform in time domain is shown in Figure 4.5 with τ_d=0.1τ.

4.1.5.3 Critical Damping Time Constant

If the damping time constant is equal to the critical value (i.e., $\tau_d = \tau/4$), the transfer function (4.40) is retained as the second-order function with two equal real poles $\sigma_1 = \sigma_2 = \sigma$ as

$$G(s) = \frac{M}{1 + s\tau + s^2 \tau \tau_d} = \frac{M / \tau \tau_d}{(s + \sigma)^2} \tag{4.47}$$

where $\sigma = 1/2\tau_d = 2/\tau$.

There are two folded real poles in the transfer function. This expression describes the characteristics of the DC/DC converter. The unit-step function response in the time domain is

$$g(t) = M\left[1 - \left(1 + \frac{2t}{\tau}\right)e^{-2t/\tau}\right] \tag{4.48}$$

The transient process is nearly 2.4 times that of the time constant τ, 2.4τ. The response process is quick without oscillation. The response waveform in time domain is shown in Figure 4.4 with $\tau_d = 0.25\tau$.

The impulse interference response is

$$\Delta g(t) = \frac{4U}{\tau} t e^{-2t/\tau} \tag{4.49}$$

where U is the interference signal. The transient process is still nearly 2.4 times that of the time constant, 2.4τ. The response waveform in time domain is shown in Figure 4.5 with $\tau_d = 0.25\tau$.

4.1.5.4 Large Damping Time Constant

If the damping time constant is large (i.e., $\tau_d > \tau/4$, $\xi > 0.25$), the transfer function (4.40) is a second-order function with a couple of conjugated complex poles $-s_1$ and $-s_2$ in the left-hand half plane (LHHP) in s-domain

$$G(s) = \frac{M}{1 + s\tau + s^2 \tau \tau_d} = \frac{M / \tau \tau_d}{(s + s_1)(s + s_2)} \tag{4.50}$$

where

$$s_1 = \sigma + j\omega$$

$$s_2 = \sigma - j\omega$$

$$\sigma = \frac{1}{2\tau_d}$$

$$\omega = \frac{\sqrt{4\tau\tau_d - \tau^2}}{2\tau\tau_d}$$

There are a couple of conjugated complex poles $-s_1$ and $-s_2$ in the transfer function. This expression describes the characteristics of the DC/DC converter. The unit-step function response in the time domain is

$$g(t) = M\left[1 - e^{-t/2\tau_d}\left(\cos \omega t - \frac{1}{\sqrt{4\tau_d/\tau - 1}}\sin \omega t\right)\right] \qquad (4.51)$$

The transient response has oscillation progress with damping factor σ and frequency ω. The corresponding waveform in time domain is shown in Figure 4.4 with $\tau_d = 0.5\tau$ and in Figure 4.6 with τ, 2τ, 5τ, and 10τ.

The impulse interference response is

$$\Delta g(t) = \frac{U}{\sqrt{(\tau_d/\tau) - (1/4)}}e^{-t/2\tau_d}\sin(\omega t) \qquad (4.52)$$

where U is the interference signal. The recovery process is a curve with damping factor σ and frequency ω. The response waveform in time domain is shown in Figure 4.5 with $\tau_d = 0.5\tau$ and in Figure 4.7 with τ, 2τ, 5τ, and 10τ.

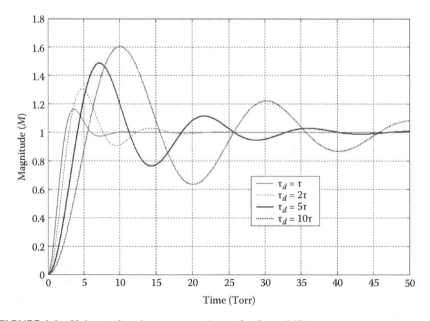

FIGURE 4.6 Unit-step function responses ($\tau_d = \tau$, 2τ, 5τ, and 10τ).

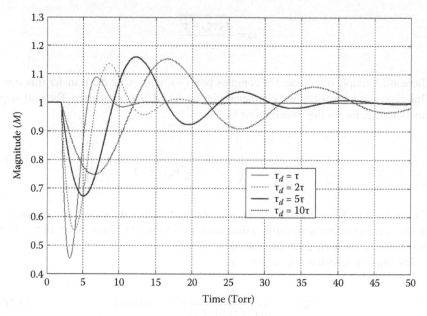

FIGURE 4.7   Impulse responses ($\tau_d = \tau$, 2τ, 5τ, and 10τ).

4.1.6   FAST FOURIER TRANSFORM

Fast Fourier Transform (FFT) [9] is a very versatile method to analyze the waveforms. A periodical function with the radian frequency ω can be represented by a series of sinusoidal functions:

$$f(t) = \frac{a_0}{2} + \sum_{n=1}^{\infty} (a_n \cos n\omega t + b_n \sin n\omega t) \tag{4.53}$$

where the Fourier coefficients are

$$a_n = \frac{1}{\pi} \int_0^{2\pi} f(t)\cos(n\omega t)d(\omega t) \quad n = 0,1,2,\dots,\infty \tag{4.54}$$

$$b_n = \frac{1}{\pi} \int_0^{2\pi} f(t)\sin(n\omega t)d(\omega t) \quad n = 1,2,\dots,\infty \tag{4.55}$$

In this case, we call the items with the radian frequency ω the fundamental harmonic and the items with the radian frequency $n\omega$ ($n > 1$) higher-order harmonics. Draw the amplitudes of all harmonics in the frequency domain. We can get the spectrum in an individual peak. The item $a_0/2$ is the DC component.

4.1.6.1   Central Symmetrical Periodical Function

If the periodical function is a central symmetrical periodical function, all items with cosine function disappear. The FFT remains as

$$f(t) = \sum_{n=1}^{\infty} b_n \sin n\omega t \qquad (4.56)$$

where

$$b_n = \frac{1}{\pi} \int_0^{2\pi} f(t)\sin(n\omega t)d(\omega t) \quad n = 1, 2, \ldots, \infty \qquad (4.57)$$

We usually call this function odd function. In this case, we call the item with the radian frequency ω the fundamental harmonic, and the items with the radian frequency $n\omega$ ($n > 1$) higher-order harmonics. Draw the amplitudes of all harmonics in the frequency domain. We can get the spectrum in individual peaks. Since it is an odd function, the DC component is zero.

4.1.6.2   Axial (Mirror) Symmetrical Periodical Function

If the periodical function is an axial symmetrical periodical function, all items with sine function disappear. The FFT remains as

$$f(t) = \frac{a_0}{2} + \sum_{n=1}^{\infty} a_n \cos n\omega t \qquad (4.58)$$

where $a_0/2$ is the DC component and

$$a_n = \frac{1}{\pi} \int_0^{2\pi} f(t)\cos(n\omega t)d(\omega t) \quad n = 0, 1, 2, \ldots, \infty \qquad (4.59)$$

The item $a_0/2$ is the DC component. We usually call this function even function. In this case, we call the item with the radian frequency ω the fundamental harmonic, and the items with the radian frequency $n\omega$ ($n > 1$) higher-order harmonics. Draw the amplitudes of all harmonics in the frequency domain. We can get the spectrum in individual peaks. Since it is an even function, the DC component is usually not zero.

4.1.6.3   Nonperiodical Function

The spectrum of a periodical function in the time domain is a discrete function in the frequency domain. If a function is a nonperiodical function in the time domain, it is possibly represented by Fourier Integration. The spectrum is a continuous function in the frequency domain.

4.1.6.4   Useful Formulae and Data

Some trigonometric formulae are useful for FFT:

$$\sin^2 x + \cos^2 x = 1 \qquad\qquad \sin x = \cos\left(\frac{\pi}{2} - x\right)$$

$$\sin x = -\sin(-x) \qquad\qquad \sin x = \sin(\pi - x)$$
$$\cos x = \cos(-x) \qquad\qquad \cos x = -\cos(\pi - x)$$

$$\frac{d}{dx}\sin x = \cos x \qquad\qquad \frac{d}{dx}\cos x = -\sin x$$

$$\int \sin x\, dx = -\cos x \qquad\qquad \int \cos x\, dx = \sin x$$

$$\sin(x \pm y) = \sin x \cos y \pm \cos x \sin y$$
$$\cos(x \pm y) = \cos x \cos y \pm \sin x \sin y$$
$$\sin 2x = 2 \sin x \cos x$$
$$\cos 2x = \cos^2 x - \sin^2 x$$

Some values corresponding to the special angles are usually used:

$$\sin\frac{\pi}{12} = \sin 15° = 0.2588 \qquad\qquad \cos\frac{\pi}{12} = \cos 15° = 0.9659$$

$$\sin\frac{\pi}{8} = \sin 22.5° = 0.3827 \qquad\qquad \cos\frac{\pi}{8} = \cos 22.5° = 0.9239$$

$$\sin\frac{\pi}{6} = \sin 30° = 0.5 \qquad\qquad \cos\frac{\pi}{6} = \cos 30° = \frac{\sqrt{3}}{2} = 0.866$$

$$\sin\frac{\pi}{4} = \sin 45° = \frac{\sqrt{2}}{2} = 0.7071 \qquad\qquad \cos\frac{\pi}{4} = \cos 45° = \frac{\sqrt{2}}{2} = 0.7071$$

$$\tan\frac{\pi}{12} = \tan 15° = 0.2679 \qquad\qquad \tan\frac{\pi}{8} = \tan 22.5° = 0.4142$$

$$\tan\frac{\pi}{6} = \tan 30° = \frac{\sqrt{3}}{3} = 0.5774 \qquad\qquad \tan\frac{\pi}{4} = \tan 45° = 1$$

$$\tan x = \frac{1}{\text{co-}\tan x} \qquad\qquad \tan x = \text{co-}\tan\left(\frac{\pi}{2} - x\right)$$

4.1.6.5   Examples of FFT Applications

**Example 4.1 An odd-square waveform is shown in Figure 4.8.
Find the FFT, HF up to seventh order, THD, and WTHD.**

Solution: The function $f(t)$ is

$$f(t) = \begin{cases} 1 & 2n\pi \le \omega t < (2n+1)\pi \\ -1 & (2n+1)\pi \le \omega t < 2(n+1)\pi \end{cases} \qquad (4.60)$$

The Fourier coefficients are

$$b_n = \frac{1}{\pi}\int_0^{2\pi} f(t)\sin(n\omega t)d(\omega t) = \frac{2}{n\pi}\int_0^{n\pi}\sin\theta d\theta = 2\frac{1-(-1)^n}{n\pi}$$

or

$$b_n = \frac{4}{n\pi} \quad n = 1,3,5,\ldots,\infty \qquad (4.61)$$

Finally, we obtain

$$F(t) = \frac{4}{\pi}\sum_{n=1}^{\infty}\frac{\sin(n\omega t)}{n} \quad n = 1,3,5,\ldots,\infty \qquad (4.62)$$

The fundamental harmonic has the amplitude $4/\pi$. If we consider the higher-order harmonics until seventh order, that is, $n = 3, 5, 7$, the HFs are

$$HF_3 = \frac{1}{3}; \quad HF_5 = \frac{1}{5}; \quad HF_7 = \frac{1}{7}$$

The THD is

$$THD = \frac{\sqrt{\sum_{n=2}^{\infty} V_n^2}}{V_1} = \sqrt{\left(\frac{1}{3}\right)^2 + \left(\frac{1}{5}\right)^2 + \left(\frac{1}{7}\right)^2} = 0.41415 \qquad (4.63)$$

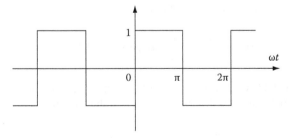

FIGURE 4.8   A waveform.

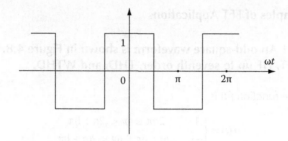

FIGURE 4.9   An even-square waveform.

The WTHD is

$$\text{WTHD} = \frac{\sqrt{\sum_{n=2}^{\infty} V_n^2/n}}{V_1} = \sqrt{\left(\frac{1}{3}\right)^3 + \left(\frac{1}{5}\right)^3 + \left(\frac{1}{7}\right)^3} = 0.219 \qquad (4.64)$$

Example 4.2 An even-square waveform is shown in Figure 4.9. Find the FFT, HF up to seventh order, THD, and WTHD.

The function $f(t)$ is

$$f(t) = \begin{cases} 1 & (2n-0.5)\pi \le \omega t < (2n+0.5)\pi \\ -1 & (2n+0.5)\pi \le \omega t < (2n+1.5)\pi \end{cases} \qquad (4.65)$$

The Fourier coefficients are

$$a_0 = 0$$

$$a_n = \frac{1}{\pi} \int_0^{2\pi} f(t)\cos(n\omega t)d(\omega t) = \frac{4}{n\pi} \int_0^{n\pi/2} \cos\theta d\theta = \frac{4\sin(n\pi/2)}{n\pi}$$

or

$$a_n = \frac{4}{n\pi}\sin\frac{n\pi}{2} \quad n = 1,3,5,\ldots,\infty \qquad (4.66)$$

The item $\sin n\pi/2$ is used to define the sign. Finally, we obtain

$$F(t) = \frac{4}{\pi}\sum_{n=1}^{\infty}\sin\frac{n\pi}{2}\cos(n\omega t) \quad n = 1,3,5,\ldots,\infty \qquad (4.67)$$

The fundamental harmonic has the amplitude $4/\pi$. If we consider the higher-order harmonics until seventh order, that is, $n = 3, 5, 7$, the HFs are

$$HF_3 = \frac{1}{3}; \quad HF_5 = \frac{1}{5}; \quad HF_7 = \frac{1}{7}$$

The THD is

$$\text{THD} = \frac{\sqrt{\sum_{n=2}^{\infty} V_n^2}}{V_1} = \sqrt{\left(\frac{1}{3}\right)^2 + \left(\frac{1}{5}\right)^2 + \left(\frac{1}{7}\right)^2} = 0.41415 \qquad (4.68)$$

The WTHD is

$$\text{WTHD} = \frac{\sqrt{\sum_{n=2}^{\infty} V_n^2/n}}{V_1} = \sqrt{\left(\frac{1}{3}\right)^3 + \left(\frac{1}{5}\right)^3 + \left(\frac{1}{7}\right)^3} = 0.219 \qquad (4.69)$$

Example 4.3 An odd-waveform pulse with pulse-width x is shown in Figure 4.10. Find the FFT, HF up to seventh order, THD, and WTHD.

The function $f(t)$ is in the period $-\pi$ to $+\pi$:

$$f(t) = \begin{cases} 1 & \dfrac{\pi - x}{2} \le \omega t < \dfrac{\pi + x}{2} \\ -1 & -\dfrac{\pi + x}{2} \le \omega t < -\dfrac{\pi - x}{2} \end{cases} \qquad (4.70)$$

The Fourier coefficients are

$$b_n = \frac{1}{\pi} \int_0^{2\pi} f(t)\sin(n\omega t)d(\omega t)$$

$$= \frac{2}{n\pi} \int_{n((\pi-x)/2)}^{n((\pi+x)/2)} \sin\theta\, d\theta = 2\frac{\cos\left(n((\pi - x)/2)\right) - \cos\left(n((\pi + x)/2)\right)}{n\pi}$$

$$= 2\frac{2\cos\left(n((\pi - x)/2)\right)}{n\pi} = \frac{4\sin(n\pi/2)\sin(nx/2)}{n\pi}$$

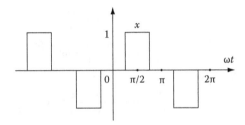

FIGURE 4.10 An odd-waveform pulse.

or

$$b_n = \frac{4}{n\pi}\sin\frac{n\pi}{2}\sin\frac{nx}{2} \quad n = 1,3,5,\ldots,\infty \tag{4.71}$$

Finally, we obtain

$$F(t) = \frac{4}{\pi}\sum_{n=1}^{\infty}\frac{\sin(n\omega t)}{n}\sin\frac{n\pi}{2}\sin\frac{nx}{2} \quad n = 1,3,5,\ldots,\infty \tag{4.72}$$

The fundamental harmonic has the amplitude $(4/\pi)\sin(x/2)$. If we consider the higher-order harmonics until seventh order, that is, $n = 3, 5, 7$, the HFs are

$$HF_3 = \frac{\sin(3x/2)}{3\sin(x/2)}; \quad HF_5 = \frac{\sin(5x/2)}{5\sin(x/2)}; \quad HF_7 = \frac{\sin(7x/2)}{7\sin(x/2)}$$

The values of the HFs should be absolute values.
   If $x = \pi$, the THD is

$$THD = \frac{\sqrt{\sum_{n=2}^{\infty}V_n^2}}{V_1} = \sqrt{\left(\frac{1}{3}\right)^2 + \left(\frac{1}{5}\right)^2 + \left(\frac{1}{7}\right)^2} = 0.41415 \tag{4.73}$$

The WTHD is

$$WTHD = \frac{\sqrt{\sum_{n=2}^{\infty}V_n^2/n}}{V_1} = \sqrt{\left(\frac{1}{3}\right)^3 + \left(\frac{1}{5}\right)^3 + \left(\frac{1}{7}\right)^3} = 0.219 \tag{4.74}$$

Example 4.4 A five-level odd waveform is shown in Figure 4.11. Find the FFT, HF up to seventh order, THD, and WTHD.

The function $f(t)$ is in the period $-\pi$ to $+\pi$:

$$f(t) = \begin{cases} 2 & \frac{\pi}{3} \leq \omega t < \frac{2\pi}{3} \\ 1 & \frac{\pi}{6} \leq \omega t < \frac{\pi}{3}, \frac{2\pi}{3} \leq \omega t < \frac{5\pi}{6} \\ 0 & \text{other} \\ -1 & -\frac{5\pi}{6} \leq \omega t < -\frac{2\pi}{3}, -\frac{\pi}{3} \leq \omega t < -\frac{\pi}{6} \\ -2 & -\frac{2\pi}{3} \leq \omega t < -\frac{\pi}{3} \end{cases} \tag{4.75}$$

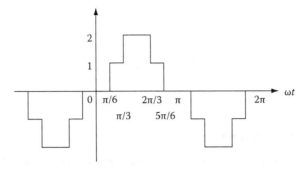

FIGURE 4.11 A five-level odd waveform.

The Fourier coefficients are

$$b_n = \frac{1}{\pi}\int_0^{2\pi} f(t)\sin(n\omega t)d(\omega t) = \frac{2}{n\pi}\left[\int_{n\pi/6}^{5n\pi/6}\sin\theta d\theta + \int_{n\pi/3}^{2n\pi/3}\sin\theta d\theta\right]$$

$$= \frac{2}{n\pi}\left[\left(\cos\frac{n\pi}{6} - \cos\frac{5n\pi}{6}\right) + \left(\cos\frac{n\pi}{3} - \cos\frac{2n\pi}{3}\right)\right] = \frac{4}{n\pi}\left(\cos\frac{n\pi}{6} + \cos\frac{n\pi}{3}\right)$$

or

$$b_n = \frac{4}{n\pi}\left(\cos\frac{n\pi}{6} + \cos\frac{n\pi}{3}\right) \quad n = 1,3,5,\ldots,\infty \qquad (4.76)$$

Finally, we obtain

$$F(t) = \frac{4}{\pi}\sum_{n=1}^{\infty}\frac{\sin(n\omega t)}{n}\left(\cos\frac{n\pi}{6} + \cos\frac{n\pi}{3}\right) \quad n = 1,3,5,\ldots,\infty \qquad (4.77)$$

The fundamental harmonic has the amplitude $(2/\pi)(1+\sqrt{3})$. If we consider the higher-order harmonics until seventh order, that is, $n = 3, 5, 7$, the HFs are

$$HF_3 = \frac{2}{3(1+\sqrt{3})} = 0.244; \quad HF_5 = \frac{\sqrt{3}-1}{5(1+\sqrt{3})} = 0.0536; \quad HF_7 = \frac{\sqrt{3}-1}{7(1+\sqrt{3})} = 0.0383$$

The values of the HFs should be absolute values.
 The THD is

$$THD = \frac{\sqrt{\sum_{n=2}^{\infty}V_n^2}}{V_1} = \sqrt{\sum_{n=2}^{\infty}HF_n^2} = \sqrt{0.244^2 + 0.0536^2 + 0.0383^2} = 0.2527 \quad (4.78)$$

The WTHD is

$$\text{WTHD} = \frac{\sqrt{\sum_{n=2}^{\infty} V_n^2/n}}{V_1} = \sqrt{\sum_{n=2}^{\infty} \frac{\text{HF}_n^2}{n}} = \sqrt{\frac{0.244^2}{3} + \frac{0.0536^2}{5} + \frac{0.0383^2}{7}} = 0.1436$$

(4.79)

4.2 AC/DC RECTIFIERS

AC/DC rectifiers [3] have been used in industrial applications for a long time. Before the 1960s most power AC/DC rectifiers were constructed by mercury-arc rectifiers. The large power silicon diode and thyristor (or SCR—silicon-controlled rectifier) were successfully developed in the 1960s. All power AC/DC rectifiers are constructed by power silicon diode and thyristor.

Using power silicon diode, we can construct uncontrolled diode rectifiers. Using power thyristor, we can construct controlled SCR rectifiers since a thyristor is usually triggered at firing angle α, which is variable. If the firing angle $\alpha=0$, the characteristics of the controlled SCR rectifier will be returned to those of the uncontrolled diode rectifier. Investigation of the characteristics of the uncontrolled diode rectifier can help designers to understand the characteristics of the controlled SCR rectifier.

A single-phase half-wave diode rectifier is shown in Figure 4.12. The load can be resistive load, inductive load, capacitive load, or back electromotive force (emf) load. The diode can be conducted when the current flows from its anode to cathode, and the corresponding voltage applied across the diode is defined positive. On the contrary, the diode is blocked when the voltage applied across the diode is negative, and no current flows through it. Therefore, the single-phase half-wave diode rectifier supplying a deferent load has a deferent output voltage waveform.

Three important points have to be emphasized in this book:

1. Clearing historic problems
2. Introducing updated circuits
3. Investigating PFC methods

4.2.1 HISTORIC PROBLEMS

Rectifier circuits are easily understood. The input power supply can be single-phase, three-phase, and multiphase sine-wave voltages. Usually, the more the phases input

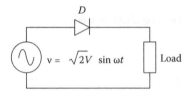

FIGURE 4.12   A single-phase half-wave diode rectifier.

power supply circuits, the simpler the circuit operation. The most difficult analysis is in the simplest circuit. Single-phase diode rectifier circuit is the simplest circuit, but it has not been deeply analyzed. Results in many papers and books recently published offer faulty ideas, for example, a single-phase diode rectifier supplies an RC circuit.

4.2.2 UPDATED CIRCUITS

Many update circuits and control methods have been developed in recent decades, especially in this century. Unfortunately, most updated circuits and control methods have not been discussed in the books recently published.

4.2.3 POWER FACTOR CORRECTION METHODS

Power factor correction (PFC) methods have attracted most attention in recent years. Many papers regarding PFC have been published but not textbooks.

4.3 DC/DC CONVERTERS

DC/DC conversion technology [5] is a significant area of study. It has developed rapidly and made great progress. According to statistics, there are more than 500 existing topologies of DC/DC converters. DC/DC converters have been widely used in industrial applications such as DC motor drives, communication equipment, mobile phones, and digital cameras. Many new topologies have been developed in the recent decade. They have to be systematically introduced in books.

Mathematical modeling is a historic problem accompanying the development of DC/DC conversion technology. Since the 1940s many scholars have devoted themselves to research in this area and offered various mathematical modeling and control methods. We will discuss these problems in detail.

Most DC/DC converters have at least one pump circuit. For example, a Buck-Boost converter shown in Figure 4.13 has the pump-circuit $S–L$. When the switch S is on, the inductor L absorbs the energy from the source V_1. When the switch S is off, the inductor L releases the SE to the load and in the mean time charges the capacitor C.

From the example we recognize that all energy obtained by the load must be a part of the energy stored in the inductor L. Theoretically, the energy transferred to the load has no limit. In a particular operation, the energy rate cannot be very high.

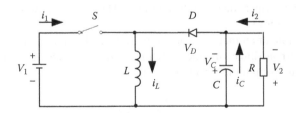

FIGURE 4.13 Buck-boost converter.

Correspondingly, the power losses will sharply increase and the power transfer efficiency will largely decrease.

The following are important points provided in this book:

- To introduce updated circuits
- To introduce new concepts and mathematical modeling
- To check the power rates

4.3.1 UPDATED CONVERTER

Voltage-lift conversion technique is widely used in electronic circuit design. Applying this technique opened a new way to design DC/DC converters. Many new topologies have been developed in the recent decade.

Furthermore, the superlift technique and ultralift technique have been created. Both techniques largely increase the voltage transfer gains of DC/DC converters. The superlift technique is the most outstanding contribution in DC/DC conversion technology.

4.3.2 NEW CONCEPTS AND MATHEMATICAL MODELING

DC/DC converters are an element in an energy control system. In order to obtain satisfactory performance of the energy control system, it is necessary to know the mathematical modeling of the used DC/DC converter. Traditionally, the modeling of power DC/DC converters is derived from the impedance voltage-division method. The clue is that the inductor impedance is sL and the inductor impedance is $1/sC$, where s is the Laplace operator. The output voltage is the voltage division by impedance calculation. Actually, it successfully solves the problem on the fundamental DC/DC converters. The transfer function of a DC/DC converter has the order number equal to the number of the energy-storage elements. A DC/DC converter with two inductors and two capacitors has a fourth-order transfer function. Moreover, a DC/DC converter with four inductors and four capacitors must have an eighth-order transfer function. It is unbelievable that it can be used for industrial applications.

4.3.3 POWER RATE CHECKING

How large power can be used in an energy system with DC/DC converters? It is a very sensitive problem for industrial applications. DC/DC converters are quite different from transformers and AC/DC rectifiers. Their output power is limited by the pump circuit power rate.

The power rate of an inductor pump circuit depends on the inductance, applying current and current ripple, and the switching frequency. The energy transferred by the inductor pump circuit in a cycle $T = 1/f$ is

$$\Delta E = \frac{L}{2}(I_{max}^2 - I_{min}^2) \qquad (4.80)$$

The maximum power can be transferred is

$$P_{max} = f \Delta E = \frac{fL}{2}(I_{max}^2 - I_{min}^2) \tag{4.81}$$

Therefore, when designing an energy system with a DC/DC converter we have to estimate the power rate.

4.4 DC/AC INVERTERS

DC/AC inverters [1,2] were not widely used in industrial applications before the 1960s because of their complexity and cost. They were used in most fractional Horsepower AC motor drives in the 1970s because the AC motors have advantages such as lower cost than DC motors, small size, and no maintenance. Because of the semiconductor development and more effective devices such as IGBT and MOSFET produced in the 1980s, DC/AC inverters started to be widely applied in industrial applications. DC/AC conversion techniques can be sorted into two categories: pulse-width modulation (PWM) and multilevel modulation (MLM). Each category has many circuits to implement the modulation. Using PWM, we can design various inverters such as voltage-source inverters (VSI), current-source inverters (CSI), impedance-source inverters (ZSI), and multistage PWM inverters.

A single-phase half-wave PWM is shown in Figure 4.14.

The pulse-width modulation (PWM) method is suitable for DC/AC conversion as the input voltage is usually a constant DC voltage (DC link). The pulse-phase modulation (PPM) method is also possible, but it is not so convenient. The pulse-amplitude modulation (PAM) method is not suitable for DC/AC conversion as the input voltage is usually a constant DC voltage. PWM operations have all pulses' leading edge starting from the beginning of the pulse period, and their trailer edge is adjustable. The PWM method is the fundamental technique for many types of PWM DC/AC inverters such as VSI, CSI, ZSI, and multistage PWM inverters.

Another group of DC/AC inverters are the multilevel inverters (MLI). MLIs were invented in the later 1970s. The early MLIs were constructed by diode-clamped and capacitor-clamped circuits. Later, various MLIs were developed.

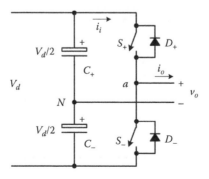

FIGURE 4.14 Single-phase half-wave PWM VSI.

The following important points have to be noted:

- To sort the existing inverters
- To introduce updated circuits
- To investigate the soft switching methods

4.4.1 SORTING EXISTING INVERTERS

Since there are a large number of inverters existing, we have to sort them systematically. Some circuits have not been defined properly, therefore, misleading the readers in understanding their function.

4.4.2 UPDATED CIRCUITS

Many updated DC/AC inverters have been developed in recent decades, but they have not been introduced in textbooks. We discuss these techniques in this book and help students in understanding the methods.

4.4.3 SOFT SWITCHING METHODS

The soft switching technique is widely used in switching circuits for a long time. It effectively reduces the power losses of the equipment and largely increases the power transfer efficiency. A few soft switching techniques are introduced in this book.

4.5 AC/AC CONVERTERS

AC/AC converters [10] were not very widely used in industrial applications before the 1960s because of their complexity and cost. They were used in heating systems for temperature control; in light dimmers in cinemas, theaters, and nightclubs; or in bedroom night dimmers for color and brightening control. Early AC/AC converters were designed by the voltage-regulation (VR) method. A typical single-phase voltage-regulation AC/AC converter is shown in Figure 4.15.

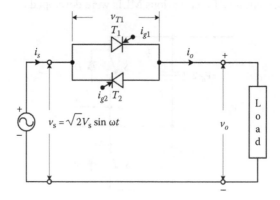

FIGURE 4.15   Single-phase voltage-regulation AC/AC converter.

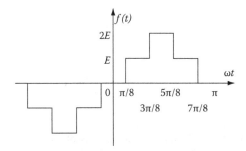

FIGURE 4.16   An five-level odd-waveform showing different angles.

Voltage-regulation AC/AC converters have been successfully used in heating and light-dimming systems. The output AC voltage of VR AC/AC converters is a heavily distorted waveform with poor THD and PF. Other disadvantages are that the output voltage is constantly lower than input voltage and the output frequency is not adjustable.

Cycloconverters and matrix converters can change the output frequency, but the output voltage is also constantly lower than input voltage. Their THD and PF are still very poor.

DC-modulated (DM) AC/AC converters can easily obtain the output voltage higher than the input voltage. They are discussed in this book. Using the DC-modulated (DM) method can successfully improve the THD and PF as indicated from Figure 4.16, by comparing with Figure 4.11.

4.6   AC/DC/AC AND DC/AC/DC CONVERTERS

AC/DC/AC and DC/AC/DC converters are for special applications. In recent years, renewable energy sources and distributed generations (DG) have developed rapidly since fossil energy sources (coal, oil, gas, and so on) are exhaustible. These sources such as solar panel, photovoltaic cells, fuel cells, and wind turbines have unstable DC and AC output voltages. They are usually formed in a microgrid. It is necessary to use the special AC/DC/AC and DC/AC/DC converters to link these sources to the general bus inside the microgrid.

Wind turbines have single-phase or multiphase AC output voltages with variable amplitude and frequency as wind speed varies from time to time. It is difficult to use these unstable AC voltages for any applications. We need use a AC/DC/AC converter to convert them to a suitable AC voltage (single-phase or multiphase) with stable amplitude and frequency.

Solar panels have DC output voltages with variable amplitude since the sunlight varies from time to time. It is difficult to use these unstable DC voltages for any applications. We need to use DC/AC/DC converters to convert them to a suitable DC voltage with a stable amplitude and frequency.

REFERENCES

1. Luo, F. L., Ye, H., and Rashid, M. H. 2005. *Digital Power Electronics and Applications.* Boston, MA: Academic Press.

2. Mohan, N., Undeland, T. M., and Robbins, W. P. 2003. *Power Electronics: Converters, Applications and Design* (3rd edn.). Hoboken, NJ: John Wiley & Sons, Inc.
3. Rashid, M. H. 2004. *Power Electronics: Circuits, Devices and Applications* (3rd edn.). Upper Saddle River, NJ: Prentice Hall.
4. Luo, F. L. and Ye, H. 2007. DC-modulated single-stage power factor correction AC/AC converters. *Proceedings of ICIEA'2007*, Harbin, China, pp. 1477–1483.
5. Luo, F. L. and Ye, H. 2004. *Advanced DC/DC Converters*. Boca Raton, FL: CRC Press.
6. Luo, F. L. and Ye, H. 2005. Energy factor and mathematical modeling for power DC/DC converters. *IEE-EPA Proceedings*, 152(2), 191–198.
7. Luo, F. L. and Ye, H. 2007. Small signal analysis of energy factor and mathematical modeling for power DC/DC converters. *IEEE-Transactions on Power Electronics*, 22(1), 69–79.
8. Luo, F. L. and Ye, H. 2006. *Synchronous and Resonant DC/DC Conversion Technology, Energy Factor and Mathematical Modeling*. Boca Raton, FL: Taylor & Francis.
9. Carlson, A. B. 2000. *Circuits*. Pacific Grove, CA: Brooks/Cole.
10. Rashid, M. H. 2007. *Power Electronics Handbook* (2nd edn.). Boston, MA: Academic Press.

5 Uncontrolled AC/DC Converters

Most of the electronic equipment and circuits require DC sources for their operation. Dry cells and rechargeable battery can be used for these applications. However, they can only offer limited power and unstable voltage. Most useful DC sources are AC/DC converters [1].

AC/DC conversion technology is a vast area in research and industrial applications. AC/DC converters are usually called rectifiers that convert an AC power supply source voltage to a DC voltage load. Usually, the uncontrolled AC/DC converters consist of diode circuits. They can be sorted into the following groups [2]:

- Single-phase half-wave (SPHW) rectifiers
- Single-phase full-wave (SPFW) rectifiers
- Three-phase (TP) rectifiers
- Multipulse (MP) rectifiers
- Power factor correction (PFC) rectifiers
- Pulse-width-modulated (PWM) boost-type rectifiers

Some historic problems are mentioned in ADVICE in order to draw readers' attention since the theoretical analysis and calculation results here are different from those of some existing papers and books in the literature.

5.1 INTRODUCTION

The input voltage of a diode rectifier is an AC voltage, which can be single-phase or three-phase voltage. They are usually a pure sinusoidal wave. A single-phase input voltage $v(t)$ can be expressed as follows:

$$v(t) = \sqrt{2}V \sin \omega t = V_m \sin \omega t \qquad (5.1)$$

where
    $v(t)$ is the instantaneous input voltage
    V is its root-mean-square (rms) value
    $V_m = \sqrt{2}V$ is its amplitude
    ω is the angular frequency
    $\omega = 2\pi f$, f is the supply frequency

Usually, the input current $i(t)$ may be a pure sinusoidal wave with phase shift angle Φ if it is not distorted, and it is expressed as follows:

$$i(t) = \sqrt{2}I \sin(\omega t - \Phi) = I_m \sin(\omega t - \Phi) \tag{5.2}$$

where
$i(t)$ is the instantaneous input current
I is its rms value
I_m is its amplitude
Φ is the phase shift angle

In this case, we define the power factor (PF) as

$$PF = \cos\Phi \tag{5.3a}$$

If the input current is distorted, it consists of harmonics. Its fundamental harmonic can be expressed as Equation 4.17, and the displacement power factor (DPF) is defined in Equation 4.18. The power factor is measured by Equation 5.3b:

$$PF = \frac{DPF}{\sqrt{1 + THD^2}} \tag{5.3b}$$

where THD is the total harmonic distortion. It is defined in Equation 4.20 [3,4].

A pure DC voltage has no ripple; it is then called ripple-free DC voltage. Otherwise, DC voltage is distorted, and its rms value is $V_{d\text{-}rms}$. For a distorted DC voltage, its rms value $V_{d\text{-}rms}$ is constantly higher than its average value V_d. The ripple factor (RF) is defined in Equation 4.24, and the form factor (FF) is defined in Equation 4.25.

5.2 SINGLE-PHASE HALF-WAVE CONVERTERS

An SPHW diode rectifier consists of a single-phase AC input voltage and one diode [5]. It is the simplest rectifier, but its analysis is most complex. This rectifier can supply various loads as described in the following sections.

5.2.1 R LOAD

An SPHW diode rectifier with R load is shown in Figure 5.1a, and the input voltage, output voltage, and current waveforms are shown in Figure 5.1b through d. The output voltage is the same as the input voltage in the positive half-cycle and zero in the negative half-cycle.

The output average voltage is

$$V_d = \frac{1}{2\pi} \int_0^\pi \sqrt{2}V \sin \omega t\, d(\omega t) = \frac{2\sqrt{2}}{2\pi}V = 0.45\,V \tag{5.4}$$

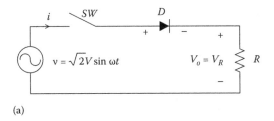

(a)

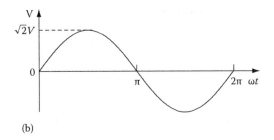

(b)

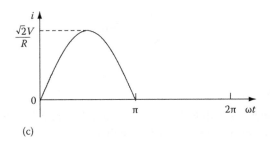

(c)

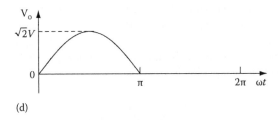

(d)

FIGURE 5.1 SPHW diode rectifier with R load: (a) circuit, (b) input voltage, (c) input current, and (d) output voltage.

The output rms voltage is

$$V_{d\text{-}rms} = \sqrt{\frac{1}{2\pi}\int_0^\pi \left(\sqrt{2}V\sin\omega t\right)^2 d(\omega t)} = V\sqrt{\frac{1}{\pi}\int_0^\pi (\sin\alpha)^2 d\alpha} = \frac{1}{\sqrt{2}}V = 0.707\,V \quad (5.5)$$

The output average and rms currents are

$$I_d = \frac{V_d}{R} = \frac{\sqrt{2}}{\pi}\frac{V}{R} = 0.45\frac{V}{R} \tag{5.6}$$

$$I_{d\text{-}rms} = \frac{V_{d\text{-}rms}}{R} = \frac{1}{\sqrt{2}}\frac{V}{R} = 0.707\frac{V}{R} \tag{5.7}$$

The FF and RF of the output voltage are

$$\text{FF} = \frac{V_{d\text{-}rms}}{V_d} = \frac{1/\sqrt{2}}{\sqrt{2}/\pi} = \frac{\pi}{2} = 1.57 \tag{5.8}$$

$$\text{RF} = \sqrt{\text{FF}^2 - 1} = \sqrt{\left(\frac{\pi}{2}\right)^2 - 1} = 1.21 \tag{5.9}$$

$$\text{PF} = \frac{1}{\sqrt{2}} = 0.707 \tag{5.10}$$

5.2.2 R–L LOAD

An SPHW diode rectifier with R–L load is shown in Figure 5.2a, while various circuit waveforms are shown in Figure 5.2b through d.

It will be seen that the load current flows not only in the positive half-cycle of the supply voltage but also in a portion of the negative half-cycle of the supply voltage [6]. The load inductor stored energy maintains the load current, and the inductor's terminal voltage changes to overcome the negative supply and keep the diode forward biased and conducting. The area A is equal to area B in Figure 5.2c. When the diode conducted, the following equation is available:

$$L\frac{di}{dt} + Ri = \sqrt{2}V\sin\omega t \tag{5.11}$$

or

$$\frac{di}{dt} + \frac{R}{L}i = \frac{\sqrt{2}V}{L}\sin\omega t$$

This is a non-normalized differential equation. The solution has two parts: The forced component is determined by

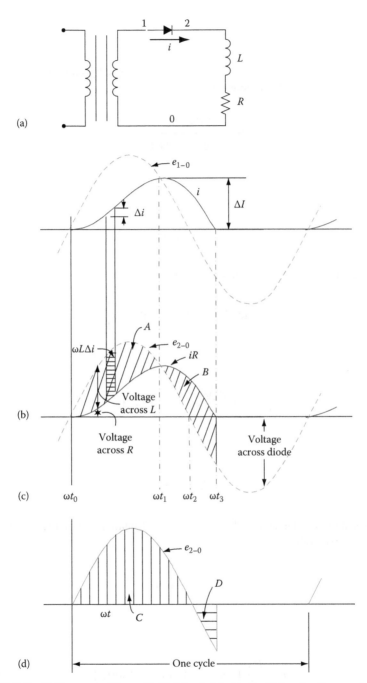

FIGURE 5.2 Half-wave rectifier with R–L load: (a) circuit, (b) input voltage and current, (c) analysis of input voltage and current, and (d) output voltage.

$$i_F = e^{-(R/L)t} \int \left(\frac{\sqrt{2}V}{L} \sin \omega t \right) e^{(R/L)t} dt \tag{5.12}$$

If the circuit is blocked during the negative half-cycle, then by sinusoidal steady-state circuit analysis the forced component of the current is

$$i_F = \frac{\sqrt{2}V \sin(\omega t - \Phi)}{\sqrt{R^2 + (\omega L)^2}} \tag{5.13}$$

where

$$\Phi = \tan^{-1} \left(\frac{\omega L}{R} \right) \tag{5.14}$$

The natural response of such a circuit is given by

$$i_N = Ae^{-(R/L)t} = Ae^{-t/\tau} \text{ with } \tau = \frac{L}{R} \tag{5.15}$$

Thus,

$$i = i_F + i_N = \frac{\sqrt{2}V}{Z} \sin(\omega t - \Phi) + Ae^{-(R/L)t} \tag{5.16}$$

where

$$Z = \sqrt{R^2 + (\omega L)^2} \tag{5.17}$$

The constant A is determined by substitution in Equation 5.16 of the initial condition $i = 0$ at $t = 0$, giving

$$A = \frac{\sqrt{2}V}{Z} \sin \Phi$$

Thus,

$$i = \frac{\sqrt{2}V}{Z} \left[\sin(\omega t - \Phi) + e^{-(R/L)t} \sin \Phi \right] \tag{5.18}$$

We define the *extinction angle* β where the current becomes zero. Therefore,

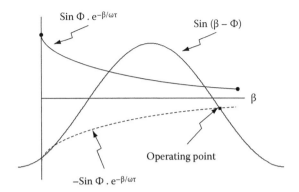

FIGURE 5.3 Determination of the extinction angle β.

$$i = 0, \ \beta \le \omega t < 2\pi \qquad (5.19)$$

The current *extinction angle* β is determined by the load impedance and can be solved from Equation 5.25 when $i=0$ and $\omega t=\beta$, that is,

$$\sin(\beta - \Phi) = -e^{-R\beta/\omega L} \sin \Phi \qquad (5.20)$$

This is a transcendental equation with the unknown value of β. It is shown in Figure 5.3. The term $\sin(\beta-\Phi)$ is a sinusoidal function. The term $e^{-R\beta/\omega L} \sin \Phi$ is an exponentially decaying function. The operating point of β is the intersection of $\sin(\beta-\Phi)$ and $e^{-R\beta/\omega L} \sin \Phi$.

The value of β can be obtained by using MATLAB® simulation, and it can be solved by numerical techniques such as iterative methods.

5.2.2.1 Graphical Method

Using MATLAB to solve Equation 5.20, the resultant values of β for the corresponding values of Φ are plotted as a graph shown in Figure 5.4.

It can be observed that the graph commences at 180° (or π radium's) on the β[X] axis, and for smaller values Φ the characteristic is linear, approximating

$$\beta \approx \pi + \Phi$$

However, for a large value of Φ, the corresponding value of β tends to be

$$\beta > \pi + \Phi$$

with a terminal value of 2π (or 360°) for a purely inductive load.

ADVICE: If $L>0$, $\beta>\pi+\Phi$. Use the graph in Figure 5.4; we cannot get accurate results. (*Historic problem*: $\beta=\pi+\Phi$.)

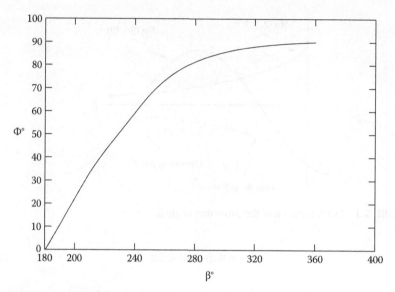

FIGURE 5.4 β versus Φ.

5.2.2.2 Iterative Method 1

The operating point of $\beta \geq \pi + \Phi$.

Let $\beta = \pi + \Phi$.

L1: Calculate $x = \sin(\beta - \Phi)$.

   Calculate $y = -e^{-R\beta/\omega L} \sin \Phi$.

   If $x = y$, then β is the correct value, *END*

   If $|x| < y$, then increment β and return to *L1*

   If $|x| > y$, then decrement β and return to *L1*

Example 5.1 An SPHW diode rectifier is operating from a supply of $V = 240\,V$, $50\,Hz$ to a load of $R = 10\,\Omega$ and $L = 0.1\,H$. Determine the extinction angle β using iterative method 1.

From Equation 5.20, $\Phi = \tan^{-1}(\omega L/R) = 72.34°$

   Let $\beta_1 = \pi + \Phi = 252.34°$.

Step	β	$x = \mathrm{Sin}\,(\beta - \Phi)$	$y = e^{-R\beta/\omega L} \sin \Phi$	$|x| : y$
1	252.34°	0	0.2345	<
2↑	260°	−0.1332	0.2248	<
3↑	270°	−0.3033	0.2126	>
4↓	265°	−0.2191	0.2186	≈
5↓	264°	−0.2020	0.2198	<
6↓	266°	−0.2360	0.2174	>

Best value of β = 265° to satisfy Equation 5.20.

5.2.2.3 Iterative Method 2

Let $\beta_n = \pi + \Phi$.

Calculate $x = \sin(\beta - \Phi)$.

L1: Calculate $y = e^{-R\beta/\omega L} \sin \Phi$.

Let $x = y$

$\beta_{(n+1)} = [\sin^{-1} y] + \pi + \Phi$.

If $\beta_{(n+1)} = \beta_n$ then *END*, else

Choose $\beta_n = \beta_{(n+1)}$ and return *L1*

From above operation, we obtain the extinction angle $\beta = 264.972°$ with high accuracy. The average value of the rectified current can be obtained by

$$v_d = v_R + v_L = \sqrt{2}V \sin \omega t$$

$$\int_0^\beta v_R d(\omega t) + \int_0^\beta v_L d(\omega t) = \int_0^\beta \sqrt{2}V \sin \omega t d(\omega t)$$

$$R \int_0^\beta i(t) d(\omega t) = \sqrt{2}V(1 - \cos \beta)$$

$$I_d = \frac{1}{2\pi} \int_0^\beta i(t) d(\omega t) = \frac{\sqrt{2}V}{2\pi R}(1 - \cos \beta) \qquad (5.21)$$

While the average output voltage is given by

$$V_d = \frac{\sqrt{2}V}{2\pi}(1 - \cos \beta) \qquad (5.22)$$

the output rms voltage is given by

$$V_{d\text{-}rms} = \sqrt{\frac{1}{2\pi} \int_0^\beta \left(\sqrt{2}V \sin \omega t\right)^2 d(\omega t)} = V\sqrt{\frac{1}{\pi} \int_0^\beta (\sin \alpha)^2 d\alpha}$$

$$= V\sqrt{\frac{1}{\pi} \int_0^\beta \left(\frac{1 - \cos 2\alpha}{2}\right) d\alpha} = V\sqrt{\frac{1}{\pi}\left(\frac{\beta}{2} - \frac{\sin 2\beta}{4}\right)} \qquad (5.23)$$

The FF and RF of the output voltage are

$$FF = \frac{V_{d\text{-}rms}}{V_d} = \frac{\sqrt{(1/\pi)\big((\beta/2)-(\sin 2\beta/4)\big)}}{(\sqrt{2}/2\pi)(1-\cos\beta)} = \sqrt{\frac{\pi}{2}}\frac{\sqrt{2\beta-\sin 2\beta}}{1-\cos\beta} \qquad (5.24)$$

$$RF = \sqrt{FF^2 - 1} = \sqrt{\frac{\pi}{2}\frac{2\beta-\sin 2\beta}{(1-\cos\beta)^2}-1} \qquad (5.25)$$

5.2.3 R–L CIRCUIT WITH FREEWHEELING DIODE

The circuit in Figure 5.2a, which has an R–L load, is characterized by discontinuous and high ripple current. Continuous load current can result when a diode is added across the load, as shown in Figure 5.5a.

This diode prevents the voltage across the load from reversing during the negative half-cycle of the supply voltage. When diode D_1 ceases to conduct at zero volts, diode D_2 provides an alternative freewheeling path, as indicated by the waveforms in Figure 5.5b.

After a large number of supply cycles, steady-state load current conditions are established, and the load current is given by

$$i_o = \frac{\sqrt{2}V}{Z}\sin(\omega t - \Phi) + Ae^{-(R/L)t} \qquad (5.26)$$

Also,

$$i_o\big|_{t=0} = I_O\big|_{t=2\pi} \qquad (5.27)$$

Substituting the initial conditions of Equation 5.27 into Equation 5.26 yields

$$i_o = \frac{\sqrt{2}V}{Z}\sin(\omega t - \Phi) + \left(I_{0-2\pi} + \frac{\sqrt{2}V}{Z}\sin\Phi\right)e^{-(R/L)t} \qquad (5.28)$$

At $\omega t = \pi$, diode D_2 begins to conduct, the input current i falls instantaneously to zero, and from Equation 5.28,

$$I_{0-\pi} = i_o\big|_{t=\pi/\omega} = \frac{\sqrt{2}V}{Z}\sin(\pi - \Phi) + \left(I_{0-2\pi} + \frac{\sqrt{2}V}{Z}\sin\Phi\right)e^{-\pi R/\omega L} \qquad (5.29)$$

During the succeeding half-cycle, v_o is zero. The stored energy in the inductor is dissipated by current i_D flowing in the R–L-D_2 mesh. Thus,

$$i_o = i_D = I_{0-\pi}e^{-(R/L)(t-\pi/\omega)} \qquad (5.30)$$

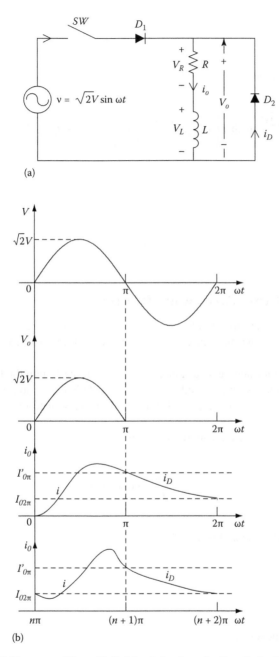

(a)

(b)

FIGURE 5.5 Half-wave rectifier with R–L load plus freewheeling diode. (a) Circuit diagram and (b) waveforms.

at $\omega t = 2\pi$, v and hence v_o becomes positive.

$$i_o\big|_{t=2\pi/\omega} = I_{0-\pi}e^{-(R/L)(t-\pi/\omega)} = I_O\big|_{\omega t=2\pi} \tag{5.31}$$

Thus, from Equations 5.29 and 5.31,

$$\frac{\sqrt{2}V}{Z}\sin\Phi + \left(I_{0-2\pi} + \frac{\sqrt{2}V}{Z}\sin\Phi\right)e^{-\pi R/\omega L} = I_{0-2\pi}e^{\pi R/\omega L} \tag{5.32}$$

so that

$$I_{0-2\pi} = \frac{\left(\sqrt{2}V/Z\right)\sin\Phi\left(1+e^{-\pi R/\omega L}\right)}{e^{\pi R/\omega L} - e^{\pi R/\omega L}} \tag{5.33}$$

5.2.4 An *R–L* Load Circuit with a Back emf

An SPHW rectifier to supply an *R–L* load with a back electromotive force (emf) V_c is shown in Figure 5.6a. The corresponding waveforms are shown in Figure 5.6b.

The effect of introducing a back electromagnetic force V_c into the load circuit of a half-wave rectifier is investigated in this section. This is the situation that would arise if such a circuit were employed to charge a battery or to excite a DC motor armature circuit.

The current component due to the AC source is

$$i_{SF} = \frac{\sqrt{2}V}{Z}\sin(\omega t - \Phi) \tag{5.34}$$

The component due to the direct emf is

$$i_{cF} = \frac{-V_c}{R} \tag{5.35}$$

The natural component is

$$i_N = Ae^{-(R/L)t} \tag{5.36}$$

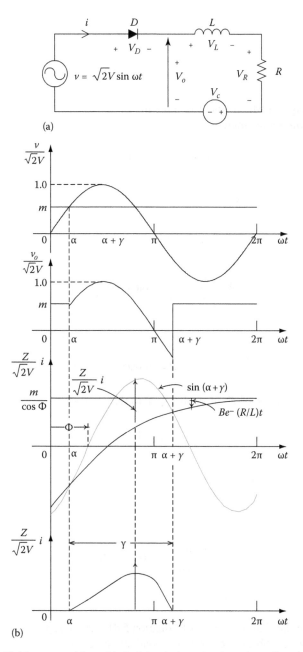

FIGURE 5.6 Half-wave rectifier with R–L load plus a back emf. (a) Circuit diagram and (b) waveforms.

The total current in the circuit is the sum of these three components:

$$i = \frac{\sqrt{2}V}{Z}\sin(\omega t - \varphi) - \frac{V_c}{R} + Ae^{-(R/L)t}$$

$$\alpha < \omega t < \alpha + \gamma \qquad\qquad (5.37)$$

where
   α is the angle at which conduction begins
   γ is the *conduction* angle

Therefore, as may be seen from the voltage curve in Figure 5.6b,

$$\sin\alpha = \frac{V_c}{\sqrt{2}V} = m \qquad\qquad (5.38)$$

At $\omega t = \alpha$, $i = 0$ so that from Equation 5.37,

$$A = \left[\frac{V_c}{R} - \frac{\sqrt{2}V}{Z}\sin(\alpha - \varphi)\right]e^{\alpha R/\omega L} \qquad\qquad (5.39)$$

Also,

$$R = Z\cos\varphi \qquad\qquad (5.40)$$

Substituting Equations 5.38 through 5.40 into (5.37) yields

$$\frac{Z}{\sqrt{2}V}i = \sin(\omega t - \varphi) - \left[\frac{m}{\cos\varphi} - Be^{-(R/L)t}\right]$$

$$\alpha < \omega t < \alpha + \gamma \qquad\qquad (5.41)$$

where

$$B = \left[\frac{m}{\cos\varphi} - \sin(\alpha - \varphi)\right]e^{\alpha R/\omega L}$$

$$\omega t = \alpha \qquad\qquad (5.42)$$

The terms on the right-hand side of Equation 5.41 may be represented separately as shown in Figure 5.6b. At the end of the conduction period,

$$i = 0, \quad \omega t = \alpha + \gamma \qquad\qquad (5.43)$$

Substituting (5.43) in Equation 5.41 yields

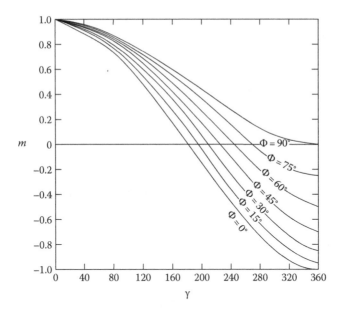

FIGURE 5.7 m versus γ referring to Φ.

$$\frac{(m/\cos\varphi) - \sin(\alpha + \gamma - \varphi)}{(m/\cos\varphi) - \sin(\alpha - \varphi)} = e^{-\gamma/\tan\varphi} \tag{5.44}$$

We get

$$e^{-\gamma/\omega\tau} = \frac{m/\cos\varphi - \sin(\eta + \gamma - \varphi)}{m/\cos\varphi - \sin(\eta - \varphi)} \tag{5.45}$$

Solve for conduction angle γ, using suitable iterative techniques. For practicing design engineers, a quick reference graph of m–φ–γ is given in Figure 5.7.

Example 5.2 An SPHW diode rectifier is operating from a supply of $V = 240\,V$, $50\,Hz$ to a load of $R = 10\,\Omega$, $L = 0.1\,H$ and an emf $V_c = 200\,V$. Determine the conduction angle γ and the total current $i(t)$.

From Equation 5.20, $\Phi = \tan^{-1}(\omega L/R) = 72.34°$
Therefore,

$$\tau = \frac{L}{R} = 10\,ms$$

$$Z = \sqrt{R^2 + (\omega L)^2} = \sqrt{100 + 986.96} = 32.969\,\Omega$$

From Equation 5.38, $m = \sin\alpha = 200/\left(240\sqrt{2}\right) = 0.589$.

Therefore, $\alpha = \sin^{-1} 0.589 = 36.1° = 0.63$ rad.

According to the graph in Figure 5.7, we obtain $\gamma = 156°$.

From Equation 5.39,

$$A = \left[\frac{V_c}{R} - \frac{\sqrt{2}V}{Z}\sin(\alpha - \Phi)\right]e^{\alpha R/\omega L} = [20 - 10.295\sin(-36.24)]e^{0.2}$$

$$= 26.086 * 1.2214 = 31.86$$

Therefore,

$$i(t) = 10.295\sin(314.16t - 72.34°) - 20 + 31.86e^{-100t} \text{ A in } 36.1° < \omega t < 192.1°$$

5.2.4.1 Negligible Load-Circuit Inductance

From Equation 5.37, if $L = 0$ we obtain

$$i = \frac{\sqrt{2}V}{R}\sin\omega t - \frac{V_c}{R} \tag{5.46}$$

or

$$\frac{R}{\sqrt{2}V}i = \sin\omega t - m \tag{5.47}$$

The current $(R/\sqrt{2}V)i$ is shown in Figure 5.8, from which it may be seen that

$$\gamma = \pi - 2\alpha \tag{5.48}$$

The average current

$$I_o = \frac{1}{2\pi}\int_\alpha^{\pi-\alpha} \frac{\sqrt{2}V}{R}(\sin\omega t - m)d(\omega t)$$

$$= \frac{\sqrt{2}V}{\pi R}[\cos\alpha - m(\pi/2 - \alpha)] = \frac{\sqrt{2}V}{\pi R}\left[\sqrt{1-m^2} - m\cos^{-1}m\right] \tag{5.49}$$

5.2.5 SINGLE-PHASE HALF-WAVE RECTIFIER WITH A CAPACITIVE FILTER

An SPHW rectifier in Figure 5.9 has a parallel R–C load. The purpose of the capacitor is to reduce the variation in the output voltage, making it more like a pure DC voltage.

Assuming that the rectifier works in steady state, the capacitor is initially charged in certain DC voltage, and the circuit is energized at $\omega t = 0$, the diode

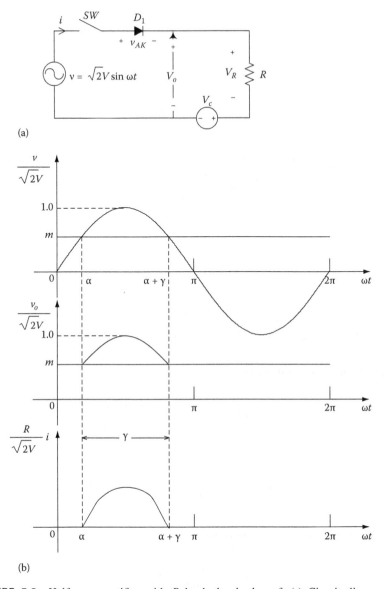

(a)

(b)

FIGURE 5.8 Half-wave rectifier with R load plus back emf. (a) Circuit diagram and (b) waveforms.

becomes forward biased at the angle $\omega t = \alpha$ as the source becomes positive. As the source decreases after $\omega t = \pi/2$, the capacitor discharges from the discharging angle θ into the load resistor. From this point, the voltage of the source becomes less than the output voltage, reverse biasing the diode and isolating the load from the source. The output voltage is a decaying exponential with time constant RC while the diode is off.

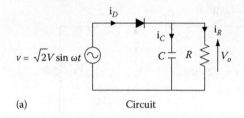

(a)   Circuit

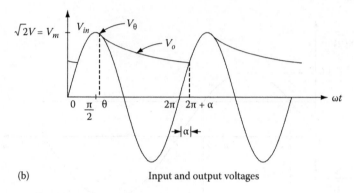

(b)   Input and output voltages

FIGURE 5.9   Half-wave rectifier with an R–C load.

The output voltage is described by

$$v_d(\omega t) = \begin{cases} \sqrt{2}V \sin \omega t & \text{diode on} \\ V_\theta e^{-(\omega t-\theta)/\omega RC} & \text{diode off} \end{cases} \tag{5.50}$$

where

$$V_\theta = \sqrt{2}V \sin \theta \tag{5.51}$$

At $\omega t = \theta$, the slopes of the voltage functions are equal to

$$\sqrt{2}V \cos \theta = \frac{\sqrt{2}V \sin \theta}{-\omega RC} e^{-(\theta-\theta)/\omega RC}$$

Hence,

$$\frac{1}{\tan \theta} = \frac{-1}{\omega RC}$$

Thus,

$$\theta = \pi - \tan^{-1}(\omega RC) \tag{5.52}$$

ADVICE: The discharging angle θ must be greater than $\pi/2$. (*Historic problem*: $\theta = \pi/2$.)

The angle at which the diode turns on in the second period, $\omega t = 2\pi + \alpha$, is the point at which the sinusoidal source reaches the same value as the decaying exponential output.

$$\sqrt{2}V \sin(2\pi + \alpha) = (\sqrt{2}V \sin\theta)e^{-(2\pi+\alpha-\theta)/\omega RC}$$

or

$$\sin\alpha - (\sin\theta)e^{-(2\pi+\alpha-\theta)/\omega RC} = 0 \qquad (5.53)$$

The preceding equation must be solved numerically.

Peak capacitor current occurs when the diode turns on at $\omega t = 2\pi + \alpha$:

$$i_{C\text{-}peak} = \omega C\sqrt{2}V \cos(2\pi + \alpha) = \omega C\sqrt{2}V \cos\alpha \qquad (5.54)$$

ADVICE: The capacitor peak current locates at $\omega t = \alpha$, which is usually much smaller than $\pi/2$. (*Historic problem*: $\alpha \approx \pi/2$.)

Resistor current $i_R(t)$ is

$$i_R(t) = \begin{cases} \dfrac{\sqrt{2}V}{R}\sin\omega t & \text{diode on} \\[2ex] \dfrac{V_\theta}{R}e^{-(\omega t-\theta)/\omega RC} & \text{diode off} \end{cases}$$

where $V_\theta = \sqrt{2}V \sin\theta$.

Its peak current at $\omega t = \pi/2$,

$$i_{R\text{-}peak} = \frac{\sqrt{2}V}{R} \qquad \omega t = \frac{\pi}{2}$$

Its current at $\omega t = 2\pi + \alpha$ (and $\omega t = \alpha$).

$$i_R(2\pi + \alpha) = \frac{\sqrt{2}V}{R}\sin(2\pi + \alpha) = \frac{\sqrt{2}V}{R}\sin\alpha \qquad (5.55)$$

Usually, the capacitive reactance is smaller than the resistance R, and the main component of the source current is capacitor current. Therefore, the peak diode (source) current is

$$i_{D\text{-}peak} = \omega C\sqrt{2}V \cos\alpha + \frac{\sqrt{2}V}{R}\sin\alpha \qquad (5.56)$$

ADVICE: The source peak current locates at $\omega t = \alpha$, which is usually much smaller than $\pi/2$. (*Historic problem*: The source peak current locates at $\omega t = \pi/2$.)

The peak-to-peak ripple of the output voltage is given by

$$\Delta V_d = \sqrt{2}V - \sqrt{2}V \sin\alpha = \sqrt{2}V(1 - \sin\alpha) \tag{5.57}$$

Example 5.3 An SPHW diode rectifier shown in Figure 5.9a is operating from a supply of $V = 240\,V$, 50 Hz to a load of $R = 100$ Ω and $C = 100\,\mu F$ in parallel. If $\alpha = 12.63°$ (see *Question* 5.5), determine the peak capacitor current and peak source current.

From Equation 5.54, the peak capacitor current at $\omega t = \alpha$ is

$$i_{C\text{-}peak} = \omega C \sqrt{2}V \cos\alpha = 100\pi * 0.0001 * 240 * \sqrt{2} * \cos 12.63° = 10.4\,A$$

From Equation 5.56, the peak source current at $\omega t = \alpha$ is

$$i_{D\text{-}peak} = \omega C \sqrt{2}V \cos\alpha + \frac{\sqrt{2}V}{R}\sin\alpha = 10.4 + \frac{240\sqrt{2}}{100}\sin 12.63° = 11.14\,A$$

In order to help readers understand the current waveforms, the simulation results are presented in the following (Figure 5.10) for reference: $V_{in} = 340\,V/50\,Hz$, $C = 100\,\mu F$, and $R = 100\,\Omega$.

5.3   SINGLE-PHASE FULL-WAVE CONVERTERS

Single-phase uncontrolled full-wave bridge circuits are shown in Figures 5.11a and 5.12a. They are called the center-tap (mid-point) rectifier and bridge (Graetz) rectifier, respectively. These two figures appear identical as far as the load is concerned. It will be seen that in Figure 5.11a too fewer diodes can be employed, but this requires a center-tapped transformer. The rectifying diodes in Figure 5.11a experience twice the reverse voltage, as the four diodes in the circuit of Figure 5.12a. Most industrial applications take the bridge (Graetz) rectifier circuit. We take the bridge rectifier for further analysis and discussion.

5.3.1   *R* LOAD

Referring to the bridge circuit shown in Figure 5.12 the load is pure resistive, *R*. Figure 5.12b shows bridge circuit voltage and current waveforms. The output average voltage is

$$V_d = \frac{1}{\pi}\int_0^\pi \sqrt{2}V \sin\omega t\, d(\omega t) = \frac{2\sqrt{2}}{\pi}V = 0.9\,V \tag{5.58}$$

The output rms voltage is

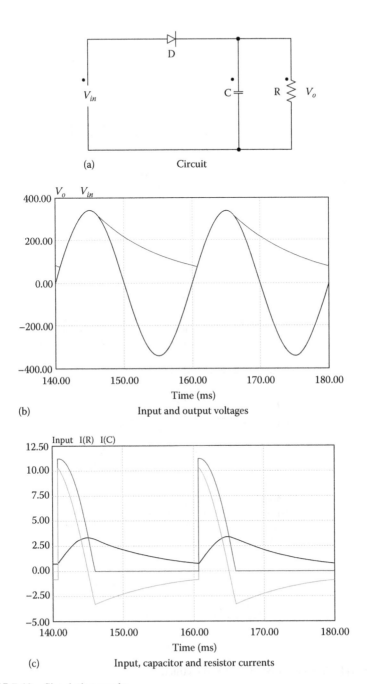

(a)   Circuit

(b)   Input and output voltages

(c)   Input, capacitor and resistor currents

FIGURE 5.10   Simulation results.

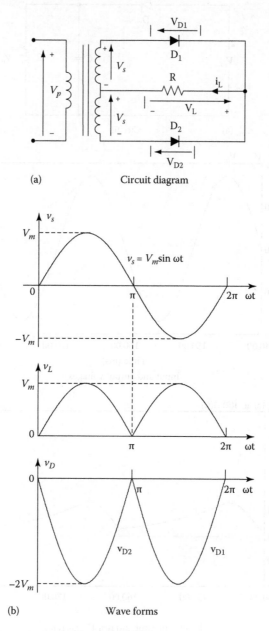

(a)   Circuit diagram

(b)   Wave forms

FIGURE 5.11 The center-tap (mid-point) rectifier.

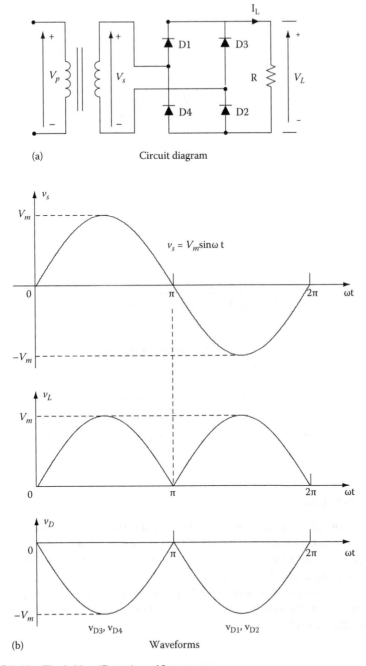

(a)   Circuit diagram

(b)   Waveforms

FIGURE 5.12 The bridge (Graetz) rectifier.

$$V_{d\text{-}rms} = \sqrt{\frac{1}{2\pi}\int_0^\pi (\sqrt{2}V\sin\omega t)^2\, d(\omega t)} = V\sqrt{\frac{1}{\pi}\int_0^\pi (\sin\alpha)^2\, d\alpha} = V \qquad (5.59)$$

The output average and rms currents are

$$I_d = \frac{V_d}{R} = \frac{2\sqrt{2}}{\pi}\frac{V}{R} = 0.9\frac{V}{R} \qquad (5.60)$$

$$I_{d\text{-}rms} = \frac{V_{d\text{-}rms}}{R} = \frac{V}{R} \qquad (5.61)$$

The FF and RF of the output voltage are

$$FF = \frac{V_{d\text{-}rms}}{V_d} = \frac{1}{2\sqrt{2}/\pi} = \frac{\pi}{2\sqrt{2}} = 1.11 \qquad (5.62)$$

$$RF = \sqrt{FF^2 - 1} = \sqrt{(1.11)^2 - 1} = 0.48 \qquad (5.63)$$

$$PF = \frac{1}{\sqrt{2}} = 0.707 \quad \text{for mid-point circuit} \qquad (5.64)$$

$$RF = 1 \quad \text{for Graetz circuit} \qquad (5.65)$$

ADVICE: For all diode rectifiers, only the Graetz (bridge) circuit has unity power factor. (*Historic problem*: MP full-wave rectifiers may have UPF.)

5.3.2 R–C LOAD

Linear and switch-mode DC power supplies require AC/DC rectification. To obtain a "smooth" output, capacitor C is connected, as shown in Figure 5.13.

Neglecting diode forward voltage drop, the peak of the output voltage is $\sqrt{2}V$. During *each half-cycle*, the capacitor undergoes cyclic changes from $v_{d\,(min)}$ to $\sqrt{2}V$ in the period between $\omega t = \alpha$ and $\omega t = \pi/2$ and discharges from $\sqrt{2}V$ to $v_{d\,(min)}$ in the period between $\omega t = \theta$ and $\omega t = \pi + \alpha$. The resultant output of the diode bridge is unipolar but time dependant.

$$V_d(\omega t) = \begin{cases} \sqrt{2}V\sin\omega t & \text{diode on} \\ V_\theta e^{-(\omega t - \theta)/\omega RC} & \text{diode off} \end{cases} \qquad (5.66)$$

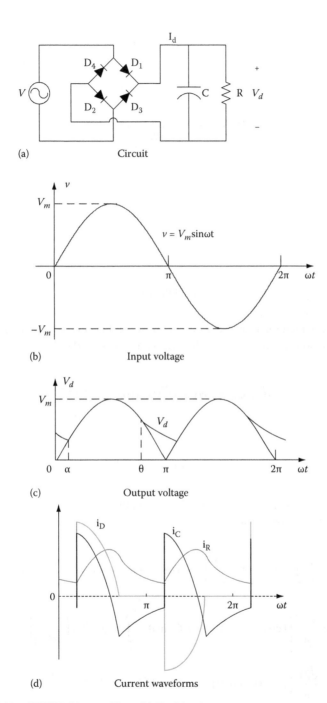

FIGURE 5.13 SPFW bridge rectifier with *R–C* load.

where

$$V_\theta = \sqrt{2}V \sin\theta \qquad (5.67)$$

At $\omega t = \theta$, the slopes of the voltage functions are equal to

$$\sqrt{2}V \cos\theta = \frac{\sqrt{2}V \sin\theta}{-\omega RC} e^{-(\theta-\theta)/\omega RC}$$

Therefore,

$$\frac{1}{\tan\theta} = \frac{-1}{\omega RC}$$

Thus,

$$\theta = \pi - \tan^{-1}(\omega RC) \qquad (5.68)$$

The angle at which the diode turns on in the second period, $\omega t = \pi + \alpha$, is the point at which the sinusoidal source reaches the same value as the decaying exponential output.

$$\sqrt{2}V \sin(\pi+\alpha) = (\sqrt{2}V \sin\theta)e^{-(\pi+\alpha-\theta)/\omega RC}$$

or

$$\sin\alpha - (\sin\theta)e^{-(\pi+\alpha-\theta)/\omega RC} = 0 \qquad (5.69)$$

The preceding equation must be solved numerically.

The output average voltage is

$$V_d = \frac{1}{\pi}\int_\alpha^{\pi+\alpha} v_d d(\omega t) = \frac{\sqrt{2}V}{\pi}\left[\int_\alpha^\theta \sin\omega t\, d(\omega t) + \int_\theta^{\pi+\alpha} \sin\theta e^{-(t-\theta/\omega)/RC} d(\omega t)\right]$$

$$= \frac{\sqrt{2}V}{\pi}\left[(\cos\alpha - \cos\theta) + \omega RC \sin\theta \int_0^{(\pi+\alpha-\theta)/\omega} e^{-t/RC} d\left(\frac{t}{RC}\right)\right]$$

$$= \frac{\sqrt{2}V}{\pi}\left[(\cos\alpha - \cos\theta) + \omega RC \sin\theta\left(1 - e^{-(\pi+\alpha-\theta)/\omega RC}\right)\right] \qquad (5.70)$$

The output rms voltage is

$$V_{d\text{-}rms} = \sqrt{\frac{1}{\pi}\int_{\alpha}^{\pi+\alpha} v_d^2 d(\omega t)} = \sqrt{\frac{2V^2}{\pi}\left[\int_{\alpha}^{\theta}(\sin\omega t)^2 d(\omega t) + \int_{\theta}^{\pi+\alpha}\sin^2\theta e^{-2(t-\theta/\omega)/RC} d(\omega t)\right]}$$

$$= \sqrt{2}V\sqrt{\frac{1}{\pi}\left[\left(\frac{\theta-\alpha}{2} - \frac{\cos 2\alpha - \cos 2\theta}{4}\right) + \omega RC\sin^2\theta\left(1 - \frac{e^{-2(\pi+\alpha-\theta/\omega RC)}}{2}\right)\right]}$$

(5.71)

Since the average capacitor current is zero, the output average current is

$$I_d = \frac{V_d}{R} = \frac{\sqrt{2}V}{\pi R}\left[(\cos\alpha - \cos\theta) + \omega RC\sin\theta(1 - e^{-(\pi+\alpha-\theta)/\omega RC})\right]$$

(5.72)

The FF and RF of the output voltage are

$$\text{FF} = \frac{V_{d\text{-}rms}}{V_d} = \frac{\sqrt{2}V\sqrt{\begin{array}{c}1/\pi\left[((\theta-\alpha)/2 - (\cos 2\alpha - \cos 2\theta)/4\right)\\ + \omega RC\sin^2\theta\left(1 - (e^{-2(\pi+\alpha-\theta)/\omega RC}/2)\right)\right]\end{array}}}{\sqrt{2}V/\pi\left[(\cos\alpha - \cos\theta) + \omega RC\sin\theta\left(1 - e^{-(\pi+\alpha-\theta)/\omega RC}\right)\right]}$$

$$= \frac{\sqrt{\pi}\sqrt{((\theta-\alpha)/2) - ((\cos 2\alpha - \cos 2\theta)/4) + \omega RC\sin^2\theta\left(1 - (e^{-2(\pi+\alpha-\theta)/\omega RC}/2)\right)}}{\cos\alpha - \cos\theta + \omega RC\sin\theta\left(1 - e^{-(\pi+\alpha-\theta)/\omega RC}\right)}$$

(5.73)

$$\text{RF} = \sqrt{\text{FF}^2 - 1}$$

5.3.3 R–L LOAD

An SPHW diode rectifier with R–L load is shown in Figure 5.14a, while various circuit waveforms are shown in Figure 5.14b and c.

If the inductance L is large enough, the load current can be considered a continuous constant current to simplify the analysis and calculations. It is accurate enough for theoretical analysis and engineering calculation. In this case, the load current is assumed a constant DC current.

The output average voltage is

$$V_d = \frac{1}{\pi}\int_0^\pi \sqrt{2}V\sin\omega t d(\omega t) = \frac{2\sqrt{2}}{\pi}V = 0.9\,\text{V}$$

(5.74)

The output rms voltage is

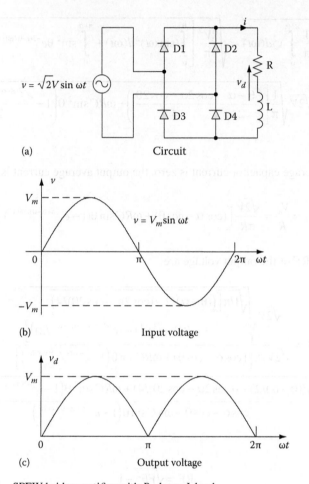

(a)   Circuit

(b)   Input voltage

(c)   Output voltage

FIGURE 5.14 SPFW bridge rectifier with $R+$ large L load.

$$V_{d\text{-}rms} = \sqrt{\frac{1}{2\pi}\int_0^\pi \left(\sqrt{2}V\sin\omega t\right)^2 d(\omega t)} = V\sqrt{\frac{1}{\pi}\int_0^\pi (\sin\alpha)^2 d\alpha} = V \qquad (5.75)$$

The output current is a constant DC value, and its average and rms currents are

$$I_d = I_{d\text{-}rms} = \frac{V_d}{R} = \frac{2\sqrt{2}}{\pi}\frac{V}{R} = 0.9\frac{V}{R} \qquad (5.76)$$

The FF and RF of the output voltage are

$$FF = \frac{V_{d\text{-}rms}}{V_d} = \frac{1}{2\sqrt{2}/\pi} = \frac{\pi}{2\sqrt{2}} = 1.11 \qquad (5.77)$$

$$RF = \sqrt{FF^2 - 1} = \sqrt{(1.11)^2 - 1} = 0.48 \tag{5.78}$$

5.4 THREE-PHASE HALF-WAVE CONVERTERS

When there is AC power supply from a transformer, four circuits can be used. The three-phase half-wave rectifiers are shown in Figure 5.15.

The first circuit is called Y/Y circuit (Figure 5.15a), the second circuit is called Δ/Y circuit (Figure 5.15b), the third circuit is called Y/Y bending circuit (Figure 5.15c), and the fourth circuit is called Δ/Y bending circuit (Figure 5.15d). Each diode is conducted in a $120°$ cycle. Some waveforms are shown in Figure 5.16 corresponding to $L=0$. The three phase voltages are balanced so that

$$v_a(t) = \sqrt{2}V \sin \omega t \tag{5.79}$$

$$v_b(t) = \sqrt{2}V \sin(\omega t - 120°) \tag{5.80}$$

$$v_c(t) = \sqrt{2}V \sin(\omega t - 240°) \tag{5.81}$$

5.4.1 R LOAD

Referring to the bridge circuit shown in Figure 5.15a the load is pure resistive, R ($L=0$). Figure 5.16 shows voltage and current waveforms. The output average voltage is

$$V_{d0} = \frac{3}{2\pi} \int_{\pi/6}^{5\pi/6} \sqrt{2}V \sin \omega t d(\omega t) = \frac{3\sqrt{3}}{\sqrt{2\pi}}V = 1.17\,V \tag{5.82}$$

The output rms voltage is

$$V_{d\text{-}rms} = \sqrt{\frac{3}{2\pi} \int_{\pi/6}^{5\pi/6} \left(\sqrt{2}V \sin \omega t\right)^2 d(\omega t)} = V\sqrt{\frac{6}{\pi}\left(\frac{\pi}{6} + \frac{\sqrt{3}}{8}\right)} = 1.1889\,V \tag{5.83}$$

The output average and rms currents are

$$I_d = \frac{V_d}{R} = 1.17\frac{V}{R} \tag{5.84}$$

$$I_{d\text{-}rms} = \frac{V_{d\text{-}rms}}{R} = 1.1889\frac{V}{R} \tag{5.85}$$

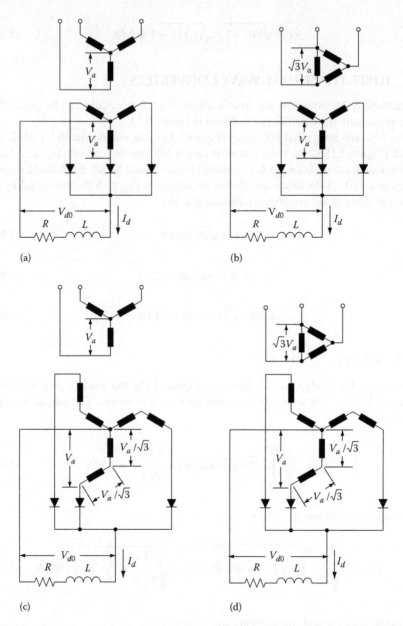

FIGURE 5.15 Three-phase half-wave diode rectifiers: (a) Y/Y circuit, (b) Δ/Y circuit, (c) Y/Y bending circuit, and (d) Δ/Y bending circuit.

The FF, RF, and PF of the output voltage are

$$\text{FF} = \frac{V_{d\text{-}rms}}{V_d} = \frac{1.1889}{1.17} = 1.016 \qquad (5.86)$$

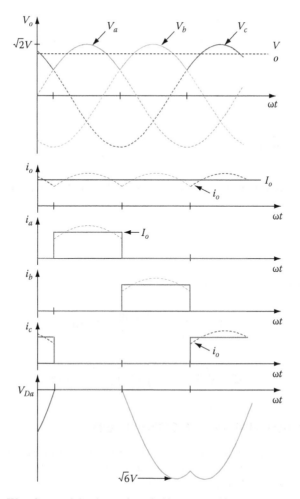

FIGURE 5.16 Waveforms of the three-phase half-wave rectifier.

$$RF = \sqrt{FF^2 - 1} = \sqrt{(1.016)^2 - 1} = 0.18 \qquad (5.87)$$

$$PF = 0.686 \qquad (5.88)$$

5.4.2 *R–L* LOAD

A three-phase half-wave diode rectifier with *R–L* load is shown in Figure 5.15a. If the inductance *L* is large enough, the load current can be considered a continuous constant current to simplify the analysis and calculations. It is accurate enough for theoretical analysis and engineering calculation. In this case, the load current is assumed a constant DC current. The output average voltage is

$$V_{d0} = \frac{3}{2\pi} \int_{\pi/6}^{5\pi/6} \sqrt{2}V \sin \omega t d(\omega t) = \frac{3\sqrt{3}}{\sqrt{2}\pi} V = 1.17\,V \tag{5.89}$$

The output rms voltage is

$$V_{d\text{-}rms} = \sqrt{\frac{3}{2\pi} \int_{\pi/6}^{5\pi/6} \left(\sqrt{2}V \sin \omega t\right)^2 d(\omega t)} = V\sqrt{\frac{6}{\pi}\left(\frac{\pi}{6} + \frac{\sqrt{3}}{8}\right)} = 1.1889\,V \tag{5.90}$$

The output current is nearly a constant DC value, and its average and rms currents are

$$I_d = I_{d\text{-}rms} = \frac{V_d}{R} = 1.17\frac{V}{R} \tag{5.91}$$

The FF and RF of the output voltage are

$$FF = \frac{V_{d\text{-}rms}}{V_d} = \frac{1.1889}{1.17} = 1.016 \tag{5.92}$$

$$RF = \sqrt{FF^2 - 1} = \sqrt{(1.016)^2 - 1} = 0.18 \tag{5.93}$$

5.5   SIX-PHASE HALF-WAVE CONVERTERS

Six-phase half-wave rectifiers have two constructions: six-phase with neutral line circuit and double antistar with balance-choke circuit. The following description is based on the R load or R plus large L load.

5.5.1   Six-Phase with Neutral Line Circuit

If the power supply from a transformer is AC, four circuits can be used. The six-phase half-wave rectifiers are shown in Figure 5.17.

The first circuit is called Y/Star circuit (Figure 5.17a), the second circuit is called Δ/Star circuit (Figure 5.17b), the third circuit is called Y/Star bending circuit (Figure 5.17c), and the fourth circuit is called Δ/Star bending circuit (Figure 5.17d). Each diode is conducted in a 60° cycle. Since the load is an R–L circuit, the output voltage average value is

$$V_{d0} = \frac{1}{\pi/3} \int_{\pi/3}^{2\pi/3} V_m \sin(\omega t) d(\omega t) = \frac{3\sqrt{2}}{\pi} V_a = 1.35 V_a \tag{5.94}$$

$$FF = 1.00088 \tag{5.95}$$

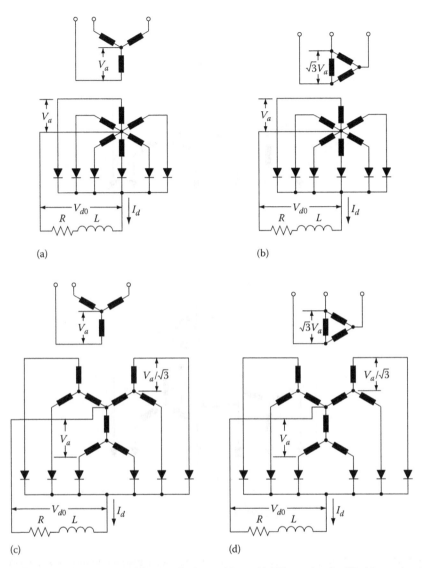

FIGURE 5.17 Six-phase half-wave diode rectifiers: (a) Y/star circuit, (b) Δ/star circuit, (c) Y/star bending circuit, and (d) Δ/star bending circuit.

$$RF = 0.042 \tag{5.96}$$

$$PF = 0.552 \tag{5.97}$$

5.5.2 Double Antistar with Balance-Choke Circuit

If there is AC power supply from a transformer, two circuits can be used. The six-phase half-wave rectifiers are shown in Figure 5.18.

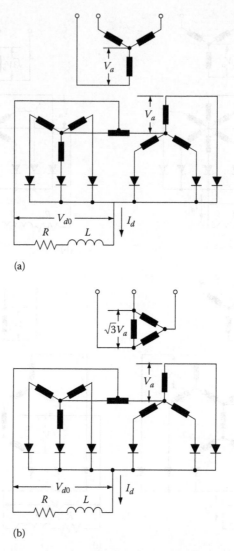

FIGURE 5.18   Three-phase double-antistar with balance-choke diode rectifiers: (a) Y/Y-Y circuit and (b) Δ/Y-Y circuit.

The three-phase double antistar with balance-choke diode rectifiers are shown in Figure 5.18. The first circuit is called *Y/Y–Y* circuit (Figure 5.18a), and the second circuit is called *Δ/Y–Y* circuit (Figure 5.18b). Each diode is conducted in a 120° cycle. Since the load is an *R–L* circuit, the output voltage average value is

$$V_{d0} = \frac{1}{2\pi/3} \int_{\pi/6}^{5\pi/6} V_m \sin(\omega t)d(\omega t) = \frac{3\sqrt{6}}{2\pi} V_a = 1.17 V_a \qquad (5.98)$$

$$FF = 1.01615 \qquad (5.99)$$

$$RF = 0.18 \tag{5.100}$$

$$PF = 0.686 \tag{5.101}$$

5.6 THREE-PHASE FULL-WAVE CONVERTERS

If there is AC power supply from a transformer, four circuits can be used. The three-phase full-wave rectifiers are shown in Figure 5.19.

The three-phase full-wave diode rectifiers shown in Figure 5.19 have four circuits all consisting of six diodes. The first circuit is called *Y/Y* circuit (Figure 5.19a), the second circuit is called Δ/*Y* circuit (Figure 5.19b), the third circuit is called *Y*/Δ circuit (Figure 5.19c), and the fourth circuit is called Δ/Δ circuit shown in Figure 5.19d. Each diode is conducted in a 120° cycle. Since the load is an *R–L* circuit, the output voltage average value is

$$V_{d0} = \frac{2}{2\pi/3} \int_{\pi/6}^{5\pi/6} V_m \sin(\omega t) d(\omega t) = \frac{3\sqrt{6}}{\pi} V_a = 2.34 V_a \tag{5.102}$$

$$FF = 1.00088 \tag{5.103}$$

$$RF = 0.042 \tag{5.104}$$

$$PF = 0.956 \tag{5.105}$$

Some waveforms are shown in Figure 5.20.

ADVICE: The three-phase full-wave bridge rectifier has high PF (although it is not unity power factor) and low RF=4.2%. It is a good circuit to be used in most industrial applications.

5.7 MULTIPHASE FULL-WAVE CONVERTERS

Usually, the more the phases the smaller the output voltage ripple. In this section, several circuits with 6-phase, 12-phase, and 18-phase supply are investigated.

5.7.1 SIX-PHASE FULL-WAVE DIODE RECTIFIERS

Figure 5.21 shows two circuits of the six-phase full-wave diode rectifiers.

The six-phase full-wave diode rectifier shown in Figure 5.21 has two configurations all consisting of 12 diodes. The first circuit is called six-phase bridge circuit (Figure 5.21a), and the second circuit is called hexagon bridge circuit (Figure 5.21b). Each diode is conducted in a 60° cycle. Since the load is an *R–L* circuit, the output voltage average value is

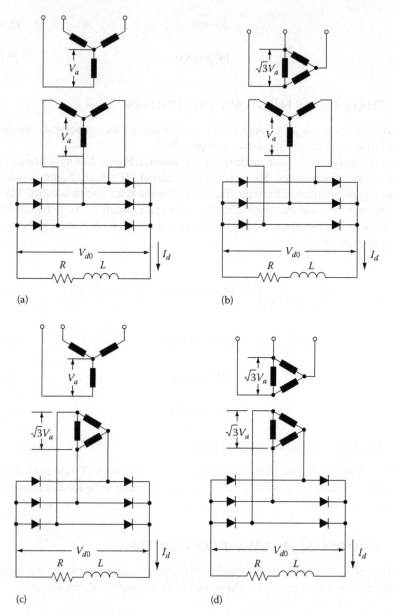

FIGURE 5.19 Three-phase full-wave diode rectifiers: (a) Y/Y circuit, (b) Δ/Y circuit, (c) Y/Δ circuit, and (d) Δ/Δ circuit.

$$V_{d0} = \frac{2}{\pi/3} \int_{\pi/3}^{2\pi/3} V_m \sin(\omega t) d(\omega t) = \frac{6\sqrt{2}}{\pi} V_a = 2.7V_a \qquad (5.106)$$

$$\text{FF} = 1.00088 \qquad (5.107)$$

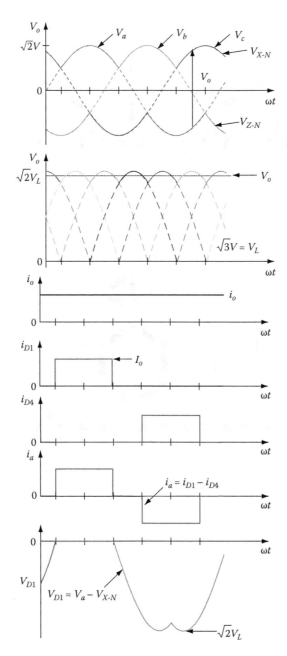

FIGURE 5.20   Waveforms of a three-phase full-wave bridge rectifier.

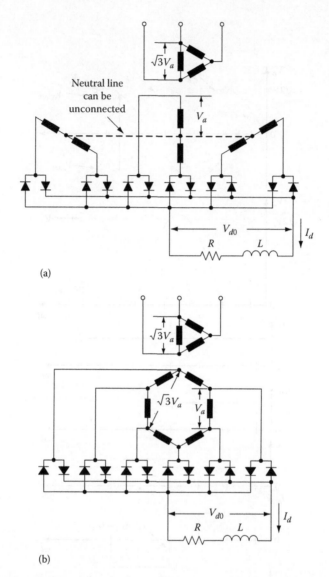

(a)

(b)

FIGURE 5.21 Six-phase full-wave diode rectifiers: (a) six-phase bridgecircuit and (b) hexagon bridge circuit.

$$RF = 0.042 \qquad (5.108)$$

$$PF = 0.956 \qquad (5.109)$$

5.7.2 Six-Phase Double-Bridge Full-Wave Diode Rectifiers

Figure 5.22 shows two circuits of the six-phase, double-bridge full-wave diode rectifiers. The first circuit is called Y/Y–Δ circuit (Figure 5.22a), and the second circuit is

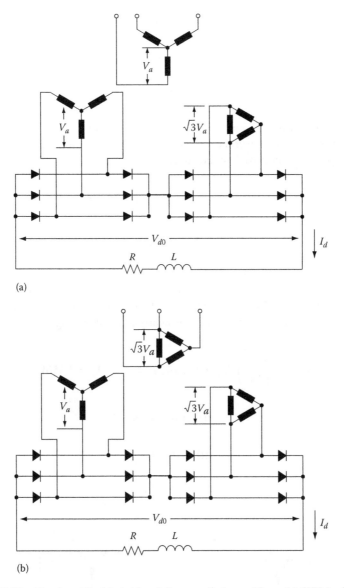

FIGURE 5.22 Six-phase double-bridge full-wave diode rectifiers: (a) Y/Y-Δ circuit and (b) Δ/Y-Δ circuit.

called Δ/Y–Δ circuit (Figure 5.22b). Each diode is conducted at a 120° cycle. There are 12 pulses a period, and phase shift is 30° each. Since the load is an R–L circuit, the output voltage V_{d0} is nearly pure DC voltage.

$$V_{d0} = \frac{4}{2\pi/3} \int_{\pi/6}^{5\pi/6} V_m \sin(\omega t) d(\omega t) = \frac{6\sqrt{6}}{\pi} V_a = 4.678 V_a \qquad (5.110)$$

$$FF = 1.0000567 \qquad (5.111)$$

$$RF = 0.0106 \qquad (5.112)$$

$$PF = 0.956 \qquad (5.113)$$

ADVICE: The six-phase double-bridge full-wave bridge rectifier has high PF (although it is not unity power factor) and lower RF = 1.06%. It is a good circuit to be used in large power industrial applications.

5.7.3 SIX-PHASE DOUBLE-TRANSFORMER DOUBLE-BRIDGE FULL-WAVE DIODE RECTIFIERS

Figure 5.23 shows the six-phase double-transformer double-bridge full-wave diode rectifier. The first transformer T_1 is called $Y/Y–\Delta$ connection transformer, and the second transformer T_2 is called bending $Y/Y–\Delta$ connection transformer with 15° phase shift. There are totally 24 diodes involved in the rectifier. Each diode is conducted at a 120° cycle. There are 24 pulses a period, and phase shift is 15° each. Since the load is an $R–L$ circuit, the output voltage V_{d0} is nearly pure DC voltage.

$$V_{d0} = \frac{8}{2\pi/3} \int_{\pi/6}^{5\pi/6} V_m \sin(\omega t)d(\omega t) = \frac{12\sqrt{6}}{\pi}V_a = 9.356V_a \qquad (5.114)$$

$$FF = 1.0000036 \qquad (5.115)$$

$$RF = 0.00267 \qquad (5.116)$$

$$PF = 0.956 \qquad (5.117)$$

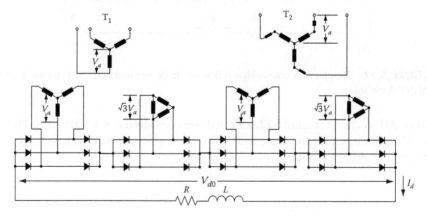

FIGURE 5.23 Six-phase double-transformer double-bridge full-wave diode rectifier.

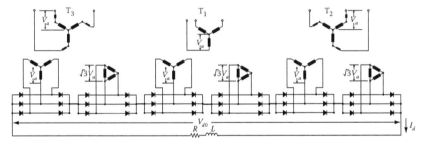

FIGURE 5.24   Six-phase triple-transformer double-bridge full-wave diode rectifier.

5.7.4   SIX-PHASE TRIPLE-TRANSFORMER DOUBLE-BRIDGE FULL-WAVE DIODE RECTIFIERS

Figure 5.24 shows the six-phase triple-transformer double-bridge full-wave diode rectifier. The first transformer T_1 is called $Y/Y–\Delta$ connection transformer, the second transformer T_2 is called positive-bending $Y/Y–\Delta$ connection transformer with $+10°$ phase shift, and the third transformer T_3 is called negative-bending $Y/Y–\Delta$ connection transformer with $-10°$ phase shift. There are totally 36 diodes involved in the rectifier. Each diode is conducted at a $120°$ cycle. There are 36 pulses a period, and phase shift is $10°$ each. Since the load is an $R–L$ circuit, the output voltage V_{d0} is nearly pure DC voltage.

$$V_{d0} = \frac{12}{2\pi/3} \int_{\pi/6}^{5\pi/6} V_m \sin(\omega t)d(\omega t) = \frac{18\sqrt{6}}{\pi}V_a = 14.035V_a \qquad (5.118)$$

$$FF = 1.0000007 \qquad (5.119)$$

$$RF = 0.00119 \qquad (5.120)$$

$$PF = 0.956 \qquad (5.121)$$

REFERENCES

1. Rashid, M. H. 2007. *Power Electronics Handbook* (2nd edn.). Boston, MA: Academic Press.
2. Luo, F. L., Ye, H., and Rashid, M. H. 2005. *Digital Power Electronics and Applications.* New York: Academic Press.
3. Dorf, R. C. 2006. *The Electrical Engineering Handbook* (3rd edn.). Boca Raton, FL: Taylor & Francis.
4. Mohan, N., Undeland, T. M., and Robbins, W. P. 2003. *Power Electronics: Converters, Applications and Design* (3rd edn.). Hoboken, NJ: John Wiley & Sons, Inc.
5. Rashid, M. H. 2003. *Power Electronics: Circuits, Devices and Applications* (3rd edn.). Upper Saddle River, NJ: Prentice Hall, Inc.
6. Keown, J. 2001. *OrCAD PSpice and Circuit Analysis* (4th edn.). Upper Saddle River, NJ: Prentice Hall, Inc.

6 Controlled AC/DC Converters

Controlled AC/DC converters are usually called controlled rectifiers that convert an AC power supply source voltage to a controlled DC load voltage [1–3]. Controlled AC/DC conversion technology is a vast area in research and industrial applications. Usually, the rectifier devices are thyristor (or SCR—silicon controlled rectifier), gate-turn-off thyristor (GTO), power transistor (BT), insulated gate bipolar transistor (IGBT), and so on. The most generally used device is thyristor (or SCR). The controlled AC/DC converters consist of thyristor/diode circuits. They can be sorted into the following groups:

- Single-phase half-wave rectifiers
- Single-phase full-wave rectifiers with half/full control
- Three-phase rectifiers with half/full control
- Multipulse rectifiers

6.1 INTRODUCTION

As in the case of the diode rectifiers discussed in Chapter 5, it must be assumed that in the controlled rectifiers the diodes are replaced by thyristors or other semiconductor devices. The controlled rectifiers are supplied from an ideal AC source. Two conditions must be met before the thyristor can be conducted [4–10]:

1. The thyristor must be forward biased.
2. A current must be applied to the gate of the thyristor.

Only one condition must be met before the thyristor can be switched off: the current that flows through it is lower than the latched value no matter whether the thyristor is forward/reverse biased.

According to these conditions, a firing pulse with variable angle is required to apply to the gate of the thyristor. We usually define the firing angle as α. If the firing angle $\alpha = 0$, the thyristor functions as a diode. The corresponding output DC voltage of the rectifier is its maximum value. We can refer to the results in Chapter 5 to design proper controlled rectifiers to satisfy the industrial applications.

6.2 SINGLE-PHASE HALF-WAVE CONTROLLED CONVERTERS

A single-phase half-wave controlled rectifier consists of a single-phase AC input voltage and one thyristor. It is the simplest rectifier. This rectifier can supply various loads as described in the following subsections.

6.2.1 R Load

A single-phase half-wave diode rectifier with R load is shown in Figure 6.1a; the input voltage, output voltage, and current waveforms are shown in Figure 6.1b through d. The output voltage is the same as the input voltage in the positive half-cycle and zero in the negative half-cycle.

The output average voltage is

$$V_d = \frac{1}{2\pi}\int_{\alpha}^{\pi}\sqrt{2}V\sin\omega t\, d(\omega t) = \frac{\sqrt{2}}{2\pi}V(1+\cos\alpha) = 0.45V\frac{1+\cos\alpha}{2} \qquad (6.1)$$

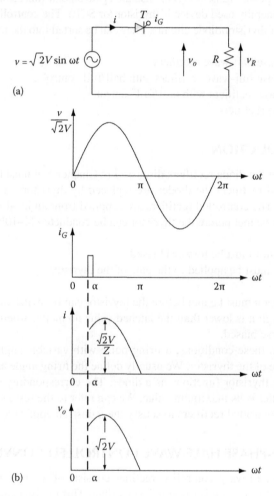

(a)

(b)

FIGURE 6.1 Single-phase half-wave controlled rectifier with R load. (a) Circuit diagram and (b) waveforms.

Using the definition,

$$V_{d0} = \frac{1}{2\pi}\int_0^\pi \sqrt{2}V\sin\omega t\, d(\omega t) = \frac{\sqrt{2}}{2\pi}V \qquad (6.2)$$

We can rewrite (6.1)

$$V_d = \frac{1}{2\pi}\int_\alpha^\pi \sqrt{2}V\sin\omega t\, d(\omega t) = \frac{1+\cos\alpha}{2}V_{d0} \qquad (6.3)$$

The output rms voltage is

$$V_{d\text{-}rms} = \sqrt{\frac{1}{2\pi}\int_\alpha^\pi (\sqrt{2}V\sin\omega t)^2\, d(\omega t)} = V\sqrt{\frac{1}{\pi}\int_\alpha^\pi (\sin x)^2\, dx} = V\sqrt{\frac{1}{\pi}\left(\frac{\pi-\alpha}{2}+\frac{\sin 2\alpha}{4}\right)}$$

$$(6.4)$$

The output average and rms currents are

$$I_d = \frac{V_d}{R} = \frac{\sqrt{2}}{\pi}\frac{V}{R}\frac{1+\cos\alpha}{2} = \frac{1+\cos\alpha}{2}\frac{V_{d0}}{R} \qquad (6.5)$$

$$I_{d\text{-}rms} = \frac{V_{d\text{-}rms}}{R} = \frac{V}{R}\sqrt{\frac{1}{\pi}\left(\frac{\pi-\alpha}{2}+\frac{\sin 2\alpha}{4}\right)} \qquad (6.6)$$

6.2.2 R–L LOAD

A single-phase half-wave diode rectifier with R–L load is shown in Figure 6.2a, while various circuit waveforms are shown in Figure 6.2b through d.

It will be seen that load current flows not only in the positive part of the supply voltage but also in the portion of the negative supply voltage [11–21]. The load inductor stored energy maintains the load current, and the inductor's terminal voltage changes to overcome the negative supply and keep the diode forward biased and conducting. The load impedance Z is

$$Z = R + j\omega L = |Z|\angle\phi \quad \text{with} \quad \phi = \tan^{-1}\frac{\omega L}{R} \qquad (6.7)$$

$$|Z| = \sqrt{R^2 + (\omega L)^2}$$

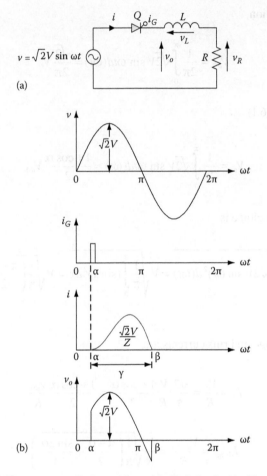

FIGURE 6.2 Half-wave controlled rectifier with R–L load. (a) Circuit diagrams and (b) waveforms.

When the thyristor is conducted, the dynamic equation is

$$L\frac{di}{dt} + Ri = \sqrt{2}V \sin \omega t \quad \text{with} \quad \alpha \le \omega t < \beta \qquad (6.8)$$

or

$$\frac{di}{dt} + \frac{R}{L}i = \frac{\sqrt{2}V}{L}\sin \omega t \quad \text{with} \quad \alpha \le \omega t < \beta$$

where
α is the firing angle
β is the extinction angle

The thyristor conducts between α and β. Equation 2.7 is a non-normalized differential equation. The solution has two parts. The forced solution is determined by

$$i_F = \frac{\sqrt{2}V}{L}\sin(\omega t - \phi) \tag{6.9}$$

The natural response of such a circuit is given by

$$i_N = Ae^{-(R/L)t} = Ae^{-t/\tau} \quad \text{with} \quad \tau = \frac{L}{R} \tag{6.10}$$

The solution of Equation 6.8 is

$$i = i_F + i_N = \frac{\sqrt{2}V}{Z}\sin(\omega t - \phi) + Ae^{-(R/L)t} \tag{6.11}$$

The constant A is determined by substitution in Equation 6.11 of the initial conditions $i = 0$ at $\omega t = \alpha$; it yields

$$i = \frac{\sqrt{2}V}{Z}\left[\sin(\omega t - \phi) - \sin(\alpha - \phi)e^{\frac{R}{L}\left(\frac{\alpha}{\omega}-t\right)}\right] \tag{6.12}$$

Also, $i = 0$, $\beta < \omega t < 2\pi$.

The current *extinction angle* β is determined by the load impedance and can be solved from Equation 2.12 when $i = 0$ and $\omega t = \beta$, that is,

$$\sin(\beta - \phi) = -e^{-R\beta/\omega L}\sin(\alpha - \phi) \tag{6.13}$$

which is a transcendental equation with the unknown value of β. The term $\sin(\beta - \phi)$ is a sinusoidal function. The term $e^{-(R\beta/\omega L)}\sin(\phi - \alpha)$ is an exponentially decaying function. The operating point of β is the intersection of $\sin(\beta - \phi)$ and $e^{-(R\beta/\omega L)}\sin(\phi - \alpha)$, and its value can be determined by iterative methods and MATLAB®. The average output voltage is

$$V_O = \frac{1}{2\pi}\int_\alpha^\beta \sqrt{2}V\sin(\omega t)d(\omega t) \tag{6.14}$$

$$V_O = \frac{V}{\sqrt{2}\pi}\left[\cos\alpha - \cos\beta\right]$$

Example 6.1 A controlled half-wave rectifier has an AC input of 240 V (rms) at 50 Hz with a load $R = 10\,\Omega$ and $L = 0.1\,H$ in series. The firing angle α is 45° as shown in Figure 6.2. Determine the extinction angle β with an accuracy of 0.01° using the iterative method 2.

Solution: Calculation of the extinction angle β using Iterative method 2

$$\frac{\omega L}{R} = \pi \approx 3.14$$

$$z = \sqrt{R^2 + \omega^2 L^2} = 33\,\Omega$$

$$\Phi = \tan^{-1}\left(\frac{\omega L}{R}\right) = 72.34°$$

$$\alpha = 45°, \quad V_m = \sqrt{2}V = 240\sqrt{2} = 340\,V$$

At $\omega t = \beta$, the current is zero:

$$\sin(\beta - \varphi) = e^{(\alpha - \beta)/\tan\varphi}\sin(\alpha - \varphi)$$

Using the iterative method 2, define

$$x = |\sin(\beta - \varphi)|$$

$$y = e^{(\alpha - \beta)/\tan\varphi}\sin(\varphi - \alpha) = \sin(72.34 - \alpha)e^{(\alpha - \beta)/\pi} = 0.46e^{(\alpha - \beta)/\pi}$$

Make a table

β (°)	x	y	$\sin^{-1}y$ (°)	$\|x\|: y$
252.34	0	0.1454	8.36	<
260.7	0.1454	0.1388	7.977	>
260.32	0.1388	0.13907	7.994	<
260.33	0.13907	0.139066	7.994	$\approx$

We can choose $\beta = 260.33°$.

6.2.3   R–L LOAD PLUS BACK EMF V_c

If the circuit involves an emf or battery V_c (refer to Figure 6.3), the minimum firing angle is requested to guarantee the thyristor is successfully fired on. If a firing angle is allowable to supply the load with an emf V_c, the minimum delay angle is

$$\alpha_{min} = \sin^{-1}\left(\frac{V_c}{\sqrt{2}V}\right) \tag{6.15}$$

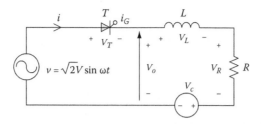

FIGURE 6.3　Half-wave controlled rectifier with R–L load plus an emf V_c.

It means that the firing pulse has to be applied to the thyristor when the supply voltage is higher than the emf V_c. Other characteristics can be derived as shown in Section 5.2.4.

Example 6.2 A controlled half-wave rectifier has an AC input of 120 V (rms) at 60 Hz, $R=2\,\Omega$, $L=20\,mH$, and an emf of $V_c=100$ V. The firing angle α is 45°. Determine

1. An expression for the current
2. The power absorbed by the dc source V_c in the load

Solution: From the parameters given,

$$z = \sqrt{R^2 + \omega^2 L^2} = 7.8\,\Omega$$

$$\phi = \tan^{-1}\left(\frac{\omega L}{R}\right) = 1.312\,\text{rad}$$

$$\frac{\omega L}{R} = 3.77$$

$$\alpha = 45°,\quad V_m = \sqrt{2}V = 120\sqrt{2} = 169.7\,\text{V}$$

1. First, use Equation 6.15 to determine the minimum delay angle, if $\alpha=45°$ is allowable. The minimum delay angle is

$$\alpha_{min} = \sin^{-1}\left(\frac{100}{120\sqrt{2}}\right) = 36°$$

which indicates that $\alpha=45°$, which is allowable. Equation

$$\frac{Z}{\sqrt{2}V}i = \sin(\omega t - \varphi) - \left[\frac{m}{\cos\varphi} - Be^{(\alpha-\omega t)/\tan\varphi}\right],\quad \alpha < \omega t \le \beta$$

$$B = \frac{m}{\cos\varphi} - \sin(\alpha - \varphi),\quad \omega t = \alpha,\quad i = 0$$

becomes

$$i = 21.8\sin(\omega t - 1.312) + 75e^{-\omega t/3.77} - 50$$

$$\text{for} \quad 0.785\,\text{rad} \le \omega t \le 3.37\,\text{rad}$$

where the extinction angle β is found numerically to be 3.37 rad from the equation $i(\beta) = 0$.

2. Power absorbed by the dc source V_c

$$P_{dc} = IV_c = V_c \frac{1}{2\pi}\int_{\alpha}^{\beta} i(\omega t)d(\omega t) = 2.19 \times 100 = 219\ \text{W}$$

6.3 SINGLE-PHASE FULL-WAVE CONTROLLED CONVERTERS

Full-wave voltage control is possible with the circuits with an R–L load shown in Figure 6.4a and b.

The circuit in Figure 6.4a uses a center-tapped transformer and two thyristors that experience a reverse bias of twice the supply. At high powers, where a transformer may not be applicable, a four thyristor configuration as in Figure 6.4b is suitable. The load current waveform becomes continuous when the (maximum) phase control angle α is given by

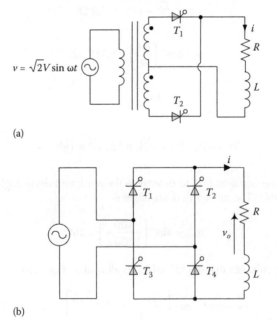

(a)

(b)

FIGURE 6.4 Full-wave voltage-controlled circuit: (a) the central-tap (mid-point) rectifier, (b) Bridge (Graetz) rectifier, (c) the discontinuous output voltage, (d) the critical output voltage, and (e) the continuous output voltage.

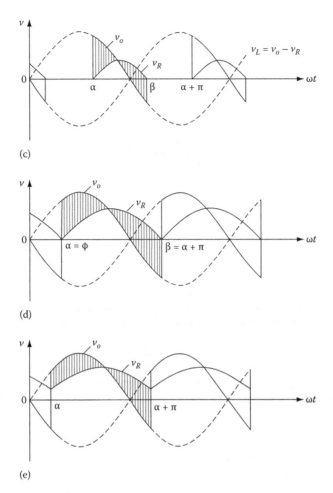

FIGURE 6.4 (continued)

$$\alpha = \tan^{-1}\left(\frac{\omega L}{R}\right) = \varphi \qquad (6.16)$$

at which the output current is a rectifier sine wave.

For $\alpha > \phi$, discontinuous load current flows as shown in Figure 6.4c. At $\alpha = \phi$, the load current becomes continuous as shown in Figure 6.4d, where $\beta = \alpha + \pi$. Further decrease in α, that is $\alpha < \phi$, results in a continuous load current that is always greater than zero, as shown in Figure 6.4e.

6.3.1 $\alpha > \phi$, Discontinuous Load Current

The load current waveform is the same as for the half-wave situation considered in Section 6.2.2, by Equation 6.15, that is,

$$i = \frac{\sqrt{2}V}{Z}\left[\sin(\omega t - \varphi) - \sin(\alpha - \varphi)e^{(R/L)(\alpha/\omega - t)}\right] \tag{6.17}$$

The average output voltage for this full-wave circuit will be twice that of the half-wave case in Section 6.2.2, by Equation 6.14.

$$V_O = \frac{1}{\pi}\int_\alpha^\beta \sqrt{2}V \sin(\omega t)d(\omega t)$$

$$V_O = \frac{\sqrt{2}V}{\pi}\left[\cos\alpha - \cos\beta\right] \tag{6.18}$$

where β has to be found numerically.

Example 6.3 A controlled full-wave rectifier, shown in Figure 6.4, has an AC input of 240 V (rms) at 50 Hz with a load $R = 10\,\Omega$ and $L = 0.1$ H in series. The firing angle α is 80°.

1. Determine whether the load current is discontinuous. If it is, find the extinction angle β within an accuracy of 0.01° using iterative method 2.
2. Derive expressions for the current i and output voltage v_O, and find the average output voltage V_O.

Solution:

1. The thyristor firing angle $\alpha = 80°$. Since the firing angle α is greater than the load phase angle $\phi = \tan^{-1}(\omega L/R) = 72.34°$, the load current is discontinuous. The extinction angle β is greater than π, but $< (\pi + \alpha) = 260°$. The output voltage goes to negative when $\pi \le \omega t \le \beta$.
   Calculation of the extinction angle β using iterative method 2

$$\left(\frac{\omega L}{R}\right) = \pi \approx 3.14$$

$$z = \sqrt{R^2 + \omega^2 L^2} = 33\,\Omega$$

$$\Phi = \tan^{-1}\left(\frac{\omega L}{R}\right) = 72.34°$$

$$\alpha = 80°, \quad V_m = \sqrt{2}V = 240\sqrt{2} = 340 \text{ V}$$

Since $\alpha > \phi$, the rectifier is working in a discontinuous current state.

We obtain the following equation, at $\omega t = \beta$, and the current is zero,

$$\sin(\beta - \varphi) = e^{(\alpha-\beta)/\tan\varphi} \sin(\alpha - \varphi)$$

Using iterative method 2, define

$$x = \sin(\beta - \varphi)$$

$$y = e^{(\alpha-\beta)/\tan\varphi} \sin(\alpha - \varphi) = \sin(\alpha - 72.34)e^{(\alpha-\beta)/\pi} = 0.1333e^{(\alpha-\beta)/\pi}$$

Make a table

β (°)	x	y	$\sin^{-1} y$ (°)	$\|x\| >, =,< y$?	Next Step
252.34	0	0.05117	2.933	<	Increase angle
255.273	0.05117	0.05034	2.886	>	Decrease angle
255.226	0.05034	0.05036	2.8864	<	Increase angle
255.2264	0.05036	0.05036		≈	~

We can choose $\beta = 255.23°$
2. The equation of the current

$$i = \frac{\sqrt{2}V}{Z}\left[\sin(\omega t - \varphi) - \sin(\alpha - \varphi)e^{(R/L)(\alpha/\omega - t)}\right]$$

becomes

$$i = \frac{\sqrt{2}V}{Z}\left[\sin(\omega t - \varphi) + 0.1333e^{(\alpha-\omega t)/\pi}\right]$$
$$= 10.29\sin(\omega t - 72.34) + 1.37e^{(\alpha-\omega t)/\pi}$$

The current expression is

$$i = 10.29\sin(\omega t - 72.34) + 2.138e^{-\omega t/\pi}$$

The output voltage expression in a period is

$$v_O(t) = \begin{cases} 240\sqrt{2}\sin\omega t & \alpha \le \omega t \le \beta, (\pi + \alpha) \le \omega t \le (\pi + \beta) \\ 0 & \text{otherwise} \end{cases}$$

The average output voltage V_O is

$$V_O = \frac{1}{\pi}\int_\alpha^\beta vd(\omega t) = \frac{240\sqrt{2}}{\pi}\int_\alpha^\beta \sin(\omega t)d(\omega t) = \frac{240\sqrt{2}}{\pi}(\cos\alpha - \cos\beta)$$

$$= \frac{240\sqrt{2}}{\pi}(0.1736 + 0.2549) = 46.3 \text{ V}$$

6.3.2 $\alpha = \phi$, VERGE OF CONTINUOUS LOAD CURRENT

When $\alpha = \phi$, the load current is given by

$$i = \frac{\sqrt{2}V}{Z} \sin(\omega t - \varphi), \quad \varphi < \omega t < \varphi + \pi \tag{6.19}$$

and the average output voltage is given by

$$V_O = \frac{2\sqrt{2}V}{\pi} \cos \alpha \tag{6.20}$$

which is independent of the load.

6.3.3 $\alpha < \phi$, CONTINUOUS LOAD CURRENT

Under these conditions, a thyristor is still conducting when another is forward biased and is turned on. The first device is instantaneously reverse biased by the second device that has been turned on. The average output voltage is

$$V_O = \frac{2\sqrt{2}V}{\pi} \cos \alpha \tag{6.21}$$

The rms output voltage is

$$V_r = V \tag{6.22}$$

6.4 THREE-PHASE HALF-WAVE CONTROLLED RECTIFIERS

A three-phase half-wave controlled rectifier is shown in Figure 6.5.

The input three-phase voltages are

$$v_a(t) = \sqrt{2}V \sin \omega t$$

$$v_b(t) = \sqrt{2}V \sin(\omega t - 120°) \tag{6.23}$$

$$v_c(t) = \sqrt{2}V \sin(\omega t + 120°)$$

Usually the load is an inductive load, that is, R–L load. If the inductance is large enough, the load current is continuous for most firing angle α; the corresponding voltage and current waveforms are shown in Figure 6.5b. Each thyristor is conducted in a 120° cycle. If the load is a pure resistive load and the firing angle is in $0 < \alpha < \pi/6$,

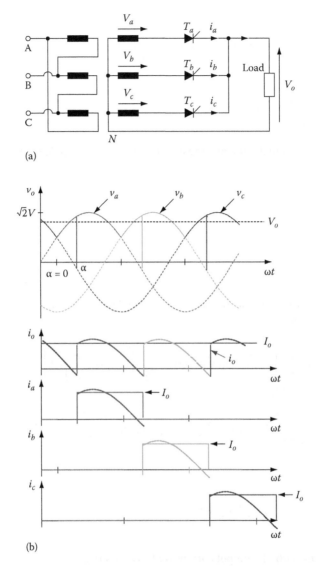

(a)

(b)

FIGURE 6.5 Three-phase half-wave controlled rectifier: (a) the circuit diagram and (b) the waveforms.

the output voltage and current are continuous and each thyristor is conducted in a 120° cycle. If the firing angle $\alpha > \pi/6$ (or 30°), the output voltage and current are discontinuous, and each thyristor is conducted in a (150°-α) cycle.

6.4.1 An *R* Load Circuit

If the load is a resistive load and the firing angle $\alpha \le \pi/6$ ($\omega t = \alpha + \pi/6$), referring to Figure 6.5, the output voltage is

$$V_O = \frac{3}{2\pi} \int_{\alpha+\frac{\pi}{6}}^{\alpha+\frac{5\pi}{6}} \sqrt{2}V \sin(\omega t)d(\omega t) = \frac{3V}{\sqrt{2\pi}}\left[\cos\left(\alpha+\frac{\pi}{6}\right)-\cos\left(\alpha+\frac{5\pi}{6}\right)\right]$$

$$= \frac{3\sqrt{3}V}{\sqrt{2\pi}}\cos\alpha = V_{d0}\cos\alpha \qquad (6.24)$$

where V_{d0} is the output voltage corresponding to the firing angle $\alpha=0$,

$$V_{d0} = \frac{3\sqrt{3}V}{\sqrt{2\pi}} = 1.17\,V \qquad (6.25)$$

The output current is ($\alpha \le \pi/6$)

$$I_O = \frac{V_O}{R} = \frac{V_{d0}}{R}\cos\alpha = 1.17\frac{V}{R}\cos\alpha \qquad (6.26)$$

If the load is a resistive load and the firing angle $\pi/6<\alpha<5\pi/6$ ($\omega t=\alpha+\pi/6$), the output voltage is

$$V_O = \frac{3}{2\pi} \int_{\alpha+\frac{\pi}{6}}^{\pi} \sqrt{2}V \sin(\omega t)d(\omega t) = \frac{3V}{\sqrt{2\pi}}\left[\cos\left(\alpha+\frac{\pi}{6}\right)+1\right]$$

$$= \frac{3V}{\sqrt{2\pi}}\left(\frac{\sqrt{3}}{2}\cos\alpha - \frac{\sin\alpha}{2}+1\right) = 0.675\,V\left(\frac{\sqrt{3}}{2}\cos\alpha - \frac{\sin\alpha}{2}+1\right) \quad (6.27)$$

The output current is

$$I_O = \frac{V_O}{R} = \frac{0.675\,V}{R}\left(\frac{\sqrt{3}}{2}\cos\alpha - \frac{\sin\alpha}{2}+1\right) \qquad (6.28)$$

Since $\pi/6<\alpha<5\pi/6$, the output current is always positive.

When $\alpha \ge 5\pi/6$, both output voltage and current are zero. In this case, all thyristors are reversely biased when the firing pulses are applied. Therefore, all thyristors cannot be conducted.

Example 6.4 A three-phase half-wave controlled rectifier, shown in Figure 6.5, has an AC input of 200 V (rms) at 50 Hz with a load $R=10\,\Omega$. The firing angle α is

   1. 20°
   2. 60°

Calculate the output voltage and current.

Solution:

1. The firing angle $\alpha = 20°$, and the output voltage and current are continuous. Referring to the formulae (6.24), (6.25), and (6.26), we have the output voltage and current as

$$V_O = 1.17V_{in}\cos\alpha = 1.17 \times 200 \times \cos 20° = 220 \text{ V}$$

$$I_O = \frac{V_O}{R} = \frac{220}{10} = 22 \text{ A}$$

2. The firing angle $\alpha = 60°$, which is greater than $\pi/6 = 30°$. The output voltage and current are discontinuous. Referring to the formulae (6.27) and (6.28), we have the output voltage and current as

$$V_O = 0.675 \, V\left(\frac{\sqrt{3}}{2}\cos\alpha - \frac{\sin\alpha}{2} + 1\right)$$

$$= 0.675 \times 200(0.433 - 0.433 + 1) = 135 \text{ V}$$

$$I_O = \frac{V_O}{R} = \frac{135}{10} = 13.5 \text{ A}$$

6.4.2 AN *R–L* LOAD CIRCUIT

Figure 6.6 shows four circuit diagrams for *R–L* load.

If the inductance is large enough and can keep the current continuity, the output voltage is

$$V_O = V_{d0}\cos\alpha = 1.17V\cos\alpha \tag{6.29}$$

The output current is for $\alpha < \pi/2$,

$$I_O = \frac{V_O}{R} = \frac{V_{d0}}{R}\cos\alpha = 1.17\frac{V}{R}\cos\alpha \tag{6.30}$$

When the firing angle $\alpha > \pi/2$, the output voltage can be a negative value, but the output current can only be positive. This situation corresponds to the regenerative state.

Example 6.5 A three-phase half-wave controlled rectifier, shown in Figure 6.5, has an AC input of 200 V (rms) at 50 Hz with a load *R* = 10 Ω plus a large inductance that can keep the continuous output current. The firing angle α is

1. 20°
2. 100°

Calculate the output voltage and current.

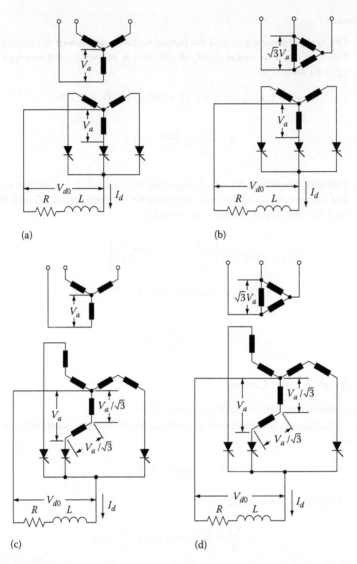

FIGURE 6.6 Three-phase half-wave controlled rectifiers. (a) *Y/Y* circuit, (b) *Δ/Y* circuit, (c) *Y/Y* bending circuit, and (d) *Δ/Y* bending circuit.

Solution:

1. The firing angle $\alpha = 20°$; the output voltage and current are continuous. Referring to the formulae (6.24), (6.25), and (6.26), we have the output voltage and current as

$$V_O = 1.17 V_{in} \cos \alpha = 1.17 \times 200 \times \cos 20° = 220 \text{ V}$$

$$I_O = \frac{V_O}{R} = \frac{220}{10} = 22 \text{ A}$$

2. The firing angle $\alpha = 100°$, but the large inductance can keep the output current to be continuous. The output voltage and current are continuous and negative values. Referring to the formulae (6.29) and (6.30), we have the output voltage and current as

$$V_O = 1.17 V_{in} \cos\alpha = 1.17 \times 200 \times \cos 100° = -40.6 \text{ V}$$

$$I_O = \frac{V_O}{R} = \frac{-40.6}{10} = -4.06 \text{ A}$$

6.5 SIX-PHASE HALF-WAVE CONTROLLED RECTIFIERS

Six-phase half-wave controlled rectifiers have two constructions: six-phase with neutral line circuit and double antistar with balance-choke circuit. The following description is based on the R load or R plus large L load.

6.5.1 SIX-PHASE WITH NEUTRAL LINE CIRCUIT

When there is AC power supply from a transformer, four circuits can be used. The six-phase half-wave rectifiers are shown in Figure 6.7.

The power supply is a six-phase balanced voltage source. Each phase is shifted by 60°.

$$v_a(t) = \sqrt{2}V \sin\omega t$$
$$v_b(t) = \sqrt{2}V \sin(\omega t - 60°)$$
$$v_c(t) = \sqrt{2}V \sin(\omega t - 120°)$$
$$v_d(t) = \sqrt{2}V \sin(\omega t - 180°) \tag{6.31}$$
$$v_e(t) = \sqrt{2}V \sin(\omega t - 240°)$$
$$v_f(t) = \sqrt{2}V \sin(\omega t - 300°)$$

The first circuit is called Y/Star circuit (shown in Figure 6.7a), the second circuit is called Δ/Star circuit (shown in Figure 6.7b), the third circuit is called Y/Star bending circuit (shown in Figure 6.7c), and the fourth circuit is called Δ/Star bending circuit (shown in Figure 6.7d). Each diode is conducted in a 60° cycle. The firing angle $\alpha = \omega t - \pi/3$ in the range of 0–2π/3. Since the load is an R–L circuit, the output voltage average value is

$$V_O = \frac{1}{\pi/3}\int_{\pi/3+\alpha}^{2\pi/3+\alpha} \sqrt{2}V \sin(\omega t)d(\omega t) = \frac{3\sqrt{2}V}{\pi}\left[\cos\left(\frac{\pi}{3}+\alpha\right) - \cos\left(\frac{2\pi}{3}+\alpha\right)\right]$$

$$= \frac{3\sqrt{2}}{\pi}V \cos\alpha = 1.35 V \cos\alpha \tag{6.32}$$

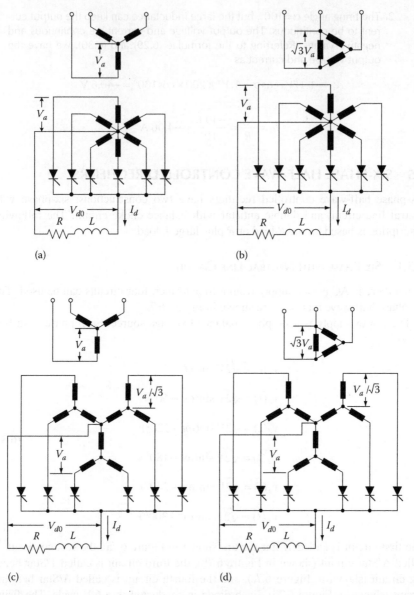

FIGURE 6.7 Six-phase half-wave controlled rectifiers. (a) Y/star circuit, (b) Δ/star circuit, (c) Y/star bending circuit, and (d) Δ/star bending circuit.

The output voltage can be positive ($\alpha < \pi/2$) and negative ($\alpha > \pi/2$) values. The output current is when $\alpha < \pi/2$.

$$I_O = \frac{V_O}{R} = \frac{3\sqrt{2}}{\pi R} V \cos\alpha = 1.35\frac{V}{R}\cos\alpha \qquad (6.33)$$

When the firing angle $\alpha > \pi/2$, the output voltage can be a negative value, but the output current can only be positive. This situation corresponds to the regenerative state.

6.5.2 DOUBLE ANTISTAR WITH BALANCE-CHOKE CIRCUIT

If there is AC power supply from a transformer, two circuits can be used. The six-phase half-wave controlled rectifiers are shown in Figure 6.8.

The three-phase double-antistar with balance-choke controlled rectifiers are shown in Figure 6.8. The first circuit is called $Y/Y–Y$ circuit (shown in Figure 6.8a),

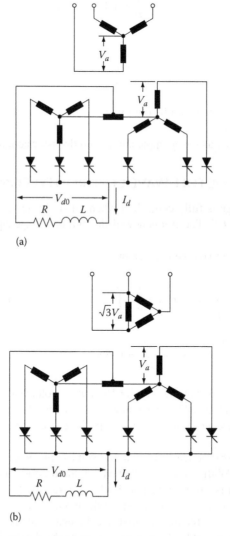

(a)

(b)

FIGURE 6.8 Three-phase double-antistar with balance-choke controlled rectifiers. (a) $Y/Y–Y$ circuit and (b) $\Delta/Y–Y$ circuit.

and the second circuit is called Δ/Y–Y circuit (shown in Figure 6.8b). Each device is conducted in a 120° cycle. The firing angle $\alpha = \omega t - \pi/6$. Since the load is an R–L circuit, the output voltage average value is

$$V_O = \frac{1}{2\pi/3} \int_{\pi/6+\alpha}^{5\pi/6+\alpha} \sqrt{2}V \sin(\omega t) d(\omega t) = \frac{3\sqrt{3}}{\sqrt{2\pi}} V \cos\alpha = 1.17\, V \cos\alpha \quad (6.34)$$

The output voltage can be ($\alpha < \pi/2$) and negative ($\alpha > \pi/2$) value. The output current is

$$I_O = \frac{V_O}{R} = 1.17 \frac{V}{R} \cos\alpha \quad (6.35)$$

When the firing angle $\alpha > \pi/2$, the output voltage can be a negative value, but the output current can only be positive. This situation corresponds to the regenerative state. These circuits have the following advantages:

- Large output current can be obtained since there are two three-phase half-wave rectifiers.
- Output voltage has lower ripple since each thyristor conducts in 120°.

6.6   THREE-PHASE FULL-WAVE CONTROLLED CONVERTERS

A three-phase bridge is fully controlled when all six bridge devices are thyristors, as shown in Figure 6.9. The frequency of the output voltage ripple is six times the supply frequency.

The average output voltage is given by

$$V_O = \frac{3}{\pi} \int_{-\pi/3+\alpha}^{\alpha} V_{ry} d(\omega t) = \frac{3}{\pi} \int_{-\pi/3+\alpha}^{\alpha} \sqrt{3}\sqrt{2}V \sin(\omega t + 2\pi/3) d(\omega t)$$

$$= \frac{3\sqrt{3}}{\pi} \sqrt{2}V \cos\alpha = 2.34\, V \cos\alpha \quad (6.36)$$

The equation illustrates that the rectifier DC output voltage V_O is positive when the firing angle α is less than $\pi/2$ and becomes negative for a firing angle α greater than $\pi/2$. However, the DC current I_O is always positive irrelevant to the polarity of the DC output voltage.

When the rectifier produces a positive DC voltage, the power is delivered from the supply to the load. With a negative DC voltage, the rectifier operates in an *inverter mode* and the power is fed from the load back to the supply. This phenomenon is usually used in electrical drives systems where the motor drive is allowed to decelerate and the kinetic energy of the motor and its mechanical load are converted to the electrical energy and then sent back to the power supply by the thyristor rectifier for fast *dynamic braking*. The power flow in the thyristor rectifier is, therefore, *bidirectional*.

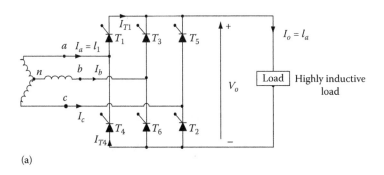

(a)

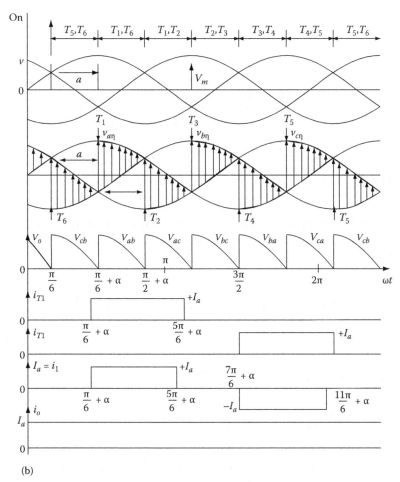

(b)

FIGURE 6.9   A three-phase bridge fully controlled rectifier. (a) Circuit and (b) waveforms.

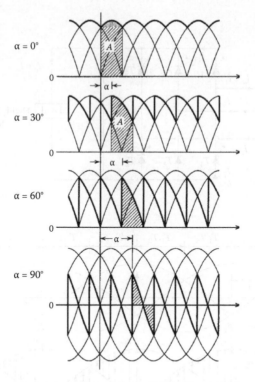

FIGURE 6.10 Rectified voltage waveforms for various firing angles.

Figure 6.10 shows some waveforms corresponding to various firing angles. The shaded area A is the device conduction period and the corresponding rectified voltage. The rms value of the output voltage is given by

$$V_{rms} = \sqrt{\frac{3}{\pi} \int_{-\pi/3+\alpha}^{\alpha} \left[\sqrt{3}\sqrt{2}V \sin(\omega t + 2\pi/3)\right]^2 d(\omega t)} = \sqrt{2}\sqrt{6}V\left[\frac{1}{4} + \frac{3\sqrt{3}}{8\pi} \cos 2\alpha\right]^{1/2}$$

(6.37)

The line current i_r can be expressed in a Fourier series as

$$i_r = \frac{2\sqrt{3}}{\pi} I_{dc}\left[\sin(\omega t - \varphi_1) - \frac{1}{5}\sin 5(\omega t - \varphi_1)\right.$$

$$\left. -\frac{1}{7}\sin 7(\omega t - \varphi_1) + \frac{1}{11}\sin 11(\omega t - \varphi_1) + \frac{1}{13}\sin 13(\omega t - \varphi_1) - \cdots\right] \quad (6.38)$$

where ϕ_1 is the phase angle between the supply voltage v_r and the fundamental frequency line current i_{r1}. The rms value of i_r can be calculated by

$$I_r = \sqrt{\frac{1}{2\pi} \int_0^{2\pi} i_r d(\omega t)} = \sqrt{\frac{1}{2\pi} \left[\int_{-60+\alpha}^{60+\alpha} I_{dc}^2 d(\omega t) + \int_{120+\alpha}^{240+\alpha} I_{dc}^2 d(\omega t) \right]}$$

$$I_r = \sqrt{\frac{2}{3}} I_{dc} = 0.816 I_{dc} \qquad (6.39)$$

from which the total harmonic distortion (THD) for the line current i_r is

$$\text{THD} = \frac{\sqrt{I_r^2 - I_{r1}^2}}{I_{r1}} = \frac{(0.816 I_{dc})^2 - (0.78 I_{dc})^2}{0.78 I_{dc}} = 0.311 \qquad (6.40)$$

where I_{r1} is the rms value of i_{r1} (i.e., $(\sqrt{6}/\pi) I_{dc}$).

Example 6.6 A three-phase full-wave controlled rectifier, shown in Figure 6.9, has an AC input of 200 V (rms) at 50 Hz with a load $R = 10\,\Omega$ plus a large inductance that can keep the continuous output current. The firing angle α is

1. 30°
2. 120°

Calculate the output voltage and current.

Solution:

1. The firing angle $\alpha = 30°$, and the output voltage and current are continuous. Referring to the formula (6.36), we have the output voltage and current as

$$V_O = 2.34\,V \cos\alpha = 2.34 \times 200 \cos 30° = 234\text{ V}$$

$$I_O = \frac{V_O}{R} = \frac{234}{10} = 23.4\text{ A}$$

2. The firing angle $\alpha = 120°$, and the output voltage and current are continuous and negative values. Referring to the formula (6.36), we have the output voltage and current as

$$V_O = 2.34\,V \cos\alpha = 2.34 \times 200 \cos 120° = -234\text{ V}$$

$$I_O = \frac{V_O}{R} = \frac{-234}{10} = -23.4\text{ A}$$

6.7 MULTI-PHASE FULL-WAVE CONTROLLED CONVERTERS

Figure 6.11 shows the typical configuration of a 12-pulse series type controlled rectifier. There are two identical ones. Two six-pulse controlled rectifiers are powered by

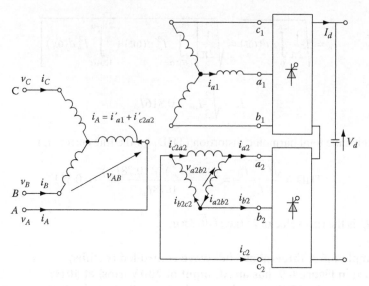

FIGURE 6.11   Twelve-pulse controlled rectifier.

a phase shifting transformer with two secondary windings in delta and star connections. Therefore, the phase-angle between both secondary windings shifts 30° each.

The DC outputs of the rectifiers are connected in series. To dominate lower-order harmonics in the line current i_a, the line-to-line voltage v_{a1b1} of the star-connected secondary winding is in phase with the primary voltage v_{AB} while the delta-connected secondary winding voltage v_{a1b1} leads the primary voltage v_{AB} by

$$\delta = \angle v_{a2b2} - \angle V_{AB} = 30° \tag{6.41}$$

The rms line-to-line voltage of each secondary winding is

$$V_{a1b1\text{-}rms} = V_{a2b2\text{-}rms} = \frac{V_{AB\text{-}rms}}{2} \tag{6.42}$$

from which the turns ratio of the transformer can be determined by

$$\frac{N_1}{N_2} = 2 \quad \text{for} \quad \frac{Y}{Y}$$

$$\frac{N_1}{N_3} = \frac{2}{\sqrt{3}} \quad \text{for} \quad \frac{Y}{\Delta} \tag{6.43}$$

Consider an idealized 12-pulse rectifier where the line inductance L_s and the total leakage inductance L_{lk} of the transformer are assumed to be zero. The current waveforms are in Figure 6.12, where i_{a1} and i_{c2a2} are the secondary line primary

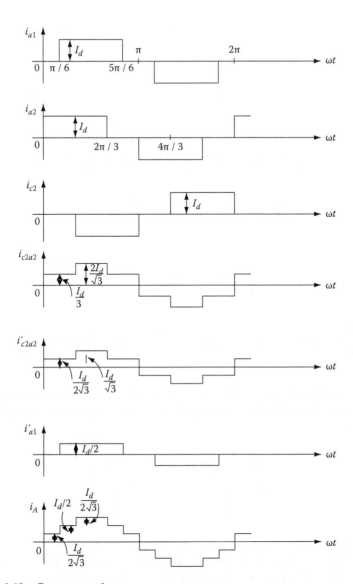

FIGURE 6.12 Current waveforms.

currents referred from the secondary side, and i_a is the primary line current given by $i_a = i'_{a1} + i'_{c2a2}$, respectively.

The secondary line current i_{a1} can be expressed as

$$i_{a1} = \frac{2\sqrt{3}}{\pi} I_d \left(\sin \omega t - \frac{1}{5} \sin 5\omega t - \frac{1}{7} \sin 7\omega t + \frac{1}{11} \sin 11\omega t + \frac{1}{13} \sin 13\omega t \cdots \right) \quad (6.44)$$

where $\omega = 2\pi f$ is the angular frequency of the supply voltage. Since the waveform of current i_{a1} is of half-wave symmetry, it does not contain any even-order harmonics. In addition, current i_a does not contain any triple harmonics either due to the balanced three-phase system.

The other secondary current such as i_{a2} leads i_{a1} by 30°, and its Fourier expression is

$$i_{a2} = \frac{2\sqrt{3}}{\pi} I_d \left[\sin(\omega t + 30°) - \frac{1}{5}\sin 5(\omega t + 30°) - \frac{1}{7}\sin 7(\omega t + 30°) \right.$$

$$\left. + \frac{1}{11}\sin 11(\omega t + 30°) + \frac{1}{13}\sin 13(\omega t + 30°)\cdots \right] \qquad (6.45)$$

The waveform for the referred current i'_{a1} in Figure 6.12 is identical to i_{a1} except that its magnitude is halved due to the turns ratio of the Y/Y-connected windings. The current i'_{a1} can be expressed in Fourier series as

$$i'_{a1} = \frac{\sqrt{3}}{\pi} I_d \left(\sin \omega t - \frac{1}{5}\sin 5\omega t - \frac{1}{7}\sin 7\omega t + \frac{1}{11}\sin 11\omega t + \frac{1}{13}\sin 13\omega t \cdots \right) \qquad (6.46)$$

The phase currents i_{b2a2}, i_{a2c2}, and i_{c2b2} can be derived from the line currents using the relationships of Equation 6.47, which apply provided the zero, a requirement fulfilled in this case

$$\begin{pmatrix} i_{a2b2} \\ i_{b2c2} \\ i_{c2a2} \end{pmatrix} = \frac{1}{3} \begin{pmatrix} -1 & 1 & 0 \\ 0 & -1 & 1 \\ 1 & 0 & -1 \end{pmatrix} \begin{pmatrix} i_{a2} \\ i_{b2} \\ i_{c2} \end{pmatrix} \qquad (6.47)$$

These currents have a stepped waveform, each step being of 60° duration and the height of the steps being $I_d/3$ and $2I_d/3$. The currents i_{a2b2}, i_{b2a2}, and i_{c2a2} need to be multiplied by $\sqrt{3}/2$ when they are referred to the primary side. Using Equation 6.45 and similar equations for i_{b2} and i_{c2}, one can derive Fourier expressions for i_{a2b2}, i_{b2c2}, and i_{c2a2}. For example,

$$i_{a2b2} = \frac{1}{3}(i_{b2} - i_{a2}) \qquad i_{b2c2} = \frac{1}{3}(i_{c2} - i_{b2})$$

and

$$i_{c2a2} = \frac{1}{3}(i_{a2} - i_{c2})$$

Therefore,

$$
i_{c2a2} = \frac{1}{3}\frac{2\sqrt{3}}{\pi}I_d\left[\sin(\omega t + 30°) - \frac{1}{5}\sin 5(\omega t + 30°) - \frac{1}{7}\sin 7(\omega t + 30°)\right.
$$

$$
+ \frac{1}{11}\sin 11(\omega t + 30°)\cdots + \sin(\omega t + 150°) - \frac{1}{5}\sin 5(\omega t + 150°)
$$

$$
\left. - \frac{1}{7}\sin 7(\omega t + 150°) + \frac{1}{11}\sin 11(\omega t + 150°)\cdots\right] \tag{6.48}
$$

By simplifying Equation 6.48 and multiplying with √3/2, we have

$$
i''_{c2a2} = \frac{\sqrt{3}}{\pi}I_d\left(\sin \omega t + \frac{1}{5}\sin 5\omega t + \frac{1}{7}\sin 7\omega t + \frac{1}{11}\sin 11\omega t\cdots\right) \tag{6.49}
$$

As it can be seen from Equation 6.48, the phase angles of some harmonics currents are altered due to the Y/Δ connected windings. As a result, the current i'_{c2a2} does not keep the same wave shape as i'_{a1}. The line current i_A can be found from

$$
i_A = i'_{a1} + i'_{c2a2} = \frac{2\sqrt{3}}{\pi}I_d\left(\sin \omega t + \frac{1}{11}\sin 11\omega t + \frac{1}{13}\sin 13\omega t\cdots\right)
$$

where the two dominant current harmonics, the 5th and 7th, are canceled in addition to the 17th and 19th.

The THD of the secondary and primary line currents i_{a1} and i_A can be determined by

$$
\mathrm{THD}(i_{a1}) = \frac{\sqrt{I_{a1}^2 - I_{a1,1}^2}}{I_{a1,1}} = \frac{\sqrt{I_{a1,5}^2 + I_{a1,7}^2 + \cdots}}{I_{a1,1}} \tag{6.50}
$$

and

$$
\mathrm{THD}(i_A) = \frac{\sqrt{I_A^2 - I_{A,1}^2}}{I_{A,1}} = \frac{\sqrt{I_{A,11}^2 + I_{A,13}^2 + \cdots}}{I_{a1,1}} \tag{6.51}
$$

The THD of the primary line current i_A in the idealized 12-pulse rectifier is reduced by nearly 50% compared with that of i_{a1}.

6.8 EFFECT OF LINE INDUCTANCE ON OUTPUT VOLTAGE (OVERLAP)

We now investigate a three-phase fully controlled rectifier as shown in Figure 6.9a. We can partially redraw the circuit in Figure 6.13 (only show phase A and phase C).

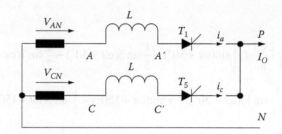

FIGURE 6.13   Effect on line inductance.

In practice, the cable lengths from phase A to A' (or C to C') have an inductance (L). The commutation process (e.g., for i_a to replace i_c) will take a certain time interval. This affects the voltage at point P to neutral point N, and the final half output voltage is V_{PN}.

During the commutation process (e.g., for i_A to replace i_C), Kirchhoff's Voltage Law (KVL) for the commutation loop and Kirchhoff's Current Law (KCL) at point P give the following: The output current I_O is filtered by a large inductance and this implies that its change is much slower than that of i_C and i_A. We can write

$$v_{AN} - v_{CN} = L\frac{di_A}{dt} - L\frac{di_C}{dt} \tag{6.52}$$

$$i_A + i_C = I_o \Rightarrow \frac{di_A}{dt} + \frac{di_C}{dt} = \frac{dI_o}{dt} \Rightarrow 0 \tag{6.53}$$

$$\frac{di_A}{dt} = -\frac{di_C}{dt} \tag{6.54}$$

From Equations 6.52 and 6.54,

$$v_{AN} - v_{CN} = L\frac{di_A}{dt} - L\frac{di_C}{dt} \Rightarrow L\frac{di_A}{dt} = \frac{v_{AN} - v_{CN}}{2} \tag{6.55}$$

This allows one to derive V_{PN}. Thus, V_{PN} takes mid-point value between V_{AN} and V_{CN} during commutation. The output voltage waveform is shown in Figure 6.14.

$$v_{PN} = v_{AN} - L\frac{di_A}{dt} = v_{AN} - \frac{v_{AN} - v_{CN}}{2} \Rightarrow v_{PN} = \frac{v_{AN} + v_{CN}}{2} \tag{6.56}$$

Thus, the integral of V_{PN} will involve two parts, one from firing angle α to $(\alpha + u)$, where u is the overlap angle, and subsequently from $(\alpha + u)$ to next phase fired, where commutation has

$$v_{PN} = \frac{3}{2\pi}\left[\int_{\pi/6+\alpha}^{\pi/6+\alpha+u} \frac{v_{AN} + v_{CN}}{2} d(\omega t) + \int_{\pi/6+\alpha+u}^{\pi/6+\alpha+2\pi/3} v_{AN} d(\omega t)\right] \tag{6.57}$$

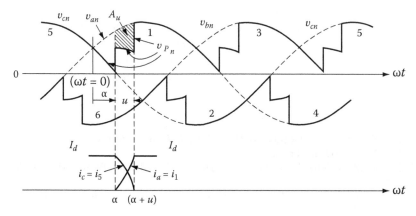

FIGURE 6.14 The waveforms affected by the line inductance.

Hence,

$$v_{PN} = \frac{3}{2\pi}\left[\int_{\pi/6+\alpha}^{\pi/6+\alpha+u} \frac{v_{AN} + v_{CN}}{2}d(\omega t) + \int_{\pi/6+\alpha+u}^{\pi/6+\alpha+2\pi/3} v_{AN}d(\omega t) \right.$$

$$+ \int_{\pi/6+\alpha}^{\pi/6+\alpha+u} \frac{v_{AN} - v_{CN}}{2}d(\omega t) - \int_{\pi/6+\alpha}^{\pi/6+\alpha+u} \frac{v_{AN} - v_{CN}}{2}d(\omega t) + \left. \right]$$

Notice that the first integral is the original integral involving V_{AN} for the full 120° interval. The second interval can be linked to the derivative of current i_A for the commutation interval.

$$\int_{\pi/6+\alpha}^{\pi/6+\alpha+u} \frac{v_{AN} - v_{CN}}{2}d(\omega t) = \int_{\pi/6+\alpha}^{\pi/6+\alpha+u} \left(L\frac{di_A}{dt} \right)\frac{\omega}{\omega}d(\omega t)$$

$$= \int_{i_{start}}^{i_{end}} L\omega d(i_A) = L\omega[I_O + 0] = L\omega I_O \qquad (6.58)$$

Therefore, by an identical analysis for the bottom three thyristors, the output voltage has

$$V_O = 2v_{PN} = \frac{3\sqrt{6}}{\pi}V\cos\alpha - \frac{3}{\pi}L\omega I_O \quad \text{for} \quad 0° < \alpha < 180° \qquad (6.59)$$

Thus, the commutation interval duration due to the line inductance modifies the output voltage waveform (finite time for current change) and this changes the average output voltage *by a reduction* of $(3/\pi)L\omega$: This can be compensated by feed forward.

The figure given earlier shows how V_{PN} is affected during the commutation interval "u." It takes the midpoint value between the incoming phase (V_{AN}) and outgoing phase (V_{CN}) voltages. The corresponding currents i_A and i_C can be seen to rise and fall at finite rates. The rate of current change will be slower for high values of line Interference (EMI), and certain standards limit this rise time.

REFERENCES

1. Luo, F. L., Ye, H., and Rashid, M. H. 2005. *Digital Power Electronics and Applications.* New York: Academic Press.
2. Rashid, M. H. 2007. *Power Electronics Handbook* (2nd edn.). Boston, MA: Academic Press.
3. Dorf, R. C. 2006. *The Electrical Engineering Handbook* (3rd edn.). Boca Raton, FL: Taylor & Francis.
4. Luo, F. L., Jackson, R. D., and Hill, R. J. 1985. Digital controller for thyristor current source. *IEE Proceedings Part B*, 132, 46–52.
5. Luo, F. L. and Hill, R. J. 1985. Disturbance response techniques for digital control systems. *IEEE Transactions on Industrial Electronics*, 32, 245–253.
6. Luo, F. L. and Hill, R. J. 1985. Minimisation of interference effects in thyristor converters by feedback feedforward control. *IEEE Transactions on Measurement and Control*, 7, 175–182.
7. Luo, F. L. and Hill, R. J. 1986. Influence of feedback filter on system stability area in digitally-controlled thyristor converters. *IEEE Transactions on Industry Applications*, 22, 18–24.
8. Luo, F. L. and Hill, R. J. 1986. Fast response and optimum regulation in digitally-controlled thyristor converters. *IEEE Transactions on Industry Applications*, 22, 10–17.
9. Luo, F. L. and Hill, R. J. 1986. System analysis of digitally-controlled thyristor converters. *IEEE Transactions on Measurement and Control*, 8, 39–45.
10. Luo, F. L. and Hill, R. J. 1986. System optimisation—Self-adaptive controller for digitally-controlled thyristor current controller. *IEEE Transactions on Industrial Electronics*, 33, 254–261.
11. Luo, F. L. and Hill, R. J. 1987. Stability analysis of thyristor current controllers. *IEEE Transactions on Industry Applications*, 23, 49–56.
12. Luo, F. L. and Hill, R. J. 1987. Current source optimisation in AC-DC GTO thyristor converters. *IEEE Transactions on Industrial Electronics*, 34, 475–482.
13. Luo, F. L. and Hill, R. J. 1989. Microprocessor-based control of steel rolling mill digital DC drives. *IEEE Transactions on Power Electronics*, 4, 289–297.
14. Luo, F. L. and Hill, R. J. 1990. Microprocessor-controlled power converter using single-bridge rectifier and GTO current switch. *IEEE Transactions on Measurement and Control*, 12, 2–8.
15. Muth, E. J. 1977. *Transform Method with Applications to Engineering and Operation Research.* Englewood Cliffs, NJ: Prentice-Hall.
16. Oliver, G., Stefanovic, R., and Jamil, A. 1979. Digitally controlled thyristor current source. *IEEE Transactions on Industrial Electronics and Control Instrumentation*, 26, 185–191.
17. Fallside, F. and Jackson, R. D. 1969. Direct digital control of thyristor amplifiers. *IEE Proceedings, Part B*, 116, 873–878.
18. Arrillaga, J., Galanos, G., and Posner, E. T. 1970. Direct digital control of HVDC converters. *IEEE Transactions on Power Apparatus and Systems*, 89, 2056–2065.
19. Daniels, A. R. and Lipczyski, R. T. 1969. Digital firing angle circuit for thyristor motor controllers. *IEE Proceedings, Part B*, 125, 245–256.

20. Dewan, S. B. and Dunford, W. G. 1983. A microprocessor-based controller for a three-phase controlled bridge rectifiers. *IEEE Transactions on Industry Applications*, 19, 113–119.
21. Cheung, W. N. 1971. The realisation of converter control using sampled-and-delay method. *IEE Proceedings, Part B*, 127, 701–705.

7 Power Factor Correction Implementing in AC/DC Converters

Power factor correction (PFC) is the capacity of generating or absorbing the reactive power produced by a load [1–3]. Power quality issues and regulations require rectifier loads connected to the utility to achieve high power factor. This means that PFC rectifier needs to draw close to a sinusoidal current in-phase with the supply voltage, unlike phase controlled rectifiers (making PFC rectifier "look like" resistive load to the utility).

7.1 INTRODUCTION

Refer to the formula [3]

$$PF = \frac{DPF}{\sqrt{1 + THD^2}}$$

where
   DPF is the displacement power factor
   THD is the total harmonic distortion

We can explain DPF that the fundamental harmonic of the current has a delay angle θ (or ϕ), DPF = cosine θ (or cosine ϕ). THD is calculated by (4.20). Most AC/DC uncontrolled and controlled rectifiers have poor PF except the single-phase full-wave uncontrolled bridge (Graetz) rectifier with R load. All three-phase uncontrolled and controlled rectifiers have the input current fundamental harmonic delaying its corresponding voltage by an angle 30° plus α, where α is the firing angle of the controlled rectifier. Consequently, AC/DC rectifiers naturally have a poor power factor. In order to maintain power quality, PFC is necessary. Implementing the PFC means

- Reducing the phase difference between the line voltage and current (DPF ⇒ 1)
- Shaping the line current to a sinusoidal waveform (THD ⇒ 0)

The first condition requires that the fundamental harmonic of the current has a delay angle $\theta \geq 0°$. The second condition requires that the harmonic components be as small as possible. In recent research there are following methods to implement PFC:

1. DC/DC converterized rectifiers
2. PWM boost-type rectifiers
3. Tapped-transformer converters
4. Single-stage power factor correction AC/DC converters
5. VIENNA rectifiers
6. Other methods

7.2 DC/DC CONVERTERIZED RECTIFIERS

A full-wave diode rectifier with R load has high PF. If this rectifier supplies an R–C load, the PF is poor. Using a DC/DC converter in this circuit can largely improve PF. The PFC rectifier circuit is shown in Figure 7.1.

The resistor emulation of the PFC rectifier is carried out by the DC/DC converter. The input to the DC/DC converter is a fully rectified sinusoidal voltage waveform. A constant DC voltage is maintained at the output of the PFC rectifier. The DC/DC converter is switched at a switching frequency f_s that is many times higher than the line frequency f. The input current waveform into the diode bridge is modified to contain a strong fundamental sinusoid at the line frequency but with harmonics at several times higher frequency than the line frequency.

Since the switching frequency f_s is very high in comparison with the line frequency f, the input and output voltages of the PFC rectifier may be considered to be constant throughout a switching period. Thus, the PFC rectifier can be analyzed like a regular DC/DC converter as given by

$$v_s = V_s \sin\theta$$

$$v_1 = V_s |\sin\theta| \quad \text{with} \quad \theta = 2\pi ft \tag{7.1}$$

The voltage transfer ratio of the PFC rectifier is required to vary with the angle θ in a half supply period. The voltage transfer ratio of the DC/DC converter is

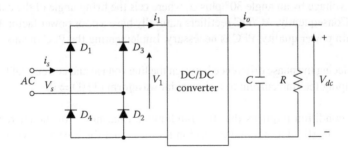

FIGURE 7.1 PFC rectifier.

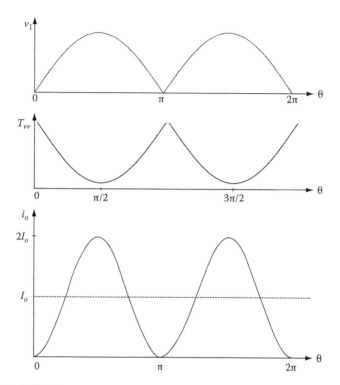

FIGURE 7.2   DC/DC converter output current required.

$$T_{vv}(\theta) = \frac{\overline{V_{dc}}}{V_1(\theta)} = \frac{\overline{V_{dc}}}{V_s \sin\theta} \quad \text{with} \quad f_s \gg f \tag{7.2}$$

where $\overline{V_{dc}}$ is the local average DC output voltage.

The T_{vv} in a supply period is shown in Figure 7.2. The high voltage transfer ratio at the vicinity of $\omega t = 0°$ and $180°$ angle can be achieved by using many converters such as boost, buck-boost, or flyback converters.

To prove this technique, a full-wave diode rectifier with $R-C$ load ($R = 100\ \Omega$ and $C = 100\,\mu\text{F}$) plus a buck-boost converter is investigated. Before applying any converter, the input voltage and current waveforms are shown in Figure 7.3. The fundamental harmonic of the input current delays the input voltage by the angle $\theta = 33.45°$.

The harmonics (FFT spectrum) of the input current are shown in Figure 7.4. The harmonic values are listed in Table 7.1.

The THD of the input current is obtained as $\text{THD}^2 = \sum_{i=2}^{\alpha}\left(I_i/I_1\right)^2 = 0.4625$

and displacement power factor is obtained to be $\cos(33.45°) = 0.834$. Therefore, the power factor can be calculated:

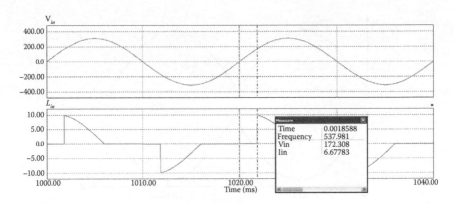

FIGURE 7.3 Input voltage and current waveforms.

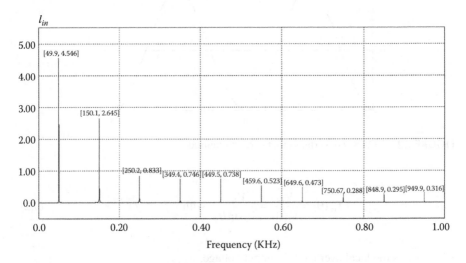

FIGURE 7.4 FFT spectrum of the input current.

$$PF = \frac{DPF}{\sqrt{1 + THD^2}} = \frac{\cos 33.45}{\sqrt{1 + 0.4625}} = 0.689 \qquad (7.3)$$

A buck-boost converter (see Figure 5.7) is used for this purpose. The circuit diagram is shown in Figure 7.5.

The input voltage is 311 V (peak)/50 Hz. The duty ratio k is calculated as 20 chopping periods for a half cycle. For one cycle, there are 40 chopping periods (keep same duty ratio) corresponding to its frequency to be 2 kHz. The inductance value was set by $L = 0.6$ mH and capacitance value was set 800 μF to maintain the output voltage as close as 200 V. Duty ratio k was calculated to set a constant DC output voltage of 200 V. For a detailed calculation of duty ratio k, see Table 7.2.

TABLE 7.1

Harmonic Current Values of Normal AC to DC Converter

Current	Frequency (Hz)	Fourier Component
I1	50	4.546
I3	150	2.645
I5	250	0.833
I7	350	0.746
I9	450	0.738
I11	550	0.523
I13	650	0.473
I15	750	0.288
I17	850	0.295
I19	950	0.316

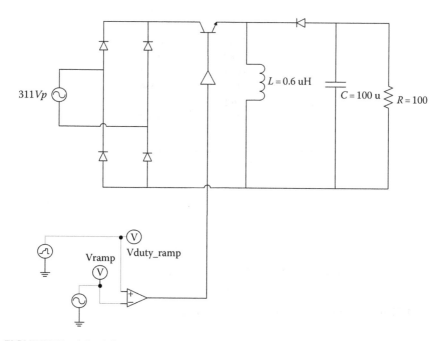

FIGURE 7.5   A buck-boost converter is used for PFC with R–C load.

The duty ratio k waveform in two and half cycles is shown in Figure 7.6 and the switch (transistor) turn-on and turn-off in half cycle is shown in Figure 7.7.

The input voltage and current waveforms are shown in Figure 7.8. From the waveform, we can see that the fundamental harmonic delay angle θ is about 3.21°. The output voltage of the buck-boost converter is 200 V as shown in Figure 7.9.

TABLE 7.2
Duty Ratio k in the 20 Chopping
Periods in Half Cycle

| | Input Voltage = 311 | |
ωt (°)	sin (ωt) (V)	k
9	48.65	0.804
18	96.1	0.676
27	141.2	0.586
36	182.8	0.522
45	219.9	0.476
54	251.6	0.443
63	277.1	0.419
72	295.8	0.403
81	307.2	0.394
90	311	0.391
99	307.2	0.394
108	295.8	0.403
117	277.1	0.419
126	251.6	0.443
135	219.9	0.476
144	182.8	0.522
153	141.2	0.586
162	96.1	0.676
171	48.65	0.804
180	0	∞

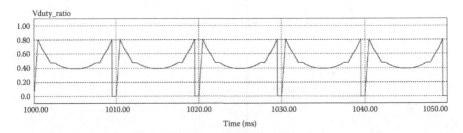

FIGURE 7.6 Duty ratio k waveform in two and half cycles.

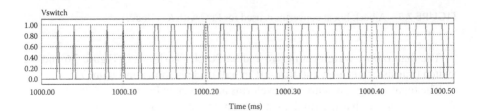

FIGURE 7.7 Switch turn-on and turn-off waveform in half cycle.

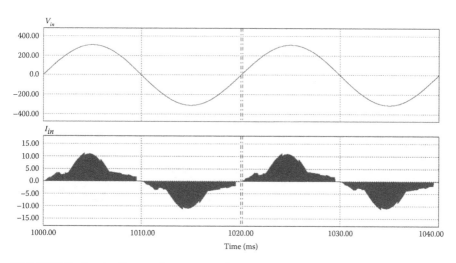

FIGURE 7.8   Input voltage and current waveforms.

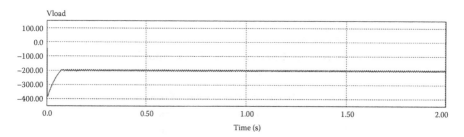

FIGURE 7.9   Output voltage of the buck-boost converter.

The FFT spectrum of the input current is shown in Figure 7.10, and the harmonic components are shown in Table 7.3.

From the data in Table 7.3, the THD of the input current is obtained as $\text{THD}^2 = \sum_{i=2}^{\alpha} (I_i / I_1)^2 = 0.110062$ and displacement power factor is obtained to be $\cos (3.21°) = 0.998431$. Therefore, the power factor can be calculated

$$PF = \frac{DPF}{\sqrt{1+\text{THD}^2}} = \frac{\cos 3.21}{\sqrt{1+0.110062}} = 0.95 \tag{7.4}$$

Using this technique, the power factor is significantly improved from 0.689 to 0.95.

From the previous investigation, we know that using a buck-boost converter to implement PFC can be successful, but the output voltage is negative polarity. If we use a positive-output Luo converter, SEPIC, or positive output buck-boost converter, we can obtain a positive-output voltage.

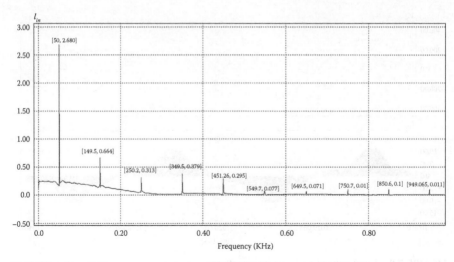

FIGURE 7.10   FFT spectrum of the input current.

TABLE 7.3
Harmonic Components of the Input Current

Current	Frequency (Hz)	Fourier Component
I1	50	2.680
I3	150	0.664
I5	250	0.313
I7	350	0.379
I9	450	0.295
I11	550	0.077
I13	650	0.071
I15	750	0.010
I17	850	0.100
I19	950	0.011

Example 7.1 A positive-output (P/O) Luo converter (see Figure 5.11) to implement PFC in a single-phase diode rectifier with R–C load. The AC supply voltage is 240 V/50 Hz, and the required output voltage is 200 V. The switching frequency is 4 kHz. Determine the duty cycle k in a half supply period (10 ms). Other component values for reference: $R = 100\ \Omega$, $C = C_O = 20\ \mu F$, and $L_1 = L_2 = 10\ mH$.

Solution: Since supply frequency is 50 Hz and switching frequency is 4 kHz, there are 40 switching periods in a half supply period (10 ms). The voltage transfer gain of the P/O Luo converter is

$$V_O = \frac{k}{1-k} V_{in}$$

$$k = \frac{V_O}{V_O + V_{in}} = \frac{200}{200 + 240\sqrt{2}\sin\omega t}$$

Duty cycle k is listed in Table 7.4.

7.3 PWM BOOST-TYPE RECTIFIERS

Using this method we can achieve unity power factor (UPF). In order to have the UPF, that is, PF = 1, the current from the diode bridge must be identical in shape and in phase with the supply voltage waveform. Hence,

$$i_1 = I_s |\sin\theta| \tag{7.5}$$

The input and output powers averaged over a switching period are given as

$$P_{in} = V_s I_s \sin^2\theta$$

$$P_o = \overline{V_{dc} i_o} \tag{7.6}$$

Assuming a lossless rectifier, the output current requirement is determined as

$$i_o = \frac{V_s I_s}{\overline{V_{dc}}} \sin^2\theta \tag{7.7}$$

The input and output powers are averaged over a supply period as

$$P_{in} = \frac{V_s I_s}{2}$$

$$P_o = \overline{V_{dc} I_o} \tag{7.8}$$

where I_o is the averaged DC output current.

The instantaneous output current is

$$i_o = \frac{V_s I_s}{\overline{V_{dc}}} \sin^2\theta = 2I_o \sin^2\theta$$

$$i_o = I_o(1 - \cos 2\theta) \tag{7.9}$$

The DC/DC converter output current required for a UPF, as a function of the angle θ, is shown in Figure 7.2.

Because the input current to the DC/DC converter is to be shaped, the DC/DC converter is operated in a current-regulated mode.

TABLE 7.4

Duty Ratio k in the 40 Chopping Periods in Half Cycle

ωt (°)	Input Voltage = $240\sqrt{2}\sin$ (ωt) (V)	k
4.5	26.6	0.88
9	53.1	0.79
13.5	79.2	0.72
18	104.9	0.66
22.5	129.9	0.61
27	154.1	0.56
31.5	177.3	0.53
36	199.5	0.5
40.5	220.4	0.48
45	240	0.45
49.5	258.1	0.44
54	274.6	0.42
58.5	289.4	0.41
63	302.4	0.4
67.5	313.6	0.39
72	322.8	0.38
76.5	330	0.377
81	335.2	0.374
85.5	338.4	0.371
90	339.4	0.37
94.5	338.4	0.371
99	335.2	0.374
103.5	330	0.377
108	322.8	0.38
112.5	313.6	0.39
117	302.4	0.4
121.5	289.4	0.41
126	274.6	0.42
130.5	258.1	0.44
135	240	0.45
139.5	220.4	0.48
144	199.5	0.5
148.5	177.3	0.53
153	154.1	0.56
157.5	129.9	0.61
162	104.9	0.66
166.5	79.2	0.72
171	53.1	0.79
175.5	26.6	0.88
180	0	∞

7.3.1   DC-SIDE PWM BOOST-TYPE RECTIFIER

The DC-side PWM boost-type rectifier is shown in Figure 7.11, where i_1^* is the reference of the desired value of the current i_1.

Here i_1^* has the same waveform shape as $|v_s|$. The amplitude of i_1^* should be such as to maintain the output voltage at a desired or reference level v_{dc}^*, in spite of the variation in load and the fluctuation of the line voltage from its nominal value. The waveform of i_1^* is obtained by measuring $|v_s|$ and multiplying it with the amplified error between the v_{dc}^* and v_{dc}. The actual current i_1 is measured. The status of the switch in the DC/DC converter is controlled by comparing the actual current with i_1^*.

Once i_1^* and i_1 are available, there are various ways to implement the current-mode control of the DC/DC converter.

7.3.1.1   Constant-Frequency Control

Here, the switching frequency f_s is kept constant. When i_1 reaches i_1^*, the switch in the DC/DC converter is turned off. The switch is turned on by a clock period at a fixed frequency f_s. This method is likely open-loop control. The operation is shown in Figure 7.12.

Example 7.2 A boost converter (see Figure 5.5) is used to implement PFC in the circuit shown in Figure 7.11a. The switching frequency is 2 kHz, $L = 10$ mH, $C_d = 20\,\mu$F, $R = 100\,\Omega$, and output voltage $V_O = 400$ V. The AC supply voltage is 240 V/50 Hz. Determine the duty cycle k in a half supply period (10 ms).

Solution: Since supply frequency is 50 Hz and switching frequency is 2 kHz, there are 20 switching periods in a half supply cycle (10 ms). The voltage transfer gain of the boost converter is

$$V_O = \frac{1}{1-k}V_{in}$$

$$k = \frac{V_O - V_{in}}{V_O} = \frac{400 - 240\sqrt{2}\sin\omega t}{400}$$

Duty ratio k is listed in Table 7.5.

7.3.1.2   Constant-Tolerance-Band (Hysteresis) Control

Here, the constant i_1 is controlled such that the peak-to-peak ripple (I_{rip}) in i_1 remains constant. With a preselected value of I_{rip}, i_1 is forced to be within the tolerance band $(i_1^* + I_{rip}/2)$ and $(i_1^* - I_{rip}/2)$ by controlling the switch status. This method is likely a closed-loop control. A current sensor is necessarily required to measure the particular current i_1 to determine switch-on and switch-off. The operation is shown in Figure 7.13.

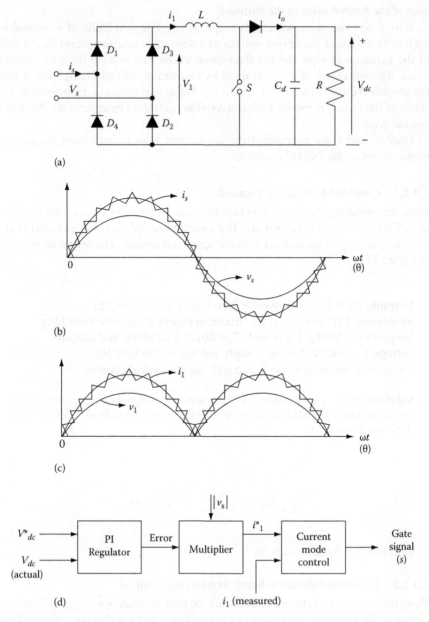

FIGURE 7.11 UPF diode rectifier with feedback control. (a) Circuit, (b) input voltage and current, (c) output voltage and current of the diode rectifier, and (d) control block diagram.

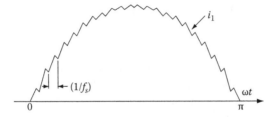

FIGURE 7.12 Operation indication of constant-frequency control.

TABLE 7.5

Duty Ratio k in the 20 Chopping Periods in Half Cycle (10 ms)

	Input Current = $240\sqrt{2}$	
ωt (°)	sin (ωt) (V)	k
9	53.1	0.867
18	104.9	0.738
27	154.1	0.615
36	199.5	0.501
45	240	0.4
54	274.6	0.314
63	302.4	0.244
72	322.8	0.193
81	335.2	0.162
90	339.4	0.152
99	335.2	0.162
108	322.8	0.193
117	302.4	0.244
126	274.6	0.314
135	240	0.4
144	199.5	0.501
153	154.1	0.615
162	104.9	0.738
171	53.1	0.867
180	0	∞

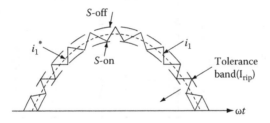

FIGURE 7.13 Operation indication of hysteresis control.

7.3.2 SOURCE-SIDE PWM BOOST-TYPE RECTIFIERS

In motor drive applications with regenerative braking, the power flow from the AC line is required to be bidirectional. A bidirectional converter can be designed using the phase angle delay control but at the penalty of poor input power factor and high waveform distortion in the line current. It is possible to overcome these limitations by using a switch-mode converter, as shown in Figure 7.14.

The rectifier being the dominant mode of operation, i_s is defined with a direction. An inductance L_s (which augments the internal inductance of the utility source) is included to reduce the ripple in i_s at a finite switching frequency. The four switching devices (IGBTs or MOSFETs) are operated in pulse-width modulation (PWM). Their switching frequency f_s is usually in kilo-Hertz. From Figure 7.14, we have

$$v_s = v_{conv} + v_L \qquad (7.10)$$

Assuming v_s to be sinusoidal, the fundamental-frequency components of v_{conv} and i_s in Figure 7.14 can be expressed as phasors $\overrightarrow{V_{conv1}}$ and $\overrightarrow{I_{s1}}$, respectively (the subscript 1 means the fundamental component). Choosing arbitrarily as the reference phasor $\overrightarrow{V_s} = V_s e^{j0^o}$, at the line frequency $\omega = 2\pi f$

$$\overrightarrow{V_s} = \overrightarrow{V_{conv1}} + \overrightarrow{V_{L1}} \qquad (7.11)$$

where

$$\overrightarrow{V_{L1}} = i\omega L_s \overrightarrow{I_{s1}} \qquad (7.12)$$

A phasor diagram corresponding to Equations 7.11 and 7.12 is shown in Figure 7.15, where $\overrightarrow{I_{s1}}$ lags $\overrightarrow{V_s}$ by an arbitrary phase angle θ.

The real power P supplied by the AC source to the converter is

$$P = V_s I_{s1} \cos\theta = \frac{V_s^2}{\omega L_s} \frac{V_{conv1}}{V_s} \sin\delta \qquad (7.13)$$

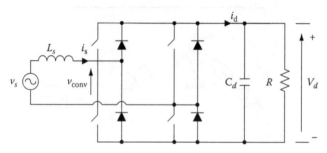

FIGURE 7.14   Switch-mode converter.

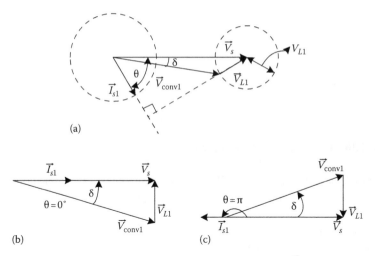

(a)

(b)   (c)

FIGURE 7.15 Phasor diagram: (a) the general vector diagram, (b) $\vec{V}_{convl}$ is delayed than $\vec{V}_s$, and (c) $\vec{V}_{convl}$ is ahead $\vec{V}_s$.

From Figure 7.15a,

$$V_{L1}\cos\theta = \omega L_s I_{s1}\cos\theta = V_{convl}\sin\delta \tag{7.14}$$

In the phasor diagram of Figure 7.15a, the reactive power Q supplied by the AC source is positive. It can be expressed as

$$Q = V_s I_{s1}\sin\theta = \frac{V_s^2}{\omega L_s}\left(1 - \frac{V_{convl}}{V_s}\cos\delta\right) \tag{7.15}$$

From Figure 7.15a, we also have

$$V_s - \omega L_s I_{s1}\sin\theta = V_{convl}\cos\delta \tag{7.16}$$

From these equations, it is clear that for a given line voltage v_s and the chosen inductance L_s, desired values of P and Q can be obtained by controlling the magnitude and the phase of v_{convl}.

Figure 7.15 shows how $\overrightarrow{V_{convl}}$ can be varied, keeping the magnitude of $\overrightarrow{I_{s1}}$ constant. The two special cases of rectification and inversion at a UPF are shown in Figure 7.15b and c. In both cases,

$$V_{convl} = \sqrt{V_s^2 + (\omega L_s I_{s1})^2} \tag{7.17}$$

In the circuit of Figure 7.14, V_d is established by charging the capacitor C_d through the switch mode converter. The value of V_d should be of a sufficiently large magnitude so that v_{convl} at the AC side of the converter is produced by a PWM (pulse-width

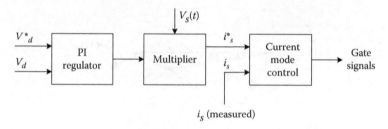

FIGURE 7.16 Block diagram of UPF operation.

modulation) that corresponds to a PWM in a linear region. The control circuit to regulate V_d in Figure 7.14 at its reference value v_d^* and to achieve a UPF of operation is shown in Figure 7.16. The amplified error between V_d and v_d^* is multiplied with the signal proportional to the input voltage v_s waveform to produce the reference signal i_s^*. A current-mode control such as a tolerance band control or a fixed-frequency control can be used to deliver i_s equal to i_s^*. The magnitude and direction of power flow are automatically controlled by regulating V_d at its desired value.

7.4 TAPPED-TRANSFORMER CONVERTERS

A simple method to improve the power factor is to use a tapped transformer. DC motor variable speed control drive-systems are widely used in industrial applications. Some applications require the DC motor running at lower speed. For example, winding machines and rolling mills mostly work in lower speed (lower than their 50% rated speed). If the DC motors are supplied by AC/DC rectifiers, the lower speed corresponds to lower armature voltage.

Assume the DC motor rated voltage corresponding to the rectifier firing angle α to be about 10°. The firing angle α will be about 60° if the motor runs at half rated speed. In the first case, the DPF is about (cos α), that is, DPF=0.98. In the second case, the DPF is about 0.48. It means that the power factor is very poor if the DC motor works in lower speed.

A tapped-transformer converter is shown in Figure 7.17a, which is a single-phase controlled rectifier. The original bridge consists of thyristors T_1–T_4. The transformer is tapped at 50% of the secondary winding. The third leg consists of thyristors T_5–T_6, which linked at the tapped point at the middle point of the secondary winding. Since the DC motor armature circuit has enough inductance, the armature current is always continuous. The motor armature voltage is

$$V_O = V_{dO} \cos \alpha \qquad (7.18)$$

If the motor works in lower speed, for example, 45% of its rated speed, the corresponding firing angle α is about 64°. The output voltage waveform from the original bridge is shown in Figure 7.17b. The fundamental harmonic component sine wave must have the delay angle $\phi_1 = \alpha = 64°$, and DPF=cos α. After Fourier Transform analysis and THD calculation, the voltage waveform in Figure 7.17b is 0.24. Therefore, the power factor is

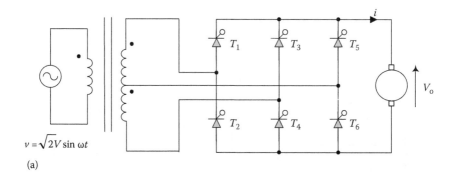

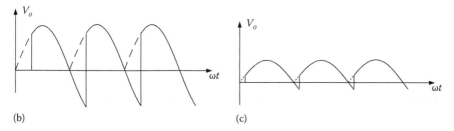

FIGURE 7.17 Tapped-transformer converter. (a) Circuit diagram, (b) output voltage waveform from original bridge, and (c) output voltage waveform from new leg.

$$PF = \frac{DPF}{\sqrt{1+THD^2}} = \frac{\cos 64°}{\sqrt{1+0.24^2}} = \frac{0.443}{1.028} = 0.43 \qquad (7.19)$$

Keeping the same armature voltage, we get the voltage from the legs 2 and 3, that is, the thyristors T_1 and T_2 are idle. It means the input voltage is reduced by half of the supplying voltage, and the firing angle α' is about 27.6°. The output voltage waveform from the legs 2 and 3 is shown in Figure 7.17c. The fundamental harmonic component sine wave must have the delay angle $\phi_1 = \alpha' = 27.6°$, and $DPF = \cos \alpha'$. After Fourier Transform analysis and THD calculation, the voltage waveform in Figure 7.17c is 0.07. Therefore, the power factor is

$$PF = \frac{DPF}{\sqrt{1+THD^2}} = \frac{\cos 27.6°}{\sqrt{1+0.07^2}} = \frac{0.8863}{1.0024} = 0.884 \qquad (7.20)$$

With comparison of the power factors in (7.19) and (7.20), it can be seen that the power factor has been significantly corrected.

This method is very simple and straightforward. The tapped point can be shifted to any other percentage (not fixed at 50%) depending on the applications.

A test rig has been constructed for collecting the measured results. The circuit is shown in Figure 7.18. The secondary voltage of the transformer is 230/115 V. The requested output voltage is set 80 V.

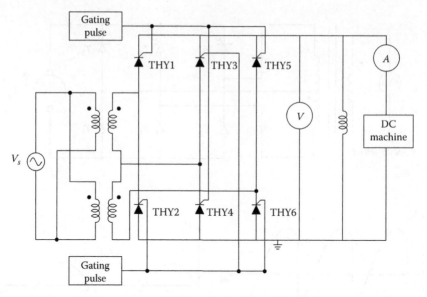

FIGURE 7.18 A single-phase controlled rectifier with tapped transformer.

If the supply voltage is 230 V, the firing angle is approximately 67°. The output voltage is shown in Figure 7.19, and the measured record is shown in Figure 7.20. The indication of the power factor in it is 0.64.

If the supply voltage is 115 V, the firing angle is approximately 39.4°. The output voltage is shown in Figure 7.21, and the measured record is shown in Figure 7.22. The indication of the power factor in it is 0.87.

If the output voltage increases to 103 V and the supply voltage remains 115 V, the firing angle is approximately 1°. The output voltage is shown in Figure 7.23, and the measured record is shown in Figure 7.24. The indication of the power factor in it is 0.98.

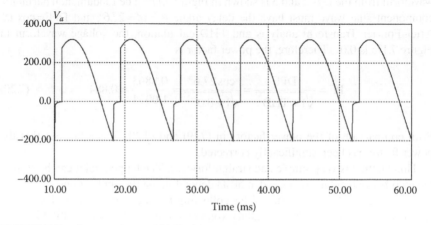

FIGURE 7.19 Output voltage of 80 V with input voltage 230 V.

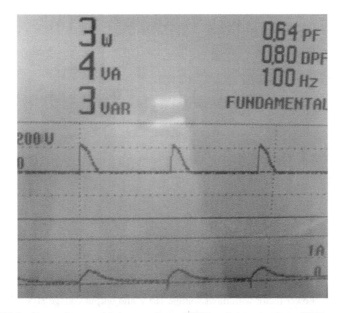

FIGURE 7.20 Power factor with input voltage 230 V and output voltage 80 V.

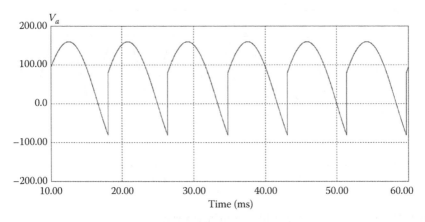

FIGURE 7.21 Output voltage of 80 V with input voltage 115 V.

7.5 SINGLE-STAGE POWER FACTOR CORRECTION AC/DC CONVERTERS

Double-current synchronous rectifier (DC-SR) converter is a popular circuit applied in computers [1,2]. Unfortunately, its power factor is not high. We introduce the single-stage PFC DC-SR converter here to improve its power factor nearly to be unity. The circuit diagram is shown in Figure 7.25.

The system consists of an AC/DC diode rectifier and a DC-SR converter [1]. Suppose that the output inductors L_1 and L_2 are equal to each other, $L_1 = L_2 = L_O$,

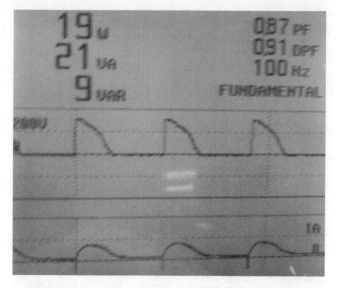

FIGURE 7.22   Power factor with input voltage 115 V and output voltage 80 V.

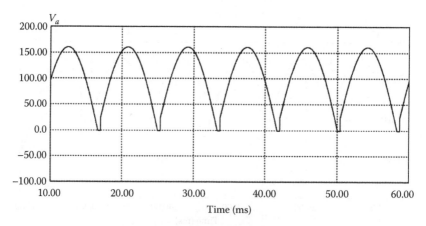

FIGURE 7.23   Output voltage of 103 V with input voltage 115 V.

which are called output inductors. There are three switches: main switch S and two auxiliary synchronous switches S_1 and S_2. It inherently exhibits high power factor because the PFC cell operates in continuous conduction mode (CCM). In addition, it is also free to suffer from high voltage stress across the bulk capacitor at light loads. In order to investigate the dynamic behaviors, the averaging method is used to drive the DC operating point and the small-signal model. A proportional-integral-differential (PID) controller is designed to achieve output voltage regulation despite variations in line voltage and load resistance.

In power electronic equipments, the PFC circuits are usually added between the bridge rectifier and the loads to eliminate high harmonics distortion of the line

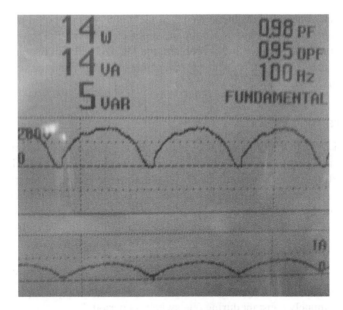

FIGURE 7.24   Power factor with input voltage 115 V and output voltage 103 V.

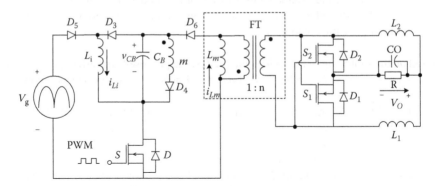

FIGURE 7.25   Proposed single-stage PFC double-current synchronous rectifier converter.

current. In general, they can be divided into two categories, the two-stage approach and the single-stage approach. In the two-stage approach, it includes a PFC stage and a DC/DC regulation stage. It has good PFC and fast output regulations, but the size and cost increase. To overcome the drawbacks, the graft scheme is proposed in Ref. [4]. Many single-stage approaches have been proposed in the literature [5–8]. It integrates a PFC cell and a DC/DC conversion cell to form a single-stage with a common switch. Therefore, the sinusoidal input current waveform and the output voltage regulation can be simultaneously achieved. It thereby meets the requirements of performances and costs.

However, there exists a high voltage stress across the bulk capacitor C_B at light loads if DC/DC cell operates in discontinuous current mode (DCM). To overcome this drawback, a negative magnetic feedback technique was proposed in the literature. However, the dead band exists in the input current and the power factor is thereby degraded. To deal with this problem, the DC/DC cell will operate in DCM. The voltage across the bulk capacitor is independent of loads, and the voltage stress is reduced effectively.

7.5.1 OPERATING PRINCIPLES

Figure 7.25 depicts the proposed forward single-stage high PFC converter topology. A physical three-winding transformer has turns ratio 1: n: m. A tertiary transformer winding, in series with diode D_4, is added to the converter for transformer flux resetting. The magnetizing inductance L_m is parallel with the ideal transformer. In the proposed converter, both PFC cell and DC/DC conversion cell are operating in CCM. To simplify the analysis of the circuit, the following assumptions are made:

1. The large-value bulk capacitor C_B and output capacitor C_o are sufficiently large so that the voltages across the bulk capacitor and output capacitor are approximately constant during one switching period T_S.
2. All switches and diodes of the converter are ideal. The switching time of the switch and the reverse recovery time of the diodes are negligible.
3. The inductors and the capacitors of the converter are considered to be ideal without parasitic components.

Based on the switching of the switch and diodes, the proposed converter operating in one switching period T_S can be divided into five linear stages described as follows:

Stage 1 $[0, t_1]$ (S: on, D_1: on, D_2: off, D_3: off, D_4: off, D_5: on, D_6: on): In the first stage, the switch S is turned on. The diodes (D_1, D_5, D_6) are turned on and the diodes (D_2, D_3, D_4) are turned off. Power is transferred from bulk capacitor C_B to the output via the transformer.

Stage 2 $[t_1, t_2]$ (S: off, D_1: off, D_2: on, D_3: on, D_4: on, D_5: off, D_6: off): The stage begins when the switch S is turned off. The diodes (D_2, D_3, D_4) are turned on, and the diodes (D_1, D_5, D_6) are turned off. The current i_{L_i} flows through the diode D_3 and charges the bulk capacitor C_B. The diode D_4 is turned on for transformer flux resetting. In this stage, the output power is provided by the inductor L_o.

Stage 3 $[t_2, t_3]$ (S: off, D_1: off, D_2: on, D_3: off, D_4: on, D_5: off, D_6: off): The stage begins at t_2 when the input current i_{L_i} falls to zero and thus diode D_3 is turned off. The switch S is still off. All diodes, except D_3, maintain their states as shown in the previous stage. During this stage, the voltages $(-v_{C_B}/m)$ and $(-v_o)$ are applied across the inductors L_m and L_o, and thus, the inductor currents continue to linearly decrease. The output power is also provided by the output inductor L_o.

Stage 4 $[t_3, t_4]$ (S: off, D_1: off, D_2: off, D_3: off, D_4: on, D_5: off, D_6: off): The stage begins when the current i_{L_o} decreases to zero and thus diode D_2 is turned off.

The switch S is still off. The diode D_4 is still turned on and the diodes (D_1, D_3, D_5, D_6) are still turned off. During this stage, the voltage $(-v_{C_B} / m)$ is applied across inductor L_m. The inductor current continues to linearly decrease. The output power is provided by the output capacitor C_o in this stage.

Stage 5 $[t_4, t_5]$ (S: off, D_1: off, D_2: off, D_3: off, D_4: off, D_5: off, D_6: off): The stage begins when the current i_{L_m} falls to zero and thus diode D_4 becomes off. The switch S is still off and all diodes are off. The output power is also provided by the output capacitor C_o. The operation of the converter returns to the first stage when the switch S is turned on again.

According to the analysis of the proposed converter, the key waveforms over one switching period T_S are schematically depicted in Figure 7.26. The slopes of the waveforms $i_{C_o}(t)$ and $i_{C_B}(t)$ are defined

$$m_{C_{o1}} = \frac{nv_{C_B} - v_{C_o}}{L_o}, \quad m_{C_{o2}} = -\frac{v_{C_o}}{L_o}$$

$$m_{C_{B1}} = -\left[\frac{v_{C_B}}{L_m} + \frac{n(nv_{C_B} - v_{C_o})}{L_o}\right], \quad m_{C_{B2}} = -\left(\frac{v_{C_B}}{L_i} + \frac{v_{C_B}}{m^2 L_m}\right), \quad m_{C_{B2}} = -\frac{v_{C_B}}{m^2 L_m} \quad (7.21)$$

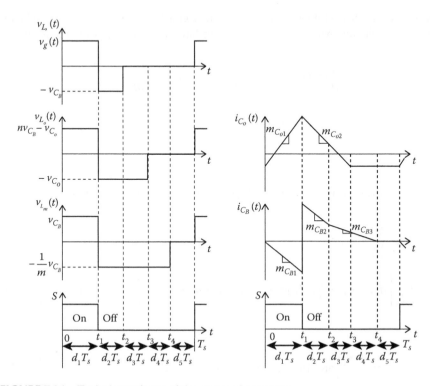

FIGURE 7.26 Typical waveforms of the proposed converter.

7.5.2   MATHEMATICAL MODEL DERIVATION

In this section, the small-signal model of the proposed converter can be derived by the averaging method. The moving average of a variable, voltage or current, over one switching period T_S is defined as the area, encompassed by its waveform and the time axis, divided by T_S.

7.5.2.1   Averaged Model over One Switching Period T_S

There are six storage elements in the proposed converter in Figure 7.25. The state variables of the converter are chosen as the current through the inductor and voltage across the capacitor. Since both PFC cell and DC/DC cells operate in DCM, the initial and final values of inductor currents vanish in each switching period T_S. From a system point of view, the inductor currents i_{L_i}, i_{L_o}, and i_{L_m} should not be considered as state variables. Only the bulk capacitor voltage v_{C_B} and the output capacitor voltage v_{C_o} are considered as state variables of the proposed converter.

For notational brevity, a variable with an upper bar denotes its moving average over one switching period T_S. With the aid of this definition, the averaged state-variable description of the converter is given by

$$C_B \frac{d\bar{v}_{C_B}}{dt} = \bar{i}_{C_B} \quad \text{and} \quad C_o \frac{d\bar{v}_{C_o}}{dt} = \bar{i}_{C_o} \tag{7.22}$$

Moreover, in discontinuous conduction, the averaged voltage across each inductor over one switching period is zero. Hence, we have three constraints of the form

$$L_i \frac{d\bar{i}_{L_i}}{dt} = \bar{v}_{L_i} = 0 \quad L_o \frac{d\bar{i}_{L_o}}{dt} = \bar{v}_{L_o} = 0 \quad L_m \frac{d\bar{i}_{L_m}}{dt} = \bar{v}_{L_m} = 0 \tag{7.23}$$

The output equation is expressed as

$$\bar{v}_o = \bar{v}_{C_o} \tag{7.24}$$

Based on the typical waveforms in Figure 7.26, the averaged variables are given by

$$\bar{i}_{C_B} = \frac{1}{T_S} \sum_{j=1}^{5} \text{area} \left[i_{C_B(j)} \right]$$

$$= \frac{1}{T_S} \left[d_1^2 T_S^2 m_{C_{B1}} + \frac{1}{2} d_2 T_S^2 \left[d_2 m_{C_{B2}} + 2(d_3 + d_4) m_{C_{B2}} \right] + \frac{1}{2} (d_3 + d_4)^2 T_S^2 m_{C_{B3}} \right] \tag{7.25}$$

$$\bar{i}_{C_o} = \frac{1}{T_S} \sum_{j=1}^{5} \text{area} \left[i_{C_o(j)} \right] = \frac{1}{T_S} \left[d_1 T_S^2 (d_1 + d_2 + d_3) \frac{(n\bar{v}_{C_B} - \bar{v}_{C_o})}{2L_o} - T_S \frac{\bar{v}_{C_o}}{R} \right] \tag{7.26}$$

where the notation area $[i_{C_B(j)}]$ denotes the area, encompassed by the waveform $i_{C_B}(t)$ and time axis, during the stage j. Similarly, we have

$$\bar{v}_{L_i} = \frac{1}{T_S}\sum_{j=1}^{5}\text{area}\,[v_{L_i(j)}] = \frac{1}{T_S}[d_1 T_S \bar{v}_g(t) + d_2 T_S(-\bar{v}_{C_B})]$$

$$\bar{v}_{L_m} = \frac{1}{T_S}\sum_{j=1}^{5}\text{area}\,[v_{L_m(j)}] = \frac{1}{T_S}\left[d_1 T_S \bar{v}_{C_B} + (d_2 + d_3 + d_4)T_S\left(-\frac{\bar{v}_{C_B}}{m}\right)\right] \quad (7.27)$$

$$\bar{v}_{L_o} = \frac{1}{T_S}\sum_{j=1}^{5}\text{area}\,[v_{L_o(j)}] = \frac{1}{T_S}[d_1 T_S(n\bar{v}_{C_B} - \bar{v}_{C_o}) + (d_2 + d_3)T_S(-\bar{v}_{C_o})]$$

Substituting (7.23) into the constraints (7.23) and performing mathematical manipulations gives

$$d_2 = \frac{\bar{v}_g(t)}{v_{C_B}}d_1 \quad d_3 = \left(\frac{n\bar{v}_{C_B}}{v_{C_o}} - 1 - \frac{\bar{v}_g(t)}{v_{C_B}}\right)d_1 \quad d_4 = \left(m + 1 - \frac{n\bar{v}_{C_B}}{v_{C_o}}\right)d_1 \quad (7.28)$$

Now, substituting (7.21) and (7.28) into (7.25) and (7.26), the averaged state equations (7.22) can be rewritten as

$$C_B\frac{d\bar{v}_{C_B}}{dt} = -d_1^2 T_S\frac{n(n\bar{v}_{C_B} - \bar{v}_{C_o})}{2L_o} + \frac{d_1^2 T_S \bar{v}_g^2(t)}{2L_i \bar{v}_{C_B}} \quad C_o\frac{d\bar{v}_{C_o}}{dt} = -\frac{\bar{v}_{C_o}}{R} + d_1^2 T_S\frac{n(n\bar{v}_{C_B} - \bar{v}_{C_o})}{2L_o \bar{v}_{C_o}}$$

$$(7.29)$$

The averaged rectified line current is given by

$$\bar{i}_g(t) = \frac{1}{T_S}\left\{\text{area}\left[i_{L_i(1)}\right]\right\} = \frac{1}{T_S}\left[\frac{1}{2}(d_1 T_S)^2\frac{\bar{v}_g(t)}{L_i}\right] \quad (7.30)$$

It reveals from (7.30) that $\bar{i}_g(t)$ is proportional to $\bar{v}_g(t)$. Thus, the proposed converter is provided with UPF.

7.5.2.2 Averaged Model over One Half Line Period T_L

Based on the derived averaged model described by (7.30) over one switching period T_S, we now proceed to develop the averaged model over one half line period T_L. Since the bulk capacitance and the output capacitance are sufficiently large, both capacitor voltages can be considered as constants over T_L. Therefore, the state equations of the averaged model over one half line period T_L can be given by

$$C_B \frac{d\langle \bar{v}_{C_B}\rangle_{T_L}}{dt} = \left\langle \frac{d_1^2 T_S}{2}\left[\frac{-n^2\bar{v}_{C_B} + n\bar{v}_{C_o})}{L_o} + \frac{\bar{v}_g^2(t)}{L_i\bar{v}_{C_B}} \right] \right\rangle_{T_L}$$

$$= \frac{1}{\pi}\int_0^\pi \frac{d_1^2 T_S}{2}\left[\frac{-n^2\bar{v}_{C_B} + n\bar{v}_{C_o})}{L_o} + \frac{v_m^2\sin^2(\omega t)}{L_i\bar{v}_{C_B}} \right] d(\omega t)$$

$$= \frac{d_1^2 T_S}{2}\left[\frac{-n^2\langle \bar{v}_{C_B}\rangle_{T_L} + n\langle \bar{v}_{C_o}\rangle_{T_L}}{L_o} + \frac{v_m^2}{2L_i\langle \bar{v}_{C_B}\rangle_{T_L}} \right] \quad (7.31)$$

$$C_o \frac{d\langle \bar{v}_{C_o}\rangle_{T_L}}{dt} = \left\langle -\frac{\bar{v}_{C_o}}{R} + d_1^2 T_S\frac{(n^2\bar{v}_{C_B}^2 - n\bar{v}_{C_B}\bar{v}_{C_o})}{2L_o\bar{v}_{C_o}} \right\rangle_{T_L}$$

$$= \frac{1}{\pi}\int_0^\pi \left[-\frac{\bar{v}_{C_o}}{R} + d_1^2 T_S\frac{(n^2\bar{v}_{C_B}^2 - n\bar{v}_{C_B}\bar{v}_{C_o})}{L_o\bar{v}_{C_o}} \right] d(\omega t)$$

$$= \frac{\langle \bar{v}_{C_o}\rangle_{T_L}}{R} + \frac{d_1^2 T_S\left[-n^2\langle \bar{v}_{C_B}\rangle_{T_L}^2 - n\langle \bar{v}_{C_B}\rangle_{T_L}\langle \bar{v}_{C_o}\rangle_{T_L} \right]}{2L_o\langle \bar{v}_{C_o}\rangle_{T_L}} \quad (7.32)$$

and the output equation is given by

$$\langle \bar{v}_o\rangle_{T_L} = \langle \bar{v}_{C_o}\rangle_{T_L} \quad (7.33)$$

Notably, (7.31) and (7.32) are nonlinear state equations that can be linearized around the DC operating point. The DC operating point can be determined by setting $d\langle \bar{v}_{C_B}\rangle_{T_L}/dt = 0$ and $d\langle \bar{v}_{C_o}\rangle_{T_L}/dt = 0$ in (7.31) and (7.32). Mathematically, we then successively compute the bulk capacitor voltage V_{C_B} and output voltage V_o as

$$V_{C_B} = \frac{1}{2n}\left(\sqrt{\frac{D_1^2 R T_S}{4L_i} + \frac{2L_o}{L_i}} + \sqrt{\frac{D_1^2 R T_S}{4L_i}} \right), \quad V_o = D_1\sqrt{\frac{R T_S}{4L_i}}V_m \quad (7.34)$$

The design specifications and component values of the proposed converter are listed in Table 7.6. In Table 7.6, it follows directly from (7.34) that $V_{C_B} = 146.6$ V and $V_o = 108$ V. Therefore, the proposed converter exhibits low-voltage stress across the bulk capacitor for a VAC 110 input voltage.

After determining the DC operating point, we proceed to derive the small-signal model linearized around the operating point. To proceed to small perturbations,

$$v_m = V_m + \tilde{v}_m, \quad d_1 = D_1 + \tilde{d}_1, \quad \langle \bar{v}_{C_B}\rangle_{T_L} = V_{C_B} + \tilde{v}_{C_B}$$

$$\langle \bar{v}_{C_o}\rangle_{T_L} = V_{C_o} + \tilde{v}_{C_o}, \quad \langle \bar{v}_o\rangle_{T_L} = V_o + \tilde{v}_o \quad (7.35)$$

TABLE 7.6

Design Specifications and Component Values of the Proposed Converter

Input peak voltage V_m	156 V	Duty ratio D_1	0.26
Input inductor L_i	75 μH	Switching period T_S	20 μs
Magnetizing inductor L_m	3.73 mH	Switching frequency f_s	50 kHz
Output inductor L_o	340 μH	Load resistance R	108 Ω
Bulk capacitor C_B	330 μF	Turns ratio $1:n:m$	1:2:1
Output capacitor C_o	1000 μF	PWM gain k_{pwm}	1/12 V^{-1}
Bulk capacitor voltage V_{C_B}	146.6 V	Output voltage V_o	108 V

and

$$V_m \gg \tilde{v}_m, \quad D_1 \gg \tilde{d}_1, \quad V_{C_B} \gg \tilde{v}_{C_B}, \quad V_{C_o} \gg \tilde{v}_{C_o}, \quad V_o \gg \tilde{v}_o \qquad (7.36)$$

are introduced into (7.31) and (7.32), and high-order terms are neglected, yielding the dynamic equations of the following form:

$$C_B \frac{d\tilde{v}_{C_B}}{dt} = \frac{D_1^2 T_S}{2}\left(-\frac{n^2}{L_o} - \frac{V_m^2}{2L_i V_{C_B}^2}\right)\tilde{v}_{C_B} + \frac{D_1^2 T_S}{2}\left(\frac{n}{L_o}\right)\tilde{v}_{C_o}$$

$$+ \frac{D_1^2 T_S}{2}\left(\frac{V_m}{L_i V_{C_B}}\right)\tilde{v}_m + D_1 T_S\left(\frac{-n^2 V_{C_B} + n V_{C_o}}{L_o} + \frac{V_m^2}{2L_i V_{C_B}}\right)\tilde{d}_1$$

$$= a_{11}\tilde{v}_{C_B} + a_{12}\tilde{v}_{C_o} + b_{11}\tilde{v}_m + b_{12}\tilde{d}_1 \qquad (7.37)$$

$$C_o \frac{d\tilde{v}_{C_o}}{dt} = \frac{D_1^2 T_S}{2}\left(\frac{2n^2 V_{C_B}}{L_o V_{C_o}} - \frac{n}{L_o}\right)\tilde{v}_{C_B} + \left(-\frac{1}{R} - \frac{D_1^2 T_S}{2}\cdot\frac{n^2 V_{C_B}^2}{L_o V_{C_o}^2}\right)\tilde{v}_{C_o}$$

$$+ 0 \cdot \tilde{v}_m + D_1 T_S\left(\frac{n^2 V_{C_B}^2}{L_o V_{C_o}} - \frac{n V_{C_B}}{L_o}\right)\tilde{d}_1$$

$$= a_{21}\tilde{v}_{C_B} + a_{22}\tilde{v}_{C_o} + b_{21}\tilde{v}_m + b_{22}\tilde{d}_1 \qquad (7.38)$$

The parameters are defined as

$$a_{11} = \frac{-D_1^2 T_S}{2}\left(\frac{n^2}{L_o} + \frac{V_m^2}{2L_i V_{C_B}^2}\right), \quad a_{12} = \frac{D_1^2 T_S}{2}\left(\frac{n}{L_o}\right)$$

$$a_{21} = \frac{D_1^2 T_S}{2}\left(\frac{2n^2 V_{CB}}{L_o V_{C_o}} - \frac{n}{L_o}\right), \quad a_{22} = -\left(\frac{1}{R} + \frac{D_1^2 T_S}{2} \cdot \frac{n^2 V_{CB}^2}{L_o V_{C_o}^2}\right)$$

$$b_{11} = \frac{D_1^2 T_S}{2}\left(\frac{V_m}{L_i V_{CB}}\right), \quad b_{12} = D_1 T_S\left(\frac{-n^2 V_{CB} + n V_{C_o}}{L_o} + \frac{V_m^2}{2 L_i V_{CB}}\right)$$

$$b_{21} = 0, \quad b_{22} = D_1 T_S\left(\frac{n^2 V_{CB}^2}{L_o V_{C_o}} - \frac{n V_{CB}}{L_o}\right)$$

Mathematically, the dynamic equations in (7.37) and (7.38) can be expressed in matrix form:

$$\begin{bmatrix} \dot{\tilde{v}}_{CB} \\ \dot{\tilde{v}}_{C_o} \end{bmatrix} = \begin{bmatrix} \dfrac{a_{11}}{C_B} & \dfrac{a_{12}}{C_B} \\ \dfrac{a_{21}}{C_o} & \dfrac{a_{22}}{C_o} \end{bmatrix}\begin{bmatrix} \tilde{v}_{CB} \\ \tilde{v}_{C_o} \end{bmatrix} + \begin{bmatrix} \dfrac{b_{11}}{C_B} & \dfrac{b_{12}}{C_B} \\ \dfrac{b_{21}}{C_o} & \dfrac{b_{22}}{C_o} \end{bmatrix}\begin{bmatrix} \tilde{v}_m \\ \tilde{d}_1 \end{bmatrix} \tag{7.39}$$

$$\tilde{v}_o = \begin{bmatrix} 0 & 1 \end{bmatrix}\begin{bmatrix} \tilde{v}_{CB} \\ \tilde{v}_{C_o} \end{bmatrix} \tag{7.40}$$

Now taking the Laplace transform for the dynamic equation, the resulting transfer functions from line to output and duty ratio to output are given by

$$\frac{\tilde{v}_o(s)}{\tilde{v}_m(s)} = \frac{\dfrac{b_{11}a_{21}}{C_B C_o}}{s^2 + \left(-\dfrac{a_{11}}{C_B} - \dfrac{a_{22}}{C_o}\right)s + \dfrac{a_{11}a_{22} - a_{12}a_{21}}{C_B C_o}}$$

$$\frac{\tilde{v}_o(s)}{\tilde{d}_1(s)} = \frac{\dfrac{b_{22}}{C_o}s + \dfrac{a_{21}b_{12} - a_{11}b_{22}}{C_B C_o}}{s^2 + \left(-\dfrac{a_{11}}{C_B} - \dfrac{a_{22}}{C_o}\right)s + \dfrac{a_{11}a_{22} - a_{12}a_{21}}{C_B C_o}} \tag{7.41}$$

7.5.3 SIMULATION RESULTS

PSpice simulation results presented in Figure 7.27 demonstrate that both PFC and DC/DC cells are operating in DCM. Both the input inductor current $i_{L_i}(t)$ and output inductor current $i_{L_o}(t)$ reach zero for the remainder of the switching period. Figure 7.28a presents the bulk capacitor voltage $V_{C_B} = 149$ V and Figure 7.28b presents the output capacitor voltage $V_{C_o} = 110$ V. They are close to the theoretical results $V_{C_B} = 146.6$ V, $V_{C_o} = 108$ V.

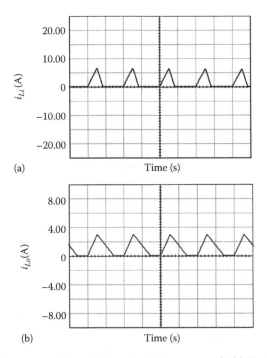

FIGURE 7.27 Current waveforms: (a) input inductor currents $i_{L_i}(t)$ (horizontal: 10 µs/div) and (b) output inductor current $i_{L_o}(t)$ (horizontal: 10 µs/div).

7.5.4 EXPERIMENTAL RESULTS

One prototype based on the topology depicted in Figure 7.25 is built and tested to verify the operating principle of the proposed converter. The experimental results are depicted in the following figures. Figure 7.29a presents the rectified line voltage and current, and Figure 7.29b presents the input line voltage and current. It reveals that the proposed converter has high power factor. According to the total harmonic distortion (THD) obtained in the simulation results, the power factor is calculated to be PF=0.999.

Figure 7.30 presents the waveform of the input inductor current $i_{L_i}(t)$ and output inductor current $i_{L_o}(t)$. Figure 7.31 presents the voltage ripples of the bulk capacitor voltage $V_{C_B}(t)$ and output capacitor voltage $V_{C_o}(t)$. Figure 7.32 presents the rectified line voltage and current and line voltage and current. The proposed converter exhibits low voltage stress and high power factor. The measured power factor of the converter is PF=0.998. The efficiency of the proposed converter is about 72%.

7.6 VIENNA RECTIFIERS

The VIENNA rectifier can be used to improve the power factor of a three-phase rectifier. However, its "critical input inductor" is calculated for the nominal load condition, and both the power factor and total harmonic distortion are degraded in a low-output power region. A novel strategy implementing reference compensation

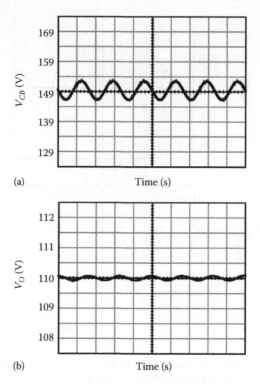

FIGURE 7.28 Ripples of (a) bulk capacitor voltage $V_{C_B}(t)$ (vertical: 5 V/div, horizontal: 5 ms/div) and (b) output capacitor voltage $V_{C_o}(t)$ (vertical: 0.5 V/div, horizontal: 5 ms/div).

current is proposed based on the operation principle of VIENNA rectifier in this section. This strategy can realize the three-phase three-level UPF rectifier. With the proposed control algorithm, the converter draws high-quality sinusoidal supply currents and maintains good DC-link-voltage regulation under wide load variation. Theoretical analysis is initially verified by digital simulation. Finally, experimental results of a 1 kVA laboratory prototype system confirm the feasibility and effectiveness of the proposed technique.

Diode rectifiers with smoothing capacitors have been widely used in many three-phase power electronic systems such as DC motor drives and switch mode power supplies. However, such topology injects large current harmonics into utilities, which result in the decrease of the power factor (PF). The expressions of the current total harmonic distortion (THD) and input power factor are given as

$$\text{THD} = 100 \times \frac{\sqrt{\sum_{h=2}^{\infty} I_{sh}^2}}{I_{s1}} \qquad (7.42)$$

$$\text{PF} = \frac{1}{\sqrt{1 + \text{THD}^2}} \cdot \text{DPF} \qquad (7.43)$$

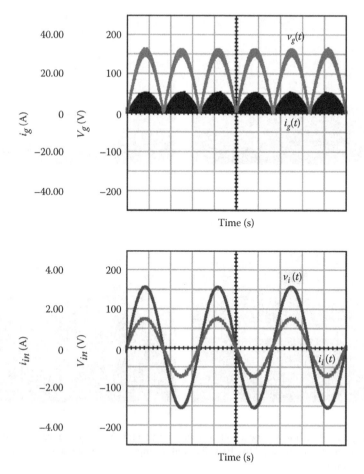

FIGURE 7.29 The line voltages and currents: (a) rectified line voltage and current (horizontal: 5 ms/div) and (b) input line voltage and current (horizontal: 5 ms/div).

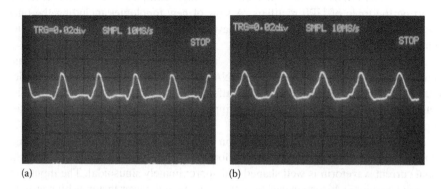

FIGURE 7.30 The inductor currents (horizontal: 10 μs/div): (a) input inductor currents $i_{L_i}(t)$ (vertical: 5 A/div) and (b) output inductor current $i_{L_o}(t)$ (vertical: 2 A/div).

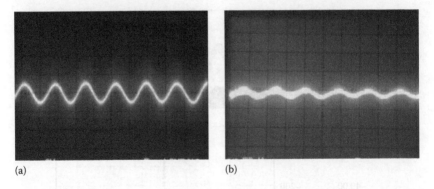

(a)   (b)

FIGURE 7.31 Ripples of (a) bulk capacitor voltage $V_{C_B}(t)$ (vertical: 5 V/div, horizontal: 5 ms/div) and (b) output capacitor voltage $V_{C_o}(t)$ (vertical: 0.5 V/div, horizontal: 5 ms/div).

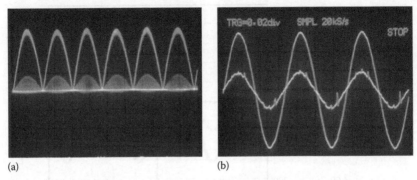

(a)   (b)

FIGURE 7.32 The line voltages and currents (horizontal: 5 ms/div): (a) rectified line voltage and current (vertical: 50 V/div, 10 A/div) and (b) input line voltage and current (vertical: 50 V/div, 2 A/div).

The international standards presented in IEC 1000-3-2 and EN61000-3-2 imposed harmonic restrictions to modern rectifiers, which stimulated a focused research effort on the topic of UPF rectifiers. A slew of new topologies including the ones based on three-level power conversion have been proposed to realize high-quality input waveforms [9–20].

Among the reported three-phase rectifier topologies, the three-phase star-connected switch three-level (VIENNA) rectifier [21–25] is an attractive choice because its switch voltage stress is one half of the total output voltage. This rectifier with three bidirectional switches, three input inductors, and two series-connected capacitors is shown in Figure 7.33.

Each bidirectional switch is turned on when the corresponding phase voltage crosses the zero-volt point and conducts for 30° of the line voltage cycle. Thus, the input current waveform is well shaped and approximately sinusoidal. The input current total harmonic distortion can be as low as 6.6% and power factor as high as 0.99. In addition, the bidirectional switches conduct at twice the line frequency; therefore, the switching losses are negligible.

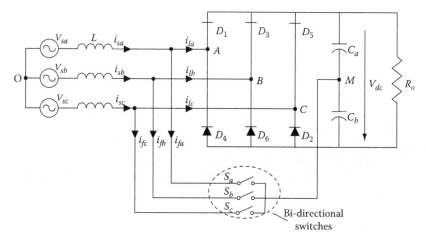

FIGURE 7.33 An AC/DC converter with bidirectional switches—VIENNA rectifier.

However, the optimal input inductance required to obtain such a result is usually large, and this technique was proposed for the rectifier operating with a fixed load and fixed optimal input inductor. So that the DC link voltage is sensitive to the load variation, and high performance is achieved within a limited output power range.

In order to overcome these drawbacks, some control strategies have been proposed [26–31]. A control strategy that takes into account the actual load level on the rectifier is proposed in Ref. [27]. With this method, high performance can be achieved within a wide output power range. The required optimal input inductance for a prototype rated at 8 kW is about 4 mH. This method is especially suitable for medium- to high-power applications. However, for low-power application (i.e., 1–5 kW), the required optimal input inductance should be larger; for example, around 24 mH for a converter with rated power 1.5 kW. This can result in a bulky and impractical structure.

The ramp comparison current control presented in Ref. [26] derives the duty cycle by comparison of the current error and the fixed-frequency carrier signal. The ripple current in the input inductor makes the current error noisy although synchronization is carefully considered. Another approach that features constant switching frequency was proposed based on integration control [28]. The input voltage sensors were eliminated in the integration control. However, significant low-frequency distortion can be observed in the input currents. Reference [29] adopted a synchronous reference-frame-based hysteresis current control as inner loop and DC-link voltage control as outer loop, but the reference-frame transformation is required, which increases the controller operation time (DSP in reference [29]). The hysteresis current controller was proposed in Refs. [30,31]. The switching signals are generated by comparison of the reference current template (sinusoidal) and the measured main currents. This approach is easy to implement, but there is a need to measure the DC current and equipment is costly.

A novel control method was proposed based on the operation principle of the VIENNA rectifier in this paper. The VIENNA rectifier is composed of two parts: an active compensation circuit and a conventional rectifier circuit. The harmonics injected by a conventional rectifier can be compensated by the active compensation circuit; thus, the input PF can be increased. The average real power consumed by load is supplied by the source, and the active compensation circuit does not provide or consume any average real power. Then the reference compensational current can be obtained. The conduction period of bidirectional switches (S_a, S_b, and S_c) is controlled by using the hysteresis current control (HCC). The idea lies in the high switching frequency, resulting in the input inductor size being effectively reduced. This control method does not need to measure the DC-link current, which results in a decrease in the equipment size and cost. The simulation and experimental results show that the input PF can be improved and input current harmonics can be effectively eliminated under wide load variation. The proposed control strategy can also maintain good DC-link voltage.

7.6.1 Circuit Analysis and Principle of Operation

The AC/DC converter topology shown in Figure 7.33 is composed of a three-phase diode rectifier with two identical series-connected capacitors and three bidirectional switches S_a, S_b, and S_c. The assembly of the switches consists of four diodes and a MOSFET to form a bidirectional switch, which is illustrated in Figure 7.34.

These bidirectional switches are controlled by using the HCC to ensure good supply current waveform, constant DC-link-voltage, and accurate voltage balance between the two capacitors. The voltage sources v_{sa}, v_{sb}, and v_{sc} in Figure 7.33 denote the three-phase AC system. The waveforms and the current of phase a i_{sa} are shown in Figure 7.35.

For the circuit analysis, in Figure 7.36, six topological stages are presented, corresponding to half cycle (0°–180°) referring to the input voltage v_{sa} shown in Figure 7.35; for simplicity, only the components where current is present were pictured at each of those intervals.

In the interval between 0° and 30° shown in Figure 7.36a and b, the polarity of the source voltages v_{sa} and v_{sc} is positive, v_{sb} negative. When the bidirectional switch S_a is on, the source current i_{sa} flows through S_a, and diodes D_5 and D_6 are on. The other diodes not shown in Figure 7.36a are off. When the bidirectional switch S_a is off, the current i_{sa} flowing through the input inductor is continued through the diode D_1, and diodes D_5 and D_6 are still on. The other diodes not shown in Figure 7.36b are off. The current commutation from S_a to D_1 is at a certain moment that is determined by HCC. Diodes D_5 and D_6 offer conventional rectifying wave. Switch S_a and diode D_1 turn on exclusively and offer the active compensation current.

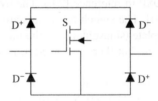

FIGURE 7.34 Construction of a bidirectional switch.

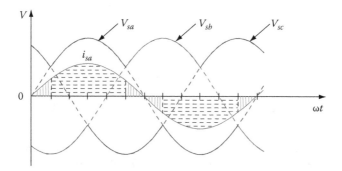

FIGURE 7.35   Waveforms of source voltages and current of phase a, i_{sa}.

In the interval between 30° and 60° shown in Figure 7.36c and d, the polarity of the source voltages v_{sa} and v_{sc} are positive, v_{sb} negative. When the bidirectional switch S_c is on, the source current i_{sc} flows through S_c, and diodes D_1 and D_6 are on. The other diodes not shown in Figure 7.36c are off. When the bidirectional switch S_c is off, the current i_{sc} flowing through the input inductor is continued through the diode D_5, and diodes D_1 and D_6 are still on. The other diodes not shown in Figure 7.36d are off. The current commutation from S_c to D_5 is at a certain moment that is determined by HCC. Diodes D_1 and D_6 offer conventional rectifying wave. Switch S_c and diode D_5 turn on exclusively and offer the active compensation current.

In the interval between 60° and 90° shown in Figure 7.36e and f, the polarity of the source voltage v_{sa} is positive, v_{sb} and v_{sc} negative. When the bidirectional switch S_c is on, the source current i_{sc} flows through S_c, diodes D_1 and D_6 are on. The other diodes not shown in Figure 7.36e are off. When the bidirectional switch S_c is off, the current i_{sc} flowing through the input inductor is continued through the diode D_2, and diodes D_1 and D_6 are still on. The other diodes not shown in Figure 7.36f are off. The current commutation from S_c to D_2 is at a certain moment that is determined by HCC. Diodes D_1 and D_6 offer a conventional rectifying wave. Switch S_c and diode D_2 turn on exclusively and offer the active compensation current.

In the interval between 90° and 120° shown in Figure 7.36g and h, the polarity of the source voltage V_a is positive, V_b and V_c negative. When the bidirectional switch S_b is on, the source current i_b flows through S_b, and diodes D_1 and D_2 are on. The other diodes not shown in Figure 7.36g are off. When the bidirectional switch S_b is off, the current i_b flowing through the input inductor is continued through the diode D_6, and diodes D_1 and D_2 are still on. The other diodes not shown in Figure 7.36h are off. The current commutation from S_b to D_6 is at certain moment that is determined by HCC. Diodes D_1 and D_2 offer a conventional rectifying wave. Switch S_b and diode D_6 turn on exclusively and offer the active compensation current.

In the interval between 120° and 150° shown in Figure 7.36i and j, the polarity of the source voltages v_{sa} and v_{sb} is positive, v_{sc} negative. When the bidirectional switch S_b is on, the source current i_{sb} flows through S_b, and diodes D_1 and D_2 are on. The other diodes not shown in Figure 7.36i are off. When the bidirectional switch S_b is off, the current i_{sb} flowing through the input inductor is continued through the diode D_3, and diodes D_1 and D_2 are still on. The other diodes not shown in Figure 7.36j

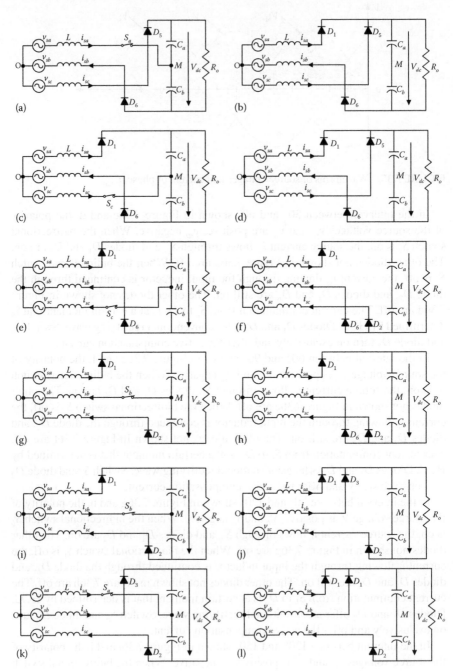

FIGURE 7.36 Topological stages for 0°–180°, referring to the input voltage V_{sa}: (a) 0°–15°, (b) 15°–30°, (c) 30°–45°, (d) 45°–60°, (e) 60°–75°, (f) 75°–90°, (g) 90°–105°, (h) 105°–120°, (i) 120°–135°, (j) 135°–150°, (k) 150°–165°, (l) 165°–180°.

are off. The current commutation from S_b to D_3 is at a certain moment that is determined by HCC. Diodes D_1 and D_2 offer conventional rectifying wave. Switch S_b and diode D_3 turn on exclusively and offer the active compensation current.

In the interval between 150° and 180° shown in Figure 7.36k and l, the polarity of the source voltages v_{sa} and v_{sb} are positive, v_{sc} negative. When the bidirectional switch S_a is on, the source current i_{sa} flows through S_a, and diodes D_3 and D_2 are on. The other diodes not shown in Figure 7.36k are off. When the bidirectional switch S_a is off, the current i_{sa} flowing through the input inductor is continued through the diode D_1, and diodes D_3 and D_2 are still on. The other diodes not shown in Figure 7.36l are off. The current commutation from S_a to D_1 is at a certain moment that is determined by HCC. Diodes D_3 and D_2 offer a conventional rectifying wave. Switch S_a and diode D_1 turn on exclusively and offer the active compensation current.

An active compensation circuit is composed of one of the bidirectional switches and an off-diode in the rectifier bridge legs, but the other legs act as conventional rectifier. Therefore, there are two circuits in the VIENNA rectifier, which are the conventional rectifier circuit and the active compensation circuit. Thus, the load average real power is supplied by the source (the same as conventional rectifier), and the active compensation circuit does not provide or consume any real power.

7.6.2 PROPOSED CONTROL ARITHMETIC

The proposed controller is based on the requirement that the source currents need to be balanced, undistorted, and in phase with the source voltages. The functions of the active compensation circuit are (1) to unitize supply power factor (2) to minimize average real power consumed or supplied by the active compensation circuit, (3) to compensate harmonics and reactive currents. To carry out the functions, the desired three-phase source currents of (7.44) must be in phase with the source voltages of (7.45):

$$\begin{cases} i_{sa} = I_m\sin(\omega t + \varphi) \\ i_{sb} = I_m\sin(\omega t + \varphi - 120°) \\ i_{sc} = I_m\sin(\omega t + \varphi + 120°) \end{cases} \qquad (7.44)$$

$$\begin{cases} v_{sa} = V_m\sin(\omega t + \varphi) \\ v_{sb} = V_m\sin(\omega t + \varphi - 120°) \\ v_{sc} = V_m\sin(\omega t + \varphi + 120°) \end{cases} \qquad (7.45)$$

where V_m and ϕ are the voltage magnitude and the phase angle of the source voltages, respectively. Under the conditions that the load active power is supplied by the source and the active compensation circuit does not provide or consume any real power, it is required to determine the current magnitude I_m from the sequential instantaneous voltage and real power components supplied to the load. According to the symmetrical-component transformation for the three-phase root mean square (rms) currents at each harmonic order, the three-phase instantaneous load currents can be expressed by

$$i_{lk} = \sum_{n=1}^{\infty} i_{lkn}^{+} + \sum_{n=1}^{\infty} i_{lkn}^{-} + \sum_{n=1}^{\infty} i_{lkn}^{0}, \quad k \in K \tag{7.46}$$

In (7.46), $K = \{a, b, c\}$; 0, +, and $-$ stand for zero-, positive-, and negative-sequence components, respectively, and n represents the fundamental (i.e., $n = 1$) and the harmonic components. Since the average real power consumed by the load over one period of time T must be supplied by the source and it requires that the active compensation circuit consumes or supplies null average real power, (7.47) through (7.51) must hold.

$$p_s = p_l + p_f \tag{7.47}$$

$$\bar{p}_s = \frac{1}{T} \int_0^T \sum_{k \in K} v_{sk} i_{sk} dt \tag{7.48}$$

$$\bar{p}_l = \frac{1}{T} \int_0^T \sum_{k \in K} v_{sk} i_{lk} dt \tag{7.49}$$

$$\bar{p}_f = 0 \tag{7.50}$$

$$\bar{p}_s = \bar{p}_l \tag{7.51}$$

Substituting (7.46) into (7.49) yields the sum of the fundamental and the harmonic power terms at the three sequential components, as given in (7.52):

$$\bar{p}_l = \bar{p}_{l1}^{+} + \bar{p}_{l1}^{-} + \bar{p}_{l1}^{0} + \bar{p}_{lh}^{+} + \bar{p}_{lh}^{-} + \bar{p}_{lh}^{0} \tag{7.52}$$

where

$$\bar{p}_{l1}^{+} = \frac{1}{T} \int_0^T \sum_{k \in K} v_{sk} i_{lk1}^{+} dt = \frac{1}{T} \int_0^T \sum_{k \in K} v_{sk} i_{sk} dt = \frac{3V_m I_m}{2} \tag{7.53}$$

and

$$\bar{p}_{l1}^{-} = \bar{p}_{l1}^{0} = \bar{p}_{lh}^{+} = \bar{p}_{lh}^{-} = \bar{p}_{lh}^{0} = 0 \tag{7.54}$$

Each power term in (7.54) is determined based on the orthogonal theorem for a periodic sinusoidal function. Then, (7.49) becomes

$$\overline{p}_s = \overline{p}_l = \overline{p}_{l1}^+ = \frac{1}{T} \int_0^T \sum_{k \in K} v_{sk} i_{sk} dt \tag{7.55}$$

By (7.51), (7.53), and (7.55), the desired source current magnitude at each phase is determined to be

$$I_m = \frac{2\overline{p}_l}{3V_m} = \frac{2 \int_0^T \sum_{k \in K} v_{sk} i_{lk} dt}{3TV_m} \tag{7.56}$$

and the source currents of (7.44) can be expressed by

$$i_{sk} = I_m \frac{v_{sk}}{V_m} = \frac{2\overline{p}_l}{3(V_m)^2} v_{sk}, \quad k \in K \tag{7.57}$$

The required current compensation at each phase by the active compensation circuit is then obtained by subtracting the desired source current from the load current as given:

$$i_{fk}^* = i_{lk} - i_{sk} = i_{lk} - \frac{2\overline{p}_l}{3(V_m)^2} v_{sk}, \quad k \in K \tag{7.58}$$

The average real power consumed or supplied by the active compensation circuit is expressed as

$$\overline{p}_f = \frac{1}{T} \int_0^T \sum_{k \in K} v_{sk} i_{fk} dt \tag{7.59}$$

Substituting (7.58) into (7.59) yields

$$\overline{p}_f = \frac{1}{T} \int_0^T \sum_{k \in K} v_{sk} i_{lk} dt - \frac{2\overline{p}_l}{3(V_m)^2} \frac{1}{T} \int_0^T \sum_{k \in K} v_{sk}^2 dt = \overline{p}_l - \frac{2\overline{p}_l}{3(V_m)^2} \frac{3(V_m)^2}{2} = \overline{p}_l - \overline{p}_l = 0$$

$$\tag{7.60}$$

Therefore, the active compensation circuit does not consume or supply average real power.

7.6.3   BLOCK DIAGRAM OF THE PROPOSED CONTROLLER FOR VIENNA RECTIFIER

Figure 7.37 depicts the block diagram of the control circuit based on the proposed approach to fulfill the function of the reference compensation current calculator. The source voltages are input to a phase-locked-loop (PLL), where the peak voltage magnitude V_m, the unity voltages (i.e., v_{sk}/V_m), and the period T are generated.

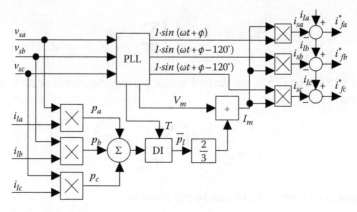

FIGURE 7.37 Block diagram of the controller.

The average real power of the load consumed is calculated by using Equation 7.55 and is input to a divider to obtain the desired source current amplitude I_m in (7.56). DI denotes the calculation of definite integral (DI). The desired source currents in (7.57) and reference compensation currents of the active compensation circuit in (7.58) are computed by using the voltage magnitude and the unity voltages.

Once the reference compensation currents are determined, they are input to a current controller to produce control signals to the bidirectional switches. The block diagram of the 1 proposed control scheme is shown in Figure 7.38. The bidirectional switches are controlled with HCC technique to ensure sinusoidal input current with UPF and DC link voltage. In addition, since the capacitor voltage must be maintained at a constant level, the power losses caused by switching and capacitor voltage variations are supplied by the source. The sum of the power losses $\bar{p}_{sw}$ is controlled via a PI controller and is then input to the reference compensation current calculator. Since the rectifier provides continuous input currents, the current

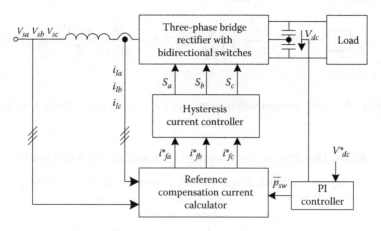

FIGURE 7.38 Block diagram of the control system.

stresses on the switching devices are smaller and the critical input inductor size can be reduced.

7.6.4 Converter Design and Simulation Result

To verify the performance of the proposed control strategy, a MATLAB®–SIMULINK® prototype of the rectifier is developed. A sinusoidal PWM (SPWM) voltage source inverter, a very popular topology in industry, is used as the DC/AC inverter for the intended rectifier–inverter AC motor drive topology, as depicted in Figure 7.39.

To illustrate the design feasibility of the proposed converter, a prototype with the following specifications is chosen:

1. Input line-to-line voltage 220 V
2. DC link reference voltage 370 V
3. Input inductance 5 mH
4. Rated output power 1 kW

A MATLAB/SIMULINK model for the proposed rectifier inverter structure is developed to perform the digital simulation. Figure 7.40 shows the converter input phase current waveform and its harmonic spectrum at rated output power operation. The same waveform for a conventional converter is shown in Figure 7.41.

Before improvement, the THD of the rectifier input current was found to be 91.5%, and the input power factor was 0.72. After the improvement, the input current THD is 3.8% and the input power factor is 0.999. Therefore, we can say that, with the proposed reference compensation current strategy, the harmonics are effectively reduced and the power factor is dramatically increased.

In order to show the performance of the converter under varying load conditions, it is operated below and above its rated value. The converter input phase current waveform and its harmonic spectrum at 50% rated output power are shown in Figure 7.42. The converter input power factor is found to be 0.996, and input current THD is 4.0%.

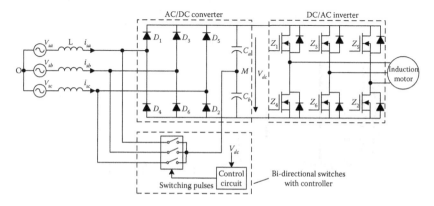

FIGURE 7.39 Complete diagram of the proposed unity power factor AC drive.

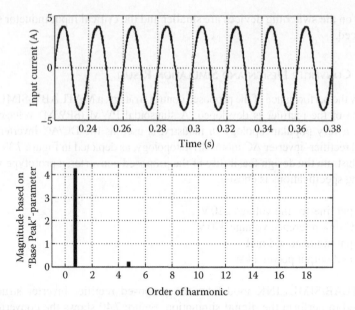

FIGURE 7.40 Input current and its spectral composition of the proposed scheme at rated load.

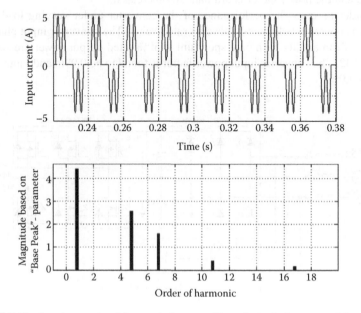

FIGURE 7.41 Input current and its spectral composition of a typical commercial converter.

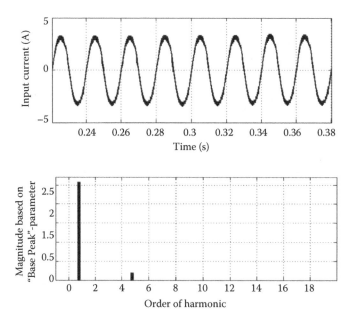

FIGURE 7.42 Input current and its spectral composition of the proposed scheme at 50% rated load.

The converter input phase current waveform and its harmonic spectrum at 150% rated output power are shown in Figure 7.43. The converter input power factor is found to be 0.999, and input current THD is 3.7%. It is evident that the proposed control strategy has a good adaptability to different load conditions. This strategy can be also used for rectifiers operating at various rated power levels.

Figure 7.44 illustrates the input phase currents and DC link voltage waveforms when the converter output power demand changes instantaneously from 50% to 100% of its rated value due to load disturbance. The load change was initiated at 0.26 s where the converter was in steady state. One can clearly see that the converter exhibits a good response to the sudden load variation. From this figure it can be seen that this proposed control technique has a good adaptability to load variation.

7.6.5 EXPERIMENTAL RESULTS

The control system is implemented using a single-board dSPACE 1102 microprocessor and developed under the integrated development of MATLAB–SIMULINK RTW provided by The MathWorks. A 1 kW hardware prototype of the rectifier–inverter structure depicted as shown in Figure 7.39 was constructed, and its performance was observed.

The rectifier input current and voltage waveforms before and after improvements are shown in Figures 7.45 and 7.46, respectively. A Fluke-43 spectrum analyzer with online numerical value illustration is used to monitor the waveforms. The input PF is shown online at the upper right-hand side of Figures 7.45 and 7.46.

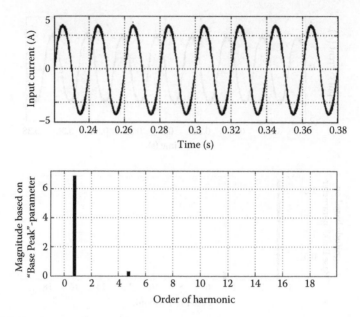

FIGURE 7.43 Input current and its spectral composition of the proposed scheme at 150% rated load.

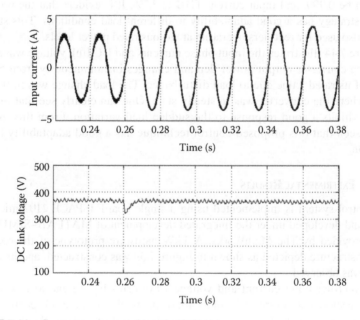

FIGURE 7.44 Converter response due to load change.

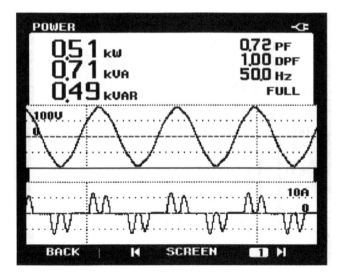

FIGURE 7.45   Input voltage and current of a typical conventional converter.

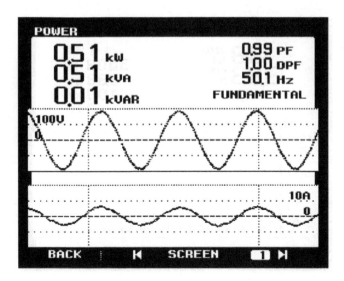

FIGURE 7.46   Input voltage and current of the proposed prototype.

Prior to improvement, the input current THD and power factor were 91.5% and 0.72, respectively.

The proposed scheme is able to improve the input current THD to 3.8% and the input power factor to 0.99. There is a remarkable improvement in power factor and THD. The experimental results are identical to the MATLAB predicted ones

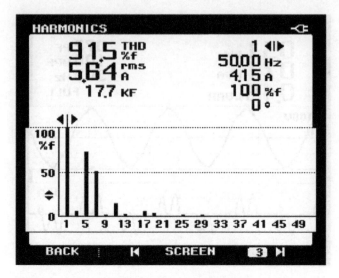

FIGURE 7.47   Input current FFT of a typical conventional converter.

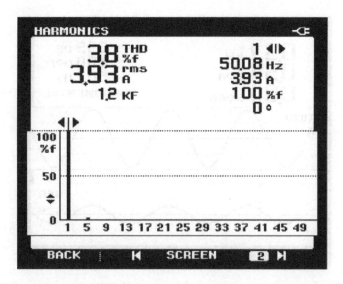

FIGURE 7.48   Input current FFT of the proposed prototype.

calculated based on the waveforms in Figures 7.40 and 7.41. Figures 7.47 and 7.48 show the experimental input current fast-Fourier transform (FFT) spectrum for a typical conventional converter and the proposed converter, respectively.

At 50% rated output power, the converter input PF is found to be 0.99 while the input current THD has increased to 4.0%, as shown in Figure 7.49. At 150% rated

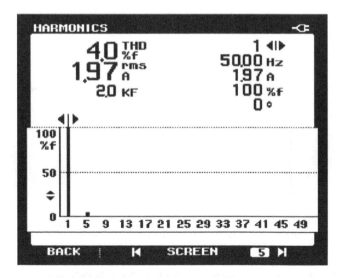

FIGURE 7.49 Input current FFT of the proposed prototype at 50% rated load.

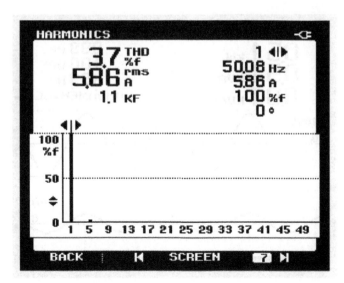

FIGURE 7.50 Input current FFT of the proposed prototype at 150% rated load.

output power, the converter input PF is found to be 0.99, and the input current THD is reduced to 3.7%, as shown in Figure 7.50.

Figure 7.51 shows the DC link voltage waveforms when the converter output power demand changes instantaneously from 50% to 100% of its rated value responding to load disturbance. One can see that with the proposed control strategy, the converter exhibits a good response to a sudden load variation.

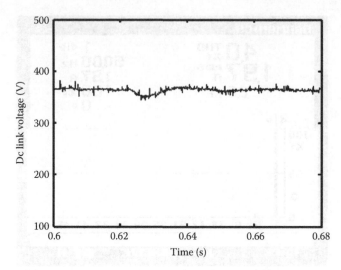

FIGURE 7.51   Converter response to a sudden load change in DC-link voltage.

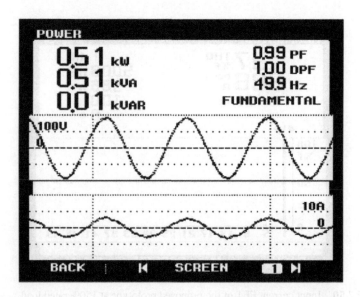

FIGURE 7.52   Converter input current and voltage for 3 mH input inductance.

To investigate the effect of input inductance, it is varied as well. Under 3 and 7 mH input inductances, the converter input currents and voltages are shown in Figures 7.52 and 7.53, respectively. These results illustrate that the proposed converter with bidirectional switches coupled with proposed strategy overcomes most of the shortcomings of the conventional converters, for example, change in the input PF due to output power, input inductance, and load torque variations.

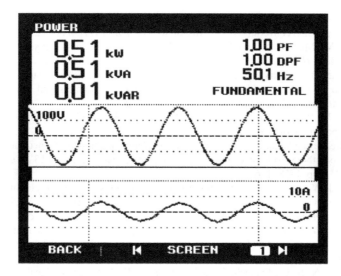

FIGURE 7.53 Converter input current and voltage for 7 mH input inductance.

REFERENCES

1. Luo, F. L. and Ye, H. 2004. *Advanced DC/DC Converters*. Boca Raton, FL: CRC Press.
2. Luo, F. L. 2005. A single-stage power factor correction AC/DC converter. *Proceedings of the International Conference IPEC'2005*, Singapore, pp. 513–518.
3. Mohan, N., Undeland, T. M., and Robbins, W. P. 2003. *Power Electronics: Converters, Applications and Design* (3rd edn.). Hoboken, NJ: John Wiley & Sons, Inc.
4. Wu, T. F. and Chen, Y. K. 1998. A systematic and unified approach to modeling PWM dc/dc converters based on the graft scheme. *IEEE Transactions on Industrial Electronics*, 45, 88–98.
5. Kheraluwala, M. H. 1991. Fast-response high power factor converter with a single power stage. *Proceedings of the IEEE-PESC*, New York, pp. 769–779.
6. Lee, Y. S. and Siu, K. W. 1996. Single-switch fast-response switching regulators with unity power factor. *Proceedings of the IEEE-APEC*, San Jose, CA, pp. 791–796.
7. Shen, M. and Qian, Z. 2002. A novel high-efficiency single-stage PFC converter with reduced voltage stress. *IEEE Transactions on Industry Applications*, 38, 507–513.
8. Qiu, M. 1999. Analysis and design of a single stage power factor corrected full-bridge converter. *Proceedings of the IEEE-APEC*, Singapore, pp. 119–125.
9. Zhang, S. and Luo, F. L. 2009. A novel reference compensation current strategy for three-phase three-level unity PF rectifier. *Proceedings of the IEEE-ICIEA'2009*, Xi'an, China, pp. 581–586.
10. Suryawanshi, H. M., Ramteke, M. R., Thakre, K. L., and Borghate, V. B. 2008. Unity-power-factor operation of three-phase AC-DC soft switched converter based on Boost active clamp topology in modular approach. *IEEE Transactions on Power Electronics*, 23, 229–236.
11. Lu, D. D., Iu, H. H., and Pjevalica, V. 2008. A single-stage AC/DC converter with high power factor, regulated bus voltage, and output voltage. *IEEE Transactions on Power Electronics*, 23, 218–228.
12. Chen, J. F., Chen, R. Y., and Liang, T. J. 2008. Study and implementation of a single-stage current-fed boost PFC converter with ZCS for high voltage applications. *IEEE Transactions on Power Electronics*, 23, 379–386.

13. Kong, P., Wang, S., and Lee, F. C. 2008. Common mode EMI noise suppression for bridgeless PFC converters. *IEEE Transactions on Power Electronics*, 23, 291–297.
14. Chen, M., Mathew, A., and Sun, J. 2007. Nonlinear current control of single-phase PFC converters. *IEEE Transactions on Power Electronics*, 22, 2187–2194.
15. Tutakne, D. R., Suryawanshi, H. M., and Tarnekar, S. G. 2007. Adaptive pulse synchronizing control for high-power-factor operation of variable speed DC-drive. *IEEE Transactions on Power Electronics*, 22, 2499–2510.
16. Greul, R., Round, S. D., and Kolar, J. W. 2007. Analysis and control of a three-phase, unity power factor Y-rectifier. *IEEE Transactions on Power Electronics*, 22, 1900–1911.
17. Bendre, A. and Venkataramanan, G. 2003. Modeling and design of a neutral point regulator for a three level diode clamped rectifier. *Proceedings of IEEE IAS'2003*, Salt Lake City, UT, pp. 1758–1765.
18. Kolar, J. W. and Drofenik, U. 1999. A new switching loss reduced discontinuous PWM scheme for a unidirectional three-phase/switch/level boost type PWM (VIENNA) rectifier. *Proceedings of the 21st INTELEC*, Copenhagen, Denmark, Paper 29-2.
19. Kolar, J. W. and Zach, F. C. 1994. A novel three-phase utility interface minimizing line current harmonics of high-power telecommunications rectifier modules. *Proceedings of the 16th INTELEC*, Vancouver, British Columbia, Canada, pp. 367–374.
20. Youssef, N. B. H., Fnaiech, F., and Al-Haddad, K. 2003. Small signal modeling and control design of a three-phase AC/DC Vienna converter. *Proceedings of the 29th IEEE IECON*, Roanoke, VA, pp. 656–661.
21. Mehl, E. L. M. and Barbi, I. 1997. An improved high power factor and low cost three-phase rectifier. *IEEE Transactions on Industry Applications*, 33, 485–492.
22. Salmon, J. 1995. Circuit topologies for PWM boost rectifiers operated from 1-phase and 3-phase ac supplies and using either single or split dc rail voltage outputs. *Proceedings of the IEEE Applied Power Electronics Conference*, Dallas, TX, pp. 473–479.
23. Kolar, J. W. and Zach, F. C. 1997. A novel three-phase utility interface minimizing line current harmonics of high power telecommunications rectifiers modules. *IEEE Transactions on Industrial Electronics*, 44, 456–467.
24. Kolar, J. W., Ertl, H., and Zach, F. C. 1996. Design and experimental investigation of a three-phase high power density high efficiency unity-powerfactor PWM (VIENNA) rectifier employing a novel integrated power semiconductor module. *Proceedings of APEC'96*, San Jose, CA, pp. 514–523.
25. Maswood, A. I., Yusop, A. K., and Rahman, M. A. 2002. A novel suppressed-link rectifier-inverter topology with unity power factor. *IEEE Transactions on Power Electronics*, 17, 692–700.
26. Drofenik, U. and Kolar, J. W. 1999. Comparison of not synchronized sawtooth carrier and synchronized triangular carrier phase current control for the VIENNA rectifier I. *Proceedings of IEEE ISIE*, New York, pp. 13–18.
27. Maswood, A. I. and Liu, F. 2005. A novel unity power factor input stage for AC drive application. *IEEE Transactions on Power Electronics*, 20, 839–846.
28. Qiao, C. and Smedley, K. M. 2003. Three-phase unity-power-factor star-connected switch (VIENNA) rectifier with unified constant-frequency integration control. *IEEE Transactions on Power Electronics*, 18, 952–957.
29. Liu, F. and Maswood, A. I. 2006. A novel variable hysteresis band current control of three-phase three-level unity PF rectifier with constant switching frequency. *IEEE Transactions on Power Electronics*, 21, 1727–1734.
30. Maswood, A. I. and Liu, F. 2006. A unity power factor front-end rectifier with hysteresis current control. *IEEE Transactions on Energy Conversion*, 21, 69–76.
31. Maswood, A. I. and Liu, F. 2007. A unity-power-factor converter using the synchronous-reference-frame-based hysteresis current control. *IEEE Transactions on Industry Applications*, 43, 593–599.

8 Classical DC/DC Converters

According to statistics, the existing DC/DC converters have more than 600 prototypes. Luo and Ye systematically sort them into six generations in their book *Advanced DC/DC Converters* [1,2]. According to the systematical categorization, the classical converters introduced in this book belong to a few generations.

8.1 INTRODUCTION

DC/DC conversion technology is a significant subject in research and industrial applications. Since the twentieth century, the DC/DC conversion technique has been greatly developed, and there are plenty of new topologies of DC/DC converters. DC/DC converters are widely used in communication equipment, hand-phone and digital cameras, computer hardware circuits, dental apparatus, and other industrial applications. Since there are numerous DC/DC converters, we have to sort them into six generations of DC/DC converters: classical/traditional converters (first generation), multiquadrant converters (second generation), switched-component converters (third generation), soft-switching converters (fourth generation), synchronous rectifier converters (fifth generation), and multielement resonant power converters (sixth generation).

The first-generation DC/DC converters are so-called classical or traditional converters. These converters perform in a single-quadrant mode and in low power range (up to 100 W). Since there are a large number of prototypes in this generation, they are sorted into six categories [1–5]:

1. Fundamental converters
2. Transformer-type converters
3. Developed converters
4. Voltage-lift converters
5. Super-lift converters
6. Ultra-lift converter

Fundamental converters such as the buck converter, boost converter, and buck-boost converter are named according to their functions. These three prototypes perform basic functions, so they are discussed in detail. Because of the effects of parasitic elements, the output voltage and power transfer efficiency of all these converters are restricted. Therefore, transformer-type and developed converters are created.

The voltage lift (VL) technique is a popular method that is widely applied in electronic circuit design. Applying this technique effectively overcomes the effects of parasitic elements and greatly increases the voltage transfer gain. Therefore, these DC/DC converters can convert the source voltage into a higher output voltage with high power efficiency, high power density, and a simple structure. Super-lift (SL) and ultra-lift (UL) techniques are even more powerful methods to increase the voltage transfer gain in power series.

The second-generation converters perform two-quadrant or four-quadrant operation with medium output power range (say hundred Watts to kilo-Watts). These converters are usually applied in industrial applications. For example, DC motor drives with multiquadrant operation. Since most second-generation converters are still made of capacitors and inductors, they are large.

The third generation converters are called switched-component DC/DC converters and are made of either inductors or capacitors, which are called switched-capacitor converters or switched-inductor converters, respectively. They usually perform two-quadrant or four-quadrant operation with high output power range (say 1000 W). Since they are made of only inductors or capacitors, they are small.

Switched-capacitor DC/DC converters are made of switched capacitors only (without any inductor). Since switched capacitors can be integrated into power semiconductor IC chips, they have limited size and work in high switching frequency. They have been successfully employed in the inductorless DC/DC converters and opened the way to build converters with high power density. Therefore, they have drawn much attention from researchers and manufacturers. However, most switched-capacitor converters in the literature perform single-quadrant operation and work in the push–pull status. In addition, their control circuit and topologies are very complex, especially, for the large difference between input and output voltages.

Switched-inductor DC/DC converters are made of only inductor and have been derived from four-quadrant choppers. They usually perform multiquadrant operation with very simple structure. The significant advantage of these converters is their simplicity and high power density. No matter how large the difference between the input and output voltages, only one inductor is required for each switched-inductor DC/DC converter. Therefore, they are widely required for industrial applications.

The fourth-generation converters are called soft-switching converters. The soft-switching technique involves many methods implementing resonance characteristics. A popular method is resonant switching. There are two main groups that are zero-current-switching (ZCS) and zero-voltage-switching (ZVS) converters. They usually perform in single-quadrant operation according to the literature.

ZCS and ZVS converters have large current and voltage stresses. In addition, the conduction duty cycle k and switching frequency f are not individually adjusted. In order to overcome these drawbacks, the zero-voltage-plus-zero-current-switching (ZV/ZCS) and zero-transition (ZT) converters are developed, which implement ZVS and ZCS technique operation. Since switches turn on and off at the moment that the voltage or current is equal to zero, the power losses during switch-on and off become zero. Consequently, these converters have high power density and transfer efficiency. Usually, the repeating frequency is not very high and the converter works in the resonance state; the components of higher order harmonics are very low. Therefore, the EMI is low, and EMS and EMC should be reasonable.

The fifth generation converters are called synchronous rectifier DC/DC converters. Corresponding to the development of the microelectronics and computer science, power supplies with low output voltage and strong current are widely required in industrial applications. These power supplies provide very low voltage (5, 3.3, 2.5, and 1.8–1.5 V) and strong current (30, 60, 100, till 200 A) with high power density and high power transfer efficiency (88%, 90%, up to 92%). Traditional diode bridge rectifiers are not available for this requirement. The new type of synchronous rectifier DC/DC converters can realize these technical features.

The sixth generation converters are called multielements resonant power converters (RPC). There are 8 topologies of 2-element RPC, 38 topologies of 3-element RPC, and 98 topologies of 4-element RPC. They are widely applied in military equipment and industrial applications.

The DC/DC converter family tree is shown in Figure 8.1.

In this book the input voltage is presented as V_1 and/or V_I (V_{in}), and output voltage is presented as V_2 and/or V_O. The input current is presented as I_1 and I_I (I_{in}); output current is presented as I_2 and/or I_O. The switching frequency is f, and the switching period is $T = 1/f$. Conduction duty cycle/ratio is k, and k is the ratio of the switching-on time over the period T. The value of k is in the range of $0 < k < 1$.

8.2 FUNDAMENTAL CONVERTERS

Fundamental converters are the buck converter, boost converter, buck-boost converter and positive-output buck-boost converter. Considering the *input current continuity*, we can divide all DC/DC converters into two main modes: continuous input current mode (CICM) and discontinuous input current mode (DICM). A boost converter operates in CICM, and buck converter and buck-boost converter operate in DICM [6–12].

8.2.1 BUCK CONVERTER

A buck converter is shown in Figure 8.2a. It converts the input voltage to output voltage that is less than the input voltage. Its switch-on and switch-off equivalent circuits are shown in Figure 8.2b and c, respectively.

8.2.1.1 Voltage Relations

When the switch S is on, the inductor current increases. For easy analysis in steady state, we assume that the capacitor C is large enough (the ripple can be negligible) $v_C = V_2$. We have

$$V_1 = v_L + v_C = L\frac{di_L}{dt} + v_C \tag{8.1}$$

$$\frac{di_L}{dt} = \frac{V_1 - v_C}{L} = \frac{V_1 - V_2}{L} \tag{8.2}$$

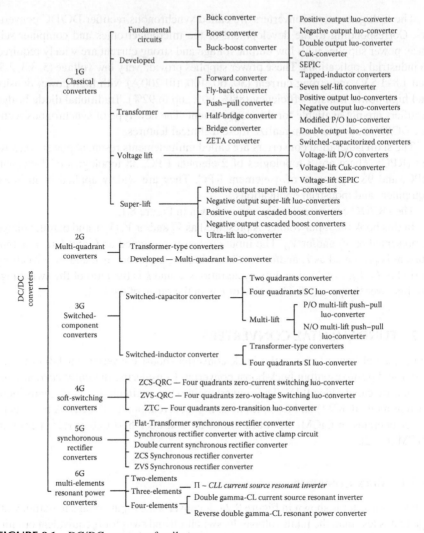

FIGURE 8.1   DC/DC converter family tree.

In this period of time length kT, the inductor current increases with a constant slope $(V_1 - V_2)/L$, which is shown in Figure 8.3. The inductor current starts at the initial value I_{min} and changes to a top value I_{max} at the end of the switch-closure period.

When the switch is off, the inductor current decreases and freewheels through the diode. We have the following equations:

$$0 = v_L + v_C \tag{8.3}$$

$$\frac{di_L}{dt} = -\frac{v_C}{L} = -\frac{V_2}{L} \tag{8.4}$$

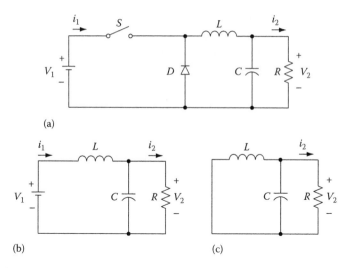

FIGURE 8.2 Buck converter and its equivalent circuits. (a) Buck converter, (b) switch-on, and (c) switch-off.

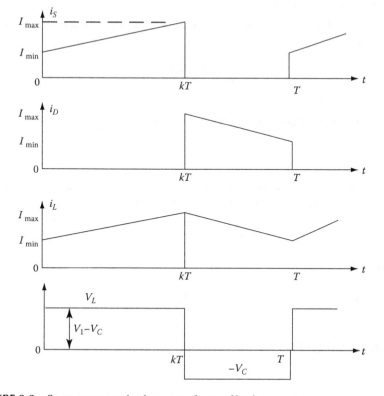

FIGURE 8.3 Some current and voltage waveforms of buck converter.

When the switch is off in the time interval $(1 - k)T$, the inductor current decreases with a constant slope $-V_2/L$ from I_{max} to I_{min}. The ending value I_{min} must be the same as that at the beginning of the period in steady state. The current increment during switch-on is equal to the current decrement during switch-off:

$$I_{max} - I_{min} = \frac{V_1 - V_2}{L} kT \tag{8.5}$$

$$I_{min} - I_{max} = \frac{-V_2}{L}(1 - k)T \tag{8.6}$$

Thus,

$$\frac{V_1 - V_2}{L} kT = \frac{V_2}{L}(1 - k)T \quad V_2 = kV_1 \tag{8.7}$$

The output voltage (capacitor voltage) depends solely on the duty cycle k and input voltage. From Figure 8.3 it can be seen that the input source current i_1 (is equal to switch current i_S) is discontinuous. Hence, buck converter operates in DICM.

8.2.1.2   Circuit Currents

From Figure 8.3, we can find the average value of inductor current easily by inspecting the waveform:

$$I_L = \frac{I_{max} + I_{min}}{2} \tag{8.8}$$

Applying Kirchhoff Current Law (KCL), we have

$$i_L = i_C + i_2 \tag{8.9}$$

Because the average capacitor current is zero in periodic operation, the result can be written by averaging values over one period of operation:

$$I_L = I_2 \tag{8.10}$$

By Ohm's law the current I_2 is given as

$$I_2 = \frac{V_2}{R} \tag{8.11}$$

Considering (8.10), (8.11), and (8.5) we get

$$I_{max} + I_{min} = 2\frac{V_2}{R} \tag{8.12}$$

$$I_{max} = kV_1 \left(\frac{1}{R} + \frac{1-k}{2L} T \right) \tag{8.13}$$

$$I_{min} = kV_1 \left(\frac{1}{R} - \frac{1-k}{2L} T \right) \tag{8.14}$$

8.2.1.3 Continuous Current Condition (Continuous Conduction Mode)

If the I_{min} is zero, it yields a relation for the minimum inductance that results in continuous inductor current:

$$L_{min} = \frac{1-k}{2} TR \tag{8.15}$$

8.2.1.4 Capacitor Voltage Ripple

The ripple-less condition in the capacitor voltage is now relaxed to allow a small ripple. This has only a second-order effect on the currents calculated in the previous section, so the previous results can be used without change.

As previously noted, the capacitor current must be entirely alternating to have periodic operation. The graph of capacitor current must be as shown in Figure 8.4 for continuous inductor current. The peak value of this triangular waveform is $+(I_{max} - I_{min})/2$. The resulting ripple in capacitor voltage depends on the area under the curve of capacitor current versus time. The charge added to the capacitor in a half-cycle is given by the triangular area above the axis:

$$\Delta Q = \frac{1}{2} \frac{I_{max} - I_{min}}{2} \frac{T}{2} = \frac{I_{max} - I_{min}}{8} T \tag{8.16}$$

The graph of capacitor voltage is also shown in the lower graph of Figure 8.4. The ripple in the voltage is exaggerated to show its effect. Minimum and maximum

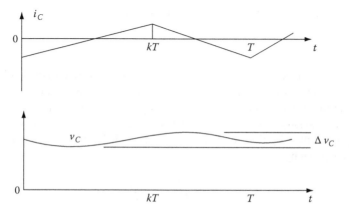

FIGURE 8.4 Waveforms of i_C and v_C.

capacitor voltage values occur at the time of capacitor current zero values. The peak-to-peak value of capacitor voltage ripple is given by

$$\Delta v_C = \frac{\Delta Q}{C} = \frac{I_{max} - I_{min}}{8C} T = \frac{k(1-k)V_1}{8CL} T^2 \tag{8.17}$$

Example 8.1 A buck converter has components: $V_1 = 20\,V$, $L = 10\,mH$, $C = 20\,\mu F$, $R = 20\,\Omega$, and switching frequency $f = 20\,kHz$ and conduction duty cycle $k = 0.6$. Calculate the output voltage and its ripple in steady state. Does this converter work in CCM or DCM?

Solution:

1. From (8.7), the output voltage is $V_2 = kV_1 = 0.6 \times 20 = 12\,V$.
2. From (8.17), the output voltage ripple is

$$\Delta v_2 = \Delta v_C = \frac{k(1-k)V_1}{8CL} T^2 = \frac{0.6 \times 0.4 \times 20}{8 \times 20\,\mu F \times 10\,mH \times (20\,kHz)^2} = 7.5\,mV$$

3. From (8.15), the inductor

$$L = 10\,mH > L_{min} = \frac{1-k}{2} TR = \frac{0.4}{2 \times 20\,kHz} 20 = 0.2\,mH$$

This converter works in CCM.

8.2.2 Boost Converter

If the three elements S, L, and D of buck converter are rearranged as shown in Figure 8.5a, a boost converter is created. Its equivalent circuits during switch-on and switch-off are shown in Figure 8.5b and c.

8.2.2.1 Voltage Relations

When the switch S is on, the inductor current increases:

$$\frac{di_L}{dt} = \frac{V_1}{L} \tag{8.18}$$

Since the diode is inversely biased, the capacitor supplies current to the load, and the capacitor current i_c is negative. Upon opening the switch, the inductor current must decrease so that the current at the end of the cycle can be the same as at the start of the cycle in steady state. For the inductor current to decrease, the value $V_c = V_2$ must be greater than V_1. For this interval with the switch open, the inductor current derivative is given by

$$\frac{di_L}{dt} = \frac{V_1 - V_C}{L} = \frac{V_1 - V_2}{L} \tag{8.19}$$

A graph of inductor current versus time is shown in Figure 8.6.

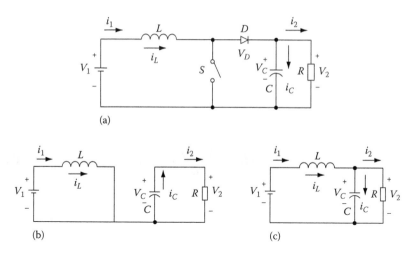

FIGURE 8.5   Boost converter. (a) Circuit, (b) switch-on, and (c) switch-off.

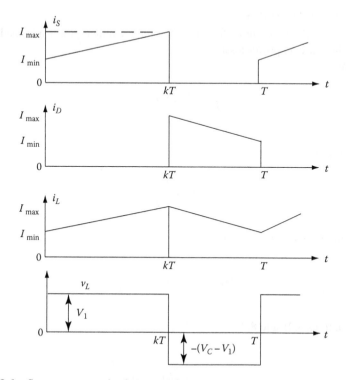

FIGURE 8.6   Some current and voltage waveforms.

The increment of i_L during switch-on must be equal to its decrement during switch-off.

$$I_{max} - I_{min} = \frac{V_1}{L} kT \qquad (8.20)$$

and

$$I_{min} - I_{max} = \frac{V_1 - V_C}{L}(1-k)T \qquad (8.21)$$

$$V_2 = V_C = \frac{V_1}{1-k} \qquad (8.22)$$

From Equation 8.22, we can see that if k is large, output voltage V_2 can be very large. In fact, as k approaches unity, the output voltage decreases rather than increases because of the effect of circuit parasitic elements. The value of k must be limited to less than certain upper limit (say 0.9) to prevent such a problem. Practical limits to this also become important for an increase in the voltage transfer gain, for example, 10. The switch may be open only for a very short time (0.1 T since $k=0.9$).

8.2.2.2 Circuit Currents

The I_{max} and I_{min} values can be found via the input average power and load average power if there are no power losses:
input power

$$P_{in} = \frac{I_{max} + I_{min}}{2} V_1 \qquad (8.23)$$

and output power

$$P_O = \frac{V_2^2}{R} \qquad (8.24)$$

Considering Equation 8.22, we have

$$I_{max} + I_{min} = 2\frac{V_1}{R(1-k)^2} \qquad (8.25)$$

From Equations 8.21 and 8.25,

$$I_{min} = \frac{V_1}{R(1-k)^2} - \frac{V_1}{2L}kT \qquad (8.26)$$

$$I_{max} = \frac{V_1}{R(1-k)^2} + \frac{V_1}{2L}kT \tag{8.27}$$

The load current value I_2 is known $I_2 = V_2/R$, and the average current flowing through the capacitor is zero. The instantaneous capacitor current is likely a triangle waveform, which is approximately $(i_L - I_2)$ during switch-off and $-I_2$ during switch-on. From Figure 8.6 the input source current $i_1 = i_S = i_L$ is continuous. Hence, buck converter operates in CICM.

8.2.2.3 Continuous Current Condition
When the I_{min} is equal to zero, the minimum inductance can be determined to ensure continuous inductor current. Using Equation 8.26 and solving it gives

$$L_{min} = \frac{k(1-k)^2}{2}TR \tag{8.28}$$

8.2.2.4 Output Voltage Ripple
The change of the charge across capacitor C is

$$\Delta Q = kTI_2 = kT\frac{V_2}{R} = \frac{kTV_1}{(1-k)R}$$

Therefore, the ripple voltage Δv_c across the capacitor C is

$$\Delta v_C = \frac{\Delta Q}{C} = \frac{kTV_2}{RC} = \frac{kTV_1}{(1-k)RC} \tag{8.29}$$

8.2.3 BUCK-BOOST CONVERTER

If the three elements S, D, and L in a boost converter are rearranged as shown in Figure 8.7a, a buck-boost type converter is created. Taking similar analysis to this converter, we can easily obtain all characteristics of a buck-boost converter in steady-state operating conditions.

8.2.3.1 Voltage and Current Relations
With the switch closed, inductor current changes:

$$\frac{di_L}{dt} = \frac{V_1}{L} \tag{8.30}$$

and

$$I_{max} - I_{min} = \frac{V_1}{L}kT \tag{8.31}$$

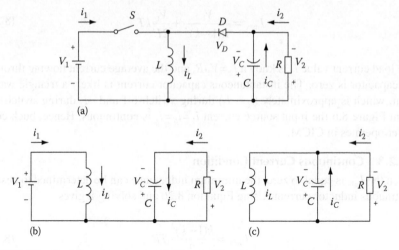

FIGURE 8.7 Buck-boost converter. (a) Circuit, (b) switch-on, and (c) switch-off.

With the switch open,

$$\frac{di_L}{dt} = -\frac{V_C}{L} \tag{8.32}$$

and

$$I_{min} - I_{max} = -\frac{V_C}{L}(1-k)T \tag{8.33}$$

Equating these two changes in i_L gives the result

$$V_2 = V_C = \frac{k}{1-k}V_1 \tag{8.34}$$

8.2.3.2 CCM Operation and Circuit Currents

Some waveforms are shown in Figure 8.8. The input source current $i_1 = i_S$ is discontinuous during switch-off. Hence, a buck-boost converter operates in DICM. Input average power then is found from input power

$$P_{in} = \frac{I_{max} + I_{min}}{2}kV_1 \tag{8.35}$$

and output power

$$P_O = \frac{V_2^{\,2}}{R} \tag{8.36}$$

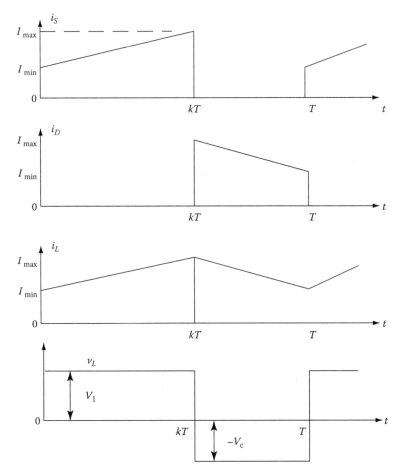

FIGURE 8.8 Some current and voltage waveforms.

Other parameters are listed as follows:

$$I_{max} + I_{min} = \frac{2kV_1}{R(1-k)^2} \tag{8.37}$$

$$I_{min} = \frac{kV_1}{R(1-k)^2} - \frac{V_1}{2L}kT \tag{8.38}$$

$$I_{max} = \frac{kV_1}{R(1-k)^2} + \frac{V_1}{2L}kT \tag{8.39}$$

The boundary for continuous current is found by setting I_{min} to zero; this defines a minimum inductance to ensure continuous inductor current. Using Equation 8.38 and solving it gives

$$L_{min} = \frac{(1-k)^2}{2} TR \qquad (8.40)$$

The ripple voltage Δv_c across the capacitor C is

$$\Delta v_C = \frac{\Delta Q}{C} = \frac{kTI_2}{C} = \frac{kTV_2}{RC} = \frac{k^2 TV_1}{(1-k)RC} \qquad (8.41)$$

Example 8.2 A buck-boost converter has components: $V_1 = 20\,V$, $L = 10\,mH$, $C = 20\,\mu F$, $R = 20\,\Omega$, and switching frequency $f = 50\,kHz$ and conduction duty cycle $k = 0.6$. Calculate the output voltage and its ripple in steady state. Does this converter work in CCM or DCM?

Solution:

1. From (8.34), the output voltage is

$$V_2 = V_C = \frac{k}{1-k} V_1 = \frac{0.6}{0.4} 20 = 30\ V$$

2. From (8.41), the output voltage ripple is

$$\Delta v_2 = \Delta v_C = \frac{kV_2}{fRC} = \frac{0.6 \times 20}{50\ kHz \times 20\ \Omega \times 20\ \mu F} = 0.6\ V$$

3. From (8.40), the inductor

$$L = 10\ mH > L_{min} = \frac{1-k}{2} TR = \frac{0.4}{2 \times 50\ kHz} 20 = 0.08\ mH$$

This converter works in CCM.

8.3   POSITIVE OUTPUT BUCK-BOOST CONVERTER

A traditional buck-boost converter has negative output voltage. In some applications, changing voltage polarity is not allowed. For example, the Li-ion battery is the common choice for most portable applications such as mobile phone and digital camera. With the increasing use of low voltage portable devices and increasing requirements of functionalities embedded into such devices, efficient power management techniques are needed for longer battery life. The voltage of a single Li-ion battery varies from 4.2 to 2.7 V. A DC–DC converter is needed to maintain the varying voltage of the Li-ion battery to a constant value of 3.3 V. This converter needs to operate in both the step-up as well as step-down conditions. Smooth transition from buck to the boost mode is the most desired criteria for a longer life of the battery. A positive output (P/O) buck-boost converter with two independent controlled switches is shown in Figure 8.9.

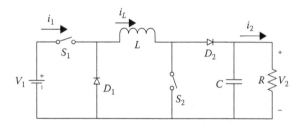

FIGURE 8.9   Circuit diagram of a positive output buck-boost converter.

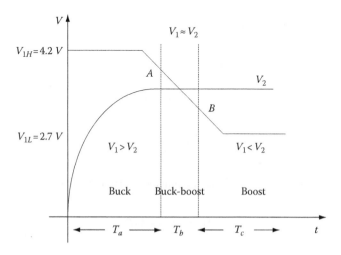

FIGURE 8.10   Input and output characteristics curves of the P/O buck-boost converter.

There are three operation modes shown in Figure 8.10:

1. Buck operation mode when V_1 is higher than V_2
2. Boost operation mode when V_1 is lower than V_2
3. Buck-boost operation mode when V_1 is similar to V_2

where $V_2 = 3.3$ V for this application.

This converter can work as a buck or boost converter depending on input–output voltages. The problem of output regulation with guaranteed transient performances for noninverting buck-boost converter topology is discussed. Various digital control techniques are addressed, which can smoothly perform the transition job. In the first two modes, the operation principles are the same as those of the buck converter and boost converter described in the previous section. The third mode of operation needs to be described here.

8.3.1   Buck Operation Mode

When the input voltage V_1 is higher than the output voltage V_2 (e.g., $V_1 > 1.03$ V_2, say 3.4 V), the positive buck-boost converter can be operated in the "buck operation mode." In this case, the switch S_2 is constantly open, and diode D_2 is constantly on. The remaining components are formed as a buck converter.

8.3.2 BOOST OPERATION MODE

When the input voltage V_1 is lower than the output voltage V_2 (e.g., $V_1 > 0.97 \, V_2$, say 3.2 V), the positive buck-boost converter can be operated in the "boost operation mode." In this case, the switch S_1 is constantly on, and diode D_1 is constantly blocked. The remaining components are formed as a boost converter.

8.3.3 BUCK-BOOST OPERATION MODE

When the input voltage V_1 is nearly equal to the output voltage V_2, for example, $3.2 \, V < V_1 < 3.4 \, V$, the positive buck-boost converter can be operated in the "buck-boost operation mode." In this case, both switches S_1 and S_2 switch on and off simultaneously. When the switches are on, we have the inductor current increase:

$$\Delta i_L = \frac{V_1}{L} kT \tag{8.42}$$

When the switches are off, we have the inductor current decrease:

$$\Delta i_L = \frac{V_2}{L} (1-k)T \tag{8.43}$$

Hence,

$$V_2 = \frac{k}{1-k} V_1 \tag{8.44}$$

The other parameters can be determined by the corresponding formulae of the normal buck-boost converter. Therefore, the positive buck-boost converter operates in "buck-boost operation mode," and the output voltage keeps positive polarity.

When this converter works in "buck operation mode" and "buck-boost operation mode," its input current is discontinuous, that is, it works in DICM.

8.3.4 OPERATION CONTROL

The general control block diagram is shown in Figure 8.11. It implements two functions: Logic control to select the operation mode and voltage closed-loop control to maintain the output voltage constant.

Refer to Figure 8.11, when the input voltage V_1 is higher than the upper limit voltage, for example, 1.03 V_{ref} (here the upper limit voltage is set 3.4 V) as the point A in Figure 8.10, the P/O buck-boost converter operates in buck mode. When the input voltage V_1 is lower than the lower limit voltage, for example, 0.97 V_{ref} (the upper limit

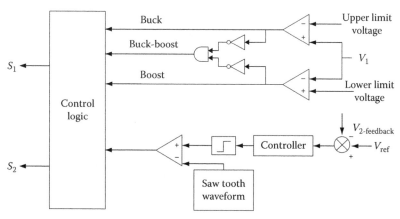

FIGURE 8.11   General control block diagram.

voltage is set to 3.2 V) as the point B in Figure 8.10, the P/O buck-boost converter operates in boost mode. When the input voltage V_1 is that between upper and lower limit voltages, for example, 0.97 $V_{ref} < V_1 < 1.03$ V_{ref}, the P/O buck-boost converter operates in buck-boost mode.

Output voltage feedback signal compares with the $V_{ref} = 3.3$ V to regulate the duty cycle k in order to keep the output voltage $V_2 = 3.3$ V. To analyze the performance of the system during the operation in the buck, boost modes and the behavior of the system in transition, the typical parameters of the converter are shown in Table 8.1. Voltage source is modeled to act as a single cell Li-ion battery, whose voltage varies from $V_{1H} = 4.2$ V when it is fully charged to $V_{1L} = 2.7$ V when it is discharged.

A PI controller is used for voltage closed-loop control. All logic operation and voltage feedback control diagram of the P/O buck-boost converter is shown in Figure 8.12.

The simulation results are shown in Figure 8.13.

TABLE 8.1

Circuit Parameters of the Positive Output Buck-Boost Converter

Variable	Parameter	Value
L	Magnetizing inductance	220 μH
C	Output filter capacitance	500 μF
V_1	Input voltage	4.2–2.7 V
	Upper limit voltage	3.4 V
V_{ref}	Output voltage reference	3.3 V
	Lower limit voltage	3.2 V
R	Load resistance	7 Ω
f	Switching frequency	20 kHz

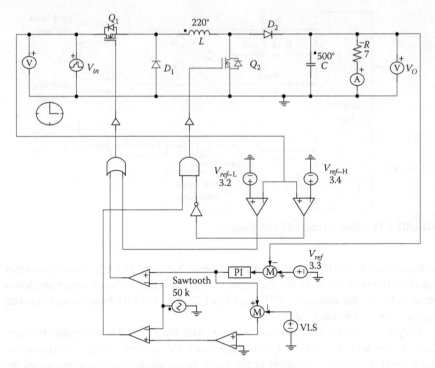

FIGURE 8.12 Simulation diagram of the P/O buck-boost converter.

A test rig is constructed for experiment testing. The measured results are shown in Table 8.2.

8.4 TRANSFORMER-TYPE CONVERTERS

Transformer-type converters consist of transformers and other parts. They can isolate the input and output circuits and have additional voltage transfer gain corresponding to the winding turns ratio n. After reviewing popular topologies, a few new circuits are introduced.

- Forward converter
- Fly-back converter
- Push–pull converters
- Half-bridge converters
- Bridge converters
- Zeta converter

8.4.1 FORWARD CONVERTER

Forward converter is the first transformer-type converter and is widely applied in industrial applications.

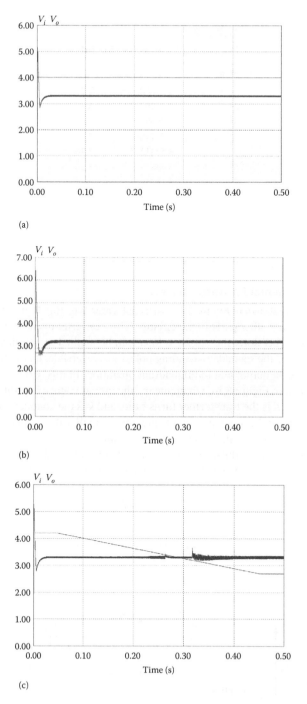

FIGURE 8.13 Simulation results. (a) Buck mode operation with $V_1 = 4.0$ V, (b) boost mode operation with $V_1 = 2.8$ V, and (c) overall operation with $V_1 = 2.7–4.2$ V.

TABLE 8.2
Measured Simulation Results

Step	V_{in}	V_{out}
1	4.20000	3.30
2	4.15909	3.30
3	3.99091	3.30
4	3.75748	3.30
5	3.54412	3.30
6	3.44875	3.30
7	3.18519	3.30
8	3.08228	3.30
9	2.95426	3.30
10	2.82877	3.30
11	2.70000	3.30

8.4.1.1 Fundamental Forward Converter

Forward converter shown in Figure 8.14 is a transformer-type topology, which consists of a transformer and other parts in the circuits. This converter insolates the input and output circuitry. Therefore, the output voltage can be applied in any floating circuit. Furthermore, since the secondary winding polarity is reversible, it is very convenient to perform negative output and multiquadrant operation. In this text explanation, the polarity is shown in Figure 8.14, which means the output voltage is positive.

In Figure 8.14, n is the transformer turns ratio, and k is the conduction duty cycle. The turns ratio n can be any value that is greater or smaller than unity, and the conduction duty cycle k is definitely is smaller than unity.

The equivalent circuits during switch-on and switch-off are shown in Figure 8.15a and b, respectively. During switch-on, we have the following equations:

$$nV_1 = v_L + v_C \quad nV_1 = L\frac{di_L}{dt} + V_C$$

$$\frac{di_L}{dt} = \frac{(nV_1 - V_C)}{L} \tag{8.45}$$

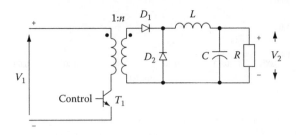

FIGURE 8.14 Forward converter.

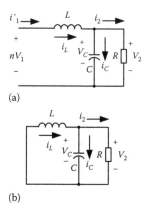

(a)

(b)

FIGURE 8.15 Equivalent circuits. (a) Switching-on and (b) switching-off.

During switch-off, we have the following equations:

$$0 = v_L + v_C \quad 0 = L\frac{di_L}{dt} + V_C$$

$$\frac{di_L}{dt} = \frac{-V_C}{L} \tag{8.46}$$

Some voltage and current waveforms are shown in Figure 8.16.

In the steady state, the current increment $(I_{max} - I_{min})$ during switch-on is equal to the current decrement $(I_{min} - I_{max})$ during switch-off. We have the equation to determine the voltage transfer gain:

$$I_{max} - I_{min} = \frac{nV_1 - V_C}{L}kT \tag{8.47}$$

$$I_{min} - I_{max} = \frac{-V_C}{L}(1-k)T \tag{8.48}$$

Thus,

$$\frac{nV_1 - V_C}{L}kT = \frac{V_C}{L}(1-k)T \tag{8.49}$$

$$(nV_1 - V_C)kT = V_C(1-k)T$$

$$V_2 = V_C = nkV_1 \tag{8.50}$$

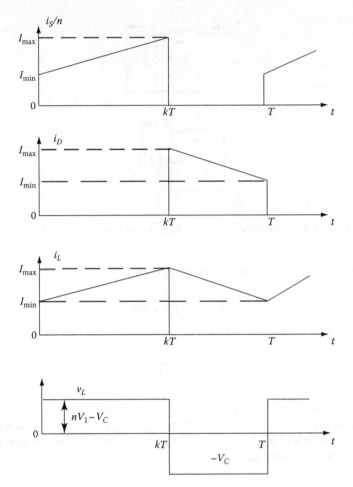

FIGURE 8.16   Some voltage and current waveforms.

From Figure 8.16, we can find the average value of inductor current easily by inspecting the waveform.

$$I_L = I_2 = \frac{V_2}{R} = \frac{I_{\max} + I_{\min}}{2} \tag{8.51}$$

The values of $I_{\max}$ and $I_{\min}$ are expressed as follows:

$$I_{\max} = V_2 \left(\frac{1}{R} + \frac{1-k}{2L} T \right) \tag{8.52}$$

$$I_{\min} = V_2 \left(\frac{1}{R} - \frac{1-k}{2L} T \right) \tag{8.53}$$

If the $I_{\min}$ is greater than zero, we call the operation the continuous conduction mode (CCM), vice versa, the discontinuous conduction mode (DCM). Solving Equation 8.53 for a zero value of $I_{\min}$ yields a relation for the minimum value of circuit inductance that results in continuous inductor current.

$$L_{\min} = \frac{1-k}{2} TR \qquad (8.54)$$

The ripple-less condition in the capacitor voltage now is relaxed to allow a small ripple. This has only a second-order effect on the currents calculated in the previous section, so the previous results can be used without change.

As previously noted, the capacitor current must be entirely alternating to have periodic operation. The graph of capacitor current must be as shown in Figure 8.17 for continuous inductor current. The peak value of this triangular waveform locates at $(I_{\max} - I_{\min})/2$. The resulting ripple in capacitor voltage depends on the area under the curve of capacitor current versus time. The charge added to the capacitor in a half-cycle is given by the triangular area above the axis:

$$\Delta Q = \frac{1}{2} \frac{I_{\max} - I_{\min}}{2} \frac{T}{2} = \frac{I_{\max} - I_{\min}}{8} T \qquad (8.55)$$

The graph of capacitor voltage is also shown as part of Figure 8.17. The ripple in the voltage is exaggerated to show its effect. Minimum and maximum capacitor voltage values occur at the time of capacitor current zero values. The peak-to-peak value of capacitor voltage ripple is given by

$$\Delta v_2 = \Delta v_C = \frac{\Delta Q}{C} = \frac{I_{\max} - I_{\min}}{8C} T = \frac{(1-k)V_2}{8CL} T^2 = \frac{nk(1-k)V_1}{8CL} T^2 \qquad (8.56)$$

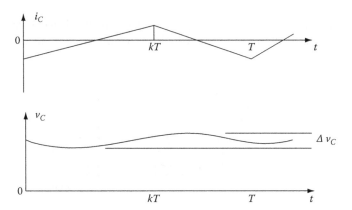

FIGURE 8.17 Waveforms of i_C and v_C.

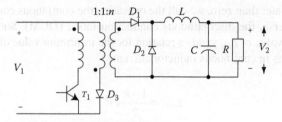

FIGURE 8.18   Forward converter with tertiary winding.

8.4.1.2   Forward Converter with Tertiary Winding

In order to exploit the magnetizing characteristic ability, a tertiary winding is applied in forward converter. The circuit diagram is shown in Figure 8.18.

The tertiary winding very much exploits the core magnetization ability and reduces the transformer size largely.

8.4.1.3   Switch Mode Power Supplies with Multiple Outputs

In many applications more than one output are required, with each output likely to have different voltage and current specifications. A forward converter with three outputs is shown in Figure 8.19. Each output voltage is determined by the turns ratio n_1, n_2, or n_3. The three output voltages are

$$V_{O1} = n_1 k V_1$$

$$V_{O2} = n_2 k V_1 \tag{8.50a}$$

$$V_{O3} = n_3 k V_1$$

However, multiple outputs can be readily obtained using any of the converters that have an isolating transformer, by employing a separate secondary winding for each output, as shown in the forward converter.

8.4.2   Fly-Back Converter

A fly-back converter is a transformer-type converter using the demagnetizing effect. Its circuit diagram is shown in Figure 8.20. The output voltage is calculated by the formula

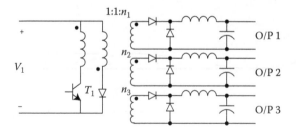

FIGURE 8.19   Forward converter with three outputs.

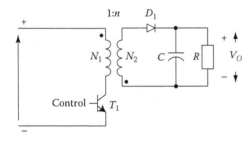

FIGURE 8.20 Fly-back converter.

$$V_O = \frac{k}{1-k} n V_{in} \tag{8.57}$$

where
  n is the transformer turns ratio
  k is the conduction duty cycle $k = t_{on}/T$

8.4.3 PUSH–PULL CONVERTER

Push–pull converter works in push–pull state, which effectively avoids the iron core saturation. Its circuit diagram is shown in Figure 8.21. Since there are two switches that work alternatively, the output voltage is doubled. The output voltage is calculated by the formula

$$V_O = 2nkV_{in} \tag{8.58}$$

where
  n is the transformer turns ratio
  k is the conduction duty cycle $k = t_{on}/T$

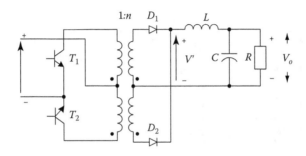

FIGURE 8.21 Push–pull converter.

8.4.4 HALF-BRIDGE CONVERTER

In order to reduce the primary side in one winding, a half-bridge converter was constructed. Its circuit diagram is shown in Figure 8.22. The output voltage is calculated by the formula

$$V_O = nkV_{in} \qquad (8.59)$$

where

n is the transformer turns ratio
k is the conduction duty cycle $k=t_{on}/T$

8.4.5 BRIDGE CONVERTER

A bridge converter is shown in Figure 8.23. The transformer has a couple of identical secondary windings. The primary circuit is a bridge inverter, so it is called bridge converter. Since the two pairs of the switches work symmetrically with 180° phase-angle shift, the transformer iron core is not saturated, and the magnetizing characteristics have been fully exploited. No tertiary winding is requested. The secondary side contains an antiparalleled diode full-wave rectifier. It is likely that two antiparalleled forward converters work altogether.

To avoid the short circuit, each pairs of the switches can only be switched on in the phase angle 0°–180°; usually it is set in 18°–162°. The corresponding conduction duty cycle k is in the range of 0.05–0.45.

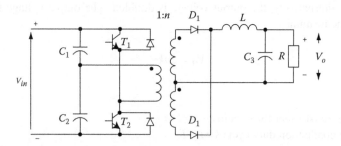

FIGURE 8.22 Half-bridge converter.

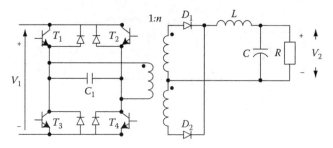

FIGURE 8.23 Bridge converter.

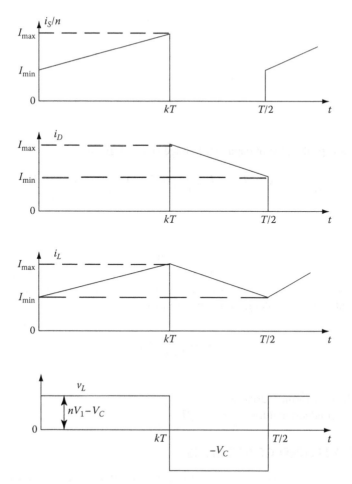

FIGURE 8.24 Some voltage and current waveforms.

The circuit analysis is also likely forward converter. Some voltage and current waveforms are shown in Figure 8.24. The repeating period is $T/2$ in bridge converter operation, which is rather than T in forward converter operation.

The voltage transfer gain is

$$V_2 = 2nkV_1 \tag{8.60}$$

Analogously, the average current

$$I_L = I_2 = \frac{V_2}{R} = \frac{I_{max} + I_{min}}{2} \tag{8.61}$$

The current I_{max} and I_{min} are

$$I_{max} = V_2 \left(\frac{1}{R} + \frac{0.5 - k}{2L} T \right) \tag{8.62}$$

$$I_{min} = V_2\left(\frac{1}{R} - \frac{0.5-k}{2L}T\right)$$ (8.63)

The minimum inductor to retain CCM is

$$L_{min} = \frac{0.5-k}{2}TR$$ (8.64)

The peak-to-peak value of capacitor voltage ripple is given by

$$\Delta v_2 = \Delta v_C = \frac{\Delta Q}{C} = \frac{I_{max} - I_{min}}{8C}T = \frac{(0.5-k)V_2}{8CL}T^2 = \frac{nk(0.5-k)V_1}{4CL}T^2$$ (8.65)

8.4.6 ZETA CONVERTER

Zeta converter is a transformer-type converter with a low-pass filter. Its circuit diagram is shown in Figure 8.25. Many people do not know its original circuit and call positive output Luo converter as Zeta converter. Its output voltage ripple is small. The output voltage is calculated by the formula

$$V_o = \frac{1}{1-k}nV_{in}$$ (8.66)

where
    n is the transformer turns ratio
    k is the conduction duty cycle $k = t_{on}/T$

8.5 DEVELOPED CONVERTERS

All developed converters are derived from fundamental converters. Since there are more components, the output voltage ripple is smaller. Five types of developed converters are introduced in this section.

1. Positive output Luo converter
2. Negative output Luo converter

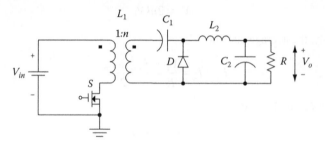

FIGURE 8.25 Zeta converter.

3. Double output Luo converter
4. Cúk converter
5. SEPIC

8.5.1 Positive Output Luo Converter (Elementary Circuit)

P/O Luo converter (elementary circuit) is shown in Figure 8.26a. Capacitor C acts as the primary means of storing and transferring energy from the input source to the output load via the pump inductor L_1. Assuming capacitor C to be sufficiently large, the variation of the voltage across capacitor C from its average value V_C can be neglected in steady state, that is, $v_C(t) \approx V_C$, even though it stores and transfers energy from the input to the output.

When switch S is on, the source current $i_I = i_{L1} + i_{L2}$. Inductor L_1 absorbs energy from the source. In the mean time inductor L_2 absorbs energy from source and capacitor C, both currents i_{L1} and i_{L2} increase. When switch S is off, source current

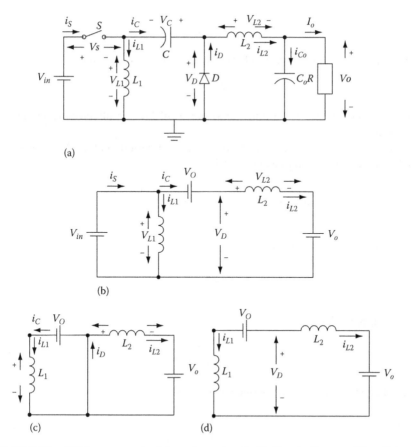

FIGURE 8.26 P/O Luo converter (elementary circuit). (a) Circuit diagram, (b) switch-on, (c) switch-off, and (d) discontinuous conduction mode (DCM).

$i_j=0$. Current i_{L1} flows through the free-wheeling diode D to charge capacitor C. Inductor L_1 transfers its stored energy to capacitor C. In the mean time, the inductor current i_{L2} flows through the $(C_O - R)$ circuit and free-wheeling diode D to keep itself continuous. Both currents i_{L1} and i_{L2} decrease. In order to analyze the circuit working procession, the equivalent circuits in switching-on and off states are shown in Figure 8.26b through d.

Actually, the variations of currents i_{L1} and i_{L2} are small so that $i_{L1} \approx I_{L1}$ and $i_{L2} \approx I_{L2}$. The charge on capacitor C increases during switch-off:

$$Q+ = (1-k)TI_{L1}$$

It decreases during switch-on:

$$Q- = kTI_{L2}$$

In the whole period of investigation, $Q+=Q-$. Thus,

$$I_{L2} = \frac{1-k}{k}I_{L1} \tag{8.67}$$

Since capacitor C_O performs as a low-pass filter, the output current

$$I_{L2} = I_O \tag{8.68}$$

These two equations (8.67) and (8.68) are available for all positive output Luo converters.

The source current $i_j=i_{L1}+i_{L2}$ during switch-on period, and $i_j=0$ during switch-off. Thus, the average source current I_j is

$$I_j = k \times i_j = k(i_{L1} + i_{L2}) = k\left(1 + \frac{1-k}{k}\right)I_{L1} = I_{L1} \tag{8.69}$$

Therefore, the output current is

$$I_O = \frac{1-k}{k}I_j \tag{8.70}$$

Hence, output voltage is

$$V_O = \frac{k}{1-k}V_j \tag{8.71}$$

The voltage transfer gain in continuous mode is

$$M_E = \frac{V_O}{V_j} = \frac{k}{1-k} \tag{8.72}$$

The curve of M_E versus k is shown in Figure 8.27.

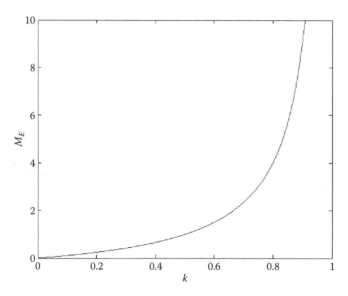

FIGURE 8.27 Voltage transfer gain M_E versus k.

Current i_{L1} increases and is supplied by V_I during switch-on. It decreases and is inversely biased by $-V_C$ during switch-off. Therefore, $kTV_I = (1 - k)TV_C$.

The average voltage across capacitor C is

$$V_C = \frac{k}{1-k}V_I = V_O \qquad (8.73)$$

Current i_{L1} increases and is supplied by V_I during switch-on. It decreases and is inversely biased by $-V_C$ during switch-off. Therefore, its peak-to-peak variation is

$$\Delta i_{L1} = \frac{kTV_I}{L_1}$$

Considering Equation 8.69, the variation ratio of the current i_{L1} is

$$\xi_1 = \frac{\Delta i_{L1}/2}{I_{L1}} = \frac{kTV_I}{2L_1 I_I} = \frac{1-k}{2M_E}\frac{R}{fL_1} \qquad (8.74)$$

Current i_{L2} increases and is supplied by the voltage $(V_I + V_C - V_O) = V_I$ during switch-on. It decreases and is inversely biased by $-V_O$ during switch-off. Therefore its peak-to-peak variation is

$$\Delta i_{L2} = \frac{kTV_I}{L_2} \qquad (8.75)$$

Considering Equation 8.67, the variation ratio of current i_{L2} is

$$\xi_2 = \frac{\Delta i_{L2}/2}{I_{L2}} = \frac{kTV_I}{2L_2 I_O} = \frac{k}{2M_E} \frac{R}{fL_2} \tag{8.76}$$

When switch is off, the free-wheeling diode current $i_D = i_{L1} + i_{L2}$ and

$$\Delta i_D = \Delta i_{L1} + \Delta i_{L2} = \frac{kTV_I}{L_1} + \frac{kTV_I}{L_2} = \frac{kTV_I}{L} = \frac{(1-k)TV_O}{L} \tag{8.77}$$

Considering (8.67) and (8.68), the average current in switch-off period is $I_D = I_{L1} + I_{L2} = I_O/(1-k)$.

The variation ratio of current i_D is

$$\zeta = \frac{\Delta i_D/2}{I_D} = \frac{(1-k)^2 TV_O}{2LI_O} = \frac{k(1-k)R}{2M_E fL} = \frac{k^2}{M_E^2} \frac{R}{2fL} \tag{8.78}$$

The peak-to-peak variation of v_C is

$$\Delta v_C = \frac{Q+}{C} = \frac{1-k}{C} TI_I$$

Considering Equation 8.73, the variation ratio of v_C is

$$\rho = \frac{\Delta v_C/2}{V_C} = \frac{(1-k)TI_I}{2CV_O} = \frac{k}{2} \frac{1}{fCR} \tag{8.79}$$

In order to investigate the variation of output voltage v_O, we have to calculate the charge variation on the output capacitor C_O, because $Q = C_O V_O$ and $\Delta Q = C_O \Delta v_O$. ΔQ is caused by Δi_{L2} and corresponds to the *area* of the triangle with the *height* of half of Δi_{L2} and the *width* of half of the repeating period $T/2$. Considering Equation 8.75,

$$\Delta Q = \frac{1}{2} \frac{\Delta i_{L2}}{2} \frac{T}{2} = \frac{T}{8} \frac{kTV_I}{L_2}$$

Thus, the half peak-to-peak variation of output voltage v_O and v_{CO} is

$$\frac{\Delta v_O}{2} = \frac{\Delta Q}{2C_O} = \frac{kT^2 V_I}{16C_O L_2}$$

The variation ratio of output voltage v_O is

$$\varepsilon = \frac{\Delta v_O/2}{V_O} = \frac{kT^2}{16C_O L_2} \frac{V_I}{V_O} = \frac{k}{16M_E} \frac{1}{f^2 C_O L_2} \tag{8.80}$$

For analysis in DCM, referring to Figure 8.26d, we can see that the diode current i_D becomes zero during switch-off before next period switch-on. The condition for discontinuous conduction mode is $\zeta \geq 1$, that is, $(k^2 / M_E^2)(R / 2fL) \geq 1$

or

$$M_E \leq k\sqrt{\frac{R}{2fL}} = k\sqrt{\frac{z_N}{2}} \qquad (8.81)$$

The graph of the boundary curve versus the normalized load $z_N = R/fL$ is shown in Figure 8.28. It can be seen that the boundary curve is a monorising function of the parameter k.

In the DCM case, the current i_D exists in the period between kT and $t_1 = [k+(1-k) m_E]T$, where m_E is the *filling efficiency* and it is defined as

$$m_E = \frac{1}{\zeta} = \frac{M_E^2}{k^2(R/2fL)} \qquad (8.82)$$

The diode current i_D decreases to zero at $t = t_1 = kT + (1-k)m_E T$; therefore, $0 < m_E < 1$ (see Figure 8.29).

For the current i_L, we have

$$kTV_I = (1-k)m_E TV_C$$

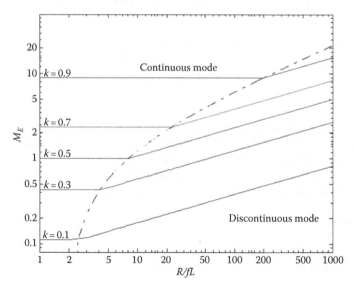

FIGURE 8.28 The boundary between continuous and discontinuous modes and the output voltage versus the normalized load $z_N = R/fL$.

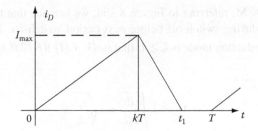

FIGURE 8.29 The discontinuous current waveform.

or

$$V_C = \frac{k}{(1-k)m_E}V_I = k(1-k)\frac{R}{2fL}V_I$$

with

$$\sqrt{\frac{R}{2fL}} \geq \frac{1}{1-k}$$

For the current i_{LO}, we have $kT(V_I+V_C-V_O)=(1-k)m_ETV_O$

Therefore, output voltage in discontinuous mode is

$$V_O = \frac{k}{(1-k)m_E}V_I = k(1-k)\frac{R}{2fL}V_I \quad \text{with} \quad \sqrt{\frac{R}{2fL}} \geq \frac{1}{1-k} \qquad (8.83)$$

The output voltage will linearly increase during load resistance R increasing. The output voltage versus the normalized load $z_N=R/fL$ is shown in Figure 8.28. It can be seen that larger load resistance R may cause higher output voltage in discontinuous conduction mode (DCM).

Example 8.3 A Positive output Luo converter has components:
$V_I=20\,V$, $L_1=L_2=10\,mH$, $C=C_O=20\,\mu F$, $R=20\,\Omega$, and switching
frequency $f=50\,kHz$ and conduction duty cycle $k=0.6$.
**Calculate the output voltage, its variation ratio, and the variation
ratio of the inductor currents i_{l1} and i_{l2} in steady state.**

Solution:

1. From (8.71), the output voltage is $V_O=kV_I/(1-k)=0.6\times20/0.4=30\,V$.
2. From (8.80), the variation ratio of v_O is

$$\varepsilon = \frac{k}{16M_E}\frac{1}{f^2C_OL_2} = \frac{0.6}{16\times1.5}\frac{1}{(50\,kHz)^2\,20\,\mu F\times10\,mH} = 0.00005$$

3. From (8.74), the variation ratio of the current i_{L1} is

$$\xi_1 = \frac{1-k}{2M_E} \frac{R}{fL_1} = \frac{0.4}{2\times1.5} \frac{20}{50\,\text{kHz}\times10\,\text{mH}} = 0.0053$$

4. From (8.76), the variation ratio of the current i_{L2} is

$$\xi_2 = \frac{k}{2M_E} \frac{R}{fL_2} = \frac{0.6}{2\times1.5} \frac{20}{50\,\text{kHz}\times10\,\text{mH}} = 0.008$$

8.5.2 Negative Output Luo Converter (Elementary Circuit)

The negative output Luo converter (elementary circuit), and its switch-on and switch-off equivalent circuits are shown in Figure 8.30. This circuit can be considered as a combination of an electronic pump S-L-D-(C) and a "Π"-type low-pass filter C-L_O-C_O. The electronic pump injects certain energy to the low-pass filter every cycle. Capacitor C in Figure 8.30 acts as the primary means of storing and transferring energy from the input source to the output load. Assuming capacitor C to be sufficiently large, the variation of the voltage across capacitor C from its average value V_C can be neglected in steady state, that is, $v_C(t) \approx V_C$, even though it stores and transfers energy from the input to the output.

The voltage transfer gain in continuous conduction mode

$$M_E = \frac{V_O}{V_I} = \frac{I_I}{I_O} = \frac{k}{1-k} \tag{8.84}$$

The transfer gain is shown in Figure 8.27. Current i_L increases and is supplied by V_I during switch-on. Thus, its peak-to-peak variation is $\Delta i_L = kTV_I/L$. The inductor current I_L is

$$I_L = I_{C\text{-}off} + I_O = \frac{I_O}{1-k} \tag{8.85}$$

Considering $R = V_O/I_O$, the variation ratio of the current i_L is

$$\zeta = \frac{\Delta i_L/2}{I_L} = \frac{k(1-k)V_I T}{2LI_O} = \frac{k(1-k)R}{2M_E fL} = \frac{k^2}{M_E^2} \frac{R}{2fL} \tag{8.86}$$

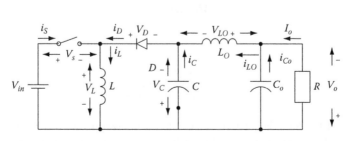

FIGURE 8.30 *N/O* Luo converter (elementary circuit).

The peak-to-peak variation of voltage v_C is

$$\Delta v_C = \frac{Q-}{C} = \frac{k}{C} T I_O \qquad (8.87)$$

The variation ratio of voltage v_C is

$$\rho = \frac{\Delta v_C / 2}{V_C} = \frac{k I_O T}{2 C V_O} = \frac{k}{2} \frac{1}{fCR} \qquad (8.88)$$

The peak-to-peak variation of current i_{LO} is

$$\Delta i_{LO} = \frac{k}{8 f^2 C L_O} I_O \qquad (8.89)$$

Considering $I_{LO} = I_O$ is

$$\xi = \frac{\Delta i_{LO} / 2}{I_{LO}} = \frac{k}{16} \frac{1}{f^2 C L_O} \qquad (8.90)$$

the variation of voltage v_{CO} is

$$\Delta v_{CO} = \frac{A}{C_O} = \frac{1}{2} \frac{T}{2} \frac{k}{16 f^2 C C_O L_O} I_O = \frac{k}{64 f^3 C C_O L_O} I_O \qquad (8.91)$$

The variation ratio of the output voltage v_{CO} is

$$\varepsilon = \frac{\Delta v_{CO} / 2}{V_{CO}} = \frac{k}{128 f^3 C C_O L_O} \frac{I_O}{V_O} = \frac{k}{128} \frac{1}{f^3 C C_O L_O R} \qquad (8.92)$$

In DCM, the diode current i_D becomes zero during switch-off before next period switch-on. The condition for discontinuous conduction mode is $\zeta \geq 1$, that is,

$$\frac{k^2}{M_E^2} \frac{R}{2 fL} \geq 1$$

or

$$M_E \leq k \sqrt{\frac{R}{2 fL}} = k \sqrt{\frac{z_N}{2}} \qquad (8.93)$$

The graph of the boundary curve versus the normalized load $z_N = R/fL$ is shown in Figure 8.28. It can be seen that the boundary curve is a monorising function of the parameter k.

In the DCM case, the current i_D exists in the period between kT and $t_1 = [k + (1-k) m_E]T$, where m_E is the *filling efficiency* and it is defined as

$$m_E = \frac{1}{\zeta} = \frac{M_E^2}{k^2(R/2fL)} \tag{8.94}$$

Considering $\zeta > 1$ for DCM operation, therefore $0 < m_E < 1$. Since the diode current i_D becomes zero at $t = t_1 = kT + (1-k)m_E T$,
for the current i_L, we have $TV_I = (1-k)m_E TV_C$ or

$$V_C = \frac{k}{(1-k)m_E} V_I = k(1-k)\frac{R}{2fL} V_I \quad \text{with} \quad \sqrt{\frac{R}{2fL}} \ge \frac{1}{1-k}$$

For the current i_{LO}, we have $kT(V_I + V_C - V_O) = (1-k)m_E TV_O$.
Therefore, output voltage in discontinuous mode is

$$V_O = \frac{k}{(1-k)m_E} V_I = k(1-k)\frac{R}{2fL} V_I \quad \text{with} \quad \sqrt{\frac{R}{2fL}} \ge \frac{1}{1-k} \tag{8.95}$$

that is, the output voltage will linearly increase during load resistance R increasing. Larger load resistance R may cause higher output voltage in discontinuous conduction mode (DCM).

8.5.3 Double Output Luo Converter (Elementary Circuit)

Combining P/O and N/O elementary Luo converters together, we obtain the double output elementary Luo converter shown in Figure 8.31. All analyses can be referred

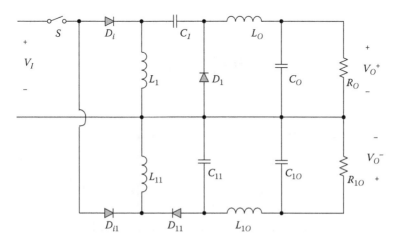

FIGURE 8.31 Double output elementary Luo converter.

to the previous two sections of P/O and N/O output elementary Luo converters. The voltage transfer gains are calculated by

$$\frac{V_O +}{V_I} = \frac{V_O -}{V_I} = \frac{k}{1-k} \tag{8.96}$$

8.5.4 CÚK CONVERTER

Cúk converter is derived from boost converter. Its circuit diagram is shown in Figure 8.32. It was published in 1977 as boost-buck converter and renamed by Cúk's students after 1986–1990.

The inductor current i_L increases with slope $+V_I/L$ during switching on and decreases with slope $-(V_C - V_I)/L$ during switching off. Thus,

$$\frac{V_I}{L} kT = \frac{V_C - V_I}{L}(1-k)T$$

$$V_C = \frac{1}{1-k}V_I$$

Since L_O–C_O is a low-pass filter, the output voltage is calculated by the formula

$$V_O = V_C - V_I = \frac{k}{1-k}V_I \tag{8.97}$$

The voltage transfer gain is

$$M = \frac{V_O}{V_I} = \frac{k}{1-k} \tag{8.98}$$

also

$$M = \frac{I_I}{I_O} = \frac{k}{1-k}$$

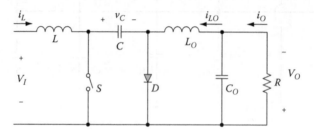

FIGURE 8.32   Cúk converter.

Since the inductor L is in series connected to the source voltage, and inductor L_O is in series connected to the output circuit $R-C_O$, we have the relations

$$I_L = I_I \quad \text{and} \quad I_{LO} = I_O$$

The variation of the current i_L is $\Delta i_L = (V_I/L)kT$.

Therefore, the variation ratio of the current i_L is

$$\xi = \frac{\Delta i_L/2}{I_L} = \frac{V_I}{2I_I L}kT = \frac{k}{2M^2}\frac{R}{fL} \tag{8.99}$$

The variation of the current i_{LO} is

$$\Delta i_{LO} = \frac{V_O}{L_O}(1-k)T$$

Therefore, the variation ratio of the current i_{LO} is

$$\xi_O = \frac{\Delta i_{LO}/2}{I_{LO}} = \frac{V_O}{2I_O L_O}(1-k)T = \frac{1-k}{2}\frac{R}{fL_O} \tag{8.100}$$

The variation of the diode current i_D is

$$\Delta i_D = \Delta i_L + \Delta i_{LO} = \left(\frac{V_O}{L} + \frac{V_O}{L_O}\right)(1-k)T$$

We can define $L_{//} = L \, // \, L_O$.

$$\Delta i_D = \Delta i_L + \Delta i_{LO} = \frac{V_O}{L_{//}}(1-k)T$$

and

$$I_D = I_L + I_{LO} = I_I + I_O = (M+1)I_O = \frac{1}{1-k}I_O$$

Therefore, the variation ratio of the diode current i_D is

$$\zeta = \frac{\Delta i_D/2}{I_D} = \frac{V_O}{2I_O L_{//}}(1-k)^2 T = \frac{(1-k)^2}{2}\frac{R}{fL_{//}} \tag{8.101}$$

The variation of the voltage v_C is

$$\Delta v_C = \frac{\Delta Q}{C} = \frac{I_I}{C}(1-k)T$$

Therefore, the variation ratio of the voltage v_C is

$$\rho = \frac{\Delta v_C / 2}{V_C} = \frac{I_I}{2CV_C}(1-k)T = \frac{k(1-k)M}{2}\frac{1}{fRC} \qquad (8.102)$$

The variation of the voltage v_{CO} is

$$\Delta v_{CO} = \frac{\Delta Q_O}{C_O} = \frac{T}{8C_O}\Delta i_{LO} = \frac{V_O}{8f^2C_OL_O}(1-k)$$

Therefore, the variation ratio of the voltage v_{CO} is

$$\varepsilon = \frac{\Delta v_{CO}/2}{V_O} = \frac{1-k}{16f^2C_OL_O} \qquad (8.103)$$

The boundary is determined by the condition:

$$\zeta = 1 \quad \text{or} \quad \zeta = \frac{(1-k)^2}{2}\frac{R}{fL_{//}} = \frac{1}{2(1+M)^2}Z_N = 1 \quad \text{with} \quad Z_N = \frac{R}{fL_{//}}$$

Therefore, the boundary between CCM and DCM is

$$M = \sqrt{\frac{Z_N}{2}} - 1 \qquad (8.104)$$

If $(M+1) > \sqrt{Z_N/2}$, the converter works in CCM; if $(M+1) < \sqrt{Z_N/2}$, the converter works in DCM.

Example 8.4 A Cúk converter has components: $V_1 = 20\,V$, $L = L_O = 10\,mH$, $C = C_O = 20\,\mu F$, $R = 20\,\Omega$, and switching frequency $f = 50\,kHz$ and conduction duty cycle $k = 0.6$. Calculate the output voltage and its ripple in steady state. Does this converter work in CCM or DCM?

Solution:

1. From (8.59), the output voltage is

$$V_2 = V_C = \frac{k}{1-k}V_1 = \frac{0.6}{0.4}20 = 30\ V$$

2. From (8.103), the output voltage ripple is

$$\varepsilon = \frac{1-k}{16f^2C_OL_O} = \frac{0.4}{16(50\ kHz)^2 \times 20\ \mu F \times 10\ mH} = 0.00005$$

3. We have $M + 1 = 2.5$, which is greater than

$$\sqrt{\frac{Z_N}{2}} = \sqrt{\frac{20}{2 \times 5\,\text{mH} \times 50\,\text{kHz}}} = 0.2$$

Referring to (8.104), we know that this converter works in CCM.

8.5.5 SINGLE-ENDED PRIMARY INDUCTANCE CONVERTER

The single-ended primary inductance converter (SEPIC) is derived from the boost converter. Its circuit diagram is shown in Figure 8.33. It was created after Cúk converter and the so-called positive output Cúk converter.

The inductor current i_{L1} increases with slope $+V_C/L_1$ during switching on and decreases with slope $-V_O/L_1$ during switch-off.

Thus,

$$\frac{V_C}{L_1} kT = \frac{V_O}{L_1}(1-k)T$$

$$V_C = \frac{1-k}{k} V_O \tag{8.105}$$

The inductor current i_L increases with slope $+V_I/L$ during switch-on and decreases with slope $-(V_C + V_O - V_I)/L$ during switch-off.

Thus,

$$\frac{V_I}{L} kT = \frac{V_C + V_O - V_I}{L}(1-k)T$$

$$V_O = \frac{k}{1-k} V_I \tag{8.106}$$

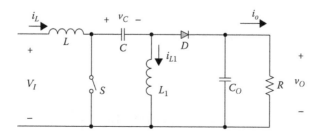

FIGURE 8.33 Single-ended primary inductance converter.

that is,

$$M = \frac{V_O}{V_I} = \frac{k}{1-k}$$

Since the inductor L is in series connected to the source voltage, the inductor average current I_L is

$$I_L = I_I$$

Since the inductor L_1 is in parallel connected to the capacitor C during switch-off, the inductor average current I_{L1} is ($I_{CO\text{-}on}=I_O$ and $I_{CO\text{-}off}=I_I$), $I_{L1}=I_O$.
    The variation of the current i_L is $\Delta i_L=(V_I/244L)kT$.
    Therefore, the variation ratio of the current i_L is

$$\xi = \frac{\Delta i_L/2}{I_L} = \frac{V_I}{2I_I L}kT = \frac{k}{2M^2}\frac{R}{fL} \tag{8.107}$$

The variation of the current i_{L1} is

$$\Delta i_{L1} = \frac{V_C}{L_1}kT$$

Therefore, the variation ratio of the current i_{L1} is

$$\xi_1 = \frac{\Delta i_{L1}/2}{I_{L1}} = \frac{V_C}{2I_O L_1}kT = \frac{1-k}{2}\frac{R}{fL_1} \tag{8.108}$$

The variation of the diode current i_D is

$$\Delta i_D = \Delta i_L + \Delta i_{L1} = \left(\frac{V_O}{L} + \frac{V_O}{L_1}\right)(1-k)T$$

We can define $L_{//}=L \text{ // } L_1$.
    Hence,

$$\Delta i_D = \Delta i_L + \Delta i_{L1} = \frac{V_O}{L_{//}}(1-k)T$$

and

$$I_D = I_L + I_{LO} = I_I + I_O = (M+1)I_O = \frac{1}{1-k}I_O$$

Therefore, the variation ratio of the diode current i_D is

$$\zeta = \frac{\Delta i_D/2}{I_D} = \frac{V_O}{2I_O L_{//}}(1-k)^2 T = \frac{(1-k)^2}{2} \frac{R}{fL_{//}} \tag{8.109}$$

The variation of the voltage v_C is

$$\Delta v_C = \frac{\Delta Q}{C} = \frac{I_I}{C}(1-k)T$$

Therefore, the variation ratio of the voltage v_C is

$$\rho = \frac{\Delta v_C/2}{V_C} = \frac{I_I}{2CV_C}(1-k)T = \frac{kM}{2} \frac{1}{fRC} \tag{8.110}$$

The variation of the voltage v_{CO} is

$$\Delta v_{CO} = \frac{\Delta Q_O}{C_O} = \frac{kTI_O}{C_O} = \frac{kI_O}{fC_O}$$

Therefore, the variation ratio of the voltage v_{CO} is

$$\varepsilon = \frac{\Delta v_{CO}/2}{V_O} = \frac{kI_O}{2fC_O V_O} = \frac{k}{2fRC_O} \tag{8.111}$$

The boundary is determined by the condition

$$\zeta = 1 \quad \text{or} \quad \zeta = \frac{(1-k)^2}{2} \frac{R}{fL_{//}} = \frac{1}{2(1+M)^2} Z_N = 1 \quad \text{with} \quad Z_N = \frac{R}{fL_{//}}$$

Therefore, the boundary between CCM and DCM is

$$M = \sqrt{\frac{Z_N}{2}} - 1 \tag{8.112}$$

8.6 TAPPED-INDUCTOR CONVERTERS

These converters have been derived from fundamental converters, whose circuit diagrams are shown in Table 8.3. The voltage transfer gains are shown in Table 8.4. Here the tapped inductor ratio is $n = n1/(n1 + n2)$.

TABLE 8.3

The Circuit Diagrams of the Tapped Inductor Fundamental Converters

TABLE 8.4

The Voltage Transfer Gains of the Tapped Inductor Fundamental Converters

Converter	No Tap	Switched to Tap	Diode to Tap	Rail to Tap
Buck	K	$\dfrac{k}{n+k(1-n)}$	$\dfrac{nk}{1+k(n-1)}$	$\dfrac{k-n}{k(1-n)}$
Boost	$\dfrac{1}{1-k}$	$\dfrac{n+k(1-n)}{n(1-k)}$	$\dfrac{1+k(n-1)}{1-k}$	$\dfrac{n-k}{n(1-k)}$
Buck-boost	$\dfrac{k}{1-k}$	$\dfrac{k}{n(1-k)}$	$\dfrac{nk}{1-k}$	$\dfrac{k}{1-k}$

REFERENCES

1. Luo, F. L. and Ye, H. 2004. *Advanced DC/DC Converters*. Boca Raton, FL: CRC Press.
2. Luo, F. L. and Ye, H. 2006. *Essential DC/DC Converters*. Boca Raton, FL: Taylor & Francis Group LLC.
3. Luo, F. L. 1999. Positive output Luo-converters: Voltage lift technique. *IEE-EPA Proceedings*, 146, 415–432.
4. Luo, F. L. 1999. Negative output Luo-converters: Voltage lift technique. *IEE-EPA Proceedings*, 146, 208–224.
5. Luo, F. L. 2000. Double output Luo-converters: Advanced voltage lift technique. *IEE-EPA Proceedings*, 147, 469–485.
6. Erickson, R. W. and Maksimovic, D. 1999. *Fundamentals of Power Electronics*. Boston, MA: Kluwer Academic Publishers.
7. Middlebrook, R. D. and Cuk, S. 1981. *Advances in Switched-Mode Power Conversion*. Pasadena, CA: TESLAco.
8. Maksimovic, D. and Cuk, S. 1991. Switching converters with wide DC conversion range. *IEEE Transactions on Power Electronics*, 6, 151–159.
9. Smedley, K. M. and Cuk, S. 1995. One-cycle control of switching converters. *IEEE Transactions on Power Electronics*, 10, 625–634.
10. Redl, R., Molnar, B., and Sokal, N. O. 1986. Class-E resonant DC-DC power converters: Analysis of operations, and experimental results at 1.5 MHz. *IEEE Transactions on Power Electronics*, 1, 111–121.
11. Kazimierczuk, M. K. and Bui, X. T. 1989. Class-E DC-DC converters with an inductive impedance inverter. *IEEE Transactions on Power Electronics*, 4, 124–133.
12. Liu, Y. and Sen, P. C. 1996. New class-E DC-DC converter topologies with constant switching frequency. *IEEE Transactions on Industry Applications*, 32, 961–972.

9 Voltage Lift Converters

Ordinary DC/DC converter has limited voltage transfer gain. Considering the component parasitic elements' effects, the conduction duty cycle k only can be $0.1 < k < 0.9$. This restriction blocks ordinary DC/DC converter voltage transfer gain increase. Voltage-lift (VL) technique is a general method used in electronics circuitry design to amplify the output voltage. Using this technique in DC/DC conversion technology, we can design VL power converters with high-voltage transfer gains in arithmetic progression stage by stage. It opens the way to largely increase the voltage transfer gain of DC/DC converters. Using this technique the following series VL converters are designed [1,2]:

- Positive-output (P/O) Luo converters
- Negative-output (N/O) Luo converters
- Double-output Luo converters
- VL Cúk converters
- VL single-ended primary inductance converter (SEPIC)
- Other VL double-output converters
- Switched-capacitorized converters

9.1 INTRODUCTION

VL technique is applied in the periodical switching circuit. Usually, a capacitor is charged during the switch-on condition by certain voltage, for example, source voltage. This charged capacitor voltage can be arranged on top-up to some parameter, for example, output voltage during switch-off. Therefore, the output voltage can be higher. Consequently, this circuit is called self-lift circuit. A typical example is the saw-tooth-wave generator with self-lift circuit.

By repeating this operation, another capacitor can be charged by certain voltage, which possibly is the input voltage or other equivalent voltage. The second capacitor-charged voltage is also possibly arranged on top-up to some parameter, especially output voltage. Therefore, the output voltage can be higher than that of self-lift circuit. As usual, this circuit is called re-lift circuit.

Analogously, this operation can be repeated many times. Consequently, the series circuits are called triple-lift circuit, quadruple-lift circuit, and so on.

Because of the effect of parasitic elements the output voltage and power transfer efficiency of DC/DC converters are limited. VL technique opens a good way to improve circuit characteristics. After long-term research, this technique has been successfully applied for DC/DC converters. Three series Luo converters are newly developed from prototypes using VL technique. These converters perform DC/DC voltage-increasing conversion with high power density, high efficiency, and cheap topology with simple structure. They are different from any other DC/DC step-up converters and possess many advantages including the high output voltage with small ripples. Therefore, these converters will be widely used in computer peripheral equipment and industrial applications, especially for high output voltage projects. This chapter contents are arranged as follows:

1. Seven types of self-lift converters
2. P/O Luo converters
3. N/O Luo converters
4. Modified P/O Luo converters
5. Double-output Luo converters

Using VL technique we can easily obtain the other series of VL converters; for example, VL Cúk converters, VL SEPICs, other types of Double-output converter, and Switched-Capacitorized converters.

9.2 SEVEN SELF-LIFT CONVERTERS

All self-lift converters introduced here are derived from developed converters such as Luo converters, Cúk converter, and SEPIC in Section 5.5. Since all circuits are simple, usually only one more capacitor and diode required, the output voltage is higher than an input voltage [3–5]. The output voltage is calculated by the formula

$$V_O = \left(\frac{k}{1-k} + 1 \right) V_{in} = \frac{1}{1-k} V_{in} \qquad (9.1)$$

Seven circuits were developed:

1. Self-lift Cúk converter
2. Self-lift P/O Luo converter
3. Reverse self-lift P/O Luo converter
4. Self-lift N/O Luo converter
5. Reverse self-lift Luo converter
6. Self-lift SEPIC
7. Enhanced self-lift P/O Luo converter

These converters perform DC–DC voltage-increasing conversion in simple structures. In these circuits, the switch S is a semiconductor device (MOSFET, BJT, IGBT, and so on). It is driven by a pulse-width-modulated (PWM) switching signal with variable frequency f and conduction duty k. For all circuits, the load is usually resistive, that is, $R = V_O/I_O$.

The normalized impedance Z_N is

$$z_N = \frac{R}{fL_{eq}} \qquad (9.2)$$

where L_{eq} is the equivalent inductance.

We concentrate on the absolute values rather than polarity in the following description and calculations. The directions of all voltages and currents are defined and shown in the corresponding figures. We also assume that the semiconductor switch and the passive components are all ideal. All capacitors are assumed to be large enough that the ripple voltage across the capacitors can be negligible in one switching cycle for the average value discussions.

For any component X (e.g., C, L, and so on), its instantaneous current and voltage are expressed as i_X and v_X. Its average current and voltage values are expressed as I_x and V_x. The input voltage and current are V_O and I_O; the output voltage and current are V_I and I_I, and T and f are the switching period and frequency.

The voltage transfer gain for the continuous conduction mode (CCM) is as follows:

$$M = \frac{V_o}{V_I} = \frac{I_I}{I_o} \tag{9.3}$$

Variation of current i_L is as follows:

$$\zeta_1 = \frac{\Delta i_L / 2}{I_L} \tag{9.4}$$

Variation of current i_{LO} is as follows:

$$\zeta_2 = \frac{\Delta i_{Lo} / 2}{I_{Lo}} \tag{9.5}$$

Variation of current i_D is as follows:

$$\xi = \frac{\Delta i_D / 2}{I_D} \tag{9.6}$$

Variation of voltage v_C is as follows:

$$\rho = \frac{\Delta v_C / 2}{V_C} \tag{9.7}$$

Variation of voltage v_{C1} is as follows:

$$\sigma_1 = \frac{\Delta v_{C1} / 2}{v_{C1}} \tag{9.8}$$

Variation of voltage v_{C2} is as follows:

$$\sigma_2 = \frac{\Delta v_{C2}/2}{v_{C2}} \tag{9.9}$$

Variation of output voltage v_O is as follows:

$$\varepsilon = \frac{\Delta V_o/2}{V_o} \tag{9.10}$$

Here, I_D refers to the average current i_D that flows through the diode D during the switch-off period, not its average current over whole period.

Detailed analysis of the seven self-lift DC/DC converters is given in the following sections. Due to the limited length of the book, only the simulation and experimental results of the self-lift Cúk converter are given. However, the results and conclusions of other self-lift converters should be quite similar to those of the self-lift Cúk converter.

9.2.1 Self-Lift Cúk Converter

Self-lift Cúk converter and its equivalent circuits during the switch-on and switch-off periods are shown in Figure 9.1. It is derived from the Cúk converter. During the switch-on period, S and D_1 are on, and D is off. During the switch-off period, D is on, and S and D_1 are off.

9.2.1.1 Continuous Conduction Mode

In steady state, the average inductor voltages over a period are zero. Thus,

$$V_{C1} = V_{CO} = V_O \tag{9.11}$$

During the switch-on period, the voltage across capacitor C and C_1 are equal. Since we assume that C and C_1 are sufficiently large, so

$$V_C = V_{C1} = V_O \tag{9.12}$$

The inductor current i_L increases during switch-on and decreases during switch-off. The corresponding voltages across L are V_I and $-(V_C - V_I)$.

Therefore,

$$kTV_I = (1-k)T(V_C - V_I)$$

Hence,

$$V_O = V_C = V_{C1} = V_{CO} = \frac{1}{1-k}V \tag{9.13}$$

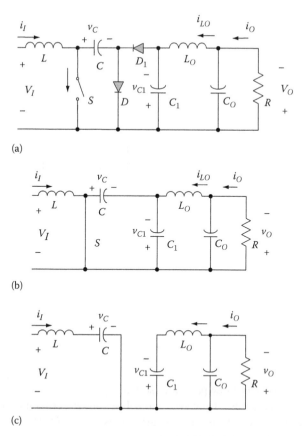

(a)

(b)

(c)

FIGURE 9.1   Self-lift Cúk converter and equivalent circuits. (a) The self-lift Cúk converter, (b) the equivalent circuit during switch-on, and (c) the equivalent circuit during switch-off.

The voltage transfer gain in the CCM is

$$M = \frac{V_O}{V_I} = \frac{I_I}{I_O} = \frac{1}{1-k} \tag{9.14}$$

The characteristics of M versus conduction duty cycle k are shown in Figure 9.2.

Since all the components are considered ideal, the power loss associated with all the circuit elements are neglected. Therefore, the output power P_O is considered to be equal to the input power P_{IN}: $V_O I_O = V_I I_I$

Thus,

$$I_L = I_I = \frac{1}{1-k} I_O$$

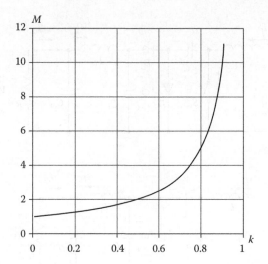

FIGURE 9.2 The voltage transfer gain M versus the normalized load at various k.

During switch-off,

$$i_D = i_L \quad I_D = \frac{1}{1-k} I_O \tag{9.15}$$

The capacitor C_O acts as a low-pass filter so that $I_{LO} = I_O$

The current i_L increases during switch-on. The voltage across it during switch-on is V_i; therefore its peak-to-peak current variation is $\Delta i_L = kTV_i/L$

The variation ratio of the current i_L is

$$\zeta_1 = \frac{\Delta i_L/2}{I_L} = \frac{kTV_I}{2I_L} = \frac{k(1-k)^2 R}{2fL} = \frac{kR}{2M^2 fL} \tag{9.16}$$

The variation of current i_D is

$$\xi = \zeta_1 = \frac{kR}{2M^2 fL} \tag{9.17}$$

The peak-to-peak variation of voltage v_C is

$$\Delta v_C = \frac{I_L(1-k)T}{C} = \frac{I_O}{fC} \tag{9.18}$$

The variation ratio of the voltage v_C is

$$\rho = \frac{\Delta v_C/2}{V_C} = \frac{I_O}{2fCV_O} = \frac{1}{2fRC} \tag{9.19}$$

The peak-to-peak variation of the voltage v_{C1} is

$$\Delta v_{C1} = \frac{I_{LO}(1-k)T}{C_1} = \frac{I_O(1-k)}{fC_1} \tag{9.20}$$

The variation ratio of the voltage v_{C1} is

$$\sigma_1 = \frac{\Delta v_{C1}/2}{V_{C1}} = \frac{I_O(1-k)}{2fC_1V_O} = \frac{1}{2MfRC_1} \tag{9.21}$$

The peak-to-peak variation of the current i_{LO} is approximately

$$\Delta i_{LO} = \frac{(1/2)(\Delta v_{C1}/2)(T/2)}{L_O} = \frac{I_O(1-k)}{8f^2L_OC_1} \tag{9.22}$$

The variation ratio of the current i_{LO} is approximately

$$\zeta_2 = \frac{\Delta i_{LO}/2}{I_{LO}} = \frac{I_O(1-k)}{16f^2L_OC_1I_O} = \frac{1}{16Mf^2L_OC_1} \tag{9.23}$$

The peak-to-peak variation of voltage v_O and v_{CO} is

$$\Delta v_O = \Delta v_{CO} = \frac{(1/2)(\Delta i_{LO}/2)(T/2)}{C_O} = \frac{I_O(1-k)}{64f^3L_OC_1C_O} \tag{9.24}$$

The variation ratio of the output voltage is

$$\varepsilon = \frac{\Delta v_O/2}{V_O} = \frac{I_O(1-k)}{128f^3L_OC_1C_OV_O} = \frac{1}{128Mf^3L_OC_1C_OR} \tag{9.25}$$

The voltage transfer gain of the self-lift Cúk converter is the same as the original Boost converter. However, the output current of the self-lift Cúk converter is continuous with small ripples.

The output voltage of the self-lift Cúk converter is higher than the corresponding Cúk converter by an input voltage. It retains one of the merits of the Cúk converter. They both have continuous input and output current in CCM. As for component stress, it can be seen that the self-lift converter has a smaller voltage and current stresses than the original Cúk converter.

9.2.1.2 Discontinuous Conduction Mode

Self-lift Cúk converter operates in the discontinuous conduction mode (DCM), if the current i_D reduces to zero during switch-off. As a special case, when i_D decreases to zero at $t=T$, then the circuit operates at the boundary of CCM and DCM. The variation ratio of the current i_D is 1 when the circuit works in the boundary state.

$$\xi = \frac{k}{2}\frac{R}{M^2 fL} = 1 \tag{9.26}$$

Therefore, the boundary between CCM and DCM is

$$M_B = \sqrt{k}\sqrt{\frac{R}{2fL}} = \sqrt{\frac{kz_N}{2}} \tag{9.27}$$

where z_N is the normalized load $R/(fL)$. The boundary between CCM and DCM is shown in Figure 9.3a. The curve that describes the relationship between M_B and z_N has the minimum value $M_B = 1.5$ and $k = 1/3$ when the normalized load z_N is 13.5.

When $M > M_B$, the circuit operates in the DCM. In this case the diode current i_D, decreases to zero at $t = t_1 = [k+(1-k)m]T$ where $kT < t_1 < T$ and $0 < m < 1$.

Define m as the current filling factor. After mathematical manipulation,

$$m = \frac{1}{\xi} = \frac{M^2}{k(R/2fL)} \tag{9.28}$$

From the previous equation we can see that the DCM is caused by the following factors:

- Switching frequency f is too low.
- Duty cycle k is too small.
- Inductance L is too small.
- Load resistor R is too big.

In the DCM, current i_L increases during switch-on and decreases in the period from kT to $(1-k)mT$. The corresponding voltages across L are V_I and $-(V_C - V_I)$. Therefore, $kTV_I = (1-k)mT(V_C - V_I)$

Hence,

$$V_C = \left[1 + \frac{k}{(1-k)m}\right]V_I \tag{9.29}$$

Since we assume that C, C_1, and C_O are large enough,

$$V_O = V_C = V_{CO} = \left[1 + \frac{k}{(1-k)m}\right]V_I \tag{9.30}$$

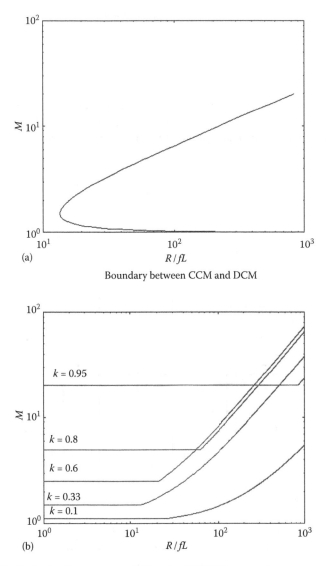

FIGURE 9.3 Output voltage characteristics of self-lift Cúk converter.

or

$$V_O = \left[1 + k^2(1-k)\frac{R}{2fL}\right]V_I \qquad (9.31)$$

The voltage transfer gain in the DCM is

$$M_{DCM} = 1 + k^2(1-k)\frac{R}{2fL} \qquad (9.32)$$

The relation between DC voltage transfer gain M and the normalized load at various k in the DCM is also shown Figure 9.3b. It can be seen that in DCM, the output voltage increases as the load resistance R increasing.

9.2.2 SELF-LIFT P/O LUO CONVERTER

Self-lift P/O Luo converter and the equivalent circuits during switch-on and switch-off periods are shown Figure 9.4. It is the self-lift circuit of the P/O Luo converter. It is derived from the elementary circuit of P/O Luo converter. During the switch-on

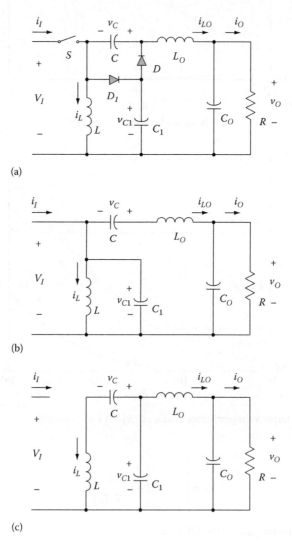

(a)

(b)

(c)

FIGURE 9.4 Self-lift P/O Luo converter and its equivalent circuits: (a) circuit, (b) the equivalent circuit during switch-on, and (c) the equivalent circuit during switch-off.

period, S and D_1 are switched on, and D is switched off. During the switch-off period, D is on, S and D_1 are off.

9.2.2.1   Continuous Conduction Mode

In steady state, the average inductor voltages over a period are zero. Thus,

$$V_C = V_{CO} = V_O$$

During the switch-on period, the voltage across capacitor C_1 is equal to the source voltage, since we assume that C and C_1 are sufficiently large, $V_{C1} = V_I$.

The inductor current i_L increases in the switch-on period and decreases in the switch-off period. The corresponding voltages across L are V_I and $-(V_C - V_{C1})$.

Therefore,

$$kTV_I = (1-k)T(V_C - V_{C1})$$

Hence,

$$V_O = \frac{1}{1-k}V_I$$

The voltage transfer gain in the CCM is

$$M = \frac{V_O}{V_I} = \frac{1}{1-k} \tag{9.33}$$

Since all the components are considered ideal, the power loss associated with all the circuit elements are neglected. Therefore, the output power P_O is considered to be equal to the input power P_{IN}, $V_O I_O = V_I I_I$

Thus,

$$I_I = \frac{1}{1-k}I_O$$

The capacitor C_O acts as a low-pass filter so that $I_{LO} = I_O$.

The charge of capacitor C increases during switch-on and decreases during switch-off.

$$Q_+ = I_{C-ON}kT = I_O kT \quad Q_- = I_{C-OFF}(1-k)T = I_L(1-k)T$$

During a switch-on period,

$$Q_+ = Q_- \quad I_L = \frac{k}{1-k}I_O$$

During switch-off period, $i_D = i_L + i_{LO}$
   Therefore,

$$I_D = I_L + I_{LO} = \frac{1}{1-k} I_O$$

For the current and voltage variations and boundary condition, we can get the following equations using similar method used in the analysis of self-lift Cúk converter.
   Current variations are

$$\zeta_1 = \frac{1}{2M^2} \frac{R}{fL} \qquad \zeta_2 = \frac{k}{2M} \frac{R}{fL_O} \qquad \xi = \frac{k}{2M^2} \frac{R}{fL_{eq}}$$

where L_{eq} refers to $L_{eq} = LL_O/(L+L_O)$
   Voltage variations are

$$\rho = \frac{k}{2} \frac{1}{fCR} \qquad \sigma_1 = \frac{M}{2} \frac{1}{fC_1R} \qquad \varepsilon = \frac{k}{8M} \frac{1}{f^2 L_O C_O}$$

9.2.2.2 Discontinuous Conduction Mode

Self-lift P/O Luo converter operates in the DCM if the current i_D reduces to zero during switch-off. In a critical case, when i_D decreases to zero at $t=T$, the circuit operates at the boundary of CCM and DCM.
   The variation ratio of the current i_D is 1 when the circuit works in the boundary state:

$$\xi = \frac{k}{2M^2} \frac{R}{fL_{eq}} = 1$$

Therefore, the boundary between CCM and DCM is

$$M_B = \sqrt{k} \sqrt{\frac{R}{2fL_{eq}}} = \sqrt{\frac{kz_N}{2}} \qquad (9.34)$$

where
   z_N is the normalized load $R/(fL_{eq})$
   L_{eq} refers to $L_{eq} = LL_O/(L+L_O)$

When $M > M_B$, the circuit operates at the DCM. In this case, the circuit is operating in the case that the diode current i_D decreases to zero at $t=t_1=[k+(1-k)\,m]\,T$, where $KT < t_1 < T$ and $0 < m < 1$, and m is the current-filling factor. We define m as

$$m = \frac{1}{\xi} = \frac{M^2}{k(R/2fL_{eq})} \qquad (9.35)$$

In the DCM, current i_L increases in switch-on kT and decreases in the period from kT to $(1-k)mT$. The corresponding voltages across L are V_I and $-(V_C - V_{C1})$. Therefore,

$$kTV_I = (1-k)mT(V_C - V_{C1})$$

and

$$V_C = V_{CO} = V_O \quad V_{C1} = V_I$$

Hence,

$$V_O = \left[1 + \frac{k}{(1-k)m}\right]V_I \quad \text{or} \quad V_O = \left[1 + k^2(1-k)\frac{R}{2fL_{eq}}\right]V_I \qquad (9.36)$$

So the real DC voltage transfer gain in the DCM is

$$M_{DCM} = 1 + k^2(1-k)\frac{R}{2fL_{eq}} \qquad (9.37)$$

In DCM, the output voltage increases as the load resistance R increasing.

Example 9.1 A P/O self-lift Luo converter has the following components: $V_I = 20\,V$, $L = L_O = 1\,mH$, $C = C_1 = C_O = 20\,\mu F$, $R = 40\,\Omega$, $f = 50\,kHz$, and $k = 0.5$. Calculate the output voltage, the variation ratios ζ_1, ζ_2, ξ, ρ, σ_1, and ε in steady state.

Solution:

1. From (9.33), the output voltage is $V_O = V_I/(1-k) = 20/0.5 = 40\,V$, that is, $M = 2$.
2. From the formulae in Section 9.2.2.1, we can get the ratios:

$$\zeta_1 = \frac{1}{2M^2}\frac{R}{fL} = \frac{1}{2 \times 2^2}\frac{40}{50\,kHz \times 1\,mH} = 0.1$$

$$\zeta_2 = \frac{k}{2M}\frac{R}{fL_O} = \frac{1}{2 \times 2^2}\frac{40}{50\,kHz \times 1\,mH} = 0.1$$

$$\xi = \frac{k}{2M^2}\frac{R}{fL_{eq}} = \frac{1}{2 \times 2^2}\frac{40}{50\,kHz \times 0.5\,mH} = 0.2$$

$$\rho = \frac{k}{2}\frac{1}{fCR} = \frac{0.5}{2}\frac{1}{50\,kHz \times 20\,\mu F \times 40} = 0.00625$$

$$\sigma_1 = \frac{M}{2} \frac{1}{fC_iR} = \frac{2}{2} \frac{1}{50\,\text{kHz} \times 20\,\mu\text{F} \times 40} = 0.025$$

$$\varepsilon = \frac{k}{8M} \frac{1}{f^2 L_O C_O} = \frac{0.5}{8 \times 2} \frac{1}{(50\,\text{kHz})^2 \times 20\,\mu\text{F} \times 1\,\text{mH}} = 0.000625$$

From the calculations, the variations of i_{L1}, i_{L2}, v_C, and v_{C1} are small. The output voltage v_O (also v_{C1}) is almost a real DC voltage with very small ripple. Because of the resistive load, the output current i_O ($i_O = v_O/R$) is almost a real DC waveform with very small ripple as well.

9.2.3   REVERSE SELF-LIFT P/O LUO CONVERTER

Reverse self-lift P/O Luo converter and the equivalent circuits during the switch-on and switch-off periods are shown in Figure 9.5. It is derived from the elementary circuit of P/O Luo converters. During switch-on period, S and D_1 are on, and D is off. During switch-off period, D is on, and S and D_1 are off.

9.2.3.1   Continuous Conduction Mode

In steady state, the average inductor voltages over a period are zero.
    Thus,

$$V_{C1} = V_{CO} = V_O$$

During switch-on period, the voltage across capacitor C is equal to the source voltage plus the voltage across C_1. Since we assume that C and C_1 are sufficiently large, $V_{C1} = V_I + V_C$.
    Therefore,

$$V_{C1} = V_I + \frac{k}{1-k}V_I = \frac{1}{1-k}V_I \quad V_O = V_{CO} = V_{C1} = \frac{1}{1-k}V_I \qquad (9.38)$$

The voltage transfer gain in the CCM is

$$M = \frac{V_O}{V_I} = \frac{1}{1-k} \qquad (9.39)$$

Since all the components are considered ideal, the power losses on all the circuit elements are neglected. Therefore, the output power P_O is considered to be equal to the input power P_{IN},

$$V_O I_O = V_I I_I$$

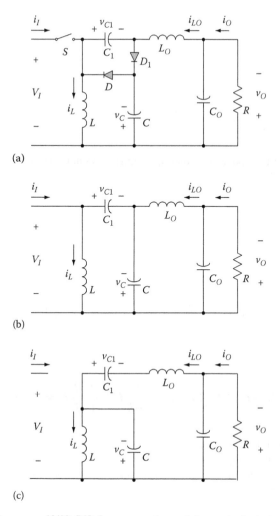

(a)

(b)

(c)

FIGURE 9.5 Reverse self-lift P/O Luo converter and its equivalent circuits: (a) circuit, (b) the equivalent circuit during switch-on, and (c) the equivalent circuit during switch-off.

Thus,

$$I_I = \frac{1}{1-k}I_O$$

The capacitor C_O acts as a low-pass filter so that $I_{LO} = I_O$

The charge of capacitor C_1 increases during switch-on and decreases during switch-off:

$$Q_+ = I_{C1-ON}kT$$

$$Q_- = I_{LO}(1-k)T = I_O(1-k)T$$

In a switching period,

$$Q_+ = Q_- \quad I_{C1-ON} = \frac{1-k}{k} I_O$$

$$I_{C-ON} = I_{LO} + I_{C1-ON} = I_O + \frac{1-k}{k} I_O = \frac{1}{k} I_O \qquad (9.40)$$

The charge on the capacitor C increases during switch-off and decreases during switch-on:

$$Q_+ = I_{C-OFF}(1-k)T \quad Q_- = I_{C-ON}kT = \frac{1}{k} I_O kT$$

In a switching period,

$$Q_+ = Q_- \quad I_{C-OFF} = \frac{1-k}{k} I_{C-ON} = \frac{1}{1-k} I_O \qquad (9.41)$$

Therefore,

$$I_L = I_{LO} + I_{C-OFF} = I_O + \frac{1}{1-k} I_O = \frac{2-k}{1-k} I_O = I_O + I_I$$

During switch-off, $i_D = i_L - i_{LO}$
    Therefore, $I_D = I_L - I_{LO} = I_O$
    The following equations can be used for current and voltage variations and boundary condition.
    Current variations is

$$\zeta_1 = \frac{k}{(2-k)M^2} \frac{R}{fL} \quad \zeta_2 = \frac{k}{2M} \frac{R}{fL_O} \quad \xi = \frac{1}{2M^2} \frac{R}{fL_{eq}}$$

where L_{eq} refers to $L_{eq} = LL_O/(L+L_O)$.
    Voltage variations are as follows:

$$\rho = \frac{1}{2k} \frac{1}{fCR} \quad \sigma_1 = \frac{1}{2M} \frac{1}{fC_1R} \quad \varepsilon = \frac{k}{16M} \frac{1}{f^2C_OL_O}$$

9.2.3.2 Discontinuous Conduction Mode

Reverse self-lift P/O Luo converter operates in the DCM if the current i_D reduces to zero during switch-off at $t=T$, then the circuit operates at the boundary of CCM and DCM. The variation ratio of the current i_D is 1 when the circuit works in the boundary state:

$$\xi = \frac{k}{2M^2} \frac{R}{fL_{eq}} = 1$$

Therefore, the boundary between CCM and DCM is

$$M_B = \sqrt{k} \sqrt{\frac{R}{2fL_{eq}}} = \sqrt{\frac{kz_N}{2}} \qquad (9.42)$$

where
z_N is the normalized load $R/(fL_{eq})$
L_{eq} refers to $L_{eq} = (LL_O/L + L_O)$

When $M > M_B$, the circuit operates in DCM. In this case, the diode current i_D, decreases to zero at $t = t_1 = [k + (1-k) m]T$, where $kT < t_1 < T$ and $0 < m < 1$, and m as the current-filling factor.

$$m = \frac{1}{\xi} = \frac{M^2}{k(R/2fL_{eq})} \qquad (9.43)$$

In the DCM, current i_L increases during switch-on and decreases during the period kT to $(1-k)mT$. The corresponding voltages across L are V_I and $-V_C$.
Therefore,

$$kTV_I = (1-k)mTV_C$$

and

$$V_{C1} = V_{CO} = V_O \qquad V_{C1} = V_I + V_C$$

Hence,

$$V_O = \left[1 + \frac{k}{(1-k)m}\right]V_I \quad \text{or} \quad V_O = \left(1 + k^2(1-k)\frac{R}{2fL_{eq}}\right)V_I \qquad (9.44)$$

So the real DC voltage transfer gain in the DCM is

$$M_{DCM} = 1 + k^2(1-k)\frac{R}{2fL} \qquad (9.45)$$

In DCM, the output voltage increases as the load resistance R increases.

9.2.4 SELF-LIFT N/O LUO CONVERTER

Self-lift N/O Luo converter and the equivalent circuits during switch-on and switch-off are shown in Figure 9.6. It is the self-lift circuit of the N/O Luo converter. The function of capacitor C_1 is to lift the voltage V_C by a source voltage V_I. S and D_1 are on, and D is off during the switch-on period. D is on, and S and D_1 are off during the switch-off period.

9.2.4.1 Continuous Conduction Mode

In steady state, the average inductor voltages over a period are zero. Thus, $V_C = V_{CO} = V_O$. During the switch-on period, the voltage across capacitor C_1 is equal to the source voltage. Since we assume that C and C_1 are sufficiently large, $V_{C1} = V_I$.

Inductor current i_L increases in the switch-on period and decreases in the switch-off period. The corresponding voltages across L are V_I and $-(V_C - V_{C1})$.

Therefore,

$$kTV_I = (1-k)T(V_C - V_{C1})$$

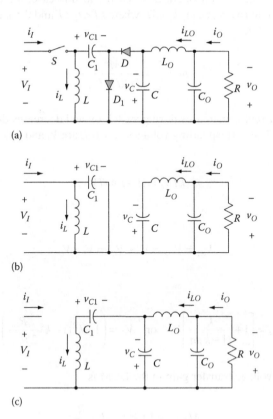

(a)

(b)

(c)

FIGURE 9.6 Self-lift N/O Luo converter and its equivalent circuits: (a) circuit, (b) the equivalent circuit during switch-on, and (c) the equivalent circuit during switch-off.

Hence,

$$V_O = V_C = V_{CO} = \frac{1}{1-k} V_I \qquad (9.46)$$

The voltage transfer gain in the CCM is

$$M = \frac{V_O}{V_I} = \frac{1}{1-k} \qquad (9.47)$$

Since all the components are considered ideal, the power loss associated with all the circuit elements are neglected. Therefore, the output power P_O is considered to be equal to the input power P_{IN}, $V_O I_O = V_I I_I$.

Thus,

$$I_I = \frac{1}{1-k} I_O$$

The capacitor C_O acts as a low-pass filter so that $I_{LO} = I_O$

For the current and voltage variations and boundary condition, the following equations can be obtained using a method similar to the one used in the analysis of self-lift Cúk converter.

Current variations are as follows:

$$\zeta_1 = \frac{k}{2M^2} \frac{R}{fL} \quad \zeta_2 = \frac{k}{16} \frac{1}{f^2 L_O C} \quad \xi = \zeta_1 = \frac{k}{2M^2} \frac{R}{fL}$$

Voltage variations are as follows:

$$\rho = \frac{k}{2} \frac{1}{fCR} \quad \sigma_1 = \frac{M}{2} \frac{1}{fC_1 R} \quad \varepsilon = \frac{k}{128} \frac{1}{f^3 L_O C C_O R}$$

9.2.4.2 Discontinuous Conduction Mode

Self-lift N/O Luo converter operates in the DCM if the current i_D reduces to zero at $t = T$, then the circuit operates at the boundary of CCM and DCM. The variation ratio of the current i_D is 1 when the circuit works at the boundary state:

$$\xi = \frac{k}{2M^2} \frac{R}{fL} = 1$$

Therefore, the boundary between CCM and DCM is

$$M_B = \sqrt{k}\sqrt{\frac{R}{2fL_{eq}}} = \sqrt{\frac{kz_N}{2}} \tag{9.48}$$

where L_{eq} refers to $L_{eq}=L$ and z_N is the normalized load $R/(fL)$.

When $M>M_B$, the circuit operates in the DCM. In this case, the diode current i_D decreases to zero at $t=t_1=[k+(1-k)\,m]T$, where $KT<t_1<T$ and $0<m<1$, and m is the current-filling factor and is defined as

$$m = \frac{1}{\xi} = \frac{M^2}{k(R/2fL)} \tag{9.49}$$

In the DCM, current i_L increases during switch-on and decreases during the period kT to $(1-k)mT$. The voltages across L are V_I and $-(V_C-V_{C1})$.

$$kTV_I = (1-k)mT(V_C - V_{C1})$$

and

$$V_{C1} = V_I \quad V_C = V_{CO} = V_O$$

Hence,

$$V_O = \left[1+\frac{k}{(1-k)m}\right]V_I \quad \text{or} \quad V_O = \left[1+k^2(1-k)\frac{R}{2fL}\right]V_I$$

So the real DC voltage transfer gain in the DCM is

$$M_{DCM} = 1+k^2(1-k)\frac{R}{2fL} \tag{9.50}$$

We can see that in DCM, the output voltage increases as the load resistance R increases.

9.2.5 REVERSE SELF-LIFT N/O LUO CONVERTER

Reverse self-lift N/O Luo converter and the equivalent circuits during the switch-on and switch-off periods are shown in Figure 9.7. During the switch-on period, S and D_1 are on, and D is off. During the switch-off period, D is on, and S and D_1 are off.

9.2.5.1 Continuous Conduction Mode
In steady state, the average inductor voltages over a period are zero. Thus,

$$V_{C1} = V_{CO} = V_O$$

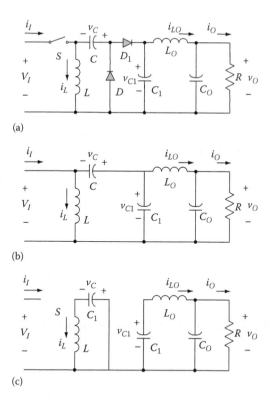

(a)

(b)

(c)

FIGURE 9.7 Reverse self-lift N/O Luo converter and its equivalent circuits: (a) circuit, (b) the equivalent circuit during switch-on, and (c) the equivalent circuit during switch-off.

The inductor current i_L increases in the switch-on period and decreases in the switch-off period. The corresponding voltages across L are V_I and $-V_C$.

Therefore,

$$kTV_I = (1-k)TV_C$$

Hence,

$$V_C = \frac{k}{1-k}V_I \qquad (9.51)$$

voltage across C. Since we assume that C and C_1 are sufficiently large, $V_{C1} = V_I + V_C$.

Therefore,

$$V_{C1} = V_I + \frac{k}{1-k}V_I = \frac{1}{1-k}V_I \quad V_O = V_{CO} = V_{C1} = \frac{1}{1-k}V_I$$

The voltage transfer gain in the CCM is

$$M = \frac{V_O}{V_I} = \frac{1}{1-k} \tag{9.52}$$

Since all the components are considered ideal, the power loss associated with all the circuit elements are neglected. Therefore, the output power P_O is considered to be equal to the input power P_{IN}, $V_O I_O = V_I I_I$

Thus,

$$I_I = \frac{1}{1-k} I_O$$

The capacitor C_O acts as a low-pass filter so that $I_{LO} = I_O$

The charge of capacitor C_1 increases during switch-on and decreases during switch-off.

$$Q_+ = I_{C1-ON} kT \quad Q_- = I_{C1-OFF}(1-k)T = I_O(1-k)T$$

In a switching period,

$$Q_+ = Q_- \quad I_{C1-ON} = \frac{1-k}{k} I_{C-OFF} = \frac{1-k}{k} I_O$$

The charge of capacitor C increases during switch-on and decreases during switch-off.

$$Q_+ = I_{C-ON} kT \quad Q_- = I_{C-OFF}(1-k)T$$

In a switching period, $Q_+ = Q_-$

$$I_{C-ON} = I_{C1-ON} + I_{LO} = \frac{1-k}{k} I_O + I_O = \frac{1}{k} I_O$$

$$I_{C-OFF} = \frac{k}{1-k} I_{C-ON} = \frac{k}{1-k} \frac{1}{k} I_O = \frac{1}{1-k} I_O$$

Therefore,

$$I_L = I_{C-OFF} = \frac{1}{1-k} I_O$$

During the switch-off period,

$$i_D = i_L \quad I_D = I_L = \frac{1}{1-k} I_O$$

For the current and voltage variations and the boundary condition, we can get the following equations using a method similar to the one used in the analysis of self-lift Cúk converter.

Current variations are as follows:

$$\zeta_1 = \frac{k}{2M^2} \frac{R}{fL} \quad \zeta_2 = \frac{1}{16M} \frac{R}{f^2 L_O C_1} \quad \xi = \frac{k}{2M^2} \frac{R}{fL}$$

Voltage variations are as follows:

$$\rho = \frac{1}{2k} \frac{1}{fCR} \quad \sigma_1 = \frac{1}{2M} \frac{1}{fC_1 R} \quad \varepsilon = \frac{1}{128M} \frac{1}{f^3 L_O C_1 C_O R}$$

9.2.5.2 Discontinuous Conduction Mode

Reverse self-lift N/O Luo converter operates in the DCM, if the current i_D reduces to zero during switch-off. As a special case, when i_D decreases to zero at $t=T$, then the circuit operates at the boundary of CCM and DCM.

The variation ratio of the current i_D is 1 when the circuit works in the boundary state:

$$\xi = \frac{k}{2M^2} \frac{R}{fL_{eq}} = 1$$

The boundary between CCM and DCM is as follows:

$$M_B = \sqrt{k}\sqrt{\frac{R}{2fL_{eq}}} = \sqrt{\frac{kz_N}{2}}$$

where
z_N is the normalized load $R/(fL_{eq})$
L_{eq} refers to $L_{eq}=L$

When $M>M_B$, the circuit operates at the DCM. In this case, diode current i_D decreases to zero at $t=t_1=[k+(1-k)m]T$ where $KT<t_1<T$ and $0<m<1$ with m as the current-filling factor.

$$m = \frac{1}{\xi} = \frac{M^2}{k(R/2fL_{eq})} \tag{9.53}$$

In the DCM, current i_L increases in the switch-on period kT and decreases during the period kT to $(1-k)mT$. The corresponding voltages across L are V_I and $-V_C$.

Therefore,

$$kTV_I = (1-k)mTV_C$$

and

$$V_{C1} = V_{CO} = V_O \quad V_{C1} = V_I + V_C$$

Hence,

$$V_O = \left[1 + \frac{k}{(1-k)m}\right]V_I \quad \text{or} \quad V_O = \left(1 + k^2(1-k)\frac{R}{2fL}\right)V_I \tag{9.54}$$

The voltage transfer gain in the DCM is

$$M_{DCM} = 1 + k^2(1-k)\frac{R}{2fL} \tag{9.55}$$

It can be seen that in DCM the output voltage increases as the load resistance R increases.

9.2.6 SELF-LIFT SEPIC

Self-lift SEPIC and the equivalent circuits during switch-on and switch-off periods are shown in Figure 9.8. It is derived from SEPIC (with output filter). S and D_1 are on, and D is off during the switch-on period. D is on, and S and D_1 are off during the switch-off period.

9.2.6.1 Continuous Conduction Mode

In steady state, the average voltage across inductor L over a period is zero. Thus, $V_C = V_I$.

During the switch-on period, the voltages across capacitor C_1 are equal to the voltages across C. Since we assume that C and C_1 are sufficiently large, $V_{C1} = V_C = V_I$.

In steady state, the average voltage across inductor L_O over a period is also zero. Thus,

$$V_{C2} = V_{CO} = V_O$$

The inductor current i_L increases in the switch-on period and decreases in the switch-off period. The corresponding voltages across L are V_I and $-(V_C - V_{C1} + V_{C2} - V_I)$.

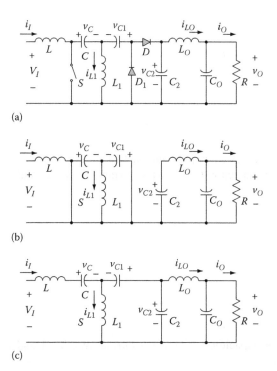

FIGURE 9.8 Self-lift SEPIC converter and its equivalent circuits: (a) self-lift SEPIC converter, (b) the equivalent circuit during switch-on, and (c) the equivalent circuit during switch-off.

Therefore,

$$kTV_I = (1-k)T(V_C - V_{C1} + V_{C2} - V_I)$$

or

$$kTV_I = (1-k)T(V_O - V_I)$$

Hence,

$$V_O = \frac{1}{1-k}V_I = V_{CO} = V_{C2} \tag{9.56}$$

The voltage transfer gain in the CCM is

$$M = \frac{V_O}{V_I} = \frac{1}{1-k} \tag{9.57}$$

Since all the components are considered ideal, the power loss associated with all the circuit elements are neglected. Therefore, the output power P_O is considered to be equal to the input power P_{IN}, $V_O I_O = V_I I_I$.
Thus,

$$I_I = \frac{1}{1-k} I_O = I_L$$

The capacitor C_O acts as a low-pass filter so that $I_{LO} = I_O$.
The charge of capacitor C increases during switch-off and decreases during switch-on.

$$Q_- = I_{C-ON} kT \quad Q_+ = I_{C-OFF}(1-k)T = I_I(1-k)T$$

In a switching period,

$$Q_+ = Q_- \quad I_{C-ON} = \frac{1-k}{k} I_{C-OFF} = \frac{1-k}{k} I_I$$

The charge of capacitor C_2 increases during switch-off and decreases during switch-on periods.

$$Q_- = I_{C2-ON} kT = I_O kT \quad Q_+ = I_{C2-OFF}(1-k)T$$

In a switching period,

$$Q_+ = Q_- \quad I_{C2-OFF} = \frac{k}{1-k} I_{C-N} = \frac{k}{1-k} I_O$$

The charge of capacitor C_1 increases during switch-on and decreases during switch-off.

$$Q_+ = I_{C1-ON} kT \quad Q_- = I_{C1-OFF}(1-k)T$$

In a switching period,

$$Q_+ = Q_- \quad I_{C1-OFF} = I_{C2-OFF} + I_{LO} = \frac{k}{1-k} I_O + I_O = \frac{1}{1-k} I_O$$

Therefore,

$$I_{C1-ON} = \frac{1-k}{k} I_{C1-OFF} = \frac{1}{k} I_O \quad I_{L1} = I_{C1-ON} - I_{C-ON} = 0$$

During switch-off,

$$i_D = i_L - i_{L1}$$

Therefore,

$$I_D = I_I = \frac{1}{1-k} I_O$$

For the current and voltage variations and the boundary condition, we can get the following equations using a method similar to the one used in the analysis of self-lift Cúk converter.

Current variations are as follows:

$$\zeta_1 = \frac{k}{2M^2} \frac{R}{fL} \quad \zeta_2 = \frac{k}{16} \frac{R}{f^2 L_O C_2} \quad \xi = \frac{k}{2M^2} \frac{R}{fL_{eq}}$$

where L_{eq} refers to $L_{eq} = LL_O/(L+L_O)$.

Voltage variations are as follows:

$$\rho = \frac{M}{2} \frac{1}{fCR} \quad \sigma_1 = \frac{M}{2} \frac{1}{fC_1 R} \quad \sigma_2 = \frac{k}{2} \frac{1}{fC_2 R} \quad \varepsilon = \frac{k}{128} \frac{1}{f^3 L_O C_2 C_O R}$$

9.2.6.2 Discontinuous Conduction Mode

Self-lift Sepic converter operates in the DCM if the current i_D reduces to zero during switch-off. As a special case, when i_D decreases to zero at $t = T$, then the circuit operates at the boundary of CCM and DCM.

The variation ratio of the current i_D is 1 when the circuit works in the boundary state:

$$\xi = \frac{k}{2M^2} \frac{R}{fL_{eq}} = 1$$

Therefore, the boundary between CCM and DCM is

$$M_B = \sqrt{k} \sqrt{\frac{R}{2fL_{eq}}} = \sqrt{\frac{kz_N}{2}} \qquad (9.58)$$

where

z_N is the normalized load $R/(fL_{eq})$
L_{eq} refers to $L_{eq} = LL_O/(L+L_O)$

When $M > M_B$, the circuit operates in the DCM. In this case the diode current i_D decreases to zero at $t = t_1 = [k + (1-k)m]T$, where $KT < t_1 < T$ and $0 < m < 1$ and m is defined as

$$m = \frac{1}{\xi} = \frac{M^2}{k(R/2\,fL_{eq})} \qquad (9.59)$$

In the DCM, current i_L increases during switch-on and decreases during the period kT to $(1-k)mT$. The corresponding voltages across L are V_I and $-(V_C - V_{C1} + V_{C2} - V_I)$. Thus,

$$kTV_I = (1-k)T(V_C - V_{C1} + V_{C2} - V_I)$$

and

$$V_C = V_I \quad V_{C1} = V_C = V_I \quad V_{C2} = V_{CO} = V_O$$

Hence,

$$V_O = \left[1 + \frac{k}{(1-k)m}\right] V_I \quad \text{or} \quad V_O = \left(1 + k^2(1-k)\frac{R}{2\,fL_{eq}}\right) V_I$$

So the real DC voltage transfer gain in the DCM is

$$M_{DCM} = 1 + k^2(1-k)\frac{R}{2\,fL_{eq}} \qquad (9.60)$$

In DCM, the output voltage increases as the load resistance R increases.

9.2.7 ENHANCED SELF-LIFT P/O LUO CONVERTER

Enhanced self-lift P/O Luo converter circuit and the equivalent circuits during the switch-on and switch-off periods are shown in Figure 9.9. It is derived from the self-lift P/O Luo converter in Figure 9.4, with the positions of switch S and inductor L swapped.

During the switch-on period, S and D_1 are on, and D is off. Thus, we obtain

$$V_C = V_{C1} \quad \text{and} \quad \Delta i_L = \frac{V_I}{L} kT$$

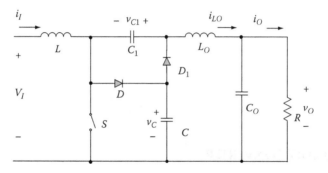

FIGURE 9.9 Enhanced self-lift P/O Luo converter.

During the switch-off period, D is on, and S and D_1 are off:

$$\Delta i_L = \frac{V_C - V_I}{L}(1-k)T$$

so that

$$V_C = \frac{1}{1-k}V_I$$

The output voltage and current and the voltage transfer gain are as follows:

$$V_O = V_I + V_{C1} = \left(1 + \frac{1}{1-k}\right)V_I \tag{9.61}$$

$$I_O = \frac{1-k}{2-k}I_I \tag{9.62}$$

$$M = 1 + \frac{1}{1-k} = \frac{2-k}{1-k} \tag{9.63}$$

Average voltages are as follows:

$$V_C = \frac{1}{1-k}V_I \tag{9.64}$$

$$V_{C1} = \frac{1}{1-k}V_I \tag{9.65}$$

Average currents are as follows:

$$I_{LO} = I_O \tag{9.66}$$

$$I_L = \frac{2-k}{1-k} I_O = I_I \qquad (9.67)$$

Therefore,

$$\frac{V_O}{V_I} = \frac{1}{1-k} + 1 = \frac{2-k}{1-k} \qquad (9.68)$$

9.3 P/O LUO CONVERTERS

P/O Luo converters perform the voltage conversion from positive to positive voltages using VL technique. They work in the first quadrant with large voltage amplification. Five circuits have been introduced in the literature [6–11]:

1. Elementary circuit
2. Self-lift circuit
3. Re-lift circuit
4. Triple-lift circuit
5. Quadruple-lift circuit

The elementary circuit was discussed in Section 5.5.1, and the self-lift circuit was discussed in Section 9.2.2.

9.3.1 Re-Lift Circuit

Re-lift circuit is derived from self-lift circuit, and its switch-on and switch-off equivalent circuits are shown in Figure 9.10. Capacitors C_1 and C_2 perform characteristics to lift the capacitor voltage V_C twice that of the source voltage V_I. L_3 performs the function as a *ladder joint* to link the two capacitors C_1 and C_2 and lift the capacitor voltage V_C up.

When switches S and S_1 turn on, the source instantaneous current $i_I = i_{L1} + i_{L2} + i_{C1} + i_{L3} + i_{C2}$. Inductors L_1 and L_3 absorb energy from the source. In the mean time, inductor L_2 absorbs energy from source and capacitor C. Three currents i_{L1}, i_{L3}, and i_{L2} increase. When switches S and S_1 are off, source current $i_I = 0$. Current i_{L1} flows through capacitor C_1, inductor L_3, capacitor C_2, and diode D to charge capacitor C. Inductor L_1 transfers its stored energy to capacitor C. In the mean time, current i_{L2} flows through the $(C_O - R)$ circuit, capacitor C_1, inductor L_3, capacitor C_2, and diode D to keep itself continuous. Both currents i_{L1} and i_{L2} decrease. In order to analyze the circuit working procession, the equivalent circuits in switch-on and switch-off states are shown in Figure 9.10b through d. Assuming capacitor C_1 and C_2 are sufficiently large, the voltages V_{C1} and V_{C2} across them are equal to V_I in steady state.

Voltage v_{L3} is equal to V_I during switch-on. The peak-to-peak variation of current i_{L3} is

$$\Delta i_{L3} = \frac{V_I k T}{L_3} \qquad (9.69)$$

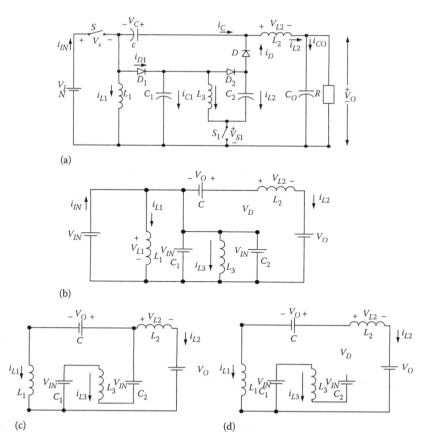

(a)

(b)

(c)   (d)

FIGURE 9.10 P/O re-lift circuit: (a) circuit diagram, (b) switch on, and (c) switch-off, (d) discontinuous mode.

This variation is equal to the current reduction in the switch-off condition. Suppose its voltage is $-V_{L3\text{-}off}$,

$$\Delta i_{L3} = \frac{V_{L3\text{-}off}(1-k)T}{L_3}$$

Thus, during the switch-off condition, the voltage drop across inductor L_3 is

$$V_{L3\text{-}off} = \frac{k}{1-k}V_I \qquad (9.70)$$

Current i_{L1} increases in the switch-on period kT, and decreases in the switch-off period $(1-k)T$. The corresponding voltages applied across L_1 are V_I and $-(V_C - 2V_I - V_{L3\text{-}off})$. Therefore,

$$kTV_I = (1-k)T(V_C - 2V_I - V_{L3\text{-}off})$$

Hence,

$$V_C = \frac{2}{1-k}V_I \tag{9.71}$$

Current i_{L2} increases in the switch-on period kT, and it decreases in switch-off period $(1-k)T$. The corresponding voltages applied across L_2 are $(V_I + V_C - V_O)$ and $-(V_O - 2V_I - V_{L3\text{-}off})$. Therefore,

$$kT(V_C + V_I - V_O) = (1-k)T(V_O - 2V_I - V_{L3\text{-}off})$$

Hence,

$$V_O = \frac{2}{1-k}V_I \tag{9.72}$$

and the output current is

$$I_O = \frac{1-k}{2}I_I \tag{9.73}$$

The voltage transfer gain in continuous mode is

$$M_R = \frac{V_O}{V_I} = \frac{2}{1-k} \tag{9.74}$$

The curve of M_R versus k is shown in Figure 9.11.

Other average currents

$$I_{L1} = \frac{k}{1-k}I_O = \frac{k}{2}I_I \tag{9.75}$$

and

$$I_{L3} = I_{L1} + I_{L2} = \frac{1}{1-k}I_O \tag{9.76}$$

Currents i_{C1} and i_{C2} equal to $i_{L1} + i_{L2}$ during the *switch-off* period, $(1-k)T$, and the charges on capacitors C_1 and C_2 decrease, that is,

$$i_{C1} = i_{C2} = (i_{L1} + i_{L2}) = \frac{1}{1-k}I_O$$

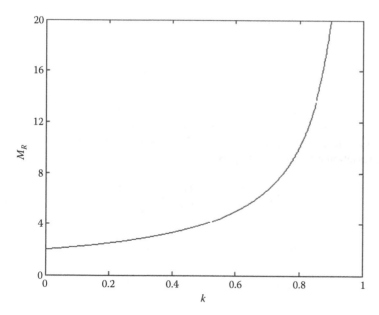

FIGURE 9.11   Voltage transfer gain M_R versus k.

The charges increase during the *switch-on* period kT, so their average currents are

$$I_{C1} = I_{C2} = \frac{1-k}{k}(I_{L1}+I_{L2}) = \frac{1-k}{k}\left(\frac{k}{1-k}+1\right)I_O = \frac{I_O}{k} \qquad (9.77)$$

During switch-off the source current i_I is 0, and in the switch-on period kT, it is

$$i_I = i_{L1}+i_{L2}+i_{C1}+i_{L3}+i_{C2}$$

Hence,

$$I_I = ki_I = k(I_{L1}+I_{L2}+I_{C1}+I_{L3}+I_{C2})$$
$$= k[2(I_{L1}+I_{L2})+2I_{C1}]$$
$$= 2k(I_{L1}+I_{L2})\left(1+\frac{1-k}{k}\right) = 2k\frac{I_{L2}}{1-k}\frac{1}{k} = \frac{2}{1-k}I_O \qquad (9.78)$$

Variations of currents and voltages: Current i_{L1} increases and is supplied by V_I during the switch-on period kT. It decreases and is inversely biased by $-(V_C-2V_I-V_{L3})$ during the switch-off period $(1-k)T$. Therefore, its peak-to-peak variation is $\Delta i_{L1} = kTV_I/L_1$.

The variation ratio of current i_{L1} is

$$\xi_1 = \frac{\Delta i_{L1}/2}{I_{L1}} = \frac{kV_IT}{kL_1I_I} = \frac{1-k}{2M_R}\frac{R}{fL_1}$$

(9.79)

Current i_{L2} increases and is supplied by the voltage $(V_I+V_C-V_O)=V_I$ during the switch-on period kT. It decreases and is inversely biased by $-(V_O-2V_I-V_{L3})$ during switch-off. Therefore, its peak-to-peak variation is $\Delta i_{L2} = kTV_I/L_2$.
  The variation ratio of current i_{L2} is

$$\xi_2 = \frac{\Delta i_{L2}/2}{I_{L2}} = \frac{kTV_I}{2L_2I_O} = \frac{k}{2M_R}\frac{R}{fL_2}$$

(9.80)

In the switch-off condition, the free-wheeling diode current $i_D=i_{L1}+i_{L2}$ and

$$\Delta i_D = \Delta i_{L3} = \Delta i_{L1} + \Delta i_{L2} = \frac{kTV_I}{L} = \frac{k(1-k)V_O}{2L}T$$

(9.81)

Since $I_D=I_{L1}+I_{L2}=I_O/(1-k)$
  the variation ratio of current i_D is

$$\zeta = \frac{\Delta i_D/2}{I_D} = \frac{k(1-k)^2TV_O}{4LI_O} = \frac{k(1-k)R}{2M_RfL} = \frac{k}{M_R^2}\frac{R}{fL}$$

(9.82)

The variation ratio of current i_{L3} is

$$\chi_1 = \frac{\Delta i_{L3}/2}{I_{L3}} = \frac{kV_IT}{2L_3(1/1-k)I_O} = \frac{k}{M_R^2}\frac{R}{fL_3}$$

(9.83)

The peak-to-peak variation of v_C is

$$\Delta v_C = \frac{Q+}{C} = \frac{1-k}{C}TI_{L1} = \frac{k(1-k)}{2C}TI_I$$

Considering Equation 9.71, the variation ratio is

$$\rho = \frac{\Delta v_C/2}{V_C} = \frac{k(1-k)TI_I}{4CV_O} = \frac{k}{2fCR}$$

(9.84)

The charges on capacitors C_1 and C_2 increase during the switch-on period kT and decrease during the switch-off period $(1-k)T$ by the current $(I_{L1}+I_{L2})$. Therefore, their peak-to-peak variations are

$$\Delta v_{C1} = \frac{(1-k)T(I_{L1}+I_{L2})}{C_1} = \frac{(1-k)I_I}{2C_1 f}$$

$$\Delta v_{C2} = \frac{(1-k)T(I_{L1}+I_{L2})}{C_2} = \frac{(1-k)I_I}{2C_2 f}$$

Considering $V_{C1} = V_{C2} = V_I$, the variation ratios of voltages v_{C1} and v_{C2} are as follows:

$$\sigma_1 = \frac{\Delta v_{C1}/2}{V_{C1}} = \frac{(1-k)I_I}{4fC_1V_I} = \frac{M_R}{2fC_1R} \tag{9.85}$$

$$\sigma_2 = \frac{\Delta v_{C2}/2}{V_{C2}} = \frac{(1-k)I_I}{4V_IC_2 f} = \frac{M_R}{2fC_2R} \tag{9.86}$$

Analogously, the variation ratio of output voltage v_O is

$$\varepsilon = \frac{\Delta v_O/2}{V_O} = \frac{kT^2}{16C_OL_2}\frac{V_I}{V_O} = \frac{k}{16M_R}\frac{1}{f^2C_OL_2} \tag{9.87}$$

Example 9.2 A P/O re-lift Luo converter has the following components: $V_I = 20$ V, $L_1 = L_2 = 1$ mH, $L_3 = 0.5$ mH, and all capacitors have $20\,\mu$F, $R = 160\,\Omega$, $f = 50$ kHz, and $k = 0.5$. Calculate the output voltage, the variation ratios ξ_1, ξ_2, ζ, χ_1, ρ, σ_1, σ_2, and ε in steady state.

Solution: From (9.72) we obtain the output voltage as follows:

$$V_O = \frac{2}{1-k}V_I = \frac{2}{1-0.5}20 = 80\,\text{V}$$

The variation ratios $\xi_1 = 0.2$, $\xi_2 = 0.2$, $\zeta = 0.1$, $\chi_1 = 0.1$, $\rho = 0.0016$, $\sigma_1 = 0.0125$, $\sigma_2 = 0.0125$, and $\varepsilon = 1.56 \times 10^{-4}$. Therefore, the variations are small.

From the example we know the variations are small. Therefore, the output voltage v_O is almost a real DC voltage with very small ripple. Because of the resistive load, the output current $i_O(t)$ is almost a real DC waveform with very small ripple as well, and $I_O = V_O/R$.

For DCM, referring to Figure 9.10d, we can see that the diode current i_D becomes zero during switch-off before the next switch-on period. The condition for DCM is $\zeta \geq 1$, that is,

$$\frac{k}{M_R^2}\frac{R}{fL} \geq 1$$

or

$$M_R \leq \sqrt{k}\sqrt{\frac{R}{fL}} = \sqrt{k}\sqrt{z_N} \qquad (9.88)$$

The graph of the boundary curve versus the normalized load $z_N = R/fL$ is shown in Figure 9.12. It can be seen that the boundary curve has a minimum value of 3.0 at $k = 1/3$.

In this case, the current i_D exists during the period between kT and $t_1 = [k + (1-k)m_R]T$, where m_R is the *filling efficiency* and is defined as

$$m_R = \frac{1}{\zeta} = \frac{M_R^2}{k(R/fL)} \qquad (9.89)$$

Therefore, $0 < m_R < 1$. Since the diode current i_D becomes zero at $t = t_1 = kT + (1-k)m_R T$, for the current i_L,

$$kTV_I = (1-k)m_R T(V_C - 2V_I - V_{L3-off})$$

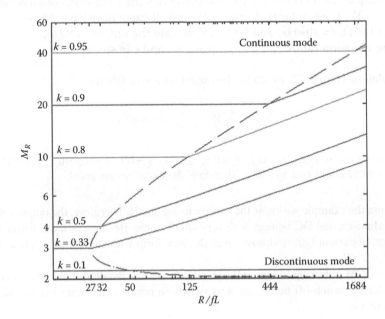

FIGURE 9.12 The boundary between continuous and discontinuous modes and the output voltage versus the normalized load $z_N = R/fL$.

or

$$V_C = \left[2 + \frac{k}{1-k} + \frac{k}{(1-k)m_R} \right]$$

$$V_I = \left[2 + \frac{k}{1-k} + k^2(1-k)\frac{R}{4fL} \right]V_I \quad \text{with} \quad \sqrt{k}\sqrt{\frac{R}{fL}} \geq \frac{2}{1-k}$$

and for the current, $i_{LO} kT(V_I + V_C - V_O) = (1-k)m_R T(V_O - 2V_I - V_{L3\text{-off}})$.

Therefore, output voltage in discontinuous mode is

$$V_O = \left[2 + \frac{k}{1-k} + \frac{k}{(1-k)m_R} \right]$$

$$V_I = \left[2 + \frac{k}{1-k} + k^2(1-k)\frac{R}{4fL} \right]V_I \quad \text{with} \quad \sqrt{k}\sqrt{\frac{R}{fL}} \geq \frac{2}{1-k} \qquad (9.90)$$

that is, the output voltage linearly increases when load resistance R increases. The output voltage versus the normalized load $z_N = R/fL$ is shown in Figure 9.12. Larger load resistance R may cause higher output voltage in discontinuous mode.

9.3.2 TRIPLE-LIFT CIRCUIT

Triple-lift circuit shown in Figure 9.13 consists of two static switches S and S_1; four inductors L_1, L_2, L_3, and L_4; five capacitors C, C_1, C_2, C_3, and C_O; and five diodes. Capacitors C_1, C_2, and C_3 perform characteristics to lift the capacitor voltage V_C by three times that of source voltage V_I. L_3 and L_4 perform the function as ladder

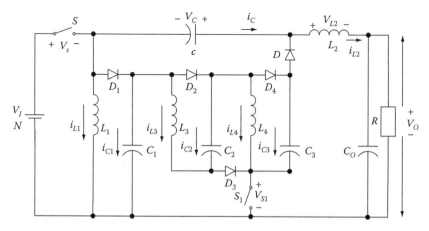

FIGURE 9.13 Triple-lift circuit.

joints to link the three capacitors C_1, C_2, and C_3 and lift the capacitor voltage V_C up. Current $i_{C1}(t)$, $i_{C2}(t)$, and $i_{C3}(t)$ are exponential functions. They have large values at the moment of power on, but they are small because $v_{C1} = v_{C2} = v_{C3} = V_I$ in steady state.

The output voltage and current are as follows:

$$V_O = \frac{3}{1-k} V_I \qquad (9.91)$$

and

$$I_O = \frac{1-k}{3} I_I \qquad (9.92)$$

The voltage transfer gain in continuous mode is

$$M_T = \frac{V_O}{V_I} = \frac{3}{1-k} \qquad (9.93)$$

The curve of M_T versus k is shown in Figure 9.14.

Other average voltages are $V_C = V_O$; $V_{C1} = V_{C2} = V_{C3} = V_I$.

Other average currents are $I_{L2} = I_O$; $I_{L1} = k/(1-k)I_O$

$$I_{L3} = I_{L4} = I_{L1} + I_{L2} = \frac{1}{1-k} I_O$$

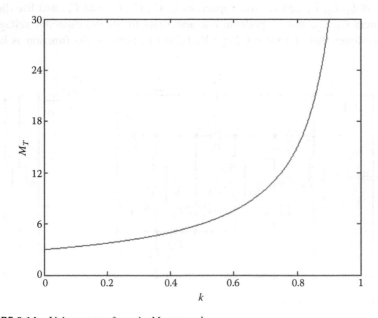

FIGURE 9.14 Voltage transfer gain M_T versus k.

Current variations are as follows:

$$\xi_1 = \frac{1-k}{2M_T} \frac{R}{fL_1} \quad \xi_2 = \frac{k}{2M_T} \frac{R}{fL_2} \quad \zeta = \frac{k(1-k)R}{2M_T fL} = \frac{k}{M_T^2} \frac{3R}{2fL}$$

$$\chi_1 = \frac{k}{M_T^2} \frac{3R}{2fL_3} \quad \chi_2 = \frac{k}{M_T^2} \frac{3R}{2fL_4}$$

Voltage variations are as follows:

$$\rho = \frac{k}{2fCR} \quad \sigma_1 = \frac{M_T}{2fC_1R} \quad \sigma_2 = \frac{M_T}{2fC_2R} \quad \sigma_3 = \frac{M_T}{2fC_3R}$$

The variation ratio of output voltage v_C is

$$\varepsilon = \frac{k}{16M_T} \frac{1}{f^2C_0L_2} \tag{9.94}$$

The output voltage ripple is very small.

The boundary between CCM and DCM is

$$M_T \le \sqrt{k}\sqrt{\frac{3R}{2fL}} = \sqrt{\frac{3kz_N}{2}} \tag{9.95}$$

This boundary curve is shown in Figure 9.15. It can be seen that the boundary curve has a minimum value of M_T that is equal to 4.5, corresponding to $k = 1/3$.

In the discontinuous mode, the current i_D exists during the period between kT and $t_1 = [k + (1-k)m_T]T$, where m_T is the filling efficiency, that is,

$$m_T = \frac{1}{\zeta} = \frac{M_T^2}{k(3R/2fL)} \tag{9.96}$$

The diode current i_D becomes zero at $t = t_1 = kT + (1-k)m_TT$, and therefore, $0 < m_T < 1$. For the current i_{L1}, $kTV_I = (1-k)m_TT(V_C - 3V_I - V_{L3\text{-}off} - V_{L4\text{-}off})$
or

$$V_C = \left[3 + \frac{2k}{1-k} + \frac{k}{(1-k)m_T}\right]$$

$$V_I = \left[3 + \frac{2k}{1-k} + k^2(1-k)\frac{R}{6fL}\right]V_I \quad \text{with} \quad \sqrt{k}\sqrt{\frac{3R}{2fL}} \ge \frac{3}{1-k}$$

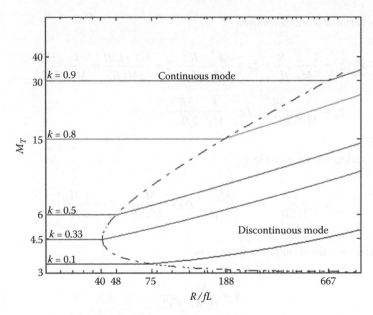

FIGURE 9.15 The boundary between continuous and discontinuous modes and the output voltage versus the normalized load $z_N = R/fL$.

and for the current i_{L2},

$$kT\left(V_I + V_C - V_O\right) = \left(1-k\right)m_T T\left(V_O - 2V_I - V_{L3-off} - V_{L4-off}\right)$$

Therefore, output voltage in discontinuous mode is

$$V_O = \left[3 + \frac{2k}{1-k} + \frac{k}{(1-k)m_T}\right]$$

$$V_I = \left[3 + \frac{2k}{1-k} + k^2(1-k)\frac{R}{6fL}\right]V_I \quad \text{with} \quad \sqrt{k}\sqrt{\frac{3R}{2fL}} \geq \frac{3}{1-k} \qquad (9.97)$$

that is, the output voltage linearly increases when load resistance R increases, as shown in Figure 9.15.

9.3.3 QUADRUPLE-LIFT CIRCUIT

Quadruple-lift circuit shown in Figure 9.16 consists of two static switches S and S_1; five inductors L_1, L_2, L_3, L_4, and L_5; and six capacitors C, C_1, C_2, C_3, C_4, and C_O and seven diodes. Capacitors C_1, C_2, C_3, and C_4 perform characteristics to lift the capacitor voltage V_C by four times that of source voltage V_I. L_3, L_4, and L_5 perform the function as ladder joints to link the four capacitors C_1, C_2, C_3, and C_4 and lift the output capacitor voltage V_C up. Currents $i_{C1}(t)$, $i_{C2}(t)$, $i_{C3}(t)$, and $i_{C4}(t)$ are exponential

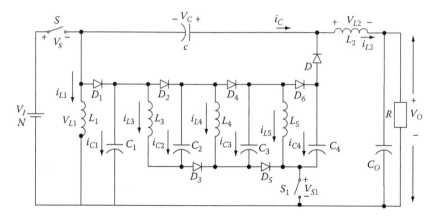

FIGURE 9.16 Quadruple-lift circuit.

functions. They have large values at the moment power is on, but they are small because $v_{C1} = v_{C2} = v_{C3} = v_{C4} = V_I$ in steady state.

The output voltage and current are

$$V_O = \frac{4}{1-k} V_I \tag{9.98}$$

and

$$I_O = \frac{1-k}{4} I_I \tag{9.99}$$

The voltage transfer gain in continuous mode is

$$M_Q = \frac{V_O}{V_I} = \frac{4}{1-k} \tag{9.100}$$

The curve of M_Q versus k is shown in Figure 9.17.

Other average voltages are $V_C = V_O$; $V_{C1} = V_{C2} = V_{C3} = V_{C4} = V_I$.

Other average currents are as follows: $I_{L2} = I_O$; $I_{L1} = (k/1-k)I_O$

$$I_{L3} = I_{L4} = L_{L5} = I_{L1} + I_{L2} = \frac{1}{1-k} I_O$$

Inductor current variations are as follows:

$$\xi_1 = \frac{1-k}{2M_Q} \frac{R}{fL_1} \quad \xi_2 = \frac{k}{2M_Q} \frac{R}{fL_2} \quad \zeta = \frac{k(1-k)R}{2M_Q fL} = \frac{k}{M_Q{}^2} \frac{2R}{fL}$$

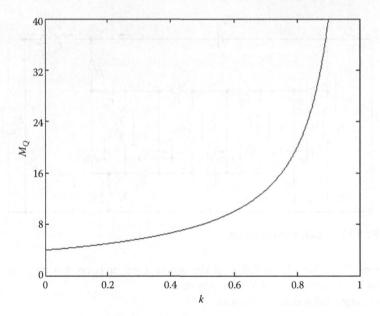

FIGURE 9.17   Voltage transfer gain M_Q versus k.

$$\chi_1 = \frac{k}{M_Q{}^2}\frac{2R}{fL_3} \qquad \chi_2 = \frac{k}{M_Q{}^2}\frac{2R}{fL_4} \qquad \chi_3 = \frac{k}{M_Q{}^2}\frac{2R}{fL_5}$$

Capacitor voltage variations are as follows:

$$\rho = \frac{k}{2fCR} \quad \sigma_1 = \frac{M_Q}{2fC_1R} \quad \sigma_2 = \frac{M_Q}{2fC_2R} \quad \sigma_3 = \frac{M_Q}{2fC_3R} \quad \sigma_4 = \frac{M_Q}{2fC_4R}$$

The variation ratio of output voltage V_C is

$$\varepsilon = \frac{k}{16M_Q}\frac{1}{f^2C_OL_2} \tag{9.101}$$

The output voltage ripple is very small.

The boundary between continuous and discontinuous modes is

$$M_Q \leq \sqrt{k}\sqrt{\frac{2R}{fL}} = \sqrt{2kz_N} \tag{9.102}$$

This boundary curve is shown in Figure 9.18. It can be seen that this boundary curve has a minimum value of M_Q that is equal to 6.0, corresponding to $k = 1/3$.

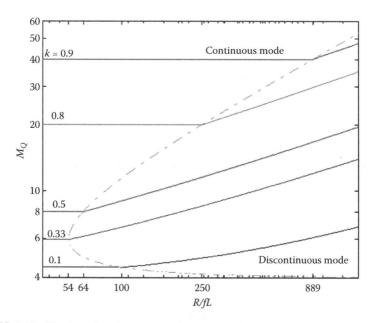

FIGURE 9.18 The boundary between continuous and discontinuous modes and the output voltage versus the normalized load $z_N = R/fL$.

In the discontinuous mode the current i_D exists during the period between kT and $t_1 = [k + (1-k)m_Q]T$, where m_Q is the filling efficiency, that is,

$$m_Q = \frac{1}{\zeta} = \frac{M_Q^2}{k(2R/fL)} \tag{9.103}$$

The current i_D becomes zero at $t = t_1 = kT + (1-k)m_QT$; therefore, $0 < m_Q < 1$. For the current i_{L1}, we have

$$kTV_I = (1-k)m_QT(V_C - 4V_I - V_{L3-off} - V_{L4-off} - V_{L5-off})$$

or

$$V_C = \left[4 + \frac{3k}{1-k} + \frac{k}{(1-k)m_Q} \right]$$

$$V_I = \left[4 + \frac{3k}{1-k} + k^2(1-k)\frac{R}{8fL} \right] V_I \quad \text{with} \quad \sqrt{k}\sqrt{\frac{2R}{fL}} \geq \frac{4}{1-k}$$

and for current i_{L2}, we have

$$kT(V_I + V_C - V_O) = (1-k)m_QT(V_O - 2V_I - V_{L3-off} - V_{L4-off} - V_{L5-off})$$

Therefore, output voltage in discontinuous mode is

$$V_O = \left[4 + \frac{3k}{1-k} + \frac{k}{(1-k)m_Q} \right]$$

$$V_I = \left[4 + \frac{3k}{1-k} + k^2(1-k)\frac{R}{8fL} \right] V_I \quad \text{with} \quad \sqrt{k}\sqrt{\frac{2R}{fL}} \geq \frac{4}{1-k} \qquad (9.104)$$

that is, the output voltage linearly increases when load resistance R increases, as shown in Figure 9.18.

9.3.4 SUMMARY

From the analysis and calculation in the previous sections, the common formulas for all circuits can be obtained as follows:

$$M = \frac{V_O}{V_I} = \frac{I_I}{I_O} \quad L = \frac{L_1 L_2}{L_1 + L_2} \quad z_N = \frac{R}{fL} \quad R = \frac{V_O}{I_O}$$

Inductor current variations are as follows:

$$\xi_1 = \frac{1-k}{2M}\frac{R}{fL_1} \quad \xi_2 = \frac{k}{2M}\frac{R}{fL_2} \quad \chi_i = \frac{k}{M^2}\frac{n}{2}\frac{R}{fL_{i+2}}$$

where
   i is the component number ($i = 1, 2, 3, \ldots, n-1$)
   n is the stage number

Capacitor voltage variations are as follows:

$$\rho = \frac{k}{2fCR} \quad \varepsilon = \frac{k}{16M}\frac{1}{f^2 C_O L_2} \quad \sigma_i = \frac{M}{2fC_i R} \quad (i = 1,2,3,4,\ldots,n)$$

In order to write common formulas for the boundaries between continuous and discontinuous modes and output voltage for all circuits, the circuits can be numbered. The definition is that subscript $n=0$ means the elementary circuit, subscript 1 means the self-lift circuit, subscript 2 means the re-lift circuit, subscript 3 means the triple-lift circuit, subscript 4 means the quadruple-lift circuit, and so on. The voltage transfer gain is

$$M_n = \frac{n + kh(n)}{1-k} \quad n = 0,1,2,3,4,\ldots \qquad (9.105)$$

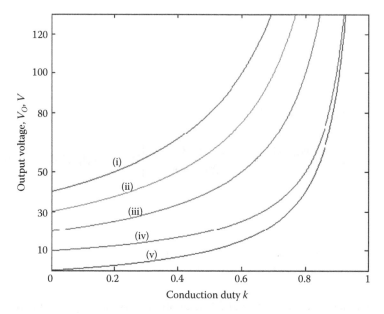

FIGURE 9.19 Output voltages of all P/O Luo converters ($V_I = 10\,\mathrm{V}$).

Assuming that $f = 50\,\mathrm{kHz}$, $L_1 = L_2 = 1\,\mathrm{mH}$, $L_2 = L_3 = L_4 = L_5 = 0.5\,\mathrm{mH}$, $C = C_1 = C_2 = C_3 = C_4 = C_O = 20\,\mu\mathrm{F}$, and the source voltage $V_I = 10\,\mathrm{V}$, the value of the output voltage V_O with various conduction duty k in continuous mode is shown in Figure 9.19. The free-wheeling diode current i_D's variation is

$$\zeta_n = \frac{k^{[1+h(n)]}}{M_n^2} \frac{n+h(n)}{2} z_N \qquad (9.106)$$

The boundaries are determined by the condition

$$\zeta_n \geq 1$$

or

$$\frac{k^{[1+h(n)]}}{M_n^2} \frac{n+h(n)}{2} z_N \geq 1 \quad n = 0,1,2,3,4,\dots \qquad (9.107)$$

Therefore, the boundaries between continuous and discontinuous modes for all circuits are as follows:

$$M_n = k^{\frac{1+h(n)}{2}} \sqrt{\frac{n+h(n)}{2} z_N} \quad n = 0,1,2,3,4,\dots \qquad (9.108)$$

The filling efficiency is

$$m_n = \frac{1}{\zeta_n} = \frac{M_n^{\,2}}{k^{[1+h(n)]}} \frac{2}{n+h(n)} \frac{1}{z_N} \tag{9.109}$$

The output voltage in the DCM for all circuits is

$$V_{O-n} = \left[n + \frac{n+h(n)-1}{1-k} + k^{[2-h(n)]} \frac{1-k}{2[n+h(n)]} z_N \right] V_I \quad n = 0,1,2,3,4,\ldots \tag{9.110}$$

where

$$h(n) = \begin{cases} 0 & if \quad n \geq 1 \\ 1 & if \quad n = 0 \end{cases} \quad \text{is the } Hong\ Function \tag{9.111}$$

The boundaries between CCM and DCM of all circuits are shown in Figure 9.20. The curves of all M versus z_N state that the continuous mode area increases from M_E via M_S, M_R, M_T to M_Q. The boundary of elementary circuit is the monorising curve, but other curves are not monorising. The minimum values of the boundaries of other circuits M_S, M_R, M_T, and M_Q are at $k = 1/3$.

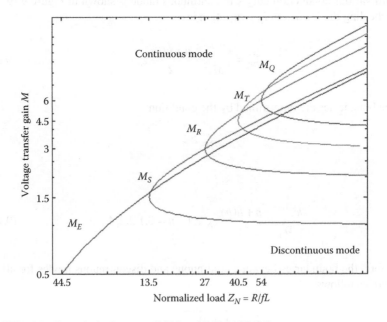

FIGURE 9.20   Boundaries between CCM and DCM of P/O Luo converters.

9.4 N/O LUO CONVERTERS

N/O Luo converters perform the voltage conversion from positive to negative voltages using VL technique. They work in the second quadrant with large voltage amplification. Five circuits have been introduced in the literature [12,13]:

1. Elementary circuit
2. Self-lift circuit
3. Re-lift circuit
4. Triple-lift circuit
5. Quadruple-lift circuit

The elementary circuit was discussed in Section 5.5.2, and the self-lift circuit was discussed in Section 9.2.4. Therefore, further circuits are discussed in this section.

9.4.1 RE-LIFT CIRCUIT

Figure 9.21 shows the N/O re-lift circuit, which is derived from the self-lift circuit. It consists of one static switch S, three inductors L, L_1, and L_O, four capacitors C, C_1, C_2, and C_O, and diodes. It can be seen that there is one capacitor C_2, one inductor L_1, and two diodes D_2, D_{11} added into the re-lift circuit. Circuit C_1-D_1-D_{11}-L_1-C_2-D_2 is the lift circuit. Capacitors C_1 and C_2 perform characteristics to lift the capacitor voltage V_C twice that of the source voltage $2V_I$. Inductor L_1 performs the function as a ladder joint to link the two capacitors C_1 and C_2 and lift the capacitor voltage V_C up. Currents $i_{C1}(t)$ and $i_{C2}(t)$ are exponential functions $\delta_1(t)$ and $\delta_2(t)$. They have large values at the moment power is on, but they are small because $v_{C1} = v_{C2} \cong V_I$ in steady state.

When switch S is on, the source current $i_I = i_L + i_{C1} + i_{C2}$. Inductor L absorbs energy from the source, and current i_L linearly increases with slope V_I/L. In the mean time, the diodes D_1, D_2 are conducted so that capacitors C_1 and C_2 are charged by the current i_{C1} and i_{C2}. Inductor L_O keeps the output current I_O continuous and transfers

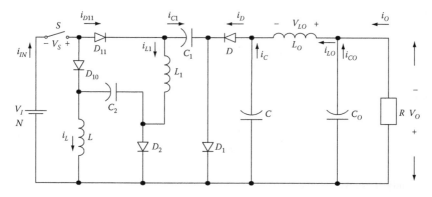

FIGURE 9.21 N/O re-lift circuit.

energy from capacitors C to the load R, that is, $i_{C\text{-}on} = i_{LO}$. When switch S is off, the source current is $i_I = 0$. Current i_L flows through the free-wheeling diode D, capacitors C_1 and C_2, and inductor L_1 to charge capacitor C, and enhances current i_{LO}. Inductor L transfers its stored energy to capacitor C and load R via inductor L_O, that is, $i_L = i_{C1\text{-}off} = i_{C2\text{-}off} = i_{L1\text{-}off} = i_{C\text{-}off} + i_{LO}$. Thus, currents i_L decreases.

The output current $I_O = I_{LO}$ because the capacitor C_O does not consume any energy in the steady state. The average output current is

$$I_O = I_{LO} = I_{C\text{-}on} \tag{9.112}$$

The charge of the capacitor C increases during the switch-off period:

$$Q_+ = (1-k)TI_{C\text{-}off}$$

And it decreases during the switch-on period: $Q_- = kTI_{C\text{-}on}$

In a whole repeating period T, $Q_+ = Q_-$. Thus,

$$I_{C\text{-}off} = \frac{k}{1-k}I_{C\text{-}on} = \frac{k}{1-k}I_O$$

Therefore, the inductor current I_L is

$$I_L = I_{C\text{-}off} + I_O = \frac{I_O}{1-k} \tag{9.113}$$

We know that

$$I_{C1\text{-}off} = I_{C2\text{-}off} = I_{L1} = I_L = \frac{1}{1-k}I_O \tag{9.114}$$

and

$$I_{C1\text{-}on} = \frac{1-k}{k}I_{C1\text{-}off} = \frac{1}{k}I_O \tag{9.115}$$

and

$$I_{C2\text{-}on} = \frac{1-k}{k}I_{C2\text{-}off} = \frac{1}{k}I_O \tag{9.116}$$

In steady state, we can use

$$V_{C1} = V_{C2} = V_I$$

and

$$V_{L1\text{-}on} = V_I \quad V_{L1\text{-}off} = \frac{k}{1-k}V_I$$

Investigate current i_L; it increases during switch-on with slope V_I/L and decreases during switch-off with slope $-(V_O-V_{C1}-V_{C2}-V_{L1\text{-}off})/L=-[V_O-2V_I-kV_I/(1-k)]/L$.
Therefore,

$$kTV_I = (1-k)T\left(V_O - 2V_I - \frac{k}{1-k}V_I\right)$$

or

$$V_O = \frac{2}{1-k}V_I \tag{9.117}$$

and

$$I_O = \frac{1-k}{2}I_I \tag{9.118}$$

The voltage transfer gain in continuous mode is

$$M_R = \frac{V_O}{V_I} = \frac{I_I}{I_O} = \frac{2}{1-k} \tag{9.119}$$

The curve of M_R versus k is shown in Figure 9.11. Circuit $(C\text{-}L_O\text{-}C_O)$ is a "Π" type low-pass filter.
Therefore,

$$V_C = V_O = \frac{2}{1-k}V_I \tag{9.120}$$

Current i_L increases and is supplied by V_I during switch-on. Thus, its peak-to-peak variation is

$$\Delta i_L = \frac{kTV_I}{L}$$

The variation ratio of the current i_L is

$$\zeta = \frac{\Delta i_L/2}{I_L} = \frac{k(1-k)V_IT}{2LI_O} = \frac{k(1-k)R}{2M_RfL} = \frac{k}{M_R^2}\frac{R}{fL} \tag{9.121}$$

The peak-to-peak variation of current i_{L1} is

$$\Delta i_{L1} = \frac{k}{L_1} TV_I$$

The variation ratio of current i_{L1} is

$$\chi_1 = \frac{\Delta i_{L1}/2}{I_{L1}} = \frac{kTV_I}{2L_1 I_O}(1-k) = \frac{k(1-k)}{2M_R}\frac{R}{fL_1} \qquad (9.122)$$

The peak-to-peak variation of voltage v_C is

$$\Delta v_C = \frac{Q-}{C} = \frac{k}{C} TI_O$$

The variation ratio of voltage v_C is

$$\rho = \frac{\Delta v_C/2}{V_C} = \frac{kI_O T}{2CV_O} = \frac{k}{2}\frac{1}{fCR} \qquad (9.123)$$

The peak-to-peak variation of voltage v_{C1} is

$$\Delta v_{C1} = \frac{kT}{C_1} I_{C1\text{-}on} = \frac{1}{fC} I_O$$

The variation ratio of voltage v_{C1} is

$$\sigma_1 = \frac{\Delta v_{C1}/2}{V_{C1}} = \frac{I_O}{2fC_1 V_I} = \frac{M_R}{2}\frac{1}{fC_1 R} \qquad (9.124)$$

Considering the same operation, variation ratio of voltage v_{C2} is

$$\sigma_2 = \frac{\Delta v_{C2}/2}{V_{C2}} = \frac{I_O}{2fC_2 V_I} = \frac{M_R}{2}\frac{1}{fC_2 R} \qquad (9.125)$$

Since

$$\Delta i_{LO} = \frac{1}{2}\frac{T}{2}\frac{k}{2CL_O} TI_O = \frac{k}{8f^2 CL_O} I_O$$

the variation ratio of current i_{LO} is

$$\xi = \frac{\Delta i_{LO}/2}{I_{LO}} = \frac{k}{16} \frac{1}{f^2 C L_O} \tag{9.126}$$

Since

$$\Delta v_{CO} = \frac{B}{C_O} = \frac{1}{2} \frac{T}{2} \frac{k}{16 f^2 C C_O L_O} I_O = \frac{k}{64 f^3 C C_O L_O} I_O$$

the variation ratio of current v_{CO} is

$$\varepsilon = \frac{\Delta v_{CO}/2}{V_{CO}} = \frac{k}{128 f^3 C C_O L_O} \frac{I_O}{V_O} = \frac{k}{128} \frac{1}{f^3 C C_O L_O R} \tag{9.127}$$

**Example 9.3 An N/O re-lift Luo converter has the following
components: $V_I = 20\,V$, $L = L_1 = L_O = 1\,mH$, and all capacitance are
$20\,\mu F$, $R = 160\,\Omega$, $f = 50\,kHz$, and $k = 0.5$. Calculate the output voltage,
and the variation ratios are ξ, ζ, χ_1, ρ, σ_1, σ_2, and ε in steady state.**

Solution: From (9.127) we obtain the output voltage as

$$V_O = \frac{2}{1-k} V_I = \frac{2}{1-0.5} 20 = 80\,V$$

The variation ratios $\xi = 6.25 \times 10^{-4}$, $\zeta = 0.04$, $\chi_1 = 0.1$, $\rho = 0.0016$, $\sigma_1 = 0.04$, $\sigma_2 = 0.04$, and $\varepsilon = 7.8 \times 10^{-5}$. Therefore, the variations are small.

In DCM, the diode current i_D becomes zero during the switch-off period before the next switch-on period. The condition for DCM is $\zeta \geq 1$, that is,

$$\frac{k}{M_R^2} \frac{R}{fL} \geq 1$$

or

$$M_R \leq \sqrt{k} \sqrt{\frac{R}{fL}} = \sqrt{k} \sqrt{z_N} \tag{9.128}$$

The graph of the boundary curve versus the normalized load $z_N = R/fL$ is shown in Figure 9.12. It can be seen that the boundary curve has a minimum value of 3.0 at $k = 1/3$.

In this case, the current i_D exists in the period between kT and $t_1 = [k + (1-k)m_R]T$, where m_R is the *filling efficiency* and is defined as

$$m_R = \frac{1}{\zeta} = \frac{M_R^2}{k(R/fL)} \qquad (9.129)$$

Therefore, $0 < m_R < 1$. Because inductor current $i_{L1} = 0$ at $t = t_1$,

$$V_{L1\text{-off}} = \frac{k}{(1-k)m_R} V_I$$

Since the current i_D becomes zero at $t = t_1 = [k + (1-k)m_R]T$, for the current i_L, $kTV_I = (1-k)m_R T(V_C - 2V_I - V_{L1\text{-off}})$ or

$$V_C = \left[2 + \frac{2k}{(1-k)m_R}\right]V_I = \left[2 + k^2(1-k)\frac{R}{2fL}\right]V_I \quad \text{with} \quad \sqrt{k}\sqrt{\frac{R}{fL}} \geq \frac{2}{1-k}$$

and for the current i_{LO} $kT(V_I + V_C - V_O) = (1-k)m_R T(V_O - 2V_I - V_{L1}\text{-off})$
Therefore, output voltage in discontinuous mode is

$$V_O = \left[2 + \frac{2k}{(1-k)m_R}\right]V_I = \left[2 + k^2(1-k)\frac{R}{2fL}\right]V_I \quad \text{with} \quad \sqrt{k}\sqrt{\frac{R}{fL}} \geq \frac{2}{1-k} \qquad (9.130)$$

that is, the output voltage linearly increases when load resistance R increases. Larger load resistance R may cause higher output voltage in discontinuous mode.

9.4.2 N/O TRIPLE-LIFT CIRCUIT

N/O triple-lift circuit is shown in Figure 9.22. It consists of one static switch S; four inductors L, L_1, L_2, and L_O; and five capacitors C, C_1, C_2, C_3, and C_O; and diodes. Circuit C_1-D_1-L_1-C_2-D_2-D_{11}-L_2-C_3-D_3-D_{12} is the lift circuit. Capacitors C_1, C_2, and C_3 perform characteristics to lift the capacitor voltage V_C by 3 times that of source voltage V_I. L_1 and L_2 perform the function as ladder joints to link the three capacitors C_1, C_2, and C_3 and lift the capacitor voltage V_C up. Current $i_{C1}(t)$, $i_{C2}(t)$, and $i_{C3}(t)$ are exponential functions. They have large values at the moment power is on, but they are small because $v_{C1} = v_{C2} = v_{C3} \cong V_I$ in steady state.

The output voltage and current are as follows:

$$V_O = \frac{3}{1-k} V_I \qquad (9.131)$$

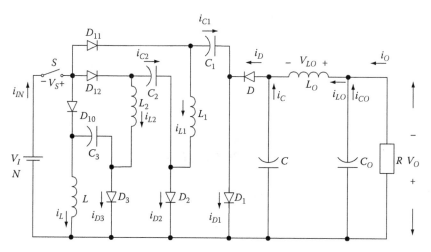

FIGURE 9.22 N/O triple-lift circuit.

and

$$I_O = \frac{1-k}{3} I_I \tag{9.132}$$

The voltage transfer gain in continuous mode is

$$M_T = \frac{V_O}{V_I} = \frac{3}{1-k} \tag{9.133}$$

The curve of M_T versus k is shown in Figure 9.14.
Other average voltages are as follows:

$$V_C = V_O; V_{C1} = V_{C2} = V_{C3} = V_I$$

Other average currents are as follows:

$$I_{LO} = I_O; \quad I_L = I_{L1} = I_{L2} = \frac{1}{1-k} I_O$$

Current variation ratios are as follows:

$$\zeta = \frac{k}{M_T^2} \frac{3R}{2fL} \quad \xi = \frac{k}{16} \frac{1}{f^2 C L_O} \quad \chi_1 = \frac{k(1-k)}{2M_T} \frac{R}{fL_1} \quad \chi_2 = \frac{k(1-k)}{2M_T} \frac{R}{fL_2}$$

Voltage variation ratios are as follows:

$$\rho = \frac{k}{2}\frac{1}{fCR} \qquad \sigma_1 = \frac{M_T}{2}\frac{1}{fC_1R} \qquad \sigma_2 = \frac{M_T}{2}\frac{1}{fC_2R} \qquad \sigma_3 = \frac{M_T}{2}\frac{1}{fC_3R}$$

The variation ratio of output voltage V_C is

$$\varepsilon = \frac{k}{128}\frac{1}{f^3CC_OL_OR} \tag{9.134}$$

The boundary between continuous and discontinuous modes is

$$M_T \le \sqrt{k}\sqrt{\frac{3R}{2fL}} = \sqrt{\frac{3kz_N}{2}} \tag{9.135}$$

It can be seen that the boundary curve has a minimum value of M_T that is equal to 4.5, corresponding to $k = 1/3$. The boundary curve versus the normalized load $z_N = R/fL$ is shown in Figure 9.15.

In discontinuous mode, the current i_D exists in the period between kT and $t_1 = [k + (1-k)m_T]T$, where m_T is the filling efficiency, that is,

$$m_T = \frac{1}{\zeta} = \frac{M_T^2}{k(3R/2fL)} \tag{9.136}$$

The inductor current $i_{L1} = i_{L2} = 0$ at $t = t_1$; therefore, $0 < m_T < 1$.

$$V_{L1\text{-}off} = V_{L2\text{-}off} = \frac{k}{(1-k)m_T}V_I$$

Since the current i_D becomes zero at $t = t_1 = [k + (1-k)m_T]T$, for the current i_L we get

$$kTV_I = (1-k)m_TT(V_C - 3V_I - V_{L1\text{-}off} - V_{L2\text{-}off})$$

or

$$V_C = \left[3 + \frac{3k}{(1-k)m_T}\right]V_I = \left[3 + k^2(1-k)\frac{R}{2fL}\right]V_I \quad \text{with} \quad \sqrt{k}\sqrt{\frac{3R}{2fL}} \ge \frac{3}{1-k}$$

and for the current i_{LO} we get

$$kT(V_I + V_C - V_O) = (1-k)m_TT(V_O - 2V_I - V_{L1\text{-}off} - V_{L2\text{-}off})$$

Therefore, output voltage in discontinuous mode is

$$V_O = \left[3 + \frac{3k}{(1-k)m_T} \right] V_I = \left[3 + k^2(1-k)\frac{R}{2fL} \right] V_I \quad \text{with} \quad \sqrt{k}\sqrt{\frac{3R}{2fL}} \geq \frac{3}{1-k} \quad (9.137)$$

that is, the output voltage linearly increases when load resistance R increases. We can see that the output voltage increases when load resistance R increases.

9.4.3 N/O Quadruple-Lift Circuit

N/O quadruple-lift circuit is shown in Figure 9.23. It consists of one static switch S; five inductors L, L_1, L_2, L_3, and L_O; and six capacitors C, C_1, C_2, C_3, C_4, and C_O. Capacitors C_1, C_2, C_3, and C_4 perform characteristics to lift the capacitor voltage V_C by four times that of source voltage V_I. L_1, L_2, and L_3 perform the function as ladder joints to link the four capacitors C_1, C_2, C_3, and C_4 and lift the output capacitor voltage V_C up. Currents $i_{C1}(t)$, $i_{C2}(t)$, $i_{C3}(t)$, and $i_{C4}(t)$ are exponential functions. They have large values at the moment power is on, but they are small because $v_{C1} = v_{C2} = v_{C3} = v_{C4} \cong V_I$ in steady state.

The output voltage and current are as follows:

$$V_O = \frac{4}{1-k}V_I \quad (9.138)$$

and

$$I_O = \frac{1-k}{4}I_I \quad (9.139)$$

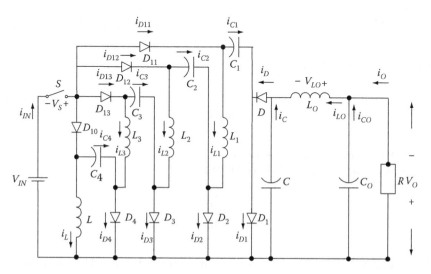

FIGURE 9.23 N/O quadruple-lift circuit.

The voltage transfer gain in continuous mode is

$$M_Q = \frac{V_O}{V_I} = \frac{4}{1-k} \qquad (9.140)$$

The curve of M_Q versus k is shown in Figure 9.44.
Other average voltages are as follows:

$$V_C = V_O; V_{C1} = V_{C2} = V_{C3} = V_{C4} = V_I.$$

Other average currents are as follows:

$$I_{LO} = I_O; \quad I_L = I_{L1} = I_{L2} = I_{L3} = \frac{1}{1-k} I_O$$

Current variation ratios are as follows:

$$\zeta = \frac{k}{M_Q^2} \frac{2R}{fL} \quad \xi = \frac{k}{16} \frac{1}{f^2 C L_O}$$

$$\chi_1 = \frac{k(1-k)}{2M_Q} \frac{R}{fL_1} \quad \chi_2 = \frac{k(1-k)}{2M_Q} \frac{R}{fL_2} \quad \chi_3 = \frac{k(1-k)}{2M_Q} \frac{R}{fL_3}$$

Voltage variation ratios are as follows:

$$\rho = \frac{k}{2} \frac{1}{fCR} \quad \sigma_1 = \frac{M_Q}{2} \frac{1}{fC_1 R} \quad \sigma_2 = \frac{M_Q}{2} \frac{1}{fC_2 R} \quad \sigma_3 = \frac{M_Q}{2} \frac{1}{fC_3 R} \quad \sigma_4 = \frac{M_Q}{2} \frac{1}{fC_4 R}$$

The variation ratio of output voltage V_C is

$$\varepsilon = \frac{k}{128} \frac{1}{f^3 C C_O L_O R} \qquad (9.141)$$

The output voltage ripple is very small.
    The boundary between CCM and DCM is

$$M_Q \leq \sqrt{k} \sqrt{\frac{2R}{fL}} = \sqrt{2k z_N} \qquad (9.142)$$

It can be seen that the boundary curve has a minimum value of M_Q that is equal to
6.0, corresponding to $k = 1/3$. The boundary curve is shown in Figure 6.18.

In the discontinuous mode, the current i_D exists during the period between kT and $t_1 = [k + (1-k)m_Q]T$, where m_Q is the filling efficiency, that is,

$$m_Q = \frac{1}{\zeta} = \frac{M_Q^2}{k(2R/fL)} \qquad (9.143)$$

The inductor current $i_{L1} = i_{L2} = i_{L3} = 0$ at $t = t_1$; therefore, $0 < m_Q < 1$.

$$V_{L1\text{-off}} = V_{L2\text{-off}} = V_{L3\text{-off}} = \frac{k}{(1-k)m_Q} V_I$$

Since the current i_D becomes zero at $t = t_1 = kT + (1-k)m_Q T$, for the current i_L we have

$$kTV_I = (1-k)m_Q T(V_C - 4V_I - V_{L1\text{-off}} - V_{L2\text{-off}} - V_{L3\text{-off}})$$

or

$$V_C = \left[4 + \frac{4k}{(1-k)m_Q}\right]V_I = \left[4 + k^2(1-k)\frac{R}{2fL}\right]V_I \quad \text{with} \quad \sqrt{k}\sqrt{\frac{2R}{fL}} \geq \frac{4}{1-k}$$

and for current i_{LO} we have

$$kT(V_I + V_C - V_O) = (1-k)m_Q T(V_O - 2V_I - V_{L1\text{-off}} - V_{L2\text{-off}} - V_{L3\text{-off}})$$

Therefore, output voltage in discontinuous mode is

$$V_O = \left[4 + \frac{4k}{(1-k)m_Q}\right]V_I = \left[4 + k^2(1-k)\frac{R}{2fL}\right]V_I \quad \text{with} \quad \sqrt{k}\sqrt{\frac{2R}{fL}} \geq \frac{4}{1-k} \qquad (9.144)$$

that is, the output voltage linearly increases when load resistance R increases. We can see that the output voltage increases when load resistance R increases.

9.4.4 SUMMARY

From the analysis and calculation in previous sections, the common formulae can be obtained for all circuits:

$$M = \frac{V_O}{V_I} = \frac{I_I}{I_O} \qquad z_N = \frac{R}{fL} \qquad R = \frac{V_O}{I_O}$$

Inductor current variation ratios are as follows:

$$\zeta = \frac{k(1-k)R}{2MfL} \quad \xi = \frac{k}{16f^2CL_O} \quad \chi_i = \frac{k(1-k)R}{2MfL_i} \quad (i=1,2,3,\ldots,n-1) \quad \text{with } n \ge 2$$

Capacitor voltage variation ratios are as follows:

$$\rho = \frac{k}{2fCR} \quad \varepsilon = \frac{k}{128f^3CC_OL_OR} \quad \sigma_i = \frac{M}{2fC_iR} \quad (i=1,2,3,4,\ldots,n) \quad \text{with } n \ge 1$$

where
   i is the component number
   n is the stage number

In order to write common formulae for the boundaries between continuous and discontinuous modes and output voltage for all circuits, the circuits can be numbered. The definition is that subscript $n=0$ means the elementary circuit, subscript 1 means the self-lift circuit, subscript 2 means the self-lift circuit, subscript 3 means the triple-lift circuit, subscript 4 means the quadruple-lift circuit, and so on. Therefore, the voltage transfer gain in continuous mode for all circuits, as shown in Figure 9.24, is

$$M_n = \frac{n+kh(n)}{1-k} \quad n=0,1,2,3,4,\ldots \tag{9.145}$$

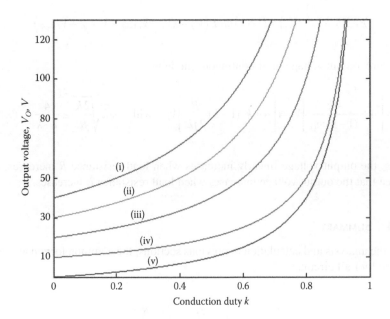

FIGURE 9.24   Output voltages of N/O Luo converters ($V_I = 10\,\text{V}$).

The variation of the free-wheeling diode current i_D is

$$\zeta_n = \frac{k^{[1+h(n)]}}{M_n^{\,2}} \frac{n+h(n)}{2} z_N \tag{9.146}$$

The boundaries are determined by the condition $\zeta_n \geq 1$ or

$$\frac{k^{[1+h(n)]}}{M_n^{\,2}} \frac{n+h(n)}{2} z_N \geq 1 \quad n = 0,1,2,3,4,\ldots \tag{9.147}$$

Therefore, the boundaries between continuous and discontinuous modes for all circuits are as follows:

$$M_n = k^{\frac{1+h(n)}{2}} \sqrt{\frac{n+h(n)}{2} z_N} \quad n = 0,1,2,3,4,\ldots \tag{9.148}$$

For DCM, the filling efficiency is

$$m_n = \frac{1}{\zeta_n} = \frac{M_n^{\,2}}{k^{[1+h(n)]}} \frac{2}{n+h(n)} \frac{1}{z_N} \tag{9.149}$$

The voltage across the capacitor C in discontinuous mode for all circuits is as follows:

$$V_{C-n} = \left[n + k^{[2-h(n)]} \frac{1-k}{2} z_N \right] V_I \quad n = 0,1,2,3,4,\ldots \tag{9.150}$$

The output voltage in the discontinuous mode for all circuits is

$$V_{O-n} = \left[n + k^{[2-h(n)]} \frac{1-k}{2} z_N \right] V_I \quad j = 0,1,2,3,4,\ldots \tag{9.151}$$

where $h(n) = \begin{cases} 0 & if & n \geq 1 \\ 1 & if & n = 0 \end{cases}$ is the Hong Function.

The voltage transfer gains in CCM for all circuits are shown in Figure 9.24. The boundaries between continuous and discontinuous modes of all circuits are shown in Figure 9.25. The curves of all M versus z_N state that the continuous mode area increases from M_E via M_S, M_R, M_T to M_Q. The boundary of elementary circuit is monorising curve, but other curves are not monorising. The minimum values of the boundaries of other circuits M_S, M_R, M_T, and M_Q are at $k = 1/3$.

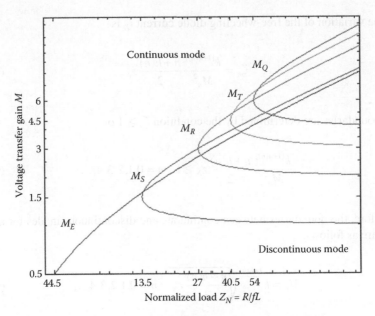

FIGURE 9.25 Boundaries between CCM and DCM of N/O Luo converters.

9.5   MODIFIED P/O LUO CONVERTERS

N/O Luo converters perform the voltage conversion from positive to negative voltages using VL technique with only one switch S. This section introduces the technique to modify P/O Luo converters that can employ only *one* switch for all circuits. Five circuits have been introduced in the literature [14]. They are as follows:

1. Elementary circuit
2. Self-lift circuit
3. Re-lift circuit
4. Triple-lift circuit
5. Quadruple-lift circuit

The elementary circuit is the original P/O Luo converter. We will introduce the self-lift circuit, re-lift circuit, and multiple-lift circuit in this section.

9.5.1   Self-Lift Circuit

Self-lift circuit is shown in Figure 9.26. It is derived from the elementary circuit of the P/O Luo converter. In steady state, the average inductor voltages in a period are zero. Thus,

$$V_{C1} = V_{CO} = V_O \qquad (9.152)$$

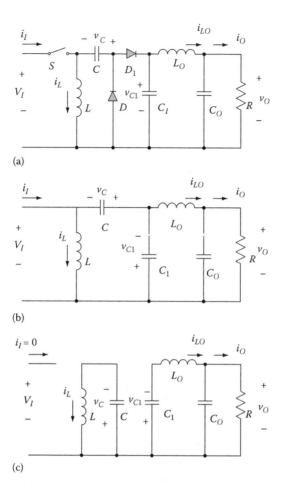

(a)

(b)

(c)

FIGURE 9.26 Self-lift circuit of modified P/O Luo converters: (a) self-lift circuit, (b) switch-on equivalent circuit, and (c) switch-off equivalent circuit.

The inductor current i_L increases in the switch-on period and decreases in the switch-off period. The corresponding voltages across L are V_I and $-V_C$.

Therefore, $kTV_I = (1-k)TV_C$.

Hence,

$$V_C = \frac{k}{1-k}V_I \tag{9.153}$$

During the switch-on period, the voltage across capacitor C_1 is equal to the source voltage plus the voltage across C. Since we assume that C and C_1 are sufficiently large,

$$V_{C1} = V_I + V_C$$

Therefore,

$$V_{C1} = V_I + \frac{k}{1-k}V_I = \frac{1}{1-k}V_I$$

$$V_O = V_{CO} = V_{C1} = \frac{1}{1-k}V_I$$

The voltage transfer gain of CCM is

$$M = \frac{V_O}{V_I} = \frac{1}{1-k}$$

The output voltage and current and the voltage transfer gain are as follows:

$$V_O = \frac{1}{1-k}V_I$$

$$I_O = (1-k)I_I$$

$$M_S = \frac{1}{1-k} \tag{9.154}$$

Average voltages are as follows:

$$V_C = kV_O$$

$$V_{C1} = V_O$$

Average currents are as follows:

$$I_{LO} = I_O$$

$$I_L = \frac{1}{1-k}I_O$$

We also implement the breadboard prototype of the proposed self-lift circuit. NMOS IRFP460 is used as the semiconductor switch. The diode is MR824. The other parameters are as follows:

$$V_I = 0 \sim 30\,\text{V},\ R = 30 \sim 340\,\Omega,\ k = 0.1 \sim 0.9,\ C = C_O = 100\,\mu\text{F, and } L = 470\,\mu\text{H}$$

9.5.2 Re-Lift Circuit

Re-lift circuit and its equivalent circuits are shown in Figure 9.27. It is derived from
the self-lift circuit. The function of capacitor C_2 is to lift the voltage v_C by source voltage V_I, the function of inductor L_1 likes a hinge of the foldable ladder (capacitor C_2)
to lift the voltage v_C in the switch-off condition.

In steady state, the average inductor voltages over a period are zero. Thus,

$$V_{C1} = V_{CO} = V_O$$

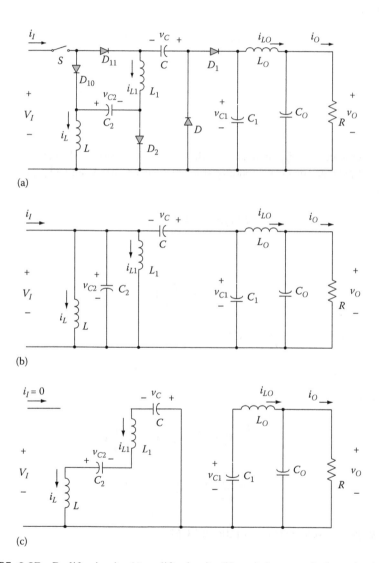

(a)

(b)

(c)

FIGURE 9.27 Re-lift circuit: (a) re-lift circuit, (b) switch-on equivalent circuit, and
(c) switch-off equivalent circuit.

We assume C_2 is large enough and C_2 is biased by the source voltage V_I during the switch-on period; thus, $V_{C2} = V_I$.

From the switch-on equivalent circuit, another capacitor voltage equation can also be derived since we assume all the capacitors to be large enough,

$$V_O = V_{C1} = V_C + V_I$$

The inductor current i_L increases in the switch-on period and decreases in the switch-off period. The corresponding voltages across L are V_I and $-V_{L\text{-}OFF}$.

Therefore,

$$kTV_I = (1-k)TV_{L\text{-}OFF}$$

Hence,

$$V_{L\text{-}OFF} = \frac{k}{1-k}V_I$$

The inductor current i_{L1} increases in the switch-on period and decreases in the switch-off period. The corresponding voltages across L_1 are V_I and $-V_{L1\text{-}OFF}$.

Therefore,

$$kTV_I = (1-k)TV_{L1\text{-}OFF}$$

Hence,

$$V_{L1\text{-}OFF} = \frac{k}{1-k}V_I$$

From the switch-off-period equivalent circuit,

$$V_C = V_{C\text{-}OFF} = V_{L\text{-}OFF} + V_{L1\text{-}OFF} + V_{C2}$$

Therefore,

$$V_C = \frac{k}{1-k}V_I + \frac{k}{1-k}V_I + V_I = \frac{1+k}{1-k}V_I \tag{9.155}$$

$$V_O = \frac{1+k}{1-k}V_I + V_I = \frac{2}{1-k}V_I$$

Then we get the voltage transfer ratio in CCM:

$$M = M_R = \frac{2}{1-k} \tag{9.156}$$

The following is a brief summary of the main equations for the re-lift circuit. The output voltage and current gain are as follows:

$$V_O = \frac{2}{1-k}V_I$$

$$I_O = \frac{1-k}{2}I_I$$

$$M_R = \frac{2}{1-k}$$

Average voltages are as follows:

$$V_C = \frac{1+k}{1-k}V_I$$

$$V_{C1} = V_{CO} = V_O$$

$$V_{C2} = V_I$$

Average currents are as follows:

$$I_{LO} = I_O$$

$$I_L = I_{L1} = \frac{1}{1-k}I_O$$

9.5.3 Multilift Circuit

Multilift circuits are derived from re-lift circuit by repeating the section of L_1-C_1-D_1 by multiple times. For example, triple-list circuit is shown in Figure 9.28. The function of capacitors C_2 and C_3 is to lift the voltage v_C across capacitor C twice that of source voltage $2V_I$, and the function of inductors L_1 and L_2 likes hinges of the foldable ladder (capacitors C_2 and C_3) to lift the voltage v_C during the switch-off period.

The output voltage and current and voltage transfer gain are as follows:

$$V_O = \frac{3}{1-k}V_I \quad \text{and} \quad I_O = \frac{1-k}{3}I_I$$

$$M_T = \frac{3}{1-k} \tag{9.157}$$

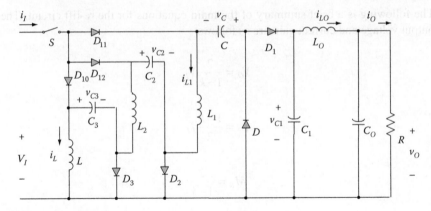

FIGURE 9.28   Triple-list circuit.

Other average voltages are as follows:

$$V_C = \frac{2+k}{1-k}V_I \quad \text{and} \quad V_{C1} = V_O$$

$$V_{C2} = V_{C3} = V_I$$

Other average currents are as follows:

$$I_{LO} = I_O$$

$$I_{L1} = I_{L2} = I_L = \frac{1}{1-k}I_O$$

The quadruple-lift circuit is shown in Figure 9.29. The function of capacitors C_2, C_3, and C_4 is to lift the voltage v_C across capacitor C by three times that of source voltage $3V_I$. The function of inductors L_1, L_2, and L_3 likes hinges of the foldable ladder (capacitors C_2, C_3, and C_4) to lift the voltage v_C during the switch-off period. The output voltage and current and voltage transfer gain are as follows:

$$V_O = \frac{4}{1-k}V_I \quad \text{and} \quad I_O = \frac{1-k}{4}I_I$$

$$M_Q = \frac{4}{1-k} \tag{9.158}$$

Average voltages are as follows:

$$V_C = \frac{3+k}{1-k}V_I \quad \text{and} \quad V_{C1} = V_O$$

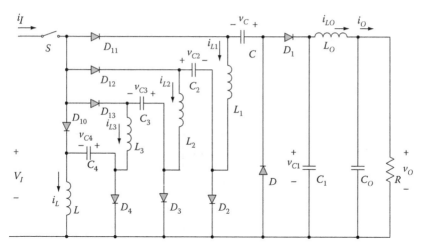

FIGURE 9.29   Quadruple-lift circuit.

$$V_{C2} = V_{C3} = V_{C4} = V_I$$

Average currents are as follows:

$$I_{LO} = I_O \quad \text{and} \quad I_L = \frac{k}{1-k} I_O$$

$$I_{L1} = I_{L2} = I_{L3} = IL + I_{LO} = \frac{1}{1-k} I_O$$

9.6   DOUBLE-OUTPUT LUO CONVERTERS

Mirror-symmetrical double-output voltages are specially required in industrial applications and computer periphery circuit. Double-output DC/DC Luo converters can convert the positive input source voltage to positive and N/O voltages. It consists of two conversion paths. Double-output Luo converters perform from positive to positive and negative DC/DC voltage-increasing conversion with high power density, high efficiency, and cheap topology with simple structure [15,16]. Like P/O and N/O Luo converters, there are five circuits in this series:

1. Elementary circuit
2. Self-lift circuit
3. Re-lift circuit
4. Triple-lift circuit
5. Quadruple-lift circuit

The elementary circuit is the original double-output Luo converter introduced in Section 5.53. We will introduce the self-lift circuit, re-lift circuit, triple-lift circuit, and quadruple-lift circuit in this section.

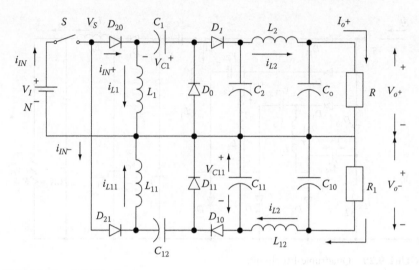

FIGURE 9.30   Self-lift circuit.

9.6.1   SELF-LIFT CIRCUIT

Self-lift circuit shown in Figure 9.30 is derived from the elementary circuit. The positive conversion path consists of a pump circuit S-L_1-D_0-C_1 and a filter (C_2)-L_2-C_O and a lift circuit D_1-C_2. The negative conversion path consists of a pump circuit S-L_{11}-D_{10}-(C_{11}) and a "Π"-type filter C_{11}-L_{12}-C_{10} and a lift circuit D_{11}-C_{12}.

9.6.1.1   Positive Conversion Path

The equivalent circuit during the switch-on period is shown in Figure 9.31a, and the equivalent circuit during the switch-off period is shown in Figure 9.31b. The voltage across inductor L_1 is equal to V_I during the switch-on period, and $-V_{C1}$ during the switch-off period. We have the following relation:

$$V_{C1} = \frac{k}{1-k} V_I$$

Hence,

$$V_O = V_{CO} = V_{C2} = V_I + V_{C1} = \frac{1}{1-k} V_I$$

and

$$V_{O+} = \frac{1}{1-k} V_I$$

The output current is $I_{O+} = (1-k)I_{I+}$.

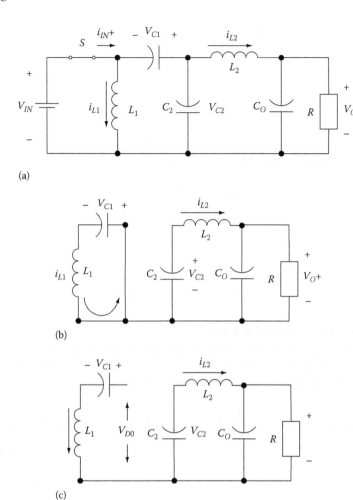

FIGURE 9.31 Equivalent circuits positive path of self-lift circuit: (a) switch-on, (b) switch-off, and (c) DCM.

Other relations are as follows:

$$I_{I+} = k i_{I+} i_{I+} = I_{L1} + i_{C1\text{-}on} \quad i_{C1\text{-}off} = \frac{k}{1-k} i_{C1\text{-}on}$$

and

$$I_{L1} = i_{C1\text{-}off} = k i_{I+} = I_{I+} \tag{9.159}$$

Therefore, the voltage transfer gain in continuous mode is

$$M_{S+} = \frac{V_{O+}}{V_I} = \frac{1}{1-k} \tag{9.160}$$

The variation ratios of the parameters are as follows:

$$\xi_{2+} = \frac{\Delta i_{L2}/2}{I_{L2}} = \frac{k}{16} \frac{1}{f^2 C_2 L_2} \quad \rho_+ = \frac{\Delta v_{C1}/2}{V_{C1}} = \frac{(1-k)I_{I+}}{2fC_1(k/1-k)V_I}$$

$$= \frac{1}{2kfC_1 R} \quad \sigma_{1+} = \frac{\Delta v_{C2}/2}{V_{C2}} = \frac{k}{2fC_2 R}$$

The variation ratio of the currents i_{D0} and i_{L1} is as follows:

$$\zeta_+ = \xi_{1+} = \frac{\Delta i_{L1}/2}{I_{L1}} = \frac{kV_I T}{2L_1 I_{I+}} = \frac{k}{M_S^2} \frac{R}{2fL_1} \tag{9.161}$$

The variation ratio of output voltage v_{O+} is as follows:

$$\varepsilon_+ = \frac{\Delta v_{O+}/2}{V_{O+}} = \frac{k}{128} \frac{1}{f^3 C_2 C_O L_2 R} \tag{9.162}$$

9.6.1.2 Negative Conversion Path

The equivalent circuit during the switch-on period is shown in Figure 9.32a, and the equivalent circuit during the switch-off period is shown in Figure 9.32b. The relations of the average currents and voltages are as follows:

$$I_{O-} = I_{L12} = I_{C11\text{-}on} \quad I_{C11\text{-}off} = \frac{k}{1-k}I_{C11\text{-}on} = \frac{k}{1-k}I_{O-} \quad I_{L11} = I_{C11\text{-}off} + I_{O-} = \frac{I_{O-}}{1-k}$$

$$\tag{9.163}$$

We know that

$$I_{C12\text{-}off} = I_{L11} = \frac{1}{1-k}I_{O-} \quad \text{and} \quad I_{C12\text{-}on} = \frac{1-k}{k}I_{C12\text{-}off} = \frac{1}{k}I_{O-}$$

so that

$$V_{O-} = \frac{1}{1-k}V_I \quad \text{and} \quad I_{O-} = (1-k)I_I$$

The voltage transfer gain in continuous mode is

$$M_{S-} = \frac{V_{O-}}{V_I} = \frac{1}{1-k} \tag{9.164}$$

Circuit $(C_{11}\text{-}L_{12}\text{-}C_{10})$ is a "Π" type low-pass filter. Therefore, $V_{C11} = V_{O-} = k/(1-k)V_I$.

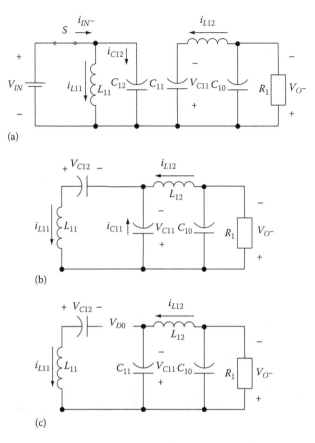

FIGURE 9.32 Equivalent circuits negative path of self-lift circuit: (a) switch-on, (b) switch-off, and (c) DCM.

From Equations 9.160 and 9.161, define $M_S = M_{S+} = M_{S-}$. The curve of M_S versus k is shown in Figure 9.33.

The variation ratios of the parameters are as follows:

$$\xi_- = \frac{\Delta i_{L12}/2}{I_{L12}} = \frac{k}{16} \frac{1}{f^2 C_{10} L_{12}} \qquad \rho_- = \frac{\Delta v_{C11}/2}{V_{C11}} = \frac{k I_{O-} T}{2 C_{11} V_{O-}} = \frac{k}{2} \frac{1}{f C_{11} R_1}$$

$$\sigma_{1-} = \frac{\Delta v_{C12}/2}{V_{C12}} = \frac{I_{O-}}{2 f C_{12} V_I} = \frac{M_S}{2} \frac{1}{f C_{12} R_1}$$

The variation ratio of currents i_{D10} and i_{L11} is

$$\zeta_- = \frac{\Delta i_{L11}/2}{I_{L11}} = \frac{k(1-k) V_I T}{2 L_{11} I_{O-}} = \frac{k(1-k) R_1}{2 M_S f L_{11}} = \frac{k}{M_S^2} \frac{R_1}{2 f L_{11}} \tag{9.165}$$

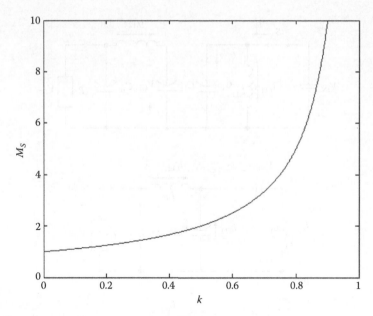

FIGURE 9.33 Voltage transfer gain M_S versus k.

The variation ratio of current v_{C10} is

$$\varepsilon_- = \frac{\Delta v_{C10}/2}{V_{C10}} = \frac{k}{128 f^3 C_{11} C_{10} L_{12}} \frac{I_{O-}}{V_{O-}} = \frac{k}{128} \frac{1}{f^3 C_{11} C_{10} L_{12} R_1} \qquad (9.166)$$

Example 9.4 A double-output self-lift Luo converter has the following components: $V_I = 20\,V$, all inductances are 1 mH, and all capacitance are 20 μF, $R = R_1 = 160\,\Omega$, $f = 50\,kHz$, and $k = 0.5$. Calculate the output voltage, the variation ratios, and ε in steady state.

Solution:

From (9.160) and (9.161), we obtain the output voltage as follows:

$$V_{O+} = V_{O-} = \frac{1}{1-k} V_I = \frac{1}{1-0.5} 20 = 40\,V$$

The variation ratios are as follows:

$$\xi_{2+} = 6.25 \times 10^{-4},\ \xi_{1+} = \zeta_{1+} = 0.2,\ \rho_+ = 0.05,\ \sigma_{1+} = 0.00625, \text{and } \varepsilon_+ = 2 \times 10^{-6}$$

$$\xi_- = 6.25 \times 10^{-4}, \zeta_- = 0.05,\ \rho_- = 0.00625,\ \sigma_{1-} = 0.025, \text{and } \varepsilon_+ = 2 \times 10^{-6}$$

Therefore, the variations are small.

9.6.1.3 Discontinuous Conduction Mode

The equivalent circuits of the DCM's operation are shown in Figures 9.31c and 9.32c. Thus, we select $z_N = z_{N+} = z_{N-}$, $M_S = M_{S+} = M_{S-}$, and $\zeta = \zeta_+ = \zeta_-$. The boundary between CCM and DCM is $\zeta \geq 1$

or

$$\frac{k}{M_S^2}\frac{z_N}{2} \geq 1$$

Hence,

$$M_S \leq \sqrt{k}\sqrt{\frac{z}{2}} = \sqrt{\frac{kz_N}{2}} \tag{9.167}$$

This boundary curve is shown in Figure 9.34. This curve has a minimum value of M_S that is equal to 1.5 at $k = 1/3$.

The filling efficiency is defined as

$$m_S = \frac{1}{\zeta} = \frac{2M_S^2}{kz_N} \tag{9.168}$$

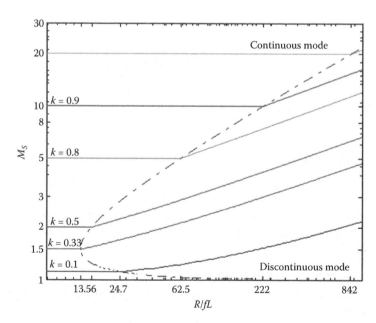

FIGURE 9.34 The boundary between continuous and discontinuous modes and the output voltage versus the normalized load $z_N = R/fL$ (self-lift circuit).

For the current i_{L1}, we have $kTV_I = (1-k)m_{S+}TV_{C1}$ or

$$V_{C1} = \frac{k}{(1-k)m_S}V_I = k^2(1-k)\frac{z_N}{2}V_I \quad \text{with} \quad \sqrt{\frac{kz_N}{2}} \geq \frac{1}{1-k} \qquad (9.169)$$

Therefore, the P/O voltage in DCM is

$$V_{O+} = V_{C1} + V_I = \left[1 + \frac{k}{(1-k)m_S}\right]V_I = \left[1 + k^2(1-k)\frac{z_N}{2}\right]V_I \quad \text{with} \quad \sqrt{\frac{kz_N}{2}} \geq \frac{1}{1-k}$$

$$(9.170)$$

For the current i_{L11}, we have $kTV_I = (1-k)m_S T(V_{C11} - V_I)$ or

$$V_{C11} = \left[1 + \frac{k}{(1-k)m_S}\right]V_I = \left[1 + k^2(1-k)\frac{z_N}{2}\right]V_I \quad \text{with} \quad \sqrt{\frac{kz_N}{2}} \geq \frac{1}{1-k} \qquad (9.171)$$

and for the current i_{L12} we have $kT(V_I + V_{C11} - V_{O-}) = (1-k)m_{S-}T(V_{O-} - V_I)$
   Therefore, the N/O voltage in the DCM is

$$V_{O-} = \left[1 + \frac{k}{(1-k)m_S}\right]V_I = \left[1 + k^2(1-k)\frac{z_N}{2}\right]V_I \quad \text{with} \quad \sqrt{\frac{kz_N}{2}} \geq \frac{1}{1-k} \qquad (9.172)$$

Then, we have

$$V_O = V_{O+} = V_{O-} = \left[1 + k^2(1-k)\frac{z_N}{2}\right]V_I$$

that is, the output voltage linearly increases when load resistance increases. Larger load resistance causes higher output voltage in the DCM as shown in Figure 9.34.

9.6.2   RE-LIFT CIRCUIT

Re-lift circuit shown in Figure 9.35 is derived from self-lift circuit. The positive conversion path consists of a pump circuit S-L_1-D_0-C_1 and a filter (C_2)-L_2-C_O, and a lift circuit D_1-C_2-D_3-L_3-D_2-C_3. The negative conversion path consists of a pump circuit S-L_{11}-D_{10}-(C_{11}) and a "Π"-type filter C_{11}-L_{12}-C_{10} and a lift circuit D_{11}-C_{12}-L_{13}-D_{22}-C_{13}-D_{12}.

9.6.2.1   Positive Conversion Path

The equivalent circuit during switch-on is shown in Figure 9.36a, and the equivalent circuit during switch-off in Figure 9.36b.
   The voltage across inductors L_1 and L_3 is equal to V_I during switch-on and $-(V_{C1} - V_I)$ during switch-off. We have the following relations:

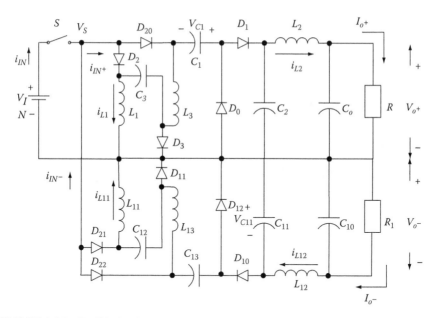

FIGURE 9.35 Re-lift circuit.

$$V_{C1} = \frac{1+k}{1-k}V_I \quad \text{and} \quad V_O = V_{CO} = V_{C2} = V_I + V_{C1} = \frac{2}{1-k}V_I$$

Thus,

$$V_{O+} = \frac{2}{1-k}V_I \quad \text{and} \quad I_{O+} = \frac{1-k}{2}I_{I+}$$

The other relations are as follows:

$$I_{I+} = ki_{I+}i_{I+} = I_{L1} + I_{L3} + i_{C3\text{-}on} + i_{C1\text{-}on} \quad i_{C1\text{-}off} = \frac{k}{1-k}i_{C1\text{-}on}$$

and

$$I_{L1} = I_{L3} = i_{C1\text{-}off} = i_{C3\text{-}off} = \frac{k}{2}i_{I+} = \frac{1}{2}I_{I+} \qquad (9.173)$$

The voltage transfer gain in continuous mode is

$$M_{R+} = \frac{V_{O+}}{V_I} = \frac{2}{1-k} \qquad (9.174)$$

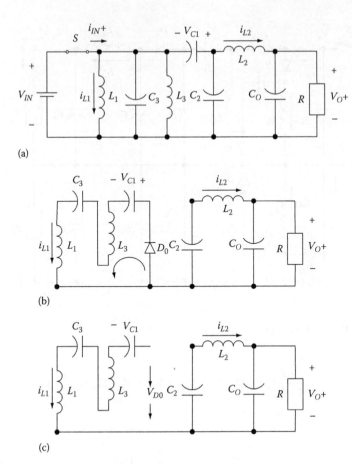

(a)

(b)

(c)

FIGURE 9.36 Equivalent circuits positive path of re-lift circuit: (a) switch-on, (b) switch-off, and (c) DCM.

The variation ratios of the parameters are as follows:

$$\xi_{2+} = \frac{\Delta i_{L2}/2}{I_{L2}} = \frac{k}{16}\frac{1}{f^2 C_2 L_2} \quad \text{and} \quad \chi_{1+} = \frac{\Delta i_{L3}/2}{I_{L3}} = \frac{kV_I T}{2L_3(1/2)I_{1+}} = \frac{k}{M_R^2}\frac{R}{fL_3}$$

$$\rho_+ = \frac{\Delta v_{C1}/2}{V_{C1}} = \frac{(1-k)TI_I}{4C_1(1+k/1-k)V_I} = \frac{1}{(1+k)fC_1 R} \qquad \sigma_{1+} = \frac{\Delta v_{C2}/2}{V_{C2}} = \frac{k}{2fC_2 R}$$

$$\sigma_{2+} = \frac{\Delta v_{C3}/2}{V_{C3}} = \frac{1-k}{4fC_3}\frac{I_{1+}}{V_I} = \frac{M_R}{2fC_3 R}$$

The variation ratio of currents i_{D0} and i_{L1} is

$$\zeta_+ = \xi_{1+} = \frac{\Delta i_{D0}/2}{I_{D0}} = \frac{kV_I T}{L_1 I_{1+}} = \frac{k}{M_R^2}\frac{R}{fL_1} \tag{9.175}$$

and the variation ratio of output voltage v_{O+} is

$$\varepsilon_+ = \frac{\Delta v_{O+}/2}{V_{O+}} = \frac{k}{128} \frac{1}{f^3 C_2 C_O L_2 R}$$ (9.176)

9.6.2.2 Negative Conversion Path

The equivalent circuit during switch-on is shown in Figure 9.37a, and the equivalent circuit during switch-off in Figure 9.37b).

The relations of the average currents and voltages are as follows:

$$I_{O-} = I_{L12} = I_{C11\text{-}on} \quad I_{C11\text{-}off} = \frac{k}{1-k} I_{C11\text{-}on} = \frac{k}{1-k} I_{O-} \quad \text{and}$$

$$I_{L11} = I_{C11\text{-}off} + I_{O-} = \frac{I_{O-}}{1-k}$$ (9.177)

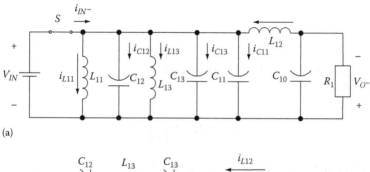

(a)

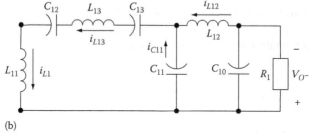

(b)

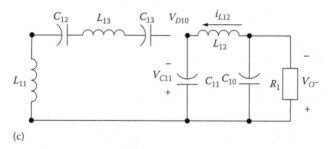

(c)

FIGURE 9.37 Equivalent circuits negative path of re-lift circuit: (a) switch-on, (b) switch-off, and (c) DCM.

$$I_{C12\text{-}off} = I_{C13\text{-}off} = I_{L11} = \frac{1}{1-k}I_{O-} \quad I_{C12\text{-}on} = \frac{1-k}{k}I_{C12\text{-}off} = \frac{1}{k}I_{O-}$$

$$I_{C13\text{-}on} = \frac{1-k}{k}I_{C13\text{-}off} = \frac{1}{k}I_{O-}$$

In steady state, we have

$$V_{C12} = V_{C13} = V_I \quad V_{L13\text{-}on} = V_I \quad \text{and} \quad V_{L13\text{-}off} = \frac{k}{1-k}V_I$$

$$V_{O-} = \frac{2}{1-k}V_I \quad \text{and} \quad I_{O-} = \frac{1-k}{2}I_{I-}$$

The voltage transfer gain in continuous mode is

$$M_{R-} = \frac{V_{O-}}{V_I} = \frac{I_{I-}}{I_{O-}} = \frac{2}{1-k} \tag{9.178}$$

Circuit $(C_{11}\text{-}L_{12}\text{-}C_{10})$ is a "Π"-type low-pass filter.
Therefore,

$$V_{C11} = V_{O-} = \frac{2}{1-k}V_I$$

From Equations 9.174 and 9.178 we define $M_R = M_{R+} = M_{R-}$. The curve of M_R versus k is shown in Figure 9.38.
The variation ratios of the parameters are as follows:

$$\xi_- = \frac{\Delta i_{L12}/2}{I_{L12}} = \frac{k}{16}\frac{1}{f^2 C_{10}L_{12}} \quad \text{and} \quad \chi_{1-} = \frac{\Delta i_{L13}/2}{I_{L13}} = \frac{kTV_I}{2L_{13}I_{O-}}(1-k) = \frac{k(1-k)}{2M_R}\frac{R_1}{fL_{13}}$$

$$\rho_- = \frac{\Delta v_{C11}/2}{V_{C11}} = \frac{kI_O T}{2C_{11}V_{O-}} = \frac{k}{2}\frac{1}{fC_{11}R_1} \quad \sigma_{1-} = \frac{\Delta v_{C12}/2}{V_{C12}} = \frac{I_{O-}}{2fC_{12}V_I} = \frac{M_R}{2}\frac{1}{fC_{12}R_1}$$

$$\sigma_{2-} = \frac{\Delta v_{C13}/2}{V_{C13}} = \frac{I_{O-}}{2fC_{13}V_I} = \frac{M_R}{2}\frac{1}{fC_{13}R_1}$$

The variation ratio of the current i_{D10} and i_{L11} is as follows:

$$\zeta_- = \frac{\Delta i_{L11}/2}{I_{L11}} = \frac{k(1-k)V_I T}{2L_{11}I_{O-}} = \frac{k(1-k)R_1}{2M_R fL_{11}} = \frac{k}{M_R^2}\frac{R_1}{fL_{11}} \tag{9.179}$$

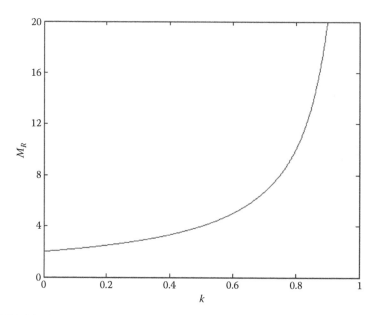

FIGURE 9.38 Voltage transfer gain M_R versus k.

The variation ratio of current v_{C10} is as follows:

$$\varepsilon_- = \frac{\Delta v_{C10}/2}{V_{C10}} = \frac{k}{128 f^3 C_{11} C_{10} L_{12}} \frac{I_{O-}}{V_{O-}} = \frac{k}{128} \frac{1}{f^3 C_{11} C_{10} L_{12} R_1} \tag{9.180}$$

9.6.2.3 Discontinuous Conduction Mode

The equivalent circuits of the DCM are shown in Figures 9.36c and 9.37c. In order to obtain the mirror-symmetrical double-output voltages, we purposely select $z_N = z_{N+} = z_{N-}$ and $\zeta = \zeta_+ = \zeta_-$. The free-wheeling diode currents i_{D0} and i_{D10} become zero during switch-off before the next switch-on period. The boundary between continuous and DCMs is $\zeta \geq 1$ or

$$\frac{k}{M_R^2} z_N \geq 1$$

Hence,

$$M_R \leq \sqrt{k z_N} \tag{9.181}$$

This boundary curve is shown in Figure 9.39. It can be seen that the boundary curve has a minimum value of M_R that is equal to 3.0, corresponding to $k = 1/3$.

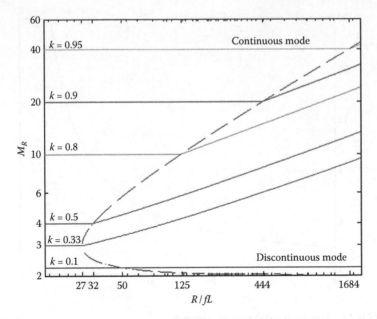

FIGURE 9.39 The boundary between continuous and discontinuous modes and the output voltage versus the normalized load $z_N = R/fL$ (re-lift circuit).

The filling efficiency m_R is

$$m_R = \frac{1}{\zeta} = \frac{M_R^2}{kz_N} \tag{9.182}$$

So,

$$V_{C1} = \left[1 + \frac{2k}{(1-k)m_R}\right]V_I = \left[1 + k^2(1-k)\frac{z_N}{2}\right]V_I \quad \text{with} \quad \sqrt{kz_N} \geq \frac{2}{1-k} \tag{9.183}$$

Therefore, the P/O voltage in the DCM is

$$V_{O+} = V_{C1} + V_I = \left[2 + \frac{2k}{(1-k)m_R}\right]V_I = \left[2 + k^2(1-k)\frac{z_N}{2}\right]V_I \quad \text{with} \quad \sqrt{kz_N} \geq \frac{2}{1-k} \tag{9.184}$$

For the current i_{L11} because inductor current $i_{L13} = 0$ at $t = t_1$,

$$V_{L13\text{-}off} = \frac{k}{(1-k)m_R}V_I$$

and for the current i_{L11}, we have $kTV_I = (1-k)m_R T(V_{C11} - 2V_I - V_{L13\text{-}off})$ or

$$V_{C11} = \left[2 + \frac{2k}{(1-k)m_R}\right]V_I = \left[2 + k^2(1-k)\frac{z_N}{2}\right]V_I \quad \text{with} \quad \sqrt{kz_N} \geq \frac{2}{1-k} \qquad (9.185)$$

and for the current i_{L12} $kT(V_I + V_{C11} - V_{O-}) = (1-k)m_R T(V_{O-} - 2V_I - V_{L13\text{-}off})$
Therefore, the N/O voltage in the DCM is

$$V_{O-} = \left[2 + \frac{2k}{(1-k)m_R}\right]V_I = \left[2 + k^2(1-k)\frac{z_N}{2}\right]V_I \quad \text{with} \quad \sqrt{kz_N} \geq \frac{2}{1-k} \qquad (9.186)$$

So

$$V_O = V_{O+} = V_{O-} = \left[2 + k^2(1-k)\frac{z_N}{2}\right]V_I$$

that is, the output voltage linearly increases when load resistance increases. Larger load resistance may cause higher output voltage in the discontinuous mode as shown in Figure 9.39.

9.6.3 TRIPLE-LIFT CIRCUIT

The triple-lift circuit is shown in Figure 9.40.

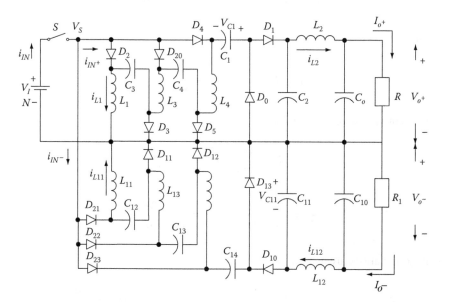

FIGURE 9.40 Triple-lift circuit.

The positive conversion path consists of a pump circuit S-L_1-D_0-C_1 and a filter (C_2)-L_2-C_O and a lift circuit D_1-C_2-D_2-C_3-D_3-L_3-D_4-C_4-D_5-L_4. The negative conversion path consists of a pump circuit S-L_{11}-D_{10}-(C_{11}) and a "Π"-type filter C_{11}-L_{12}-C_{10} and a lift circuit D_{11}-C_{12}-D_{22}-C_{13}-L_{13}-D_{12}-D_{23}-L_{14}-C_{14}-D_{13}.

9.6.3.1 Positive Conversion Path

The lift circuit is D_1-C_2-D_2-C_3-D_3-L_3-D_4-C_4-D_5-L_4. Capacitors C_2, C_3, and C_4 perform characteristics to lift the capacitor voltage V_{C1} by three times that of source voltage V_I. L_3 and L_4 perform the function as ladder joints to link the three capacitors C_3 and C_4 and lift the capacitor voltage V_{C1} up. Currents $i_{C2}(t)$, $i_{C3}(t)$, and $i_{C4}(t)$ are exponential functions. They have large values at the moment power is on, but they are small because $v_{C3} = v_{C4} = V_I$ and $v_{C2} = V_{O+}$ in steady state.

The output voltage and current are as follows:

$$V_{O+} = \frac{3}{1-k}V_I \quad \text{and} \quad I_{O+} = \frac{1-k}{3}I_{I+}$$

The voltage transfer gain in continuous mode is

$$M_{T+} = \frac{V_{O+}}{V_I} = \frac{3}{1-k} \tag{9.187}$$

Other average voltages are as follows:

$$V_{C1} = \frac{2+k}{1-k}V_I \quad V_{C3} = V_{C4} = V_I \quad V_{CO} = V_{C2} = V_{O+}$$

Other average currents are as follows:

$$I_{L2} = I_{O+}; \quad I_{L1} = I_{L3} = I_{L4} = \frac{1}{3}I_{I+} = \frac{1}{1-k}I_{O+}$$

Current variations are as follows:

$$\xi_{1+} = \zeta_{+} = \frac{k(1-k)R}{2M_T fL} = \frac{k}{M_T{}^2}\frac{3R}{2fL}$$

$$\chi_{1+} = \frac{k}{M_T{}^2}\frac{3R}{2fL_3} \quad \chi_{2+} = \frac{k}{M_T{}^2}\frac{3R}{2fL_4}$$

Voltage variations are as follows:

$$\rho_{+} = \frac{3}{2(2+k)fC_1R} \quad \sigma_{1+} = \frac{k}{2fC_2R} \quad \sigma_{2+} = \frac{M_T}{2fC_3R} \quad \sigma_{3+} = \frac{M_T}{2fC_4R}$$

The variation ratio of output voltage V_{CO} is

$$\varepsilon_+ = \frac{k}{128} \frac{1}{f^3 C_2 C_O L_2 R} \tag{9.188}$$

9.6.3.2 Negative Conversion Path

Circuit C_{12}-D_{11}-L_{13}-D_{22}-C_{13}-D_{12}-L_{14}-D_{23}-C_{14}-D_{13} is the lift circuit. Capacitors C_{12}, C_{13}, and C_{14} perform characteristics to lift the capacitor voltage V_{C11} by three times that of source voltage V_I. L_{13} and L_{14} perform the function as ladder joints to link the three capacitors C_{12}, C_{13}, and C_{14} and lift the capacitor voltage V_{C11} up. Currents $i_{C12}(t)$, $i_{C13}(t)$, and $i_{C14}(t)$ are exponential functions. They have large values at the moment power is on, but they are small because $v_{C12} = v_{C13} = v_{C14} \cong V_I$ in steady state.

The output voltage and current are

$$V_{O-} = \frac{3}{1-k} V_I \quad \text{and} \quad I_{O-} = \frac{1-k}{3} I_{I-}$$

The voltage transfer gain in the continuous mode is

$$M_{T-} = \frac{V_{O-}}{V_I} = \frac{3}{1-k} \tag{9.189}$$

From Equations 9.187 and 9.189 we define $M_T = M_{T+} = M_{T-}$. The curve of M_T versus k is shown in Figure 9.41.

Other average voltages are as follows: $V_{C11} = V_{O-}$; $V_{C12} = V_{C13} = V_{C14} = V_I$.

Other average currents are as follows:

$$I_{L12} = I_{O-}; \quad I_{L11} = I_{L13} = I_{L14} = \frac{1}{1-k} I_{O-}$$

Current variation ratios are as follows:

$$\zeta_- = \frac{k}{M_T^2} \frac{3R_1}{2fL_{11}} \quad \xi_{2-} = \frac{k}{16} \frac{1}{f^2 C_{10} L_{12}}$$

$$\chi_{1-} = \frac{k(1-k)}{2M_T} \frac{R_1}{fL_{13}} \quad \chi_{2-} = \frac{k(1-k)}{2M_T} \frac{R_1}{fL_{14}}$$

Voltage variation ratios are as follows:

$$\rho_- = \frac{k}{2} \frac{1}{fC_{11}R_1} \quad \sigma_{1-} = \frac{M_T}{2} \frac{1}{fC_{12}R_1} \quad \sigma_{2-} = \frac{M_T}{2} \frac{1}{fC_{13}R_1} \quad \sigma_{3-} = \frac{M_T}{2} \frac{1}{fC_{14}R_1}$$

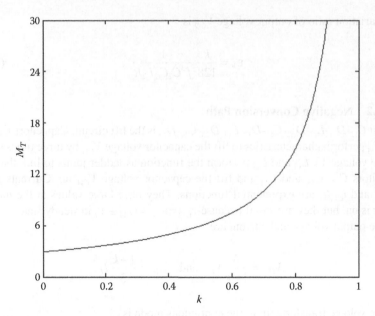

FIGURE 9.41 Voltage transfer gain M_T versus k.

The variation ratio of output voltage V_{C10} is

$$\varepsilon_- = \frac{k}{128} \frac{1}{f^3 C_{11} C_{10} L_{12} R_1} \tag{9.190}$$

9.6.3.3 Discontinuous Mode
To obtain the mirror-symmetrical double-output voltages, we purposely select $L_1 = L_{11}$ and $R = R_1$.

Define

$$V_O = V_{O+} = V_{O-} \quad M_T = M_{T+} = M_{T-}$$

$$= \frac{V_O}{V_I} = \frac{3}{1-k} \quad z_N = z_{N+} = z_{N-} \quad \text{and} \quad z = z_+ = z_-$$

The free-wheeling diode currents i_{D0} and i_{D10} become zero during the switch-off period before the next switch-on period. The boundary between continuous and discontinuous modes is $\zeta \geq 1$

The boundary between continuous and discontinuous modes is

$$M_T \leq \sqrt{\frac{3k z_N}{2}} \tag{9.191}$$

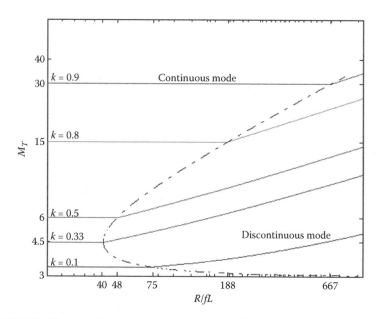

FIGURE 9.42 The boundary between continuous and discontinuous modes and the output voltage versus the normalized load $z_N = R/fL$ (triple-lift circuit).

This boundary curve is shown in Figure 9.42. It can be seen that the boundary curve has a minimum value of M_T that is equal to 4.5, corresponding to $k = 1/3$.

In discontinuous mode, the currents i_{D0} and i_{D10} exist during the period between kT and $[k+(1-k)m_T]T$, where m_T is the filling efficiency, that is,

$$m_T = \frac{1}{\zeta} = \frac{2M_T^2}{3kz_N} \tag{9.192}$$

Considering Equation 9.191, we have $0 < m_T < 1$. Since the current i_{D0} becomes zero at $t = t_1 = [k+(1-k)m_T]T$, for the current i_{L1}, i_{L3}, and i_{L4} $3kTV_I = (1-k)m_T T(V_{C1} - 2V_I)$ or

$$V_{C1} = \left[2 + \frac{3k}{(1-k)m_T}\right]V_I = \left[2 + k^2(1-k)\frac{z_N}{2}\right]V_I \quad \text{with} \quad \sqrt{\frac{3kz_N}{2}} \ge \frac{3}{1-k} \tag{9.193}$$

Therefore, the P/O voltage in discontinuous mode is

$$V_{O+} = V_{C1} + V_I = \left[3 + \frac{3k}{(1-k)m_T}\right]V_I = \left[3 + k^2(1-k)\frac{z_N}{2}\right]V_I \quad \text{with} \quad \sqrt{\frac{3kz_N}{2}} \ge \frac{3}{1-k} \tag{9.194}$$

Because inductor current $i_{L11} = 0$ at $t = t_1$, $V_{L13\text{-}off} = V_{L14\text{-}off} = (k/(1-k)m_T)V_I$

Since i_{D10} becomes 0 at $t_1 = [k + (1-k)m_T]T$, for the current i_{L11}, $kTV_I = (1-k)$ $m_T T(V_{C11} - 3V_I - V_{L13\text{-}off} - V_{L14\text{-}off})$, we obtain

$$V_{C11} = \left[3 + \frac{3k}{(1-k)m_T}\right]V_I = \left[3 + k^2(1-k)\frac{z_N}{2}\right]V_I \quad \text{with} \quad \sqrt{\frac{3kz_N}{2}} \geq \frac{3}{1-k} \qquad (9.195)$$

For the current i_{L12} $kT(V_I + V_{C14} - V_{O-}) = (1-k)m_T T(V_{O-} - 2V_I - V_{L13\text{-}off} - V_{L14\text{-}off})$ Therefore, the N/O voltage in discontinuous mode is

$$V_{O-} = \left[3 + \frac{3k}{(1-k)m_T}\right]V_I = \left[3 + k^2(1-k)\frac{z_N}{2}\right]V_I \quad \text{with} \quad \sqrt{\frac{3kz_N}{2}} \geq \frac{3}{1-k} \qquad (9.196)$$

So

$$V_O = V_{O+} = V_{O-} = \left[3 + k^2(1-k)\frac{z_N}{2}\right]V_I$$

that is, the output voltage linearly increases when load resistance increases. The output voltage increases when load resistance increases, as shown in Figure 9.42.

9.6.4 QUADRUPLE-LIFT CIRCUIT

Quadruple-lift circuit is shown in Figure 9.43.

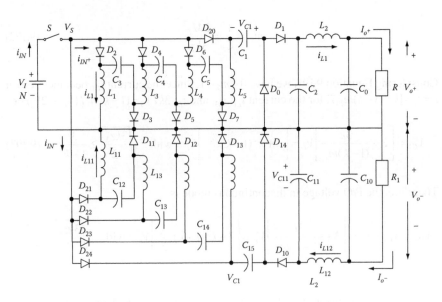

FIGURE 9.43 Quadruple-lift circuit.

The positive conversion path consists of a pump circuit S-L_1-D_0-C_1 and a filter (C_2)-L_2-C_O, and a lift circuit D_1-C_2-L_3-D_2-C_3-D_3-L_4-D_4-C_4-D_5-L_5-D_6-C_5-S_1. The negative conversion path consists of a pump circuit S-L_{11}-D_{10}-(C_{11}) and a "Π"-type filter C_{11}-L_{12}-C_{10} and a lift circuit D_{11}-C_{12}-D_{22}-L_{13}-C_{13}-D_{12}-D_{23}-L_{14}-C_{14}-D_{13}-D_{24}-L_{15}-C_{15}-D_{14}.

9.6.4.1 Positive Conversion Path

Capacitors C_2, C_3, C_4 and C_5 perform characteristics to lift the capacitor voltage V_{C1} by four times that of source voltage V_I. L_3, L_4, and L_5 perform the function as ladder joints to link the four capacitors C_2, C_3, C_4, and C_5 and lift the output capacitor voltage V_{C1} up. Current $i_{C2}(t)$, $i_{C3}(t)$, $i_{C4}(t)$, and $i_{C5}(t)$ are exponential functions. They have large values at the moment power is on, but they are small because $v_{C3}=v_{C4}=v_{C5}=V_I$ and $v_{C2}=V_{O+}$ in steady state.

The output voltage and current are as follows:

$$V_{O+} = \frac{4}{1-k}V_I \quad \text{and} \quad I_{O+} = \frac{1-k}{4}I_{I+}$$

The voltage transfer gain in continuous mode is

$$M_{Q+} = \frac{V_{O+}}{V_I} = \frac{4}{1-k} \tag{9.197}$$

Other average voltages are as follows:

$$V_{C1} = \frac{3+k}{1-k}V_I \quad V_{C3} = V_{C4} = V_{C5} = V_I \quad V_{CO} = V_{C2} = V_O$$

Other average currents are as follows:

$$I_{L2} = I_{O+}; \quad I_{L1} = I_{L3} = I_{L4} = I_{L5} = \frac{1}{4}I_{I+} = \frac{1}{1-k}I_{O+}$$

Current variations are as follows:

$$\xi_{1+} = \zeta_+ = \frac{k(1-k)R}{2M_Q fL} = \frac{k}{M_Q^2}\frac{2R}{fL} \quad \xi_{2+} = \frac{k}{16}\frac{1}{f^2C_2L_2}$$

$$\chi_{1+} = \frac{k}{M_Q^2}\frac{2R}{fL_3} \quad \chi_{2+} = \frac{k}{M_Q^2}\frac{2R}{fL_4} \quad \chi_{3+} = \frac{k}{M_Q^2}\frac{2R}{fL_5}$$

Voltage variations are as follows:

$$\rho_+ = \frac{2}{(3+2k)fC_1R} \quad \sigma_{1+} = \frac{M_Q}{2fC_2R} \quad \sigma_{2+} = \frac{M_Q}{2fC_3R} \quad \sigma_{3+} = \frac{M_Q}{2fC_4R} \quad \sigma_{4+} = \frac{M_Q}{2fC_5R}$$

The variation ratio of output voltage V_{CO} is

$$\varepsilon_+ = \frac{k}{128}\frac{1}{f^3 C_2 C_0 L_2 R} \tag{9.198}$$

9.6.4.2 Negative Conversion Path

Capacitors C_{12}, C_{13}, C_{14}, and C_{15} perform characteristics to lift the capacitor voltage V_{C11} by four times that of source voltage V_I. L_{13}, L_{14}, and L_{15} perform the function as ladder joints to link the four capacitors C_{12}, C_{13}, C_{14}, and C_{15} and lift the output capacitor voltage V_{C11} up. Current $i_{C12}(t)$, $i_{C13}(t)$, $i_{C14}(t)$, and $i_{C15}(t)$ are exponential functions. They have large values at the moment power is on, but they are small because $v_{C12}=v_{C13}=v_{C14}=v_{C15} \cong V_I$ in steady state.

The output voltage and current are as follows:

$$V_{O-} = \frac{4}{1-k}V_I \quad \text{and} \quad I_{O-} = \frac{1-k}{4}I_{I-}$$

The voltage transfer gain in continuous mode is

$$M_{Q-} = \frac{V_{O-}}{V_I} = \frac{4}{1-k} \tag{9.199}$$

From Equations 9.197 and 9.199, we define $M_Q = M_{Q+} = M_{Q-}$. The curve of M_Q versus k is shown in Figure 9.44.

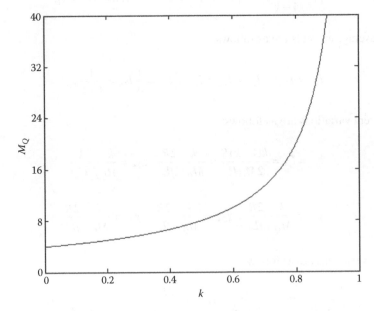

FIGURE 9.44 Voltage transfer gain M_Q versus k.

Other average voltages are as follows: $V_{C10} = V_{O-}$; $V_{C12} = V_{C13} = V_{C14} = V_{C15} = V_I$. Other average currents are as follows:

$$I_{L12} = I_{O-}; \quad I_{L11} = I_{L13} = I_{L14} = I_{L15} = \frac{1}{1-k} I_{O-}$$

Current variation ratios are as follows:

$$\zeta_- = \frac{k}{M_Q^2} \frac{2R_1}{fL_{11}} \quad \xi_- = \frac{k}{16} \frac{1}{f^2 CL_{12}}$$

$$\chi_{1-} = \frac{k(1-k)}{2M_Q} \frac{R_1}{fL_{13}} \quad \chi_{2-} = \frac{k(1-k)}{2M_Q} \frac{R_1}{fL_{14}} \quad \chi_{3-} = \frac{k(1-k)}{2M_Q} \frac{R_1}{fL_{15}}$$

Voltage variation ratios are as follows:

$$\rho_- = \frac{k}{2} \frac{1}{fC_{11}R_1} \quad \sigma_{1-} = \frac{M_Q}{2} \frac{1}{fC_{12}R_1} \quad \sigma_{2-} = \frac{M_Q}{2} \frac{1}{fC_{13}R_1}$$

$$\sigma_{3-} = \frac{M_Q}{2} \frac{1}{fC_{14}R_1} \quad \sigma_{4-} = \frac{M_Q}{2} \frac{1}{fC_{15}R_1}$$

The variation ratio of output voltage V_{C10} is

$$\varepsilon_- = \frac{k}{128} \frac{1}{f^3 C_{11}C_{10}L_{12}R_1} \tag{9.200}$$

9.6.4.3 Discontinuous Conduction Mode

In order to obtain the mirror-symmetrical double-output voltages, we purposely select $L_1 = L_{11}$ and $R = R_1$. Therefore, we may define

$$V_O = V_{O+} = V_{O-} \quad M_Q = M_{Q+} = M_{Q-}$$

$$= \frac{V_O}{V_I} = \frac{4}{1-k} \quad z_N = z_{N+} = z_{N-} \quad \text{and} \quad z = z_+ = z_-$$

The free-wheeling diode currents i_{D0} and i_{D10} become zero during the switch-off period before the next switch–on period. The boundary between CCM and DCM is $\zeta \geq 1$ or

$$M_Q \leq \sqrt{2kz_N} \tag{9.201}$$

This boundary curve is shown in Figure 9.45. It can be seen that this boundary curve has a minimum value of M_Q that is equal to 6.0, corresponding to $k = 1/3$.

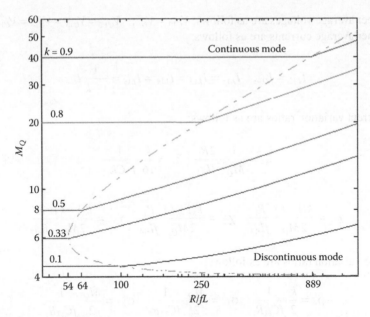

FIGURE 9.45 The boundary between continuous and discontinuous modes and the output voltage versus the normalized load $z_N = R/fL$ (quadruple-lift circuit).

In the discontinuous mode the currents i_{D0} and i_{D10} exist during the period between kT and $[k+(1-k)m_Q]T$, where m_Q is the filling efficiency, that is,

$$m_Q = \frac{1}{\zeta} = \frac{M_Q^2}{2kz_N} \tag{9.202}$$

where $0 < m_Q < 1$. Since the current i_{D0} becomes zero at $t = t_1 = kT + (1-k)m_Q T$, for the current i_{L1}, i_{L3}, i_{L4}, and i_{L5}

$$4kTV_I = (1-k)m_Q T(V_{C1} - 3V_I)$$

$$V_{C1} = \left[3 + \frac{4k}{(1-k)m_Q}\right]V_I = \left[3 + k^2(1-k)\frac{z_N}{2}\right]V_I \quad \text{with} \quad \sqrt{2kz_N} \geq \frac{4}{1-k} \tag{9.203}$$

Therefore, the P/O voltage in the DCM is

$$V_{O+} = V_{C1} + V_I = \left[4 + \frac{4k}{(1-k)m_Q}\right]V_I = \left[4 + k^2(1-k)\frac{z_N}{2}\right]V_I \quad \text{with} \quad \sqrt{2kz_N} \geq \frac{4}{1-k} \tag{9.204}$$

Because inductor current $i_{L11}=0$ at $t=t_1$,

$$V_{L13\text{-}off} = V_{L14\text{-}off} = V_{L15\text{-}off} = \frac{k}{(1-k)m_Q}V_I$$

Since the current i_{D10} becomes zero at $t=t_1=kT+(1-k)m_QT$, for the current i_{L11} we have

$$kTV_I=(1-k)m_{Q\text{-}}T(V_{C11}-4V_I-V_{L13\text{-}off}-V_{L14\text{-}off}-V_{L15\text{-}off})$$

So

$$V_{C11} = \left[4+\frac{4k}{(1-k)m_Q}\right]V_I = \left[4+k^2(1-k)\frac{z_N}{2}\right]V_I \quad \text{with} \quad \sqrt{2kz_N} \geq \frac{4}{1-k} \quad (9.205)$$

For the current $i_{L12}kT(V_I+V_{C15}-V_{O\text{-}})=(1-k)m_QT(V_{O\text{-}}-2V_I-V_{L13\text{-}off}-V_{L14\text{-}off}-V_{L15\text{-}off})$, therefore, the N/O voltage in DCM is

$$V_{O\text{-}} = \left[4+\frac{4k}{(1-k)m_Q}\right]V_I = \left[4+k^2(1-k)\frac{z_N}{2}\right]V_I \quad \text{with} \quad \sqrt{2kz_N} \geq \frac{4}{1-k} \quad (9.206)$$

So,

$$V_O = V_{O+} = V_{O\text{-}} = \left[4+k^2(1-k)\frac{z_N}{2}\right]V_I$$

Thus, the output voltage linearly increases when load resistance increases. It can be seen that the output voltage increases when load resistance increases, as shown in Figure 9.45.

9.6.5 SUMMARY

9.6.5.1 Positive Conversion Path

From the analysis and calculation in previous sections, the common formulae can be obtained for all circuits:

$$M = \frac{V_{O+}}{V_I} = \frac{I_{I+}}{I_{O+}} \quad z_N = \frac{R}{fL} \quad R = \frac{V_{O+}}{I_{O+}}$$

$L = L_1L_2/(L_1+L_2)$ for elementary circuit only
$L=L_1$ for other lift circuits

Current variations are as follows:

$$\xi_{1+} = \frac{1-k}{2M_E}\frac{R}{fL_1} \quad \text{and} \quad \xi_{2+} = \frac{k}{2M_E}\frac{R}{fL_2} \quad \text{for elementary circuit only}$$

$$\xi_{1+} = \zeta_+ = \frac{k(1-k)R}{2MfL} \quad \text{and} \quad \xi_{2+} = \frac{k}{16}\frac{1}{f^2C_2L_2} \quad \text{for other lift circuits}$$

$$\zeta_+ = \frac{k(1-k)R}{2MfL} \quad \chi_{j+} = \frac{k}{M^2}\frac{R}{fL_{j+2}} \quad (j=1,2,3,...)$$

Voltage variations are as follows:

$$\rho_+ = \frac{k}{2fC_1R} \quad \varepsilon_+ = \frac{k}{8M_E}\frac{1}{f^2C_0L_2} \quad \text{for elementary circuit only}$$

$$\rho_+ = \frac{M}{M-1}\frac{1}{2fC_1R} \quad \varepsilon_+ = \frac{k}{128}\frac{1}{f^3C_2C_0L_2R} \quad \text{for other lift circuits}$$

$$\sigma_{1+} = \frac{k}{2fC_2R} \quad \sigma_{j+} = \frac{M}{2fC_{j+1}R} \quad (j=2,3,4,...)$$

9.6.5.2 Negative Conversion Path
From the analysis and calculation in previous sections, the common formulae can be obtained for all circuits:

$$M = \frac{V_{O-}}{V_I} = \frac{I_{I-}}{I_{O-}} \quad Z_{N-} = \frac{R_1}{fL_{11}} \quad R_1 = \frac{V_{O-}}{I_{O-}}$$

Current variation ratios are as follows:

$$\zeta_- = \frac{k(1-k)R_1}{2MfL_{11}} \quad \xi_- = \frac{k}{16f^2C_{11}L_{12}} \quad \chi_{j-} = \frac{k(1-k)R_1}{2MfL_{j+2}} \quad (j=1,2,3,...)$$

Voltage variation ratios are as follows:

$$\rho_- = \frac{k}{2fC_{11}R_1} \quad \varepsilon_- = \frac{k}{128f^3C_{11}C_{10}L_{12}R_1} \quad \sigma_{j-} = \frac{M}{2fC_{j+11}R_1} \quad (j=1,2,3,4,...)$$

9.6.5.3 Common Parameters

Usually, we select the loads $R=R_1$, $L=L_{11}$, so that we get $z_N=z_{N+}=z_{N-}$. In order to write common formulas for the boundaries between continuous and discontinuous modes and output voltage for all circuits, the circuits can be numbered. The definition is that subscript 0 means the elementary circuit, subscript 1 means the self-lift circuit, subscript 2 means the self-lift circuit, subscript 3 means the triple-lift circuit, subscript 4 means the quadruple-lift circuit, and so on.

The voltage transfer gain is

$$M_j = \frac{k^{h(j)}[j+h(j)]}{1-k} \quad j=0,1,2,3,4,\ldots$$

The characteristics of output voltage of all circuits are shown in Figure 9.46. The free-wheeling diode current's variation is

$$\zeta_j = \frac{k^{[1+h(j)]}}{M_j^2} \frac{j+h(j)}{2} z_N$$

The boundaries are determined by the condition $\zeta_j \geq 1$ or

$$\frac{k^{[1+h(j)]}}{M_j^2} \frac{j+h(j)}{2} z_N \geq 1 \quad j=0,1,2,3,4,\ldots$$

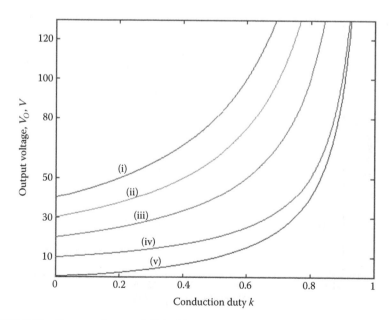

FIGURE 9.46 Output voltages of all double-output Luo converters ($V_I = 10\,\text{V}$).

Therefore, the boundaries between continuous and discontinuous modes for all circuits are as follows:

$$M_j = k^{1+h(j)/2}\sqrt{\frac{j+h(j)}{2}z_N} \quad j = 0,1,2,3,4,\dots$$

The filling efficiency is

$$m_j = \frac{1}{\zeta_j} = \frac{M_j^2}{k^{[1+h(j)]}}\frac{2}{j+h(j)}\frac{1}{z_N} \quad j = 0,1,2,3,4,\dots$$

The output voltage in discontinuous mode for all circuits is

$$V_{O-j} = \left[j + k^{[2-h(j)]}\frac{1-k}{2}z_N\right]V_I$$

where $h(j) = \begin{cases} 0 & if & j \ge 1 \\ 1 & if & j = 0 \end{cases}$ $j = 0,1,2,3,4,\dots$ $h(j)$ is the Hong Function.

The boundaries between continuous and discontinuous modes of all circuits are shown in Figure 9.47. The curves of all M versus z_N state that the continuous mode area increases from M_E via M_S, M_R, M_T to M_Q. The boundary of elementary circuit

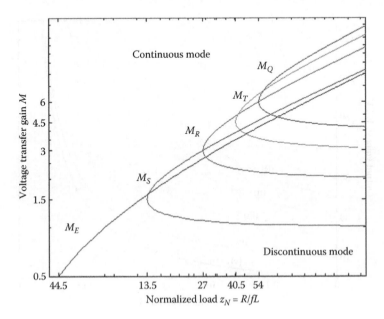

FIGURE 9.47 Boundaries between continuous and discontinuous modes of all double-output Luo converters.

is a monorising curve, while others not. The minimum values of the boundaries of other circuits M_S, M_R, M_T, and M_Q are at $k = 1/3$.

9.7 VOLTAGE-LIFT CÚK CONVERTERS

The proposed N/O Cúk converters are developed from Cúk converter as shown in Figure 5.32. They are as follows:

- Elementary self-lift circuit
- Developed self-lift circuit
- Re-lift circuit
- Multiple-lift circuits (e.g., triple-lift and quadruple-lift circuit)

These converters perform positive to negative DC/DC voltage-increasing conversion with higher voltage transfer gains, power density, small ripples, high efficiency, and cheap topology with simple structure [17–19].

9.7.1 ELEMENTARY SELF-LIFT CÚK CIRCUIT

The elementary self-lift circuit is derived from Cúk converter by adding the components (D_1-C_1). The circuit diagram is shown in Figure 9.48. The lift circuit consists of L_1-D_1-C_1, and it is a basic VL cell. When switches S turn on, D_1 is on, and D_0 is off. When S turns off, D_1 is off, and D_0 is on. The capacitor C_1 performs characteristics to lift the output capacitor voltage V_{Co} by the capacitor voltage V_{Cs}.

In the steady state, the average voltage across inductor L_1 over a period is zero. Thus,

$$V_{C1} = V_{Co} = V_o$$

During the switch-on period, the voltages across capacitor C_1 is equal to the voltage across C_s. Since C_s and C_1 are sufficiently large, we have

$$V_{C1} = V_{Cs} = V_o$$

The inductor current i_L increases during the switch-on period and decreases during the switch-off period. The corresponding voltages across L are V_{in} and $-(V_{Cs} - V_{in})$. Therefore, $kTV_{in} = (1-k)T(V_{Cs} - V_{in})$.

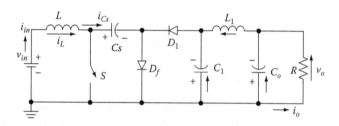

FIGURE 9.48 Elementary self-lift Cúk converter.

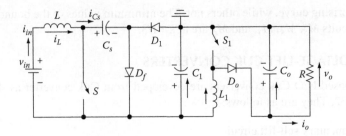

FIGURE 9.49 Developed self-lift Cúk circuit.

Hence, the voltage transfer gain of the elementary self-lift circuit is

$$M_S = \frac{V_O}{V_{in}} = \frac{1}{1-k} \tag{9.207}$$

9.7.2 Developed Self-Lift Cúk Circuit

The developed self-lift circuit is derived from the elementary self-lift Cúk circuit by adding the components $(D_0\text{-}S_1)$ and redesigning the connection of L_1. Static switches S and S_1 are switched simultaneously. The circuit diagram is shown in Figure 9.49. The lift circuit consists of $C_1\text{-}L_1\text{-}S_1\text{-}D_1$. When switches S and S_1 turn on, D_1 is on, and D_f and D_0 are off. When S and S_1 turn off, D_1 is off, and D_f and D_0 are on. Capacitor C_1 performs characteristics to lift the output capacitor voltage V_{Co} by the capacitor voltage V_{Cs}.

During the switch-on period, the voltages across capacitor C_1 are equal to the voltage across C_s. Since C_s and C_1 are sufficiently large, we have

$$V_{C1} = V_{Cs} = \frac{1}{1-k}V_{in}$$

The inductor current i_{L1} increases during the switch-on period and decreases during the switch-off period. The corresponding voltages across L are V_{Cs} and $-(V_o - V_{C1})$. Therefore,

$$kTV_{Cs} = (1-k)T(V_o - V_{C1})$$

Hence, the voltage transfer gain of the developed self-lift circuit is

$$M_S' = \frac{V_O}{V_{in}} = \frac{1}{(1-k)^2} \tag{9.208}$$

9.7.3 Re-Lift Cúk Circuit

The re-lift circuit is derived from the developed self-lift Cúk circuit by adding the components $(D_2\text{-}C_2\text{-}L_2\text{-}D_3)$. Static switches S and S_1 are switched simultaneously. The circuit diagram is shown in Figure 9.50. The lift circuit consists of

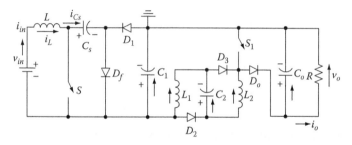

FIGURE 9.50 Re-lift Cúk circuit.

L_1-D_1-C_1-D_2-C_2-L_2-D_3-S_1, and it can be divided into two basic VL cells. When switches S and S_1 turn on, D_1, D_2, and D_3 are on, and D_0 is off. When S and S_1 turn off, D_1, D_2, and D_3 are off, and D_0 is on. Capacitors C_1 and C_2 perform characteristics to lift the output capacitor voltage V_{C_0} by twice that of the capacitor voltage V_{C_s}. L_2 performs the function of a ladder joint to link the two capacitors C_1 and C_2 and lift V_{C_0}. To avoid the abnormal phenomena of diodes during the switch-off period, it is assumed that L_1 and L_2 are the same, which simplifies the theoretical analysis.

During the switch-on period, both the voltages across capacitor C_1 and C_2 are equal to the voltage across C_s. Since C_s, C_1, and C_2 are sufficiently large, we have

$$V_{C1} = V_{C2} = V_{Cs} = V_{in} = \frac{1}{1-k}V_{in}$$

The voltage across L_1 is equal to V_{Cs} during the switch-on period. With the second voltage balance, we have

$$V_{L1-off} = \frac{k}{1-k}V_{Cs}$$

The inductor current i_{L2} increases during the switch-on period and decreases during the switch-off period. The corresponding voltages across L_2 are V_{Cs} and $-(V_o - V_{C1} - V_{C2} - V_{L1-off})$. Therefore,

$$kTV_{Cs} = (1-k)T(V_o - V_{C1} - V_{C2} - V_{L1-off})$$

Hence, the voltage transfer gain of the re-lift circuit is

$$M_R = \frac{V_o}{V_{in}} = \frac{2}{(1-k)^2} \qquad (9.209)$$

9.7.4 MULTIPLE-LIFT CÚK CIRCUIT

It is possible to construct multiple-lift circuit by adding the components D_2-C_2-L_2-D_3. Assuming that there are n VL cells, the generalized representation of

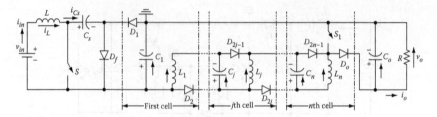

FIGURE 9.51   Generalized representation of N/O Cúk converters.

multiple-lift circuits is shown in Figure 9.51. Only two synchronous switches S and S_1 are required for each complex multiple-lift circuits, which simplify the control scheme and decrease the cost significantly. Hence, each circuit has two switches, $(n+1)$ inductors, $(n+1)$ capacitors, and $(2n-1)$ diodes. It is noted that all inductors existing in the VL cells are the same here because of the same reasons explained for the re-lift circuit. All the capacitors are sufficiently large. From the foregoing analysis and calculation, the general formulas for all multiple-lift circuits can be obtained using similar steps.

The generalized voltage transfer gain is

$$M = \frac{n}{(1-k)^{h(n)}} \quad n = 1,2,3,4,\ldots \tag{9.210}$$

where $h(n) = \begin{cases} 1 & \text{self-lift} \\ 2 & \text{others} \end{cases}$

If the generalized circuit possesses three VL cells, it is termed the triple-lift circuit. If the generalized circuit possesses four VL cells, it is termed the quadruple-lift circuit.

9.7.5   SIMULATION AND EXPERIMENTAL VERIFICATION OF ELEMENTARY AND DEVELOPED SELF-LIFT CIRCUITS

Referring to Figures 9.48 and 9.49, we set these two circuits to have the same conditions: $V_{in} = 10\,\text{V}$, $R = 100\,\Omega$, $L = 1\,\text{mH}$, $L_1 = 500\,\mu\text{H}$, $C_s = 110\,\mu\text{F}$, $C_1 = 22\,\mu\text{F}$, $C_o = 47\,\mu\text{F}$, $k = 0.5$, and $f = 100\,\text{kHz}$. According to (9.207), the theoretical value V_O of the elementary self-lift circuit is equal to 20 V. According to (9.208), the theoretical value V_O of the developed self-lift circuit is equal to 40 V. The simulation results in Psim are shown in Figure 9.52, where curve 1 is for v_O of the elementary self-lift circuit and curve 2 is for v_O of the developed self-lift circuit. The steady-state values in the simulation are identical to that given in theoretical analysis.

The same parameters are chosen to construct the corresponding testing hardware circuits. A single n-channel MOSFET is used in the elementary self-lift circuit. Two n-channel MOSFETs are used in the developed self-lift circuit. The corresponding

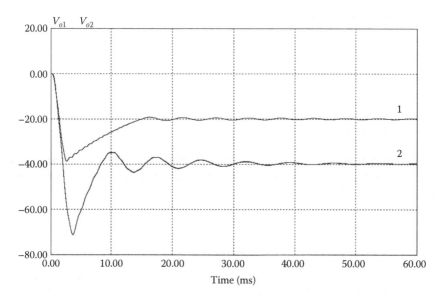

FIGURE 9.52 Simulation results of the elementary and developed self-lift circuit.

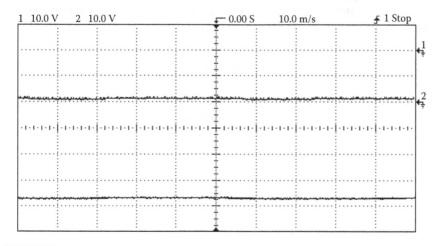

FIGURE 9.53 Experimental results of the elementary and developed self-lift circuit.

experimental curves in the steady state are shown in Figure 9.53, respectively. The curve shown in Channel 1 with 10 V/Div corresponds to the output voltage of the elementary self-lift circuit, which is about 19 V. The curve shown in Channel 2 with 10 V/Div corresponds to the output voltage of the developed self-lift circuit, which is about 37 V. Considering the effects caused by the parasitic parameters, we can see that the measured results are very close to the theoretical analysis and simulation results.

9.8 VOLTAGE-LIFT SEPICs

The proposed P/O SEPICs are developed from SEPIC as shown in Figure 5.33. They are as follows:

- Self-lift circuit
- Re-lift circuit
- Multiple circuits (e.g., triple-lift and quadruple-lift circuit)

These converters perform positive-to-positive DC/DC voltage-increasing conversion with higher voltage transfer gains, power density, small ripples, high efficiency, and cheap topology with simple structure [18–21].

9.8.1 SELF-LIFT SEPIC

The self-lift circuit is derived from SEPIC converter by adding the components $(D_1\text{-}C_1)$. The circuit diagram is shown in Figure 9.54. The lift circuit consists of $L_1\text{-}D_1\text{-}C_1$, and it is a basic VL cell. When switches S turn on, D_1 is on, and D_0 is off. When S turn off, D_1 is off, and D_0 is on. The capacitor C_1 performs characteristics to lift the output capacitor voltage V_{Co} by the capacitor voltage V_{Cs}.

In the steady state, the average voltage across inductor L over a period is zero. Thus,

$$V_{Cs} = V_{in}$$

During the switch-on period, the voltages across capacitor C_1 is equal to the voltage across C_s. Since C, and C_1 are sufficiently large, we have $V_{C1} = V_{Cs} = V_{in}$.

The inductor current i_L increases during the switch-on period and decreases during the switch-off period. The corresponding voltages across L are V_{Cs} and $-(V_{Co} - V_{C1} - V_{in} + V_{Cs})$. Therefore,

$$kTV_{Cs} = (1-k)T(V_{Co} - V_{C1} - V_{in} + V_{Cs})$$

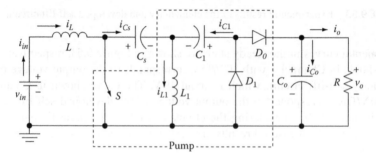

FIGURE 9.54   Self-lift SEPIC.

Hence, the voltage transfer gain of the self-lift circuit is

$$M_S = \frac{V_o}{V_{in}} = \frac{1}{1-k}$$ (9.211)

9.8.2 RE-LIFT SEPIC

The re-lift circuit is derived from the self-lift circuit by adding the components (L_2-D_2-C_2-S_1). Static switches S and S_1 are switched simultaneously. The circuit diagram and equivalent circuits during the switch-on and switch-off periods are shown in Figure 9.55. The lift circuit consists of L_1-D_1-C_1-L_2-D_2-C_2-S_1, and it can be divided into two basic VL cells. When switches S and S_1 turn on, D_1 and D_2 are on, and D_0 is off. When S and S_1 turn off, D_1 and D_2 are off, and D_0 is on. Capacitors C_1 and C_2 perform characteristics to lift the output capacitor voltage V_{Co} by twice that of the capacitor voltage V_{Cs}. L_2 performs the function of a ladder joint to link the two capacitors C_1 and C_2 and lift V_{Co}. To avoid the abnormal phenomena of diodes during the switch-off period [11], it is assumed that L_1 and L_2 are the same, which simplifies the theoretical analysis.

In the steady state, both the average voltages across inductor L and L_1 over a period are zero. Thus, $V_{Cs} = V_{in}$.

During the switch-on period, both the voltages across capacitor C_1 and C_2 are equal to the voltage across C_s. Since C, C_1, and C_2 are sufficiently large, we have

$$V_{C1} = V_{C2} = V_{Cs} = V_{in}$$

The voltage across L_1 is equal to V_{Cs} during the switch-on period. With the second voltage balance, we have

$$V_{L1\text{-}off} = \frac{k}{1-k} V_{in}$$

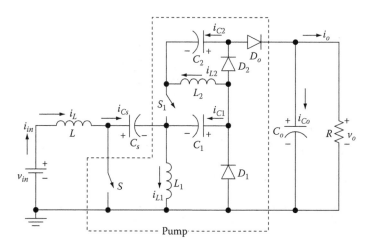

FIGURE 9.55   Re-lift SEPIC.

The inductor current i_{L2} increases during the switch-on period and decreases during the switch-off period. The corresponding voltages across L_2 are V_{Cs} and $-(V_{Co} - V_{C1} - V_{C2} - V_{L1\text{-}off})$. Therefore,

$$kTV_{Cs} = (1-k)T(V_{Co} - V_{C1} - V_{C2} - V_{L1\text{-}off})$$

Hence, the voltage transfer gain of the re-lift circuit is

$$M_R = \frac{V_o}{V_{in}} = \frac{2}{1-k} \tag{9.212}$$

9.8.3 MULTIPLE-LIFT SEPICs

It is possible to construct multiple-lift circuit by adding the components (L_2-D_2-C_2-S_1). Assuming that there are n VL cells, the generalized representation of multiple-lift circuits is shown in Figure 9.56.

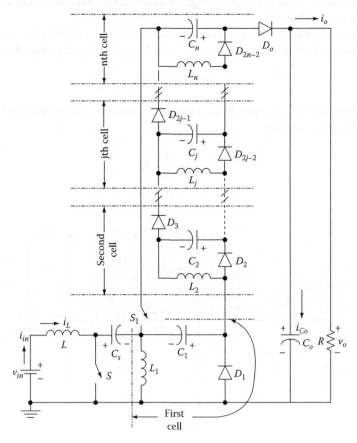

FIGURE 9.56 Multilift SEPIC.

All future active switches can be replaced by passive diodes. According to this principle, only two synchronous switches S and S_1 are required for each complex multiple-lift circuits, which simplifies the control scheme and decreases the cost significantly. Hence, each circuit has two switches, $(n+1)$ inductors, $(n+1)$ capacitors, and $(2n-1)$ diodes. It is noted that all inductors existing in the VL cells are the same here because of the reasons explained for the re-lift circuit. All the capacitors are sufficiently large. From the foregoing analysis and calculation, the general formulas for all multiple-lift circuits can be obtained using similar steps. The generalized voltage transfer gain is

$$M = \frac{n}{1-k} \quad n = 1,2,3,4,... \tag{9.213}$$

If the generalized circuit possesses three VL cells, it is termed the triple-lift circuit. If the generalized circuit possesses four VL cells, it is termed the quadruple-lift circuit.

9.8.4 SIMULATION AND EXPERIMENTAL RESULTS OF A RE-LIFT SEPIC

The circuit parameters for simulation are as follows: $V_{in}=10\,\text{V}$, $R=100\,\Omega$, $L=1\,\text{mH}$, $L_1=L_2=500\,\mu\text{H}$, $C_s=110\,\mu\text{F}$, $C_1=C_2=22\,\mu\text{F}$, $C_o=110\,\mu\text{F}$, and $k=0.6$. The switching frequency f is $100\,\text{kHz}$. According to (9.212), we obtain a theoretical value V_o, which is equal to $50\,\text{V}$. The simulation results in Psim are shown in Figure 9.57, where curves 1–3 are for v_o, i_{L2}, and i_{L1}, respectively. The steady-state performance in the simulation is identical to that given in theoretical analysis.

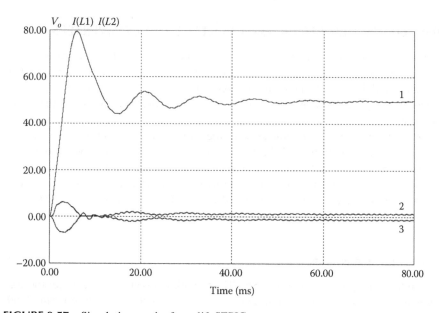

FIGURE 9.57   Simulation result of a re-lift SEPIC.

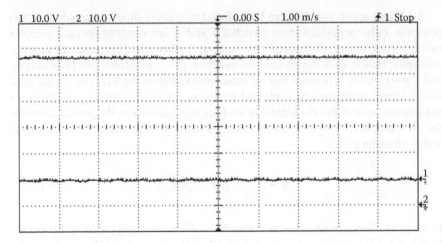

FIGURE 9.58 Experimental result of a re-lift SEPIC.

The same parameters are chosen to construct a testing hardware circuit. Two n-channel MOSFETs 2SK2267 are selected. We obtained the output voltage value of $V_o = 46.2\,\text{V}$ (shown in Channel 1 with 10 V/Div) and capacitor value of $V_{Cs} = 9.9\,\text{V}$ (shown in Channel 1 with 10 V/Div). The corresponding experimental curves in the steady-state are shown in Figure 9.58. The practical output voltage is smaller than the theoretical values because of the effects caused by parasitic parameters. It is seen that the measured results are very close to the theoretical analysis and simulation results.

9.9 OTHER DOUBLE-OUTPUT VOLTAGE-LIFT CONVERTERS

For all converters mentioned previously, each topology is divided into two sections, one is the source section including the voltage source, the inductor L, and the active switch S, and the other is the pump section consisting of the rest of the components. Each topology can be considered as a special, cascaded connection of these two sections.

We compare the SEPIC converter to the Cúk converter; both converters have the same source sections and the voltage transfer gains with opposite polarities. Hence, a series of novel, double-output converters based on the SEPIC and Cúk converters can be constructed by combining two converters at the input side. They are as follows: elementary circuit, self-lift circuit, and corresponding enhanced series [18,19].

9.9.1 ELEMENTARY CIRCUIT

Combining the prototypes of the SEPIC and Cúk converters, we can get the elementary circuit of novel double-output converters, which is shown in Figure 9.59.

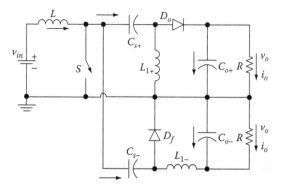

FIGURE 9.59 The novel elementary double-output converter.

The positive conversion path is the same as that of the SEPIC converter. The negative conversion path is the same as that of the Cúk converter. Hence, from the foregoing analysis and calculation, the voltage transfer gains of are obtained as follows:

$$
\begin{cases}
M_{E+} = \dfrac{V_{O+}}{V_{in}} = \dfrac{k}{1-k} \\[3mm]
M_{E-} = \dfrac{V_{O-}}{V_{in}} = -\dfrac{k}{1-k}
\end{cases}
\tag{9.214}
$$

9.9.2 SELF-LIFT DOUBLE-OUTPUT CIRCUIT

The self-lift circuit is a derivative of the elementary circuit that is shown in Figure 9.60.

The positive conversion path is the same as that of the self-lift SEPIC converter. The negative conversion path is the same as that of the self-lift Cúk converter.

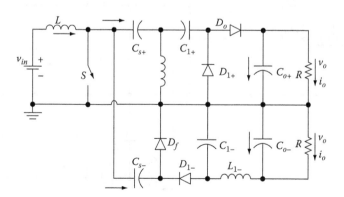

FIGURE 9.60 The novel self-lift double-output converter.

Hence, from the foregoing analysis and calculation, the voltage transfer gains are obtained as follows:

$$\begin{cases} M_{S+} = \dfrac{V_{O+}}{V_{in}} = \dfrac{1}{1-k} \\ M_{S-} = \dfrac{V_{O-}}{V_{in}} = -\dfrac{1}{1-k} \end{cases}$$ (9.215)

9.9.3 Enhanced Series Double-Output Circuits

Since the positive and negative conversion paths share a common source section that can be regarded as a boost converter circuit, we can construct the corresponding enhanced series using the VL technique. A series of novel boost circuits are added to the source section, which transfers much more energy to C_{s+} and C_{s-} in each cycle and increases V_{Cs+} and V_{Cs-} stage by stage along the geometric progression.

As shown in Figure 9.61, the source section is redesigned by adding the components (L_{s1}-D_{s1}-D_{s2}-C_{s1}), which form a basic VL cell and is expressed by *boost*[1]. The newly derived topology provides a single boost circuit enhancement using the supplementary components. When switches S turn on, D_{s2} is on, and D_{s1} is off. When S turn off, D_{s2} is off, and D_{s1} is on. The capacitor C_{s1} performs characteristics to lift the source voltage V_{in}. The energy is transferred to C_{s+} and C_{s-} in each cycle from C_{s1} and increase V_{Cs+} and V_{Cs-}. Thus, we get

$$\begin{cases} V_{Cs+} = V_{Cs1} = \dfrac{1}{1-k} V_{in} \\ V_{Cs-} = \dfrac{1}{1-k} V_{Cs1} = \dfrac{1}{(1-k)^2} V_{in} \end{cases}$$ (9.216)

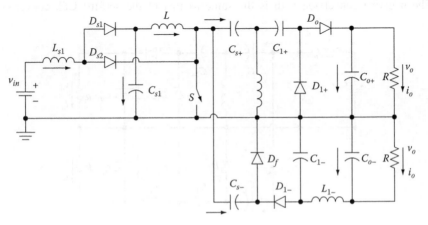

FIGURE 9.61 Enhanced double-output self-lift DC/DC converter (single boost circuit enhancement).

Therefore, from the foregoing analysis and calculation, it is apparent that the voltage transfer gains of this enhanced double-output self-lift DC/DC converters are as follows:

$$\begin{cases} M_{boost^1-S+} = \dfrac{V_{O+}}{V_{in}} = \dfrac{1}{(1-k)^2} \\[3mm] M_{boost^1-S-} = \dfrac{V_{O-}}{V_{in}} = -\dfrac{1}{(1-k)^2} \end{cases} \tag{9.217}$$

Referring to Figure 9.61, it is possible to realize the multiple boost circuits enhancement in the source section by repeating the components $(L_{s1}\text{-}D_{s1}\text{-}D_{s2}\text{-}C_{s1})$ stage by stage. Assuming that there are n VL cells (denoted by $boost^M$), the generalized representation of enhanced series for the double-output self-lift DC/DC converter is shown in Figure 9.62. All circuits share the same power switch S, which simplifies the control scheme and decreases the cost significantly. Hence, each circuit has one switch, $(n+3)$ inductors, $(n+5)$ capacitors, and $(2n+4)$ diodes. It is noted that all inductors existing in the VL cells are the same here because of the reasons explained in the foregoing sections. All the capacitors are sufficiently large. The energy is transferred to C_{s+} and C_{s-} in each cycle from C_{sn} and increases V_{Cs+} and V_{Cs-}. Thus, we get

$$\begin{cases} V_{Cs+} = V_{Csn} = \dfrac{1}{(1-k)^n} V_{in} \\[3mm] V_{Cs-} = \dfrac{1}{1-k} V_{Csn} = \dfrac{1}{(1-k)^{n+1}} V_{in} \end{cases} \tag{9.218}$$

Therefore, from the foregoing analysis and calculation, the general voltage transfer gains of enhanced double-output self-lift DC/DC converters are as follows:

$$\begin{cases} M_{boost^M-S+} = \dfrac{V_{O+}}{V_{in}} = \dfrac{1}{(1-k)^{n+1}} \\[3mm] M_{boost^M-S-} = \dfrac{V_{O-}}{V_{in}} = -\dfrac{1}{(1-k)^{n+1}} \end{cases} \tag{9.219}$$

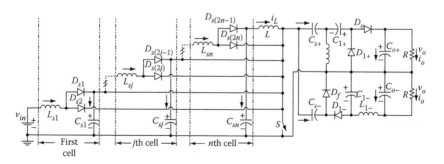

FIGURE 9.62 Generalized representation of enhanced double-output self-lift DC/DC converters (multiple boost circuits enhancement).

Analogically, we also can develop a series of enhanced double-output elementary circuits using the same source section. The general voltage transfer gains of enhanced double-output elementary DC/DC converters are also given here for ready reference.

$$
\begin{cases}
M_{boost^M-E+} = \dfrac{V_{o+}}{V_{in}} = \dfrac{k}{(1-k)^{n+1}} \\[3mm]
M_{boost^M-E-} = \dfrac{V_{o-}}{V_{in}} = -\dfrac{k}{(1-k)^{n+1}}
\end{cases}
\tag{9.220}
$$

9.9.4 SIMULATION AND EXPERIMENTAL VERIFICATION OF AN ENHANCED DOUBLE-OUTPUT SELF-LIFT CIRCUIT

Referring to Figure 9.61, the circuit parameters for simulation are $V_{in}=10$ V, $R=100\ \Omega$, $L_{s1}=L=1\,\mathrm{mH}$, $C_{1+}=C_{1-}=C_{s1}=22\,\mu\mathrm{F}$, $C_{s+}=C_{s-}=110\,\mu\mathrm{F}$, $C_{o+}=C_{o-}=47\,\mu\mathrm{F}$, $C_o=110\,\mu\mathrm{F}$, $k=0.5$, and $f=100\,\mathrm{kHz}$. According to (9.219), we obtain the theoretical values of double-output voltage V_{o+} and V_{o-}, which are equal to 40 V and −40 V, respectively. The simulation results in Psim are shown in Figure 9.63, where curve 1 is for v_{o+} of the positive conversion path and curve 2 is for v_{O-} of the negative conversion path. The steady-state values in the simulation are identical to that given in theoretical analysis.

The same parameters are chosen to construct the testing hardware circuit. Only a single n-channel MOSFET is used in the circuit. The corresponding experimental curves in the steady state are shown in Figure 9.64 respectively. The curve shown in Channel 1 with 20 V/Div corresponds to positive output v_{o+}, which is about 37 V. The curve shown in Channel 2 with 20 V/Div corresponds to the N/O v_{o-}, which is also

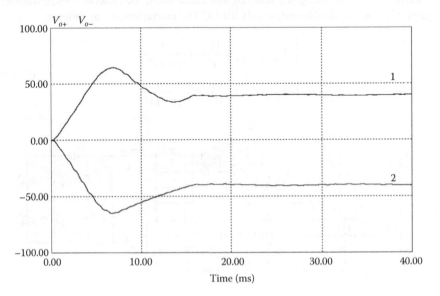

FIGURE 9.63 Simulation result for an enhanced double-output self-lift circuit (single boost circuit enhancement).

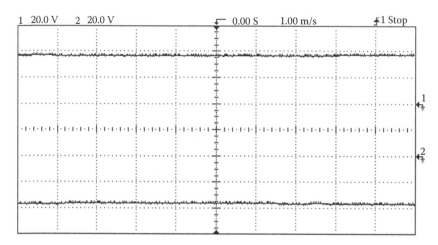

FIGURE 9.64 Experimental result for an enhanced double-output self-lift circuit (single boost circuit enhancement).

about 37 V. Considering the effects caused by the parasitic parameters, we can see that the measured results are very close to the theoretical analysis and simulation results.

9.10 SWITCHED-CAPACITORIZED CONVERTERS

Switched-capacitorized capacitor is an updated component used in power electronics. Switched-capacitorized capacitors can be used to construct new types of DC/DC converters, which are called the switched-capacitorized capacitor DC/DC converters. Switched-capacitorized capacitors can be integrated into a power IC chip. Taking this manufacturing technology, we gain advantages of small size and low power losses. Consequently, switched capacitor DC/DC converters have small size, high power density, high power transfer efficiency, and high voltage transfer gain [22–27].

DC/DC converters are supplied by a DC voltage source. The input source current can be continuous and discontinuous. With some converters such as buck converter and buck-boost converters, the input current is discontinuous. We call them working in the discontinuous input current mode (DICM). With other converters such as boost converter, the input current is continuous. We call them working in the continuous input current mode (CICM). Switched-capacitorized capacitor can be used in VL technique to construct DC/DC converters. The clue is that for the converters operating in DICM, the switched capacitors can be charged to the source voltage, with the energy stored during the input current discontinuous period (the main switch is off). They join the conversion operation when the main switch is on, and their stored energy will be delivered through the DICM converters to the load. These converters are called the switched-capacitorized (SC) DC/DC converters.

It is easy to construct switched-capacitorized DC/DC converters. Depending on how many switched-capacitors to be used, we call them one-stage SC converters, two-stage SC converters, three-stage SC converters, and n-stage SC converters in general. The corresponding circuits are shown in Figures 9.65 through 9.67.

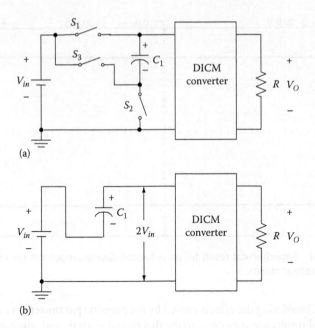

(a)

(b)

FIGURE 9.65   One-stage SC converter: (a) circuit diagram and (b) equivalent circuit when the main switch is on.

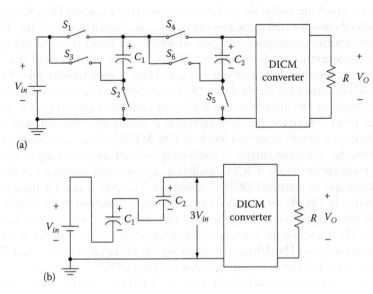

(a)

(b)

FIGURE 9.66   Two-stage SC converter: (a) circuit diagram and (b) equivalent circuit when the main switch is on.

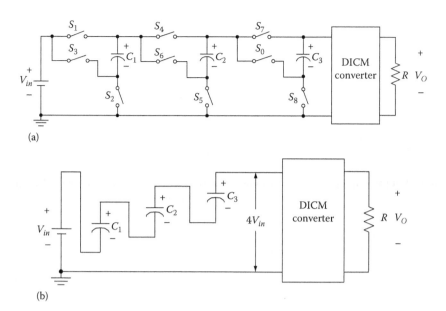

FIGURE 9.67 Three-stage SC converter: (a) two-stage SC converter and (b) equivalent circuit when the main switch is on.

The one-stage SC converter circuit is shown in Figure 9.65a. The input source voltage is V_{in}, and the output voltage is V_O. To simplify the description, we assume the load is resistive load R. The auxiliary switches S_1 and S_2 are switched on (the auxiliary switch S_3 is off) during the switch-off period. The switched capacitor C_1 is changed to the source voltage V_{in}. The auxiliary switches S_1 and S_2 are switched off, and the auxiliary switch S_3 is on during the switch-on period. The equivalent circuit is shown in Figure 9.65b. Therefore, the equivalent input voltage supplied to the ICDM converter is $2V_{in}$ [28–32]. In other words, the equivalent input voltage has been lifted by using the switched capacitor.

Analogously, the circuit diagram of the two-stage SC converter is shown in Figure 9.66a, and the corresponding equivalent circuit during the switch-on period is shown in Figure 9.66b. It supplies $3V_{in}$ to the DICM converter. The equivalent input voltage is lifted twice that of the supplying voltage V_{in}.

The circuit diagram of the three-stage SC converter is shown in Figure 9.67a, and the corresponding equivalent circuit when main switch is on as shown in Figure 9.67b. It supplies $4V_{in}$ to the DICM converter. The equivalent input voltage is lifted three times that of the supplying voltage V_{in}.

Several circuits will be introduced in this chapter:

- Switched-capacitorized buck converter
- Switched-capacitorized buck-boost converter
- Switched-capacitorized P/O Luo converter
- Switched-capacitorized N/O Luo converter

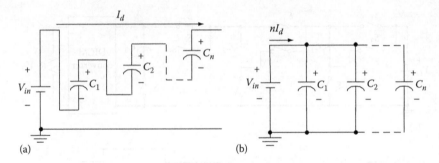

FIGURE 9.68 Discharging and charging currents of switched-capacitors: (a) discharging current when the main switch is on and (b) charging currents during switch-off.

We assume the stage number is n and the voltage transfer gain of the DICM converter is M. In the ideal condition, the output voltage is

$$V_O = (n+1)MV_{in} \qquad (9.221)$$

The ideal condition means that the voltage drop across all switches and diodes is zero, and there is no drop in voltage across the all SCs when the main switch is off. This assumption is reasonable for the investigation. We will discuss the nonideal condition operation in Section 9.10.5 [33–38].

Another advantage is that the input current becoming continuous. The input current of the original DICM converter is zero when the main switch is off. For example, the input current of the one-stage SC DC/DC converter flows through the auxiliary switches S_1 and S_2 to charge capacitor C_1 when the main switch is off. For the n-stage SC DC/DC converter, each switched capacitor is discharged by the discharging current I_d as shown in Figure 9.68a. The charging current of each switched capacitor should be I_d in the switch-off period since the average current of each switched capacitor is zero in the steady state. Therefore, the source input average current should be

$$I_{in} = (n+1)I_d \qquad (9.222)$$

9.10.1 ONE-STAGE SWITCHED-CAPACITORIZED BUCK CONVERTER

The one-stage SC buck converter is shown in Figure 9.69. The main switch S and the auxiliary switch S_3 are on and off simultaneously. The auxiliary switches S_1 and S_2 are off and on exclusively.

9.10.1.1 Operation Analysis

We assume the converter works in the steady state and that the switched-capacitor C_1 is fully charged. The main switch S is on during the switch-on period, and the auxiliary switch S_3 is on simultaneously. The voltage V_1 is about $2V_{in}$ when the main switch S is on. This is an equivalent input voltage $2V_{in}$ to supply to the buck converter.

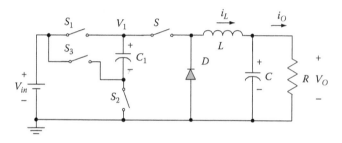

FIGURE 9.69 One-stage SC buck converter.

Refer to the buck converter voltage transfer gain $M=k$; thus, we can easily obtain the output voltage

$$V_O = 2kV_{in} \tag{9.223}$$

Using this technique, we can obtain an output voltage that is higher than the input voltage if the conduction duty cycle k is greater than 0.5. The output voltage of the original buck converter is always lower than the input voltage.

9.10.1.2 Simulation and Experimental Results

In order to verify the design and the analysis, the simulation result is shown in Figure 9.70. The simulation condition is that $V_{in}=20\,\text{V}$, $L=10\,\text{mH}$, $C=C_1=20\,\mu\text{F}$,

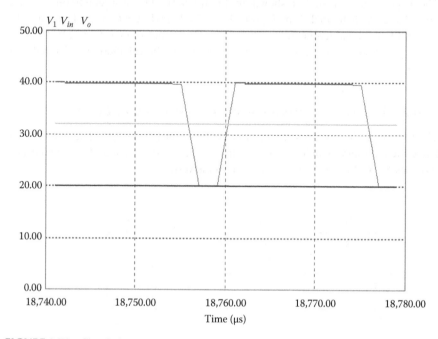

FIGURE 9.70 Simulation result.

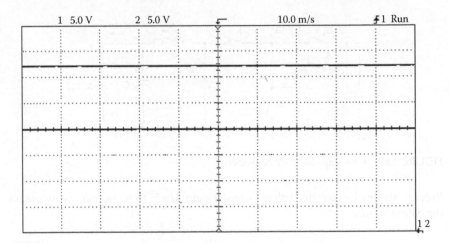

FIGURE 9.71 Experimental result.

$f=50\,\text{kHz}$, $R=100\,\Omega$, and conduction duty cycle $k=0.8$. The voltage on the top-end of the switched capacitor C_1 varies from 20 to 40 V. The output voltage $V_O=32\,\text{V}$, which is same as the calculation result.

$$V_O = 2kV_{in} = 2\times0.8\times20 = 32\,\text{V} \tag{9.224}$$

The experimental result is shown in Figure 9.71. The test condition is same that $V_{in}=20\,\text{V}$ (Channel 1 in Figure 9.71), $L=10\,\text{mH}$, $C=C_1=20\,\mu\text{F}$, $f=50\,\text{kHz}$, $R=100\,\Omega$, and conduction duty cycle $k=0.8$. The output voltage $V_O=32\,\text{V}$ (Channel 2 in Figure 9.71) is similar to that obtained by using the calculation and simulation results.

9.10.2 TWO-STAGE SWITCHED-CAPACITORIZED BUCK-BOOST CONVERTER

The two-stage SC buck-boost converter is shown in Figure 9.72. The main switch S and the auxiliary switches S_3 and S_6 are on and off simultaneously. The auxiliary switches S_1, S_2, S_4, and S_5 are off and on exclusively.

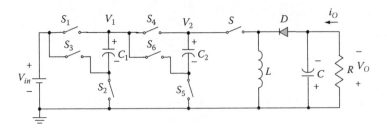

FIGURE 9.72 Two-stage SC buck-boost converter.

9.10.2.1 Operation Analysis

We assume the converter works in the steady state and that the switched capacitors C_1 and C_2 are fully charged. The main switch S is on during the switch-on period, and the auxiliary switches S_3 and S_6 are on simultaneously. The voltage V_1 is about $2V_{in}$ and the voltage V_2 is about $3V_{in}$ when the main switch S is on. This is equivalent input voltage $3V_{in}$ to supply to the buck-boost converter. Refer to the buck-boost converter voltage transfer gain $M = -k/(1-k)$; we can easily obtain the output voltage as follows

$$V_O = -\frac{3k}{1-k}V_{in} \qquad (9.225)$$

Using this technique, we can easily obtain a higher output voltage. For example, if $k=0.5$, the output voltage of the original buck-boost converter is equal to the input source voltage V_{in}. The output voltage of the two-stage SC buck-boost converter is equal to six times that of the source voltage.

9.10.2.2 Simulation and Experimental Results

In order to verify the design, the simulation result is shown in Figure 9.73. The simulation condition is that $V_{in}=20\,V$, $L=10\,mH$, $C=C_1=C_2=20\,\mu F$, $f=50\,kHz$, $R=200\,\Omega$,

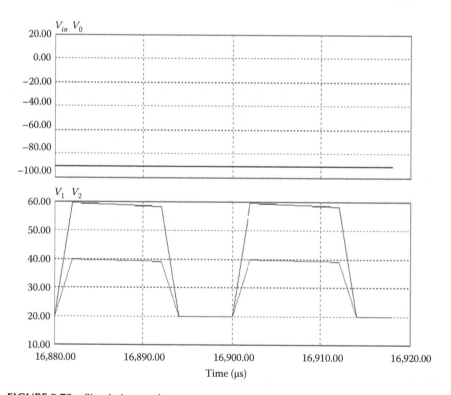

FIGURE 9.73 Simulation result.

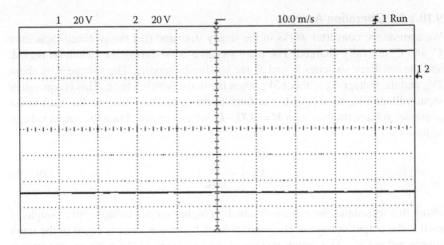

FIGURE 9.74 Experimental result.

and conduction duty cycle $k=0.6$. The voltage on the top end of the switched capacitor C_1 varies from 20 to 40 V. The voltage on the top end of the switched capacitor C_2 varies from 20 to 60 V. The output voltage $V_O=-90$ V, which is same as the calculation result.

$$V_O = -\frac{3k}{1-k}V_{in} = -\frac{3\times0.6}{1-0.6}\times20 = -90\,\text{V} \qquad (9.226)$$

The experimental result is shown in Figure 9.74. The test condition is that $V_{in}=20$ V (Channel 1 in Figure 9.74), $L=10$ mH, $C=C_1=C_2=20\,\mu\text{F}$, $f=50$ kHz, $R=200\,\Omega$, and conduction duty cycle $k=0.6$. The output voltage is $V_O=-90$ V (Channel 2 in Figure 9.74), which is same as the simulation and calculation results.

9.10.3 THREE-STAGE SWITCHED-CAPACITORIZED P/O LUO CONVERTER

The three-stage SC P/O Luo converter is shown in Figure 9.75. The main switch S and the auxiliary switches S_3, S_6, and S_9 are on and off simultaneously. The auxiliary switches S_1, S_2, S_4, S_5, S_7, and S_8 are off and on exclusively.

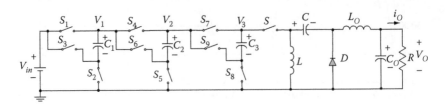

FIGURE 9.75 Three-stage SC P/O Luo converter.

9.10.3.1 Operation Analysis

We assume the converter works in the steady state and that the switched capacitors C_1, C_2, and C_3 are fully charged. The main switch S is on during the switch-on period, and the auxiliary switches S_3, S_6, and S_9 are on simultaneously. The voltage V_1 is about $2V_{in}$, the voltage V_2 is about $3V_{in}$, and the voltage V_3 is about $4V_{in}$ when the main switch S is on. This is equivalent input voltage $4V_{in}$ to supply to the P/O Luo converter. Refer to the P/O Luo converter voltage transfer gain $M = k/(1-k)$; we can easily obtain the output voltage as

$$V_O = \frac{4k}{1-k} V_{in} \tag{9.227}$$

9.10.3.2 Simulation and Experimental Results

In order to verify the design, the simulation result is shown in Figure 9.76. The simulation condition is $V_{in} = 20\,\text{V}$, $L = L_O = 10\,\text{mH}$, $C = C_1 = C_2 = C_3 = 20\,\mu\text{F}$, $f = 50\,\text{kHz}$, $R = 400\,\Omega$, and conduction duty cycle $k = 0.6$. The voltage on the top end of the switched capacitor C_1 varies from 20 to 40 V. The voltage on the top end of the switched capacitor C_2 varies from 20 to 60 V. The voltage on the top end of the switched capacitor

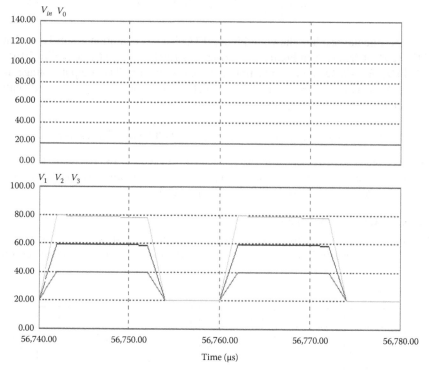

FIGURE 9.76 Simulation result.

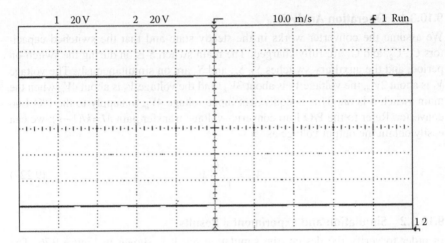

FIGURE 9.77 Experimental result.

C_3 varies from 20 to 80 V. The output voltage $V_O = 120$ V is the same as the one shown by calculation result.

$$V_O = \frac{4k}{1-k}V_{in} = \frac{4 \times 0.6}{1-0.6} \times 20 = 120 \text{ V} \tag{9.228}$$

The experimental result is shown in Figure 9.77. The test condition is similar to that of $V_{in} = 20$ V (Channel 1 in Figure 9.77), $L = L_O = 10$ mH, $C = C_O = C_1 = C_2 = C_3 = 20 \mu$F, $f = 50$ kHz, $R = 400 \Omega$, and conduction duty cycle $k = 0.6$. The output voltage is $V_O = 120$ V (Channel 2 in Figure 9.77), which is the same as the observed with simulation and calculation results.

9.10.4 THREE-STAGE SWITCHED-CAPACITORIZED N/O LUO CONVERTER

The three-stage SC N/O Luo converter is shown in Figure 9.78. The main switch S and the auxiliary switches S_3, S_6, and S_9 are on and off simultaneously. The auxiliary switches S_1, S_2, S_4, S_5 S_7, and S_8 are off and on exclusively.

9.10.4.1 Operation Analysis

We assume the converter works in the steady state and that the switched-capacitors C_1, C_2, and C_3 are fully charged. The main switch S is on during the switch-on

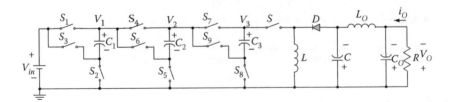

FIGURE 9.78 Three-stage SC N/O Luo converter.

period, and the auxiliary switches S_3, S_6, and S_9 are on simultaneously. The voltage V_1 is about $2V_{in}$, the voltage V_2 is about $3V_{in}$, and the voltage V_3 is about $4V_{in}$ when the main switch S is on. This is equivalent input voltage $4V_{in}$ to supply to the N/O Luo converter. Refer to the N/O Luo converter voltage transfer gain $M = -k/(1-k)$; we can easily obtain the output voltage as

$$V_O = -\frac{4k}{1-k}V_{in} \tag{9.229}$$

9.10.4.2 Simulation and Experimental Results

In order to verify the design, the simulation result is shown in Figure 9.79. The simulation condition is that $V_{in} = 20\,\text{V}$, $L = L_O = 10\,\text{mH}$, $C = C_1 = C_2 = C_3 = 20\,\mu\text{F}$, $f = 50\,\text{kHz}$, $R = 400\,\Omega$, and conduction duty cycle $k = 0.6$. The voltage on the top end of the switched capacitor C_1 varies from 20 to 40 V. The voltage on the top end of the switched capacitor C_2 varies from 20 to 60 V. The voltage on the top end of the switched capacitor C_3 varies from 20 to 80 V. The output voltage $V_O = -120\,\text{V}$, which is the same as that observed with calculation results.

$$V_O = -\frac{4k}{1-k}V_{in} = -\frac{4 \times 0.6}{1-0.6} \times 20 = -120\,\text{V} \tag{9.230}$$

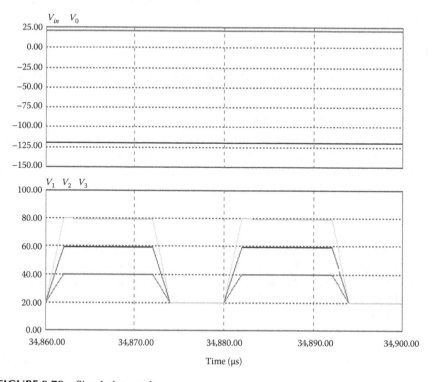

FIGURE 9.79 Simulation result.

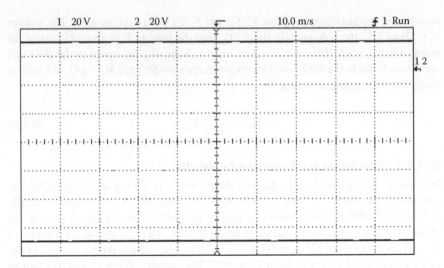

FIGURE 9.80 Experimental result.

The experimental result is shown in Figure 9.80. The test condition is identical: $V_{in}=20\,\text{V}$ (Channel 1 in Figure 9.80), $L=L_O=10\,\text{mH}$, $C=C_O=C_1=C_2=C_3=20\,\mu\text{F}$, $f=50\,\text{kHz}$, $R=400\,\Omega$, and conduction duty cycle $k=0.6$. The output voltage $V_O=120\,\text{V}$ (Channel 2 in Figure 9.80), which is the same as the one observed in simulation and calculation results.

9.10.5 Discussion

There are several factors of this technique that have to be discussed for the converter design consideration and industrial applications.

9.10.5.1 Voltage Drop across the Switched Capacitors

Refer to the waveform in Figures 9.72, 9.75, and 9.78, we can obviously see the voltage drop across the switched-capacitors. For an n-stage SC converter, there are n switched-capacitors to be used. In an ideal condition, the total voltage across all switched-capacitors should be

$$V_n = nV_{in} \tag{9.231}$$

If all switched-capacitors have same capacitance C, the equivalent capacitance in the switch-on period is C/n. The discharging current during the switch-on period can be assumed a constant value I_d, the conduction duty cycle is k, the switching frequency is f and the switch-on period is $kT=k/f$. We can calculate the voltage drop of the last switched capacitor as

$$\Delta V_n = \frac{1}{C/n} \int_0^{kT} i_d dt = \frac{nkT}{C} I_d \tag{9.232}$$

Thus, the average current flowing through switched-capacitors in a period T is zero in the steady state. The average input current from the source is $I_{in} = (n+1)I_d$. The current I_d is the input current to the DICM converter. If there are no energy losses inside the DICM converter, we can obtain it as

$$I_{in}V_{in} = (n+1)I_dV_{in} = V_OI_O = \frac{V_o^2}{R}$$ (9.233)

Considering (9.221), we have

$$I_d = \frac{V_O}{(n+1)V_{in}}I_O = MI_O = M\frac{V_O}{R}$$ (9.234)

$$\Delta V_n = \frac{nkT}{C}I_d = \frac{nk}{fC}MI_O = \frac{nkM}{fC}\frac{V_O}{R}$$ (9.235)

From Equation 9.235, we can see that the voltage drop is linearly proportional to the stages n, the duty cycle k, and the output voltage V_O. It is reversely proportional to the switching frequency f, the capacitance C of the used switched-capacitors, and the load R. In order to reduce the voltage drop, we can carry out the following steps:

- Increasing switching frequency f
- Increasing the capacitance C
- Increasing the load R
- Decreasing the duty cycle k

Correspondingly, the voltage drops across each switched capacitor is

$$\Delta V_{each} = \frac{\Delta V_n}{n} = \frac{k}{fC}I_d = \frac{kM}{fC}\frac{V_O}{R}$$ (9.236)

9.10.5.2 Necessity of the Voltage Drop across the Switched Capacitors and Energy Transfer

The voltage drops across the switched-capacitors are necessary for the energy transfer from the source to the DICM converter. The switched-capacitors absorb the energy from supply source during the switch-off period and release the stored energy to the DICM converter during the switch-on period. In the steady state, the energy transferred by the switched-capacitors in a period T is

$$\Delta E = \frac{1}{2}\frac{C}{n}[V_n^2 - (V_n - \Delta V_n)^2] = \frac{C}{2n}(2V_n\Delta V_n - \Delta V_n^2) = \frac{C}{2n}(2V_n - \Delta V_n)\Delta V_n$$ (9.237)

Considering the $2V_n \gg \Delta V_n$, Equation 9.237 can be rewritten as

$$\Delta E \approx \frac{C}{n} V_n \Delta V_n \qquad (9.238)$$

Substituting (9.231) and (9.235) into (9.238), the total power transferred by the switched-capacitors is

$$P = f\Delta E = \frac{fC}{n} V_n \Delta V_n = \frac{fC}{n} (nV_{in})\left(\frac{nkM}{fC} I_O\right) = nkMV_{in}I_O \qquad (9.239)$$

If we would like to obtain the power transferred to the DICM converter to be high, increasing the switching frequency f and capacitance C does not help. From (9.239), the following methods are helpful:

- Increasing the duty cycle k
- Increasing the stage number n
- Increasing the transfer gain M

9.10.5.3 Inrush Input Current

Inrush input current is large for all SC DC/DC converters since the charging current to the switched capacitors is high during the main switch-off period. For example, we can show the simulation result of the inrush input current of a three-stage SC P/O Luo converter in Figure 9.81.

The load current is very small $I = 120/400 = 0.3\,A$, but the peak value of the input inrush current is about 27.3 A. Another phenomenon is that the input inrush current does not usually fully occupy the switch-off period. We will discuss how to overcome this phenomenon in Section 9.8.5.5.

9.10.5.4 Power Switch-on Process

Surge input current is large for all switched-capacitorized DC/DC converters during power switch-on process, since all switched-capacitors were not precharged. For example, we can show the simulation result of power-on surge input current of a three-stage SC P/O Luo converter in Figure 9.82.

The peak value of the power-on surge input current is very high, at about 262 A.

9.10.5.5 Suppression of the Inrush and Surge Input Current

From Figures 9.81 and 9.82, we can see that the peak inrush input current can be 90 times that of the normal load current, and the peak power-on surge input current can be about 880 times that of the normal load current. This is a big trouble for industrial applications of the SC DC/DC converters. In order to suppress the large inrush input current and the peak power-on surge input current, we set a small resistor (the so-called suppression resistor R_S) in series to each switched-capacitor. The circuit of such three-stage SC P/O Luo converter is shown in Figure 9.83.

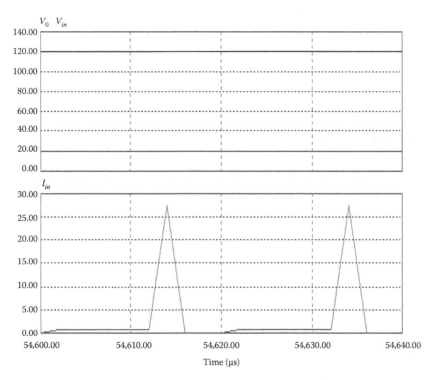

FIGURE 9.81   Simulation result (inrush input current).

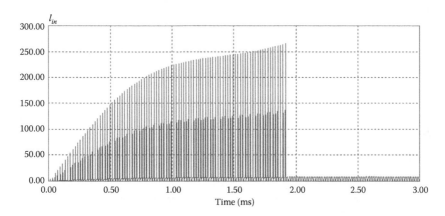

FIGURE 9.82   Simulation result (power-on surge input current).

The resistance R_S is designed to have the time constant of the RC circuit to complete to the switch-off period.

$$R_S = \frac{1-k}{C}T = \frac{1-k}{fC} \qquad (9.240)$$

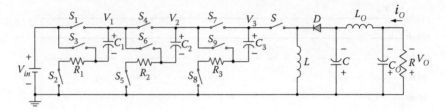

FIGURE 9.83   Improved three-stage SC P/O Luo converter.

As the data used in the previous section $f=50\,\text{kHz}$, all $C=20\,\mu\text{F}$, and conduction duty cycle $k=0.6$. We can choose $R_1=R_2=R_3=0.4\,\Omega$. The inrush input current and the load current are shown in Figure 9.84.

In comparison to the Figure 9.81, we can see that the peak inrush input current is largely reduced to 4.8 A and the input current becomes continuous during the switch-off period.

The power-on surge input current waveform is shown in Figure 9.85. The peak power-on surge input current is about 138 A, which is largely reduced.

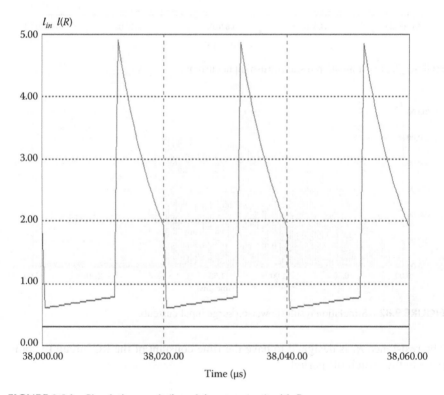

FIGURE 9.84   Simulation result (inrush input current) with R_S.

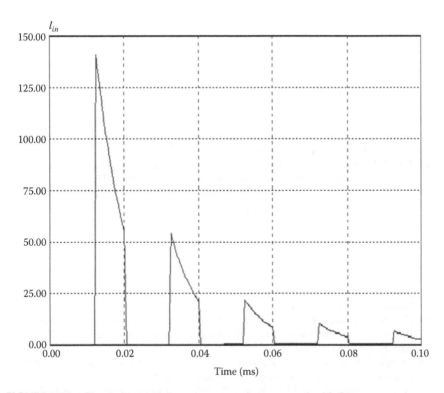

FIGURE 9.85 Simulation result (power-on surge input current) with R_S.

REFERENCES

1. Luo, F. L. and Ye, H. 2004. *Advanced DC/DC Converters*. Boca Raton, FL: CRC Press.
2. Luo, F. L. and Ye, H. 2006. *Essential DC/DC Converters*. Boca Raton, FL: Taylor & Francis Group LLC.
3. Luo, F. L. 2001. Seven self-lift DC/DC converters: Voltage lift technique. *IEE Proceedings on EPA*, 148(4), 329–338.
4. Luo, F. L. 2001. Six self-lift DC/DC converters: Voltage lift technique. *IEEE Transactions on Industrial Electronics*, 48(6), 1268–1272.
5. Luo, F. L. and Chen X. F. 1998. Self-lift DC-DC converters. *Proceedings of the Second IEEE International Conference PEDES'98*, Perth, Western Australia, Australia, pp. 441–446.
6. Luo, F. L. 1999. Positive output Luo-converters: Voltage lift technique. *IEE EPA Proceedings*, 146(4), 415–432.
7. Luo, F. L. 1998. Re-lift converter: Design, test, simulation and stability analysis. *IEE EPA Proceedings*, 145(4), 315–325.
8. Luo, F. L. 1997. Re-lift circuit: A new DC-DC step-up (boost) converter. *IEE Electronics Letters*, 33(1), 5–7.
9. Luo, F. L. 1998. Luo-converters—Voltage lift technique. *Proceedings of the IEEE Power Electronics Special Conference IEEE-PESC'98*, Fukuoka, Japan, pp. 1783–1789.

10. Luo, F. L. 1997. Luo-converters, a series of new DC-DC step-up (boost) conversion circuits. *Proceedings of the IEEE International Conference on Power Electronics & Drive Systems, 1997 (PEDS'97)*, Singapore, pp. 882–888.

11. Massey, R. P. and Snyder, E. C. 1977. High voltage single ended DC-DC converter. *Record of IEEE PESC*, Palo Alto, CA, pp. 156–159.

12. Luo, F. L. 1999. Negative output Luo-converters: Voltage lift technique. *IEE EPA Proceedings*, 146(2), 208–224.

13. Luo, F. L. 1998. Negative output Luo-converters, implementing the voltage lift technique. *Proceedings of the Second World Energy System International Conference WES'98*, Toronto, Ontario, Canada, pp. 253–260.

14. Luo, F. L. and Ye, H. 1999. Modified positive output Luo converters. *Proceedings of the IEEE International Conference PEDS'99*, Hong Kong, pp. 450–455.

15. Luo, F. L. 2000. Double output Luo-converters: Advanced voltage lift technique. *Proceedings of IEE-EPA*, 147(6), 469–485.

16. Luo, F. L. 1999. Double output Luo-converters. *Proceedings of the International Conference IEE-IPEC'99*, Singapore, pp. 647–652.

17. Cuk, S. and Middlebrook, R. D. 1977. A new optimum topology switching dc-to-dc converter. *Proceedings of IEEE PESC*, 160–179.

18. Zhu, M. and Luo, F. L. 2007. Implementing of developed voltage lift technique on SEPIC, Cúk and double-output DC/DC converters. *Proceedings of IEEE-ICIEA'2007*, Harbin, China, pp. 674–681.

19. Zhu, M. and Luo, F. L. 2007. Implementing of development of voltage lift technique on double-output transformerless DC-DC converters. *Proceedings of IECON'2007*, Taipei, Taiwan, pp. 1983–1988.

20. Jozwik, J. J. and Kazimerczuk, M. K. 1989. Dual sepic PWM switching-mode dc/dc power converter. *IEEE Transactions on Industrial Electronics*, 36(1), 64–70.

21. Adar, D., Rahav, G., and Ben-Yaakov, S. 1996. Behavioural average model of SEPIC converters with coupled inductors. *IEE Electronics Letters*, 32(17), 1525–1526.

22. Luo, F. L. 2009. Switched-capacitorized DC-DC converters. *Proceedings of IEEE-ICIEA'2009*, Xi'an, China, pp. 385–389.

23. Luo, F. L. 2009. Investigation of switched-capacitorized DC-DC converters. *Proceedings of IEEE-IPEMC'2009*, Wuhan, China, pp. 1283–1288.

24. Luo, F. L. and Ye, H. 2004. Positive output multiple-lift push-pull switched-capacitor Luo-converters. *IEEE Transactions on Industrial Electronics*, 51(3), 594–602.

25. Gao, Y. and Luo, F. L. 2001. Theoretical analysis on performance of a 5V/12V push-pull switched capacitor DC/DC converter. *Proceedings of the International Conference IPEC'2001*, Singapore, pp. 711–715.

26. Luo, F. L. and Ye, H. 2003. Negative output multiple-lift push-pull switched-capacitor Luo-converters. *Proceedings of IEEE International Conference PESC'2003*, Acapulco, Mexico, pp. 1571–1576.

27. Luo, F. L., Ye, H., and Rashid, M. H. 1999. Switched capacitor four -quadrant Luo-converter. *Proceedings of the IEEE-IAS Annual Meeting*, Phoenix, AZ, pp. 1653–1660.

28. Makowski, M. S. 1997. Realizability conditions and bounds on synthesis of switched capacitor DC-DC voltage multiplier circuits. *IEEE Transactions on CAS*, 44(8), 684–691.

29. Cheong, S. V., Chung, H., and Ioinovici, A. 1994. Inductorless DC-DC converter with high power density. *IEEE Transactions on Industrial Electronics*, 41(2), 208–215.

30. Midgley, D. and Sigger, M. 1974. Switched-capacitors in power control. *IEE Proceedings*, 121, 703–704.

31. Mak, O. C., Wong, Y. C., and Ioinovici, A. 1995. Step-up DC power supply based on a switched-capacitor circuit. *IEEE Transactions on IE*, 42(1), 90–97.

32. Chung, H. S., Hui, S. Y. R., Tang, S. C., and Wu, A. 2000. On the use of current control scheme for switched-capacitor DC/DC converters. *IEEE Transactions on Industrial Electronics*, 47(2), 238–244.

33. Pan, C. T. and Liao, Y. H. 2007. Modeling and coordinate control of circulating currents in parallel three-phase boost rectifiers. *IEEE Transactions on Industrial Electronics*, 55(7), 825–838.

34. Mazumder, S. K., Tahir, M., and Acharya, K. 2008. Master–slave current-sharing control of a parallel DC–DC converter system over an RF communication interface. *IEEE Transactions on Industrial Electronics*, 55, 59–66.

35. Asiminoaei, L., Aeloiza, E., Enjeti, P., and Blaabjerg, F. 2008. Shunt active-power-filter topology based on parallel interleaved inverters. *IEEE Transactions on Industrial Electronics*, 55(3), 1175–1189.

36. Chen, W. and Ruan, X. 2008. Zero-voltage-switching PWM hybrid full-bridge three-level converter with secondary-voltage clamping scheme. *IEEE Transactions on Industrial Electronics*, 55(2), 644–654.

37. Wang, C. M. 2006. New family of zero-current-switching PWM converters using a new zero-current-switching PWM auxiliary circuit. *IEEE Transactions on Industrial Electronics*, 53(3), 768–777.

38. Ye, Z., Jain, P. K., and Sen, P. C. 2007. Circulating current minimization in high-frequency AC power distribution architecture with multiple inverter modules operated in parallel. *IEEE Transactions on Industrial Electronics*, 54(5), 2673–2687.

10 Super-Lift Converters and Ultra-Lift Converters

Voltage lift (VL) technique has been successfully employed in the design of DC/DC converters and effectively enlarges the voltage transfer gains of the VL converters. However, the output voltage increases in arithmetic progression stage by stage. Super-lift (SL) technique is more powerful than VL technique; its voltage transfer gain can be a very large number. SL technique facilitates the output voltage increase, in geometric progression, stage by stage. It effectively enhances the voltage transfer gain in power series [1–6].

10.1 INTRODUCTION

SL technique is the most outstanding contribution in DC/DC conversion technology. Applying this technique, we can design a great number of SL converters. The following series of VL converters are introduced in this chapter:

- Positive-output (P/O) SL Luo converters
- Negative-output (N/O) SL Luo converters
- P/O cascaded boost converters
- N/O cascaded boost converters
- Ultra-lift Luo converter

Each series of converters has several subseries. For example, the P/O SL Luo converters have five subseries: Main series, additional series, enhanced series, re-enhanced series, and multiple (j)-enhanced series.

In order to concentrate the voltage enlargement, assume the converters are working in steady state with continuous conduction mode (CCM). The conduction duty ratio is k, switching frequency is f, switching period is $T = 1/f$, and the load is resistive load R. The input voltage and current are V_{in} and I_{in}, and output voltage and current are V_O and I_O. Assume there are no power losses during the conversion process, $V_{in} \times I_{in} = V_O \times I_O$. The voltage transfer gain is G: $G = V_O/V_{in}$.

10.2 P/O SL LUO CONVERTERS

We introduce only three circuits in each subseries. Once the readers catch the clue, you can design further circuits easily [1–4].

10.2.1 MAIN SERIES

The first three stages of P/O SL Luo converters and the main series are shown in Figures 10.1 through 10.3. For the sake of convenience, they are called elementary circuit, re-lift circuit, and triple-lift circuit, respectively, and are numbered $n = 1, 2,$ and 3.

10.2.1.1 Elementary Circuit

The elementary circuit and its equivalent circuits during switch-on and switch-off are shown in Figure 10.1.

The voltage across capacitor C_1 is charged to V_{in}. The current i_{L1} flowing through inductor L_1 increases with voltage V_{in} during the switch-on period kT and decreases with voltage $-(V_O - 2V_{in})$ during the switch-off period $(1 - k)T$. Therefore, the ripple of the inductor current i_{L1} is

$$\Delta i_{L1} = \frac{V_{in}}{L_1} kT = \frac{V_O - 2V_{in}}{L_1}(1 - k)T \tag{10.1}$$

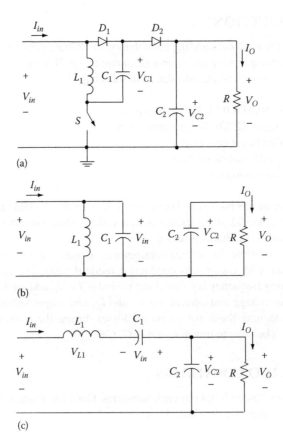

(a)

(b)

(c)

FIGURE 10.1 Elementary circuit of P/O SL Luo converters—main series. (a) Circuit diagram. (b) Equivalent circuit during the switch-on period. (c) Equivalent circuit during the switch-off period.

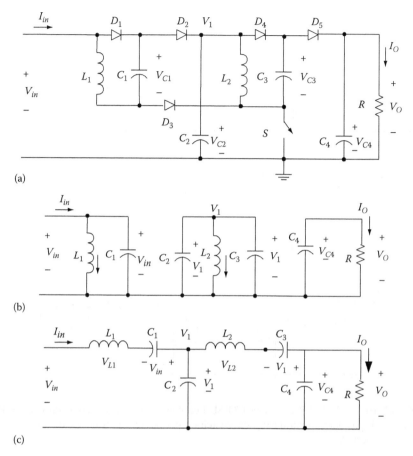

(a)

(b)

(c)

FIGURE 10.2 Re-lift circuit of P/O SL Luo converters—main series. (a) Circuit diagram. (b) Equivalent circuit during the switch-on period. (c) Equivalent circuit during the switch-off period.

$$V_O = \frac{2-k}{1-k} V_{in} \tag{10.2}$$

The voltage transfer gain is

$$G = \frac{V_O}{V_{in}} = \frac{2-k}{1-k} \tag{10.3}$$

The average input current is

$$I_{in} = k i_{in\text{-}on} + (1-k) i_{in\text{-}off} = I_{L1} + (1-k) I_{L1} = (2-k) I_{L1} \tag{10.4}$$

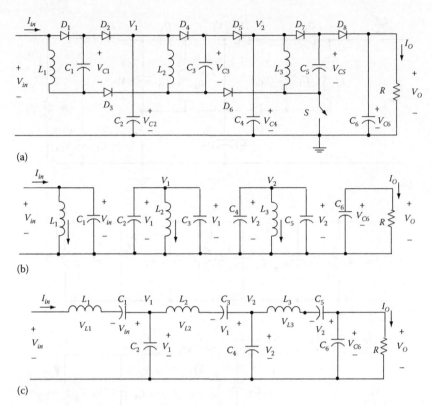

FIGURE 10.3 Triple-lift circuit of P/O SL Luo converters—main series. (a) Circuit diagram. (b) Equivalent circuit during the switch-on period. (c) Equivalent circuit during the switch-off period.

The variation ratio of current i_{L1} through inductor L_1 is

$$\xi_1 = \frac{\Delta i_{L1}/2}{I_{L1}} = \frac{k(2-k)TV_{in}}{2L_1 I_{in}} = \frac{k(1-k)^2}{2(2-k)} \frac{R}{fL_1} \qquad (10.5)$$

Usually ξ_1 is small (much lower than unity), which means this converter normally works in the continuous mode.

The ripple voltage of output voltage v_O is

$$\Delta v_O = \frac{\Delta Q}{C_2} = \frac{I_O kT}{C_2} = \frac{k}{fC_2} \frac{V_O}{R}$$

Therefore, the variation ratio of output voltage v_O is

$$\varepsilon = \frac{\Delta v_O/2}{V_O} = \frac{k}{2RfC_2} \qquad (10.6)$$

Example 10.1 A P/O SL Luo converter in Figure 10.1a has $V_{in}=20\,V$, $L_1=10\,mH$, $C_1=C_2=20\,\mu F$, $R=100\,\Omega$, $f=50\,kHz$, and conduction duty cycle $k=0.6$. Calculate the variation ratio of current i_{L1}, the output voltage, and its variation ratio.

Solution: From formula (10.5), we can get the variation ratio of current i_{L1},

$$\xi_1 = \frac{k(1-k)^2}{2(2-k)}\frac{R}{fL_1} = \frac{0.6(1-0.6)^2}{2(2-0.6)}\frac{100}{50k\times10m} = 0.00686$$

From formula (10.2), we can get the output voltage

$$V_O = \frac{2-k}{1-k}V_{in} = \frac{2-0.6}{1-0.6}20 = 70\,V$$

From (10.6), its variation ratio is

$$\varepsilon = \frac{k}{2RfC_2} = \frac{0.6}{2\times100\times50k\times20\mu} = 0.003$$

10.2.1.2 Re-Lift Circuit

The re-lift circuit is derived from elementary circuit by adding the parts L_2-D_3-D_4-D_5-C_3-C_4. Its circuit diagram and equivalent circuits during the switch-on and switch-off periods are shown in Figure 3.2. The voltage across capacitor C_1 is charged to V_{in}. As described in the previous section, the voltage V_1 across capacitor C_2 is $V_1 = (2-k)/(1-k)V_{in}$.

The voltage across capacitor C_3 is charged to V_1. The current flowing through inductor L_2 increases with voltage V_1 during the switch-on period kT and decreases with voltage $-(V_O - 2V_1)$ during the switch-off period $(1-k)T$. Therefore, the ripple of the inductor current i_{L2} is

$$\Delta i_{L2} = \frac{V_1}{L_2}kT = \frac{V_O-2V_1}{L_2}(1-k)T \tag{10.7}$$

$$V_O = \frac{2-k}{1-k}V_1 = \left(\frac{2-k}{1-k}\right)^2 V_{in} \tag{10.8}$$

The voltage transfer gain is

$$G = \frac{V_O}{V_{in}} = \left(\frac{2-k}{1-k}\right)^2 \tag{10.9}$$

The variation ratio of current i_{L1} through inductor L_1 is

$$\xi_1 = \frac{\Delta i_{L1}/2}{I_{L1}} = \frac{k(2-k)TV_{in}}{2L_1I_{in}} = \frac{k(1-k)^4}{2(2-k)^3}\frac{R}{fL_1} \tag{10.10}$$

The variation ratio of current i_{L2} through inductor L_2 is

$$\xi_2 = \frac{\Delta i_{L2}/2}{I_{L2}} = \frac{k(1-k)TV_1}{2L_2I_O} = \frac{k(1-k)^2TV_O}{2(2-k)L_2I_O} = \frac{k(1-k)^2}{2(2-k)}\frac{R}{fL_2} \tag{10.11}$$

and the variation ratio of output voltage v_O is

$$\varepsilon = \frac{\Delta v_O/2}{V_O} = \frac{k}{2RfC_4} \tag{10.12}$$

10.2.1.3 Triple-Lift Circuit

Triple-lift circuit is derived from re-lift circuit by secondly readding the parts L_2-D_3-D_4-D_5-C_3-C_4. Its circuit diagram and equivalent circuits during the switch-on and switch-off periods are shown in Figure 3.3. The voltage across capacitor C_1 is charged to V_{in}. As described in the previous section, the voltage V_1 across capacitor C_2 is $V_1 = ((2-k)/(1-k))V_{in}$ and the voltage V_2 across capacitor C_4 is $V_2 = ((2-k)/(1-k))^2 V_{in}$.

The voltage across capacitor C_5 is charged to V_2. The current flowing through inductor L_3 increases with voltage V_2 during the switch-on period kT and decreases with voltage $-(V_O - 2V_2)$ during the switch-off period $(1-k)T$. Therefore, the ripple of the inductor current i_{L2} is

$$\Delta i_{L3} = \frac{V_2}{L_3}kT = \frac{V_O - 2V_2}{L_3}(1-k)T \tag{10.13}$$

$$V_O = \frac{2-k}{1-k}V_2 = \left(\frac{2-k}{1-k}\right)^2 V_1 = \left(\frac{2-k}{1-k}\right)^3 V_{in} \tag{10.14}$$

The voltage transfer gain is

$$G = \frac{V_O}{V_{in}} = \left(\frac{2-k}{1-k}\right)^3 \tag{10.15}$$

The variation ratio of current i_{L1} through inductor L_1 is

$$\xi_1 = \frac{\Delta i_{L1}/2}{I_{L1}} = \frac{k(2-k)TV_{in}}{2L_1I_{in}} = \frac{k(1-k)^6}{2(2-k)^5}\frac{R}{fL_1} \tag{10.16}$$

The variation ratio of current i_{L2} through inductor L_2 is

$$\xi_2 = \frac{\Delta i_{L2}/2}{I_{L2}} = \frac{k(1-k)^2TV_1}{2(2-k)L_2I_O} = \frac{kT(2-k)^4V_O}{2(1-k)^3L_2I_O} = \frac{k(2-k)^4}{2(1-k)^3}\frac{R}{fL_2} \tag{10.17}$$

The variation ratio of current i_{L3} through inductor L_3 is

$$\xi_3 = \frac{\Delta i_{L3}/2}{I_{L3}} = \frac{k(1-k)TV_2}{2L_3I_O} = \frac{k(1-k)^2TV_O}{2(2-k)L_2I_O} = \frac{k(1-k)^2}{2(2-k)}\frac{R}{fL_3} \tag{10.18}$$

and the variation ratio of output voltage v_O is

$$\varepsilon = \frac{\Delta v_O/2}{V_O} = \frac{k}{2RfC_6} \tag{10.19}$$

Example 10.2 A triple-lift circuit of P/O SL Luo converter in Figure 10.3a has $V_{in}=20\,V$, all inductors are $10\,mH$, all capacitors are $20\,\mu F$, $R=1000\,\Omega$, $f=50\,kHz$, and conduction duty cycle $k=0.6$. Calculate the variation ratio of current i_{L1}, the output voltage, and its variation ratio.

Solution: From formula (10.16), we can get the variation ratio of current i_{L1}:

$$\xi_1 = \frac{k(1-k)^6}{2(2-k)^5}\frac{R}{fL_1} = \frac{0.6(1-0.6)^6}{2(2-0.6)^5}\frac{1000}{50k \times 10m} = 0.00046$$

From formula (10.14), we can get the output voltage

$$V_O = \left(\frac{2-k}{1-k}\right)^3 V_{in} = \left(\frac{2-0.6}{1-0.6}\right)^3 20 = 857.5\,V$$

From (10.19), its variation ratio is

$$\varepsilon = \frac{k}{2RfC_6} = \frac{0.6}{2 \times 1000 \times 50k \times 20\mu} = 0.0003$$

10.2.1.4 Higher-Order Lift-Circuit

Higher-order lift circuit can be designed by just readding the parts L_2-D_3-D_4-D_5-C_3-C_4 multiple times. For the nth-order lift circuit, the final output voltage across capacitor C_{2n} is

$$V_O = \left(\frac{2-k}{1-k}\right)^n V_{in}$$

The voltage transfer gain is

$$G = \frac{V_O}{V_{in}} = \left(\frac{2-k}{1-k}\right)^n \tag{10.20}$$

The variation ratio of current i_{Li} through inductor L_i ($i = 1, 2, 3, \ldots, n$) is

$$\xi_i = \frac{\Delta i_{Li}/2}{I_{Li}} = \frac{k(1-k)^{2(n-i+1)}}{2(2-k)^{2(n-i)+1}}\frac{R}{fL_i} \tag{10.21}$$

and the variation ratio of output voltage v_O is

$$\varepsilon = \frac{\Delta v_O/2}{V_O} = \frac{1-k}{2RfC_{2n}} \tag{10.22}$$

10.2.2 Additional Series

Using two diodes and two capacitors (D_{11}-D_{12}-C_{11}-C_{12}), a circuit, namely, "double/ enhanced circuit" (DEC), can be constructed, which is shown in Figure 10.4. If the input voltage is V_{in}, the output voltage V_O can be $2V_{in}$, or any other value that is higher than V_{in}. The DEC is very versatile in that it enhances DC/DC converter's voltage transfer gain.

All circuits of P/O SL Luo converters and their additional series are derived from the corresponding circuits of the main series by adding a DEC. The first three stages of this series are shown in Figures 10.5 through 10.7. For the sake of convenience, they are named elementary additional circuit, re-lift additional circuit, and triple-lift additional circuit, respectively, and numbered $n = 1, 2,$ and 3.

10.2.2.1 Elementary Additional Circuit

The elementary additional circuit is derived from elementary circuit by adding a DEC. Its circuit diagram and switch-on and switch-off equivalent circuits are shown in Figure 10.5.

The voltage across capacitor C_1 is charged to V_{in}, and the voltage across capacitor C_2 and C_{11} is charged to V_1. The current i_{L1} flowing through inductor L_1 increases with voltage V_{in} during the switch-on period kT and decreases with voltage $-(V_O - 2V_{in})$ during the switch-off period $(1 - k)T$. Therefore,

$$V_1 = \frac{2-k}{1-k}V_{in} \tag{10.23}$$

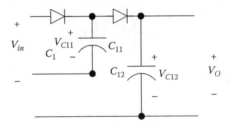

FIGURE 10.4 Double/enhance circuit.

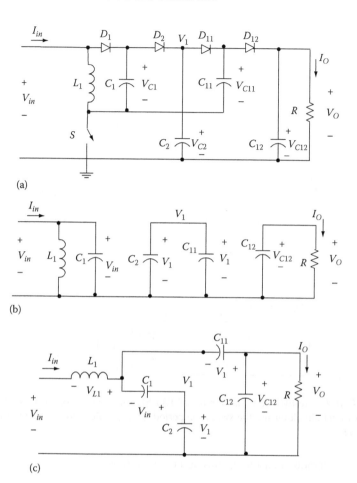

FIGURE 10.5 Elementary additional circuit of P/O SL Luo converters. (a) Circuit diagram. (b) Equivalent circuit during the switch-on period. (c) Equivalent circuit during the switch-off period.

and

$$V_{L1} = \frac{k}{1-k} V_{in} \qquad (10.24)$$

The output voltage is

$$V_O = V_{in} + V_{L1} + V_1 = \frac{3-k}{1-k} V_{in} \qquad (10.25)$$

The voltage transfer gain is

$$G = \frac{V_O}{V_{in}} = \frac{3-k}{1-k} \qquad (10.26)$$

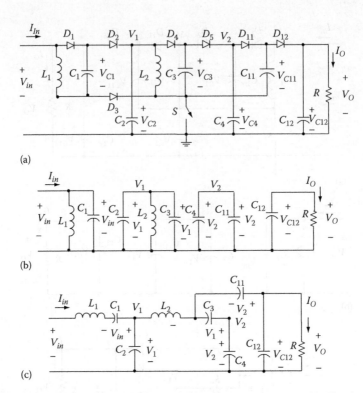

FIGURE 10.6 Re-lift additional circuit of P/O SL Luo converters. (a) Circuit diagram. (b) Equivalent circuit during the switch-on period. (c) Equivalent circuit during the switch-off period.

The variation ratio of current i_{L1} through inductor L_1 is

$$\xi_1 = \frac{\Delta i_{L1}/2}{I_{L1}} = \frac{k(1-k)TV_{in}}{4L_1I_O} = \frac{k(1-k)^2}{4(3-k)}\frac{R}{fL_1} \qquad (10.27)$$

The ripple voltage of output voltage v_O is

$$\Delta v_O = \frac{\Delta Q}{C_{12}} = \frac{I_O kT}{C_{12}} = \frac{k}{fC_{12}}\frac{V_O}{R}$$

Therefore, the variation ratio of output voltage v_O is

$$\varepsilon = \frac{\Delta v_O/2}{V_O} = \frac{k}{2RfC_{12}} \qquad (10.28)$$

10.2.2.2 Re-Lift Additional Circuit

This circuit is derived from re-lift circuit by adding a DEC. Its circuit diagram and switch-on and switch -off equivalent circuits are shown in Figure 10.6. The voltage

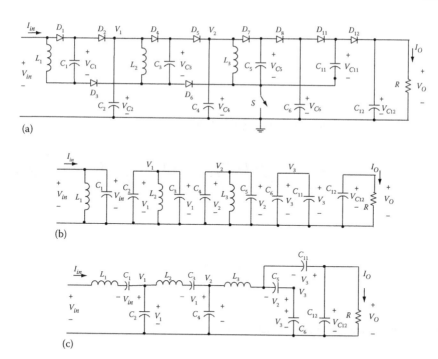

(a)

(b)

(c)

FIGURE 10.7 Triple-lift additional circuit of P/O SL Luo converters. (a) Circuit diagram. (b) Equivalent circuit during the switch-on period. (c) Equivalent circuit during the switch-off period.

across capacitor C_1 is charged to V_{in}. As described in the previous section, the voltage across C_2 is $V_1 = (2-k)/(1-k)V_{in}$.

The voltage across capacitor C_3 is charged to V_1, and the voltage across capacitor C_4 and C_{11} is charged to V_2. The current flowing through inductor L_2 increases with voltage V_1 during the switch-on period kT and decreases with voltage $-(V_O - 2V_1)$ during the switch-off period $(1-k)T$. Therefore,

$$V_2 = \frac{2-k}{1-k}V_1 = \left(\frac{2-k}{1-k}\right)^2 V_{in} \tag{10.29}$$

and

$$V_{L2} = \frac{k}{1-k}V_1 \tag{10.30}$$

The output voltage is

$$V_O = V_1 + V_{L2} + V_2 = \frac{2-k}{1-k}\frac{3-k}{1-k}V_{in} \tag{10.31}$$

The voltage transfer gain is

$$G = \frac{V_O}{V_{in}} = \frac{2-k}{1-k}\frac{3-k}{1-k} \tag{10.32}$$

The variation ratio of current i_{L1} through inductor L_1 is

$$\xi_1 = \frac{\Delta i_{L1}/2}{I_{L1}} = \frac{k(1-k)^2 TV_{in}}{2(3-k)L_1 I_O} = \frac{k(1-k)^4}{2(2-k)(3-k)^2}\frac{R}{fL_1} \tag{10.33}$$

and the variation ratio of current i_{L2} through inductor L_2 is

$$\xi_2 = \frac{\Delta i_{L2}/2}{I_{L2}} = \frac{k(1-k)TV_1}{4L_2 I_O} = \frac{k(1-k)^2}{4(3-k)}\frac{R}{fL_2} \tag{10.34}$$

The ripple voltage of output voltage v_O is

$$\Delta v_O = \frac{\Delta Q}{C_{12}} = \frac{I_O kT}{C_{12}} = \frac{k}{fC_{12}}\frac{V_O}{R}$$

Therefore, the variation ratio of output voltage v_O is

$$\varepsilon = \frac{\Delta v_O/2}{V_O} = \frac{k}{2RfC_{12}} \tag{10.35}$$

10.2.2.3 Triple-Lift Additional Circuit

This circuit is derived from triple-lift circuit by adding a DEC. Its circuit diagram and equivalent circuits during the switch-on and switch-off periods are shown in Figure 10.7. The voltage across capacitor C_1 is charged to V_{in}. As described in the previous section, the voltage across C_2 is $V_1 = ((2-k)/(1-k))V_{in}$ and the voltage across C_4 is $V_2 = ((2-k)/(1-k))V_1 = (2-k/1-k)^2 V_{in}$.

The voltage across capacitor C_5 is charged to V_2 and the voltage across capacitor C_6 and C_{11} is charged to V_3. The current flowing through inductor L_3 increases with voltage V_2 during the switch-on period kT and decreases with voltage $-(V_O - 2V_2)$ during the switch-off period $(1-k)T$. Therefore,

$$V_3 = \frac{2-k}{1-k}V_2 = \left(\frac{2-k}{1-k}\right)^2 V_1 = \left(\frac{2-k}{1-k}\right)^3 V_{in} \tag{10.36}$$

and

$$V_{L3} = \frac{k}{1-k}V_2 \qquad (10.37)$$

The output voltage is

$$V_O = V_2 + V_{L3} + V_3 = \left(\frac{2-k}{1-k}\right)^2 \frac{3-k}{1-k}V_{in} \qquad (10.38)$$

The voltage transfer gain is

$$G = \frac{V_O}{V_{in}} = \left(\frac{2-k}{1-k}\right)^2 \frac{3-k}{1-k} \qquad (10.39)$$

The variation ratio of current i_{L1} through inductor L_1 is

$$\xi_1 = \frac{\Delta i_{L1}/2}{I_{L1}} = \frac{k(1-k)^3 TV_{in}}{2(2-k)(3-k)L_1 I_O} = \frac{k(1-k)^3 T}{2(2-k)(3-k)L_1 I_O} \frac{(1-k)^3}{(2-k)^2(3-k)}$$

$$V_O = \frac{k(1-k)^6}{2(2-k)^3(3-k)^2} \frac{R}{fL_1} \qquad (10.40)$$

and the variation ratio of current i_{L2} through inductor L_2 is

$$\xi_2 = \frac{\Delta i_{L2}/2}{I_{L2}} = \frac{k(1-k)^2 TV_1}{2(3-k)L_2 I_O} = \frac{k(1-k)^2 T}{2(3-k)L_2 I_O} \frac{(1-k)^2}{(2-k)(3-k)}V_O = \frac{k(1-k)^4}{2(2-k)(3-k)^2} \frac{R}{fL_2}$$

$$(10.41)$$

and the variation ratio of current i_{L3} through inductor L_3 is

$$\xi_3 = \frac{\Delta i_{L3}/2}{I_{L3}} = \frac{k(1-k)TV_2}{4L_3 I_O} = \frac{k(1-k)T}{4L_3 I_O} \frac{1-k}{3-k}V_O = \frac{k(1-k)^2}{4(3-k)} \frac{R}{fL_3} \qquad (10.42)$$

The ripple voltage of output voltage v_O is

$$\Delta v_O = \frac{\Delta Q}{C_{12}} = \frac{I_O kT}{C_{12}} = \frac{k}{fC_{12}} \frac{V_O}{R}$$

Therefore, the variation ratio of output voltage v_O is

$$\varepsilon = \frac{\Delta v_O / 2}{V_O} = \frac{k}{2RfC_{12}} \tag{10.43}$$

10.2.2.4 Higher-Order-Lift Additional Circuit

Higher-order-lift additional circuit is derived from the corresponding circuit of the main series by adding a DEC. For the nth-order-lift additional circuit, the final output voltage is

$$V_O = \left(\frac{2-k}{1-k}\right)^{n-1} \frac{3-k}{1-k} V_{in}$$

The voltage transfer gain is

$$G = \frac{V_O}{V_{in}} = \left(\frac{2-k}{1-k}\right)^{n-1} \frac{3-k}{1-k} \tag{10.44}$$

Analogously, the variation ratio of current i_{Li} through inductor L_i $(i = 1, 2, 3, \ldots, n)$ is

$$\xi_i = \frac{\Delta i_{Li} / 2}{I_{Li}} = \frac{k(1-k)^{2(n-i+1)}}{2[2(2-k)]^{h(n-i)}(2-k)^{2(n-i)+1}(3-k)^{2u(n-i-1)}} \frac{R}{fL_i} \tag{10.45}$$

where

$$h(x) = \begin{cases} 0 & x > 0 \\ 1 & x \le 0 \end{cases} \text{ is the } \textit{Hong function}$$

$$u(x) = \begin{cases} 1 & x \ge 0 \\ 0 & x < 0 \end{cases} \text{ is the } \textit{unit-step function}$$

the variation ratio of output voltage v_O is

$$\varepsilon = \frac{\Delta v_O / 2}{V_O} = \frac{k}{2RfC_{12}} \tag{10.46}$$

10.2.3 ENHANCED SERIES

All circuits of P/O SL Luo converters and their enhanced series are derived from the corresponding circuits of the main series by adding a DEC in each stage circuit. The first three stages of this series are shown in Figures 10.5, 10.8, and 10.9. For the sake

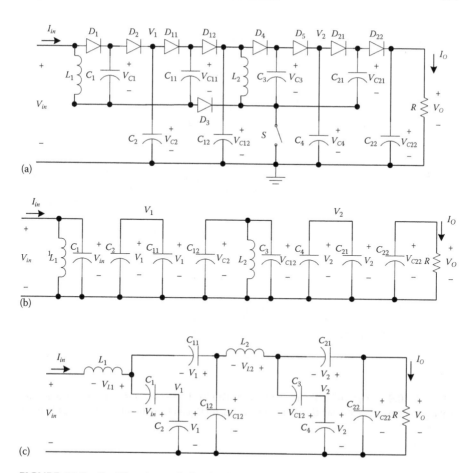

FIGURE 10.8 Re-lift enhanced circuit of P/O SL Luo converters. (a) Circuit diagram. (b) Equivalent circuit during the switch-on period. (c) Equivalent circuit during the switch-off period.

of convenience, they are called elementary enhanced circuit, re-lift enhanced circuit, and triple-lift enhanced circuit, respectively, and numbered $n = 1$, 2, and 3.

10.2.3.1 Elementary Enhanced Circuit

This circuit is shared in additional and enhanced series as in Figure 10.16, and is the same to the elementary additional circuit shown in Figure 10.5. Analysis refers to Section 10.2.2.1.

10.2.3.2 Re-Lift Enhanced Circuit

The re-lift enhanced circuit is derived from the re-lift circuit of the main series by adding the DEC in each stage circuit. Its circuit diagram and switch-on and

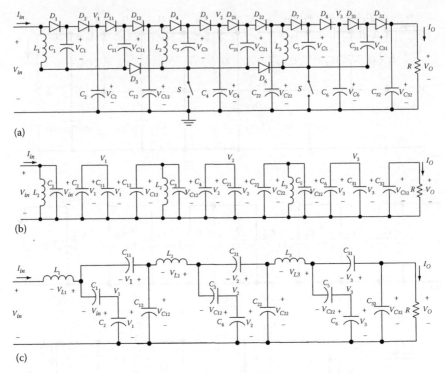

(a)

(b)

(c)

FIGURE 10.9 Triple-lift enhanced circuit of P/O SL Luo converters. (a) Circuit diagram.
(b) Equivalent circuit during the switch-on period. (c) Equivalent circuit during the switch-off
period.

switch-off equivalent circuits are shown in Figure 10.8. As described in the previous
section, the voltage across capacitor C_{12} is charged to $V_{C12} = ((3 - k)/(1 - k))V_{in}$.

The voltage across capacitor C_3 is charged to V_{C12} and the voltage across capacitor
C_4 and C_{21} is charged to V_{C4},

$$V_{C4} = \frac{2-k}{1-k}V_{C12} = \frac{2-k}{1-k}\frac{3-k}{1-k}V_{in} \qquad (10.47)$$

The current flowing through inductor L_2 increases with voltage V_{C12} during the
switch-on period kT and decreases with voltage $-(V_O - V_{C4} - V_{C12})$ during the switch-
off period $(1 - k)T$. Therefore,

$$\Delta i_{L2} = \frac{k}{L_2}V_{C12} = \frac{1-k}{L_2}(V_O - V_{C4} - V_{C12}) \qquad (10.48)$$

$$V_O = \frac{3-k}{1-k}V_{C12} = \left(\frac{3-k}{1-k}\right)^2 V_{in} \qquad (10.49)$$

The voltage transfer gain is

$$G = \frac{V_O}{V_{in}} = \left(\frac{3-k}{1-k}\right)^2 \qquad (10.50)$$

The variation ratio of current i_{L1} through inductor L_1 is

$$\xi_1 = \frac{\Delta i_{L1}/2}{I_{L1}} = \frac{k(1-k)^2 TV_{in}}{2(3-k)L_1 I_O} = \frac{k(1-k)^4}{2(2-k)(3-k)^2}\frac{R}{fL_1} \qquad (10.51)$$

and the variation ratio of current i_{L2} through inductor L_2 is

$$\xi_2 = \frac{\Delta i_{L2}/2}{I_{L2}} = \frac{k(1-k)TV_1}{4L_2 I_O} = \frac{k(1-k)^2}{4(3-k)}\frac{R}{fL_2} \qquad (10.52)$$

The ripple voltage of output voltage v_O is

$$\Delta v_O = \frac{\Delta Q}{C_{22}} = \frac{I_O kT}{C_{22}} = \frac{k}{fC_{22}}\frac{V_O}{R}$$

Therefore, the variation ratio of output voltage v_O is

$$\varepsilon = \frac{\Delta v_O/2}{V_O} = \frac{k}{2RfC_{22}} \qquad (10.53)$$

10.2.3.3 Triple-Lift Enhanced Circuit

The triple-lift enhanced circuit is derived from triple-lift circuit of the main series by adding DEC in each stage circuit. Its circuit diagram and equivalent circuits during the switch-on and switch-off periods are shown in Figure 10.9. As described in the previous section, the voltage across capacitor C_{12} is charged to $V_{C12}=((3-k)/(1-k))V_{in}$ and the voltage across capacitor C_{22} is charged to $V_{C22}=((3-k)/(1-k))^2 V_{in}$.

The voltage across capacitor C_5 is charged to V_{C22}, and the voltage across capacitor C_6 and C_{31} is charged to V_{C6}.

$$V_{C6} = \frac{2-k}{1-k}V_{C22} = \frac{2-k}{1-k}\left(\frac{3-k}{1-k}\right)^2 V_{in} \qquad (10.54)$$

The current flowing through inductor L_3 increases with voltage V_{C22} during the switch-on period kT and decreases with voltage $-(V_O - V_{C6} - V_{C22})$ during the switch-off period $(1 - k)T$. Therefore,

$$\Delta i_{L3} = \frac{k}{L_3} V_{C22} = \frac{1-k}{L_3} (V_O - V_{C6} - V_{C22}) \tag{10.55}$$

$$V_O = \frac{3-k}{1-k} V_{C22} = \left(\frac{3-k}{1-k}\right)^3 V_{in} \tag{10.56}$$

The voltage transfer gain is

$$G = \frac{V_O}{V_{in}} = \left(\frac{3-k}{1-k}\right)^3 \tag{10.57}$$

The variation ratio of current i_{L1} through inductor L_1 is

$$\xi_1 = \frac{\Delta i_{L1}/2}{I_{L1}} = \frac{k(1-k)^3 T V_{in}}{2(2-k)(3-k)L_1 I_O} = \frac{k(1-k)^3 T}{2(2-k)(3-k)L_1 I_O} \frac{(1-k)^3}{(2-k)^2(3-k)} V_O$$

$$= \frac{k(1-k)^6}{2(2-k)^3(3-k)^2} \frac{R}{fL_1} \tag{10.58}$$

The variation ratio of current i_{L2} through inductor L_2 is

$$\xi_2 = \frac{\Delta i_{L2}/2}{I_{L2}} = \frac{k(1-k)^2 T V_1}{2(3-k)L_2 I_O} = \frac{k(1-k)^2 T}{2(3-k)L_2 I_O} \frac{(1-k)^2}{(2-k)(3-k)} V_O$$

$$= \frac{k(1-k)^4}{2(2-k)(3-k)^2} \frac{R}{fL_2} \tag{10.59}$$

and the variation ratio of current i_{L3} through inductor L_3 is

$$\xi_3 = \frac{\Delta i_{L3}/2}{I_{L3}} = \frac{k(1-k)T V_2}{4L_3 I_O} = \frac{k(1-k)T}{4L_3 I_O} \frac{1-k}{3-k} V_O = \frac{k(1-k)^2}{4(3-k)} \frac{R}{fL_3} \tag{10.60}$$

The ripple voltage of output voltage v_O is

$$\Delta v_O = \frac{\Delta Q}{C_{32}} = \frac{I_O kT}{C_{32}} = \frac{k}{fC_{32}} \frac{V_O}{R}$$

Therefore, the variation ratio of output voltage v_O is

$$\varepsilon = \frac{\Delta v_O / 2}{V_O} = \frac{k}{2RfC_{32}} \tag{10.61}$$

10.2.3.4 Higher-Order-Lift Enhanced Circuit

Higher-order-lift enhanced circuit is derived from the corresponding circuit of the main series by adding the DEC in each stage circuit. For the nth-order-lift enhanced circuit, the final output voltage is $V_O = ((3 - k)/(1 - k))^n V_{in}$.

The voltage transfer gain is

$$G = \frac{V_O}{V_{in}} = \left(\frac{3-k}{1-k}\right)^n \tag{10.62}$$

Analogously, the variation ratio of current i_{Li} through inductor L_i ($i = 1, 2, 3, \ldots, n$) is

$$\xi_i = \frac{\Delta i_{Li} / 2}{I_{Li}} = \frac{k(1-k)^{2(n-i+1)}}{2[2(2-k)]^{h(n-i)}(2-k)^{2(n-i)+1}(3-k)^{2u(n-i-1)}} \frac{R}{fL_i} \tag{10.63}$$

where

$$h(x) = \begin{cases} 0 & x > 0 \\ 1 & x \leq 0 \end{cases} \text{ is the } Hong \ function$$

$$u(x) = \begin{cases} 1 & x \geq 0 \\ 0 & x < 0 \end{cases} \text{ is the } unit\text{-}step \ function$$

the variation ratio of output voltage v_O is

$$\varepsilon = \frac{\Delta v_O / 2}{V_O} = \frac{k}{2RfC_{n2}} \tag{10.64}$$

10.2.4 RE-ENHANCED SERIES

All circuits of P/O SL Luo converters and their re-enhanced series are derived from the corresponding circuits of the main series by adding the DEC twice in each stage circuit.

The first three stages of this series are shown in Figures 10.10 through 10.12. For the sake of convenience, they are named elementary re-enhanced circuit, re-lift re-enhanced circuit, and triple-lift re-enhanced circuit, respectively, and numbered $n = 1, 2,$ and 3.

10.2.4.1   Elementary Re-Enhanced Circuit

This circuit is derived from elementary circuit by adding the DEC twice. Its circuit diagram and switch-on and switch-off equivalent circuits are shown in Figure 10.10.
    The output voltage is

$$V_O = V_{in} + V_{L1} + V_{C12} = \frac{4-k}{1-k} V_{in} \qquad (10.65)$$

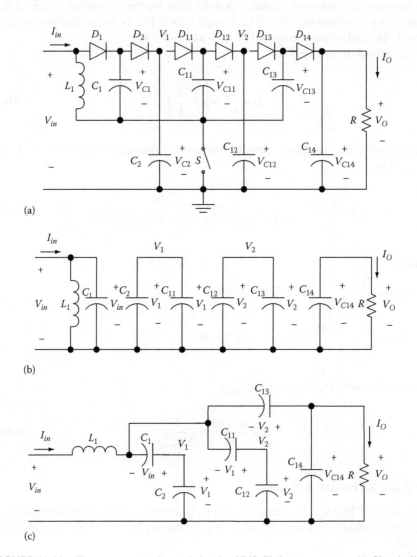

(a)

(b)

(c)

FIGURE 10.10   Elementary re-enhanced circuit of P/O SL Luo converters. (a) Circuit diagram. (b) Equivalent circuit during the switch-on period. (c) Equivalent circuit during the switch-off period.

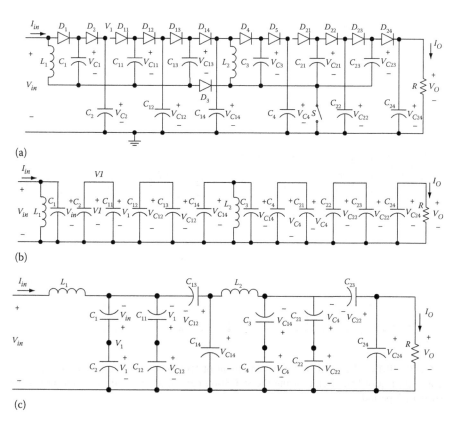

(a)

(b)

(c)

FIGURE 10.11 Re-lift re-enhanced circuit of P/O SL Luo converters. (a) Circuit diagram. (b) Equivalent circuit during the switch-on period. (c) Equivalent circuit during the switch-off period.

The voltage transfer gain is

$$G = \frac{V_O}{V_{in}} = \frac{4-k}{1-k} \tag{10.66}$$

where

$$V_{C2} = \frac{2-k}{1-k} V_{in} \tag{10.67}$$

$$V_{C12} = \frac{3-k}{1-k} V_{in} \tag{10.68}$$

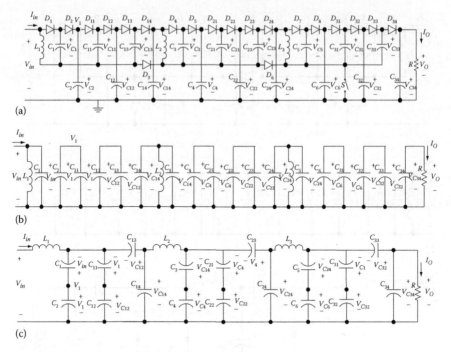

FIGURE 10.12 Triple-lift re-enhanced circuit of P/O SL Luo converters. (a) Circuit diagram. (b) Equivalent circuit during the switch-on period. (c) Equivalent circuit during the switch-off period.

and

$$V_{L1} = \frac{k}{1-k} V_{in} \tag{10.69}$$

The variation ratio of current i_{L1} through inductor L_1 is

$$\xi_1 = \frac{\Delta i_{L1}/2}{I_{L1}} = \frac{k(1-k)TV_{in}}{4L_1 I_O} = \frac{k(1-k)^2}{4(3-k)} \frac{R}{fL_1} \tag{10.70}$$

The ripple voltage of output voltage v_O is

$$\Delta v_O = \frac{\Delta Q}{C_{14}} = \frac{I_O kT}{C_{14}} = \frac{k}{fC_{14}} \frac{V_O}{R}$$

Therefore, the variation ratio of output voltage v_O is

$$\varepsilon = \frac{\Delta v_O/2}{V_O} = \frac{k}{2RfC_{14}} \tag{10.71}$$

10.2.4.2   Re-Lift Re-Enhanced Circuit

This circuit is derived from re-lift circuit of the main series by adding the DEC twice in each stage circuit. Its circuit diagram and switch-on and switch-off equivalent circuits are shown in Figure 10.11.

The voltage across capacitor C_{14} is

$$V_{C14} = \frac{4-k}{1-k} V_{in} \tag{10.72}$$

By the same analysis

$$V_O = \frac{4-k}{1-k} V_{C14} = \left(\frac{4-k}{1-k}\right)^2 V_{in} \tag{10.73}$$

The voltage transfer gain is

$$G = \frac{V_O}{V_{in}} = \left(\frac{4-k}{1-k}\right)^2 \tag{10.74}$$

The variation ratio of current i_{L1} through inductor L_1 is

$$\xi_1 = \frac{\Delta i_{L1}/2}{I_{L1}} = \frac{k(1-k)^2 T V_{in}}{2(3-k)L_1 I_O} = \frac{k(1-k)^4}{2(2-k)(3-k)^2} \frac{R}{fL_1} \tag{10.75}$$

The variation ratio of current i_{L2} through inductor L_2 is

$$\xi_2 = \frac{\Delta i_{L2}/2}{I_{L2}} = \frac{k(1-k)T V_1}{4 L_2 I_O} = \frac{k(1-k)^2}{4(3-k)} \frac{R}{fL_2} \tag{10.76}$$

The ripple voltage of output voltage v_O is

$$\Delta v_O = \frac{\Delta Q}{C_{24}} = \frac{I_O kT}{C_{24}} = \frac{k}{fC_{24}} \frac{V_O}{R}$$

Therefore, the variation ratio of output voltage v_O is

$$\varepsilon = \frac{\Delta v_O/2}{V_O} = \frac{k}{2RfC_{24}} \tag{10.77}$$

10.2.4.3 Triple-Lift Re-Enhanced Circuit

This circuit is derived from triple-lift circuit of the main series by adding the DEC twice in each stage circuit. Its circuit diagram and switch-on and switch-off equivalent circuits are shown in Figure 10.12.

The voltage across capacitor C_{14} is

$$V_{C14} = \frac{4-k}{1-k} V_{in} \tag{10.78}$$

The voltage across capacitor C_{24} is

$$V_{C24} = \left(\frac{4-k}{1-k} \right)^2 V_{in} \tag{10.79}$$

By the same analysis

$$V_O = \frac{4-k}{1-k} V_{C24} = \left(\frac{4-k}{1-k} \right)^3 V_{in} \tag{10.80}$$

The voltage transfer gain is

$$G = \frac{V_O}{V_{in}} = \left(\frac{4-k}{1-k} \right)^3 \tag{10.81}$$

The variation ratio of current i_{L1} through inductor L_1 is

$$\xi_1 = \frac{\Delta i_{L1}/2}{I_{L1}} = \frac{k(1-k)^3 T V_{in}}{2(2-k)(3-k)L_1 I_O} = \frac{k(1-k)^3 T}{2(2-k)(3-k)L_1 I_O} \frac{(1-k)^3}{(2-k)^2(3-k)} V_O$$

$$= \frac{k(1-k)^6}{2(2-k)^3(3-k)^2} \frac{R}{f L_1} \tag{10.82}$$

The variation ratio of current i_{L2} through inductor L_2 is

$$\xi_2 = \frac{\Delta i_{L2}/2}{I_{L2}} = \frac{k(1-k)^2 T V_1}{2(3-k)L_2 I_O} = \frac{k(1-k)^2 T}{2(3-k)L_2 I_O} \frac{(1-k)^2}{(2-k)(3-k)} V_O$$

$$= \frac{k(1-k)^4}{2(2-k)(3-k)^2} \frac{R}{f L_2} \tag{10.83}$$

The variation ratio of current i_{L3} through inductor L_3 is

$$\xi_3 = \frac{\Delta i_{L3}/2}{I_{L3}} = \frac{k(1-k)TV_2}{4L_3I_O} = \frac{k(1-k)T}{4L_3I_O}\frac{1-k}{3-k}V_O = \frac{k(1-k)^2}{4(3-k)}\frac{R}{fL_3} \quad (10.84)$$

The ripple voltage of output voltage v_O is

$$\Delta v_O = \frac{\Delta Q}{C_{34}} = \frac{I_O kT}{C_{34}} = \frac{k}{fC_{34}}\frac{V_O}{R}$$

Therefore, the variation ratio of output voltage v_O is

$$\varepsilon = \frac{\Delta v_O/2}{V_O} = \frac{k}{2RfC_{34}} \quad (10.85)$$

10.2.4.4 Higher-Order-Lift Re-Enhanced Circuit
Higher-order-lift additional circuit is derived from the corresponding circuit of the main series by adding DEC twice in each stage circuit. For the nth-order-lift additional circuit, the final output voltage is

$$V_O = \left(\frac{4-k}{1-k}\right)^n V_{in}$$

The voltage transfer gain is

$$G = \frac{V_O}{V_{in}} = \left(\frac{4-k}{1-k}\right)^n \quad (10.86)$$

Analogously, the variation ratio of current i_{Li} through inductor L_i ($i = 1, 2, 3, \ldots, n$) is

$$\xi_i = \frac{\Delta i_{Li}/2}{I_{Li}} = \frac{k(1-k)^{2(n-i+1)}}{2[2(2-k)]^{h(n-i)}(2-k)^{2(n-i)+1}(3-k)^{2u(n-i-1)}}\frac{R}{fL_i} \quad (10.87)$$

where

$$h(x) = \begin{cases} 0 & x > 0 \\ 1 & x \le 0 \end{cases} \text{ is the } Hong\ function$$

$$u(x) = \begin{cases} 1 & x \ge 0 \\ 0 & x < 0 \end{cases} \text{ is the } unit\text{-}step\ function$$

the variation ratio of output voltage v_O is

$$\varepsilon = \frac{\Delta v_O / 2}{V_O} = \frac{k}{2 R f C_{n4}} \qquad (10.88)$$

10.2.5 Multiple-(j)Enhanced Series

All circuits of P/O SL Luo converters and their multiple-enhanced series are derived from the corresponding circuits of the main series by adding the DEC multiple (j) times in each stage circuit. The first three stages of this series are shown in Figures 10.13 through 10.15. For the sake of convenience, they are called elementary

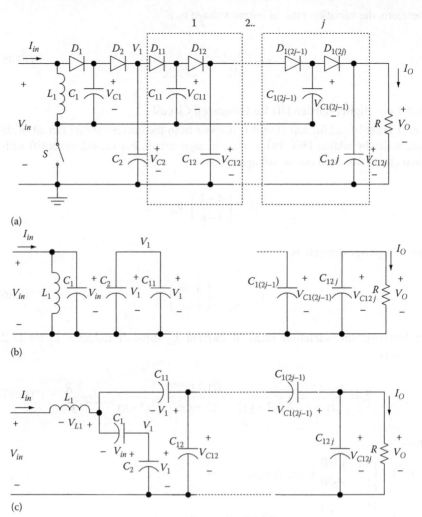

FIGURE 10.13 Elementary multiple-enhanced circuit of P/O SL Luo converters. (a) Circuit diagram. (b) Equivalent circuit during the switch-on period. (c) Equivalent circuit during the switch-off period.

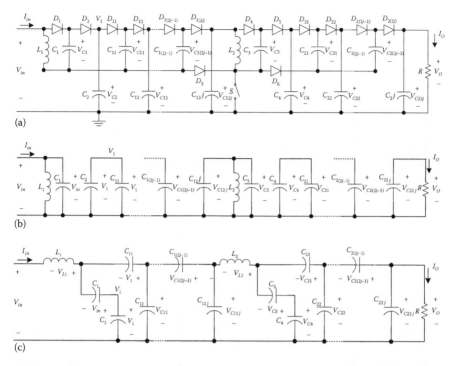

(a)

(b)

(c)

FIGURE 10.14 Re-lift multiple-enhanced circuit of P/O SL Luo converters. (a) Circuit diagram. (b) Equivalent circuit during the switch-on period. (c) Equivalent circuit during the switch-off period.

multiple-enhanced circuit, re-lift multiple-enhanced circuit, and triple-lift multiple-enhanced circuit, respectively, and numbered $n = 1$, 2, and 3.

10.2.5.1 Elementary Multiple-Enhanced Circuit

This circuit is derived from the elementary circuit of the main series by adding the DEC multiple (j) times. Its circuit diagram and switch-on and switch-off equivalent circuits are shown in Figure 10.13.

The output voltage is

$$V_O = \frac{j+2-k}{1-k} V_{in} \tag{10.89}$$

The voltage transfer gain is

$$G = \frac{V_O}{V_{in}} = \frac{j+2-k}{1-k} \tag{10.90}$$

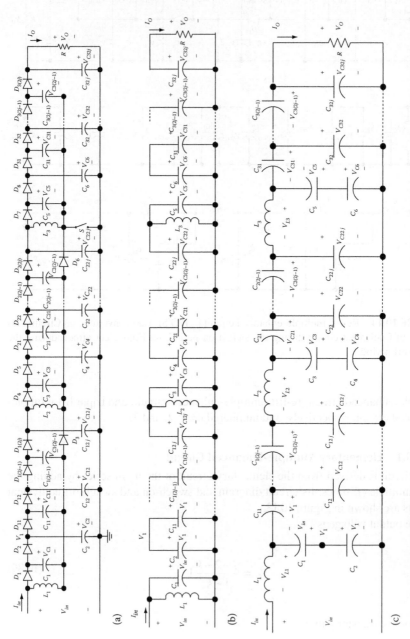

FIGURE 10.15 Triple-lift multiple-enhanced circuit of P/O SL Luo converters. (a) Circuit diagram. (b) Equivalent circuit during the switch-on period. (c) Equivalent circuit during the switch-off period.

The variation ratio of current i_{L1} through inductor L_1 is

$$\xi_1 = \frac{\Delta i_{L1}/2}{I_{L1}} = \frac{k(1-k)TV_{in}}{4L_1 I_O} = \frac{k(1-k)^2}{4(3-k)} \frac{R}{fL_1} \qquad (10.91)$$

The ripple voltage of output voltage v_O is

$$\Delta v_O = \frac{\Delta Q}{C_{12j}} = \frac{I_O kT}{C_{12j}} = \frac{k}{fC_{12j}} \frac{V_O}{R} \qquad$$

Therefore, the variation ratio of output voltage v_O is

$$\varepsilon = \frac{\Delta v_O/2}{V_O} = \frac{k}{2RfC_{12j}} \qquad (10.92)$$

10.2.5.2 Re-Lift Multiple-(j)Enhanced Circuit

This circuit is derived from the re-lift circuit of the main series by adding the DEC multiple (j) times in each stage circuit. Its circuit diagram and switch-on and switch-off equivalent circuits are shown in Figure 10.14.

The voltage across capacitor C_{12j} is

$$V_{C12j} = \frac{j+2-k}{1-k} V_{in} \qquad (10.93)$$

The output voltage across capacitor C_{22j} is

$$V_O = V_{C22j} = \left(\frac{j+2-k}{1-k}\right)^2 V_{in} \qquad (10.94)$$

The voltage transfer gain is

$$G = \frac{V_O}{V_{in}} = \left(\frac{j+2-k}{1-k}\right)^2 \qquad (10.95)$$

The variation ratio of current i_{L1} through inductor L_1 is

$$\xi_1 = \frac{\Delta i_{L1}/2}{I_{L1}} = \frac{k(1-k)^2 TV_{in}}{2(3-k)L_1 I_O} = \frac{k(1-k)^4}{2(2-k)(3-k)^2} \frac{R}{fL_1} \qquad (10.96)$$

and the variation ratio of current i_{L2} through inductor L_2 is

$$\xi_2 = \frac{\Delta i_{L2}/2}{I_{L2}} = \frac{k(1-k)TV_1}{4L_2 I_O} = \frac{k(1-k)^2}{4(3-k)} \frac{R}{fL_2} \qquad (10.97)$$

The ripple voltage of output voltage v_O is

$$\Delta v_O = \frac{\Delta Q}{C_{22j}} = \frac{I_O kT}{C_{22j}} = \frac{k}{fC_{22j}} \frac{V_O}{R}$$

Therefore, the variation ratio of output voltage v_O is

$$\varepsilon = \frac{\Delta v_O / 2}{V_O} = \frac{k}{2RfC_{22j}} \tag{10.98}$$

10.2.5.3 Triple-Lift Multiple(j)-Enhanced Circuit

This circuit is derived from triple-lift circuit of the main series by adding the DEC multiple (j) times in each stage circuit. Its circuit diagram and switch-on and switch-off equivalent circuits are shown in Figure 10.15.

The voltage across capacitor C_{12j} is

$$V_{C12j} = \frac{j+2-k}{1-k} V_{in} \tag{10.99}$$

The voltage across capacitor C_{22j} is

$$V_{C22j} = \left(\frac{j+2-k}{1-k}\right)^2 V_{in} \tag{10.100}$$

By the same analysis

$$V_O = \frac{j+2-k}{1-k} V_{C22j} = \left(\frac{j+2-k}{1-k}\right)^3 V_{in} \tag{10.101}$$

The voltage transfer gain is

$$G = \frac{V_O}{V_{in}} = \left(\frac{j+2-k}{1-k}\right)^3 \tag{10.102}$$

The variation ratio of current i_{L1} through inductor L_1 is

$$\xi_1 = \frac{\Delta i_{L1}/2}{I_{L1}} = \frac{k(1-k)^3 T V_{in}}{2(2-k)(3-k)L_1 I_O} = \frac{k(1-k)^3 T}{2(2-k)(3-k)L_1 I_O} = \frac{(1-k)^3}{(2-k)^2(3-k)} V_O$$

$$= \frac{k(1-k)^6}{2(2-k)^3(3-k)^2} \frac{R}{fL_1} \tag{10.103}$$

The variation ratio of current i_{L2} through inductor L_2 is

$$\xi_2 = \frac{\Delta i_{L2}/2}{I_{L2}} = \frac{k(1-k)^2 T V_1}{2(3-k)L_2 I_O} = \frac{k(1-k)^2 T}{2(3-k)L_2 I_O}\frac{(1-k)^2}{(2-k)(3-k)}V_O$$

$$= \frac{k(1-k)^4}{2(2-k)(3-k)^2}\frac{R}{fL_2} \tag{10.104}$$

The variation ratio of current i_{L3} through inductor L_3 is

$$\xi_3 = \frac{\Delta i_{L3}/2}{I_{L3}} = \frac{k(1-k)T V_2}{4L_3 I_O} = \frac{k(1-k)T}{4L_3 I_O}\frac{1-k}{3-k}V_O = \frac{k(1-k)^2}{4(3-k)}\frac{R}{fL_3} \tag{10.105}$$

The ripple voltage of output voltage v_O is

$$\Delta v_O = \frac{\Delta Q}{C_{32j}} = \frac{I_O kT}{C_{32j}} = \frac{k}{fC_{32j}}\frac{V_O}{R}$$

Therefore, the variation ratio of output voltage v_O is

$$\varepsilon = \frac{\Delta v_O/2}{V_O} = \frac{k}{2RfC_{32j}} \tag{10.106}$$

10.2.5.4 Higher-Order-Lift Multiple-Enhanced Circuit

Higher–order-lift multiple-enhanced circuit can be derived from the corresponding circuit of the main series converters by adding the DEC multiple (j) times in each stage circuit. For the nth-order-lift additional circuit, the final output voltage is

$$V_O = \left(\frac{j+2-k}{1-k}\right)^n V_{in}$$

The voltage transfer gain is

$$G = \frac{V_O}{V_{in}} = \left(\frac{j+2-k}{1-k}\right)^n \tag{10.107}$$

Analogously, the variation ratio of current i_{Li} through inductor L_i $(i = 1, 2, 3, \ldots, n)$ is

$$\xi_i = \frac{\Delta i_{Li}/2}{I_{Li}} = \frac{k(1-k)^{2(n-i+1)}}{2[2(2-k)]^{h(n-i)}(2-k)^{2(n-i)+1}(3-k)^{2u(n-i-1)}} \frac{R}{fL_i} \quad (10.108)$$

where

$$h(x) = \begin{cases} 0 & x > 0 \\ 1 & x \le 0 \end{cases} \text{ is the } \textit{Hong function}$$

$$u(x) = \begin{cases} 1 & x \ge 0 \\ 0 & x < 0 \end{cases} \text{ is the } \textit{unit-step function}$$

The variation ratio of output voltage v_O is

$$\varepsilon = \frac{\Delta v_O/2}{V_O} = \frac{k}{2RfC_{n2j}} \quad (10.109)$$

10.2.6 Summary of P/O SL Luo Converters

A family tree of all circuits of P/O SL Luo converters are shown in Figure 10.16.

From the analysis in previous sections, the common formula to calculate the output voltage is presented as follows:

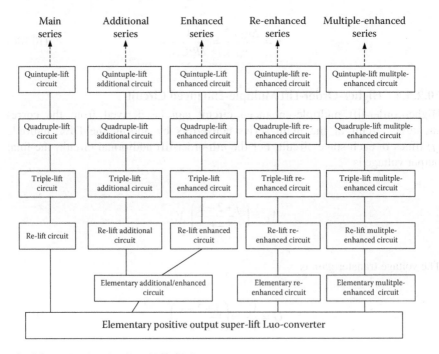

FIGURE 10.16 The family of P/O SL Luo converters.

$$V_O = \begin{cases} \left(\dfrac{2-k}{1-k}\right)^n V_{in} & \text{Main\_series} \\[2mm] \left(\dfrac{2-k}{1-k}\right)^{n-1}\left(\dfrac{3-k}{1-k}\right)V_{in} & \text{Additional\_series} \\[2mm] \left(\dfrac{3-k}{1-k}\right)^n V_{in} & \text{Enhanced\_series} \\[2mm] \left(\dfrac{4-k}{1-k}\right)^n V_{in} & \text{Re-enhanced\_series} \\[2mm] \left(\dfrac{j+2-k}{1-k}\right)^n V_{in} & \text{Multiple-enhanced\_series} \end{cases} \qquad (10.110)$$

The voltage transfer gain is

$$G = \frac{V_O}{V_{in}} = \begin{cases} \left(\dfrac{2-k}{1-k}\right)^n & \text{Main\_series} \\[2mm] \left(\dfrac{2-k}{1-k}\right)^{n-1}\left(\dfrac{3-k}{1-k}\right) & \text{Additional\_series} \\[2mm] \left(\dfrac{3-k}{1-k}\right)^n & \text{Enhanced\_series} \\[2mm] \left(\dfrac{4-k}{1-k}\right)^n & \text{Re-enhanced\_series} \\[2mm] \left(\dfrac{j+2-k}{1-k}\right)^n & \text{Multiple-enhanced\_series} \end{cases} \qquad (10.111)$$

In order to show the advantages of SL Luo converters, their voltage transfer gains can be compared to that of buck converter, $G = V_O/V_{in} = k$.

Forward converter, $G = V_O/V_{in} = kN$, N is the transformer turns ratio
Cúk converter, $G = V_O/V_{in} = k/1-k$
Fly-back converter, $G = V_O/V_{in} = (k/1-k)N$, N is the transformer turns ratio
Boost converter, $G = V_O/V_{in} = 1/1-k$
and P/O Luo converters.

$$G = \frac{V_O}{V_{in}} = \frac{n}{1-k} \qquad (10.112)$$

Assume the conduction duty k is 0.2; the output voltage transfer gains are listed in Table 10.1.

TABLE 10.1
Voltage Transfer Gains of Converters in the Condition $k = 0.2$

Stage No. (n)	1	2	3	4	5	n
Buck converter				0.2		
Forward converter			0.2N (N is the transformer turns ratio)			
Cuk converter				0.25		
Fly-back converter			0.25N (N is the transformer turns ratio)			
Boost converter				1.25		
P/O Luo converters	1.25	2.5	3.75	5	6.25	1.25^n
P/O SL Luo converters—main series	2.25	5.06	11.39	25.63	57.67	2.25^n
P/O SL Luo converters—additional series	3.5	7.88	17.72	39.87	89.7	$3.5^*2.25^{(n-1)}$
P/O SL Luo converters—enhanced series	3.5	12.25	42.88	150	525	3.5^n
P/O SL Luo converters—re-enhanced series	4.75	22.56	107.2	509	2,418	4.75^n
P/O SL Luo converters—multiple ($j=4$)-enhanced series	7.25	52.56	381	2,762	20,030	7.25^n

TABLE 10.2
Voltage Transfer Gains of Converters in the Condition $k = 0.5$

Stage No. (n)	1	2	3	4	5	n
Buck converter				0.5		
Forward converter			0.5N (N is the transformer turns ratio)			
Cuk converter				1		
Fly-back converter			N (N is the transformer turns ratio)			
Boost converter				2		
P/O Luo converters	2	4	6	8	10	2^n
P/O SL Luo converters—main series	3	9	27	81	243	3^n
P/O SL Luo converters—additional series	5	15	45	135	405	$5^*3^{(n-1)}$
P/O SL Luo converters—enhanced series	5	25	125	625	3,125	5^n
P/O SL Luo converters—re-enhanced series	7	49	343	2,401	16,807	7^n
P/O SL Luo converters—multiple ($j=4$)-enhanced series	11	121	1,331	14,641	16^*10^4	11^n

Assume the conduction duty k is 0.5; the output voltage transfer gains are listed in Table 10.2.

Assume the conduction duty k is 0.8; the output voltage transfer gains are listed in Table 10.3.

10.3 N/O SL LUO CONVERTERS

As the P/O SL Luo converters, the N/O SL Luo converters have been developed simultaneously. They perform SL technique as well. We still only introduce three circuits in each subseries [1,2,5,6].

TABLE 10.3

Voltage Transfer Gains of Converters in the Condition $k=0.8$

Stage No. (n)	1	2	3	4	5	n
Buck converter				0.8		
Forward converter			0.8N (N is the transformer turns ratio)			
Cuk converter				4		
Fly-back converter			4N (N is the transformer turns ratio)			
Boost converter				5		
P/O Luo converters	5	10	15	20	25	5^n
P/O SL Luo converters—main series	6	36	216	1296	7,776	6^n
P/O SL Luo converters—additional series	11	66	396	2,376	14,256	$11*6^{(n-1)}$
P/O SL Luo converters—enhanced series	11	121	1,331	14,641	$16*10^4$	11^n
P/O SL Luo converters—re-enhanced series	16	256	4,096	65,536	$104*10^4$	16^n
P/O SL Luo converters—multiple ($j=4$)-enhanced series	26	676	17,576	$46*10^4$	$12*10^6$	26^n

10.3.1 Main Series

The first three stages of N/O SL Luo converters and their main series are shown in Figures 10.17 through 10.19. For the sake of convenience, they are called elementary circuit, re-lift circuit, and triple-lift circuit, respectively, and numbered $n=1, 2,$ and 3.

10.3.1.1 N/O Elementary Circuit

N/O elementary circuit and its equivalent circuits during the switch-on and switch-off equivalent periods are shown in Figure 10.17.

The voltage across capacitor C_1 is charged to V_{in}. The current flowing through inductor L_1 increases with slope V_{in}/L_1 during the switch-on period kT and decreases with slope $-(V_O - V_{in})/L_1$ during the switch-off period $(1 - k)T$. Therefore, the variation of current i_{L1} is

$$\Delta i_{L1} = \frac{V_{in}}{L_1} kT = \frac{V_O - V_{in}}{L_1} (1-k)T \tag{10.113}$$

$$V_O = \frac{1}{1-k} V_{in} = \left(\frac{2-k}{1-k} - 1\right) V_{in} \tag{10.114}$$

The voltage transfer gain is

$$G_1 = \frac{V_O}{V_{in}} = \frac{2-k}{1-k} - 1 \tag{10.115}$$

If inductance L_1 is large enough, i_{L1} is nearly equal to its average current I_{L1}.

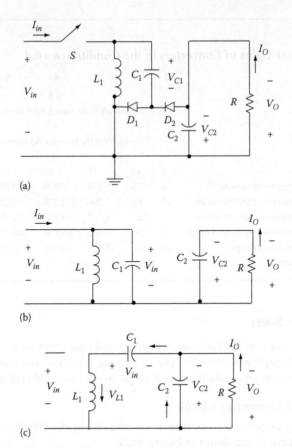

FIGURE 10.17 Elementary circuit of N/O SL Luo converters—main series. (a) Circuit diagram. (b) Equivalent circuit during the switch-on period. (c) Equivalent circuit during the switch-off period.

Therefore,

$$i_{in\text{-}on} = i_{L1\text{-}on} + i_{C1\text{-}on} = i_{L1\text{-}on} + \frac{1-k}{k} i_{C1\text{-}off} = \left(1 + \frac{1-k}{k}\right) I_{L1} = \frac{1}{k} I_{L1}$$

and

$$I_{in} = k i_{in\text{-}on} = I_{L1} \qquad\qquad (10.116)$$

Variation ratio of inductor current i_{L1} is

$$\xi_1 = \frac{\Delta i_{L1}/2}{I_{L1}} = \frac{k(1-k)TV_{in}}{2L_1 I_O} = \frac{k(1-k)}{G_1} \frac{R}{2fL_1} \qquad\qquad (10.117)$$

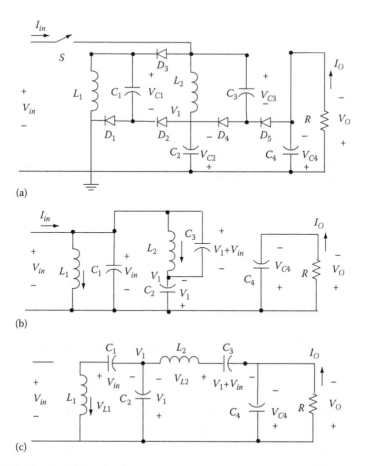

(a)

(b)

(c)

FIGURE 10.18 Re-lift circuit of N/O SL Luo converters—main series. (a) Circuit diagram. (b) Equivalent circuit during the switch-on period. (c) Equivalent circuit during the switch-off period.

Usually ξ_1 is small (much lower than unity), which means this converter works in the CCM.

The ripple voltage of output voltage v_O is

$$\Delta v_O = \frac{\Delta Q}{C_2} = \frac{I_O kT}{C_2} = \frac{k}{fC_2}\frac{V_O}{R}$$

Therefore, the variation ratio of output voltage v_O is

$$\varepsilon = \frac{\Delta v_O / 2}{V_O} = \frac{k}{2RfC_2} \tag{10.118}$$

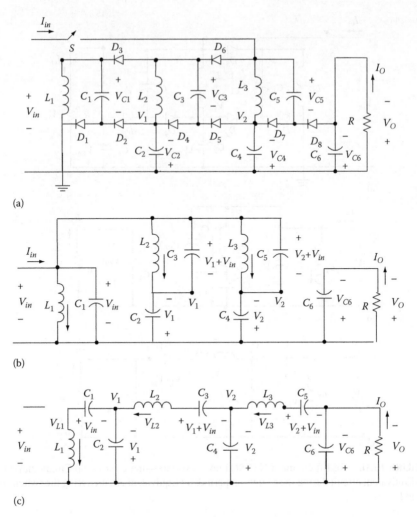

(a)

(b)

(c)

FIGURE 10.19 Triple-lift circuit of N/O SL Luo converters—main series. (a) Circuit diagram. (b) Equivalent circuit during the switch-on period. (c) Equivalent circuit during the switch-off period.

10.3.1.2 N/O Re-Lift Circuit

N/O re-lift circuit is derived from N/O elementary circuit by adding the parts L_2-D_3-D_4-D_5-C_3-C_4. Its circuit diagram and equivalent circuits during the switch-on and switch-off periods are shown in Figure 10.18.

The voltage across capacitor C_1 is charged to V_{in}. As described in the previous section, the voltage V_1 across capacitor C_2 is $V_1 = (1/1 - k)V_{in}$.

The voltage across capacitor C_3 is charged to $(V_1 + V_{in})$. The current flowing through inductor L_2 increases with slope $(V_1 + V_{in})/L_2$ during the switch-on period kT and decreases with slope $-(V_O - 2V_1 - V_{in})/L_2$ during the switch-off period $(1 - k)T$. Therefore, the variation of current i_{L2} is

$$\Delta i_{L2} = \frac{V_1 + V_{in}}{L_2} kT = \frac{V_O - 2V_1 - V_{in}}{L_2}(1-k)T \qquad (10.119)$$

$$V_O = \frac{(2-k)V_1 + V_{in}}{1-k} = \left[\left(\frac{2-k}{1-k}\right)^2 - 1\right]V_{in} \qquad (10.120)$$

The voltage transfer gain is

$$G_2 = \frac{V_O}{V_{in}} = \left(\frac{2-k}{1-k}\right)^2 - 1 \qquad (10.121)$$

The variation ratio of current i_{L1} through inductor L_1 is

$$\xi_1 = \frac{\Delta i_{L1}/2}{I_{L1}} = \frac{kTV_{in}}{\dfrac{2-k}{(1-k)^2}2L_1 I_O} = \frac{k(1-k)^2}{(2-k)G_2}\frac{R}{2fL_1} \qquad (10.122)$$

The variation ratio of current i_{L2} through inductor L_2 is

$$\xi_2 = \frac{\Delta i_{L2}/2}{I_{L2}} = \frac{k(2-k)TV_{in}}{2L_2 I_O} = \frac{k(2-k)}{G_2}\frac{R}{2fL_2} \qquad (10.123)$$

The ripple voltage of output voltage v_O is

$$\Delta v_O = \frac{\Delta Q}{C_4} = \frac{I_O kT}{C_4} = \frac{k}{fC_4}\frac{V_O}{R}$$

Therefore, the variation ratio of output voltage v_O is

$$\varepsilon = \frac{\Delta v_O/2}{V_O} = \frac{k}{2RfC_4} \qquad (10.124)$$

Example 10.3 A N/O re-lift circuit in Figure 10.18a has $V_{in} = 20\,V$, all inductors are 10 mH, all capacitors are 20 µF, $R = 200\,\Omega$, $f = 50\,kHz$, and conduction duty cycle $k = 0.6$. Calculate the variation ratio of current i_{L1}, the output voltage, and its variation ratio.

Solution: From formula (10.122), we can get the variation ratio of current i_{L1}:

$$\xi_1 = \frac{k(1-k)^3}{(2-k)^2}\frac{R}{2fL_1} = \frac{0.6(1-0.6)^3}{(2-0.6)^2}\frac{200}{2\times 50k \times 10m} = 0.0039$$

From formula (10.120), we can get the output voltage:

$$V_O = \left[\left(\frac{2-k}{1-k}\right)^2 - 1\right]V_{in} = \left[\left(\frac{2-0.6}{1-0.6}\right)^2 - 1\right] \times 20 = 225\,\text{V}$$

From (10.124), its variation ratio is

$$\varepsilon = \frac{k}{2RfC_4} = \frac{0.6}{2 \times 200 \times 50k \times 20\mu} = 0.0015$$

10.3.1.3 N/O Triple-Lift Circuit

N/O triple-lift circuit is derived from N/O re-lift circuit by readding the parts L_2-D_3-D_4-D_5-C_3-C_4. Its circuit diagram and equivalent circuits during the switch-on and switch-off periods are shown in Figure 10.19.

The voltage across capacitor C_1 is charged to V_{in}. As described in the previous section, the voltage V_1 across capacitor C_2 is $V_1 = [(2 - k/1 - k) - 1]V_{in} = (1/1 - k)V_{in}$, and the voltage V_2 across capacitor C_4 is $V_2 = [((2 - k)/(1 - k))^2 - 1]V_{in} = [(3 - 2k)/(1 - k)^2]V_{in}$.

The voltage across capacitor C_5 is charged to $(V_2 + V_{in})$. The current flowing through inductor L_3 increases with slope $(V_2 + V_{in})/L_3$ during the switch-on period kT and decreases with slope $-(V_O - 2V_2 - V_{in})/L_3$ during the switch-off period $(1 - k)T$. Therefore, the variation of current i_{L3} is

$$\Delta i_{L3} = \frac{V_2 + V_{in}}{L_3}kT = \frac{V_O - 2V_2 - V_{in}}{L_3}(1-k)T \tag{10.125}$$

$$V_O = \frac{(2-k)V_2 + V_{in}}{1-k} = \left[\left(\frac{2-k}{1-k}\right)^3 - 1\right]V_{in} \tag{10.126}$$

The voltage transfer gain is

$$G_3 = \frac{V_O}{V_{in}} = \left(\frac{2-k}{1-k}\right)^3 - 1 \tag{10.127}$$

The variation ratio of current i_{L1} through inductor L_1 is

$$\xi_1 = \frac{\Delta i_{L1}/2}{I_{L1}} = \frac{k(1-k)^3 TV_{in}}{2(2-k)^2 L_1 I_O} = \frac{k(1-k)^3}{(2-k)^2 G_3}\frac{R}{2fL_1} \tag{10.128}$$

The variation ratio of current i_{L2} through inductor L_2 is

$$\xi_2 = \frac{\Delta i_{L2}/2}{I_{L2}} = \frac{k(1-k)TV_{in}}{2L_2 I_O} = \frac{k(1-k)}{G_3}\frac{R}{2fL_2} \tag{10.129}$$

The variation ratio of current i_{L3} through inductor L_3 is

$$\xi_3 = \frac{\Delta i_{L3}/2}{I_{L3}} = \frac{k(2-k)^2 T V_{in}}{2(1-k)L_3 I_O} = \frac{k(2-k)^2}{(1-k)G_3} \frac{R}{2fL_3} \tag{10.130}$$

The ripple voltage of output voltage v_O is

$$\Delta v_O = \frac{\Delta Q}{C_6} = \frac{I_O kT}{C_6} = \frac{k}{fC_6} \frac{V_O}{R}$$

Therefore, the variation ratio of output voltage v_O is

$$\varepsilon = \frac{\Delta v_O/2}{V_O} = \frac{k}{2RfC_6} \tag{10.131}$$

10.3.1.4 N/O Higher-Order-Lift Circuit

N/O higher-order-lift circuit can be designed by reading the parts L_2-D_3-D_4-D_5-C_3-C_4 multiple times. For the nth-order-lift circuit, the final output voltage across capacitor C_{2n} is

$$V_O = \left[\left(\frac{2-k}{1-k}\right)^n - 1\right] V_{in} \tag{10.132}$$

The voltage transfer gain is

$$G_n = \frac{V_O}{V_{in}} = \left(\frac{2-k}{1-k}\right)^n - 1 \tag{10.133}$$

The variation ratio of current i_{Li} through inductor L_i ($i = 1, 2, 3, \ldots, n$) is

$$\xi_1 = \frac{\Delta i_{L1}/2}{I_{L1}} = \frac{k(1-k)^n}{(2-k)^{(n-1)}G_n} \frac{R}{2fL_i} \tag{10.134}$$

$$\xi_2 = \frac{\Delta i_{L2}/2}{I_{L2}} = \frac{k(2-k)^{(3-n)}}{(1-k)^{(n-3)}G_n} \frac{R}{2fL_i} \tag{10.135}$$

$$\xi_3 = \frac{\Delta i_{L3}/2}{I_{L3}} = \frac{k(2-k)^{(n-i+2)}}{(1-k)^{(n-i+1)}G_n} \frac{R}{2fL_3} \tag{10.136}$$

The variation ratio of output voltage v_O is

$$\varepsilon = \frac{\Delta v_O/2}{V_O} = \frac{k}{2RfC_{2n}} \tag{10.137}$$

10.3.2 N/O Additional Series

All circuits of N/O SL Luo converters and their additional series are derived from the corresponding circuits of the main series by adding a DEC. The first three stages of this series are shown in Figures 10.20 through 10.22. For the sake of convenience, they are called elementary additional circuit, re-lift additional circuit, and triple-lift additional circuit, respectively, and numbered $n = 1$, 2, and 3.

10.3.2.1 N/O Elementary Additional Circuit

This circuit is derived from N/O elementary circuit by adding a DEC. Its circuit diagram and switch-on and switch-off equivalent circuits are shown in Figure 10.20. The voltage across capacitor C_1 is charged to V_{in}. The voltage across capacitor C_2 is charged to V_1 and C_{11} is charged to $(V_1 + V_{in})$. The current i_{L1} flowing through

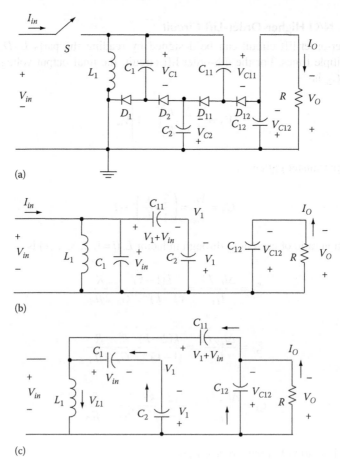

(a)

(b)

(c)

FIGURE 10.20 Elementary additional circuit of N/O SL Luo converters. (a) Circuit diagram. (b) Equivalent circuit during the switch-on period. (c) Equivalent circuit during the switch-off.

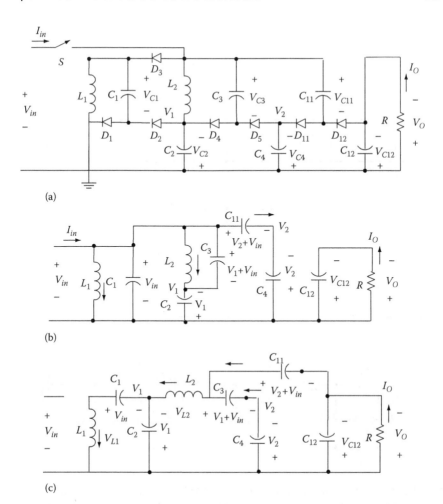

(a)

(b)

(c)

FIGURE 10.21 Re-lift additional circuit of N/O SL Luo converters. (a) Circuit diagram. (b) Equivalent circuit during the switch-on period. (c) Equivalent circuit during the switch-off period.

inductor L_1 increases with slope V_{in}/L_1 during the switch-on period kT and decreases with slope $-(V_1 - V_{in})/L_1$ during the switch-off period $(1 - k)T$.

Therefore,

$$\Delta i_{L1} = \frac{V_{in}}{L_1} kT = \frac{V_1 - V_{in}}{L_1}(1-k)T \tag{10.138}$$

$$V_1 = \frac{1}{1-k} V_{in} = \left(\frac{2-k}{1-k} - 1\right) V_{in}$$

$$V_{L1\text{-off}} = \frac{k}{1-k} V_{in}$$

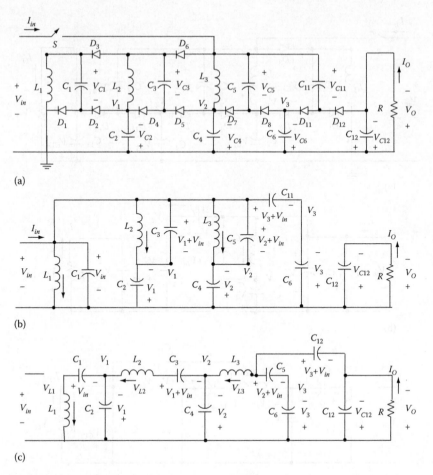

FIGURE 10.22 Triple-lift additional circuit of N/O SL Luo converters. (a) Circuit diagram. (b) Equivalent circuit during the switch-on period. (c) Equivalent circuit during the switch-off period.

The output voltage is

$$V_O = V_{in} + V_{L1} + V_1 = \frac{2}{1-k}V_{in} = \left[\frac{3-k}{1-k} - 1\right]V_{in} \tag{10.139}$$

The voltage transfer gain is

$$G_1 = \frac{V_O}{V_{in}} = \frac{3-k}{1-k} - 1 \tag{10.140}$$

The variation ratio of current i_{L1} through inductor L_1 is

$$\xi_1 = \frac{\Delta i_{L1}/2}{I_{L1}} = \frac{k(1-k)TV_{in}}{4L_1 I_O} = \frac{k(1-k)}{2G_1} \frac{R}{2fL_1} \qquad (10.141)$$

The ripple voltage of output voltage v_O is

$$\Delta v_O = \frac{\Delta Q}{C_{12}} = \frac{I_O kT}{C_{12}} = \frac{k}{fC_{12}} \frac{V_O}{R}$$

Therefore, the variation ratio of output voltage v_O is

$$\varepsilon = \frac{\Delta v_O/2}{V_O} = \frac{k}{2RfC_{12}} \qquad (10.142)$$

10.3.2.2 N/O Re-Lift Additional Circuit

The N/O re-lift additional circuit is derived from N/O re-lift circuit by adding a DEC. Its circuit diagram and switch-on and switch-off equivalent circuits are shown in Figure 10.21.

The voltage across capacitor C_1 is charged to V_{in}. As described in the previous section, the voltage across C_2 is

$$V_1 = \frac{1}{1-k} V_{in}$$

The voltage across capacitor C_3 is charged to $(V_1 + V_{in})$, voltage across capacitor C_4 is charged to V_2, and the voltage across capacitor C_{11} is charged to $(V_2 + V_{in})$. The current flowing through inductor L_2 increases with voltage $(V_1 + V_{in})$ during the switch-on period kT and decreases with voltage $-(V_2 - 2V_1 - V_{in})$ during the switch-off period $(1 - k)T$. Therefore,

$$\Delta i_{L2} = \frac{V_1 + V_{in}}{L_2} kT = \frac{V_2 - 2V_1 - V_{in}}{L_2}(1-k)T \qquad (10.143)$$

$$V_2 = \frac{(2-k)V_1 + V_{in}}{1-k} = \frac{3-2k}{(1-k)^2} = \left[\left(\frac{2-k}{1-k} \right)^2 - 1 \right] V_{in}$$

and

$$V_{L2\text{-}off} = V_2 - 2V_1 - V_{in} = \frac{k(2-k)}{(1-k)^2} V_{in} \qquad (10.144)$$

The output voltage is

$$V_O = V_2 + V_{in} + V_{L2} + V_1 = \frac{5-3k}{(1-k)^2} V_{in} = \left[\frac{3-k}{1-k} \frac{2-k}{1-k} - 1 \right] V_{in} \qquad (10.145)$$

The voltage transfer gain is

$$G_2 = \frac{V_O}{V_{in}} = \frac{2-k}{1-k} \frac{3-k}{1-k} - 1 \qquad (10.146)$$

The variation ratio of current i_{L1} through inductor L_1 is

$$\xi_1 = \frac{\Delta i_{L1}/2}{I_{L1}} = \frac{k(1-k)^2 T V_{in}}{2(3-k)L_1 I_O} = \frac{k(1-k)^2}{(3-k)G_2} \frac{R}{2fL_1} \qquad (10.147)$$

The variation ratio of current i_{L2} through inductor L_2 is

$$\xi_2 = \frac{\Delta i_{L2}/2}{I_{L2}} = \frac{k(2-k)T V_{in}}{4L_2 I_O} = \frac{k(2-k)}{2G_2} \frac{R}{2fL_2} \qquad (10.148)$$

The ripple voltage of output voltage v_O is

$$\Delta v_O = \frac{\Delta Q}{C_{12}} = \frac{I_O kT}{C_{12}} = \frac{k}{fC_{12}} \frac{V_O}{R}$$

Therefore, the variation ratio of output voltage v_O is

$$\varepsilon = \frac{\Delta v_O/2}{V_O} = \frac{k}{2RfC_{12}} \qquad (10.149)$$

10.3.2.3 Triple-Lift Additional Circuit

This circuit is derived from N/O triple-lift circuit by adding a DEC. Its circuit diagram and equivalent circuits during the switch-on and switch-off periods are shown in Figure 10.22.

The voltage across capacitor C_1 is charged to V_{in}. As described in the previous section, the voltage across C_2 is $V_1 = (1/1-k)V_{in}$ and the voltage across C_4 is

$$V_2 = \frac{3-2k}{1-k} V_1 = \frac{3-2k}{(1-k)^2} V_{in}$$

The voltage across capacitor C_5 is charged to $(V_2 + V_{in})$, the voltage across capacitor C_6 is charged to V_3, and the voltage across capacitor C_{11} is charged to $(V_3 + V_{in})$.

The current flowing through inductor L_3 increases with voltage $(V_2 + V_{in})$ during the switch-on period kT and decreases with voltage $-(V_3 - 2V_2 - V_{in})$ during the switch-off period $(1 - k)T$. Therefore,

$$\Delta i_{L3} = \frac{V_2 + V_{in}}{L_3} kT = \frac{V_3 - 2V_2 - V_{in}}{L_3}(1-k)T \qquad (10.150)$$

$$V_3 = \frac{(2-k)V_2 + V_{in}}{1-k} = \frac{7 - 9k + 3k^2}{(1-k)^3}V_{in} = \left[\left(\frac{2-k}{1-k}\right)^3 - 1\right]V_{in}$$

and

$$V_{L3\text{-}off} = V_3 - 2V_2 - V_{in} = \frac{k(2-k)^2}{(1-k)^3}V_{in} \qquad (10.151)$$

The output voltage is

$$V_O = V_3 + V_{in} + V_{L3} + V_2 = \frac{11 - 13k + 4k^2}{(1-k)^3}V_{in} = \left[\frac{3-k}{1-k}\left(\frac{2-k}{1-k}\right)^2 - 1\right]V_{in} \qquad (10.152)$$

The voltage transfer gain is

$$G_3 = \frac{V_O}{V_{in}} = \left(\frac{2-k}{1-k}\right)^2 \frac{3-k}{1-k} - 1 \qquad (10.153)$$

The variation ratio of current i_{L1} through inductor L_1 is

$$\xi_1 = \frac{\Delta i_{L1}/2}{I_{L1}} = \frac{k(1-k)^3 TV_{in}}{2(2-k)(3-k)L_1 I_O} = \frac{k(1-k)^3}{(2-k)(3-k)G_3}\frac{R}{2fL_1} \qquad (10.154)$$

and the variation ratio of current i_{L2} through inductor L_2 is

$$\xi_2 = \frac{\Delta i_{L2}/2}{I_{L2}} = \frac{k(1-k)(2-k)TV_1}{2(3-k)L_2 I_O} = \frac{k(1-k)(2-k)}{(3-k)G_3}\frac{R}{2fL_2} \qquad (10.155)$$

and the variation ratio of current i_{L3} through inductor L_3 is

$$\xi_3 = \frac{\Delta i_{L3}/2}{I_{L3}} = \frac{k(2-k)^2 TV_{in}}{4(1-k)L_3 I_O} = \frac{k(2-k)^2}{2(1-k)G_3}\frac{R}{2fL_3} \qquad (10.156)$$

The ripple voltage of output voltage v_O is

$$\Delta v_O = \frac{\Delta Q}{C_{12}} = \frac{I_O kT}{C_{12}} = \frac{k}{fC_{12}} \frac{V_O}{R}$$

Therefore, the variation ratio of output voltage v_O is

$$\varepsilon = \frac{\Delta v_O / 2}{V_O} = \frac{k}{2RfC_{12}} \qquad (10.157)$$

10.3.2.4   N/O Higher-Order-Lift Additional Circuit

N/O higher-order-lift additional circuit can be derived from the corresponding circuit of the main series by adding a DEC. Each stage voltage V_i $(i = 1, 2, ..., n)$ is

$$V_i = \left[\left(\frac{2-k}{1-k} \right)^i - 1 \right] V_{in} \qquad (10.158)$$

It means V_1 is the voltage across capacitor C_2, V_2 is the voltage across capacitor C_4, and so on.

For the nth-order-lift additional circuit, the final output voltage is

$$V_O = \left[\frac{3-k}{1-k} \left(\frac{2-k}{1-k} \right)^{n-1} - 1 \right] V_{in} \qquad (10.159)$$

The voltage transfer gain is

$$G_n = \frac{V_O}{V_{in}} = \frac{3-k}{1-k} \left(\frac{2-k}{1-k} \right)^{n-1} - 1 \qquad (10.160)$$

Analogously, the variation ratio of current i_{Li} through inductor L_i $(i = 1, 2, 3, ..., n)$ is

$$\xi_1 = \frac{\Delta i_{L1} / 2}{I_{L1}} = \frac{k(1-k)^n}{2^{h(1-n)}[(2-k)^{(n-2)}(3-k)]^{u(n-2)} G_n} \frac{R}{fL_1} \qquad (10.161)$$

$$\xi_2 = \frac{\Delta i_{L2} / 2}{I_{L2}} = \frac{k(1-k)^{(n-2)}(2-k)}{2^{h(n-2)}(3-k)^{(n-2)} G_n} \frac{R}{2fL_2} \qquad (10.162)$$

$$\xi_3 = \frac{\Delta i_{L3}/2}{I_{L3}} = \frac{k(2-k)^{(n-1)}}{2^{h(n-3)}(1-k)^{(n-2)}G_n}\frac{R}{2fL_3} \qquad (10.163)$$

where

$$h(x) = \begin{cases} 0 & x > 0 \\ 1 & x \le 0 \end{cases} \quad \text{is the } Hong \text{ function}$$

$$u(x) = \begin{cases} 1 & x \ge 0 \\ 0 & x < 0 \end{cases} \quad \text{is the } unit\text{-}step \text{ function}$$

the variation ratio of output voltage v_O is

$$\varepsilon = \frac{\Delta v_O/2}{V_O} = \frac{k}{2RfC_{12}} \qquad (10.164)$$

10.3.3 ENHANCED SERIES

All circuits of N/O SL Luo converters and their enhanced series are derived from the corresponding circuits of the main series by adding the DEC into each stage circuit of all series of converters.

The first three stages of this series are shown in Figures 10.20, 10.23, and 10.24. For the sake of convenience, they are called elementary enhanced circuit, re-lift enhanced circuit, and triple-lift enhanced circuit, respectively, and numbered $n=1$, 2, and 3.

10.3.3.1 N/O Elementary Enhanced Circuit

The circuit is shared in additional and enhanced series of N/O elementary SL Luo-converters as in Figure 10.31, and is the same to the N/O elementary additional circuit shown in Figure 10.20. Analysis can be referred to Section 10.3.2.1.

10.3.3.2 N/O Re-Lift Enhanced Circuit

The N/O re-lift enhanced circuit is derived from N/O re-lift circuit of the main series by adding the DEC into each stage. Its circuit diagram and switch-on and switch-off equivalent circuits are shown in Figure 10.23.

The voltage across capacitor C_{12} is charged to

$$V_{C12} = \frac{3}{1-k}V_{in} \qquad (10.165)$$

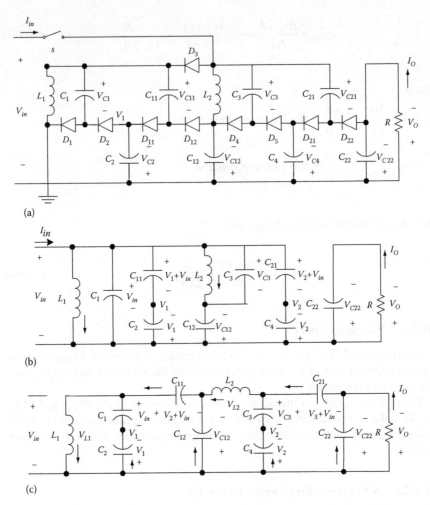

(a)

(b)

(c)

FIGURE 10.23 Re-lift enhanced circuit of N/O SL Luo converters. (a) Circuit diagram. (b) Equivalent circuit during the switch-on period. (c) Equivalent circuit during the switch-off period.

The voltage across capacitor C_3 is charged to V_{C12}, and the voltage across capacitors C_4 and C_{12} is charged to V_{C4}:

$$V_{C4} = \frac{2-k}{1-k}V_{C12} = \frac{2-k}{1-k}\frac{3-k}{1-k}V_{in} \qquad (10.166)$$

The current flowing through inductor L_2 increases with voltage V_{C12} during the switch-on period kT and decreases with voltage $-(V_{C21} - V_{C4} - V_{C12})$ during the switch-off period $(1 - k)T$.

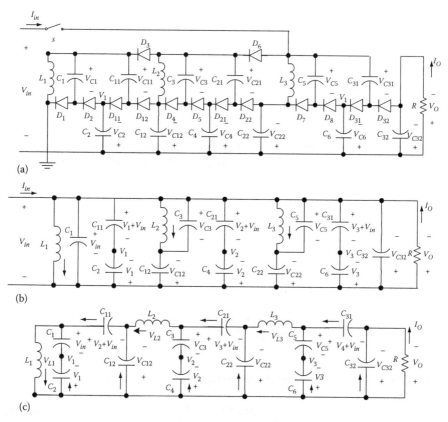

FIGURE 10.24 Triple-lift enhanced circuit of N/O SL Luo converters. (a) Circuit diagram. (b) Equivalent circuit during the switch-on period. (c) Equivalent circuit during the switch-off period.

Therefore,

$$\Delta i_{L2} = \frac{kT}{L_2}(V_{C12} - V_{in}) = \frac{V_{C21} - V_{C4} - V_{C12}}{L_2}(1-k)T \tag{10.167}$$

$$V_{C21} = \left(\frac{3-k}{1-k}\right)^2 V_{in}$$

The output voltage is

$$V_O = V_{C21} - V_{in} = \left[\left(\frac{3-k}{1-k}\right)^2 - 1\right] V_{in} \tag{10.168}$$

The voltage transfer gain is

$$G_2 = \frac{V_O}{V_{in}} = \left(\frac{3-k}{1-k}\right)^2 - 1 \qquad (10.169)$$

The variation ratio of current i_{L1} through inductor L_1 is

$$\xi_1 = \frac{\Delta i_{L1}/2}{I_{L1}} = \frac{k(1-k)^2 T V_{in}}{4(3-k)L_1 I_O} = \frac{k(1-k)^2}{2(3-k)G_2} \frac{R}{2fL_1} \qquad (10.170)$$

The variation ratio of current i_{L2} through inductor L_2 is

$$\xi_2 = \frac{\Delta i_{L2}/2}{I_{L2}} = \frac{k(2+k)T V_{in}}{4L_2 I_O} = \frac{k(2+k)}{2G_2} \frac{R}{2fL_2} \qquad (10.171)$$

The ripple voltage of output voltage v_O is

$$\Delta v_O = \frac{\Delta Q}{C_{22}} = \frac{I_O k T}{C_{22}} = \frac{k}{fC_{22}} \frac{V_O}{R}$$

Therefore, the variation ratio of output voltage v_O is

$$\varepsilon = \frac{\Delta v_O/2}{V_O} = \frac{k}{2RfC_{22}} \qquad (10.172)$$

10.3.3.3 N/O Triple-Lift Enhanced Circuit

This circuit is derived from N/O triple-lift circuit of main series by adding the DEC into each stage. Its circuit diagram and equivalent circuits during the switch-on and switch-off periods are shown in Figure 10.24.

The voltage across capacitor C_{12} is charged to V_{C12}. As described in the previous section the voltage across C_{C12} is $V_{C12} = ((3-k)/(1-k))V_{in}$ and the voltage across C_4 and C_{C22} is

$$V_{C22} = \frac{3-k}{1-k} V_{C12} = \left(\frac{3-k}{1-k}\right)^2 V_{in}$$

The voltage across capacitor C_5 is charged to V_{C22}, and the voltage across capacitor C_6 is charged to V_{C6}:

$$V_{C6} = \frac{2-k}{1-k} V_{C22} = \frac{2-k}{1-k}\left(\frac{3-k}{1-k}\right)^2 V_{in}$$

The current flowing through inductor L_3 increases with voltage V_{C22} during the switch-on period kT and decreases with voltage $-(V_{C32} - V_{C6} - V_{C22})$ during the switch-off period $(1 - k)T$.

Therefore,

$$\Delta i_{L3} = \frac{kT}{L_3}(V_{C22} - V_{in}) = \frac{V_{C31} - V_{C6} - V_{C22}}{L_3}(1-k)T \tag{10.173}$$

$$V_{C31} = \left(\frac{3-k}{1-k}\right)^3 V_{in}$$

and

$$V_O = V_{C31} - V_{in} = \left[\left(\frac{3-k}{1-k}\right)^3 - 1\right]V_{in} \tag{10.174}$$

The voltage transfer gain is

$$G_3 = \frac{V_O}{V_{in}} = \left(\frac{3-k}{1-k}\right)^2 - 1 \tag{10.175}$$

The variation ratio of current i_{L1} through inductor L_1 is

$$\xi_1 = \frac{\Delta i_{L1}/2}{I_{L1}} = \frac{k(1-k)^3 T V_{in}}{4(4-k)(3-k)L_1 I_O} = \frac{k(1-k)^3}{2(4-k)(3-k)G_3} \frac{R}{2fL_1} \tag{10.176}$$

and the variation ratio of current i_{L2} through inductor L_2 is

$$\xi_2 = \frac{\Delta i_{L2}/2}{I_{L2}} = \frac{k(1-k)(2-k)T V_1}{4(3-k)L_2 I_O} = \frac{k(1-k)(2-k)}{2(3-k)G_3} \frac{R}{2fL_2} \tag{10.177}$$

and the variation ratio of current i_{L3} through inductor L_3 is

$$\xi_3 = \frac{\Delta i_{L3}/2}{I_{L3}} = \frac{k(2-k)^2 T V_{in}}{4(1-k)L_3 I_O} = \frac{k(2-k)^2}{2(1-k)G_3} \frac{R}{2fL_3} \tag{10.178}$$

The ripple voltage of output voltage v_O is

$$\Delta v_O = \frac{\Delta Q}{C_{32}} = \frac{I_O kT}{C_{32}} = \frac{k}{fC_{32}} \frac{V_O}{R}$$

Therefore, the variation ratio of output voltage v_O is

$$\varepsilon = \frac{\Delta v_O / 2}{V_O} = \frac{k}{2 R f C_{32}} \tag{10.179}$$

10.3.3.4 N/O Higher-Order-Lift Enhanced Circuit

N/O higher-order-lift enhanced circuit is derived from the corresponding circuit of the main series by adding the DEC in each stage. Each stage final voltage V_{Ci1} ($i = 1, 2, \ldots, n$) is

$$V_{Ci1} = \left(\frac{3-k}{1-k} \right)^i V_{in} \tag{10.180}$$

For the nth-order-lift enhanced circuit, the final output voltage is

$$V_O = \left[\left(\frac{3-k}{1-k} \right)^n - 1 \right] V_{in} \tag{10.181}$$

The voltage transfer gain is

$$G_n = \frac{V_O}{V_{in}} = \left(\frac{3-k}{1-k} \right)^n - 1 \tag{10.182}$$

The variation ratio of output voltage v_O is

$$\varepsilon = \frac{\Delta v_O / 2}{V_O} = \frac{k}{2 R f C_{n2}} \tag{10.183}$$

10.3.4 RE-ENHANCED SERIES

All circuits of N/O SL Luo converters and their re-enhanced series are derived from the corresponding circuits of the main series by adding the DEC *twice* in each stage circuit.

The first three stages of this series are shown in Figures 10.25 through 10.27. For the sake of convenience, they are called elementary re-enhanced circuit, re-lift re-enhanced circuit, and triple-lift re-enhanced circuit, respectively, and numbered $n = 1, 2$, and 3.

10.3.4.1 N/O Elementary Re-Enhanced Circuit

This circuit is derived from N/O elementary circuit by adding the DEC twice. Its circuit diagram and switch-on and switch-off equivalent circuits are shown in Figure 10.25.

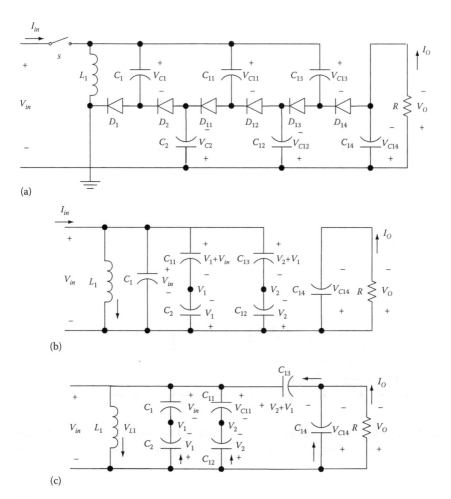

FIGURE 10.25 Elementary re-enhanced circuit of N/O SL Luo converters. (a) Circuit diagram. (b) Equivalent circuit during the switch-on period. (c) Equivalent circuit during the switch-off period.

The voltage across capacitor C_1 is charged to V_{in}. The voltage across capacitor C_{12} is charged to V_{C12}.

The voltage across capacitor C_{13} is charged to V_{C13}.

$$V_{C13} = \frac{4-k}{1-k}V_{in} \tag{10.184}$$

The output voltage is

$$V_O = V_{C13} - V_{in} = \left[\frac{4-k}{1-k} - 1\right]V_{in} \tag{10.185}$$

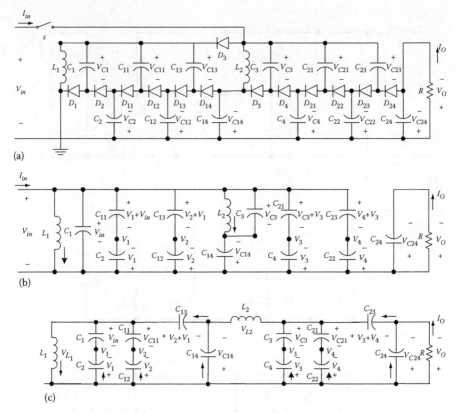

FIGURE 10.26 Re-lift re-enhanced circuit of N/O SL Luo converters. (a) Circuit diagram. (b) Equivalent circuit during the switch-on period. (c) Equivalent circuit during the switch-off period.

The voltage transfer gain is

$$G_1 = \frac{V_O}{V_{in}} = \frac{4-k}{1-k} - 1 \tag{10.186}$$

The ripple voltage of output voltage v_O is

$$\Delta v_O = \frac{\Delta Q}{C_{14}} = \frac{I_O kT}{C_{14}} = \frac{k}{fC_{14}} \frac{V_O}{R}$$

Therefore, the variation ratio of output voltage v_O is

$$\varepsilon = \frac{\Delta v_O / 2}{V_O} = \frac{k}{2RfC_{14}} \tag{10.187}$$

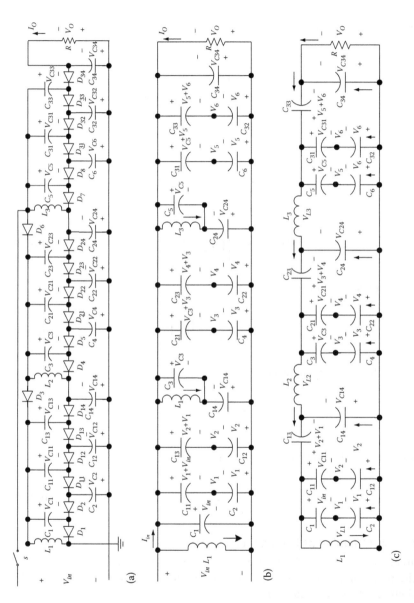

FIGURE 10.27 Triple-lift re-enhanced circuit of N/O SL Luo converters. (a) Circuit diagram. (b) Equivalent circuit during the switch-on period. (c) Equivalent circuit during the switch-off period.

10.3.4.2 N/O Re-Lift Re-Enhanced Circuit

The N/O re-lift re-enhanced circuit is derived from N/O re-lift circuit by adding the DEC twice in each stage. Its circuit diagram and switch-on and switch-off equivalent circuits are shown in Figure 10.26.

The voltage across capacitor C_{13} is charged to V_{C13}. As described in the previous section, the voltage across C_{13} is

$$V_{C13} = \frac{4-k}{1-k} V_{in}$$

Analogously,

$$V_{C23} = \left(\frac{4-k}{1-k}\right)^2 V_{in} \tag{10.188}$$

The output voltage is

$$V_O = V_{C23} - V_{in} = \left[\left(\frac{4-k}{1-k}\right)^2 - 1\right] V_{in} \tag{10.189}$$

The voltage transfer gain is

$$G_2 = \frac{V_O}{V_{in}} = \left(\frac{4-k}{1-k}\right)^2 - 1 \tag{10.190}$$

The ripple voltage of output voltage v_O is

$$\Delta v_O = \frac{\Delta Q}{C_{24}} = \frac{I_O kT}{C_{24}} = \frac{k}{fC_{24}} \frac{V_O}{R}$$

Therefore, the variation ratio of output voltage v_O is

$$\varepsilon = \frac{\Delta v_O / 2}{V_O} = \frac{k}{2RfC_{24}} \tag{10.191}$$

10.3.4.3 N/O Triple-Lift Re-Enhanced Circuit

This circuit is derived from N/O triple-lift circuit by adding the DEC twice in each stage circuit. Its circuit diagram and equivalent circuits during switch-on and switch-off periods are shown in Figure 10.27.

The voltage across capacitor C_{13} is

$$V_{C13} = \frac{4-k}{1-k} V_{in}$$

The voltage across capacitor C_{23} is

$$V_{C23} = \left(\frac{4-k}{1-k}\right)^2 V_{in}$$

Analogously, the voltage across capacitor C_{33} is

$$V_{C33} = \left(\frac{4-k}{1-k}\right)^3 V_{in} \tag{10.192}$$

The output voltage is

$$V_O = V_{C33} - V_{in} = \left[\left(\frac{4-k}{1-k}\right)^3 - 1\right] V_{in} \tag{10.193}$$

The voltage transfer gain is

$$G_3 = \frac{V_O}{V_{in}} = \left(\frac{4-k}{1-k}\right)^3 - 1 \tag{10.194}$$

The ripple voltage of output voltage v_O is

$$\Delta v_O = \frac{\Delta Q}{C_{34}} = \frac{I_O kT}{C_{34}} = \frac{k}{fC_{34}} \frac{V_O}{R}$$

Therefore, the variation ratio of output voltage v_O is

$$\varepsilon = \frac{\Delta v_O / 2}{V_O} = \frac{k}{2RfC_{34}} \tag{10.195}$$

10.3.4.4 N/O Higher-Order-Lift Re-Enhanced Circuit

N/O higher-order-lift re-enhanced circuit can be derived from the corresponding circuit of the main series by adding the DEC twice in each stage circuit. Each stage final voltage V_{Ci3} ($i = 1, 2, \ldots, n$) is

$$V_{Ci3} = \left(\frac{4-k}{1-k}\right)^i V_{in} \tag{10.196}$$

For the nth-order-lift additional circuit, the final output voltage is

$$V_O = V_{Cn3} - V_{in} = \left[\left(\frac{4-k}{1-k}\right)^n - 1\right] V_{in} \tag{10.197}$$

The voltage transfer gain is

$$G_n = \frac{V_O}{V_{in}} = \left(\frac{4-k}{1-k}\right)^n - 1 \tag{10.198}$$

The variation ratio of output voltage v_O is

$$\varepsilon = \frac{\Delta v_O / 2}{V_O} = \frac{k}{2RfC_{n4}} \tag{10.199}$$

10.3.5 N/O MULTIPLE-ENHANCED SERIES

All circuits of N/O SL Luo converters and their multiple-enhanced series are derived from the corresponding circuits of the main series by adding the DEC multiple (j) times in each stage circuit.

The first three stages of this series are shown in Figures 10.28 through 10.30. For the sake of convenience, they are called elementary multiple-enhanced circuit, re-lift multiple-enhanced circuit, and triple-lift multiple-enhanced circuit, respectively, and numbered $n = 1$, 2, and 3.

10.3.5.1 N/O Elementary Multiple-Enhanced Circuit

This circuit is derived from N/O elementary circuit by adding the DEC multiple (j) times. Its circuit diagram and switch-on and switch-off equivalent circuits are shown in Figure 10.28.

The voltage across capacitor C_{12j-1} is

$$V_{C12j-1} = \frac{j+2-k}{1-k} V_{in} \tag{10.200}$$

The output voltage is

$$V_O = V_{C12j-1} - V_{in} = \left[\frac{j+2-k}{1-k} - 1\right] V_{in} \tag{10.201}$$

The voltage transfer gain is

$$G_1 = \frac{V_O}{V_{in}} = \frac{j+2-k}{1-k} - 1 \tag{10.202}$$

The ripple voltage of output voltage v_O is

$$\Delta v_O = \frac{\Delta Q}{C_{12j}} = \frac{I_O kT}{C_{12j}} = \frac{k}{fC_{12j}} \frac{V_O}{R}$$

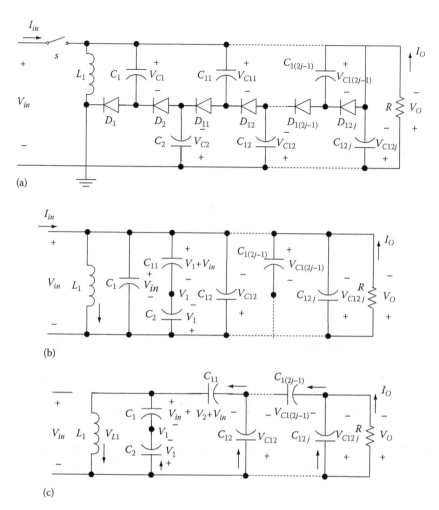

FIGURE 10.28 Elementary multiple-enhanced circuit of N/O SL Luo converters. (a) Circuit diagram. (b) Equivalent circuit during the switch-on period. (c) Equivalent circuit during the switch-off period.

Therefore, the variation ratio of output voltage v_O is

$$\varepsilon = \frac{\Delta v_O / 2}{V_O} = \frac{k}{2RfC_{12j}} \qquad (10.203)$$

10.3.5.2 N/O Re-Lift Multiple-Enhanced Circuit

The N/O re-lift multiple-enhanced circuit is derived from N/O re-lift circuit by adding the DEC multiple (j) times into each stage. Its circuit diagram and switch-on and switch-off equivalent circuits are shown in Figure 10.29.

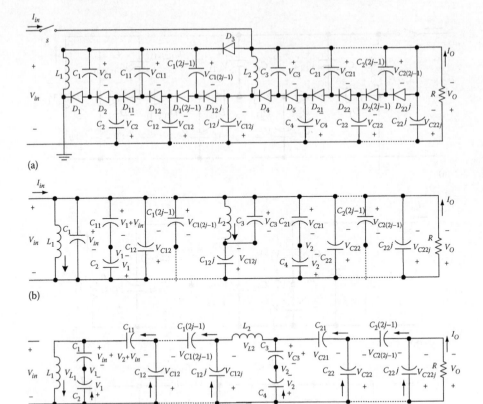

(a)

(b)

(c)

FIGURE 10.29 Re-lift multiple-enhanced circuit of the N/O SL Luo converters. (a) Circuit diagram. (b) Equivalent circuit during the switch-on period. (c) Equivalent circuit during the switch-off period.

The voltage across capacitor C_{22j-1} is

$$V_{C22j-1} = \left(\frac{j+2-k}{1-k} \right)^2 V_{in} \qquad (10.204)$$

The output voltage is

$$V_O = V_{C22j-1} - V_{in} = \left[\left(\frac{j+2-k}{1-k} \right)^2 - 1 \right] V_{in} \qquad (10.205)$$

The voltage transfer gain is

$$G_2 = \frac{V_O}{V_{in}} = \left(\frac{j+2-k}{1-k} \right)^2 - 1 \qquad (10.206)$$

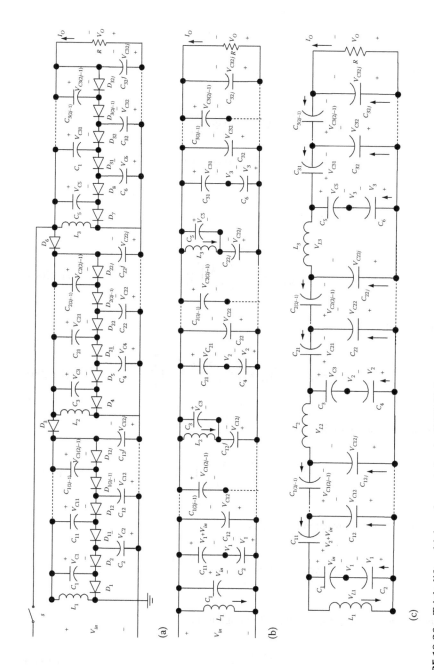

FIGURE 10.30 Triple-lift multiple-enhanced circuit of N/O SL Luo converters. (a) Circuit diagram. (b) Equivalent circuit during the switch-on period. (c) Equivalent circuit during the switch-off period.

The ripple voltage of output voltage v_O is

$$\Delta v_O = \frac{\Delta Q}{C_{22j}} = \frac{I_O kT}{C_{22j}} = \frac{k}{fC_{22j}} \frac{V_O}{R}$$

Therefore, the variation ratio of output voltage v_O is

$$\varepsilon = \frac{\Delta v_O / 2}{V_O} = \frac{k}{2RfC_{22j}} \tag{10.207}$$

10.3.5.3 N/O Triple-Lift Multiple-Enhanced Circuit

This circuit is derived from N/O triple-lift circuit by adding the DEC multiple (j) times in each stage circuit. Its circuit diagram and equivalent circuits during the switch-on and switch-off periods are shown in Figure 10.30.

The voltage across capacitor $C_{32}j - 1$ is

$$V_{C32j-1} = \left(\frac{j+2-k}{1-k}\right)^3 V_{in} \tag{10.208}$$

The output voltage is

$$V_O = V_{C32j-1} - V_{in} = \left[\left(\frac{j+2-k}{1-k}\right)^3 - 1\right] V_{in} \tag{10.209}$$

The voltage transfer gain is

$$G_3 = \frac{V_O}{V_{in}} = \left(\frac{j+2-k}{1-k}\right)^3 - 1 \tag{10.210}$$

The ripple voltage of output voltage v_O is

$$\Delta v_O = \frac{\Delta Q}{C_{32j}} = \frac{I_O kT}{C_{32j}} = \frac{k}{fC_{32j}} \frac{V_O}{R}$$

Therefore, the variation ratio of output voltage v_O is

$$\varepsilon = \frac{\Delta v_O / 2}{V_O} = \frac{k}{2RfC_{32j}} \tag{10.211}$$

10.3.5.4 N/O Higher-Order-Lift Multiple-Enhanced Circuit

N/O higher-order-lift multiple-enhanced circuit is derived from the corresponding circuit of the main series by adding the DEC multiple (j) times in each stage circuit. Each stage final voltage V_{Ci2j-1} ($i = 1, 2, \ldots, n$) is

$$V_{Ci2j-1} = \left(\frac{j+2-k}{1-k} \right)^i V_{in} \tag{10.212}$$

For the nth-order-lift multiple-enhanced circuit, the final output voltage is

$$V_O = \left[\left(\frac{j+2-k}{1-k} \right)^n - 1 \right] V_{in} \tag{10.213}$$

The voltage transfer gain is

$$G_n = \frac{V_O}{V_{in}} = \left(\frac{j+2-k}{1-k} \right)^n - 1 \tag{10.214}$$

The variation ratio of output voltage v_O is

$$\varepsilon = \frac{\Delta v_O / 2}{V_O} = \frac{k}{2RfC_{n2j}} \tag{10.215}$$

10.3.6 Summary of N/O SL Luo Converters

A family tree of all circuits of N/O SL Luo converters are shown in Figure 10.31.

Based on the analyses provided in the previous sections, the common formula to calculate the output voltage can be presented as follows:

$$V_O = \begin{cases} \left[\left(\dfrac{2-k}{1-k} \right)^n - 1 \right] V_{in} & \text{Main\_series} \\[3ex] \left[\left(\dfrac{2-k}{1-k} \right)^{n-1} \left(\dfrac{3-k}{1-k} \right) - 1 \right] V_{in} & \text{Additional\_series} \\[3ex] \left[\left(\dfrac{3-k}{1-k} \right)^n - 1 \right] V_{in} & \text{Enhanced\_series} \\[3ex] \left[\left(\dfrac{4-k}{1-k} \right)^n - 1 \right] V_{in} & \text{Re-enhanced\_series} \\[3ex] \left[\left(\dfrac{j+2-k}{1-k} \right)^n - 1 \right] V_{in} & \text{Multiple-enhanced\_series} \end{cases} \tag{10.216}$$

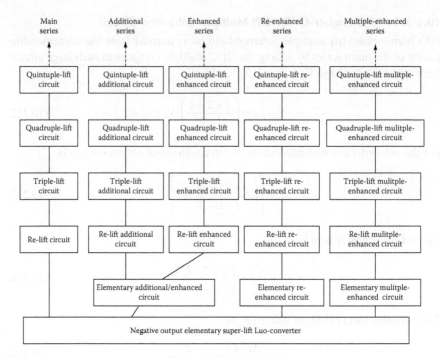

FIGURE 10.31 The family of N/O SL Luo converters.

The corresponding voltage transfer gain is

$$
G = \frac{V_O}{V_{in}} =
\begin{cases}
\left(\dfrac{2-k}{1-k}\right)^{n} - 1 & \text{Main\_series} \\[2.2ex]
\left(\dfrac{2-k}{1-k}\right)^{n-1}\left(\dfrac{3-k}{1-k}\right) - 1 & \text{Additional\_series} \\[2.2ex]
\left(\dfrac{3-k}{1-k}\right)^{n} - 1 & \text{Enhanced\_series} \\[2.2ex]
\left(\dfrac{4-k}{1-k}\right)^{n} - 1 & \text{Re-enhanced\_series} \\[2.2ex]
\left(\dfrac{j+2-k}{1-k}\right)^{n} - 1 & \text{Multiple-enhanced\_series}
\end{cases}
\tag{10.217}
$$

In order to show the advantages of N/O SL converters, their voltage transfer gains can be compared to that of buck converter, $G = V_O/V_{in} = k$.

For the forward converter $G = V_O/V_{in} = kN$, N is the transformer turns ratio.

Cuk converter, $G = \dfrac{V_O}{V_{in}} = \dfrac{k}{1-k}$

TABLE 10.4

Voltage Transfer Gains of Converters in the Condition $k = 0.2$

Stage No. (n)	1	2	3	4	5	n
Buck converter	0.2					
Forward converter	0.2N (N is the transformer turns ratio)					
Cuk converter	0.25					
Fly-back converter	0.25N (N is the transformer turns ratio)					
Boost converter	1.25					
N/O Luo converters	1.25	2.5	3.75	5	6.25	1.25^n
N/O SL converters—main series	1.25	4.06	10.39	24.63	56.67	2.25^{n-1}
N/O SL converters—additional series	2.5	6.88	16.72	38.87	88.7	$3.5*2.25^{(n-1)} - 1$

Fly-back converter, $G = V_O/V_{in} = kN/1 = k$ (N is the transformer turns ratio)
Boost converter, $G = V_O/V_{in} = 1/1 - k$
N/O Luo converter,

$$G = \frac{V_O}{V_{in}} = \frac{n}{1-k} \tag{10.218}$$

Assume the conduction duty k is 0.2; the output voltage transfer gains are listed in Table 10.4.

Assume the conduction duty k is 0.5; the output voltage transfer gains are listed in Table 10.5.

Assume the conduction duty k is 0.8; the output voltage transfer gains are listed in Table 10.6.

10.4 P/O CASCADED BOOST CONVERTERS

SL Luo converters largely increase the voltage transfer gain in geometric progression. However, their circuits are a bit complex. We introduce a novel approach—P/O cascaded boost converters that facilitates an increase in the output voltage increasing in geometric progression as well, but simpler structure. They also effectively enhance the voltage transfer gain in power-law. There are several subseries. As shown in the previous sections, only three circuits of each subseries are introduced [1,2,7–9].

10.4.1 MAIN SERIES

The first three stages of P/O cascaded boost converters and their main series are shown in Figures 8.5, 10.32, and 10.33. For the sake of convenience, they are called elementary boost converter, two-stage circuit, and three-stage circuit, respectively, and numbered $n = 1$, 2, and 3.

TABLE 10.5
Voltage Transfer Gains of Converters in the Condition $k = 0.5$

Stage No. (n)	1	2	3	4	5	n
Buck converter	0.5					
Forward converter	0.5N (N is the transformer turns ratio)					
Cuk converter	1					
Fly-back converter	N (N is the transformer turns ratio)					
Boost converter	2					
N/O Luo converters	2	4	6	8	10	2^n
–N/O SL converters—main series	2	8	26	80	242	3^{n-1}
N/O SL converters—additional series	4	14	44	134	404	$5^*3^{(n-1)} - 1$

TABLE 10.6
Voltage Transfer Gains of Converters in the Condition $k = 0.8$

Stage No. (n)	1	2	3	4	5	n
Buck converter	0.8					
Forward converter	0.8N (N is the transformer turns ratio)					
Cuk converter	4					
Fly-back converter	4N (N is the transformer turns ratio)					
Boost converter	5					
N/O Luo converters	5	10	15	20	25	5^n
N/O SL converters—main series	5	35	215	1295	7775	6^{n-1}
N/O SL converters—additional series	10	65	395	2,375	14,255	$11^*6^{(n-1)} - 1$

10.4.1.1 Elementary Boost Circuit

The elementary boost converter is the fundamental boost converter; it has already been introduced in Section 5.2.2. Its circuit diagram and its equivalent circuits during the switch-on and switch-off periods are shown in Figure 5.5. The output voltage is

$$V_O = \frac{1}{1-k} V_{in}$$

The voltage transfer gain is

$$G = \frac{V_O}{V_{in}} = \frac{1}{1-k}$$

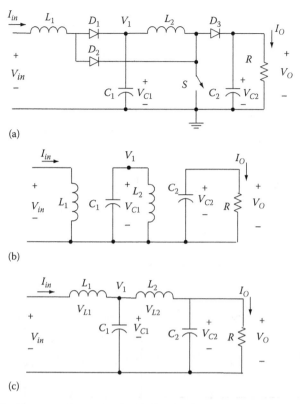

(a)

(b)

(c)

FIGURE 10.32 Two-stage boost circuit. (a) Circuit diagram. (b) Equivalent circuit during the switch-on period. (c) Equivalent circuit during the switch-off period.

Therefore, the variation ratio of output voltage v_O is

$$\varepsilon = \frac{\Delta v_O / 2}{V_O} = \frac{k}{2RfC}$$

10.4.1.2 Two-Stage Boost Circuit

The two-stage boost circuit is derived from the elementary boost converter by adding the parts L_2-D_2-D_3-C_2. Its circuit diagram and equivalent circuits during the switch-on and switch-off periods are shown in Figure 10.32.

The voltage across capacitor C_1 is charged to V_1. As described in the previous section, the voltage V_1 across capacitor C_1 is

$$V_1 = \frac{1}{1-k} V_{in}$$

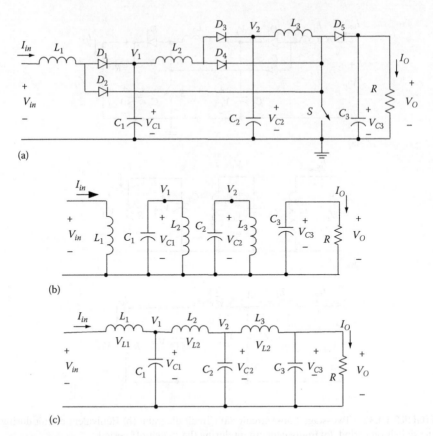

(a)

(b)

(c)

FIGURE 10.33   Three-stage boost circuit. (a) Circuit diagram. (b) Equivalent circuit during the switch-on period. (c) Equivalent circuit during the switch-off period

The voltage across capacitor C_2 is charged to V_O. The current flowing through inductor L_2 increases with voltage V_1 during the switch-on period kT and decreases with voltage $-(V_O - V_1)$ during the switch-off period $(1 - k)T$. Therefore, the ripple of the inductor current i_{L2} is

$$\Delta i_{L2} = \frac{V_1}{L_2} kT = \frac{V_O - V_1}{L_2}(1 - k)T \qquad (10.219)$$

$$V_O = \frac{1}{1-k} V_1 = \left(\frac{1}{1-k}\right)^2 V_{in} \qquad (10.220)$$

The voltage transfer gain is

$$G = \frac{V_O}{V_{in}} = \left(\frac{1}{1-k}\right)^2 \tag{10.221}$$

The variation ratio of current i_{L1} through inductor L_1 is

$$\xi_1 = \frac{\Delta i_{L1}/2}{I_{L1}} = \frac{k(1-k)^2 TV_{in}}{2L_1 I_O} = \frac{k(1-k)^4}{2} \frac{R}{fL_1} \tag{10.222}$$

The variation ratio of current i_{L2} through inductor L_2 is

$$\xi_2 = \frac{\Delta i_{L2}/2}{I_{L2}} = \frac{k(1-k)TV_1}{2L_2 I_O} = \frac{k(1-k)^2}{2} \frac{R}{fL_2} \tag{10.223}$$

The variation ratio of output voltage v_O is

$$\varepsilon = \frac{\Delta v_O/2}{V_O} = \frac{k}{2RfC_2} \tag{10.224}$$

10.4.1.3 Three-Stage Boost Circuit

The three-stage boost circuit is derived from the two-stage boost circuit by readding the parts L_2-D_2-D_3-C_2. Its circuit diagram and equivalent circuits during the switch-on and switch-off periods are shown in Figure 10.33.

The voltage across capacitor C_1 is charged to V_1. As described in the previous section, the voltage V_1 across capacitor C_1 is $V_1 = (1/1 - k)V_{in}$, and the voltage V_2 across capacitor C_2 is

$$V_2 = \left(\frac{1}{1-k}\right)^2 V_{in}$$

The voltage across capacitor C_3 is charged to V_O. The current flowing through inductor L_3 increases with voltage V_2 during the switch-on period kT and decreases with voltage $-(V_O - V_2)$ during the switch-off period $(1 - k)T$. Therefore, the ripple of the inductor current i_{L3} is

$$\Delta i_{L3} = \frac{V_2}{L_3} kT = \frac{V_O - V_2}{L_3}(1-k)T \tag{10.225}$$

$$V_O = \frac{1}{1-k}V_2 = \left(\frac{1}{1-k}\right)^2 V_1 = \left(\frac{1}{1-k}\right)^3 V_{in} \tag{10.226}$$

The voltage transfer gain is

$$G = \frac{V_O}{V_{in}} = \left(\frac{1}{1-k}\right)^3 \qquad (10.227)$$

The variation ratio of current i_{L1} through inductor L_1 is

$$\xi_1 = \frac{\Delta i_{L1}/2}{I_{L1}} = \frac{k(1-k)^3 TV_{in}}{2L_1 I_O} = \frac{k(1-k)^6}{2} \frac{R}{fL_1} \qquad (10.228)$$

The variation ratio of current i_{L2} through inductor L_2 is

$$\xi_2 = \frac{\Delta i_{L2}/2}{I_{L2}} = \frac{k(1-k)^2 TV_1}{2L_2 I_O} = \frac{k(1-k)^4}{2} \frac{R}{fL_2} \qquad (10.229)$$

The variation ratio of current i_{L3} through inductor L_3 is

$$\xi_3 = \frac{\Delta i_{L3}/2}{I_{L3}} = \frac{k(1-k)TV_2}{2L_3 I_O} = \frac{k(1-k)^2}{2} \frac{R}{fL_3} \qquad (10.230)$$

The variation ratio of output voltage v_O is

$$\varepsilon = \frac{\Delta v_O/2}{V_O} = \frac{k}{2RfC_3} \qquad (10.231)$$

Example 10.4 A three-stage boost converter in Figure 10.33a has $V_{in} = 20\,V$, all inductors are $10\,mH$, all capacitors are $20\,\mu F$, $R = 400\,\Omega$, $f = 50\,kHz$, and conduction duty cycle $k = 0.6$. Calculate the variation ratio of current i_{L1}, the output voltage, and its variation ratio.

Solution: From formula (10.228), we can get the variation ratio of current i_{L1}:

$$\xi_1 = \frac{k(1-k)^6}{2} \frac{R}{fL_1} = \frac{0.6(1-0.6)^6}{2} \frac{400}{50k \times 10m} = 0.00098$$

From formula (10.226), we can get the output voltage:

$$V_O = \left(\frac{1}{1-k}\right)^3 V_{in} = \left(\frac{1}{1-0.6}\right)^3 \times 20 = 312.5\,V$$

From (10.231), its variation ratio is

$$\varepsilon = \frac{k}{2RfC_3} = \frac{0.6}{2 \times 400 \times 50k \times 20\mu} = 0.00075$$

10.4.1.4 Higher-Stage Boost Circuit

Higher-stage boost circuit can be designed by just readding the parts L_2-D_2-D_3-C_2 multiple times. For the nth-stage boost circuit, the final output voltage across capacitor C_n is

$$V_O = \left(\frac{1}{1-k}\right)^n V_{in}$$

The voltage transfer gain is

$$G = \frac{V_O}{V_{in}} = \left(\frac{1}{1-k}\right)^n \tag{10.232}$$

The variation ratio of current i_{Li} through inductor L_i ($i = 1, 2, 3, \ldots, n$) is

$$\xi_i = \frac{\Delta i_{Li}/2}{I_{Li}} = \frac{k(1-k)^{2(n-i+1)}}{2} \frac{R}{fL_i} \tag{10.233}$$

The variation ratio of output voltage v_O is

$$\varepsilon = \frac{\Delta v_O/2}{V_O} = \frac{k}{2RfC_n} \tag{10.234}$$

10.4.2 ADDITIONAL SERIES

All circuits of P/O cascaded boost converters and their additional series are derived from the corresponding circuits of the main series by adding a DEC.

The first three stages of this series are shown in Figures 10.34 through 10.36. For the sake of convenience, they are called elementary additional circuit, two-stage additional circuit, and three-stage additional circuit, respectively, and numbered $n = 1, 2,$ and 3.

10.4.2.1 Elementary Boost Additional (Double) Circuit

This elementary boost additional circuit is derived from elementary boost converter by adding a DEC. Its circuit diagram and switch-on and switch-off equivalent circuits are shown in Figure 10.34.

The voltage across capacitors C_1 and C_{11} is charged to V_1, and the voltage across capacitor C_{12} is charged to $V_O = 2 V_1$. The current i_{L1} flowing through inductor

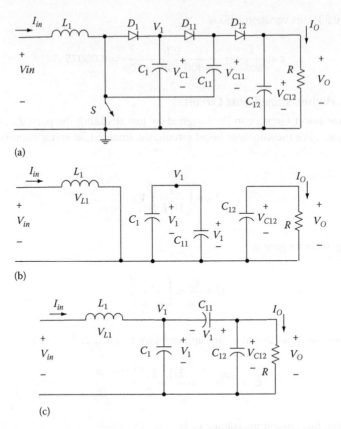

(a)

(b)

(c)

FIGURE 10.34 Elementary boost additional circuit. (a) Circuit diagram. (b) Equivalent circuit during the switch-on period. (c) Equivalent circuit during the switch-off period.

L_1 increases with voltage V_{in} during the switch-on period kT and decreases with voltage $-(V_1 - V_{in})$ during the switch-off period $(1 - k)T$. Therefore,

$$\Delta i_{L1} = \frac{V_{in}}{L_1} kT = \frac{V_1 - V_{in}}{L_1} (1 - k)T \qquad (10.235)$$

$$V_1 = \frac{1}{1-k} V_{in}$$

The output voltage is

$$V_O = 2V_1 = \frac{2}{1-k} V_{in} \qquad (10.236)$$

The voltage transfer gain is

$$G = \frac{V_O}{V_{in}} = \frac{2}{1-k} \qquad (10.237)$$

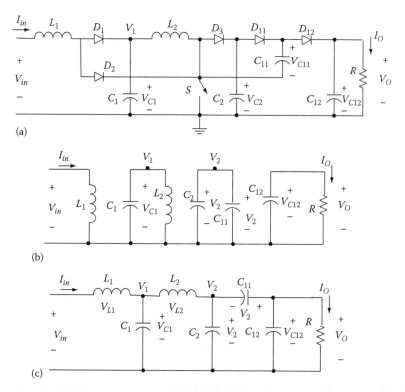

(a)

(b)

(c)

FIGURE 10.35 Two-stage additional boost circuit. (a) Circuit diagram. (b) Equivalent circuit during the switch-on period. (c) Equivalent circuit during the switch-off period.

and

$$i_{in} = I_{L1} = \frac{2}{1-k} I_O \qquad (10.238)$$

The variation ratio of current i_{L1} through inductor L_1 is

$$\xi_1 = \frac{\Delta i_{L1}/2}{I_{L1}} = \frac{k(1-k)TV_{in}}{4L_1 I_O} = \frac{k(1-k)^2}{8} \frac{R}{fL_1} \qquad (10.239)$$

The ripple voltage of output voltage v_O is

$$\Delta v_O = \frac{\Delta Q}{C_{12}} = \frac{I_O kT}{C_{12}} = \frac{k}{fC_{12}} \frac{V_O}{R}$$

Therefore, the variation ratio of output voltage v_O is

$$\varepsilon = \frac{\Delta v_O/2}{V_O} = \frac{k}{2RfC_{12}} \qquad (10.240)$$

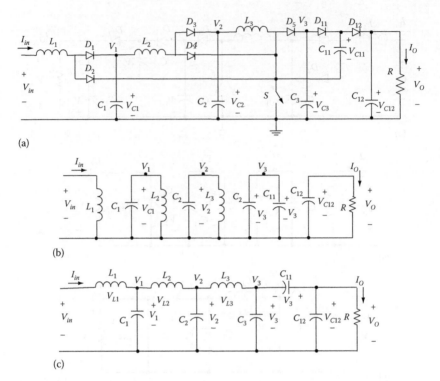

(a)

(b)

(c)

FIGURE 10.36 Three-stage additional boost circuit. (a) Circuit diagram. (b) Equivalent circuit during the switch-on period. (c) Equivalent circuit during the switch-off period.

10.4.2.2 Two-Stage Boost Additional Circuit

The two-stage boost additional circuit is derived from the two-stage boost circuit by adding a DEC. Its circuit diagram and switch-on and switch-off equivalent circuits are shown in Figure 10.35.

The voltage across capacitor C_1 is charged to V_1. As described in the previous section, the voltage V_1 across capacitor C_1 is

$$V_1 = \frac{1}{1-k}V_{in}$$

The voltage across capacitor C_2 and capacitor C_{11} is charged to V_2, and the voltage across capacitor C_{12} is charged to V_0. The current flowing through inductor L_2 increases with voltage V_1 during the switch-on period kT and decreases with voltage $-(V_2 - V_1)$ during the switch-off period $(1 - k)T$. Therefore, the ripple of the inductor current i_{L2} is

$$\Delta i_{L2} = \frac{V_1}{L_2}kT = \frac{V_2 - V_1}{L_2}(1-k)T \qquad (10.241)$$

$$V_2 = \frac{1}{1-k}V_1 = \left(\frac{1}{1-k}\right)^2 V_{in} \qquad (10.242)$$

The output voltage is

$$V_O = 2V_2 = \frac{2}{1-k}V_1 = 2\left(\frac{1}{1-k}\right)^2 V_{in} \qquad (10.243)$$

The voltage transfer gain is

$$G = \frac{V_O}{V_{in}} = 2\left(\frac{1}{1-k}\right)^2 \qquad (10.244)$$

The variation ratio of current i_{L1} through inductor L_1 is

$$\xi_1 = \frac{\Delta i_{L1}/2}{I_{L1}} = \frac{k(1-k)^2 TV_{in}}{4L_1 I_O} = \frac{k(1-k)^4}{8}\frac{R}{fL_1} \qquad (10.245)$$

The variation ratio of current i_{L2} through inductor L_2 is

$$\xi_2 = \frac{\Delta i_{L2}/2}{I_{L2}} = \frac{k(1-k)TV_1}{4L_2 I_O} = \frac{k(1-k)^2}{8}\frac{R}{fL_2} \qquad (10.246)$$

The ripple voltage of output voltage v_O is

$$\Delta v_O = \frac{\Delta Q}{C_{12}} = \frac{I_O kT}{C_{12}} = \frac{k}{fC_{12}}\frac{V_O}{R}$$

Therefore, the variation ratio of output voltage v_O is

$$\varepsilon = \frac{\Delta v_O/2}{V_O} = \frac{k}{2RfC_{12}} \qquad (10.247)$$

10.4.2.3 Three-Stage Boost Additional Circuit

This circuit is derived from the three-stage boost circuit by adding a DEC. Its circuit diagram and equivalent circuits during the switch-on and switch-off periods are shown in Figure 10.36.

The voltage across capacitor C_1 is charged to V_1. As described in the previous section, the voltage V_1 across capacitor C_1 is $V_1 = 1/(1-k)V_{in}$ and the voltage V_2 across capacitor C_2 is

$$V_2 = \left(\frac{1}{1-k}\right)^2 V_{in}$$

The voltage across capacitor C_3 and capacitor C_{11} is charged to V_3. The voltage across capacitor C_{12} is charged to V_O. The current flowing through inductor L_3 increases with voltage V_2 during the switch-on period kT and decreases with voltage $-(V_3 - V_2)$ during the switch-off period $(1 - k)T$. Therefore,

$$\Delta i_{L3} = \frac{V_2}{L_3} kT = \frac{V_3 - V_2}{L_3} (1 - k)T \tag{10.248}$$

and

$$V_3 = \frac{1}{1-k} V_2 = \left(\frac{1}{1-k}\right)^2 V_1 = \left(\frac{1}{1-k}\right)^3 V_{in} \tag{10.249}$$

The output voltage is

$$V_O = 2V_3 = 2\left(\frac{1}{1-k}\right)^3 V_{in} \tag{10.250}$$

The voltage transfer gain is

$$G = \frac{V_O}{V_{in}} = 2\left(\frac{1}{1-k}\right)^3 \tag{10.251}$$

The variation ratio of current i_{L1} through inductor L_1 is

$$\xi_1 = \frac{\Delta i_{L1}/2}{I_{L1}} = \frac{k(1-k)^3 TV_{in}}{4L_1 I_O} = \frac{k(1-k)^6}{8} \frac{R}{fL_1} \tag{10.252}$$

The variation ratio of current i_{L2} through inductor L_2 is

$$\xi_2 = \frac{\Delta i_{L2}/2}{I_{L2}} = \frac{k(1-k)^2 TV_1}{4L_2 I_O} = \frac{k(1-k)^4}{8} \frac{R}{fL_2} \tag{10.253}$$

The variation ratio of current i_{L3} through inductor L_3 is

$$\xi_3 = \frac{\Delta i_{L3}/2}{I_{L3}} = \frac{k(1-k)TV_2}{4L_3 I_O} = \frac{k(1-k)^2}{8} \frac{R}{fL_3} \tag{10.254}$$

The ripple voltage of output voltage v_O is

$$\Delta v_O = \frac{\Delta Q}{C_{12}} = \frac{I_O kT}{C_{12}} = \frac{k}{fC_{12}} \frac{V_O}{R}$$

Therefore, the variation ratio of output voltage v_O is

$$\varepsilon = \frac{\Delta v_O / 2}{V_O} = \frac{k}{2RfC_{12}} \tag{10.255}$$

10.4.2.4 Higher-Stage Boost Additional Circuit

Higher-stage boost additional circuit can be designed by just readding the parts L_2-D_2-D_3-C_2 multiple times. For the nth-stage additional circuit, the final output voltage is

$$V_O = 2\left(\frac{1}{1-k}\right)^n V_{in}$$

The voltage transfer gain is

$$G = \frac{V_O}{V_{in}} = 2\left(\frac{1}{1-k}\right)^n \tag{10.256}$$

Analogously, the variation ratio of current i_{Li} through inductor L_i ($i = 1, 2, 3, ..., n$) is

$$\xi_i = \frac{\Delta i_{Li} / 2}{I_{Li}} = \frac{k(1-k)^{2(n-i+1)}}{8} \frac{R}{fL_i} \tag{10.257}$$

and the variation ratio of output voltage v_O is

$$\varepsilon = \frac{\Delta v_O / 2}{V_O} = \frac{k}{2RfC_{12}} \tag{10.258}$$

10.4.3 DOUBLE SERIES

All circuits of P/O cascaded boost converters and their double series are derived from the corresponding circuits of the main series by adding a DEC in each stage circuit. The first three stages of this series are shown in Figures 10.34, 10.37, and 10.38. For the sake of convenience, they are called elementary double circuit, two-stage double circuit, and three-stage double circuit, respectively, and numbered $n = 1$, 2, and 3.

10.4.3.1 Elementary Double-Boost Circuit

From the construction principle, the elementary double-boost circuit is derived from the elementary boost converter by adding a DEC. Its circuit diagram and switch-on and switch-off equivalent circuits are shown in Figure 10.34, which is the same as that of elementary boost additional circuit.

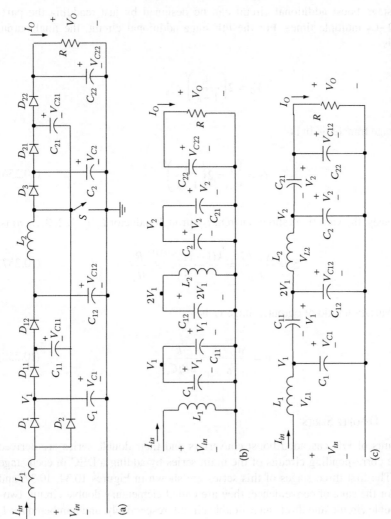

FIGURE 10.37 Two-stage double-boost circuit. (a) Circuit diagram. (b) Equivalent circuit during the switch-on period. (c) Equivalent circuit during the switch-off period.

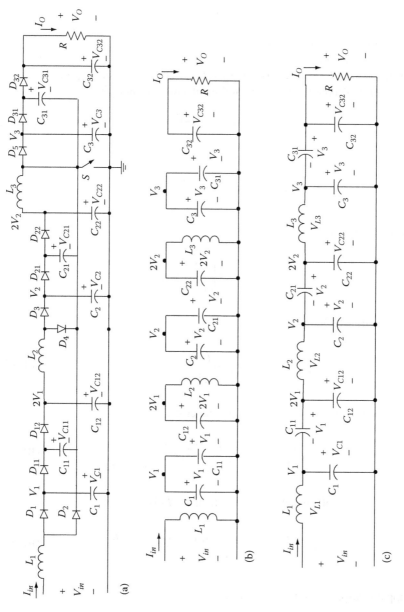

FIGURE 10.38 Three-stage double-boost circuit. (a) Circuit diagram. (b) Equivalent circuit during the switch-on period. (c) Equivalent circuit during the switch-off period.

10.4.3.2 Two-Stage Double-Boost Circuit

The two-stage double-boost circuit is derived from the two-stage boost circuit by adding a DEC in each stage circuit. Its circuit diagram and switch-on and switch-off equivalent circuits are shown in Figure 10.37.

The voltage across capacitor C_1 and capacitor C_{11} is charged to V_1. As described in the previous section, the voltage V_1 across capacitor C_1 and capacitor C_{11} is $V_1 = (1/1 - k)V_{in}$. The voltage across capacitor C_{12} is charged to $2V_1$.

The current flowing through inductor L_2 increases with voltage $2V_1$ during the switch-on period kT and decreases with voltage $-(V_2 - 2V_1)$ during the switch-off period $(1 - k)T$. Therefore, the ripple of the inductor current i_{L2} is

$$\Delta i_{L2} = \frac{2V_1}{L_2}kT = \frac{V_2 - 2V_1}{L_2}(1-k)T \tag{10.259}$$

$$V_2 = \frac{2}{1-k}V_1 = 2\left(\frac{1}{1-k}\right)^2 V_{in} \tag{10.260}$$

The output voltage is

$$V_O = 2V_2 = \left(\frac{2}{1-k}\right)^2 V_{in} \tag{10.261}$$

The voltage transfer gain is

$$G = \frac{V_O}{V_{in}} = \left(\frac{2}{1-k}\right)^2 \tag{10.262}$$

The variation ratio of current i_{L1} through inductor L_1 is

$$\xi_1 = \frac{\Delta i_{L1}/2}{I_{L1}} = \frac{k(1-k)^2 TV_{in}}{8L_1 I_O} = \frac{k(1-k)^4}{16}\frac{R}{fL_1} \tag{10.263}$$

The variation ratio of current i_{L2} through inductor L_2 is

$$\xi_2 = \frac{\Delta i_{L2}/2}{I_{L2}} = \frac{k(1-k)TV_1}{4L_2 I_O} = \frac{k(1-k)^2}{8}\frac{R}{fL_2} \tag{10.264}$$

The ripple voltage of output voltage v_O is

$$\Delta v_O = \frac{\Delta Q}{C_{22}} = \frac{I_O kT}{C_{22}} = \frac{k}{fC_{22}}\frac{V_O}{R}$$

Therefore, the variation ratio of output voltage v_O is

$$\varepsilon = \frac{\Delta v_O / 2}{V_O} = \frac{k}{2 R f C_{22}}$$

(10.265)

10.4.3.3 Three-Stage Double-Boost Circuit

This circuit is derived from the three-stage boost circuit by adding a DEC in each stage circuit. Its circuit diagram and equivalent circuits during the switch-on and switch-off periods are shown in Figure 10.38.

The voltage across capacitor C_1 and capacitor C_{11} is charged to V_1. As described in the previous section, the voltage V_1 across capacitor C_1 and capacitor C_{11} is $V_1 = (1/1 - k)V_{in}$ and the voltage V_2 across capacitor C_2 and capacitor C_{12} is $V_2 = 2(1/1 - k)^2 V_{in}$.

The voltage across capacitor C_{22} is $2V_2 = (2/1 - k)^2 V_{in}$. The voltage across capacitor C_3 and capacitor C_{31} is charged to V_3. The voltage across capacitor C_{12} is charged to V_O. The current flowing through inductor L_3 increases with voltage V_2 during the switch-on period kT and decreases with voltage $-(V_3 - 2V_2)$ during the switch-off period $(1 - k)T$. Therefore,

$$\Delta i_{L3} = \frac{2V_2}{L_3} kT = \frac{V_3 - 2V_2}{L_3} (1 - k)T$$

(10.266)

and

$$V_3 = \frac{2V_2}{(1-k)} = \frac{4}{(1-k)^3} V_{in}$$

(10.267)

The output voltage is

$$V_O = 2V_3 = \left(\frac{2}{1-k} \right)^3 V_{in}$$

(10.268)

The voltage transfer gain is

$$G = \frac{V_O}{V_{in}} = \left(\frac{2}{1-k} \right)^3$$

(10.269)

The variation ratio of current i_{L1} through inductor L_1 is

$$\xi_1 = \frac{\Delta i_{L1} / 2}{I_{L1}} = \frac{k(1-k)^3 T V_{in}}{16 L_1 I_O} = \frac{k(1-k)^6}{128} \frac{R}{f L_1}$$

(10.270)

The variation ratio of current i_{L2} through inductor L_2 is

$$\xi_2 = \frac{\Delta i_{L2}/2}{I_{L2}} = \frac{k(1-k)^2 TV_1}{8L_2 I_O} = \frac{k(1-k)^4}{32}\frac{R}{fL_2} \quad (10.271)$$

The variation ratio of current i_{L3} through inductor L_3 is

$$\xi_3 = \frac{\Delta i_{L3}/2}{I_{L3}} = \frac{k(1-k)TV_2}{4L_3 I_O} = \frac{k(1-k)^2}{8}\frac{R}{fL_3} \quad (10.272)$$

The ripple voltage of output voltage v_O is

$$\Delta v_O = \frac{\Delta Q}{C_{32}} = \frac{I_O kT}{C_{32}} = \frac{k}{fC_{32}}\frac{V_O}{R}$$

Therefore, the variation ratio of output voltage v_O is

$$\varepsilon = \frac{\Delta v_O/2}{V_O} = \frac{k}{2RfC_{32}} \quad (10.273)$$

10.4.3.4　Higher-Stage Double-Boost Circuit

Higher-stage double-boost circuit can be derived from the corresponding main series circuit by adding a DEC in each stage circuit. For the nth-stage additional circuit, the final output voltage is

$$V_O = \left(\frac{2}{1-k}\right)^n V_{in}$$

The voltage transfer gain is

$$G = \frac{V_O}{V_{in}} = \left(\frac{2}{1-k}\right)^n \quad (10.274)$$

Analogously, the variation ratio of current i_{Li} through inductor L_i ($i=1,2,3,\ldots,n$) is

$$\xi_i = \frac{\Delta i_{Li}/2}{I_{Li}} = \frac{k(1-k)^{2(n-i+1)}}{2*2^{2n}}\frac{R}{fL_i} \quad (10.275)$$

The variation ratio of output voltage v_O is

$$\varepsilon = \frac{\Delta v_O/2}{V_O} = \frac{k}{2RfC_{n2}} \quad (10.276)$$

10.4.4 TRIPLE SERIES

All circuits of P/O cascaded boost converters and their triple series are derived from the corresponding circuits of the double series by adding the DEC twice in each stage circuit. The first three stages of this series are shown in Figures 10.39 through 10.41. For the sake of convenience, they are called elementary triple-boost circuit, two-stage triple-boost circuit, and three-stage triple-boost circuit, respectively, and numbered $n = 1$, 2, and 3.

10.4.4.1 Elementary Triple-Boost Circuit

From the construction principle, the elementary triple-boost circuit is derived from elementary double-boost circuit by adding another DEC. Its circuit diagram and switch-on and switch-off equivalent circuits are shown in Figure 10.39.

The output voltage of the first-stage boost circuit is V_1, $V_1 = V_{in}/(1 - k)$.

The voltage across capacitors C_1 and C_{11} is charged to V_1, and the voltage across capacitors C_{12} and C_{13} is charged to $V_{C13} = 2 V_1$. The current i_{L1} flowing through

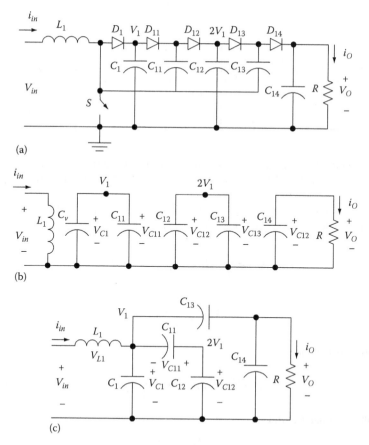

FIGURE 10.39 Elementary triple-boost circuit. (a) Circuit diagram. (b) Equivalent circuit during the switch on period. (c) Equivalent circuit during the switch-off period.

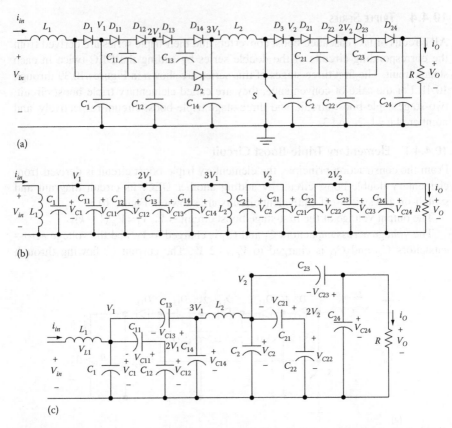

FIGURE 10.40 Two-stage triple-boost circuit. (a) Circuit diagram. (b) Equivalent circuit during the switch-on period. (c) Equivalent circuit during the switch-off period.

inductor L_1 increases with voltage V_{in} during the switch-on period kT and decreases with voltage $-(V_1 - V_{in})$ during the switch-off period $(1 - k)T$. Therefore,

$$\Delta i_{L1} = \frac{V_{in}}{L_1} kT = \frac{V_1 - V_{in}}{L_1} (1-k)T \qquad (10.277)$$

$$V_1 = \frac{1}{1-k} V_{in}$$

The output voltage is

$$V_O = V_{C1} + V_{C13} = 3V_1 = \frac{3}{1-k} V_{in} \qquad (10.278)$$

The voltage transfer gain is

$$G = \frac{V_O}{V_{in}} = \frac{3}{1-k} \qquad (10.279)$$

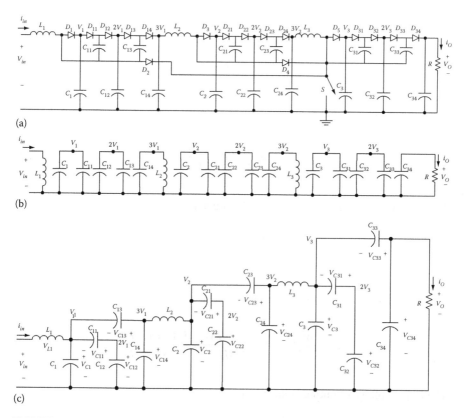

FIGURE 10.41 Three-stage triple-boost circuit. (a) Circuit diagram. (b) Equivalent circuit during the switch-on period. (c) Equivalent circuit during the switch-off period.

10.4.4.2 Two-Stage Triple-Boost Circuit

The two-stage triple-boost circuit is derived from the two-stage double-boost circuit by adding another DEC in each stage circuit. Its circuit diagram and switch-on and switch-off equivalent circuits are shown in Figure 10.40.

As described in the previous section, the voltage V_1 across capacitors C_1 and C_{11} is $V_1 = (1/1 - k)V_{in}$. The voltage across capacitor C_{14} is charged to $3V_1$.

The voltage across capacitors C_2 and C_{21} is charged to V_2, and the voltage across capacitors C_{22} and C_{23} is charged to $V_{C23} = 2V_2$. The current flowing through inductor L_2 increases with voltage $3V_1$ during the switch-on period kT and decreases with voltage $-(V_2 - 3V_1)$ during the switch-off period $(1 - k)T$. Therefore, the ripple of the inductor current i_{L2} is

$$\Delta i_{L2} = \frac{3V_1}{L_2}kT = \frac{V_2 - 3V_1}{L_2}(1-k)T \qquad (10.280)$$

$$V_2 = \frac{3}{1-k}V_1 = 3\left(\frac{1}{1-k}\right)^2 V_{in} \qquad (10.281)$$

The output voltage is

$$V_O = V_{C2} + V_{C23} = 3V_2 = \left(\frac{3}{1-k}\right)^2 V_{in} \qquad (10.282)$$

The voltage transfer gain is

$$G = \frac{V_O}{V_{in}} = \left(\frac{3}{1-k}\right)^2 \qquad (10.283)$$

The variation ratio of current i_{L1} through inductor L_1 is

$$\xi_1 = \frac{\Delta i_{L1}/2}{I_{L1}} = \frac{k(1-k)^2 T V_{in}}{8L_1 I_O} = \frac{k(1-k)^4}{16} \frac{R}{fL_1} \qquad (10.284)$$

The variation ratio of current i_{L2} through inductor L_2 is

$$\xi_2 = \frac{\Delta i_{L2}/2}{I_{L2}} = \frac{k(1-k)T V_1}{4L_2 I_O} = \frac{k(1-k)^2}{8} \frac{R}{fL_2} \qquad (10.285)$$

The ripple voltage of output voltage v_O is

$$\Delta v_O = \frac{\Delta Q}{C_{22}} = \frac{I_O kT}{C_{22}} = \frac{k}{fC_{22}} \frac{V_O}{R}$$

Therefore, the variation ratio of output voltage v_O is

$$\varepsilon = \frac{\Delta v_O/2}{V_O} = \frac{k}{2RfC_{22}} \qquad (10.286)$$

10.4.4.3 Three-Stage Triple-Boost Circuit

This circuit is derived from the three-stage double-boost circuit by adding another DEC in each stage circuit. Its circuit diagram and equivalent circuits during the switch-on and switch-off periods are shown in Figure 10.41.

As described in the previous section, the voltage V_2 across capacitors C_2 and C_{11} is $V_2 = 3V_1 = (3/1 - k)V_{in}$ and the voltage across capacitor C_{24} is charged to $3V_2$.

The voltage across capacitors C_3 and C_{31} is charged to V_3 and the voltage across capacitors C_{32} and C_{33} is charged to $V_{C33} = 2V_3$. The current flowing through inductor L_3 increases with voltage $3V_2$ during the switch-on period kT and decreases with voltage $-(V_3 - 3V_2)$ during the switch-off period $(1 - k)T$. Therefore, the ripple of the inductor current i_{L3} is

$$\Delta i_{L3} = \frac{3V_2}{L_3}kT = \frac{V_3 - 3V_2}{L_3}(1-k)T \qquad (10.287)$$

and

$$V_3 = \frac{3}{1-k}V_2 = 9\left(\frac{1}{1-k}\right)^3 V_{in} \qquad (10.288)$$

The output voltage is

$$V_O = V_{C3} + V_{C33} = 3V_3 = \left(\frac{3}{1-k}\right)^3 V_{in} \qquad (10.289)$$

The voltage transfer gain is

$$G = \frac{V_O}{V_{in}} = \left(\frac{3}{1-k}\right)^3 \qquad (10.290)$$

The variation ratio of current i_{L1} through inductor L_1 is

$$\xi_1 = \frac{\Delta i_{L1}/2}{I_{L1}} = \frac{k(1-k)^3 TV_{in}}{64L_1 I_O} = \frac{k(1-k)^6}{12^3}\frac{R}{fL_1} \qquad (10.291)$$

The variation ratio of current i_{L2} through inductor L_2 is

$$\xi_2 = \frac{\Delta i_{L2}/2}{I_{L2}} = \frac{k(1-k)^2 TV_1}{16L_2 I_O} = \frac{k(1-k)^4}{12^2}\frac{R}{fL_2} \qquad (10.292)$$

The variation ratio of current i_{L3} through inductor L_3 is

$$\xi_3 = \frac{\Delta i_{L3}/2}{I_{L3}} = \frac{k(1-k)TV_2}{4L_3 I_O} = \frac{k(1-k)^2}{12}\frac{R}{fL_3} \qquad (10.293)$$

The ripple voltage of output voltage v_O is

$$\Delta v_O = \frac{\Delta Q}{C_{32}} = \frac{I_O kT}{C_{32}} = \frac{k}{fC_{32}}\frac{V_O}{R}$$

Therefore, the variation ratio of output voltage v_O is

$$\varepsilon = \frac{\Delta v_O/2}{V_O} = \frac{k}{2RfC_{32}} \qquad (10.294)$$

10.4.4.4 Higher-Stage Triple-Boost Circuit

Higher-stage triple-boost circuit can be derived from the corresponding circuit of the double-boost series by adding another DEC in each stage circuit. For the nth-stage additional circuit, the final output voltage is

$$V_O = \left(\frac{3}{1-k}\right)^n V_{in}$$

The voltage transfer gain is

$$G = \frac{V_O}{V_{in}} = \left(\frac{3}{1-k}\right)^n \tag{10.295}$$

Analogously, the variation ratio of current i_{Li} through inductor L_i $(i = 1, 2, 3, \ldots, n)$ is

$$\xi_i = \frac{\Delta i_{Li}/2}{I_{Li}} = \frac{k(1-k)^{2(n-i+1)}}{12^{(n-i+1)}} \frac{R}{fL_i} \tag{10.296}$$

and the variation ratio of output voltage v_O is

$$\varepsilon = \frac{\Delta v_O/2}{V_O} = \frac{k}{2RfC_{n2}} \tag{10.297}$$

10.4.5 MULTIPLE SERIES

All circuits of P/O cascaded boost converters and their multiple series are derived from the corresponding circuits of the main series by adding DEC multiple (j) times in each stage circuit. The first three stages of this series are shown in Figures 10.42 through 10.44. For the sake of convenience, they are called elementary multiple-boost circuit, two-stage multiple-boost circuit, and three-stage multiple-boost circuit, respectively, and numbered $n = 1$, 2, and 3.

10.4.5.1 Elementary Multiple-Boost Circuit

From the construction principle, the elementary multiple-boost circuit is derived from the elementary boost converter by adding DEC multiple (j) times in the circuit. Its circuit diagram and switch-on and switch-off equivalent circuits are shown in Figure 10.42.

The voltage across capacitors C_1 and C_{11} is charged to V_1, and the voltage across capacitors C_{12} and C_{13} is charged to $V_{C13} = 2 V_1$. The voltage across capacitors $C_{1(2j-2)}$ and $C_{1(2j-1)}$ is charged to $V_{C1(2j-1)} = jV_1$. The current i_{L1} flowing through inductor L_1 increases with voltage V_{in} during the switch-on period kT and decreases with voltage $-(V_1 - V_{in})$ during the switch-off period $(1 - k)T$. Therefore,

$$\Delta i_{L1} = \frac{V_{in}}{L_1} kT = \frac{V_1 - V_{in}}{L_1}(1-k)T \tag{10.298}$$

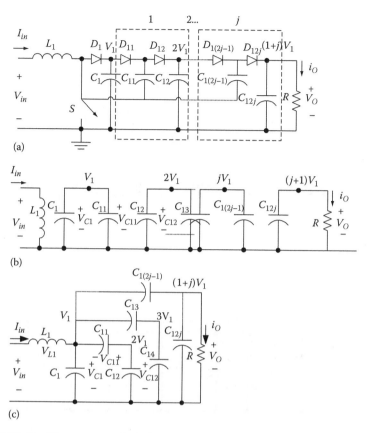

FIGURE 10.42 Elementary multiple-boost circuit. (a) Circuit diagram. (b) Equivalent circuit during the switch-on period. (c) Equivalent circuit during the switch-off period.

$$V_1 = \frac{1}{1-k} V_{in} \qquad (10.299)$$

The output voltage is

$$V_O = V_{C1} + V_{C1(2j-1)} = (1+j)V_1 = \frac{1+j}{1-k} V_{in} \qquad (10.300)$$

The voltage transfer gain is

$$G = \frac{V_O}{V_{in}} = \frac{1+j}{1-k} \qquad (10.301)$$

10.4.5.2 Two-Stage Multiple-Boost Circuit

The two-stage multiple-boost circuit is derived from the two-stage boost circuit by adding multiple (j) DECs in each stage circuit. Its circuit diagram and the switch-on and switch-off equivalent circuits are shown in Figure 10.43.

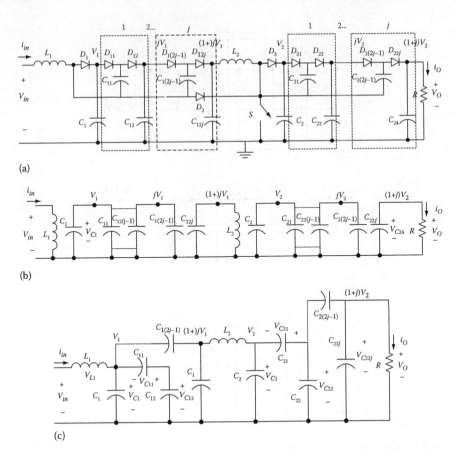

FIGURE 10.43 Two-stage multiple-boost circuit. (a) Circuit diagram. (b) Equivalent circuit during the switch-on period. (c) Equivalent circuit during the switch-off period.

The voltage across capacitor C_1 and capacitor C_{11} is charged to $V_1 = (1/1 - k)V_{in}$. The voltage across capacitor $C_{1(2j)}$ is charged to $(1+j)V_1$.

The current flowing through inductor L_2 increases with voltage $(1+j)V_1$ during the switch-on period kT and decreases with voltage $-[V_2 - (1+j)V_1]$ during the switch-off period $(1-k)T$. Therefore, the ripple of the inductor current i_{L2} is

$$\Delta i_{L2} = \frac{1+j}{L_2}kTV_1 = \frac{V_2 - (1+j)V_1}{L_2}(1-k)T \qquad (10.302)$$

$$V_2 = \frac{1+j}{1-k}V_1 = (1+j)\left(\frac{1}{1-k}\right)^2 V_{in} \qquad (10.303)$$

The output voltage is

$$V_O = V_{C1} + V_{C1(2j-1)} = (1+j)V_2 = \left(\frac{1+j}{1-k}\right)^2 V_{in} \qquad (10.304)$$

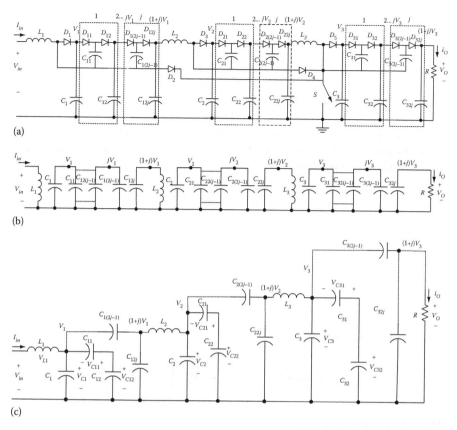

(a)

(b)

(c)

FIGURE 10.44 Three-stage multiple-boost circuit. (a) Circuit diagram. (b) Equivalent circuit during the switch-on period. (c) Equivalent circuit during the switch-off period.

The voltage transfer gain is

$$G = \frac{V_O}{V_{in}} = \left(\frac{1+j}{1-k}\right)^2 \tag{10.305}$$

The ripple voltage of output voltage v_O is

$$\Delta v_O = \frac{\Delta Q}{C_{22j}} = \frac{I_O kT}{C_{22j}} = \frac{k}{fC_{22j}}\frac{V_O}{R}$$

Therefore, the variation ratio of output voltage v_O is

$$\varepsilon = \frac{\Delta v_O / 2}{V_O} = \frac{k}{2RfC_{22j}} \tag{10.306}$$

10.4.5.3 Three-Stage Multiple-Boost Circuit

This circuit is derived from the three-stage boost circuit by adding multiple (j) DECs in each stage circuit. Its circuit diagram and equivalent circuits during the switch-on and switch-off periods are shown in Figure 10.44.

The voltage across capacitor C_1 and capacitor C_{11} is charged to $V_1 = (1/1 - k)V_{in}$. The voltage across capacitor $C_{1(2j)}$ is charged to $(1+j)V_1$. The voltage V_2 across capacitor C_2 and capacitor $C_{2(2j)}$ is charged to $(1+j)V_2$.

The current flowing through inductor L_3 increases with voltage $(1+j)V_2$ during the switch-on period kT and decreases with voltage $-[V_3 - (1+j)V_2]$ during the switch-off period $(1-k)T$. Therefore,

$$\Delta i_{L3} = \frac{1+j}{L_3} kTV_2 = \frac{V_3 - (1+j)V_2}{L_3}(1-k)T \tag{10.307}$$

and

$$V_3 = \frac{(1+j)V_2}{(1-k)} = \frac{(1+j)^2}{(1-k)^3}V_{in} \tag{10.308}$$

The output voltage is

$$V_O = V_{C3} + V_{C3(2j-1)} = (1+j)V_3 = \left(\frac{1+j}{1-k}\right)^3 V_{in} \tag{10.309}$$

The voltage transfer gain is

$$G = \frac{V_O}{V_{in}} = \left(\frac{1+j}{1-k}\right)^3 \tag{10.310}$$

The ripple voltage of output voltage v_O is

$$\Delta v_O = \frac{\Delta Q}{C_{32j}} = \frac{I_O kT}{C_{32j}} = \frac{k}{fC_{32j}}\frac{V_O}{R}$$

Therefore, the variation ratio of output voltage v_O is

$$\varepsilon = \frac{\Delta v_O/2}{V_O} = \frac{k}{2RfC_{32j}} \tag{10.311}$$

10.4.5.4 Higher-Stage Multiple-Boost Circuit

Higher-stage multiple-boost circuit is derived from the corresponding circuit of the main series by adding multiple (j) DECs in each stage circuit. For the nth-stage additional circuit, the final output voltage is

$$V_O = \left(\frac{1+j}{1-k}\right)^n V_{in}$$

The voltage transfer gain is

$$G = \frac{V_O}{V_{in}} = \left(\frac{1+j}{1-k}\right)^n \qquad (10.312)$$

Analogously, the variation ratio of output voltage v_O is

$$\varepsilon = \frac{\Delta v_O / 2}{V_O} = \frac{k}{2RfC_{n2j}} \qquad (10.313)$$

10.4.6 SUMMARY OF P/O CASCADED BOOST CONVERTERS

A family tree of all circuits of P/O cascaded boost converters are shown in Figure 10.45.

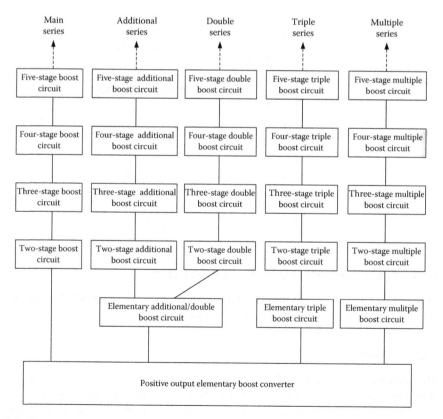

FIGURE 10.45 The family of P/O cascaded boost converters.

From the analysis of the previous two sections, we can have the common formula to calculate the output voltage:

$$V_O = \begin{cases} \left(\dfrac{1}{1-k}\right)^n V_{in} & \text{Main\_series} \\[3mm] 2*\left(\dfrac{1}{1-k}\right)^n V_{in} & \text{Additional\_series} \\[3mm] \left(\dfrac{2}{1-k}\right)^n V_{in} & \text{Double\_series} \\[3mm] \left(\dfrac{3}{1-k}\right)^n V_{in} & \text{Triple\_series} \\[3mm] \left(\dfrac{j+1}{1-k}\right)^n V_{in} & \text{Multiple}(j)\text{\_series} \end{cases}$$ (10.314)

The voltage transfer gain is

$$G = \frac{V_O}{V_{in}} = \begin{cases} \left(\dfrac{1}{1-k}\right)^n & \text{Main\_series} \\[3mm] 2*\left(\dfrac{1}{1-k}\right)^n & \text{Additional\_series} \\[3mm] \left(\dfrac{2}{1-k}\right)^n & \text{Double\_series} \\[3mm] \left(\dfrac{3}{1-k}\right)^n & \text{Triple\_series} \\[3mm] \left(\dfrac{j+1}{1-k}\right)^n & \text{Multiple}(j)\text{\_series} \end{cases}$$ (10.315)

10.5 N/O CASCADED BOOST CONVERTERS

This section introduces N/O cascaded boost converters. As P/O cascaded boost converters, these converters execute the SL technique [1,2].

10.5.1 MAIN SERIES

The first three stages of the N/O cascaded boost converters and their main series are shown in Figures 10.46 through 10.48. For the sake of convenience, they are called elementary boost converter, two-stage boost circuit, and three-stage boost circuit, respectively, and numbered $n = 1$, 2, and 3.

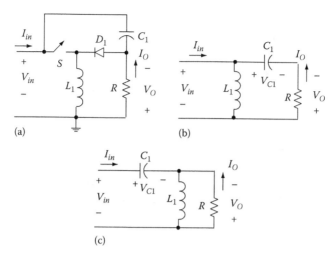

FIGURE 10.46 N/O elementary boost converter. (a) Circuit diagram. (b) Equivalent circuit during the switch-on period. (c) Equivalent circuit during the switch-off period.

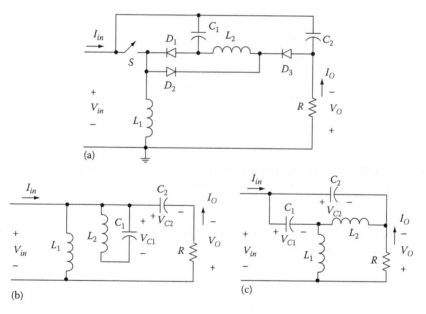

FIGURE 10.47 N/O two-stage boost circuit. (a) Circuit diagram. (b) Equivalent circuit during the switch-on period. (c) Equivalent circuit during the switch-off period.

10.5.1.1 N/O Elementary Boost Circuit

The N/O elementary boost converter and its equivalent circuits during the switch-on and switch-off periods are shown in Figure 10.46.

The voltage across capacitor C_1 is charged to V_{C1}. The current i_{L1} flowing through inductor L_1 increases with voltage V_{in} during the switch-on period kT and decreases

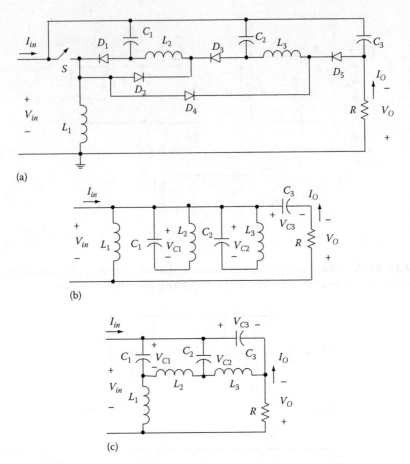

(a)

(b)

(c)

FIGURE 10.48 N/O three-stage boost circuit. (a) Circuit diagram. (b) Equivalent circuit during the switch-on period. (c) Equivalent circuit during the switch-off period.

with voltage $-(V_{C1} - V_{in})$ during the switch-off period $(1 - k)T$. Therefore, the ripple of the inductor current i_{L1} is

$$\Delta i_{L1} = \frac{V_{in}}{L_1} kT = \frac{V_{C1} - V_{in}}{L_1}(1-k)T \qquad (10.316)$$

$$V_{C1} = \frac{1}{1-k} V_{in}$$

$$V_O = V_{C1} - V_{in} = \frac{k}{1-k} V_{in} \qquad (10.317)$$

The voltage transfer gain is

$$G = \frac{V_O}{V_{in}} = \frac{1}{1-k} - 1 \tag{10.318}$$

The inductor average current

$$I_{L1} = \frac{1}{1-k} \frac{V_O}{R} \tag{10.319}$$

The variation ratio of current i_{L1} through inductor L_1 is

$$\xi_1 = \frac{\Delta i_{L1}/2}{I_{L1}} = \frac{kTV_{in}}{2L_1V_O/(1-k)R} = \frac{k(1-k)}{2G} \frac{R}{fL_1} = \frac{(1-k)^2}{2} \frac{R}{fL_1} \tag{10.320}$$

Usually ξ_1 is small (much lower than unity), which means this converter works in the continuous mode.

The ripple voltage of output voltage v_O is

$$\Delta v_O = \frac{\Delta Q}{C_1} = \frac{I_O kT}{C_1} = \frac{k}{fC_1} \frac{V_O}{R}$$

since

$$\Delta Q = I_O kT$$

Therefore, the variation ratio of output voltage v_O is

$$\varepsilon = \frac{\Delta v_O/2}{V_O} = \frac{k}{2RfC_1} \tag{10.321}$$

10.5.1.2 N/O Two-Stage Boost Circuit

The N/O two-stage boost circuit is derived from N/O elementary boost converter by adding the parts L_2-D_2-D_3-C_2. Its circuit diagram and equivalent circuits during the switch-on and switch-off periods are shown in Figure 10.47.

The voltage across capacitor C_1 is charged to V_1. As described in the previous section, the voltage V_1 across capacitor C_1 is

$$V_1 = \frac{1}{1-k} V_{in}$$

The voltage across capacitor C_2 is charged to V_{C2}. The current flowing through inductor L_2 increases with voltage V_1 during the switch-on period kT and decreases with voltage $-(V_{C2} - V_1)$ during switch-off period $(1 - k)T$. Therefore, the ripple of the inductor current i_{L2} is

$$\Delta i_{L2} = \frac{V_1}{L_2}kT = \frac{V_{C2} - V_1}{L_2}(1-k)T \qquad (10.322)$$

$$V_{C2} = \frac{1}{1-k}V_1 = \left(\frac{1}{1-k}\right)^2 V_{in}$$

$$V_O = V_{C2} - V_{in} = \left[\left(\frac{1}{1-k}\right)^2 - 1\right]V_{in} \qquad (10.323)$$

The voltage transfer gain is

$$G = \frac{V_O}{V_{in}} = \left(\frac{1}{1-k}\right)^2 - 1 \qquad (10.324)$$

Analogously,

$$\Delta i_{L1} = \frac{V_{in}}{L_1}kT \quad I_{L1} = \frac{I_O}{(1-k)^2}$$

$$\Delta i_{L2} = \frac{V_1}{L_2}kT \quad I_{L2} = \frac{I_O}{1-k}$$

Therefore, the variation ratio of current i_{L1} through inductor L_1 is

$$\xi_1 = \frac{\Delta i_{L1}/2}{I_{L1}} = \frac{k(1-k)^2 TV_{in}}{2L_1 I_O} = \frac{k(1-k)^4}{2}\frac{R}{fL_1} \qquad (10.325)$$

The variation ratio of current i_{L2} through inductor L_2 is

$$\xi_2 = \frac{\Delta i_{L2}/2}{I_{L2}} = \frac{k(1-k)TV_1}{2L_2 I_O} = \frac{k(1-k)^2}{2}\frac{R}{fL_2} \qquad (10.326)$$

The variation ratio of output voltage v_O is

$$\varepsilon = \frac{\Delta v_O/2}{V_O} = \frac{k}{2RfC_2} \qquad (10.327)$$

10.5.1.3 N/O Three-Stage Boost Circuit

The N/O three-stage boost circuit is derived from the N/O two-stage boost circuit by readding the parts L_2-D_2-D_3-C_2 the second time. Its circuit diagram and equivalent circuits during the switch-on and switch-off periods are shown in Figure 10.48.

The voltage across capacitor C_1 is charged to V_1. As described in the previous section, the voltage V_{C1} across capacitor C_1 is $V_{C1} = (1/1 - k)V_{in}$ and the voltage V_{C2} across capacitor C_2 is $V_{C2} = (1/1 - k)^2 V_{in}$.

The voltage across capacitor C_3 is charged to V_O. The current flowing through inductor L_3 increases with voltage V_{C2} during the switch-on period kT and decreases with voltage $-(V_{C3} - V_{C2})$ during the switch-off period $(1 - k)T$. Therefore, the ripple of the inductor current i_{L3} is

$$\Delta i_{L3} = \frac{V_{C2}}{L_3}kT = \frac{V_{C3} - V_{C2}}{L_3}(1-k)T \tag{10.328}$$

$$V_{C3} = \frac{1}{1-k}V_{C2} = \left(\frac{1}{1-k}\right)^2 V_{C1} = \left(\frac{1}{1-k}\right)^3 V_{in}$$

$$V_O = V_{C3} - V_{in} = \left[\left(\frac{1}{1-k}\right)^3 - 1\right]V_{in} \tag{10.329}$$

The voltage transfer gain is

$$G = \frac{V_O}{V_{in}} = \left(\frac{1}{1-k}\right)^3 - 1 \tag{10.330}$$

The variation ratio of current i_{L1} through inductor L_1 is

$$\xi_1 = \frac{\Delta i_{L1}/2}{I_{L1}} = \frac{k(1-k)^3 TV_{in}}{2L_1 I_O} = \frac{k(1-k)^6}{2}\frac{R}{fL_1} \tag{10.331}$$

The variation ratio of current i_{L2} through inductor L_2 is

$$\xi_2 = \frac{\Delta i_{L2}/2}{I_{L2}} = \frac{k(1-k)^2 TV_1}{2L_2 I_O} = \frac{k(1-k)^4}{2}\frac{R}{fL_2} \tag{10.332}$$

The variation ratio of current i_{L3} through inductor L_3 is

$$\xi_3 = \frac{\Delta i_{L3}/2}{I_{L3}} = \frac{k(1-k)TV_2}{2L_3 I_O} = \frac{k(1-k)^2}{2}\frac{R}{fL_3} \tag{10.333}$$

The variation ratio of output voltage v_O is

$$\varepsilon = \frac{\Delta v_O / 2}{V_O} = \frac{k}{2RfC_3} \tag{10.334}$$

10.5.1.4 N/O Higher-Stage Boost Circuit

N/O higher-stage boost circuit can be designed by just adding multiple times the parts L_2-D_2-D_3-C_2. For the nth-stage boost circuit, the final output voltage across capacitor C_n is

$$V_O = \left[\left(\frac{1}{1-k}\right)^n - 1\right]V_{in}$$

The voltage transfer gain is

$$G = \frac{V_O}{V_{in}} = \left(\frac{1}{1-k}\right)^n - 1 \tag{10.335}$$

The variation ratio of current i_{Li} through inductor L_i ($i = 1, 2, 3, ..., n$) is

$$\xi_i = \frac{\Delta i_{Li} / 2}{I_{Li}} = \frac{k(1-k)^{2(n-i+1)}}{2} \frac{R}{fL_i} \tag{10.336}$$

and the variation ratio of output voltage v_O is

$$\varepsilon = \frac{\Delta v_O / 2}{V_O} = \frac{k}{2RfC_n} \tag{10.337}$$

10.5.2 N/O ADDITIONAL SERIES

All circuits of N/O cascaded boost converters and their additional series are derived from the corresponding circuits of the main series by adding a DEC.

The first three stages of this series are shown in Figures 10.49 through 10.51. For the sake of convenience, they are called elementary additional boost circuit, two-stage additional boost circuit, and three-stage additional boost circuit, respectively, and numbered $n = 1, 2,$ and 3.

10.5.2.1 N/O Elementary Additional Boost Circuit

This N/O elementary additional boost circuit is derived from the N/O elementary boost converter by adding a DEC. Its circuit diagram and switch-on and switch-off equivalent circuits are shown in Figure 10.49.

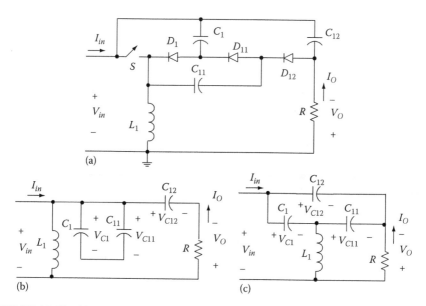

FIGURE 10.49 Negative-output elementary double-boost additional circuit. (a) Circuit diagram. (b) Equivalent circuit during the switch-on period. (c) Equivalent circuit during the switch-off additional.

The voltage across capacitor C_1 and C_{11} is charged to V_{C1}, and the voltage across capacitor C_{12} is charged to $V_{C12} = 2 V_{C1}$. The current i_{L1} flowing through inductor L_1 increases with voltage V_{in} during the switch-on period kT and decreases with voltage $-(V_{C1} - V_{in})$ during the switch-off period $(1 - k)T$. Therefore,

$$\Delta i_{L1} = \frac{V_{in}}{L_1}kT = \frac{V_{C1} - V_{in}}{L_1}(1 - k)T \qquad (10.338)$$

$$V_{C1} = \frac{1}{1-k}V_{in}$$

The voltage V_{C12} is

$$V_{C12} = 2V_{C1} = \frac{2}{1-k}V_{in} \qquad (10.339)$$

The output voltage is

$$V_O = V_{C12} - V_{in} = \left[\frac{2}{1-k} - 1\right]V_{in} \qquad (10.340)$$

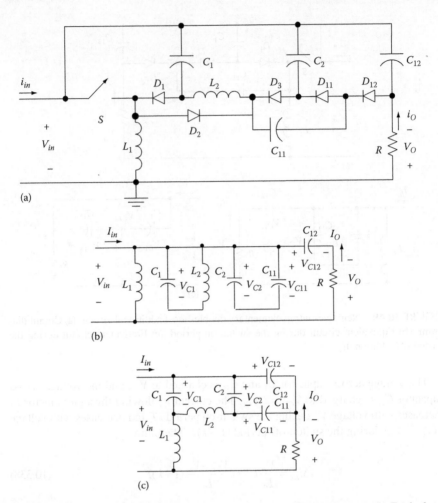

FIGURE 10.50 N/O two-stage additional boost circuit. (a) Circuit diagram. (b) Equivalent circuit during the switch-on period. (c) Equivalent circuit during the switch-off period.

The voltage transfer gain is

$$G = \frac{V_O}{V_{in}} = \frac{2}{1-k} - 1 \qquad (10.341)$$

The variation ratio of current i_{L1} through inductor L_1 is

$$\xi_1 = \frac{\Delta i_{L1}/2}{I_{L1}} = \frac{k(1-k)TV_{in}}{4L_1 I_O} = \frac{k(1-k)^2}{8}\frac{R}{fL_1} \qquad (10.342)$$

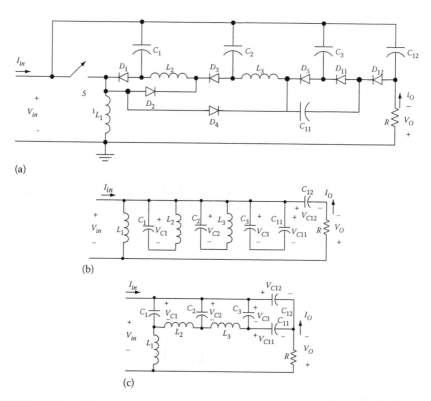

(a)

(b)

(c)

FIGURE 10.51 N/O three-stage additional boost circuit. (a) Circuit diagram. (b) Equivalent circuit during the switch-on period. (c) Equivalent circuit during the switch-off period.

The ripple voltage of output voltage v_O is

$$\Delta v_O = \frac{\Delta Q}{C_{12}} = \frac{I_O kT}{C_{12}} = \frac{k}{fC_{12}} \frac{V_O}{R}$$

Therefore, the variation ratio of output voltage v_O is

$$\varepsilon = \frac{\Delta v_O / 2}{V_O} = \frac{k}{2RfC_{12}} \tag{10.343}$$

10.5.2.2 N/O Two-Stage Additional Boost Circuit

The N/O two-stage additional boost circuit is derived from the N/O two-stage boost circuit by adding a DEC. Its circuit diagram and switch-on and switch-off equivalent circuits are shown in Figure 10.50.

The voltage across capacitor C_1 is charged to V_{C1}. As described in the previous section, the voltage V_{C1} across capacitor C_1 is $V_{C1} = (1/1 - k)V_{in}$.

The voltage across capacitor C_2 and capacitor C_{11} is charged to V_{C2}, and the voltage across capacitor C_{12} is charged to V_{C12}. The current flowing through inductor L_2 increases with voltage V_{C1} during the switch-on period kT and decreases with voltage $-(V_{C2} - V_{C1})$ during the switch-off period $(1 - k)T$. Therefore, the ripple of the inductor current i_{L2} is

$$\Delta i_{L2} = \frac{V_{C1}}{L_2} kT = \frac{V_{C2} - V_{C1}}{L_2}(1-k)T \tag{10.344}$$

$$V_{C2} = \frac{1}{1-k} V_{C1} = \left(\frac{1}{1-k}\right)^2 V_{in} \tag{10.345}$$

$$V_{C12} = 2V_{C2} = \frac{2}{1-k} V_{C1} = 2\left(\frac{1}{1-k}\right)^2 V_{in}$$

The output voltage is

$$V_O = V_{C12} - V_{in} = \left[2\left(\frac{1}{1-k}\right)^2 - 1\right]V_{in} \tag{10.346}$$

The voltage transfer gain is

$$G = \frac{V_O}{V_{in}} = 2\left(\frac{1}{1-k}\right)^2 - 1 \tag{10.347}$$

The variation ratio of current i_{L1} through inductor L_1 is

$$\xi_1 = \frac{\Delta i_{L1}/2}{I_{L1}} = \frac{k(1-k)^2 TV_{in}}{4L_1 I_O} = \frac{k(1-k)^4}{8} \frac{R}{fL_1} \tag{10.348}$$

and the variation ratio of current i_{L2} through inductor L_2 is

$$\xi_2 = \frac{\Delta i_{L2}/2}{I_{L2}} = \frac{k(1-k)TV_1}{4L_2 I_O} = \frac{k(1-k)^2}{8} \frac{R}{fL_2} \tag{10.349}$$

The ripple voltage of output voltage v_O is

$$\Delta v_O = \frac{\Delta Q}{C_{12}} = \frac{I_O kT}{C_{12}} = \frac{k}{fC_{12}} \frac{V_O}{R}$$

Therefore, the variation ratio of output voltage v_O is

$$\varepsilon = \frac{\Delta v_O / 2}{V_O} = \frac{k}{2RfC_{12}} \tag{10.350}$$

10.5.2.3 N/O Three-Stage Additional Boost Circuit

This N/O three-stage additional boost circuit is derived from the three-stage boost circuit by adding a DEC. Its circuit diagram and equivalent circuits during the switch-on and switch-off periods are shown in Figure 10.51.

The voltage across capacitor C_1 is charged to V_{C1}. As described in the previous section, the voltage V_{C1} across capacitor C_1 is $V_{C1} = (1/1 - k)V_{in}$, and the voltage V_2 across capacitor C_2 is

$$V_{C2} = \left(\frac{1}{1-k}\right)^2 V_{in}.$$

The voltage across capacitor C_3 and capacitor C_{11} is charged to V_{C3}. The voltage across capacitor C_{12} is charged to V_{C12}. The current flowing through inductor L_3 increases with voltage V_{C2} during the switch-on period kT and decreases with voltage $-(V_{C3} - V_{C2})$ during the switch-off period $(1 - k)T$. Therefore,

$$\Delta i_{L3} = \frac{V_{C2}}{L_3}kT = \frac{V_{C3} - V_{C2}}{L_3}(1-k)T \tag{10.351}$$

and

$$V_{C3} = \frac{1}{1-k}V_{C2} = \left(\frac{1}{1-k}\right)^2 V_{C1} = \left(\frac{1}{1-k}\right)^3 V_{in} \tag{10.352}$$

The voltage V_{C12} is

$$V_{C12} = 2V_{C3} = 2\left(\frac{1}{1-k}\right)^3 V_{in}$$

The output voltage is

$$V_O = V_{C12} - V_{in} = \left[2\left(\frac{1}{1-k}\right)^3 - 1\right]V_{in} \tag{10.353}$$

The voltage transfer gain is

$$G = \frac{V_O}{V_{in}} = 2\left(\frac{1}{1-k}\right)^3 - 1 \tag{10.354}$$

The variation ratio of current i_{L1} through inductor L_1 is

$$\xi_1 = \frac{\Delta i_{L1}/2}{I_{L1}} = \frac{k(1-k)^3 TV_{in}}{4L_1 I_O} = \frac{k(1-k)^6}{8} \frac{R}{fL_1} \qquad (10.355)$$

The variation ratio of current i_{L2} through inductor L_2 is

$$\xi_2 = \frac{\Delta i_{L2}/2}{I_{L2}} = \frac{k(1-k)^2 TV_1}{4L_2 I_O} = \frac{k(1-k)^4}{8} \frac{R}{fL_2} \qquad (10.356)$$

The variation ratio of current i_{L3} through inductor L_3 is

$$\xi_3 = \frac{\Delta i_{L3}/2}{I_{L3}} = \frac{k(1-k)TV_2}{4L_3 I_O} = \frac{k(1-k)^2}{8} \frac{R}{fL_3} \qquad (10.357)$$

The ripple voltage of output voltage v_O is

$$\Delta v_O = \frac{\Delta Q}{C_{12}} = \frac{I_O kT}{C_{12}} = \frac{k}{fC_{12}} \frac{V_O}{R}$$

Therefore, the variation ratio of output voltage v_O is

$$\varepsilon = \frac{\Delta v_O/2}{V_O} = \frac{k}{2RfC_{12}} \qquad (10.358)$$

10.5.2.4 N/O Higher-Stage Additional Boost Circuit

The N/O higher-stage additional boost circuit is derived from the corresponding circuit of the main series by adding a DEC. For the nth-stage additional circuit, the final output voltage is

$$V_O = \left[2\left(\frac{1}{1-k}\right)^n - 1 \right] V_{in}$$

The voltage transfer gain is

$$G = \frac{V_O}{V_{in}} = 2\left(\frac{1}{1-k}\right)^n - 1 \qquad (10.359)$$

Analogously, the variation ratio of current i_{Li} through inductor L_i $(i = 1, 2, 3, \ldots, n)$ is

$$\xi_i = \frac{\Delta i_{Li}/2}{I_{Li}} = \frac{k(1-k)^{2(n-i+1)}}{8} \frac{R}{fL_i} \qquad (10.360)$$

and the variation ratio of output voltage v_O is

$$\varepsilon = \frac{\Delta v_O / 2}{V_O} = \frac{k}{2RfC_{12}} \qquad (10.361)$$

10.5.3 DOUBLE SERIES

All circuits of N/O cascaded boost converters and their double series are derived from the corresponding circuits of the main series by adding a DEC in each stage circuit. The first three stages of this series are shown in Figures 10.49, 10.52, and 10.53. For the sake of convenience, they are called elementary double circuit, two-stage double circuit, and three-stage double-boost circuit, respectively, and numbered $n = 1$, 2, and 3.

10.5.3.1 N/O Elementary Double-Boost Circuit

This N/O elementary double-boost circuit is derived from elementary boost converter by adding a DEC. Its circuit diagram and switch-on and switch-off equivalent circuits are shown in Figure 10.49, which are the same as that of the elementary boost additional circuit.

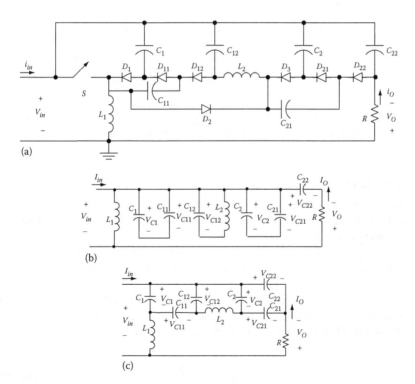

FIGURE 10.52 N/O two-stage double-boost circuit. (a) Circuit diagram. (b) Equivalent circuit during the switch-on period. (c) Equivalent circuit during the switch-off period.

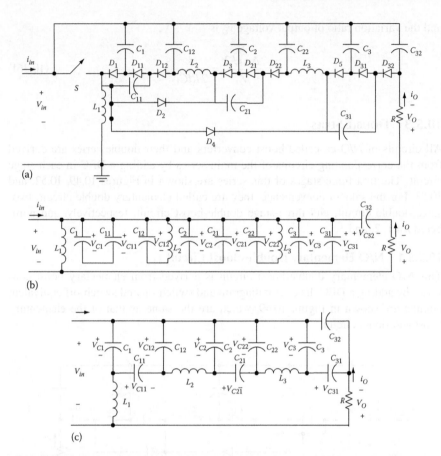

FIGURE 10.53 N/O three-stage double-boost circuit. (a) Circuit diagram. (b) Equivalent circuit during the switch-on period. (c) Equivalent circuit during the switch-off period.

10.5.3.2 N/O Two-Stage Double-Boost Circuit

The N/O two-stage double-boost circuit is derived from the two-stage boost circuit by adding a DEC in each stage circuit. Its circuit diagram and switch-on and switch-off equivalent circuits are shown in Figure 10.52.

The voltage across capacitor C_1 and capacitor C_{11} is charged to V_1. As described in the previous section, the voltage V_{C1} across capacitor C_1 and capacitor C_{11} is $V_{C1} = (1/1 - k)V_{in}$. The voltage across capacitor C_{12} is charged to $2V_{C1}$.

The current flowing through inductor L_2 increases with voltage $2V_{C1}$ during the switch-on period kT and decreases with voltage $-(V_{C2} - 2V_{C1})$ during the switch-off period $(1 - k)T$. Therefore, the ripple of the inductor current i_{L2} is

$$\Delta i_{L2} = \frac{2V_{C1}}{L_2}kT = \frac{V_{C2} - 2V_{C1}}{L_2}(1-k)T \qquad (10.362)$$

$$V_{C2} = \frac{2}{1-k}V_{C1} = 2\left(\frac{1}{1-k}\right)^2 V_{in} \qquad (10.363)$$

The voltage V_{C22} is

$$V_{C22} = 2V_{C2} = \left(\frac{2}{1-k}\right)^2 V_{in}$$

The output voltage is

$$V_O = V_{C22} - V_{in} = \left[\left(\frac{2}{1-k}\right)^2 - 1\right]V_{in} \qquad (10.364)$$

The voltage transfer gain is

$$G = \frac{V_O}{V_{in}} = \left(\frac{2}{1-k}\right)^2 - 1 \qquad (10.365)$$

The variation ratio of current i_{L1} through inductor L_1 is

$$\xi_1 = \frac{\Delta i_{L1}/2}{I_{L1}} = \frac{k(1-k)^2 T V_{in}}{8L_1 I_O} = \frac{k(1-k)^4}{16}\frac{R}{fL_1} \qquad (10.366)$$

and the variation ratio of current i_{L2} through inductor L_2 is

$$\xi_2 = \frac{\Delta i_{L2}/2}{I_{L2}} = \frac{k(1-k)T V_1}{4L_2 I_O} = \frac{k(1-k)^2}{8}\frac{R}{fL_2} \qquad (10.367)$$

The ripple voltage of output voltage v_O is

$$\Delta v_O = \frac{\Delta Q}{C_{22}} = \frac{I_O k T}{C_{22}} = \frac{k}{fC_{22}}\frac{V_O}{R}$$

Therefore, the variation ratio of output voltage v_O is

$$\varepsilon = \frac{\Delta v_O/2}{V_O} = \frac{k}{2RfC_{22}} \qquad (10.368)$$

10.5.3.3 N/O Three-Stage Double-Boost Circuit

This N/O three-stage double-boost circuit is derived from the three-stage boost circuit by adding a DEC in each stage circuit. Its circuit diagram and equivalent circuits during the switch-on and switch-off periods are shown in Figure 10.53.

The voltage across capacitor C_1 and capacitor C_{11} is charged to V_{C1}. As described in the previous section, the voltage V_{C1} across capacitor C_1 and capacitor C_{11} is $V_{C1} = (1/1 - k)V_{in}$ and the voltage V_{C2} across capacitor C_2 and capacitor C_{12} is $V_{C2} = 2(1/1 - k)^2 V_{in}$.

The voltage across capacitor C_{22} is $2V_{C2} = (2/1 - k)^2 V_{in}$. The voltage across capacitor C_3 and capacitor C_{31} is charged to V_3. The voltage across capacitor C_{12} is charged to V_O. The current flowing through inductor L_3 increases with voltage V_2 during the switch-on period kT and decreases with voltage $-(V_{C3} - 2V_{C2})$ during the switch-off period $(1 - k)T$. Therefore,

$$\Delta i_{L3} = \frac{2V_{C2}}{L_3} kT = \frac{V_{C3} - 2V_{C2}}{L_3}(1-k)T \tag{10.369}$$

and

$$V_{C3} = \frac{2V_{C2}}{(1-k)} = \frac{4}{(1-k)^3} V_{in} \tag{10.370}$$

The voltage V_{C32} is

$$V_{C32} = 2V_{C3} = \left(\frac{2}{1-k}\right)^3 V_{in}$$

The output voltage is

$$V_O = V_{C32} - V_{in} = \left[\left(\frac{2}{1-k}\right)^3 - 1\right] V_{in} \tag{10.371}$$

The voltage transfer gain is

$$G = \frac{V_O}{V_{in}} = \left(\frac{2}{1-k}\right)^3 - 1 \tag{10.372}$$

The variation ratio of current i_{L1} through inductor L_1 is

$$\xi_1 = \frac{\Delta i_{L1}/2}{I_{L1}} = \frac{k(1-k)^3 TV_{in}}{16L_1 I_O} = \frac{k(1-k)^6}{128} \frac{R}{fL_1} \tag{10.373}$$

The variation ratio of current i_{L2} through inductor L_2 is

$$\xi_2 = \frac{\Delta i_{L2}/2}{I_{L2}} = \frac{k(1-k)^2 TV_1}{8L_2 I_O} = \frac{k(1-k)^4}{32}\frac{R}{fL_2} \tag{10.374}$$

The variation ratio of current i_{L3} through inductor L_3 is

$$\xi_3 = \frac{\Delta i_{L3}/2}{I_{L3}} = \frac{k(1-k)TV_2}{4L_3 I_O} = \frac{k(1-k)^2}{8}\frac{R}{fL_3} \tag{10.375}$$

The ripple voltage of output voltage v_O is

$$\Delta v_O = \frac{\Delta Q}{C_{32}} = \frac{I_O kT}{C_{32}} = \frac{k}{fC_{32}}\frac{V_O}{R}$$

Therefore, the variation ratio of output voltage v_O is

$$\varepsilon = \frac{\Delta v_O/2}{V_O} = \frac{k}{2RfC_{32}} \tag{10.376}$$

10.5.3.4 N/O Higher-Stage Double-Boost Circuit

The N/O higher-stage double-boost circuit is derived from the corresponding circuit of the main series by adding a DEC in each stage circuit. For the nth-stage additional circuit, the final output voltage is

$$V_O = \left[\left(\frac{2}{1-k}\right)^n - 1\right]V_{in}$$

The voltage transfer gain is

$$G = \frac{V_O}{V_{in}} = \left(\frac{2}{1-k}\right)^n - 1 \tag{10.377}$$

Analogously, the variation ratio of current i_{Li} through inductor L_i ($i=1, 2, 3,\ldots, n$) is

$$\xi_i = \frac{\Delta i_{Li}/2}{I_{Li}} = \frac{k(1-k)^{2(n-i+1)}}{2*2^{2n}}\frac{R}{fL_i} \tag{10.378}$$

and the variation ratio of output voltage v_O is

$$\varepsilon = \frac{\Delta v_O/2}{V_O} = \frac{k}{2RfC_{n2}} \tag{10.379}$$

10.5.4 TRIPLE SERIES

All circuits of N/O cascaded boost converters and their triple series are derived from the corresponding circuits of the main series by adding DEC twice in each stage circuit. The first three stages of this series are shown in Figures 10.54 through 10.56. For the sake of convenience, they are called elementary double (or additional) circuit, two-stage double circuit, and three-stage double circuit, respectively, and numbered $n = 1, 2$, and 3.

10.5.4.1 N/O Elementary Triple-Boost Circuit

This N/O elementary triple-boost circuit is derived from the elementary boost converter by adding DEC twice. Its circuit diagram and switch-on and switch-off

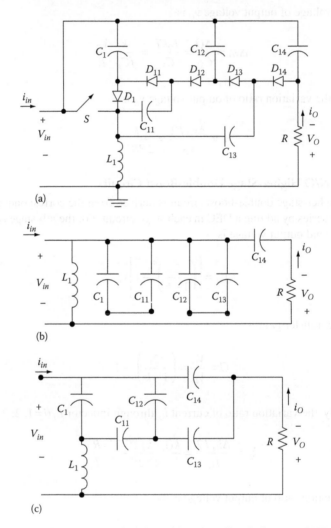

FIGURE 10.54 N/O elementary triple-boost circuit. (a) Circuit diagram. (b) Equivalent circuit during the switch-on period. (c) Equivalent circuit during the switch-off period.

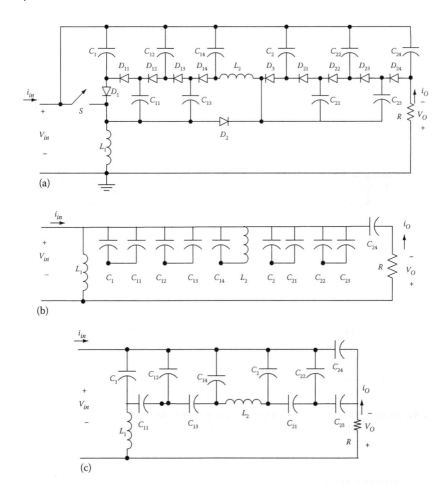

FIGURE 10.55 N/O two-stage triple-boost circuit. (a) Circuit diagram. (b) Equivalent circuit during the switch-on period. (c) Equivalent circuit during the switch-off period.

equivalent circuits are shown in Figure 10.54. The output voltage of the first-stage boost circuit is V_{C1}, $V_{C1} = V_{in}/(1 - k)$.

After the first DEC, the voltage (across capacitor C_{12}) increases to

$$V_{C12} = 2V_{C1} = \frac{2}{1-k}V_{in} \qquad (10.380)$$

After the second DEC, the voltage (across capacitor C_{14}) increases to

$$V_{C14} = V_{C12} + V_{C1} = \frac{3}{1-k}V_{in} \qquad (10.381)$$

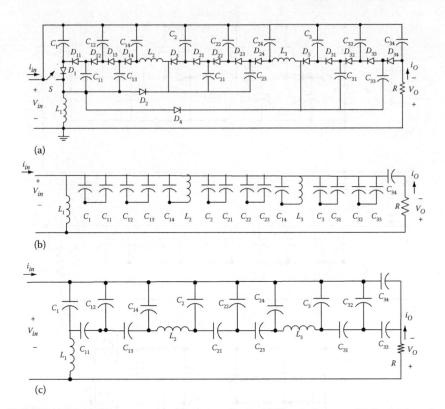

FIGURE 10.56 N/O three-stage triple-boost circuit. (a) Circuit diagram. (b) Equivalent circuit during the switch-on period. (c) Equivalent circuit during the switch-off period.

The final output voltage V_O is equal to

$$V_O = V_{C14} - V_{in} = \left[\frac{3}{1-k} - 1 \right] V_{in} \qquad (10.382)$$

The voltage transfer gain is

$$G = \frac{V_O}{V_{in}} = \frac{3}{1-k} - 1 \qquad (10.383)$$

10.5.4.2 N/O Two-Stage Triple-Boost Circuit

The N/O two-stage triple-boost circuit is derived from the two-stage boost circuit by adding DEC twice in each stage circuit. Its circuit diagram and switch-on and switch-off equivalent circuits are shown in Figure 10.55.

As described in the previous section, the voltage across capacitor C_{14} is $V_{C14} = (3/1-k)V_{in}$. Analogously, the voltage across capacitor C_{24} is

$$V_{C24} = \left(\frac{3}{1-k}\right)^2 V_{in} \qquad (10.384)$$

The final output voltage V_O is equal to

$$V_O = V_{C24} - V_{in} = \left[\left(\frac{3}{1-k}\right)^2 - 1\right]V_{in} \qquad (10.385)$$

The voltage transfer gain is

$$G = \frac{V_O}{V_{in}} = \left(\frac{3}{1-k}\right)^2 - 1 \qquad (10.386)$$

The variation ratio of current i_{L1} through inductor L_1 is

$$\xi_1 = \frac{\Delta i_{L1}/2}{I_{L1}} = \frac{k(1-k)^2 TV_{in}}{8L_1 I_O} = \frac{k(1-k)^4}{16}\frac{R}{fL_1} \qquad (10.387)$$

and the variation ratio of current i_{L2} through inductor L_2 is

$$\xi_2 = \frac{\Delta i_{L2}/2}{I_{L2}} = \frac{k(1-k)TV_1}{4L_2 I_O} = \frac{k(1-k)^2}{8}\frac{R}{fL_2} \qquad (10.388)$$

The ripple voltage of output voltage v_O is

$$\Delta v_O = \frac{\Delta Q}{C_{22}} = \frac{I_O kT}{C_{22}} = \frac{k}{fC_{22}}\frac{V_O}{R} \qquad$$

Therefore, the variation ratio of output voltage v_O is

$$\varepsilon = \frac{\Delta v_O/2}{V_O} = \frac{k}{2RfC_{22}} \qquad (10.389)$$

10.5.4.3 N/O Three-Stage Triple-Boost Circuit

This N/O three-stage triple-boost circuit is derived from the three-stage boost circuit by adding DEC twice in each stage circuit. Its circuit diagram and equivalent circuits during the switch-on and switch-off periods are shown in Figure 10.56.

As described in the previous section, the voltage across capacitor C_{14} is $V_{C14} = (3/1 - k)V_{in}$ and the voltage across capacitor C_{24} is $V_{C24} = (3/1 - k)^2 V_{in}$. Analogously, the voltage across capacitor C_{34} is

$$V_{C34} = \left(\frac{3}{1-k}\right)^3 V_{in} \qquad (10.390)$$

The final output voltage V_O is equal to

$$V_O = V_{C34} - V_{in} = \left[\left(\frac{3}{1-k}\right)^3 - 1\right] V_{in} \qquad (10.391)$$

The voltage transfer gain is

$$G = \frac{V_O}{V_{in}} = \left(\frac{3}{1-k}\right)^3 - 1 \qquad (10.392)$$

The variation ratio of current i_{L1} through inductor L_1 is

$$\xi_1 = \frac{\Delta i_{L1}/2}{I_{L1}} = \frac{k(1-k)^3 TV_{in}}{64L_1 I_O} = \frac{k(1-k)^6}{12^3}\frac{R}{fL_1} \qquad (10.393)$$

The variation ratio of current i_{L2} through inductor L_2 is

$$\xi_2 = \frac{\Delta i_{L2}/2}{I_{L2}} = \frac{k(1-k)^2 TV_1}{16L_2 I_O} = \frac{k(1-k)^4}{12^2}\frac{R}{fL_2} \qquad (10.394)$$

The variation ratio of current i_{L3} through inductor L_3 is

$$\xi_3 = \frac{\Delta i_{L3}/2}{I_{L3}} = \frac{k(1-k)TV_2}{4L_3 I_O} = \frac{k(1-k)^2}{12}\frac{R}{fL_3} \qquad (10.395)$$

Usually ξ_1, ξ_2, and ξ_3 are small, which means this converter works in the continuous mode.

The ripple voltage of output voltage v_O is

$$\Delta v_O = \frac{\Delta Q}{C_{32}} = \frac{I_O kT}{C_{32}} = \frac{k}{fC_{32}}\frac{V_O}{R}$$

Therefore, the variation ratio of output voltage v_O is

$$\varepsilon = \frac{\Delta v_O / 2}{V_O} = \frac{k}{2RfC_{32}}$$

(10.396)

Usually R is in kΩ, f is in 10 kHz, and C_{34} is in μF; thus, this ripple is very small, smaller than 1%.

10.5.4.4 N/O Higher-Stage Triple-Boost Circuit

The N/O higher-stage triple-boost circuit is derived from the corresponding circuit of the main series by adding DEC twice in each stage circuit. For the nth-stage additional circuit, the voltage across capacitor C_{n4} is

$$V_{Cn4} = \left(\frac{3}{1-k}\right)^n V_{in}$$

The output voltage is

$$V_O = V_{Cn4} - V_{in} = \left[\left(\frac{3}{1-k}\right)^n - 1\right] V_{in}$$

(10.397)

The voltage transfer gain is

$$G = \frac{V_O}{V_{in}} = \left(\frac{3}{1-k}\right)^n - 1$$

(10.398)

Analogously, the variation ratio of current i_{Li} through inductor L_i ($i=1, 2, 3, ..., n$) is

$$\xi_i = \frac{\Delta i_{Li}/2}{I_{Li}} = \frac{k(1-k)^{2(n-i+1)}}{12^{(n-i+1)}} \frac{R}{fL_i}$$

(10.399)

and the variation ratio of output voltage v_O is

$$\varepsilon = \frac{\Delta v_O/2}{V_O} = \frac{k}{2RfC_{n2}}$$

(10.400)

10.5.5 MULTIPLE SERIES

All circuits of N/O cascaded boost converters and their multiple series are derived from the corresponding circuits of the main series by adding DEC multiple (j) times in each stage circuit. The first three stages of this series are shown in

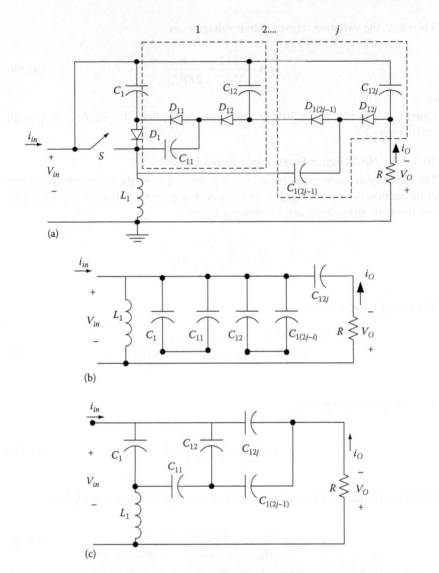

FIGURE 10.57 N/O elementary multiple-boost circuit. (a) Circuit diagram. (b) Equivalent circuit during the switch-on period. (c) Equivalent circuit during the switch-off period.

Figures 10.57 through 10.59. For the sake of convenience, they are called elementary multiple-boost circuit, two-stage multiple-boost circuit, and three-stage multiple-boost circuit, respectively, and numbered $n = 1$, 2, and 3.

10.5.5.1 N/O Elementary Multiple-Boost Circuit

This N/O elementary multiple-boost circuit is derived from the elementary boost converter by adding DEC multiple (j) times. Its circuit diagram and switch-on and switch-off equivalent circuits are shown in Figure 10.57.

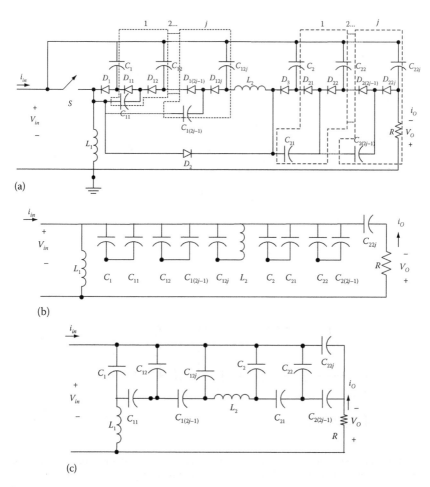

(a)

(b)

(c)

FIGURE 10.58 N/O two-stage multiple-boost circuit. (a) Circuit diagram. (b) Equivalent circuit during the switch-on period. (c) Equivalent circuit during the switch-off period.

The output voltage of the first DEC (across capacitor C_{12j}) increases to

$$V_{C12j} = \frac{j+1}{1-k}V_{in} \qquad (10.401)$$

The final output voltage V_O is equal to

$$V_O = V_{C12j} - V_{in} = \left[\frac{j+1}{1-k}-1\right]V_{in} \qquad (10.402)$$

The voltage transfer gain is

$$G = \frac{V_O}{V_{in}} = \frac{j+1}{1-k}-1 \qquad (10.403)$$

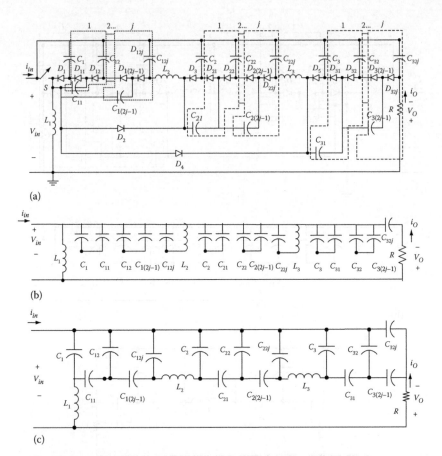

(a)

(b)

(c)

FIGURE 10.59 N/O three-stage multiple-boost circuit. (a) Circuit diagram. (b) Equivalent circuit during the switch-on period. (c) Equivalent circuit during the switch-off period.

10.5.5.2 N/O Two-Stage Multiple-Boost Circuit

The N/O two-stage multiple-boost circuit is derived from the two-stage boost circuit by adding DEC multiple (j) times in each stage circuit. Its circuit diagram and the switch-on and switch-off equivalent circuits are shown in Figure 10.58.

As described in the previous section, the voltage across capacitor C_{12j} is $V_{C12j} = (j + 1/1 - k)V_{in}$. Analogously, the voltage across capacitor C_{22j} is

$$V_{C22j} = \left(\frac{j+1}{1-k}\right)^2 V_{in} \tag{10.404}$$

The final output voltage V_O is equal to

$$V_O = V_{C22j} - V_{in} = \left[\left(\frac{j+1}{1-k}\right)^2 - 1\right]V_{in} \tag{10.405}$$

The voltage transfer gain is

$$G = \frac{V_O}{V_{in}} = \left(\frac{j+1}{1-k}\right)^2 - 1 \tag{10.406}$$

The ripple voltage of output voltage v_O is

$$\Delta v_O = \frac{\Delta Q}{C_{22j}} = \frac{I_O kT}{C_{22j}} = \frac{k}{fC_{22j}} \frac{V_O}{R}$$

Therefore, the variation ratio of output voltage v_O is

$$\varepsilon = \frac{\Delta v_O / 2}{V_O} = \frac{k}{2 R f C_{22j}} \tag{10.407}$$

Example 10.5 An N/O two-stage multiple-boost (j=4) converter in Figure 10.58a has V_{in}=20 V, all inductors are 10 mH, all capacitors are 20 μF, R=10 kΩ, f=50 kHz, and conduction duty cycle k=0.6. Calculate the output voltage and its variation ratio.

Solution: From formula (10.405), we can get the output voltage:

$$V_O = \left[\left(\frac{j+1}{1-k}\right)^2 - 1\right]V_{in} = \left[\left(\frac{4+1}{1-0.6}\right)^2 - 1\right] \times 20 = 605\,\text{V}$$

From (10.407), its variation ratio is

$$\varepsilon = \frac{k}{2RfC_{28}} = \frac{0.6}{2 \times 10,000 \times 50k \times 20\mu} = 0.00003$$

10.5.5.3 N/O Three-Stage Multiple-Boost Circuit

This N/O three-stage multiple-boost circuit is derived from the three-stage boost circuit by adding DEC multiple (j) times in each stage circuit. Its circuit diagram and equivalent circuits during the switch-on and switch-off periods are shown in Figure 10.59.

As described in the previous section, the voltage across capacitor C_{12j} is $V_{C12j}=(j+1/1-k)V_{in}$ and the voltage across capacitor C_{22j} is $V_{C22j}=(j+1/1-k)^2V_{in}$. Analogously, the voltage across capacitor C_{32j} is

$$V_{C32j} = \left(\frac{j+1}{1-k}\right)^3 V_{in} \tag{10.408}$$

The final output voltage V_O is equal to

$$V_O = V_{C32j} - V_{in} = \left[\left(\frac{j+1}{1-k} \right)^3 - 1 \right] V_{in} \tag{10.409}$$

The voltage transfer gain is

$$G = \frac{V_O}{V_{in}} = \left(\frac{j+1}{1-k} \right)^3 - 1 \tag{10.410}$$

The ripple voltage of output voltage v_O is

$$\Delta v_O = \frac{\Delta Q}{C_{32j}} = \frac{I_O k T}{C_{32j}} = \frac{k}{f C_{32j}} \frac{V_O}{R}$$

Therefore, the variation ratio of output voltage v_O is

$$\varepsilon = \frac{\Delta v_O / 2}{V_O} = \frac{k}{2 R f C_{32j}} \tag{10.411}$$

10.5.5.4 N/O Higher-Stage Multiple-Boost Circuit

The N/O higher-stage multiple-boost circuit is derived from the corresponding circuit of the main series by adding DEC multiple (j) times in each stage circuit. For the nth-stage multiple-boost circuit, the voltage across capacitor C_{n2j} is

$$V_{Cn2j} = \left(\frac{j+1}{1-k} \right)^n V_{in}$$

The output voltage is

$$V_O = V_{Cn2j} - V_{in} = \left[\left(\frac{j+1}{1-k} \right)^n - 1 \right] V_{in} \tag{10.412}$$

The voltage transfer gain is

$$G = \frac{V_O}{V_{in}} = \left(\frac{j+1}{1-k} \right)^n - 1 \tag{10.413}$$

The variation ratio of output voltage v_O is

$$\varepsilon = \frac{\Delta v_O / 2}{V_O} = \frac{k}{2 R f C_{n2j}} \tag{10.414}$$

10.5.6 SUMMARY OF N/O CASCADED BOOST CONVERTERS

A family tree of all circuits of the N/O cascaded boost converters are shown in Figure 10.60.

On the basis of the analyses provided in the previous two sections, we can have the common formula to calculate the output voltage:

$$V_O = \begin{cases} \left[\left(\dfrac{1}{1-k}\right)^n - 1\right]V_{in} & \text{Main\_series} \\[3mm] \left[2*\left(\dfrac{1}{1-k}\right)^n - 1\right]V_{in} & \text{Additional\_series} \\[3mm] \left[\left(\dfrac{2}{1-k}\right)^n - 1\right]V_{in} & \text{Double\_series} \\[3mm] \left[\left(\dfrac{3}{1-k}\right)^n - 1\right]V_{in} & \text{Triple\_series} \\[3mm] \left[\left(\dfrac{j+1}{1-k}\right)^n - 1\right]V_{in} & \text{Multiple}(j)\_\text{series} \end{cases} \qquad (10.415)$$

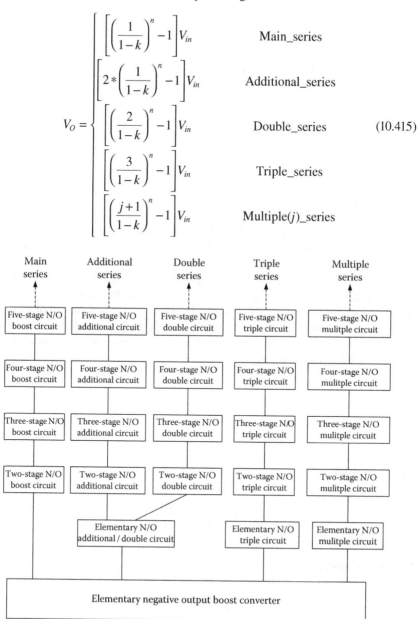

FIGURE 10.60 The family of N/O cascaded boost converters.

The voltage transfer gain is

$$G = \frac{V_O}{V_{in}} = \begin{cases} \left(\dfrac{1}{1-k}\right)^n - 1 & \text{Main\_series} \\[2ex] 2*\left(\dfrac{1}{1-k}\right)^n - 1 & \text{Additional\_series} \\[2ex] \left(\dfrac{2}{1-k}\right)^n - 1 & \text{Double\_series} \\[2ex] \left(\dfrac{3}{1-k}\right)^n - 1 & \text{Triple\_series} \\[2ex] \left(\dfrac{j+1}{1-k}\right)^n - 1 & \text{Multiple}(j)\text{\_series} \end{cases} \tag{10.416}$$

10.6 ULTRA-LIFT LUO CONVERTER

VL technique has been widely applied in the electronic circuit design. Since the past century it has been successfully applied in the design of power DC/DC converters. Good examples include the three-series Luo converters. VL technique can be used to obtain the converter's voltage transfer gain stage by stage in arithmetical series, which is higher than that of classical converters such as buck converter, boost converter, and buck-boost converter. Assume the input voltage and current of a DC/DC converter are V_1 and I_1, the output voltage and current are V_2 and I_2, and the conduction duty cycle is k. In order to compare the transfer gains of these converters, we provide the following formulae:

Buck converter

$$G = \frac{V_2}{V_1} = k$$

Boost converter

$$G = \frac{V_2}{V_1} = \frac{1}{1-k}$$

Buck-boost converter

$$G = \frac{V_2}{V_1} = \frac{k}{1-k}$$

Luo converters

$$G = \frac{V_2}{V_1} = \frac{k^{h(n)}[n+h(n)]}{1-k} \tag{10.417}$$

where
  n is the stage number
  $h(n)$ is the Hong function

$$h(n) = \begin{cases} 1 & n=0 \\ 0 & n>0 \end{cases}$$

and $n=0$ for the elementary circuit with the voltage transfer gain

$$G = \frac{V_2}{V_1} = \frac{k}{1-k} \tag{10.418}$$

SL technique has been given more attention since it yields higher voltage transfer gain. A good example is SL Luo converter. Using this technique, we can obtain the converter's voltage transfer gain stage by stage in geometrical series. The gain calculation formulae are as follows:

$$G = \frac{V_2}{V_1} = \left(\frac{j+2-k}{1-k} \right)^n \tag{10.419}$$

where
  n is the stage number
  j is the multiple-enhanced number

and $n=1$ and $j=0$ for the elementary circuit with

$$G = \frac{V_2}{V_1} = \frac{2-k}{1-k} \tag{10.420}$$

We introduce the ultra-lift Luo converter as a new type converter that has been designed using a novel approach—ultra-lift (UL) technique; thus, the ultra-lift Luo converter produces even higher voltage transfer gains [1,2,10,11]. Simulation results verified our analysis and calculation and illustrated the advanced features of this converter.

10.6.1 Operation of Ultra-Lift Luo Converter

The circuit diagram is shown in Figure 10.61a, which consists of one switch S, two inductors L_1 and L_2, two capacitors C_1 and C_2, three diodes, and the load R.

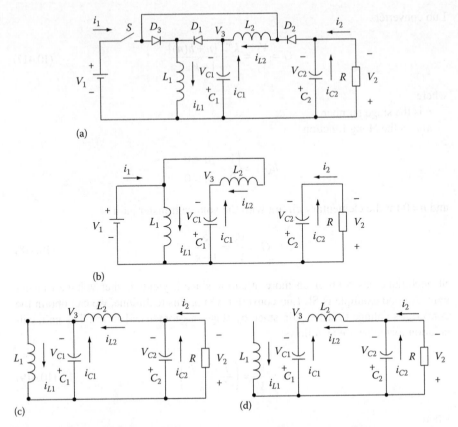

FIGURE 10.61 Ultra-lift Luo converter. (a) Circuit diagram. (b) Equivalent circuit during the switch-on period. (c) Equivalent circuit during the switch-off period (CCM). (d) Equivalent circuit during the switch-off period (DCM).

Its switch-on equivalent circuit is shown in Figure 10.61b. Its switch-off equivalent circuit for the CCM is shown in Figure 10.61c and switch-off equivalent circuit for the discontinuous conduction mode (DCM) is shown in Figure 10.61d.

It is a converter with a very simple structure, compared to other types of converters. As usual, the input voltage and current of the ultra-lift Luo converter are V_1 and I_1, the output voltage and current are V_2 and I_2, the conduction duty cycle is k, and the switching frequency is f. Consequently, the repeating period is $T = 1/f$, switch-on period is kT, and switch-off period is $(1 - k)T$. In order to concentrate on the operation process, we assume that all components, except load R, are ideal ones. Therefore, there are no power losses to be anticipated during the power transformation, that is, $P_{in} = P_O$ or $V_1 \times I_1 = V_2 \times I_2$.

10.6.1.1 Continuous Conduction Mode

Refer to the Figure 10.61b and c; we have got the current i_{L1} increasing with the slope $+V_1/L_1$ during the switch-on period and decreasing with the slope $-V_3/L_1$ during the switch-off period. In the steady state, the current increment is equal to the decrement in a whole period T. Thus, the following relation is obtained:

$$kT \frac{V_1}{L_1} = (1-k)T \frac{V_3}{L_1}$$

Thus,

$$V_{C1} = V_3 = \frac{k}{1-k} V_1 \qquad (10.421)$$

The current i_{L2} increases with the slope $+(V_1 - V_3)/L_2$ during the switch-on period and decreases with the slope $-(V_3 - V_2)/L_2$ during the switch-off period. In the steady state, the current increment is equal to the decrement in a whole period T. Thus, we obtain the following relation:

$$kT \frac{V_1 + V_3}{L_2} = (1-k)T \frac{V_2 - V_3}{L_2}$$

$$V_2 = V_{C2} = \frac{2-k}{1-k} V_3 = \frac{k}{1-k} \frac{2-k}{1-k} V_1 = \frac{k(2-k)}{(1-k)^2} V_1 \qquad (10.422)$$

The voltage transfer gain is

$$G = \frac{V_2}{V_1} = \frac{k}{1-k} \frac{2-k}{1-k} = \frac{k(2-k)}{(1-k)^2} \qquad (10.423)$$

It is much higher than the voltage transfer gains of the VL Luo converter and SL Luo converter, as shown in Equations 10.418 and 10.420. Actually, the gain in (10.423) is the production of those in (10.418) and (10.420). Another advantage is the starting output voltage, from 0 V. The curve of the voltage transfer gain M versus the conduction duty cycle k is shown in Figure 10.62.

The relation between the input and output average currents is presented as follows:

$$I_2 = \frac{(1-k)^2}{k(2-k)} I_1 \qquad (10.424)$$

The relation between the average currents I_{L2} and I_{L1} is as follows:

$$I_{L2} = (1-k)I_{L1} \qquad (10.425)$$

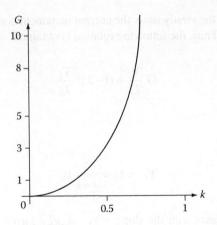

FIGURE 10.62   The voltage transfer gain G versus conduction duty cycle k.

Other relations are as follows:

$$I_{L2} = \left(1 + \frac{k}{1-k}\right)I_2 = \frac{1}{1-k}I_2 \tag{10.426}$$

$$I_{L1} = \frac{1}{1-k}I_{L2} = \left(\frac{1}{1-k}\right)^2 I_2 \tag{10.427}$$

The variation of inductor current i_{L1} is

$$\Delta i_{L1} = kT\frac{V_1}{L_1} \tag{10.428}$$

and its variation ratio is

$$\xi_1 = \frac{\Delta i_{L1}/2}{I_{L1}} = \frac{k(1-k)^2 TV_1}{2L_1 I_2} = \frac{k(1-k)^2 TR}{2L_1 M} = \frac{(1-k)^4 R}{2(2-k)fL_1} \tag{10.429}$$

The diode current i_{D1} is the same as the inductor current i_{L1} during the switch-off period. For the CCM operation, both currents do not descend to zero, that is,

$$\xi_1 \le 1$$

The variation of inductor current i_{L2} is

$$\Delta i_{L2} = \frac{kTV_1}{(1-k)L_2} \tag{10.430}$$

and its variation ratio is

$$\xi_2 = \frac{\Delta i_{L2}/2}{I_{L2}} = \frac{kTV_1}{2L_2I_2} = \frac{kTR}{2L_2M} = \frac{(1-k)^2 R}{2(2-k)fL_2} \tag{10.431}$$

The variation of capacitor voltage v_{C1} is

$$\Delta v_{C1} = \frac{\Delta Q_{C1}}{C_1} = \frac{kTI_{L2}}{C_1} = \frac{kTI_2}{(1-k)C_1} \tag{10.432}$$

and its variation ratio is

$$\sigma_1 = \frac{\Delta v_{C1}/2}{V_{C1}} = \frac{kTI_2}{2(1-k)V_3C_1} = \frac{k(2-k)}{2(1-k)^2 fC_1R} \tag{10.433}$$

The variation of capacitor voltage v_{C2} is

$$\Delta v_{C2} = \frac{\Delta Q_{C2}}{C_2} = \frac{kTI_2}{C_2} \tag{10.434}$$

and its variation ratio is

$$\varepsilon = \sigma_2 = \frac{\Delta v_{C2}/2}{V_{C2}} = \frac{kTI_2}{2V_2C_2} = \frac{k}{2fC_2R} \tag{10.435}$$

Example 10.6 An ultra-lift Luo converter in Figure 10.61a has $V_1 = 20\,V$, all inductors are $10\,mH$, all capacitors are $20\,\mu F$, $R = 500\,\Omega$, $f = 50\,kHz$, and conduction duty cycle $k = 0.6$. Calculate the variation ratios of current i_{L1}, voltage v_{C1}, the output voltage, and its variation ratio.

Solution: From formula (10.429), we can get the variation ratio of current i_{L1}:

$$\xi_1 = \frac{(1-k)^4 R}{2(2-k)fL_1} = \frac{(1-0.6)^4 \times 500}{2(2-0.6)\times 50k \times 10m} = 0.0091$$

From formula (10.431), we can get the variation ratio of current i_{L2}:

$$\xi_2 = \frac{(1-k)^2 R}{2(2-k)fL_2} = \frac{(1-0.6)^2 \times 500}{2(2-0.6)\times 50k \times 10m} = 0.057$$

From formula (10.433), we can get the variation ratio of voltage v_{C1}:

$$\sigma_1 = \frac{k(2-k)}{2(1-k)^2 fC_1R} = \frac{0.6(2-0.6)}{2(1-0.6)^2 \times 50k \times 20\mu \times 500} = 0.00525$$

This converter works in CCM. From formula (10.422), we can get the output voltage:

$$V_2 = \frac{k(2-k)}{(1-k)^2}V_1 = \frac{0.6(2-0.6)}{(1-0.6)^2}20 = 105\,\text{V}$$

From (10.435), its variation ratio can be obtained as follows:

$$\varepsilon = \frac{k}{2fC_2R} = \frac{0.6}{2\times50k\times20\mu\times500} = 0.0006$$

10.6.1.2 Discontinuous Conduction Mode

Refer to Figure 10.61b through d; we have got the current i_{L1} increasing with the slope $+V_1/L_1$ during the switch-on period and decreasing with the slope $-V_3/L_1$ during the switch-off period. The inductor current i_{L1} decreases to zero before $t=T$; that is, the current becomes zero before the next time the switch turns on.

The current waveform is shown in Figure 10.63. The DCM operation condition is defined as

$$\xi_1 \leq 1$$

or

$$\xi_1 = \frac{k(1-k)^2 TR}{2L_1M} = \frac{(1-k)^4 R}{2(2-k)fL_1} \geq 1 \tag{10.436}$$

Taking the equal mark, we obtain the boundary between CCM and DCM operations. Here we define the normalized impedance Z_n:

$$Z_n = \frac{R}{fL_1} \tag{10.437}$$

The boundary equation is

$$G = \frac{k(1-k)^2}{2}Z_N \tag{10.438}$$

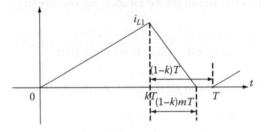

FIGURE 10.63 Discontinuous inductor current i_{L1}.

or

$$\frac{G}{Z_N} = \frac{k(1-k)^2}{2}$$

The corresponding Z_N is

$$Z_N = \frac{k(2-k)/(1-k)^2}{k(1-k)^2/2} = \frac{2(2-k)}{(1-k)^4} \tag{10.439}$$

The curve is shown in Figure 10.64 and Table 10.7.

We define the filling factor m to describe the current existing at a particular time. For DCM operation,

$$0 < m \le 1$$

In the steady state, the current increment is equal to the decrement in a whole period T. Thus, the following relation is obtained:

$$kT\frac{V_1}{L_1} = (1-k)mT\frac{V_3}{L_1}$$

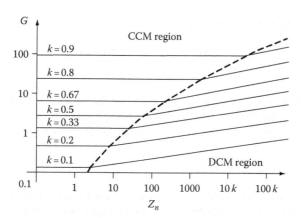

FIGURE 10.64 Boundary between CCM and DCM.

TABLE 10.7
Boundary between CCM and DCM

k	0.2	0.33	0.5	0.67	0.8	0.9
G	0.5625	1.25	3	8	24	99
G/Z_n	0.064	2/27	1/16	1/27	0.016	0.0045
Z_n	8.8	16.9	48	216	1500	22000

Thus,

$$V_{C1} = V_3 = \frac{k}{(1-k)m}V_1 \tag{10.440}$$

Comparing between Equations 10.421 and 10.440, we found that the voltage V_3 is higher during the DCM operation since the filling factor $m < 1$. Its expression is

$$m = \frac{1}{\xi_1} = \frac{2L_1 G}{k(1-k)^2 TR} = \frac{2(2-k)}{(1-k)^4 Z_N} \tag{10.441}$$

The current i_{L2} increases with the slope $+(V_1 - V_3)/L_2$ during the switch-on period and decreases with the slope $-(V_3 - V_2)/L_2$ during the switch-on period. In the steady state, the current increment is equal to the decrement in a whole period T. Thus, we obtain the following relation:

$$kT\frac{V_1 + V_3}{L_2} = (1-k)T\frac{V_2 - V_3}{L_2}$$

$$V_2 = V_{C2} = \frac{2-k}{1-k}V_3 = \frac{k(2-k)}{m(1-k)^2}V_1 \tag{10.442}$$

The voltage transfer gain in DCM is

$$G_{DCM} = \frac{V_2}{V_1} = \frac{k(2-k)}{m(1-k)^2} = \frac{k(1-k)^2}{2}Z_N \tag{10.443}$$

It is higher than the voltage transfer gain achieved during the CCM operation, since $m < 1$. We can see that the voltage transfer gain G_{DCM} linearly increases, in proportion to the normalized impedance Z_N, and this relation is shown in Figure 10.64.

10.6.2 Instantaneous Values

Instantaneous values of the voltage and current of each component are an important aspect to be covered while describing the converter operations. Refer to Figure 10.61; we can get the components' values in CCM and DCM operations.

10.6.2.1 Continuous Conduction Mode

Refer to Figure 10.61b and c; we have got the instantaneous values of the voltage and current of each component in CCM operations as shown in the following:

$$i_{L1}(t) = \begin{cases} I_{L1\text{-min}} + \frac{V_1}{L_1}t & 0 \le t \le kT \\ I_{L1\text{-max}} - \frac{V_3}{L_1}t & kT \le t \le T \end{cases} \tag{10.444}$$

$$i_{L2}(t) = \begin{cases} I_{L2\text{-min}} + \dfrac{V_1 - V_3}{L_2}t & 0 \le t \le kT \\[4mm] I_{L2\text{-max}} - \dfrac{V_2 - V_1}{L_2}t & kT \le t \le T \end{cases} \tag{10.445}$$

$$i_1(t) = i_s = \begin{cases} I_{1\text{-min}} + \left(\dfrac{V_1}{L_1} + \dfrac{V_1 - V_3}{L_2}\right)t & 0 \le t \le kT \\[4mm] 0 & kT \le t \le T \end{cases} \tag{10.446}$$

$$i_{D1}(t) = \begin{cases} 0 & 0 \le t \le kT \\[4mm] I_{L1\text{-max}} - \dfrac{V_3}{L_1}t & kT \le t \le T \end{cases} \tag{10.447}$$

$$i_{C1}(t) = \begin{cases} -\left(I_{L2\text{-min}} + \dfrac{V_1 - V_3}{L_2}t\right) & 0 \le t \le kT \\[4mm] I_{C1} & kT \le t \le T \end{cases} \tag{10.448}$$

$$i_{C2}(t) = \begin{cases} -I_2 & 0 \le t \le kT \\[2mm] I_{C2} & kT \le t \le T \end{cases} \tag{10.449}$$

$$v_{L1}(t) = \begin{cases} V_1 & 0 \le t \le kT \\[2mm] V_3 & kT \le t \le T \end{cases} \tag{10.450}$$

$$v_{L2}(t) = \begin{cases} V_1 - V_3 & 0 \le t \le kT \\[2mm] V_2 - V_3 & kT \le t \le T \end{cases} \tag{10.451}$$

$$v_s = \begin{cases} 0 & 0 \le t \le kT \\[2mm] V_1 - V_3 & kT \le t \le T \end{cases} \tag{10.452}$$

$$v_{D1}(t) = \begin{cases} V_1 - V_3 & 0 \le t \le kT \\[2mm] 0 & kT \le t \le T \end{cases} \tag{10.453}$$

$$v_{C1}(t) = \begin{cases} V_3 - \dfrac{I_{L2}}{C_1}t & 0 \le t \le kT \\[4mm] V_3 + \dfrac{I_{C1}}{C_1}t & kT \le t \le T \end{cases} \tag{10.454}$$

$$v_{C2}(t) = \begin{cases} V_2 - \dfrac{I_2}{C_2}t & 0 \leq t \leq kT \\[3mm] V_2 + \dfrac{I_{C2}}{C_2}t & kT \leq t \leq T \end{cases} \tag{10.455}$$

10.6.2.2 Discontinuous Conduction Mode

Refer to Figure 10.61b through d; we have got the instantaneous values of the voltage and current of each component in DCM operation. Since inductor current i_{L1} is discontinuous, some parameters have three states, with $T' = kT + (1 - k)mT < T$.

$$i_{L1}(t) = \begin{cases} \dfrac{V_1}{L_1}t & 0 \leq t \leq kT \\[3mm] I_{L1\text{-max}} - \dfrac{V_3}{L_1}t & kT \leq t \leq T' \\[3mm] 0 & T' \leq t \leq kT \end{cases} \tag{10.456}$$

$$i_{L2}(t) = \begin{cases} I_{L2\text{-min}} + \dfrac{V_1 - V_3}{L_2}t & 0 \leq t \leq kT \\[3mm] I_{L2\text{-max}} - \dfrac{V_2 - V_1}{L_2}t & kT \leq t \leq T \end{cases} \tag{10.457}$$

$$i_1(t) = i_s = \begin{cases} I_{1\text{-min}} + \left(\dfrac{V_1}{L_1} + \dfrac{V_1 - V_3}{L_2} \right)t & 0 \leq t \leq kT \\[3mm] 0 & kT \leq t \leq T \end{cases} \tag{10.458}$$

$$i_{D1}(t) = \begin{cases} 0 & 0 \leq t \leq kT \\[3mm] I_{L1\text{-max}} - \dfrac{V_3}{L_1}t & kT \leq t \leq T' \\[3mm] 0 & T' \leq t \leq kT \end{cases} \tag{10.459}$$

$$i_{C1}(t) = \begin{cases} -\left(I_{L2\text{-min}} + \dfrac{V_1 - V_3}{L_2}t \right) & 0 \leq t \leq kT \\[3mm] I_{C1} & kT \leq t \leq T \end{cases} \tag{10.460}$$

$$i_{C2}(t) = \begin{cases} -I_2 & 0 \leq t \leq kT \\[3mm] I_{C2} & kT \leq t \leq T \end{cases} \tag{10.461}$$

$$v_{L1}(t) = \begin{cases} V_1 & 0 \leq t \leq kT \\ V_3 & kT \leq t \leq T' \\ 0 & T' \leq t \leq kT \end{cases} \tag{10.462}$$

$$v_{L2}(t) = \begin{cases} V_1 - V_3 & 0 \leq t \leq kT \\ V_2 - V_3 & kT \leq t \leq T \end{cases} \tag{10.463}$$

$$v_s(t) = \begin{cases} 0 & 0 \leq t \leq kT \\ V_1 - V_3 & kT \leq t \leq T' \\ V_1 & T' \leq t \leq kT \end{cases} \tag{10.464}$$

$$v_{D1}(t) = \begin{cases} V_1 - V_3 & 0 \leq t \leq kT \\ 0 & kT \leq t \leq T' \\ -V_3 & T' \leq t \leq kT \end{cases} \tag{10.465}$$

$$v_{C1}(t) = \begin{cases} V_3 - \dfrac{I_{L2}}{C_1} t & 0 \leq t \leq kT \\ V_3 + \dfrac{I_{C1}}{C_1} t & kT \leq t \leq T \end{cases} \tag{10.466}$$

$$v_{C2}(t) = \begin{cases} V_2 - \dfrac{I_2}{C_2} t & 0 \leq t \leq kT \\ V_2 + \dfrac{I_{C2}}{C_2} t & kT \leq t \leq T \end{cases} \tag{10.467}$$

10.6.3 COMPARISON OF THE GAINS BETWEEN ULTRA-LIFT LUO CONVERTER AND OTHER CONVERTERS

The ultra-lift Luo converter has been successfully developed using the novel approach of a new technology—ultra-lift (UL) technique. Table 10.8 lists the voltage transfer gains of various converters at $k=0.2$, 0.33, 0.5, 0.67, 0.8, and 0.9. The outstanding characteristics of the ultra-lift Luo converter are very well presented. From the comparison, we can obviously see that the ultra-lift Luo converter has very high voltage transfer gains: $G(k)|_{k=0.5}=3$, $G(k)|_{k=0.667}=8$, $G(k)|_{k=0.8}=24$, and $G(k)|_{k=0.9}=99$.

10.6.4 SIMULATION RESULTS

To verify the advantages of the ultra-lift Luo converter, a PSpice simulation method is applied. We choose the parameter's values as follows: $V_1=10\,\text{V}$, $L_1=L_2=1\,\text{mH}$, $C_1=C_2=1\,\mu\text{F}$, $R=3\,\text{k}\Omega$, $f=50\,\text{kHz}$, and conduction duty cycles $k=0.6$ and 0.66, and

TABLE 10.8
Comparison of Various Converters' Gains

k	0.2	0.33	0.5	0.67	0.8	0.9
Buck	0.2	0.33	0.5	0.67	0.8	0.9
Boost	1.25	1.5	2	3	5	10
Buck-boost	0.25	0.5	1	2	4	9
P/O Luo converter	0.25	0.5	1	2	4	9
P/O SL Luo converter	2.25	2.5	3	4	6	11
Ultra-lift Luo converter	**0.56**	**1.25**	**3**	**8**	**24**	**99**

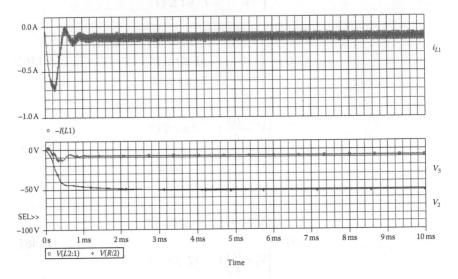

FIGURE 10.65   Simulation results for $k=0.6$.

the output voltage $V_2=52.5$ and 78 V, correspondingly. The first waveform is the inductor's current i_{L1}, which flows through the inductor L_1. The second and third waveforms are the input and output voltages V_3 and V_2. These simulation results are identical to the calculation results. The results are shown in Figures 10.65 and 10.66, respectively.

10.6.5   EXPERIMENTAL RESULTS

To verify the advantages and design of the ultra-lift Luo converter and compare them with the simulation results, we constructed a test rig with the following components: $V_1=10$ V, $L_1=L_2=1$ mH, $C_1=C_2=1\,\mu$F, $R=3$ kΩ, $f=50$ kHz, and conduction duty cycle $k=0.6$ and 0.66, and the output voltage, $V_2=52$ and 78 V, correspondingly. The first waveform is the inductor's current i_{L1}, which flows through the inductor L_1. The second waveform is the output voltage V_2. The experimental results are shown in Figures 10.67 and 10.68, respectively. The test results are identical to those of the

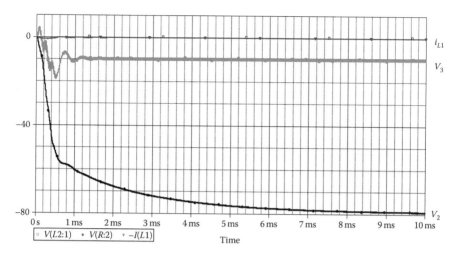

FIGURE 10.66 Simulation results for $k=0.66$.

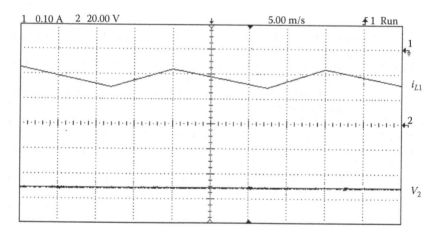

FIGURE 10.67 Experimental results for $k=0.6$.

simulation results shown in Figures 10.65 and 10.66 and verify the calculation results and our design.

10.6.6 SUMMARY

The ultra-lift Luo converter has been successfully developed using a novel approach of the new technology—ultra-lift (UL) technique, which produces even higher voltage transfer gains, much higher than that of VL Luo converters and SL Luo converters. This chapter introduced the operation and characteristics of this converter in detail. This converter will be mainly used in industrial applications with high output voltages.

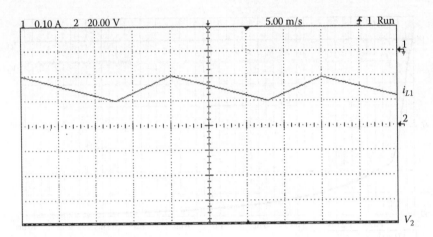

FIGURE 10.68 Experimental results for $k = 0.66$.

REFERENCES

1. Luo, F. L. and Ye, H. 2004. *Advanced DC/DC Converters*. Boca Raton, FL: CRC Press.
2. Luo, F. L. and Ye, H. 2006. *Essential DC/DC Converters*. Boca Raton, FL: Taylor & Francis Group Press.
3. Luo, F. L. and Ye, H. 2002. Super-lift Luo-converters. *Proceedings of the IEEE International Conference PESC'2002*, Cairns, Queensland, Australia, pp. 425–430.
4. Luo, F. L. and Ye, H. 2003. Positive output super-lift converters. *IEEE-Transactions on Power Electronics*, 18, 105–113.
5. Luo, F. L. and Ye, H. 2003. Negative output super-lift Luo-converters. *Proceedings of the IEEE International Conference PESC'03*, Acapulco, Mexico, pp. 1361–1366.
6. Luo, F. L. and Ye, H. 2003. Negative output Super-lift Converters. *IEEE-Transactions on Power Electronics*, 18, 1113–1121.
7. Luo, F. L. and Ye, H. 2004. Positive output cascaded boost converters. *IEE-Proceedings on Electric Power Applications*, 151, 590–606.
8. Zhu, M. and Luo, F. L. 2006. Steady-state performance analysis of cascaded boost converters. *Proceedings of IEEE Asia Pacific Conference on Circuits and Systems*, Singapore, pp. 659–662.
9. Zhu, M. and Luo, F. L. 2006. Generalized steady-state analysis on developed series of cascaded boost converters. *Proceedings of IEEE Asia Pacific Conference on Circuits and Systems, APCCAS'2006*, Singapore, pp. 1399–1402.
10. Luo, F. L. and Ye, H. 2005. Ultra-lift Luo-converter. *IEE- Proceedings on Electric Power Applications*, 152, 27–32.
11. Luo, F. L. and Ye, H. 2004. Investigation of Ultra-lift Luo-converter. *Proceedings of the IEEE International Conference POWERCON'2004*, Singapore, pp. 13–18.

11 Split-Capacitor and Split-Inductor Techniques and Their Application in Positive-Output Super-Lift Luo Converters

Voltage lift technique has been successfully employed in the design of DC/DC converters, e.g., three series Luo converters [1–3]. However, the output voltage increases in arithmetic progression. Super-lift (SL) technique is the most significant contribution in Power Electronics, e.g., four series SL converters. Their output voltage increases in geometric progression. This chapter introduces a novel approach, SL technique armed with split capacitors and split inductors, that implements the output voltage increasing in higher geometric progression. It also effectively enhances the voltage transfer gain in power series.

11.1 INTRODUCTION

SL technique is the most significant contribution in power electronics, e.g., five series SL converters [4,5]. Their output voltage increases in geometric progression. SL technique is the most outstanding contribution in DC/DC conversion technology. Applying this technique, we can design a great number of SL converters. The following series voltage-lift converters are introduced in the previous chapter:

- Positive-output (P/O) SL Luo converters
- Negative-output SL Luo converters
- P/O cascaded boost converters
- Negative-output cascaded boost converters
- Ultralift Luo converter

Each series converter has several subseries. For example, the P/O SL Luo converters have five subseries:

1. Main series
2. Additional series

3. Enhanced series
4. Re-enhanced series
5. Multiple (j)-enhanced series

In order to concentrate the voltage enlargement, assume the converters are working in steady state with continuous conduction mode (CCM). The conduction duty ratio is k, switching frequency is f, switching period is $T = 1/f$, and the load is resistive load R. The input voltage and current are V_{in} and I_{in}, output voltage and current are V_O and I_O. Assume no power losses during the conversion process, $V_{in} \times I_{in} = V_O \times I_O$. The voltage transfer gain is G: $G = V_O/V_{in}$.

This chapter introduces a novel approach, SL technique armed with split capacitors and split inductors, that implements the output voltage increasing stage by stage along higher geometric progression. It effectively enhances the voltage transfer gain in power series as well [6–8].

In order to sort these converters different from existing VL converters, we entitle them "P/O SL Luo converters armed by split capacitors." There are a few subseries, but we introduce only two subseries, *main series* and *additional series*, in detail in this chapter and summarize other series. Each circuit of the main series and additional series has one switch S, n inductors (where n is the stage number), other capacitors, and diodes.

11.2 SPLIT CAPACITORS

A capacitor C_1 as shown in Figure 11.1a can be split into two parts: two capacitors C_1 and C_2 as shown in Figure 11.1b; and three parts: three capacitors C_1, C_2, and C_3 as shown in Figure 11.1c. Furthermore, it can be split into α parts as shown in Figure 11.1d [9,10].

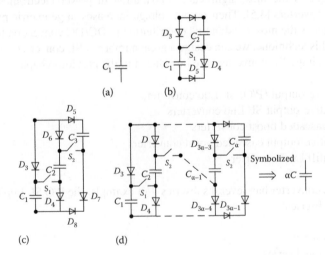

FIGURE 11.1 Single capacitor and α-split capacitors. (a) One capacitor, (b) two-split capacitors, (c) three-split capacitors, and (d) α-split capacitors.

The split stage can be defined α-times. Now, we define the single capacitor to be in $\alpha = 1$ split stage as shown in Figure 11.1a. We define the two-split capacitors to be in $\alpha = 2$ split stage as shown in Figure 11.1b; the slave switch S_1 is exclusively switched with the Main switch S. We define the three-split capacitors to be in $\alpha = 3$ split stage as shown in Figure 11.1c, and the slave switches S_1 and S_2 are exclusively switched with the main switch S. We define the α-split capacitors to be symbolized by αC. These capacitors can be charged by a DC voltage V_{in}. In the steady state, each capacitor is assumed to be charged to the source voltage V_{in}. All split capacitors are charged by source voltage V_{in} in parallel. When the capacitors are discharged, all split capacitors are discharged by an external voltage in series.

11.3 SPLIT INDUCTORS

An inductor L_1 as shown in Figure 11.2a can be split into two parts: two inductors L_1 and L_2, as shown in Figure 11.2b; and three parts: three capacitors L_1, L_2, and L_3, as shown in Figure 11.2c [11,12]. Furthermore, it can be split into β parts, as shown in Figure 11.2d with βL.

The split stage can be defined β-times. Now, we define the single capacitor to be in $\beta = 1$ split stage as shown in Figure 11.2a. We define the two-split inductors to be in $\beta = 2$ split stage as shown in Figure 11.2b, and the slave switch S_1 is exclusively switched with the main switch S; We define the three-split inductors to be in $\beta = 3$ split stage as shown in Figure 11.2c, and the slave switches S_1 and S_2 are exclusively switched with the main switch S. We define the β-split inductors to be symbolized by βL. These inductors can be charged by a DC voltage V_{in} in parallel during switch-on. When the inductors are discharged, all split inductors are discharged in series.

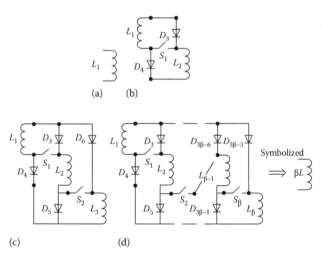

FIGURE 11.2 Single inductor and β-split inductors. (a) One inductor, (b) two-split inductors, (c) three-split inductors, and (d) β-split inductors.

11.4 SPLIT CAPACITORS AND SPLIT INDUCTORS APPLIED IN THE POSITIVE-OUTPUT ELEMENTARY SUPER-LIFT LUO CONVERTER

A P/O elementary SL Luo converter is shown in Figure 11.3. Its circuit diagram is shown in Figure 11.3a, and its equivalent circuits in switch-on and switch-off are shown in Figure 11.3b and c, respectively.

The elementary circuit and its equivalent circuits during switch-on and switch-off are shown in Figure 11.3. The voltage across capacitor C_1 is charged to V_{in} in the steady state. The current i_L flowing through inductor L increases with voltage V_{in} during the switch-on period kT and decreases with voltage $-(V_O - 2V_{in})$ during the switch-off period $(1 - k)T$. Therefore, the ripple of the inductor current i_L is

$$\Delta i_L = \frac{V_{in}}{L}kT = \frac{V_O - 2V_{in}}{L}(1-k)T \tag{11.1}$$

$$V_O = \frac{2-k}{1-k}V_{in} \tag{11.2}$$

The voltage transfer gain is

$$G = \frac{V_O}{V_{in}} = \frac{2-k}{1-k} \tag{11.3}$$

11.4.1 Two-Split Capacitors ($\alpha = 2$) Applied in the P/O Elementary SL Circuit

If the capacitor C_1 is split into two capacitors C_1 and C_2, the circuit and its equivalent circuits during switch-on and switch-off are as shown in Figure 11.4 (as mentioned, the slave switch S_1 is an exclusive switch with the main switch S).

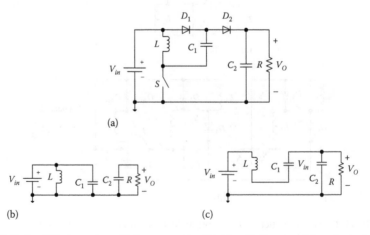

(a)

(b)   (c)

FIGURE 11.3 P/O elementary SL Luo converter. (a) Circuit diagram, (b) equivalent circuit during switch-on, and (c) equivalent circuit during switch-off.

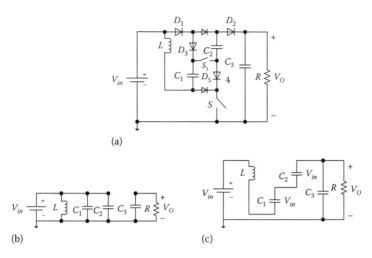

FIGURE 11.4 Two-split capacitors applied in the elementary P/O SL Luo converter. (a) Circuit diagram, (b) equivalent circuit during switch-on, and (c) equivalent circuit during switch-off.

The voltage across capacitors C_1 and C_2 is charged to V_{in} in the steady state. The current i_L flowing through inductor L increases with voltage V_{in} during the switch-on period kT and decreases with voltage $-(V_O - 3V_{in})$ during the switch-off period $(1 - k)T$. Therefore, the ripple of the inductor current i_L is

$$\Delta i_L = \frac{V_{in}}{L} kT = \frac{V_O - 3V_{in}}{L}(1-k)T \tag{11.4}$$

$$V_O = \frac{3-2k}{1-k} V_{in} \tag{11.5}$$

The voltage transfer gain is

$$G = \frac{V_O}{V_{in}} = \frac{3-2k}{1-k} \tag{11.6}$$

11.4.2 Two Split Inductors ($\beta = 2$) Applied in the Elementary P/O SL Circuit

If the inductor L is split into two inductors L_1 and L_2, the circuit and its equivalent circuits during switch-on and switch-off are as shown in Figure 11.5 (as mentioned, the slave switch S_1 is an exclusive switch with the main switch S).

The inductors L_1 and L_2 are charged to V_{in} in the steady state. The current i_L flowing through each inductor increases with voltage V_{in} during the switch-on period kT and decreases with voltage $-(V_O - 2V_{in})$ during the switch-off period $(1 - k)T$. Therefore, the ripple of the inductor current i_L is

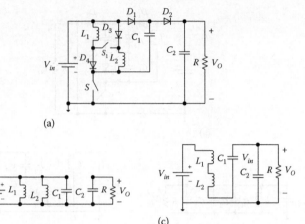

(a)

(b)   (c)

FIGURE 11.5 Two-split inductors applied in the elementary P/O SL Luo converter. (a) Circuit diagram, (b) equivalent circuit during switch-on, and (c) equivalent circuit during switch-off.

$$\Delta i_L = \frac{V_{in}}{L}kT = \frac{V_O - 2V_{in}}{2L}(1-k)T \tag{11.7}$$

$$V_O = \frac{2}{1-k}V_{in} \tag{11.8}$$

The voltage transfer gain is

$$G = \frac{V_O}{V_{in}} = \frac{2}{1-k} \tag{11.9}$$

11.4.3 α-Split Capacitors and β-Split Inductors Applied in the Elementary P/O SL Circuit

If the capacitor C_1 is split into α capacitors, the circuit is as shown in Figure 11.6.

The voltage across the α capacitors is charged to V_{in} in parallel during switch-on in the steady state. The current i_L flowing through inductor L increases with voltage V_{in} during the switch-on period kT and decreases with voltage $-[V_O - (\alpha+1)V_{in}]$ during the switch-off period $(1 - k)T$. Therefore, the ripple of the inductor current i_L is

$$\Delta i_L = \frac{V_{in}}{L}kT = \frac{V_O - (\alpha+1)V_{in}}{\beta L}(1-k)T \tag{11.10}$$

$$V_O = \left(\alpha+1+\frac{\beta k}{1-k}\right)V_{in} = \frac{(\alpha+1)+(\beta-\alpha-1)k}{1-k}V_{in} \tag{11.11}$$

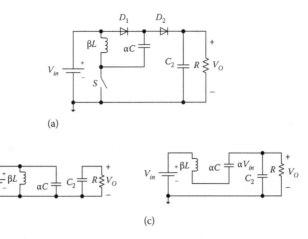

FIGURE 11.6 α-split capacitors and β-split inductors applied in the elementary P/O SL Luo converter. (a) Circuit diagram, (b) equivalent circuit during switch-on, and (c) equivalent circuit during switch-off.

The voltage transfer gain is

$$G_E = \frac{V_O}{V_{in}} = \frac{\alpha + 1 + (\beta - \alpha - 1)k}{1 - k} \qquad (11.12)$$

We define a new factor as

$$A = \frac{\alpha + 1 + (\beta - \alpha - 1)k}{1 - k} \qquad (11.12a)$$

The voltage transfer gain is $G_E = (V_O/V_{in}) = A$.

11.5 MAIN SERIES

The main series has several circuits such as the re-lift circuit, triple-lift circuit, and high-order-lift circuit. We define the stage as $n = 1$, meaning an elementary circuit; $n = 2$ means a re-lift circuit, $n = 3$ means a triple-lift circuit, and n can be higher number, meaning a high-order-lift circuit. Figure 11.6 shows the elementary circuit. Figure 11.7 shows the re-lift circuit, and Figure 11.8 shows the triple-lift circuit.

The voltage transfer gain of the re-lift circuit is

$$G_R = \frac{V_O}{V_{in}} = \left[\frac{\alpha + 1 + (\beta - \alpha - 1)k}{1 - k}\right]^2 = A^2 \qquad (11.13)$$

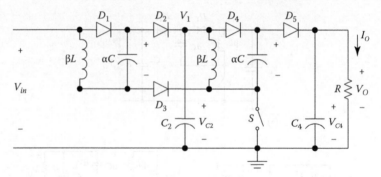

FIGURE 11.7   Re-lift circuit.

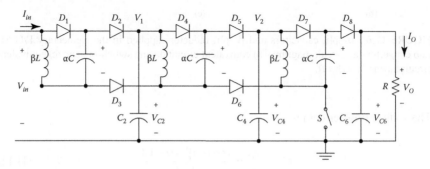

FIGURE 11.8   Triple-lift circuit.

The voltage transfer gain of the triple-lift circuit is

$$G_T = \frac{V_O}{V_{in}} = \left[\frac{\alpha + 1 + (\beta - \alpha - 1)k}{1 - k} \right]^3 = A^3 \qquad (11.14)$$

The voltage transfer gain of the nth-order-lift circuit, if each stage uses split capacitor αC and split inductor βL, is

$$G_n = \frac{V_O}{V_{in}} = \left[\frac{\alpha + 1 + (\beta - \alpha - 1)k}{1 - k} \right]^n = A^n \qquad (11.15)$$

It is a very high-voltage transfer gain. For example, if $V_{in} = 20\,\text{V}$, $n = 3$, $\alpha = 3$, $\beta = 3$, and $k = 0.5$. Hence $A = 7$; the voltage transfer gain G_T is equal to 343. The output voltage will be 6860 V!

11.6   MEC, SPLIT CAPACITORS USED IN DOUBLE/ENHANCED CIRCUIT

The original double/enhanced circuit (DEC) is shown in Figure 11.9, which consists of two diodes (D_{11} and D_{12}) and two capacitors (C_{11} and C_{12}).

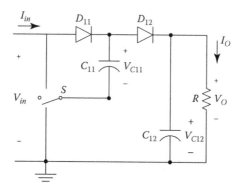

FIGURE 11.9 Double/enhanced circuit.

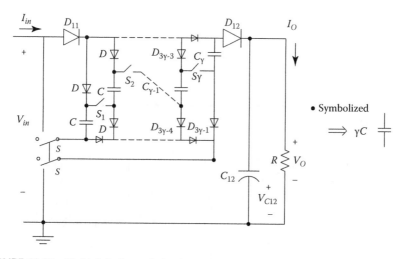

FIGURE 11.10 Multiple/enhanced circuit.

The output voltage is

$$V_O = 2V_{in} \tag{11.16}$$

If the capacitor C_{11} is replaced by a γ-split capacitor γC as shown in Figure 11.10, we can call it multiple/enhanced circuit (MEC).

The output voltage will be

$$V_O = (\gamma + 1)V_{in} \tag{11.17}$$

11.7 ADDITIONAL SERIES

All additional series of P/O SL converters are derived from the corresponding circuits of the main series. We can add an MEC following all circuits to obtain the circuits of the additional series.

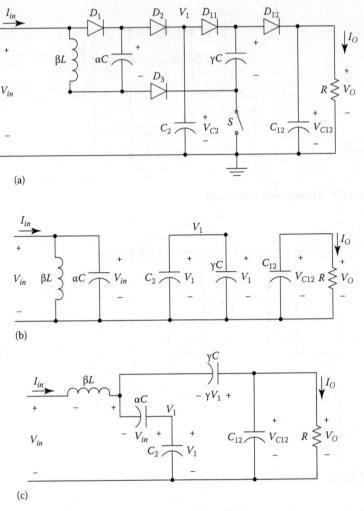

FIGURE 11.11 Elementary additional circuit. (a) Circuit diagram, (b) equivalent circuit during switch-on, and (c) equivalent circuit during switch-off.

The first three stages of this series are shown in Figures 11.11 through 11.13. For convenience, we call them elementary additional circuit, re-lift additional circuit, and triple-lift additional circuit, respectively. We can number them as $n = 1$, 2, and 3.

11.7.1 Elementary Additional Circuit

This circuit is derived from an elementary circuit by adding an MEC (D_{11}-D_{12}-γC-C_{12}). Its circuit and switch-on and switch-off equivalent circuits are shown in Figure 11.11.

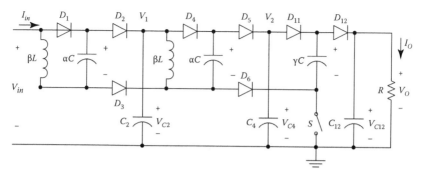

FIGURE 11.12 Re-lift additional circuit.

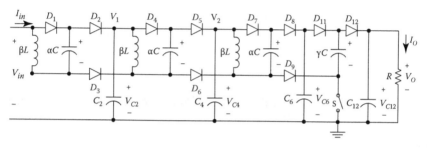

FIGURE 11.13 Triple-lift additional circuit.

The voltage across all capacitor γC is V_1 in steady state as shown in Equation 11.11. The inductor current increases when switch S is on and decreases when switch S is off. The current variation is

$$\frac{V_{in}}{L}kT = \frac{V_O - \gamma V_1 - V_{in}}{\beta L}(1-k)T \tag{11.18}$$

and

$$V_1 = \frac{\alpha + 1 + (\beta - \alpha - 1)k}{1-k}V_{in} = AV_{in}$$

Therefore, the output voltage V_O is

$$V_O = \left[\frac{\beta k}{1-k} + 1 + \gamma \frac{\alpha + 1 + (\beta - \alpha - 1)k}{1-k}\right]V_{in} \tag{11.19}$$

We define another new factor as

$$B = \left[\frac{\beta k}{1-k} + 1 + \gamma \frac{\alpha + 1 + (\beta - \alpha - 1)k}{1-k} \right] \qquad (11.12b)$$

The voltage transfer gain G is

$$G_{E-A} = \frac{V_O}{V_{in}} = B$$

11.7.2   Re-Lift Additional Circuit

This circuit is derived from an elementary additional circuit by adding the parts $(\beta L\text{-}D_3\text{-}D_4\text{-}D_5\text{-}\alpha C\text{-}C_4)$. Its circuit diagram is shown in Figure 11.12.

The voltage across capacitor C_2 is charged to V_1, and the voltage across capacitor C_4 is charged to V_2. All γC is charged to V_2. The V_2 is shown in Equation 11.13. The second inductor current increases when switch S is on and decreases when switch S is off. The current variation is

$$\frac{V_1}{L_2} kT = \frac{V_O - \gamma V_2 - V_I}{\beta L_2}(1-k)T \qquad (11.20)$$

and

$$V_2 = \frac{\alpha + 1 + (\beta - \alpha - 1)k}{1-k} V_1 = AV_1$$

Therefore, the output voltage V_O is

$$V_O = \frac{V_1}{1-k} + \gamma V_2 = \left[\frac{\beta k}{1-k} + 1 + \gamma \frac{\alpha + 1 + (\beta - \alpha - 1)k}{1-k} \right] V_1$$

$$= \left[\frac{\beta k}{1-k} + 1 + \gamma \frac{\alpha + 1 + (\beta - \alpha - 1)k}{1-k} \right] \left[\frac{\alpha + 1 + (\beta - \alpha - 1)k}{1-k} \right] V_{in} = BAV_{in} \qquad (11.21)$$

The voltage transfer gain G is

$$G_{R-A} = \frac{V_O}{V_{in}} = BA$$

11.7.3   Triple-Lift Additional Circuit

This circuit is derived from a re-lift additional circuit by adding the parts $(\beta L\text{-}D_6\text{-}D_7\text{-}D_8\text{-}\alpha C\text{-}C_6)$. Its circuit diagram is shown in Figure 11.13.

The voltage across capacitor C_2 is charged to V_1. The voltage across capacitor C_4 is charged to V_2, and the voltage across capacitor C_6 is charged to V_3. All γC

is charged to V_3 as shown in Equation 11.15. The inductor current increases when switch S is on and decreases when switch S is off. The current variation is

$$\frac{V_2}{L_3}kT = \frac{V_O - \gamma V_3 - V_2}{\beta L_3}(1-k)T \qquad (11.22)$$

and

$$V_3 = \frac{\alpha+1+(\beta-\alpha-1)k}{1-k}V_2 = AV_2$$

$$V_2 = \left[\frac{\alpha+1+(\beta-\alpha-1)k}{1-k}\right]^2 V_{in} = A^2 V_{in}$$

Therefore, the output voltage V_O is

$$V_O = \frac{\beta k}{1-k}V_1 + V_2 + \gamma V_3 = \left[\frac{\beta k}{1-k}+1+\gamma\left(\frac{\alpha+1-\alpha k}{1-k}\right)\right]V_2 = BA^2 V_{in} \qquad (11.23)$$

The voltage transfer gain G is

$$G_{T-A} = \frac{V_O}{V_{in}} = BA^2$$

11.7.4 Higher-Order Lift Additional Circuits

A higher-order lift additional circuit can be designed by just repeating the parts $(\beta L\text{-}D_3\text{-}D_4\text{-}D_5\text{-}\alpha C\text{-}C_4)$. For an nth-order lift additional circuit, the final voltage transfer gain G_{n-A} is

$$G_{n-A} = \frac{V_O}{V_{in}} = BA^{n-1} \qquad (11.24)$$

11.8 HIGHER-ORDER SERIES

Higher-order series are similar to the design /methods introduced in the previous chapter. The following series can be designed:

- Main series
- Additional series
- Enhanced series
- Re-enhanced series
- Multiple (j)-enhanced series

We have introduced the main series and additional series in previous sections. We introduce a further three series in the following sections.

11.8.1 ENHANCED SERIES

Refer to Figure 11.11; there is an MEC added in the elementary circuit of the main series. We then obtain the elementary additional circuit. By adding an MEC in every stage circuit of the main series circuit, we obtain the enhanced series. Therefore, the output voltage V_O is

$$V_O = B^n V_{in} \tag{11.25}$$

The voltage transfer gain G_{n-E} is

$$G_{n-E} = \frac{V_O}{V_{in}} = B^n \tag{11.26}$$

where n is the stage number as explained previously.

11.8.2 RE-ENHANCED SERIES

Refer to Figure 11.11; there is an MEC added in the elementary circuit of the main series. We then obtain the elementary additional circuit. By adding two MECs in the elementary circuit of the main series, we can obtain the elementary re-enhanced circuit, which is shown in Figure 11.14.

The voltage across all capacitors across the first γC is V_1 in steady state as shown in Equation 11.11. The voltage across all capacitors across the second γC is V_{11}. The V_{11} is

$$V_{11} = \gamma V_1 = \gamma \frac{(\alpha+1)+(\beta-\alpha-1)k}{1-k} V_{in} \tag{11.27}$$

Therefore, the output voltage is

$$V_O = \gamma V_{11} = \gamma^2 V_1 = \gamma^2 \frac{(\alpha+1)+(\beta-\alpha-1)k}{1-k} V_{in} \tag{11.28}$$

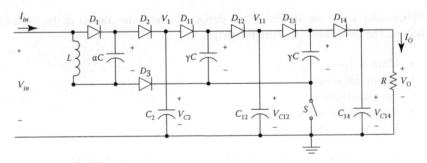

FIGURE 11.14   Elementary re-enhanced circuit.

We define another new factor as

$$C = \gamma^2 = \frac{\alpha + 1 + (\beta - \alpha - 1)k}{1 - k} \tag{11.12c}$$

The voltage transfer gain G_{E-RE} is

$$G_{E-RE} = \frac{V_O}{V_{in}} = C$$

By adding two MECs in every stage circuit of the main series circuit, we obtain the re-enhanced series. Therefore, the output voltage V_O is

$$V_O = C^n V_{in} \tag{11.29}$$

The voltage transfer gain G_{n-RE} is

$$G_{n-RE} = \frac{V_O}{V_{in}} = C^n \tag{11.30}$$

where n is the stage number as explained previously.

11.8.3 MULTIPLE (j)-ENHANCED SERIES

Refer to Figure 11.11; there is an MEC added in the elementary circuit of the main series. We then obtain the elementary additional circuit. If we add j MECs in the elementary circuit of the main series, we can obtain the elementary re-enhanced circuit, which is shown in Figure 11.15.

The voltage across all capacitors across the first γC is V_1 in steady state as shown in Equation 11.11. The voltage across all capacitors across the second γC is V_{11}. Furthermore, V_{1j} is

$$V_{1j} = \gamma^{j-1} V_{11} = \gamma^j \frac{(\alpha + 1) + (\beta - \alpha - 1)k}{1 - k} V_{in} \tag{11.31}$$

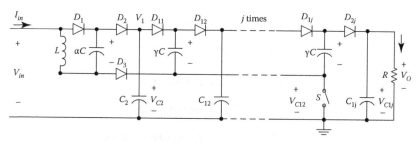

FIGURE 11.15 Elementary (j)-enhanced circuit.

Therefore, the output voltage is

$$V_o = V_{1j} = \gamma^j \frac{(\alpha+1)+(\beta-\alpha-1)k}{1-k} V_{in} \qquad (11.32)$$

We define another new factor

$$J = \gamma^j \frac{\alpha+1+(\beta-\alpha-1)k}{1-k} \qquad (11.12j)$$

The voltage transfer gain G_{E-JE} is

$$G_{E-JE} = \frac{V_O}{V_{in}} = J$$

If we add j MECs in every stage circuit of the main series circuit, we obtain the (j)-enhanced series. Therefore, the output voltage V_O is

$$V_O = C^n V_{in} \qquad (11.33)$$

The voltage transfer gain G_{n-JE} is

$$G_{n-JE} = \frac{V_O}{V_{in}} = J^n \qquad (11.34)$$

where n is the stage number as explained earlier.

11.9 SUMMARY OF P/O SUPER-LIFT LUO CONVERTERS APPLYING SPLIT CAPACITORS AND SPLIT INDUCTORS

All circuits of P/O SL Luo converters using split capacitors αC and split inductors βL and MEC (with γC) as a family are shown in Figure 11.16 (the family tree). From the analysis of previous two sections, we can obtain a common formula to calculate the output voltage:

$$V_O = \begin{cases} A^n V_{in} & \text{Main\_series} \\ BA^{n-1} V_{in} & \text{Additional\_series} \\ B^n V_{in} & \text{Enhanced\_series} \\ C^n V_{in} & \text{Re-enhanced\_series} \\ J^n V_{in} & \text{Multiple-enhanced\_series} \end{cases} \qquad (11.35)$$

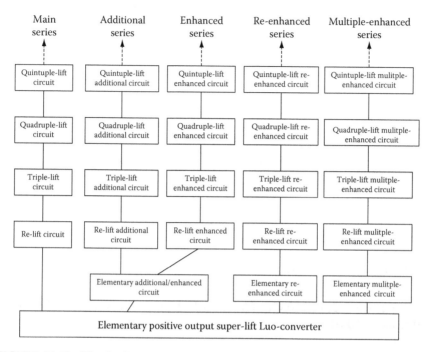

FIGURE 11.16 The family tree of P/O SL Luo converters applying capacitors and split inductors.

The voltage transfer gain is

$$
G = \frac{V_O}{V_{in}} =
\begin{cases}
A^n & \text{Main\_series} \\[4pt]
BA^{n-1} & \text{Additional\_series} \\[4pt]
B^n & \text{Enhanced\_series} \\[4pt]
C^n & \text{Re-enhanced\_series} \\[4pt]
J^n & \text{Multiple-enhanced\_series}
\end{cases}
\tag{11.36}
$$

where A, B, C, and J are presented in Equations 11.12a through 11.12c and 11.12j, respectively. These techniques can be used not only in P/O SL Luo converters but also in other DC/DC converters.

11.10 SIMULATION RESULTS

To verify the design and calculation results, the PSim simulation package was applied to these converters. Simulation conditions are as follows: re-lift circuit with $n=2$,

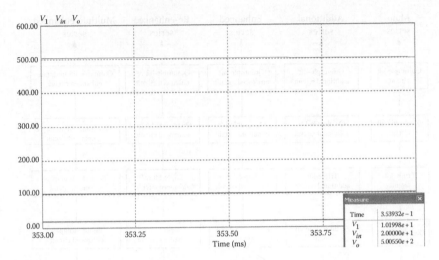

FIGURE 11.17   Simulation results of a re-lift circuit.

$\alpha=2$, $\beta=2$, and $\gamma=2$; $V_{in}=20$V, $L_1=L_2=10$mH, all capacitances are 2μF, $R=30$kΩ, $k=0.5$, and $f=50$kHz.

11.10.1   SIMULATION RESULTS OF A RE-LIFT CIRCUIT

We obtain the voltage values V_{in} and V_O of a re-lift circuit (with $n=2$, $\alpha=2$, and $\beta=2$) as 20 and 500 V, respectively. The simulation results are shown in Figure 11.17. The voltage values match the calculated results.

11.10.2   SIMULATION RESULTS OF A RE-LIFT ADDITIONAL CIRCUIT

We obtain the voltage values V_{in} and V_O of a re-lift additional circuit (see Figure 11.11) (with $n=2$, $\alpha=2$, $\beta=2$, and $\gamma=2$) as 20 and 1300 V, respectively. The simulation results are shown in Figure 11.18. The voltage values match the calculated results.

11.11   EXPERIMENTAL RESULTS

A test rig was constructed to verify the design and calculation results and compare with PSim simulation results. We still choose $V_{in}=20$V, $L_1=L_2=10$mH, and all capacitances are 2μF and $R=30$kΩ, using $k=0.5$ and $f=100$kHz. The component of the switch is a MOSFET device IRF950 with the rates 1000 V/5 A/2 MHz. We measured the values of the input and output voltages in the following converters.

11.11.1   EXPERIMENTAL RESULTS OF A RE-LIFT CIRCUIT

After carefully measuring a re-lift circuit (with $n=2$, $\alpha=2$, and $\beta=2$), we obtained the input voltage value of $V_{in}=20$V (shown in Channel 1 with 5.0 V/Div) and output

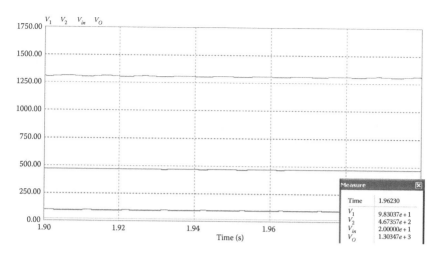

FIGURE 11.18 Simulation results of a re-lift additional circuit.

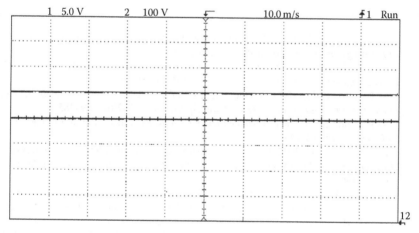

FIGURE 11.19 Experimental results of a re-lift circuit.

voltage value of $V_O = 500$ V (shown in Channel 2 with 50 V/Div). The experimental results are shown in Figure 11.19, which match the calculated and simulation results, which are $V_{in} = 20$ V and $V_O = 320$ V, as shown in Figure 11.17.

11.11.2 EXPERIMENTAL RESULTS OF A RE-LIFT ADDITIONAL CIRCUIT

After carefully measuring a re-lift additional circuit (with $n = 2$, $\alpha = 2$, $\beta = 2$, and $\gamma = 2$), we obtained the input voltage value of $V_{in} = 20$ V (shown in Channel 1 with 5.0 V/Div) and output voltage value of $V_O = 1300$ V (shown in Channel 2 with 200 V/Div). The experimental results are shown in Figure 11.20, which identically match the calculated and simulation results, which are $V_{in} = 20$ V and $V_O = 1300$ V, as shown in Figure 11.18.

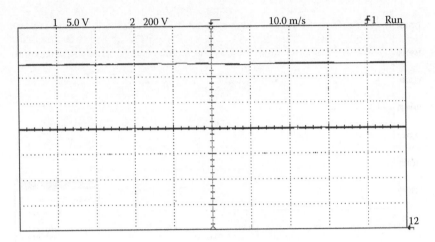

1 5.0 V	2 200 V		10.0 m/s	⨍1 Run

FIGURE 11.20   Experimental results of a re-lift additional circuit.

REFERENCES

1. Luo, F. L. July 1999. Positive output Luo-Converters, voltage lift technique. *IEE Proceedings on Electric Power Applications*, 146(4), 415–432.
2. Luo, F. L. March 1999. Negative output Luo-Converters, voltage lift technique. *IEE Proceedings on Electric Power Applications*, 146(2), 208–224.
3. Luo, F. L. November 2000. Double output Luo-Converters, advanced voltage lift technique. *IEE Proceedings on Electric Power Applications*, 147(6), 469–485.
4. Luo, F. L. and Ye, H. 2004. *Advanced DC/DC Converters*. Boca Raton, FL: CRC Press LLC, ISBN: 0-8493-1956-0.
5. Luo, F. L. and Ye, H. 2002. Positive output super-lift Luo-converters. *Proceedings of the IEEE International Conference PESC'2002*, June 23–27, Cairns, Queensland, Australia, pp. 425–430.
6. Yang, L. S., Liang, T. J., and Chen, J. F. Transformerless DC-DC converters with high step-up voltage gain. *IEEE Transactions on Industrial Electronics*, 56(8), 3144–3152.
7. Axelrod, B., Berkovich, Y., and Ioinovici, A. 2008. Switched-capacitor/switched-inductor structures for getting transformerless hybrid DC-DC PWM converters. *IEEE Transactions on Circuits and Systems I: Regular Papers*, 55(2), 687–696.
8. Jiao, Y., Luo, F. L., and Bose, B. 2010. Analysis, modeling and control of voltage–lift split-inductor-type boost converters. *IET Power Electronics*, 3(6), 845–854.
9. Luo, F. L. 2011. Study on split-capacitors applied in positive output super-lift Luo-Converters (227-fang). *Proceedings (CD-ROM) of the IEEE-ICREPQ'2011*, April 13–15, Canary Islands, Spain, pp. 60–65.
10. Luo, F. L. 2011. Investigation on split capacitors applied in positive output super-lift Luo-Converters. *Proceedings of the International Conference CCDC'2010*, May 23–25, Mianyang, China, pp. 2797–2802.
11. Luo, F. L. 2011. Investigation on split inductors applied in positive output super-lift Luo-Converters. *Proceedings of the International Conference CCDC'2010*, May 23–25, Mianyang, China, pp. 2808–2813.
12. Luo, F. L. 2011. Investigation on hybrid split-capacitors and split-inductors applied in positive output super-lift Luo-Converters. *Proceedings of the IEEE International Conference IEEE-ICIEA'2011*, June 21–23, Beijing, China, pp. 322–328.

12 Pulse-Width-Modulated DC/AC Inverters

DC/AC inverters are quickly developed converter branch of the power switching circuits used in industrial applications in comparison with other switching circuits. In recent decades, plenty of topologies of DC/AC inverters have been created. Generally say, the DC/AC inverters are mainly used in AC motor adjustable speed drive (ASD). Power DC/AC inverters were widely used in other industrial applications since the late 1980s. Semiconductor manufacture development brought power devices such as GTO, Triac, BT, IGBT, MOSFET, and so on in higher switching frequency (say from tens of kHz up to a few MHz). Because of the devices such as thyristor (SCR) with low switching frequency and high power rate, the previously mentioned devices have low power rate and high switching frequency [1,2].

Square-waveform DC/AC inverters were used before the 1980s. In those equipment, thyristor, GTO, and triac could be used in low-frequency switching operations. High-frequency equipment such as power BT and IGBT was produced. The corresponding equipment implementing the pulse-width modulation (PWM) technique has a wide range of output voltage and frequency and low THD.

Nowadays, two DC/AC inversion techniques are popular in this area: PWM technique and multilevel-modulation (MLM) technique. Most DC/AC inverters are still PWM DC/AC inverters in different prototypes. We introduce the PWM inverters in this chapter and MLM inverters in the next Chapter.

12.1 INTRODUCTION

DC/AC inverters are used for converting a DC power source into AC power applications. They are generally used in the following applications:

1. Variable-voltage/variable-frequency AC supplies in ASD, such as induction motor drives, synchronous machine drives, and so on
2. Constant regulated voltage AC power supplies such as uninterruptible power supplies (UPS's)
3. Static var (reactive power) compensations
4. Passive/active series/parallel filters
5. Flexible AC transmission systems (FACTS's)
6. Voltage compensations

Adjustable-speed induction motor drive systems are widely applied in industrial applications. These system require DC/AC power supply with variable frequency usually from 0 to 400 Hz in fractional horse-power (HP) to hundreds of HP.

537

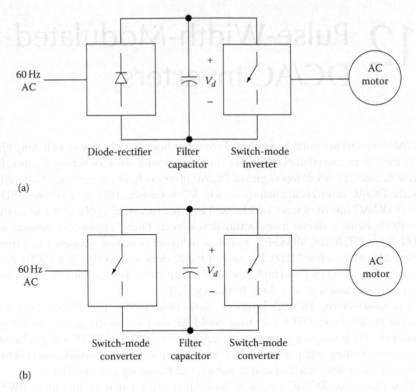

FIGURE 12.1 Standard ASD scheme. (a) Switch-mode inverter in AC motor drive. (b) Switch-mode converters for motoring/regenerative braking.

A large number of DC/AC inverters are available in the world market. The typical block circuit of an ASD is shown in Figure 12.1. From this block diagram, we can see that the power DC/AC inverter produces variable frequency and voltage to implement ASD.

The PWM technique is deferent from pulse-amplitude modulation (PAM) and pulse-phase modulation (PPM) technique. Implementing this technique, all pulses have adjustable pulse width with constant amplitude and phase. The corresponding circuit is called the pulse-width modulator. Typical input and output waveforms of a pulse-width modulator are shown in Figure 12.2. The output pulse train has pulses with the same amplitude and different widths, which corresponds to the input signal at the sampling instants.

12.2 PARAMETERS USED IN PWM OPERATION

Some parameters are specially used in PWM operation.

12.2.1 MODULATION RATIOS

The modulation ratio is usually yielded by a uniformed-amplitude triangle (carrier) signal with the amplitude V_{tri-m}. The maximum amplitude of the input signal is

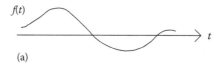

(a)

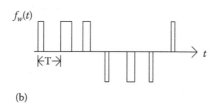

(b)

FIGURE 12.2 Typical input and output waveforms of a pulse-width modulator. (a) Input signal. (b) Output signal.

assumed to be $V_{in\text{-}m}$. We define the amplitude modulation ratio m_a for a single-phase inverter as follows:

$$m_a = \frac{V_{in-m}}{V_{tri-m}} \tag{12.1}$$

We also define the frequency modulation ratio m_f as follows:

$$m_f = \frac{f_{tri\text{-}m}}{f_{in\text{-}m}} \tag{12.2}$$

A one-leg switch-mode inverter is shown in Figure 12.3. The DC-link voltage is V_d. Two large capacitors are used to establish the neutral point N. The AC output voltage from point a to N is V_{AO}, and its fundamental component is $(V_{AO})_1$. We mark $(\hat{V}_{AO})_1$ to show the maximum amplitude of $(V_{AO})_1$. The waveforms of the input (control) signal and triangle signal and the spectrum of the PWM pulse train are shown in Figure 12.4.

If the maximum amplitude $(\hat{V}_{AO})_1$ of the input signal is smaller than or equal to half of the DC-link voltage $V_d/2$, the modulation ratio m_a is smaller than or equal to unity.

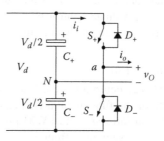

FIGURE 12.3 One-leg switch-mode inverter.

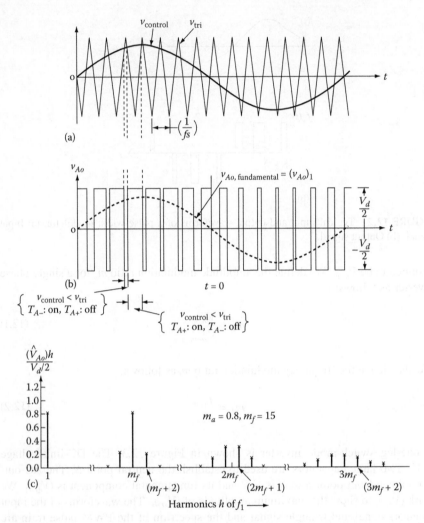

FIGURE 12.4 Pulse-width modulation, (a) control signals, (b) output pulse-train and its fundamental sine-wave, and (c) the harmonics of the output pulse-train.

In this case, the fundamental component $(V_{AO})_1$ of the output AC voltage is proportional to the input voltage. The voltage control by varying m_a for a single-phase PWM is split into three areas, which are shown in Figure 12.5.

12.2.1.1 Linear Range ($m_a \leq 1.0$)

The condition $(\hat{V}_{Ao})_1 = m_a \cdot (V_d/2)$ determines the linear region. It is a sinusoidal PWM, where the amplitude of the fundamental frequency voltage varies linearly with the amplitude modulation ratio m_a. The PWM pushes the harmonics into a high-frequency range around the switching frequency and its multiples. However,

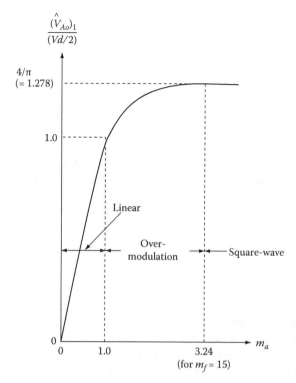

FIGURE 12.5   Voltage control by varying m_a.

the maximum available amplitude of the fundamental frequency component may not be as high as desired.

12.2.1.2   Overmodulation ($1.0 < m_a \leq 1.27$)

The condition $V_d/2 < (\hat{V}_{Ao})_1 \leq (4/\pi)(V_d/2)$ determines the overmodulation region. When the amplitude of the fundamental frequency component in the output voltage increases beyond 1.0, it reaches overmodulation. In overmodulation range, the amplitude of the fundamental frequency voltage no longer varies linearly with m_a.

Overmodulation causes the output voltage to contain many more harmonics in the sidebands as compared with the linear range. The harmonics with dominant amplitudes in the linear range may not be dominant during overmodulation.

12.2.1.3   Square Wave (Sufficiently Large $m_a > 1.27$)

The condition $(\hat{V}_{Ao})_1 > (4/\pi)(V_d/2)$ determines the square-wave region. The inverter voltage waveform degenerates from a pulse-width modulated waveform into a square wave. Each switch of the inverter leg in Figure 12.3 is on for one half-cycle (180°) of the desired output frequency.

12.2.1.4   Small m_f ($m_f \leq 21$)

Usually the triangle waveform frequency is much larger than the input signal frequency to obtain small THD. For the situation with a small $m_f \leq 21$, two points have to be mentioned:

- *Synchronous PWM*: For small value of m_f, the triangle waveform signal and the input signal should be synchronized to each other (synchronous PWM). This synchronous PWM requires that m_f be an integer. The reason for using the synchronous PWM is that the asynchronous PWM (where m_f is not an integer) results in subharmonics (of the fundamental frequency), which are very undesirable in most applications. This implies that the triangle waveform frequency varies with the desired inverter frequency (e.g., if the inverter output frequency and hence the input signal frequency is 65.42 Hz and $m_f = 5$, the triangle wave frequency should be exactly $15 \times 65.42 = 981.3$ Hz).
- m_f *should be an odd integer*: As discussed previously, m_f should be an odd integer except in single-phase inverters with PWM unipolar voltage switching, which is discussed in Section 12.7.1.

12.2.1.5   Large m_f ($m_f > 21$)

The amplitudes of subharmonics due to asynchronous PWM are small at a large value of m_f. Therefore, at large value of m_f, the asynchronous PWM can be used where the frequency of the triangle waveform is kept constant, whereas the input signal frequency varies, resulting in noninteger values of m_f (so long as they are large). However, if the inverter is supplying a load such as an AC motor, the subharmonics at zero or close to zero frequency, even though small in amplitude, will result in a large current, which will be highly undesirable. Therefore, the asynchronous PWM should be avoided.

It is a very important work to determine the harmonic components of the output voltage. Referring to Figure 12.4c, we have the FFT spectrum and the harmonics. Choosing frequency modulation ratio m_f as an odd integer and amplitude modulation ratio $m_a < 1$, we have the generalized harmonics of the output voltage shown in Table 12.1.

The rms voltages of the output voltage harmonics are calculated by the following formula:

$$(V_O)_h = \frac{V_d}{\sqrt{2}} \frac{(\hat{V}_{AO})_h}{V_d/2} \tag{12.3}$$

where
$(V_O)_h$ is the hth harmonic rms voltage of the output voltage
V_d is the DC link voltage
$(\hat{V}_{AO})_h/(V_d/2)$ is tabulated as a function of m_a

If the input (control) signal is a sinusoidal wave, we usually call this inversion sinusoidal pulse-width modulation (SPWM). The typical waveforms of an SPWM are also shown in Figure 12.4a and b.

TABLE 12.1

Generalized Harmonics of V_O (or V_{AO}) for a Large m_f

h	m_a				
	0.2	0.4	0.6	0.8	1.0
1 (Fundamental)	0.2	0.4	0.6	0.8	1.0
m_f	1.242	1.15	1.006	0.818	0.601
$m_f \pm 2$	0.016	0.061	0.131	0.220	0.318
$m_f \pm 4$					0.018
$2m_f \pm 1$	0.190	0.326	0.370	0.314	0.181
$2m_f \pm 3$		0.024	0.071	0.139	0.212
$2m_f \pm 5$				0.013	0.033
$3m_f$	0.335	0.123	0.083	0.171	0.113
$3m_f \pm 2$	0.044	0.139	0.203	0.176	0.062
$3m_f \pm 4$		0.012	0.047	0.104	0.157
$3m_f \pm 6$				0.016	0.044
$4m_f \pm 1$	0.163	0.157	0.008	0.105	0.068
$4m_f \pm 3$	0.012	0.070	0.132	0.115	0.009
$4m_f \pm 5$			0.034	0.084	0.119
$4m_f \pm 7$				0.017	0.050

Note: $(\hat{V}_{AO})_h / (V_d/2)$ is tabulated as a function of m_a.

Example 12.1 A single-phase half-bridge DC/AC inverter is shown in Figure 12.3 to implement SPWM with $V_d = 200\,V$, $m_a = 0.8$, and $m_f = 27$. The fundamental frequency is 50 Hz. Determine the rms value of the fundamental frequency and some of the harmonics in output voltage using Table 12.1.

Solution: From Equation 12.3, we have the general rms values:

$$(V_O)_h = \frac{V_d}{\sqrt{2}} \frac{(\hat{V}_{AO})_h}{V_d/2} = \frac{200}{\sqrt{2}} \frac{(\hat{V}_{AO})_h}{V_d/2} = 141.42 \frac{(\hat{V}_{AO})_h}{V_d/2} V \quad (12.4)$$

Checking the data from Table 12.1 we can get the rms values as follows:

Fundamental:

$(V_O)_1 = 141.42 \times 0.8 = 113.14\,V$	at 50 Hz
$(V_O)_{23} = 141.42 \times 0.818 = 115.68\,V$	at 1150 Hz
$(V_O)_{25} = 141.42 \times 0.22 = 31.11\,V$	at 1250 Hz
$(V_O)_{27} = 141.42 \times 0.818 = 115.68\,V$	at 1350 Hz
$(V_O)_{51} = 141.42 \times 0.139 = 19.66\,V$	at 2550 Hz
$(V_O)_{53} = 141.42 \times 0.314 = 44.41\,V$	at 2650 Hz
$(V_O)_{55} = 141.42 \times 0.314 = 44.41\,V$	at 2750 Hz
$(V_O)_{57} = 141.42 \times 0.139 = 19.66\,V$	at 2850 Hz

etc.

12.2.2 Harmonic Parameters

Refer to Figure 12.4c; various harmonic parameters were introduced in Chapter 4, which are used in PWM operation.
Harmonic factor (HF) is

$$HF_n = \frac{V_n}{V_1} \tag{4.21}$$

Total harmonic distortion (THD) is

$$THD = \sqrt{\sum_{n=2}^{\infty} HF_n^2} \text{ or } THD = \frac{\sqrt{\sum_{n=2}^{\infty} V_n^2}}{V_1} \tag{4.22}$$

Weighted total harmonic distortion (WTHD) is

$$WTHD = \frac{\sqrt{\sum_{n=2}^{\infty} V_n^2/n}}{V_1} \tag{4.23}$$

12.3 TYPICAL PWM INVERTERS

DC/AC inverters have three typical supply methods:

1. Voltage source inverter (VSI)
2. Current source inverter (CSI)
3. Impedance source inverter (z-source inverter or ZSI)

Generally, the circuits of various PWM inverters can be same. The difference between them is the type of the power supply sources or network, which are voltage source, current source, or impedance source.

12.3.1 Voltage Source Inverter

A VSI is supplied by a voltage source since the source is a DC voltage power supply. In an ASD, the DC source is usually an AC/DC rectifier. There is a large capacitor used to keep the DC-Link voltage stable. Usually, a VSI has buck operation function. Its output voltage peak value is lower than the DC link voltage.

It is necessary to avoid *short-circuit* across the DC voltage source during operation. If a VSI takes a bipolar operation, that is, the upper switch and lower switch in a legwork to provide a PWM output waveform, the control circuit and interface have to be designed to leave small gaps between switching signals to the upper switch and lower switch in the same leg. For example, the output voltage frequency is in the range 0–400 Hz, and the PWM carrying frequency is in 2–20 kHz; the gaps are usually set in 20–100 ns. This requirement is not very convenient for the control circuit and interface design. Therefore, a unipolar operation is implemented in most industrial applications.

12.3.2 Current Source Inverter

A CSI is supplied by a DC current source. In an ASD, the DC current source is usually an AC/DC rectifier with a large inductor to keep the supplying current stable. Usually, a CSI has boost operation function. Its output voltage peak value can be higher than the DC link voltage.

Since the source is a DC current source, it is necessary to avoid *open-circuit* of the inverter during operation. The control circuit and interface have to be designed to have small overlaps between switching signals to the upper switches and lower switches at least in one leg. For example, the output voltage frequency is in the range 0–400 Hz, and the PWM carrying frequency is in 2–20 kHz; the overlaps are usually set in 20–100 ns. This requirement is easy for the control circuit and interface design.

12.3.3 Impedance Source Inverter (z-Source Inverter)

An ZSI is supplied by a voltage source or current source via an "X"-shaped impedance network formed by two capacitors and two inductors, which is called Z-network. In an ASD, the DC impedance source is usually an AC/DC rectifier. A Z-network is located between the rectifier and the inverter. Since there are two inductors and two capacitors to be set in front of the chopping legs, there is no restriction to avoid the legs opened or short-circuited. A ZSI has buck-boost operation function. Its output voltage peak value can be higher or lower than the DC link voltage.

12.3.4 Circuits of DC/AC Inverters

The generally used DC/AC inverters are introduced here.

1. Single-phase half-bridge VSI
2. Single-phase full-bridge VSI
3. Three-phase full-bridge VSI
4. Three-phase full-bridge CSI
5. Multistage PWM inverters
6. Soft-switching inverters
7. ZSI

12.4 SINGLE-PHASE VOLTAGE SOURCE INVERTER

Single-phase VSIs can be implemented in the half-bridge circuit and full-bridge circuit.

12.4.1 Single-Phase Half-Bridge VSI

A single-phase half-bridge VSI is shown in Figure 12.6. The carrier-based PWM technique is applied in this inverter. Two large capacitors are required to provide a neutral point N; therefore, each capacitor keeps half of the input DC voltage. Since the output voltage refers to the neutral point N, the maximum output voltage is smaller than half of the DC-link voltage if it is operating in linear modulation.

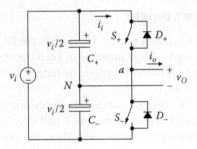

FIGURE 12.6 Single-phase half-bridge VSI.

The modulation operations are shown in Figure 12.5. Two switches S_+ and S_- in one chopping leg are switched by the PWM signal. Two switches S_+ and S_- operate in an exclusive state with small dead time to avoid short-circuit.

In general, linear modulation operation is considered so that m_a is usually smaller than unity, for example, $m_a = 0.8$. Generally, in order to obtain low THD, the m_f is usually taken large number. For the convenience of description, we choose $m_f = 9$. In order to better understand each inverter, we show some typical waveforms in Figure 12.7.

How to determine the pulse width is the clue of the PWM. If the control signal v_C is a sine-wave function as shown in Figure 12.7a, we call the modulation sinusoidal pulse-width modulation. Figure 12.7b offers the switching signal. When it is positive switch on the upper switch S_+, and the lower switch S off; vice-versa, switch off the upper switch S_+, and the lower switch S on.

Assume that the amplitude of the triangle wave is unity; the amplitude of the sine wave is 0.8. Refer to Figure 12.7a; the sine-wave function is

$$f(t) = m_a \sin \omega t = 0.8 \sin 100\pi t \qquad (12.5)$$

where
$$\omega = 2\pi f$$
$$f = 50\,\text{Hz}$$

The triangle functions are lines:

$$f_{\Delta 1}(t) = -4fm_f t = -1800t \quad f_{\Delta 2}(t) = 4fm_f t - 2 = 1800t - 2$$

$$f_{\Delta 3}(t) = 4 - 4fm_f t = 4 - 1800t \quad f_{\Delta 4}(t) = 4fm_f t - 6 = 1800t - 6$$

$\cdots$

$$f_{\Delta(2n-1)}(t) = 4(n-1) - 4fm_f t \quad f_{\Delta 2n}(t) = 4fm_f t - (4n - 2) \qquad (12.6)$$

$\cdots$

$$f_{\Delta 17}(t) = 32 - 1800t \quad f_{\Delta 18}(t) = 1800t - 34$$

$$f_{\Delta 19}(t) = 36 - 1800t$$

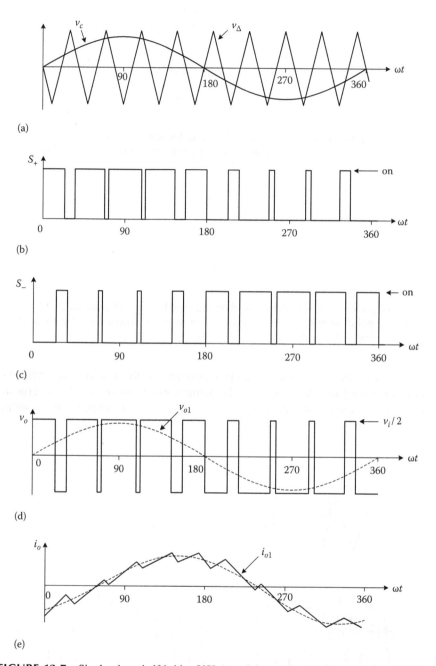

(a)

(b)

(c)

(d)

(e)

FIGURE 12.7 Single-phase half-bridge VSI ($m_a=0.8$, $m_f=9$). (a) Carrier and modulating signals. (b) Switch S_+ state. (c) Switch S_- state. (d) AC output voltage. (e) AC output current.

Example 12.2 A single-phase half-bridge DC/AC inverter is shown in Figure 12.6 to implement SPWM with $m_a = 0.8$ and $m_f = 9$. Determine the first pulse width of the pulse shown in Figure 12.7a.

Solution: The leading age of the first pulse is at $t = 0$. Refer to the triangle formulae; the first pulse-width (time or degree) is determined by the equation:

$$0.8\sin 100\pi t = 1800t - 2 \qquad (12.7)$$

This is a transcendental equation with the unknown parameter t. Using iterative method to solve the equation, let $x = 0.8\sin 100\pi t$ and $y = 1800t - 2$. We can choose the initial $t_0 = 1.38889\,\text{ms} = 25°$,

t (ms/degree)	x	Y	$\lvert x\rvert{:}y$	Remarks
1.38889/25°	0.338	0.5	<	Decrease t
1.27778/23°	0.3126	0.3	>	Increase t
1.2889/23.2°	0.3152	0.32	<	Decrease t
1.2861/23.15°	0.3145	0.315	≈	

The first pulse width to switch on and off the switch S_+ is 1.2861 ms (or 23.15°).
    Other pulse widths can be determined from other equations using iterative method. For a PWM operation with large mf, readers can refer to Figure 12.8.

Figure 12.7 shows the ideal waveforms associated with the half-bridge VSI. We can find out the phase delay between the output current and voltage. For a large m_f we can see the cross points demonstrated in Figure 12.8 with a smaller phase delay between the output current and voltage.

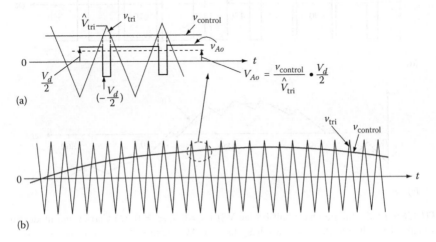

FIGURE 12.8 Sinusoidal PWM (a) partially enlarged waveform and (b) original waveform.

12.4.2 Single-Phase Full-Bridge VSI

A single-phase full-bridge VSI is shown in Figure 12.9. The carrier-based PWM technique is applied in this inverter. Two large capacitors may be used to provide a neutral point N, but not necessarily. Since the output voltage does not refer to the neutral point N, the maximum output voltage is possibly greater than half of the DC-link voltage. If it is operating in linear modulation, the output voltage is smaller than the DC-link voltage. The modulation operation is different from that of single-phase half-bridge VSI in the previous section. It is shown in Figure 12.13. Four switches S_{1+}/S_{1-} and S_{2+}/S_{2-} in two legs are applied and switched by the PWM signal.

Figure 12.10 shows the ideal waveforms associated with the full-bridge VSI. There are two sine waves used in Figure 12.10a corresponding to the two legs operation. We can find out the phase delay between the output current and voltage.

The method to determine the pulse widths is the same as that introduced in the previous section. Refer to Figure 12.10a; we can find that there are two sine-wave functions:

$$f_+(t) = m_a \sin \omega t = 0.8 \sin 100\pi t \tag{12.8}$$

and

$$f_-(t) = -m_a \sin \omega t = -0.8 \sin 100\pi t \tag{12.9}$$

The triangle functions are

$$f_{\Delta 1}(t) = -4 fm_f t = -1600t \quad f_{\Delta 2}(t) = 4 fm_f t - 2 = 1600t - 2$$

$$f_{\Delta 3}(t) = 4 - 4 fm_f t = 4 - 1600t \quad f_{\Delta 4}(t) = 4 fm_f t - 6 = 1600t - 6$$

...

$$f_{\Delta(2n-1)}(t) = 4(n-1) - 4 fm_f t \quad f_{\Delta 2n}(t) = 4 fm_f t - (4n-2) \tag{12.10}$$

...

$$f_{\Delta 15}(t) = 28 - 1600t \quad f_{\Delta 16}(t) = 1600t - 30$$

$$f_{\Delta 17}(t) = 32 - 1600t$$

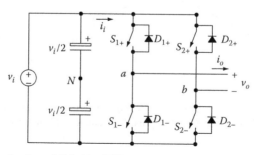

FIGURE 12.9 Single-phase full-bridge VSI.

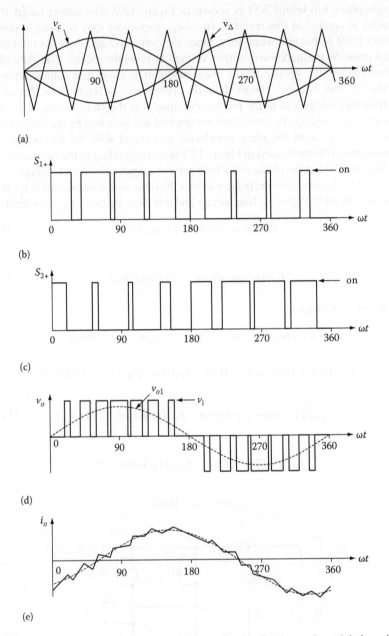

FIGURE 12.10 Full-bridge VSI (m_a=0.8, m_f=8). (a) Carrier and modulating signals. (b) Switch S_{1+} and S_{1-} state. (c) Switch S_{2+} and S_{2-} state. (d) AC output voltage. (e) AC output current.

The first pulse width to switch on and switch off the switches S_{1+} and S_{1-} is determined by the following equation:

$$0.8 \sin 100\pi t = 1600t - 2 \tag{12.11}$$

The first pulse width to switch on and switch off the switches S_{2+} and S_{2-} is determined by the following equation:

$$-0.8 \sin 100\pi t = 1600t - 2$$

or

$$0.8 \sin 100\pi t = 2 - 1600t \tag{12.12}$$

Since the output voltage is between leg and leg, the rms voltages of the output voltage harmonics are calculated by the following formula:

$$(V_O)_h = \frac{2V_d}{\sqrt{2}} \frac{(\hat{V}_{AO})_h}{V_d/2} \tag{12.13}$$

where
  $(V_O)_h$ is the hth harmonic rms voltage of the output voltage
  V_d is the DC link voltage
  $(\hat{V}_{AO})_h/(V_d/2)$ is tabulated as a function of m_a, which can be got from Table 12.1.

Example 12.3 A single-phase full-bridge DC/AC inverter is shown in Figure 12.9 to implement SPWM with $V_d = 300\,V$, $m_a = 1.0$, and $m_f = 31$. The fundamental frequency is 50 Hz. Determine the rms value of the fundamental frequency and some of the harmonics in output voltage using Table 12.1.

Solution: From Equation 12.13, we have the general rms values:

$$(V_O)_h = \frac{2\hat{V}_d}{\sqrt{2}} \frac{(V_{AO})_h}{V_d/2} = \frac{600}{\sqrt{2}} \frac{(\hat{V}_{AO})_h}{V_d/2} = 424.26 \frac{(\hat{V}_{AO})_h}{V_d/2} V$$

Checking the data from Table 12.1, we can get the rms values as follows:

Fundamental

$(V_O)_1 = 424.26 \times 1.0 = 424.26\,V$ at 50 Hz
$(V_O)_{27} = 424.26 \times 0.018 = 7.64\,V$ at 1350 Hz
$(V_O)_{29} = 424.26 \times 0.318 = 134.92\,V$ at 1450 Hz
$(V_O)_{31} = 424.26 \times 0.601 = 254.98\,V$ at 1550 Hz
$(V_O)_{33} = 424.26 \times 0.318 = 134.92\,V$ at 1650 Hz
$(V_O)_{35} = 424.26 \times 0.018 = 7.64\,V$ at 1750 Hz
$(V_O)_{57} = 424.26 \times 0.033 = 14\,V$ at 2850 Hz
$(V_O)_{59} = 424.26 \times 0.212 = 89.94\,V$ at 2950 Hz
$(V_O)_{61} = 424.26 \times 0.181 = 76.79\,V$ at 3050 Hz
$(V_O)_{63} = 424.26 \times 0.181 = 76.79\,V$ at 3150 Hz
$(V_O)_{65} = 424.26 \times 0.212 = 89.94\,V$ at 3250 Hz
$(V_O)_{67} = 424.26 \times 0.033 = 14\,V$ at 3350 Hz

12.5   THREE-PHASE FULL-BRIDGE VOLTAGE SOURCE INVERTER

A three-phase full-bridge VSI is shown in Figure 12.11. The carrier-based PWM technique is applied in this single-phase full-bridge VSI. Two large capacitors may be used to provide a neutral point N, but not necessarily. Six switches S_1–S_6 in three legs are applied and switched by the PWM signal.

Figure 12.12 shows the ideal waveforms associated with the full-bridge VSI. We can find out the phase delay between the output current and voltage.

Since the three-phase waveform in Figure 12.12a is not referring to the neutral point N, the operation conditions are different from single-phase half-bridge VSI. The maximum output line-to-line voltage is possibly greater than half of the DC-link voltage. If it is operating in linear modulation, the output voltage is smaller than the DC-link voltage. The modulation indication of a three-phase VSI is different from that of a single-phase half-bridge VSI (Section 12.4.1). It is shown in Figure 12.13.

12.6   THREE-PHASE FULL-BRIDGE CURRENT SOURCE INVERTER

A three-phase full-bridge CSI is shown in Figure 12.14.

The carrier-based PWM technique is applied in this three-phase full-bridge CSI. The main objective of these static power converters is to produce AC output current waveforms from a DC current power supply. Six switches S_1–S_6 are applied and switched by the PWM signal. Figure 12.15 shows the ideal waveforms associated with the full-bridge CSI.

The CSI has boost function. Usually, the output voltage can be higher than the input voltage. We can find out the phase ahead between the output voltage and current.

12.7   MULTISTAGE PWM INVERTER

Multistage PWM inverters can be constructed by two methods: multicell and multi-level. Unipolar modulation PWM inverters can be considered as multistage inverters.

12.7.1   UNIPOLAR PWM VSI

In Section 12.4, we introduced the single-phase source inverter operating in the *bipolar modulation*. Refer to the circuit in Figure 12.6; both upper switch S_+ and

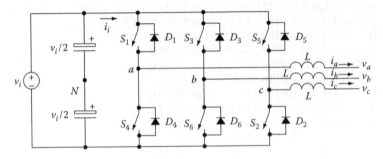

FIGURE 12.11   Three-phase full-bridge VSI.

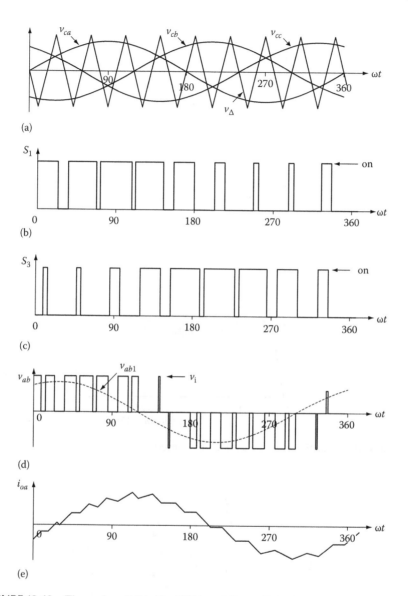

FIGURE 12.12 Three-phase full-bridge VSI ($m_a = 0.8$, $m_f = 9$). (a) Carrier and modulating signals. (b) Switch S_1/S_4 state. (c) Switch S_3/S_4 state. (d) AC output voltage. (e) AC output current.

lower switch S_- work together. The carrier and modulating signals are shown in Figure 12.7a, and the switching signals for upper switch S_+ and lower switch S_- are shown in Figure 12.7b and c. The output voltage of the inverter is the pulse train with both polarities shown in Figure 12.7d.

There are some drawbacks using bipolar modulation: (1) If the inverter is VSI, a dead time has to be set to avoid the short-circuit; (2) the zero output voltage corresponds to the equal-pulse width of positive and negative pulses; (3) power losses

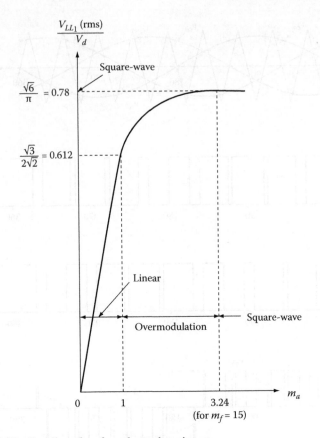

FIGURE 12.13   Function of m_a for a three-phase inverter.

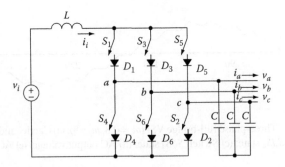

FIGURE 12.14   Three-phase CSI.

are high since two devices work, and hence the efficiency is lower; (4) two devices should be controlled simultaneously.

In most industrial applications, unipolar modulation is widely applied. The regulation and corresponding waveforms are shown in Figure 12.16 with $m_a = 0.8$ and $m_f = 9$. For unipolar regulation, the m_a is measured by

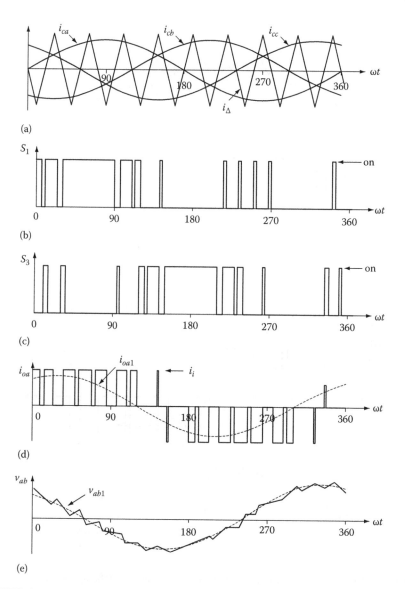

FIGURE 12.15 Three-phase CSI (m_a=0.8, m_f=9). (a) Carrier and modulating signals. (b) Switch S_{1+} state. (c) Switch S_3 state. (d) AC output current. (e) AC output voltage.

$$m_a = \frac{V_{in\text{-}m}}{2V_{tri\text{-}m}} \qquad (12.14)$$

This regulation method is likely a two-stage PWM inverter. If the output voltage is positive, only the upper device works and the lower device idles. Therefore, the output voltage remains the positive polarity pulse train. On the contrary, if the output voltage is negative, only the lower device works and the upper device idles.

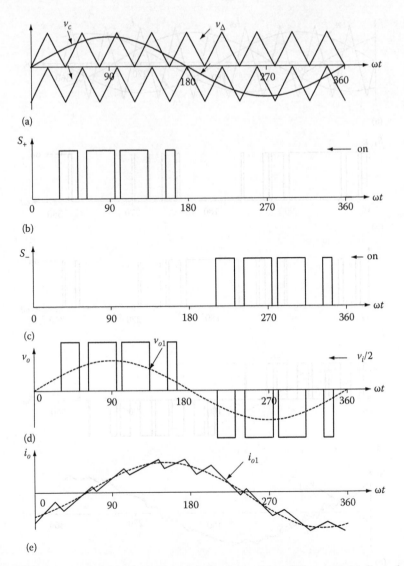

FIGURE 12.16 Three-phase unipolar regulation inverter ($m_a=0.8$, $m_f=9$). (a) Control signals, (b) upper switch control signal, (c) lower switch control signal, (d) output pulse-train and its fundamental sine-wave, and (e) the output current waveform.

Therefore, the output voltage remains the negative polarity pulse train. The advantages of implementing unipolar regulation are as follows:

- No need to set dead time.
- The pulses are narrow, for example, zero output voltage requires zero pulse width.
- Power losses are low and hence the efficiency is high.
- Only one device should be controlled in half-cycle.

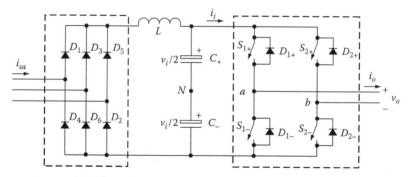

FIGURE 12.17   Three-phase input single-phase output cell.

12.7.2 MULTICELL PWM VSI

Multistage PWM inverters can consist of many cells. Each cell can be a single-phase or three-phase input plus single-phase output VSI, which is shown in Figure 12.17. If the three-phase AC supply is a secondary winding of a main transformer, it is floating and isolated from other cells and common ground point. Therefore, all cells can be linked in series or in a parallel manner.

A three-stage PWM inverter is shown in Figure 12.18. Each phase consists of three cells with different phase-angle shift by 20° from one another.

The carrier-based PWM technique is applied in this three-phase multistage PWM inverter. Figure 12.19 shows the ideal waveforms associated with the full-bridge VSI. We can find out the output, the phase delayed between the output current and voltage.

12.7.3 MULTILEVEL PWM INVERTER

A three-level PWM inverter is shown in Figure 12.20. The carrier-based PWM technique is applied in this multilevel PWM inverter. Figure 12.21 shows the ideal waveforms associated with the multilevel PWM inverter. We can find out the output, the phase delayed between the output current and voltage.

12.8   IMPEDANCE-SOURCE INVERTERS

ZSI is a new approach of DC/AC conversion technology. It was published by Peng in 2003 [3–5]. The ZSI circuit diagram shown in Figure 12.22 consists of an "X"-shaped impedance network formed by two capacitors and two inductors, and it provides unique buck-boost characteristics. Moreover, unlike VSI, the need for dead time would not arise with this topology. Due to these attractive features, it has found use in numerous industrial applications including variable-speed drives and distributed generation (DG). However, it has not been widely researched as a DG topology. Moreover, all these industrial applications require proper closed-loop controlling to adjust their operating conditions subject to changes in both input and output conditions. On the other hand, the presence of "X" shaped impedance network and the need of short-circuiting of inverter arm to boost the voltage would complicate the controlling of ZSI.

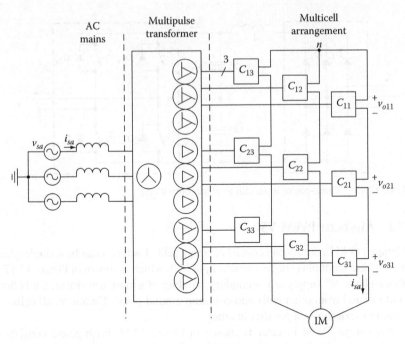

FIGURE 12.18 Multistage converter based on a multicell arrangement. (a) Carrier and modulating signals. (b) Cell c_{11} AC output voltage. (c) Cell c_{21} AC output voltage. (d) Cell c_{31}. AC output voltage. (e) Phase a load voltage.

12.8.1 COMPARISON WITH VSI AND CSI

ZSI is a new inverter different from traditional VSI and CSI. In order to express ZSI's advantages, it is necessary to compare it with VSI and CSI.

A three-phase VSI is shown in Figure 12.11. A DC voltage source supported by a relatively large capacitor feeds the main converter circuit, a three-phase bridge. The V-source inverter has the following conceptual and theoretical barriers and limitations.

1. The AC output voltage is limited and cannot exceed the DC link. Therefore, the VSI is a buck (step-down) inverter for DC/AC power conversion. For applications where overdrive is desirable and the available DC voltage is limited, an additional DC/DC boost converter is needed to obtain the desired AC output. The additional power converter stage increases system cost and lowers efficiency.
2. The upper and lower devices of each phase leg cannot be gated on simultaneously either by purpose or by EMI noise. Otherwise, a shoot-through would occur and destroy the devices. The shoot-through problem by electromagnetic interference (EMI) noise's misgating-on is a major killer to the converter's reliability. Dead time to block both upper and lower devices has to be provided in the VSI, which causes waveform distortion, etc.
3. An output LC filter is needed for providing a sinusoidal voltage compared with the CSI, which causes additional power loss and control complexity.

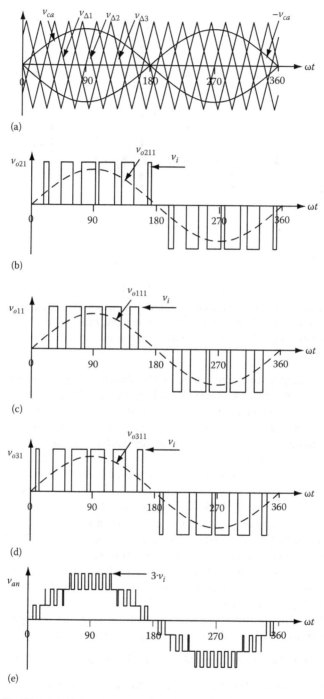

FIGURE 12.19 Multicell PWM inverter (three stages, $m_a=0.8$, $m_f=6$).

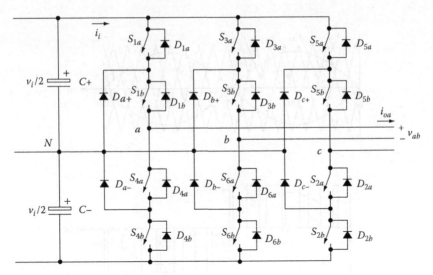

FIGURE 12.20   Three-phase three-level VSI.

A three-phase CSI is shown in Figure 12.14. A DC voltage source feeds the main inverter circuit, a three-phase bridge. The DC current source can be a relatively large DC inductor fed by a voltage source such as a battery, fuel-cell stack, diode rectifier, or thyristor converter. The CSI has the following conceptual and theoretical barriers and limitations.

1. The AC output voltage has to be greater than the original DC voltage that feeds the DC inductor, or the DC voltage produced is always smaller than the AC input voltage. Therefore, the CSI is a boost inverter for DC/AC power conversion. For applications where a wide voltage range is desirable, an additional DC–DC buck (or boost) converter is needed. The additional power conversion stage increases system cost and lowers efficiency.
2. At least one of the upper devices and one of the lower devices have to be gated on and maintained at any time. Otherwise, an open circuit of the DC inductor would occur and destroy the devices. The open-circuit problem by EMI noise's misgating-off is a major concern of the converter's reliability. Overlap time for safe current commutation is needed in the I-source converter, which also causes waveform distortion, etc.
3. The main switches of the I-source converter have to block reverse voltage, which requires a series diode to be used in combination with high-speed and high-performance transistors such as insulated gate bipolar transistors (IGBTs). This prevents the direct use of low-cost and high-performance performance IGBT modules and intelligent power modules (IPMs).

In addition, both the VSI and the CSI have the following common problems.

1. They are either a boost or a buck converter and cannot be a buck-boost converter. That is, their obtainable output voltage range is limited to either greater or smaller than the input voltage.

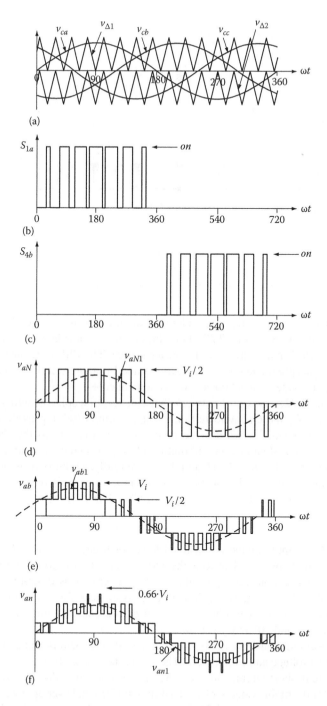

FIGURE 12.21 Three-level VSI (three levels, $m_a = 0.8$, $m_f = 15$). (a) Carrier and modulating signals. (b) Switch S_{1a} status. (c) Switch S_{4b} status. (d) Inverter phase a-N voltage. (e) AC output line voltage. (f) AC output phase voltage.

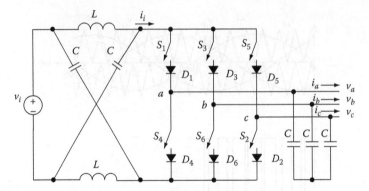

FIGURE 12.22 Impedance source inverter (ZSI).

2. Their main circuits cannot be interchanged. In other words, neither the VSI main circuit can be used for the CSI nor vice versa.
3. They are vulnerable to EMI noise in terms of reliability.

To overcome the aforementioned problems of the traditional VSI and CSI, ZST was designed as shown in Figure 12.22. It employs a unique impedance network to couple the converter's main circuit to the power source. The ZSI overcomes the aforementioned conceptual and theoretical barriers and limitations of the traditional VSI and CSI and provides a novel power conversion concept.

In Figure 12.22, a two-port network that consists of a split inductors L_1 and L_2 and capacitors C_1 and C_2 connected in X shape is employed to provide an impedance source (z-source) coupling the converter (or inverter) to the DC source. Switches used in ZSI can be a combination of switching devices and diodes such as those shown in Figures 12.11 and 12.14. Actually, if the two inductors have zero inductance, the ZSI becomes a VSI. In other words, if the two capacitors have zero capacitance, The ZSI becomes a CSI. The advantages of the ZSI are listed here:

1. The AC output voltage is not fixed lower or higher than the DC-link (or DC source) voltage. Therefore, the ZSI is a buck-boost inverter for DC/AC power conversion. For applications where overdrive is desirable and the available DC voltage is not limited, there is no need for an additional DC/DC boost converter to obtain a desired AC output. Therefore, the system cost is low and efficiency is high.
2. The z-circuit consists of two inductors and two capacitors and can restrict the overvoltage and overcurrent. Therefore, the legs in the main bridge can operate in short-circuit and open circuit in a short time. There is no restriction for the main bridge such as dead time for VSI and overlap time for CSI.
3. ZSI has a function as anti-EMI noise. The shoot-through problem by electromagnetic interference (EMI) noise's misgating-on will not damage the devices and the converter's reliability.

12.8.2   Equivalent Circuit and Operation

A three-phase ZSI used for fuel-cell application is shown in Figure 12.23. It has nine permissible switching states (vectors): six active vectors as a traditional VSI has plus three 0 vectors when the load terminals are shorted through both the upper and lower devices of any one phase leg (i.e., both devices are gated on), any two phase legs, or all three phase legs. This shoot-through zero state (or vector) is forbidden in the traditional VSI, because it would cause a shoot-through. We call this third zero state (vector) the shoot-through zero state (or vector), which can be generated by seven different ways: shoot-through via any one phase leg, combinations of any two phase legs, and all three phase legs. The z-source network makes the shoot-through zero state possible. This shoot-through zero state provides the unique buck-boost feature to the inverter.

Figure 12.24 shows the equivalent circuit of the ZSI shown in Figure 12.23 when viewed from the DC link. The inverter bridge is equivalent to a short circuit when the inverter bridge is in the shoot-through zero state, as shown in Figure 12.25, whereas the inverter bridge becomes an equivalent current source as shown in Figure 12.26 when in one of the six active states. Note that the inverter bridge can also be represented by a current source with zero value (i.e., an open circuit) when it is in one of the two traditional zero states. Therefore, Figure 12.26 shows the equivalent circuit of the z-source inverter viewed from the DC link when the inverter bridge is in one of the eight non-shoot-through switching states.

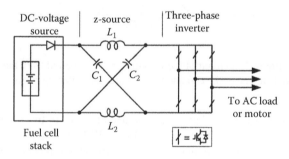

FIGURE 12.23   z-source inverter for fuel-cell applications.

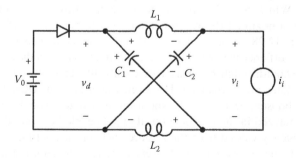

FIGURE 12.24   Equivalent circuit of the z-source inverter viewed from the DC link.

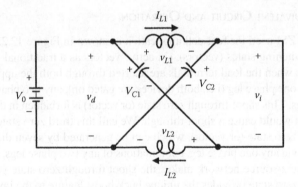

FIGURE 12.25   Equivalent circuit of the z-source inverter viewed from the DC link when the inverter bridge is in the shoot-through zero state.

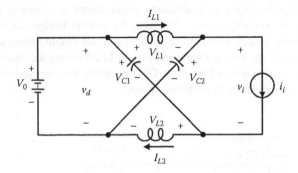

FIGURE 12.26   Equivalent circuit of the z-source inverter viewed from the DC link when the inverter bridge is in one of the eight non-shoot-through switching states.

All the traditional PWM schemes can be used to control the z-source inverter, and their theoretical input–output relationships still hold. Figure 12.27 shows the traditional PWM switching sequence based on the triangular carrier method. In every switching cycle, the two non-shoot-through zero states are used along with two adjacent active states to synthesize the desired voltage. When the DC voltage is high enough to generate the desired AC voltage, the traditional PWM of Figure 12.27 is used. While the DC voltage is not enough to directly generate a desired output voltage, a modified PWM with shoot-through zero states will be used as shown in Figure 12.28 to boost voltage. It should be noted that each phase leg still switches on and off once per switching cycle. Without change in the total zero-state time interval, shoot-through zero states are evenly allocated into each phase. That is, the active states are unchanged. However, the equivalent DC-link voltage to the inverter is boosted because of the shoot-through states. The detailed relationship will be analyzed in the next section. It is noticeable here that the equivalent switching frequency viewed from the z-source network is six times the switching frequency of the main inverter, which greatly reduces the required inductance of the z-source network.

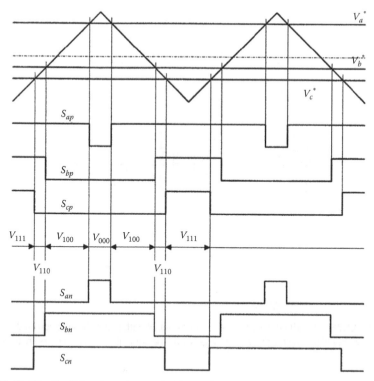

FIGURE 12.27 Traditional carrier-based PWM control without shoot-through zero states, where the traditional zero states (vectors) V111 and V000 are generated every switching cycle and determined by the references.

12.8.3 Circuit Analysis and Calculations

Assuming that the inductors L_1 and L_2 and capacitors C_1 and C_2 have the same inductance L and capacitance C, respectively, the z-source network becomes symmetrical. From the symmetry and the equivalent circuits, we have

$$V_{C1} = V_{C2} = V_C \quad v_{L1} = v_{L2} = v_L \tag{12.15}$$

Given that the inverter bridge is in the shoot-through *zero state* for an interval of T_0 during a switching cycle T. From the equivalent circuit in Figure 12.25, one has

$$v_L = V_C \quad v_d = 2V_C \quad v_i = 0 \tag{12.16}$$

Now consider that the inverter bridge is in one of the eight non-shoot-through states for an interval of T_1 during the switching cycle T. From the equivalent circuit in Figure 12.25, one has

$$v_L = V_0 - V_C \quad v_d = V_0 \quad v_i = V_C - v_L = 2V_C - V_0 \tag{12.17}$$

where
 V_0 is the DC source voltage

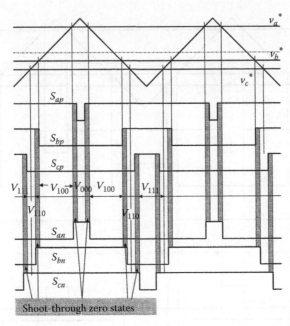

FIGURE 12.28 Modified carrier-based PWM control with shoot-through zero states that are evenly distributed among the three phase legs, while the equivalent active vectors are unchanged.

The switching duty cycle $k = T_1/T$ and $T = T_0 + T_1$.

The average voltage of the inductors over one switching period should be zero in steady state, from (12.16) to (12.17); thus, we have

$$V_L = \bar{v}_L = \frac{T_0 V_C + T_1(V_0 - V_C)}{T} = 0 \qquad (12.18)$$

or

$$\frac{V_C}{V_0} = \frac{T_1}{T_1 - T_0} \qquad (12.19)$$

Similarly, the average DC-link voltage across the inverter bridge can be found as follows:

$$V_i = \bar{v}_i = \frac{T_0 \times 0 + T_1(2V_C - V_0)}{T} = \frac{T_1}{T_1 - T_0} V_0 = V_C \qquad (12.20)$$

The peak DC-link voltage across the inverter bridge is expressed in (12.17) and can be rewritten as

$$\hat{v}_i = V_C - v_L = 2V_C - V_0 = \frac{T}{T_1 - T_0} V_0 = B \cdot V_0 \qquad (12.21)$$

where

$$B = \frac{T}{T_1 - T_0} = \frac{1}{1 - 2(T_0 / T)} \geq 1 \qquad (12.22)$$

B is the boost factor resulting from the shoot-through zero state. Usually, T_1 is greater than T_0, that is., $T_0 < T/2$. The peak DC-link voltage $\hat{V}_i$ is the equivalent DC-link voltage of the inverter. On the other side, the output peak-phase voltage from the inverter can be expressed as

$$\hat{v}_{ac} = M \cdot \frac{\hat{v}_i}{2} \qquad (12.23)$$

where M is the modulation index. Using (12.21), (12.23) can be further expressed as

$$\hat{v}_{ac} = M \cdot B \cdot \frac{V_0}{2} \qquad (12.24)$$

For the traditional VSI, we have the well-known relationship $\hat{V}_{ac} = M \cdot (V_0/2)$. Equation 12.24 shows that the output voltage can be stepped up and down by choosing an appropriate buck-boost factor MB.

$$MB = \frac{T}{T_1 - T_0} M \qquad (12.25)$$

MB is changeable from 0 to ∞. From (12.15), (12.19), and (12.22), the capacitor voltage can be expressed as

$$V_C = \frac{1 - (T_1 / T)}{1 - 2(T_0 / T)} V_0 \qquad (12.26)$$

The buck-boost factor MB is determined by the modulation index M and boost factor B. The boost factor B as expressed in (12.22) can be controlled by duty cycle (i.e., interval ratio) of the shoot-through zero state over the non-shoot-through states of the inverter PWM.

Note that the shoot-through zero state does not affect the PWM control of the inverter, because it equivalently produces the same zero voltage to the load terminal. The available shoot-through period is limited by the zero-state period that is determined by the modulation index.

12.9   EXTENDED BOOST z-SOURCE INVERTERS

In recent years many researchers have focused in many directions to develop ZSI to achieve different objectives [6–13]. Some have worked on developing different kinds of topological variations where others have worked on developing ZSI into deferent applications where controller design, modeling and analyzing its operating modes, and developing modulation methods are addressed. Theoretically, ZSI can produce infinite gain like many other DC–DC boosting topologies; however it is well known that this cannot be achieved due to effects of parasitic components where the gain tends to drop drastically [6]. Conversely, high boost could increase power losses and instability. On the other hand, shoot-through interval, the variable that is responsible for increasing the gain, is interdependent with the other variable modulation index that controls the output of the ZSI and also imposes limitation on variability, thereby boosting output voltage. That is, increase in boosting factor would compromise the modulation index and result in lower modulation index [7]. In addition, the voltage stress on the switches would be high due to the pulsating nature of the output voltage.

Unlike the DC–DC converters, so far researchers of ZSIs have not given their focus to improve the gain of the converter. This opens a significant research gap in the field of ZSI development. Particularly, some applications such as solar and fuel cells where generated power is integrated into the grid may require high voltage gain to match the voltage difference and also to compensate for the voltage variations. Its effect is significant when such sources are connected to 415 V three-phase systems. In the case of fuel and solar cells, although it is possible to increase the number of cells to increase the voltage, there are other influencing factors that need to be taken into account. Sometimes the available number of cells is limited or environmental factors could come into play due to shading of light to some cells that could result in poor overall energy catchment. Then with fuel cells some manufactures produce fuel cell with lower voltage to achieve a faster response. Such factors could demand power converters with larger boosting ratio. This cannot be realized with a single ZSI. Hence, this paper focuses on developing a new family of ZSI that would realize extended boosting capability.

12.9.1 INTRODUCTION TO ZSI AND BASIC TOPOLOGIES

The basic topology of ZSI is originally proposed in Ref. [3]. This is a single-stage buck-boost topology due to the presence of the X shaped impedance network as shown in Figure 12.29a, which allows the safe shoot-through of inverter arms avoiding the need of dead time that was needed in traditional VSI. However, unlike the VSI, the original ZSI does not share the ground point of DC source with the converter and also the current drawn from the source will be discontinuous and these would be a disadvantage in some applications, and it may be required to have a decoupling capacitor bank at the front end to avoid current discontinuity. Subsequently, the ZSI has been modified as shown in Figure 12.29b and c where now a impedance network is placed at the bottom or top arm of the inverter. The advantage of this topology is that in one topology ground point can be shared and in both cases the voltage stress on the component is much lower than that of the traditional ZSI. However, the current discontinuity still prevails, an alternative continuous current quasi ZSI (qZSI) is proposed in reference, but this continuous current circuit is not considered in developing new converters. In terms of topology, the quasi ZSI has no disadvantage over the traditional topology. In this paper, a discontinuous current quasi ZSI inverter is used to extend the boosting capability.

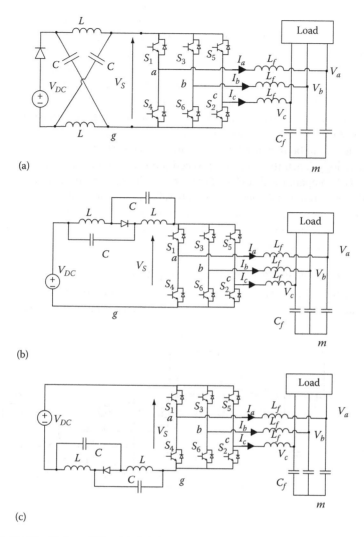

(a)

(b)

(c)

FIGURE 12.29 Various ZSIs. (a) Original ZSI. (b) Discontinuous current quasi z-source inverter with shared ground. (c) Discontinuous current quasi ZSI with low voltage level at components.

In summary, the proposed qZSIs operate similar to the original ZSI and the same modulation schemes can be applied.

12.9.2 Extended Boost qZSI Topologies

In this chapter, four new converter topologies are proposed. These topologies can be categorized mainly into diode-assisted boost or capacitor-assisted boost. They can be further divided into continuous current and discontinuous current topologies. Their operation is extensively described in the following sections. All these topologies can be modulated using the modulation methods proposed for the original ZSI. In this context, the modulation method proposed is used. The other advantage of the proposed new topologies is their expandability. This was not possible with the original ZSI, that is, if one needs additional boosting another stage can be cascaded at the front end. The new topology would operate with the same number of active switches. The only addition would be one inductor, one capacitor, and two diodes for the diode-assisted case and one inductor, two capacitor, and one diode for the capacitor-assisted case for each added new stage. By defining shoot-through duty ratio (D_S) for each added new stage boosting factor can be increased with a factor of $1/(1 - D_S)$ in the case of the diode-assisted topology. Then the capacitor-assisted topology would have a boosting factor of $1/(1 - 3D_S)$ compared with $1/(1 - 2D_S)$ in the traditional topology. However, similar to the other boosting topologies it is not advisable to operate with very high or very low shoot-through values. In addition, a careful consideration is needed in selecting the boosting factor modulation index for a suitable topology to achieve high efficiency. These aspects need further research and they will be addressed in a future study.

12.9.2.1 Diode-Assisted Extended Boost qZSI Topologies

In this category, two new families of topologies are proposed, namely the continuous current and the discontinuous current type topology. Figure 12.30 shows

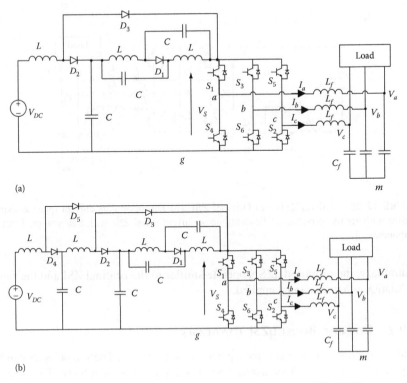

(a)

(b)

FIGURE 12.30 Diode-assisted extended boost continuous current qZSI. (a) First extension. (b) Second extension.

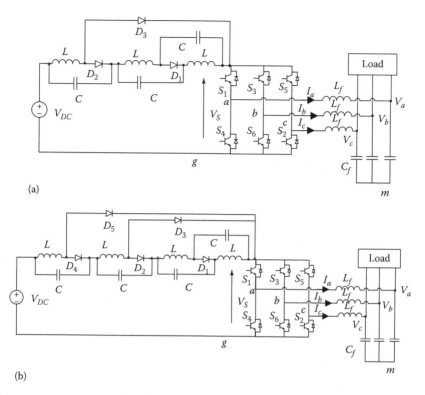

FIGURE 12.31 Diode-assisted extended boost discontinuous current qZSI. (a) First extension. (b) Second extension.

the continuous current type topology, and it can be extended to have high boost by cascading more stages as shown in Figure 12.31. This new topology comprises an additional inductor, a capacitor, and two diodes. The operating principle of this additional impedance networks is similar to that found in cascaded boost and Luo converters [9–12]. The added impedance network provides the boosting function without disturbing the operation inverter.

First, considering the continuous current topology and its steady-stage operation, we know that this converter has three operating states similar to those of the traditional ZSI topology. It can be simplified into shoot-through and non-shoot-through states. Then the inverter's action is replaced by a current source plus a single switch. First, consider the non-shoot-through state, which is represented with an open switch. In addition, diodes D_1 and D_2 are conducting and D_3 is in blocking state; therefore, the inductors discharge and the capacitors get charged. Figure 12.32b shows the equivalent circuit diagram for non-shoot-through state.

By applying KVL, the following steady-state relationships can be observed: $V_{DC} + v_{L3} = V_{c3}$, $v_{L1} = V_{c1}$, $V_{L2} = V_{c2}$, and $V_S = V_{c3} + V_{c2} + V_{L1}$. Figure 12.32c shows the equivalent circuit diagram for the shoot-through state where it is represented with the closed switch, D_3 is conducting, and D_1 and D_2 diodes are in blocking state where all the inductors get charged. Energy is transferred from the source to the inductor or the

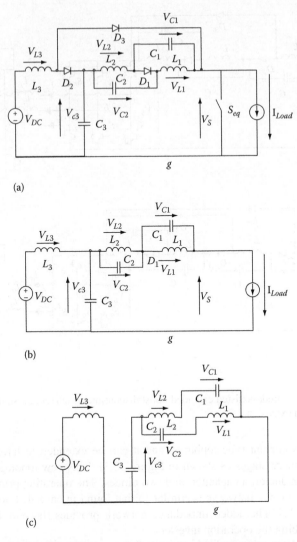

(a)

(b)

(c)

FIGURE 12.32 Simplified diagram of diode-assisted extended boost continuous current qZSI. (a) Simplified circuit. (b) Non-shoot-through state. (c) Shoot-through state.

capacitor to the inductor while capacitors get discharged. Similar relationships can be derived as $V_{DC}+v_{L3}=0$, $V_{c3}+V_{L2}+V_{c1}=0$, and $V_{c3}+V_{c2}+V_{L1}=0$, and $V_S=0$, $V_{c3}+V_{c2}=V_{L1}$. Considering the fact that the average voltage across the inductors is zero and by defining the shoot-through duty ratio as D_S and non-shoot-through duty ratio as D_A where $D_A+D_S=1$, the following relations can be derived.

$$V_{C3} = \frac{1}{1-D_S}V_{DC} \quad \text{and} \quad V_{C1} = V_{C2} = \frac{D_S}{1-2D_S}V_{C3} = \frac{D_S}{(1-2D_S)(1-D_S)}V_{DC} \quad (12.27)$$

From these equations, the peak voltage across the inverter $\hat{V}_S$ and the peak AC output voltage $\hat{V}_x$ can be obtained as follows:

$$\hat{v}_S = \frac{1}{(1-2D_S)(1-D_S)} V_{DC} \quad \text{and} \quad \hat{v}_x = M \frac{\hat{v}_S}{2} \tag{12.28}$$

Define $B = 1/(1 - 2D_S)(1 - D_S)$ and the boost factor in the DC side; then the peak AC side can be written as

$$\hat{v}_x = B\left(M \frac{V_{DC}}{2}\right) \tag{12.29}$$

Now the boosting factor has increased by a factor of $1/(1 - D_S)$ compared with that of the original ZSI. Similarly, the steady-state equations can be derived for the diode-assisted extended boost discontinuous current qZSI. Then it is possible to prove that this converter also has the same boosting factor as that of continuous current topology. In addition, the voltage stress on the capacitors is similar except the voltage across capacitor 3. This can be shown as $V_{c3} = D_S/(1 - D_S)*V_{DC}$. By studying these two topologies, it can be noted that with the discontinuous current topology, the capacitors are subjected to a small voltage stress and if there is no boosting then the voltage across them is zero. Also it is possible to derive the boost factor for topologies shown in Figures 12.30b and 12.31b as $B = 1/((1 - 2D_S)(1 - D_S)^2)$.

12.9.2.2 Capacitor-Assisted Extended Boost qZSI Topologies

Similar to the previous family of extended boost qZSI, this section proposes another family of converters. The difference is that now a much higher boost is achieved with only a simple structural change to the previous topology. Now the D_3 is replaced with a capacitor as shown in Figure 12.32. In this context, two topological variations are derived as continuous current or discontinuous current forms as shown in Figure 12.33.

In the previous scenario, the steady-state relations are derived using the continuous current topology. Therefore, in this context the discontinuous current topology is considered. In this case also, the converter's three operating states are simplified into shoot-through and non-shoot-through states.

The simplified circuit diagram is shown in Figure 12.34a. First, consider the non-shoot-through state shown in Figure 12.34b, which is represented with an open switch. As diodes D_1 and D_2 are conducting, the inductors discharge and capacitors get charged. Then by applying KVL, the following steady-state relationships can be observed. $V_{DC} + V_{c3} + V_{c2} + V_{c1} = V_S$ and $V_{DC} + V_{c3} + V_{c4} + V_{c1} = V_S$, $V_{c1} = v_{L1}$, $V_{c2} = v_{L2}$, $V_{c3} = v_{L3}$, $V_{DC} + V_{c3} = V_d$, $V_{c2} = V_{c4}$. Figure 12.34c shows the equivalent circuit diagram for the shoot-through state, where it is represented with the closed switch. Both diodes D_1 and D_2 are in blocking state, where all the inductors get charged and energy is transferred from the source to inductors or the capacitor to inductors while capacitors get discharged. Similar relationships can be derived as $V_{DC} + v_{L3} + V_{c4} + V_{c1} = 0$, $V_{DC} + V_{c3} = V_d$, $V_d + V_{L1} + V_{c2} = 0$, $V_d + V_{L2} + V_{c1} = 0$, and $V_S = 0$.

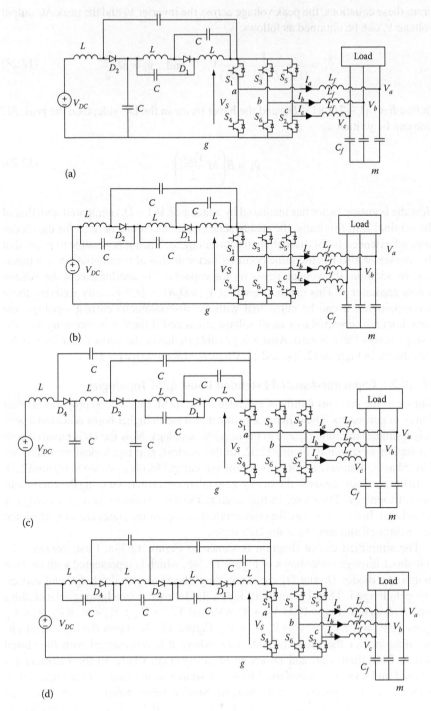

FIGURE 12.33 Capacitor-assisted extended boost qZSIs. (a) Continuous current. (b) Discontinuous current. (c) High extended continuous current. (d) Discontinuous current.

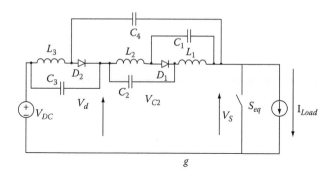

(a)

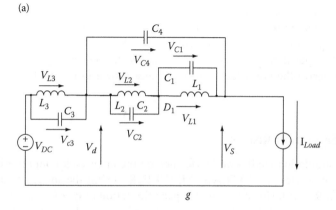

(b)

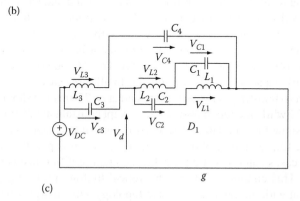

(c)

FIGURE 12.34 Simplified diagram of capacitor-assisted extended boost continuous current qZSI. (a) Simplified circuit. (b) Non-shoot-through state. (c) Shoot-through state.

Considering the fact that the average voltage across the inductors is zero, the following relations can be derived:

$$V_d = \frac{1-2D_S}{1-3D_S}V_{DC} \quad \text{and} \quad V_{C1} = V_{C2} = V_{C3} = V_{C4} = \frac{D_S}{1-2D_S}V_d = \frac{D_S}{1-3D_S}V_{DC}$$

$$(12.30)$$

Then from these equations the peak voltage across the inverter $\hat{V}_S$ can be obtained as follows:

$$\hat{v}_S = \frac{1}{1-3D_S}V_{DC} \tag{12.31}$$

Similar equations can be derived for the continuous current topology. Now the difference would be the continuity of source current and the difference in voltage across the capacitor C_3, which was derived as $V_{C3} = V_d$. The voltage across the capacitor is much larger than that of the discontinuous current topology. Similarly, it is possible to derive the boost factor for topologies shown in Figure 12.33c and d as $B = 1/1 - 4D_S$.

12.9.3 SIMULATION RESULTS

Extensive simulation studies are performed on the open-loop configuration of all proposed topologies in MATLAB®/SIMULINK® using the modulation method proposed in Ref. [6]. However, due to space limitation only a few results are presented. This would validate the operation of diode-assisted and capacitor-assisted topologies as well as continuous current and discontinuous current topologies. Here three cases are simulated. In all three cases, the input voltage is kept constant at 240 V and a three-phase load of 9.7 Ω resistor bank is used. All DC side capacitors are 1000 μF and inductors are 3.5 mH. The AC side second-order filter is used with 10 μF capacitor and 7 mH inductor. In all three cases, the converter is operated with zero boosting in the beginning and at $t = 250$ ms the shoot through is increased to 0.25 while the modulation index is kept constant at 0.7. Figures 12.35 through 12.37 show the simulation results corresponding to topologies shown in Figures 12.30a, 12.31a, and 12.32b. From these figures, it is possible to note that in the first two cases equal boosting is achieved and the difference is the voltage across the V_{C3}. This complies with the theoretical finding. From Figure 12.37 it can be noted that with the capacitor-assisted topology a much higher boosting can be achieved with the same shoot-through value, and also voltage across all four capacitors are equal and comply with the equations derived in Section 12.9.2. A comprehensive set of simulation results will be presented in the further chapters of this book.

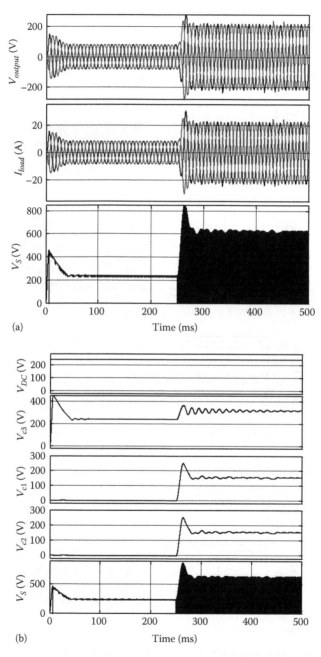

FIGURE 12.35 Simulation results for diode-assisted extended boost continuous current qZSI. (a) Waveforms of V_o, I_{load}, and V_s. (b) Waveforms of V_{dc}, V_{c3}, V_{o1}, V_{o2}, and V_s.

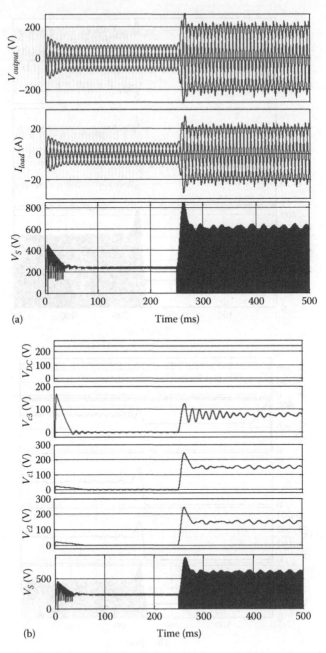

FIGURE 12.36 Simulation results for diode-assisted extended boost discontinues current qZSI. (a) Waveforms of V_o, I_{load}, and V_s. (b) Waveforms of V_{dc}, V_{c3}, V_{o1}, V_{o2}, and V_s.

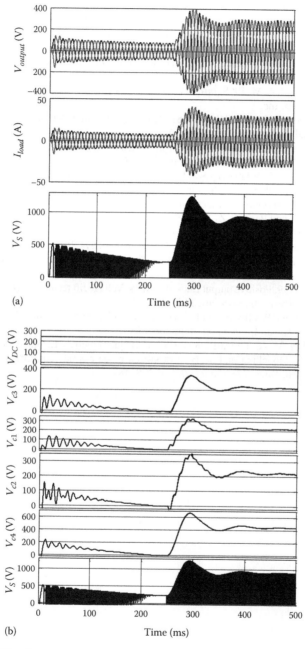

FIGURE 12.37 Simulation results for capacitor-assisted extended boost discontinuous current qZSI. (a) Waveforms of V_o, I_{load}, and V_s. (b) Waveforms of V_{dc}, V_{c3}, V_{o1}, V_{o2}, and V_s.

REFERENCES

1. Mohan, N., Undeland, T. M., and Robbins, W. P. 2003. *Power Electronics: Converters, Applications and Design* (3rd edn.). Hoboken, NJ: John Wiley & Sons, Inc.
2. Holtz, J. 1992. Pulsewidth modulation—A survey. *IEEE-Transactions on Industrial Electronics*, 39, 410–420.
3. Peng, F. Z. 2003. Z-source inverter. *IEEE-Transactions on Industry Applications*, 39, 504–510.
4. Trzynadlowski, A. M. 1998. *Introduction to Modern Power Electronics*. New York: John Wiley & Sons, Inc.
5. Middlebrook, R. D. and Cúk, S. 1981. *Advances in Switched-Mode Power Conversion* (Vols. I and II). Pasadena, CA: TESLAco.
6. Gajanayake, C. J. and Luo, F. L. 2009. Extended boost Z-source inverters. *Proceedings of IEEE ECCE'2009*, San Jose, CA, pp. 368–373.
7. Gajanayake, C. J., Vilathgamuwa, D. M., and Loh, P. C. 2007. Development of a comprehensive model and a multiloop controller for Z-source inverter DG systems. *IEEE Transactions on Industrial Electronics*, 54, 2352–2359.
8. Anderson, J. and Peng, F. Z. 2008. Four quasi-Z-Source inverters. *Proceedings of IEEE PESC'2008*, Rhodes, Greece, pp. 2743–2749.
9. Luo, F. L. and Ye, H. 2005. *Advanced DC/DC Converters*. Boca Raton, FL: CRC Press.
10. Luo, F. L. and Ye, H. 2005. *Essential DC/DC Converters*. Boca Raton, FL: Taylor & Francis Group LLC.
11. Luo, F. L. 1999. Positive output Luo-converters: Voltage lift technique. *IEE-Proceedings on Electric Power Applications*, 146, 415–432.
12. Luo, F. L. 1999. Positive output Luo-converters: Voltage lift technique. *IEE-Proceedings on Electric Power Applications*, 146, 208–224.
13. Ortiz-Lopez, M. G., Leyva-Ramos, J. E., Carbajal-Gutierrez, E., and Morales-Saldana, J. A. 2008. Modelling and analysis of switch-mode cascade converters with a single active switch. *Power Electronics, IET*, 1, 478–487.

13 Multilevel and Soft-Switching DC/AC Inverters

Multilevel inverters are a different method to construct DC/AC inverter. This idea was published by Nabae in 1980 in an IEEE international conference *IEEE APEC'80* [1], and the same idea was again published in 1981 in *IEEE Transactions on Industry Applications* [2]. Actually, multilevel inverters are a different technique from PWM method, which requires vertically chopping a reference waveform to achieve the similar output waveform (e.g., sine-wave). Multilevel inverting technique involves horizontally accumulating the levels to achieve the waveform (e.g., sine-wave).

Soft-switching technique was implemented in DC/DC conversion for more than 20 years. We also would like to introduce soft-switching technique in DC/AC inverters in this chapter.

13.1 INTRODUCTION

Although the PWM inverters have been used in industrial applications, they have many drawbacks:

1. The carrier frequency must be very high. Mohan nominated $m_f > 21$, which means $f_\Delta > 1\,\text{kHz}$ if the output waveform with the frequency is 50 Hz. Usually, in order to keep the THD is small, f_Δ is selected in $2 \sim 20\,\text{kHz}$ [3].
2. The pulse height is very high. In a normal PWM waveform (not multistage PWM) all pulse height is the DC linkage voltage. Output voltage of this PWM inverter has large jumping span. For example, if the DC linkage voltage is 400 V, all pulses have the peak value to be 400 V. Usually it causes large *dv/dt* and strong electromagnetic interference (EMI).
3. The pulse width would be very narrow when the output voltage has low value. For example, if the DC linkage voltage is 400 V, the output is 10 V, the corresponding pulse width should be 2.5% of the pulse full period.
4. Items 2 and 3 cause plenty of harmonics to produce poor THD.
5. Items 2 and 3 offer very rigorous switching condition. The switching devices have large switching power losses.
6. Inverter control circuitry is complex and the devices are costly. Therefore, the whole inverter is costly.

Multilevel inverter accumulates the output voltage in horizontal levels (layers). Therefore, using this technique overcomes the said drawbacks of the PWM technique:

1. The switching frequencies of most switching devices are low, which are same or only few times of the output signal frequency.
2. The pulse heights are quite low. For an m-level inverter with output amplitude V_m, the pulse heights are V_m/m or only few times of it. Usually it causes low dv/dt and ignorable EMI.
3. The pulse widths of all pulses have reasonable values to be comparable with the output signal.
4. Items 2 and 3 cannot cause plenty of harmonics to produce lower THD.
5. Items 2 and 3 offer smooth switching condition. The switching devices have small switching power losses.
6. Inverter control circuitry is comparably simple, and the devices are not costly. Therefore, the whole inverter is inexpensive.

Multilevel inverters contain several power switches and capacitors [4]. Output voltages of multilevel inverters are the addition of the voltages due to the commutation of the switches. Figure 13.1 shows a schematic diagram of one-phase leg of inverters with different level numbers. A two-level inverter, as shown in Figure 13.1a, generates an output voltage with two levels with respect to the negative terminal of the capacitor. The three-level inverter shown in Figure 13.1b generates three-level voltage, and an m-level inverter shown in Figure 13.1c generates m-level voltage. Thus, the output voltages of multilevel inverters have several levels. Moreover, they can also reach high voltage, and the power semiconductors must withstand only reduced voltages.

Multilevel inverters have gained increasing attention in recent decades, since multilevel inverters have many attractive features as described earlier. Various kinds of multilevel inverters have been proposed, tested, and installed.

- Diode-clamped (neutral-clamped) multilevel inverters
- Capacitor-clamped (flying capacitors) multilevel inverters
- Cascaded multilevel inverters with separate DC sources
- Hybrid multilevel inverters

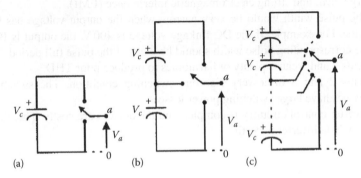

FIGURE 13.1 One-phase leg of an inverter: (a) two levels, (b) three levels, and (c) m levels.

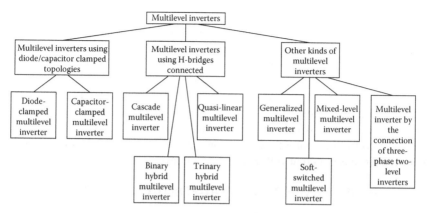

FIGURE 13.2 Family tree of multilevel inverters.

- Generalized multilevel inverters (GMI)
- Mixed-level multilevel inverters
- Multilevel inverters by the connection of three-phase two-level inverters
- Soft-switched multilevel inverters

The family tree of multilevel level inverters is shown in Figure 13.2.

The family of multilevel inverters has emerged as a solution for high-power applications, since it is hard to be implemented via single power semiconductor switch directly in medium-voltage network. Multilevel inverters have been applied to different high-power applications, such as large motor drives, railway traction applications, high-voltage DC transmissions (HVDC), unified power flow controllers (UPFC), static var compensators (SVC), and static synchronous compensators (STATCOM). The output voltage of the multilevel inverter has many levels synthesized from several DC voltage sources. The quality of the output voltage is improved as the number of voltage levels increases, so the effort of output filters can be decreased. The transformers can be eliminated due to reduced voltage that the switch endures. Moreover, as cost-effective solutions, the applications of multilevel inverters are also extended to medium- and low-power applications, such as electrical vehicle propulsion systems, active power filters (APF), voltage sag compensations, photovoltaic systems, and distributed power systems.

Multilevel inverter circuits have been studied for nearly 30 years. They separate DC-sourced power cells in series to synthesize a staircase AC output voltage. The diode-clamped inverter, also called the neutral-point-clamped (NPC) inverter, was presented in 1980 by Nabae. Because the NPC inverter effectively doubles the device voltage level without requiring precise voltage matching, the circuit topology prevailed in 1980s. The capacitor-clamped multilevel inverter came in the 1990s. Although the cascaded multilevel inverter was invented earlier, its application did not gain prominence until the mid-1990s. The advantages of cascaded multilevel inverters were indicated for motor drives and utility applications. The cascaded inverter has drawn great interest due to the great demand of medium-voltage high-power inverters.

The cascaded inverter is also used in regenerative-type motor drive application. Recently, some new topologies of multilevel inverters have emerged, such as GMI, mixed multilevel inverters, hybrid multilevel inverters, and soft-switched multilevel inverters. Today, multilevel inverters are extensively used in high-power applications with medium-voltage levels, such as laminators, mills, conveyors, pumps, fans, blowers, compressors, and so on. Moreover, as a cost-effective solution, the applications of multilevel inverters are also extended to low-power applications, such as photovoltaic systems, hybrid electrical vehicles and voltage sag compensation, in which the effort of output filter components can be decreased much due to low harmonic distortions of output voltages of the multilevel inverters.

13.2 DIODE-CLAMPED (NEUTRAL-POINT-CLAMPED) MULTILEVEL INVERTERS

Diode-clamped multilevel inverters are also called neutral-clamped multilevel inverters. In this category, the switching devices are connected in series to make up the desired voltage rating and output levels. The inner voltage points are clamped by either two extra diodes or one high-frequency capacitor. The switching devices of an m-level inverter are required to block a voltage level of $V_{dc}/(m-1)$. The clamping diode needs to have different voltage rating for different inner voltage levels. To sum up, an m-level diode-clamped inverter has

- Number of power electronic switch $= 2(m-1)$
- Number of DC link capacitor $= (m-1)$
- Number of clamped diode $= 2(m-2)$
- The voltage across each DC link capacitor $= V_{dc}/(m-1)$

where V_{dc} is the DC link voltage. A three-level diode-clamped inverter is shown in Figure 13.3a with $V_{dc} = 2E$. In this circuit, the DC-bus voltage is split into three levels

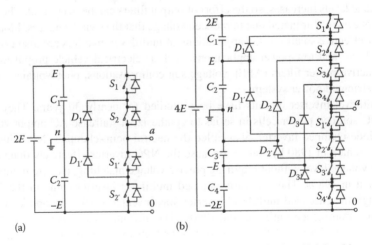

(a)   (b)

FIGURE 13.3 Diode-clamped multilevel inverter circuit topologies. (a) 3-level inverter and (b) 5-level inverter.

by two series-connected bulk capacitors, C_1 and C_2. The middle point of the two capacitors, n, can be defined as the neutral point. The output voltage v_{an} has three states: E, 0, and $-E$. For voltage level E, switches S_1 and S_2 need to be turned on; for $-E$, switches $S_{1'}$ and $S_{2'}$ need to be turned on; and for the 0 level, S_2 and $S_{2'}$ need to be turned on.

The key components that distinguish this circuit from a conventional two-level inverter are D_1 and $D_{1'}$. These two diodes clamp the switch voltage to half the level of the DC-bus voltage. When both S_1 and S_2 turn on, the voltage across a and 0 is $2E$, that is, $v_{a0} = 2E$. In this case, $D_{1'}$ balances out the voltage sharing between $S_{1'}$ and $S_{2'}$ with $S_{1'}$ blocking the voltage across C_1 and $S_{2'}$ blocking the voltage across C_2. Notice that output voltage v_{an} is AC, and v_{a0} is DC. The difference between v_{an} and v_{a0} is voltage across C_2, which is E. If the output is removed out between a and 0, then the circuit becomes a DC/DC converter, which has three output voltage levels: E, 0, and $-E$. The simulation waveform is shown in Figure 13.4.

Usually, when the number of levels increases, the corresponding THD of the output voltage is lower. The switching angle decides the THD of the output voltage as well. The three-level diode-clamped inverter has the THD as shown in Table 13.1.

Figure 13.3b shows a five-level diode-clamped converter in which the DC bus consists of four capacitors, C_1, C_2, C_3, and C_4. For DC-bus voltage $4E$, the voltage across each capacitor is E, and each device voltage stress will be limited to one capacitor voltage level E through clamping diodes.

To explain how the staircase voltage is synthesized, the neutral point n is considered as the output-phase voltage reference point. There are five switch combination to synthesize five levels of voltages across a and n.

- For voltage level $v_{an} = 2E$, turn on all upper switches S_1–S_4.
- For voltage level $v_{an} = E$, turn on three upper switches S_2–S_4 and one lower switch $S_{1'}$.
- For voltage level $v_{an} = 0$, turn on two upper switches S_3 and S_4 and two lower switches $S_{1'}$ and $S_{2'}$.

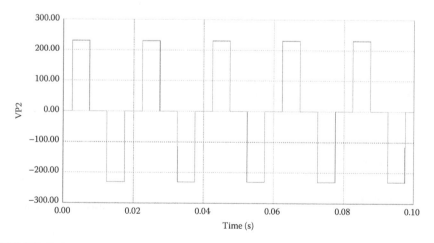

FIGURE 13.4 Output waveform of a three-level inverter.

TABLE 13.1
THD Content for Different Switching Angle

Switching Angle	THD (%)
15	31.76
30	30.9

- For voltage level $v_{an}=-E$, turn on one upper switch S_4 and three lower switches S_1–S_3.
- For voltage level $v_{an}=-2E$, turn on all lower switches S_1–S_4.

For diode-clamped inverter, each output level has only one combination to implement its output voltage. Four complementary switch pairs exist in each phase. The complementary switch pair is defined such that turning on one of the switches will exclude the other from being turned on. In this example, the four complementary pairs are $(S_1, S_{1'})$, $(S_2, S_{2'})$, $(S_3, S_{3'})$, and $(S_4, S_{4'})$. Although each active switching device is only required to block a voltage level of E, the clamping diodes must have different voltage ratings for reverse voltage blocking. Using $D_{1'}$ of Figure 13.3b as an example, when lower devices S_2–$S_{4'}$ are turned on, $D_{1'}$ needs to block three capacitor voltages, or $3E$. Similarly, D_2 and $D_{2'}$ need to block $2E$, and D_1 needs to block $3E$.

The simulation waveform is shown in Figure 13.5.

A seven-level diode-clamped inverter has the waveform as shown in Figure 13.6.

From Figures 13.4 through 13.6, the THD is reduced when the number of level of inverter is increased. Hence, a higher level of inverter will be considered to produce

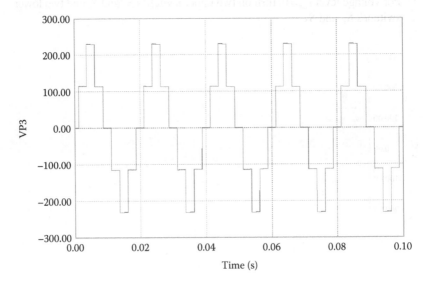

FIGURE 13.5 Output waveform of a five-level inverter.

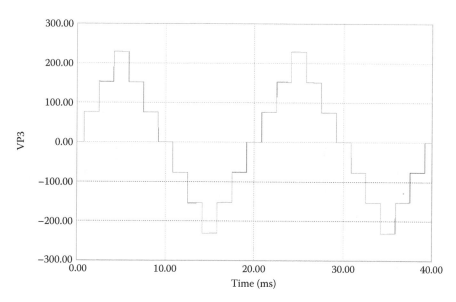

FIGURE 13.6 Output waveform of a seven-level inverter.

the output with less harmonic content. For each inverter, carefully setting the firing angles can obtain best THD. Table 13.2 shows various inverters' best firing angle to produce lowest THD.

By applying MATLAB® graph fitting tool, the relationship between lowest THD and the number of level of inverter can be estimated as the following equation:

$$\text{THD}_{Lowest} = 72.42e^{-0.4503m} + 11.86e^{-0.05273m} \tag{13.1}$$

where the m is the level number of the inverter. The corresponding figure of the THD versus m is shown in Figure 13.7.

Example 13.1 A diode-clamped three-level inverter shown in Figure 13.3a operates in the state with best THD. Determine the corresponding switching angles and switch status and THD.

Solution: Refer to Table 13.2; we have the best switching angles in a cycle to be

$\alpha_1 = 0.2332 \text{ rad} = 13.36°$
$\alpha_2 = \pi - \alpha_1 = 180° - 13.36° = 166.64°$
$\alpha_3 = \pi + \alpha_1 = 180° + 13.36° = 193.36°$
$\alpha_4 = 2\pi - \alpha_1 = 360° - 13.36° = 346.64°$

The switches referring to Figure 13.3a operate in a cycle (0°–360°) as follows:

Turn on the upper switches S_2 and the lower switches $S_{1'}$ in 0°–α_1.
Turn on all upper switches S_1 and S_2 in α_1–α_2.
Turn on the upper switches S_2 and the lower switches $S_{1'}$ in α_2–α_3.

TABLE 13.2
Best Switching Angle to Obtain Best THD

Number of Level (m)	α_1	α_2	α_3	α_4	α_5	α_6	α_7	α_8	α_9	α_{10}	α_{11}	α_{12}	α_{13}	α_{14}	α_{15}	THD (%)
3	0.2332	—	—	—	—	—	—	—	—	—	—	—	—	—	—	28.96
5	0.2242	0.7301	—	—	—	—	—	—	—	—	—	—	—	—	—	16.42
7	0.155	0.4817	0.8821	—	—	—	—	—	—	—	—	—	—	—	—	11.53
9	0.1185	0.3625	0.6323	0.9744	—	—	—	—	—	—	—	—	—	—	—	8.90
11	0.0958	0.2912	0.4989	0.7341	1.3078	—	—	—	—	—	—	—	—	—	—	7.26
13	0.0804	0.2436	0.4136	0.5976	0.8088	1.0848	—	—	—	—	—	—	—	—	—	6.13
15	0.0693	0.2094	0.3538	0.5064	0.6733	0.8666	1.1214	—	—	—	—	—	—	—	—	5.31
17	0.0609	0.1836	0.3093	0.4402	0.5798	0.7337	0.913	1.1509	—	—	—	—	—	—	—	4.68
19	0.0544	0.1635	0.275	0.3897	0.5105	0.6400	0.7834	0.9513	1.1754	—	—	—	—	—	—	4.19
21	0.049	0.1475	0.2474	0.3500	0.4565	0.569	0.6902	0.8252	0.9839	1.1961	—	—	—	—	—	3.79
23	0.0466	0.1342	0.225	0.3176	0.4132	0.513	0.6187	0.7331	0.8609	1.0116	1.2137	—	—	—	—	3.46
25	0.0412	0.1233	0.2063	0.2909	0.3777	0.4675	0.5616	0.6619	0.7705	0.8921	1.0359	1.2294	—	—	—	3.18
27	0.0379	0.1138	0.1905	0.2683	0.3478	0.4297	0.5147	0.6042	0.6995	0.8032	0.9195	1.0573	1.243	—	—	2.95
29	0.0353	0.1058	0.1769	0.2491	0.3224	0.3977	0.4754	0.5563	0.6416	0.7328	0.8320	0.9437	1.0761	1.2551	—	2.74
31	0.0329	0.0988	0.1652	0.2324	0.3005	0.3703	0.4419	0.516	0.5934	0.6751	0.7625	0.8580	0.9655	1.0933	1.266	2.57

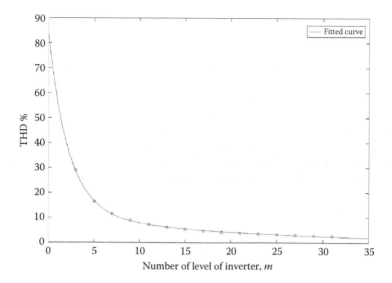

FIGURE 13.7   THD versus m.

Turn on all lower switches $S_{1'}$–$S_{2'}$ in α_3–α_4.

Turn on the upper switches S_2 and the lower switches $S_{1'}$ in α_4–360°.

The best THD = 28.96%.

13.3   CAPACITOR-CLAMPED (FLYING CAPACITOR) MULTILEVEL INVERTERS

Figure 13.8 illustrates the fundamental building block of a phase-leg capacitor-clamped inverter. The circuit has been called the flying capacitor inverter with

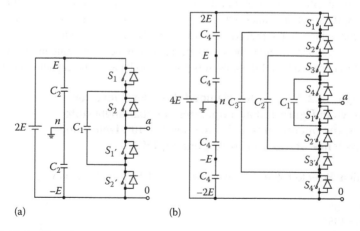

FIGURE 13.8   Capacitor-clamped multilevel inverter circuit topologies. (a) 3-level inverter and (b) 5-level inverter.

dependent capacitors clamping the device voltage to one capacitor voltage level. The inverter in Figure 13.8a provides a three-level output across a and n, that is, $v_{an}=E$, 0, or $-E$. For the voltage level E, switches S_1 and S_2 need to be turned on; for $-E$, switches $S_{1'}$ and $S_{2'}$ need to be turned on; and for the 0 level, either pair (S_1, $S_{1'}$ or S_2, $S_{2'}$) needs to be turned on. Clamping capacitor C_1 is charged when S_1 and $S_{1'}$ are turned on and is discharged when S_2 and $S_{2'}$ are turned on. The charge of C_1 can be balanced by proper selection of the zero-level switch combination.

The voltage synthesis in a five-level capacitor-clamped converter has more flexibility than a diode-clamped converter. Using Figure 13.8b as the example, the voltage of the five-level phase-leg a output with respect to the neutral point n, v_{an}, can be synthesized by the following switching combinations.

1. For voltage level $v_{an}=2E$, turn on all upper switches S_1–S_4
2. For voltage level $v_{an}=E$, there are three combinations:
   a. $S_1, S_2, S_3, S_{1'}$: $v_{an}=2E$ (upper C_4) – E (C_1)
   b. $S_2, S_3, S_4, S_{4'}$: $v_{an}=3E$ (C_3) – $2E$ (lower C_4)
   c. $S_1, S_3, S_4, S_{3'}$: $v_{an}=2E$ (upper C_4) – $3E$ (C_3) + $2E$ (C_2)
3. For voltage level $v_{an}=0$, there are six combinations:
   a. $S_1, S_2, S_{1'}, S_{4'}$: $v_{an}=2E$ (upper C_4) – $2E$ (C_2)
   b. $S_3, S_4, S_{3'}, S_{4'}$: $v_{an}=2E$ (C_2) – $2E$ (lower C_4)
   c. $S_1, S_3, S_{1'}, S_{3'}$: $v_{an}=2E$ (upper C_4) – $3E$ (C_3) + $2E$ (C_2) – E (C_1)
   d. $S_1, S_4, S_{2'}, S_{3'}$: $v_{an}=2E$ (upper C_4) – $3E$ (C_3) + E (C_1)
   e. $S_2, S_4, S_{2'}, S_{4'}$: $v_{an}=3E$ (C_3) – $2E$ (C_2) + E (C_1) – $2E$ (lower C_4)
   f. $S_2, S_3, S_{1'}, S_{4'}$: $v_{an}=3E$ (C_3) – E (C_1) – $2E$ (lower C_4)
4. For voltage level $V_{an}=-E$, there are three combinations:
   a. $S_1, S_{1'}, S_{2'}, S_{3'}$: $v_{an}=2E$ (upper C_4) – $3E$ (C_3)
   b. $S_4, S_{2'}, S_{3'}, S_{4'}$: $v_{an}=E$ (C_1) – $2E$ (lower C_4)
   c. $S_3, S_{1'}, S_{3'}, S_{4'}$: $v_{an}=2E$ (C_2) – E (C_1) – $2E$ (lower C_4)
5. For voltage level $v_{an}=-2E$, turn on all lower switches, $S_{1'}$–$S_{4'}$

Usually the positive top level and negative top level have only one combination to implement their output values. Other levels have various combinations to implement their output values. In the preceding description, the capacitors with positive signs are in discharging mode, whereas those with negative sign are in charging mode. By proper selection of capacitor combinations, it is possible to balance the capacitor charge.

Example 13.2 A capacitor-clamped three-level inverter is shown in Figure 13.8a. It operates in equal-angle state; that is, the operation time in each level is 90°. Determine the switches' status and corresponding THD.

Solution: Refer to Figure 1.10; we have the switching angles in a cycle to be

$\alpha_1 = 45°$
$\alpha_2 = 135°$
$\alpha_3 = 225°$
$\alpha_4 = 315°$

The switches referring to Figure 13.8a operate in a cycle (0°–360°) as follows:

Turn on the upper switches S_2 and the lower switches $S_{2'}$ in 0°–α_1.
(Or turn on the upper switches S_1 and the lower switches $S_{1'}$ in 0°–α_1).
Turn on all upper switches S_1 and S_2 in $\alpha_1 - \alpha_2$.
Turn on the upper switches S_2 and the lower switches $S_{2'}$ in α_2–α_3.
(Or turn on the upper switches S_1 and the lower switches $S_{1'}$ in α_2–α_3).
Turn on all lower switches $S_{1'}$–$S_{2'}$ in α_3–α_4.
Turn on the upper switches S_2 and the lower switches $S_{2'}$ in α_4–360°.
(Or turn on the upper switches S_1 and the lower switches $S_{1'}$ in α_4–360°.)

Refer to Example 4.6, the fundamental harmonic has the amplitude $4/\pi \sin x/2$, where $x = 90°$ in this example. Therefore, $4/\pi \sin x/2 = 0.9$. If we consider the higher-order harmonics until seventh order, that is, $n = 3, 5, 7$. The HFs are

$$\text{HF}_3 = \frac{\sin 3x/2}{3 \sin x/2} = \frac{1}{3}; \quad \text{HF}_5 = \frac{\sin 5x/2}{5 \sin x/2} = -\frac{1}{5}; \quad \text{HF}_7 = \frac{\sin 7x/2}{7 \sin x/2} = -\frac{1}{7}$$

The values of the HFs should be absolute values:

$$\text{THD} = \frac{\sqrt{\sum_{n=2}^{\infty} V_n^2}}{V_1} = \sqrt{\left(\frac{1}{3}\right)^2 + \left(\frac{1}{5}\right)^2 + \left(\frac{1}{7}\right)^2} = 0.41415$$

13.4 MULTILEVEL INVERTERS USING H-BRIDGES CONVERTERS

The basic structure is based on the connection of H-bridges (HBs). Figure 13.9 shows the power circuit for one-phase leg of a multilevel inverter with three HBs

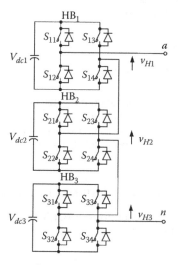

FIGURE 13.9 Multilevel inverter based on the connection of HBs.

(HB_1, HB_2, and HB_3) in each phase. Each HB is supplied by a separate DC source. The resulting phase voltage is synthesized by the addition of the voltages generated by the different HBs. If the DC link voltages of HBs are identical, the multilevel inverter is called the cascaded multilevel inverter. However, it is possible to have different values among the DC link voltages of HBs, and the circuit can be called the hybrid multilevel inverter.

Example 13.3 A three-HB multilevel inverter is shown in Figure 13.9. The output voltage is v_{an}. It is implemented as a binary Hybrid Multilevel Inverter (BHMI). Explain the inverter working operation and draw the corresponding waveforms and indicate the source voltages arrangement and how many levels can be implemented.

Solution: The DC link voltages of HB_i (the ith HB), V_{dci}, is $2^{i-1}E$. In a three-HB one-phase leg,

$$V_{dc1} = E, \quad V_{dc2} = 2E, \quad V_{dc3} = 4E$$

The operation is listed as follows:

+0: $v_{H1}=0$, $v_{H2}=0$, $v_{H3}=0$
+1E: $v_{H1}=E$, $v_{H2}=0$, $v_{H3}=0$
+2E: $v_{H1}=0$, $v_{H2}=2E$, $v_{H3}=0$
+3E: $v_{H1}=E$, $v_{H2}=2E$, $v_{H3}=0$
+4E: $v_{H1}=0$, $v_{H2}=0$, $v_{H3}=4E$
+5E: $v_{H1}=E$, $v_{H2}=0$, $v_{H3}=4E$
+6E: $v_{H1}=0$, $v_{H2}=2E$, $v_{H3}=4E$
+7E: $v_{H1}=E$, $v_{H2}=2E$, $v_{H3}=4E$
−E: $v_{H1}=-E$, $v_{H2}=0$, $v_{H3}=0$
−2E: $v_{H1}=0$, $v_{H2}=-2E$, $v_{H3}=0$
−3E: $v_{H1}=-E$, $v_{H2}=-2E$, $v_{H3}=0$
−4E: $v_{H1}=0$, $v_{H2}=0$, $v_{H3}=-4E$
−5E: $v_{H1}=-E$, $v_{H2}=0$, $v_{H3}=-4E$
−6E: $v_{H1}=0$, $v_{H2}=-2E$, $v_{H3}=-4E$
−7E: $v_{H1}=-E$, $v_{H2}=-2E$, $v_{H3}=-4E$

As shown in the foregoing figure, the output waveform, v_{an}, has 15 levels. One of the advantages is the HB with higher DC link voltage has lower number of commutations, thereby reducing the associated switching losses. The higher switching frequency components, for example, IGBT, are used to construct the HB with lower DC link voltage.

13.4.1 Cascaded Equalvoltage Multilevel Inverters

In cascaded equalvoltage multilevel inverter (CEMI), the DC link voltages of HBs are identical as shown in Figure 13.9.

$$V_{dc1} = V_{dc2} = V_{dc3} = E \tag{13.2}$$

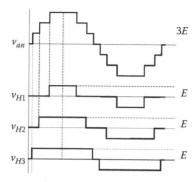

FIGURE 13.10 Waveforms of cascaded multilevel inverter.

where E is unit voltage. Each HB generates three voltages at the output: $+E$, 0, and $-E$. This is made possible by connecting the capacitors sequentially to the AC side via the three power switches. The resulting output AC voltage swings from $-3E$ to $3E$ with seven levels as shown in Figure 13.10.

13.4.2 BINARY HYBRID MULTILEVEL INVERTER

In BHMI, the DC link voltages of HB_i (the ith HB), V_{dci}, is $2i{-}1E$. In a three-HB one-phase leg,

$$V_{dc1} = E, \quad V_{dc2} = 2E, \quad V_{dc3} = 4E \tag{13.3}$$

As shown in Figure 13.11, the output waveform, v_{an}, has 15 levels. One of the advantages is that the HB with higher DC link voltage has lower number of commutations, thereby reducing the associated switching losses. The BHMI illustrates a seven-level (in half-cycle) inverter using this hybrid topology. The HB with higher DC link voltage consists of lower-switching-frequency component, for example, IGBT. The higher-switching-frequency components, for example, IGBT, are used to construct the HB with lower DC link voltage.

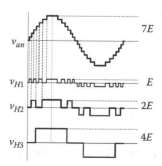

FIGURE 13.11 Waveforms of BHMI.

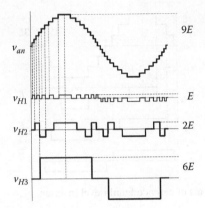

FIGURE 13.12 Waveforms of quasi-linear multilevel inverter.

13.4.3 QUASI-LINEAR MULTILEVEL INVERTER

In quasi-linear multilevel inverter (QLMI), the DC link voltages of HB_i, V_{dci}, can be expressed as

$$V_{dci} = \begin{cases} E & i=1 \\ 2 \times 3^{i-2} E & i \geq 2 \end{cases} \tag{13.4}$$

In a three-HB one-phase leg,

$$V_{dc1} = E, \quad V_{dc2} = 2E, \quad V_{dc3} = 6E \tag{13.5}$$

As shown in Figure 13.12, the output waveform, v_{an}, has 19 levels.

13.4.4 TRINARY HYBRID MULTILEVEL INVERTER

In trinary hybrid multilevel inverter (THMI), the DC link voltages of HB_i, V_{dci}, is $3^{i-1}E$. In a three-HB one-phase leg,

$$V_{dc1} = E, \quad V_{dc2} = 3E, \quad V_{dc3} = 9E \tag{13.6}$$

As shown in Figure 13.13, the output waveform, v_{an}, has 27 levels. To the best of author's knowledge, this circuit has the greatest level number for a given number of HBs among existing multilevel inverters.

13.5 OTHER KINDS OF MULTILEVEL INVERTERS

There are several other kinds of multilevel inverters introduced in this section [5].

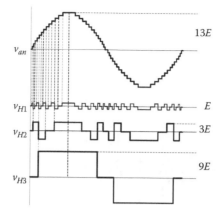

FIGURE 13.13 Waveforms of a 27-level THMI.

13.5.1 GENERALIZED MULTILEVEL INVERTERS

A GMI topology has previously been presented. The existing multilevel inverters, such as diode-clamped and capacitor-clamped multilevel inverters, can be derived from this GMI topology. Moreover, the GMI topology can balance each voltage level by itself regardless of load characteristics. Therefore, the GMI topology provides a true multilevel structure that can balance each DC voltage level automatically at any number of levels, regardless of active or reactive power conversion, and without any assistance from other circuits. Thus, in principle, it provides a complete multilevel topology that embraces the existing multilevel inverters.

Figure 13.14 shows the GMI structure per phase leg. Each switching device, diode, or capacitor's voltage is E, that is, $1/(m-1)$ of the DC link voltage. Any inverter with any number of levels, including the conventional two-level inverter, can be obtained using this generalized topology.

As an application example, a four-level bidirectional DC/DC converter, shown in Figure 13.15, is suitable for the dual-voltage system to be adopted in future automobiles. The four-level DC/DC converter has a unique feature—no magnetic components are needed. From this GMI topology, several new multilevel inverter structures can be derived.

13.5.2 MIXED-LEVEL MULTILEVEL INVERTER TOPOLOGIES

For high-voltage high-power applications, it is possible to adopt multilevel diode-clamped or capacitor-clamped inverters to replace the full-bridge cell in a cascaded multilevel inverter. The reason for doing so is to reduce the amount of separate DC sources. The 9-level cascaded inverter requires 4 separate DC sources for 1-phase leg and 12 for a 3-phase inverter. If a three-level inverter replaces the full-bridge cell, the voltage level is effectively doubled for each cell. Thus, to achieve the same nine voltage levels for each phase, only two separate DC sources are needed for one-phase leg and six for a three-phase inverter. The configuration can be considered as having

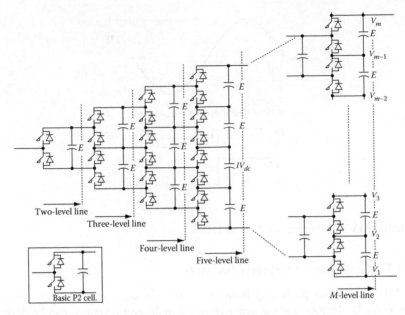

FIGURE 13.14   Generalized multilevel inverter structure.

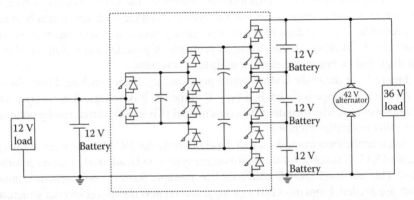

FIGURE 13.15   Application example: a four-level inverter for the dual-voltage system in automobiles.

mixed-level multilevel cells because it embeds multilevel cells as the building block of the cascaded multilevel inverter.

13.5.3 MULTILEVEL INVERTERS BY CONNECTION OF THREE-PHASE TWO-LEVEL INVERTERS

Standard three-phase two-level inverters are connected by transformers as shown in Figure 13.16. In order for the inverter output voltages to be added up, the inverter outputs of the three modules need to be synchronized with a separation of 120°

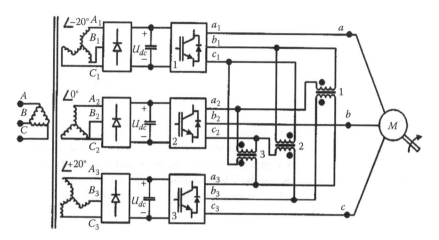

FIGURE 13.16 Cascaded inverter with three-phase cells.

between each phase. For example, obtaining a three-level voltage between outputs a and b, the voltage is synthesized by $V_{ab} = V_{a1-b1} + V_{a1-b1} + V_{a1-b1}$. The phase between b_1 and a_2 is provided by a_3 and b_3 through an isolated transformer. With three inverters synchronized, the voltages V_{a1-b1}, V_{a1-b1}, and V_{a1-b1} are all in phase; thus, the output level is simply tripled.

13.6 SOFT-SWITCHING MULTILEVEL INVERTERS

There are numerous ways of implementing soft-switching methods, such as zero-voltage switching (ZVS) and zero-current switching (ZCS), to reduce the switching losses and to increase efficiency for different multilevel inverters. For the cascaded multilevel inverter, because each inverter cell is a two-level circuit, the implementation of soft switching is not at all different from that of conventional two-level inverters. For capacitor- or diode-clamped inverters, however, the choices of soft-switching circuit can be found with different circuit combinations. Although ZCS is possible, most literatures proposed ZVS types, including auxiliary resonant commutated pole (ARCP), coupled inductor with zero-voltage transition (ZVT), and their combinations.

13.6.1 NOTCHED DC LINK INVERTERS FOR BRUSHLESS DC MOTOR DRIVE

Brushless DC motor (BDCM) has been widely used in industrial applications because of its low inertia, fast response, high power density, high reliability, and less maintenance. It exhibits the operating characteristics of a conventional commutated DC permanent magnet motor but eliminates the mechanical commutator and brushes. Hence, many problems associated with brushes are eliminated, such as radio-frequency interference, and sparking, which is the potential source of ignition in inflammable atmosphere. It is usually supplied by a hard-switching PWM inverter, which normally has low efficiency since the power losses across the switching devices are high. In order to reduce the losses, many soft-switching inverters have been designed [6].

Soft-switching operation of power inverter has attracted much attention in recent decade. In electric motor drive applications, soft-switching inverters are usually classified into three categories, namely, resonant pole inverters, resonant DC link inverters, and resonant AC link inverters [7]. Resonant pole inverter has the disadvantage of containing a considerably large number of additional components, in comparison to other hard- and soft-switching inverter topologies. Resonant AC link inverter is not suitable to BDCM drivers.

In medium-power applications, the resonant DC link concept [8] offered a first practical and reliable way to reduce commutation losses and to eliminate individual snubbers. Thus, it allows for high operating frequencies and improved efficiency. The inverter is quite simple to get the ZVS condition of the six main switches only by adding one auxiliary switch. However, the inverter has drawbacks such as high-voltage stress of the switches, high-voltage ripple of the DC link, and the frequency of the inverter related to the resonant frequency. Furthermore, the inductor power losses of the inverter are also considerable as the inductor flows current always. In order to overcome the drawback of high-voltage stress of the switches, actively clamped resonant DC link inverter was introduced [9–12]. The control scheme of the inverter is too complex, and the output contains subharmonic that, in some cases, cannot be accepted. These inverters still do not overcome the drawback of high inductor power losses.

In order to generate voltage notches of the DC link at controllable instants and reduce the power losses of the inductor, several quasi-parallel resonant schemes were proposed [13–15]. As a dwell time is generally required after every notch, severe interferences occur, mainly in multiphase inverters, substantially worsening the modulation quality. A novel DC-rail parallel resonant ZVT voltage source inverter [16] is introduced; it overcomes many drawbacks mentioned earlier. However, it requires two ZVT per PWM cycle, and it would worsen the output and limit the switch frequency of the inverter.

On the other hand, the majority of soft-switching inverters proposed in the recent years have been used in induction motor drive applications. So it is necessary to study the novel topology of soft-switching inverter and special control circuit for BDCM drive systems. This paper proposed a novel resonant DC link inverter for BDCM drive system that can generate voltage notches of the DC link at controllable instant and width. And the inverter possesses the advantage of low switching power loss, low inductor power loss, low-voltage ripple of the DC link, low device voltage stress, and simple control scheme.

The construction of the soft-switching inverter is shown in Figure 13.17. There is a front uncontrolled rectifier to obtain DC supply. The input AC supply can be single phase for low/medium power or three phases for medium/high power. It contains a resonant circuit and a conventional and control circuit. The resonant circuit contains three auxiliary switches (one IGBT and two fast-switching thyristors), a resonant inductor, and a resonant capacitor. All auxiliary switches work under ZVS or ZCS condition. It generates voltage notches of the DC link to guarantee the main switches S_1–S_6 of the inverter operating in ZVS condition. The fast-switching thyristor is the proper device to be used as auxiliary switch. We need not control the turning-off of the thyristor, and it has higher surge current capability than any other power semiconductor switches.

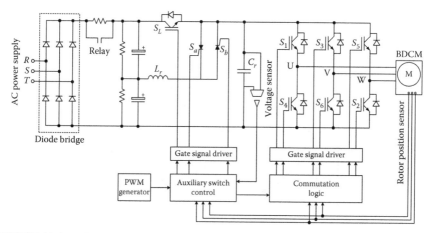

FIGURE 13.17 The construction of soft switching for BDCM drive system.

13.6.1.1 Resonant Circuit

The resonant circuit consists of three auxiliary switches, one resonant inductor, and one resonant capacitor. The auxiliary switches are controlled at certain instant to obtain the resonance between inductor and capacitor. Thus, the voltage of DC link reaches zero temporarily (voltage notch) and the main switches of the inverter get ZVS condition for commutation.

Since the resonant process is very short, the load current can be supposed constant. The equivalent circuit is shown in Figure 13.18. The corresponding waveforms of the auxiliary switches gate signal, resonant capacitor voltage (u_{Cr}), inductor current (i_{Lr}), and current of switch S_L (i_{SL}) are illustrated in Figure 13.19. The operation of the ZVT process can be divided into six modes.

Mode 0 (as shown in Figure 13.20a) $0 < t < t_0$: Its operation is the same as conventional inverter. Current flows from DC source through S_L to the load. The voltage across C_r (u_{Cr}) is equal to the voltage of the supply (V_s). The auxiliary switches S_a and S_b are in off state.

Mode 1 (as shown in Figure 13.20b) $t_0 < t < t_1$: When it is the instant for phase current commutation or PWM signal is flopped from "1" to "0," thyristor S_a is fired

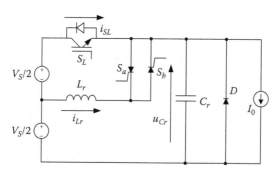

FIGURE 13.18 The equivalent circuit.

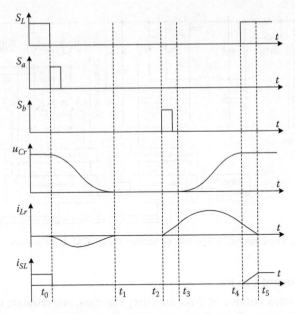

FIGURE 13.19 Some waveforms of the equivalent circuit.

(ZCS turned on due to L_r) and IGBT S_L is turned off (ZVS turned off due to C_r) at the same time. Capacitor C_r resonates with inductor L_r, and the voltage across capacitor C_r is decreased. Redefining the initial time, we have the equation

$$\begin{cases} u_{Cr}(t) + R_{Lr}i_{Lr}(t) + L_r\dfrac{di_{Lr}(t)}{dt} = \dfrac{V_S}{2} \\[4mm] I_0 - i_{Lr}(t) + C_r\dfrac{du_{cr}(t)}{dt} = 0 \end{cases} \tag{13.7}$$

where
  R_{Lr} is the resistance of inductor L_r
  I_0 is load current
  V_S is the DC power supply voltage, with initial condition $u_{cr}(0) = V_S, i_{Lr}(0) = 0$

and solving the Equation 13.7, we get

$$\begin{cases} u_{Cr}(t) = \left(\dfrac{V_S}{2} - R_{Lr}I_0\right) + \left(\dfrac{V_S}{2} - R_{Lr}I_0\right)e^{-t/\tau}\cos(\omega t) \\[4mm] \qquad + \dfrac{1}{L_r C_r \omega}e^{-t/\tau}\left(\dfrac{1}{4}R_{Lr}C_r V_S - L_r I_0 + \dfrac{1}{2}R_{Lr}^2 C_r I_0\right)\sin(\omega t) \\[4mm] i_{Lr}(t) = I_0 - I_0 e^{-t/\tau}\cos(\omega t) - \dfrac{V_S + R_{Lr}I_0}{2L_r\omega}e^{-t/\tau}\sin(\omega t) \end{cases} \tag{13.8}$$

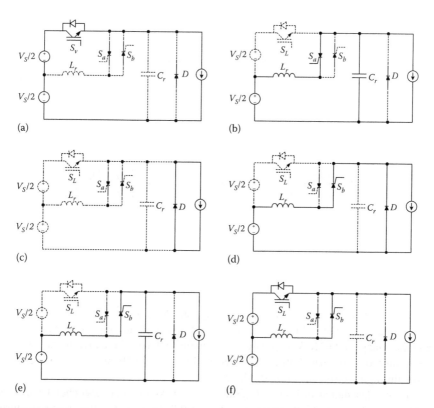

FIGURE 13.20 Operation mode of the ZVS process: (a) mode 0, (b) mode 1, (c) mode 2, (d) mode 3, (e) mode 4, and (f) mode 5.

where

$$\tau = \frac{2L_r}{R_{Lr}}, \quad \omega = \sqrt{\frac{1}{L_r C_r} - \frac{1}{\tau^2}}$$

As the resonant frequency is very high (several hundred kHz), $\omega L_r \gg R_{Lr}$, resonant inductor resistance R_{Lr} can be neglected. Then Equation 13.8 can be simplified as

$$
\begin{cases}
u_{Cr}(t) = \dfrac{V_S}{2} - I_0 \sqrt{\dfrac{L_r}{C_r}} \sin\left(\dfrac{1}{\sqrt{L_r C_r}} t\right) + \dfrac{V_S}{2}\cos\left(\dfrac{1}{\sqrt{L_r C_r}} t\right) \\[4mm]
i_{Lr}(t) = I_0 - I_0 \cos\left(\dfrac{1}{\sqrt{L_r C_r}} t\right) - \dfrac{V_S}{2}\sqrt{\dfrac{C_r}{L_r}} \sin\left(\dfrac{1}{\sqrt{L_r C_r}} t\right)
\end{cases}
\tag{13.9}
$$

that is,

$$\begin{cases} u_{Cr}(t) = \dfrac{V_S}{2} + K\cos(\omega_r t + \alpha) \\[3mm] i_{Lr}(t) = I_0 - K\sqrt{\dfrac{C_r}{L_r}}\sin(\omega_r t + \alpha) \end{cases}$$

(13.10)

where

$$K = \sqrt{\dfrac{V_S^2}{4} + \dfrac{I_0^2 L_r}{C_r}}, \quad \omega_r = \sqrt{\dfrac{1}{L_r C_r}}, \quad \alpha = tg^{-1}\left(\dfrac{2I_0}{V_S}\sqrt{\dfrac{L_r}{C_r}}\right)$$

Let $u_{Cr}(t)=0$, we get

$$\Delta T_1 = t_1 - t_0 = \dfrac{\pi - 2\alpha}{\omega_r}$$

(13.11)

$i_{Lr}(t_1)$ is zero at the same time. Then the thyristor S_a is self–turned off.

Mode 2 (as shown in Figure 13.20c) $t_1<t<t_2$: None of auxiliary switches is fired, and the voltage of DC link (u_{Cr}) is zero. The main switches of the inverter can now be either turned on or turned off under ZVS condition during the interval. Load current flows through the freewheeling diode D.

Mode 3 (as shown in Figure 13.20d) $t_2<t<t_3$: As the main switches have been turned on or turned off, thyristor S_b is fired (ZCS turn on due to L_r) and i_{Lr} starts to build up linearly in the auxiliary branch. The current in the freewheeling diode D begins to fall linearly. The load current is slowly diverted from the freewheeling diodes to the resonant branch. But u_{Cr} is still equal to zero. We have

$$\Delta T_2 = t_3 - t_2 = \dfrac{2I_0 L_r}{V_S}$$

(13.12)

At t_3, i_{Lr} equals the load current I_0 and the current through the diode becomes zero. Thus, the freewheeling diode turns off under zero-current condition.

Mode 4 (as shown in Figure 13.20e) $t_3<t<t_4$: i_{Lr} is increased continuously from I_0 and u_{Cr} is increased from zero when the freewheeling diode D is turned off. Redefining the initial time, we can get the same equation as Equation 13.7. The initial condition is $u_{Cr}(0) = 0$, $i_{Lr}(0) = I_0$; thus, by neglecting the inductor resistance and solving the equation, we get

$$\begin{cases} u_{Cr}(t) = \dfrac{V_S}{2} - \dfrac{V_S}{2}\cos\left(\dfrac{1}{\sqrt{L_r C_r}}t\right) \\[3mm] i_{Lr}(t) = I_0 + \dfrac{V_S}{2}\sqrt{\dfrac{C_r}{L_r}}\sin\left(\dfrac{1}{\sqrt{L_r C_r}}t\right) \end{cases}$$

(13.13)

that is,

$$
\begin{cases}
u_{Cr}(t) = \dfrac{V_S}{2}[1 - \cos(\omega_r t)] \\[4mm]
i_{Lr}(t) = I_0 + \dfrac{V_S}{2}\sqrt{\dfrac{C_r}{L_r}}\,\sin(\omega_r t)
\end{cases}
\tag{13.14}
$$

When

$$
\Delta T = t_4 - t_3 = \frac{\pi}{\omega_r}
\tag{13.15}
$$

$u_{Cr}=E$, IGBT S_L is fired (ZVS turned on), $i_{Lr}=I_0$ again. The peak inductor current can be derived from Equation 13.14, and then

$$
i_{Lr-m} = I_0 + \frac{V_S}{2}\sqrt{\frac{C_r}{L_r}}
\tag{13.16}
$$

Mode 5 (as shown in Figure 13.20f) $t_4 < t < t_5$: When the DC link voltage is equal to the supply voltage, auxiliary switch S_L is turned on (ZVS turned on due to C_r); i_{lr} is decreased linear from I_0 to zero at t_5 and the thyristor S_b is self turned off.

Then go back to mode 0 again. The operation principle of the other procedure is the same as conventional inverter.

13.6.1.2 Design Consideration

The design of the resonant circuit is to determine the resonant capacitor C_r, resonant inductor L_r, and the switching instants of auxiliary switches S_a, S_b, S_L. It is assumed that the inductance of BDCM is much higher than resonant inductance L_r. From the analysis presented previously, the design considerations can be summarized as follows:

Auxiliary switch S_L works under ZVS condition, and the voltage stress is DC power supply voltage V_S. The current flow through it is load current. Auxiliary switches S_a and S_b work under ZCS condition, the voltage stress is $V_S/2$, and the peak current flow through them is i_{Lr-m}. As the resonant auxiliary switches S_a and S_b carry the peak current only during switch transitions, they can be rated lower continuous current rating.

The resonant period is expressed as $T_r = 1/f_r = 2\pi\sqrt{L_r C_r}$; for high-switching-frequency inverter, T_r should be as short as possible. For getting the expected T_r, the resonant inductor and capacitor values have to be selected. The first component to be designed is the resonant inductor. Small inductance value can get small T_r, but the rise slop of the inductor current $di_{Lr}/dt = V_S/2L_r$ should be small to guarantee that freewheeling diode is turned off. For 600–1200 V power diode, the reverse recovery time is about 50–200 ns, and the rule to select an inductor is [11]

$$\frac{di_{Lr}}{dt} = \frac{V_S}{2L_r} = 75 - 150 \, A/\mu s \tag{13.17}$$

Certainly inductance is as high as possible. This implies that high inductance value is necessary. Thus, an optimum value of the inductance has to be chosen, which would reduce the inductor current rise slope, and T_r would be small enough.

The capacitance value is inversely proportional to the ascend or descend slope of the DC link voltage. It means the capacitance is as high as possible for switch S_L to get ZVS condition, but as the capacitance increases, there is more and more energy stored in it. This energy should be charged or discharged via resonant inductor; with high capacitance, the peak value of inductor current will be high. The peak value of i_{Lr} should be limited to twice that of the peak load current. From Equations 13.9 through 13.16, we obtain

$$\sqrt{\frac{C_r}{L_r}} \le \frac{2I_{0\max}}{V_S} \tag{13.18}$$

Thus, an optimum value of the capacitance has to be chosen that would limit the peak inductor current, whereas the ascend or descend slope of the DC link voltage is low enough.

13.6.1.3 Control Scheme

When the duty of PWM is 100%, that is, there is no PWM, the main switches of inverter work under the commutation frequency. When it is the instant to commutate the phase current of the BDCM, we control the auxiliary switches S_a, S_b, and S_L, and resonant occurs between L_r and C_r. The voltage of DC link reach zero temporarily; thus, ZVS condition of the main switches is obtained. When the duty of PWM is less than 100%, the auxiliary switch S_L works as chop. The main switches of the inverter do not switch within a PWM cycle when the phase current need not commutate. It has the benefit of reducing phase current drop when the PWM is turned off. The phase current is commutated when the DC link voltage becomes zero. So there is only one DC link voltage notch per PWM cycle. It is very important especially for very low or very high duty of PWM where the interval between two voltage notches is very short, even overlapped, which will limit the tuning range.

The commutation logical circuit of the system is shown in Figure 13.21. It is similar to conventional BDCM commutation logical circuit, except adding six D flip-flops to the output. Thus, the gate signal of the main switches is controlled by the synchronous pulse CK, which will be mentioned later, and the commutation can be synchronized with auxiliary switches control circuit. The operation of the inverter can be divided into PWM operation and non-PWM operation.

1. *Non-PWM operation*: When the duty of PWM is 100%, that is, there is no PWM, and the whole ZVT process (modes 1–5) occurs when the phase current commutation is ongoing. The control scheme for the auxiliary switches

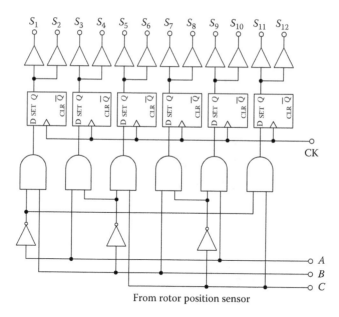

FIGURE 13.21 Commutation logical circuit for main switches.

in this operation is illustrated in Figure 13.22a. When mode 1 begins, pulse signal for thyristor S_a is generated by a monostable flip-flop and gate signal for IGBT S_L drops to low level (i.e., S_L is turned off) at the same time. Then, pulse signal for thyristor S_b and the synchronous pulse CK can be obtained after two short delays (delay 1 and delay 2, respectively). Obviously delay 1 is longer than delay 2. Pulse CK is generated during mode 2 when the voltage of DC link is zero and the main switches of the inverter get ZVS condition. Then modes 3–5 occur, and the voltage of DC link increases proportionate to supply again.

2. *PWM operation:* In this operation the auxiliary switch S_L works as chop, but the main switches of the inverter do not turn on or turn off within a single PWM cycle when the phase current need not commutate. The load current is commutated when the DC link voltage becomes zero; that is, when PWM signal is "0" (as the PWM cycle is very short, it does not affect the operation of the motor). The control scheme for the auxiliary switches in PWM operation is illustrated in Figure 13.22b.

   a. When PWM signal is flopped from "1" to "0," mode 1 begins, pulse signal for thyristor S_a is generated, and gate signal for IGBT S_L drops to low level. But the voltage of DC link does not increase until PWM signal is flipped from "0" to "1." Pulse CK is generated during mode 2.

   b. When PWM signal is flipped from "0" to "1," mode 3 begins, and pulse signal for thyristor S_b is generated at the moment (mode 3). Then when the voltage of the DC link is increased to E (voltage of supply), the gate signal for IGBT S_L is flipped to high level (modes 4 and 5).

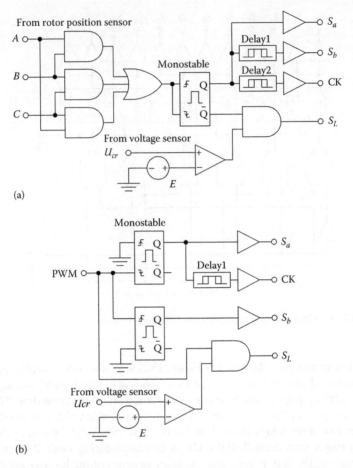

(a)

(b)

FIGURE 13.22 Control scheme for the auxiliary switches in (a) non-PWM operation and (b) PWM operation.

Thus, only one ZVT occurs per PWM cycle: modes 1 and 2 for PWM are turned off and modes 3–5 for PWM are turned on. And the switching frequency would be not greater than PWM frequency.

Normally, a drive system requires a speed or position feedback signal to get high speed or position precision and be less susceptible to disturbance of load and power supply. Speed feedback signal can be derived from a tachometer generator, optical encoder, resolver or derived from rotor position sensor. Quadrature Encoder Pulse (QEP) is a standard digital speed or position signal and can be inputted to many devices (e.g., special DSP for drive system TMS320C24x has QEP receive circuit). The QEP can be derived from rotor position sensor of a BDCM easily. The converter digital circuit and interesting waveforms are shown in Figure 13.23. Some single-chip computer has digital counter and may require only direction and pulse signal; thus, the converter circuit can be simpler. The circuit can be implemented by Complex Programmer Logical Device and only occupy partial resource of one chip.

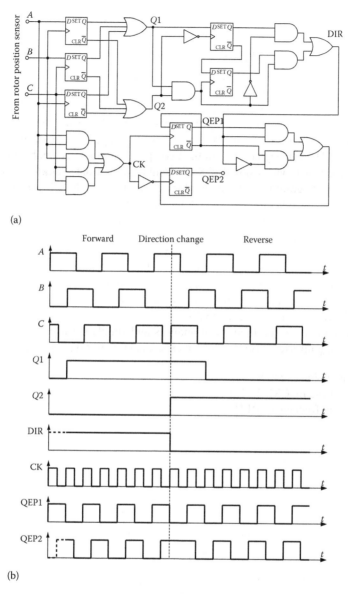

(a)

(b)

FIGURE 13.23 Circuit of derived QEP from Hall signal and waveforms. (a) Circuit diagram and (b) signals of the circuit.

The circuit can also be implemented by one Gate Array Logic (GAL) IC (e.g., 16V8) and some D flip-flop IC (e.g., 74LS74). The circuit high precision speed or position signal can be obtained when the motor speed is high or the drive system has high ratio speed reduction mechanism. In high-performance systems the rotor position sensor may be a resolver or optical encoder, with special-purpose decoding circuitry. At this level of control sophistication, it is possible to fine-tune the firing angles and the PWM control as a function of speed and load, to improve various aspects of performance such as efficiency, dynamic performance, or speed range.

13.6.1.4 Simulation and Experimental Results

The proposed topology is verified by Psim simulation software. The schematic circuit of the soft-switching inverter is shown in Figure 13.24. The left bottom of the figure is the auxiliary switches gate signal generator circuit (reference from Figure 13.22), which is made up of monostable flip-flop, delay, and logical gate. The gate signals of auxiliary switches S_a and S_b in PWM and non-PWM operation mode are combined by OR gate. The gate signal of S_L in the two operation modes is combined by AND gate and the synchronous signal (CK) is combined by a date selector. The middle bottom of the diagram is commutation logical circuit of the BDCM (reference from Figure 13.21); it is synchronized (by CK) with auxiliary switches control circuit.

Waveforms of DC link voltage u_{Cr}, resonant inductor current i_{Lr}, BDCM phase current, inverter output line–line voltage, and gate signal of the auxiliary switches are shown in Figure 13.25. The value of the resonant inductor L_r is $10\,\mu H$ and the resonant capacitor C_r is $0.047\,\mu F$, so the period of the resonant circuit is about $4\,\mu s$. The frequency of the PWM is $20\,kHz$. From the figure, we can see that the output of the simulation matches the theoretical analysis. The waveforms in Figure 13.25b through h are the same as Figure 13.26.

In order to verify the theoretical analysis and simulation results, the proposed soft-switching inverter was tested on an experimental prototype rated as

DC link voltage: 240 V
Power of the BDCM: 3.3 Hp
Switching frequency 10 kHz

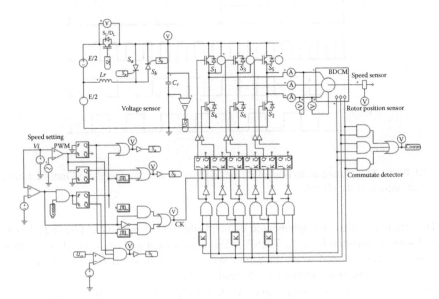

FIGURE 13.24 Schematic circuit of the drive system for PSim simulation.

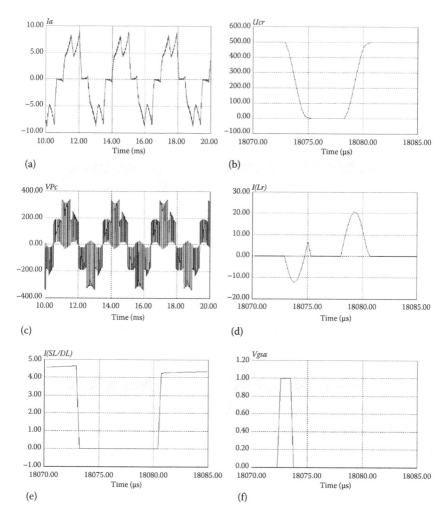

FIGURE 13.25 Simulation results: (a) current of phase a, (b) resonant capacitor voltage u_{Cr}, (c) voltage of phase a, (d) resonant inductor current (i_{Lr}), (e) current of S_L, (f) S_a gate signal, (g) S_L gate signal, and (h) S_b gate signal.

A polyester capacitor of 47 nF, 1500 V was adopted as DC link resonant capacitor C_r. The resonant inductor was of 6 μH/20A with ferrite core. The design of the auxiliary switches control circuit was referenced from Figure 13.24. The monostable flip-flop can be implemented via IC 74LS123, the delay can be implemented via Schmitt Trigger and RC circuit, and the logical gate can be replaced by programmable logical device to reduce the number of IC.

The waveforms of voltage across switch and current under hard switching and soft switching are shown in Figure 13.26a and b, respectively. All the voltage signals come from differential probe, and there is a gain of 20. For voltage waveform, 5.00 V/div = 100 V/div, which is the same for Figure 13.27. It can also be seen that there

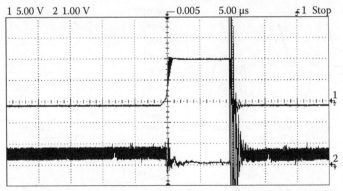

(a) Waveform of switch voltage and current with hard switching (10 A/div)

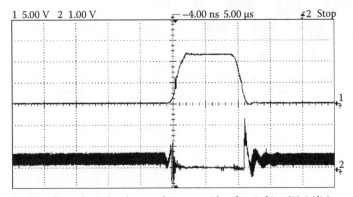

(b) Waveform of switch voltage and current with soft switching (10 A/div)

FIGURE 13.26   (a) Voltage and current waveforms of switch S_L in hard-switching and (b) soft-switching inverter.

is a considerable overlap between the voltage and current waveforms during hard switching. The overlap is much less with soft switching.

A series of key waveforms with soft-switching inverter is shown in Figure 13.27. The default scale is DC link voltage: 100 V/div, current: 20 A/div. The default switching frequency is 10 kHz. The DC link voltage is fixed to 240 V. These experimental waveforms are similar to the simulation waveforms in Figure 13.25.

13.6.2   RESONANT POLE INVERTER

Resonant pole inverter is a soft-switching DC/AC inverter circuit, which is shown in Figure 13.28. Each resonant pole comprises a resonant inductor and a pair of resonant capacitors at each phase leg. These capacitors are directly connected in parallel to the main inverter switches in order to achieve ZVS condition. In contrast to the resonant DC link inverter, the DC link voltage remains unaffected during the resonant transitions. The resonant transitions occur separately at each resonant pole when the corresponding main inverter switch needs switching. Therefore, the

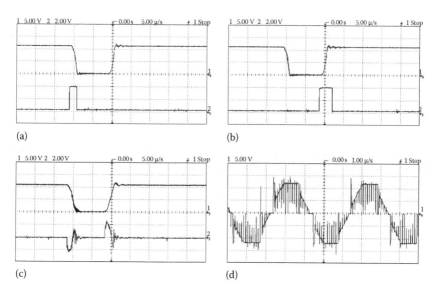

(a)   (b)

(c)   (d)

FIGURE 13.27 Experiment waveforms: (a) waveform of u_{Cr} and S_a gate signal, (b) waveform of u_{Cr} and S_b gate signal, (c) waveform of u_{Cr} and i_{Lr}, and (d) waveform of phase voltage $(L - L)$.

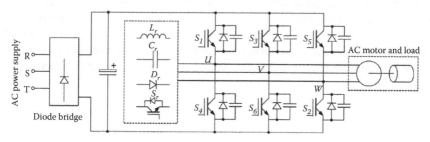

FIGURE 13.28 Resonant pole inverter.

main switches in the inverter phase legs can switch independently from each other and choose the commutation instant freely. Moreover, there is no additional main conduction path switch. Thus, the normal operation of the resonant pole inverter is entirely the same as that of the conventional hard-switching inverter [17].

The ARCP inverter [18] and the ordinary resonant snubber inverter [19] provide a ZVS condition without increasing the device voltage and current stress. These inverters are able to achieve real PWM control. However, they require a stiff DC link capacitor bank that is center-taped to accomplish commutation. The center voltage of DC link is susceptible to drift that may affect the operation of the resonant circuit. The resonant transition inverter [20,21] only uses one auxiliary switch, whose switching frequency is much higher than when applying to the main switches. Thus, it will limit the switching frequency of the inverter. Furthermore, the three resonant branches of the inverter work together and are affected by each other. A Y-configured resonant snubber inverter [22] has a floating neutral voltage that may cause overvoltage failure of the auxiliary switches. A delta (Δ) configured resonant snubber inverter [23] avoids the floating neutral voltage and is suitable for multiphase operation without circulating currents between the off-state branch and its corresponding output load. However, the inverter requires three inductors and six auxiliary switches.

Moreover, resonant pole inverters have been applied in induction motor drive applications. They are usually required to change two-phase switch states at the same time to obtain a resonant path. It is not suitable for a BDCM drive system, as only one switch is needed to change the switching state in a PWM cycle. The switching frequency of three upper switches (S_1, S_3, S_5) is different than that of three lower switches (S_2, S_4, S_6) in an inverter for a BDCM drive system. All the switches have the same switching frequency in a conventional inverter for induction motor applications. Therefore, it is necessary to develop a novel topology of soft-switching inverter and special control circuit for BDCM drive systems. This chapter proposes a specially designed resonant pole inverter that is suitable for BDCM drive systems and is easy to apply in industry. In addition, this inverter possesses the following advantages: low switching power losses, low inductor power losses, low switching noise, and simple control scheme.

13.6.2.1 Topology of the Resonant Pole Inverter

A typical controller for BDCM drive system [24] is shown in Figure 13.29.

The rotor position can be sensed by a Hall-effect sensor or a slotted optical disk, providing three square waves with phase shift in 120°. These signals are decoded by a combinatorial logic to provide the firing signals for 120° conduction on each of the three phases. The basic forward control loop is voltage control implemented by PWM (voltage reference signal compared with triangular wave or generated by

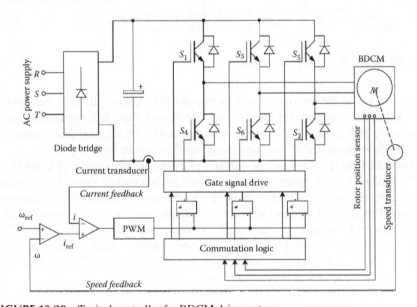

FIGURE 13.29   Typical controller for BDCM drive system.

microprocessor). The PWM is applied only to the lower switches. This not only reduces the current ripple but also avoids the need for wide bandwidth in the level-shifting circuit that feeds the upper switches. The three upper switches work under commutation frequency (typically, several hundred Hz) and the three lower switches work under PWM frequency (typically, tens of kHz). So it is not important that the three upper switches work under soft-switching condition. The switching power losses can be reduced significantly and the auxiliary circuit would be simpler if only three lower switches work under soft-switching condition. Thus, a special design resonant pole inverter for BDCM drive system is introduced for this purpose. The structure of the proposed inverter is shown in Figure 13.30.

The system contains a diode bridge rectifier, a resonant circuit, a conventional three-phase inverter, and control circuitry. The resonant circuit consists of three auxiliary switches (S_a, S_b, S_c), one transformer with turns ratio 1:n, and two diodes D_{fp}, D_r. Diode D_{fp} is connected in parallel to the primary winding of the transformer, and diode D_r is serially connected with secondary winding across the DC link. There is one snubber capacitor connected in parallel to each lower switch of phase leg. The snubber capacitor resonates with the primary winding of the transformer. The emitters of the three auxiliary switches are connected together. Thus, the gate drive of these auxiliary switches can use one common output DC power supply.

In a whole PWM cycle, the three lower switches (S_2, S_4, S_6) can be turned off in the ZVS condition as the snubber capacitors (C_{ra}, C_{rb}, C_{rc}) can slow down the voltage rise rate. The turn-off power losses can be reduced, and the turn-off voltage spike is eliminated. Before turning on the lower switch, the corresponding auxiliary switch (S_a, S_b, S_c) must be turned on ahead. The snubber capacitor is then discharged, and the lower switches get the ZVS condition. During phase current commutation, the switching state is changed from one lower switch to another; for example, by turning off S_6 and turning on S_2, S_6 can be turned off directly in the ZVS condition, and by turning on the auxiliary switch S_c to discharge the snubber capacitor C_{rc}, switch S_2 can get the ZVS condition. During phase current commutation, if the switching state

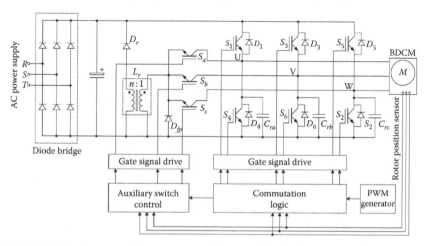

FIGURE 13.30 Structure of the resonant pole inverter for BDCM drive system.

is changed from one upper switch to another upper switch, the operation is the same as that of the hard-switching inverter, as the switching power losses of the upper switches is much smaller than that of the lower switches.

13.6.2.2 Operation Principle

For convenience sake, to describe the operation principle, we investigate the period of time when the switch S_1 is always turned on, when switch S_6 works under PWM frequency, and when other main inverter switches are tuned off. Since the resonant transition is very short, it can be assumed that the load current is constant. The equivalent circuit is shown in Figure 13.31. Where V_S is the DC link voltage, i_{Lr} is the transformer primary winding current, u_{S6} is the voltage drop across the switch S_6 (i.e., snubber capacitor C_{rb} voltage), and I_O is the load current. The waveforms of the switches (S_6, S_b) gate signal, PWM signal, main switch S_6 voltage drop (u_{S6}), and the transformer primary winding current (i_{Lr}) are illustrated in Figure 13.32, and details will be explained as follows. Accordingly, at the instant t_0–t_6, the operation of one switching cycle can be divided into seven modes.

Mode 0 (shown in Figure 13.33a) $0<t<t_0$: After the lower switch S_6 is turned off, load current flows through the upper freewheeling diode D_3, and the voltage drop u_{S6} (i.e., snubber capacitor C_{rb} voltage) across the switch S_6 is the same as that of the DC link voltage. The auxiliary resonant circuit does not operate.

Mode 1 (shown in Figure 13.33b) $t_0<t<t_1$: If the switch S_6 is turned on directly, the capacitor discharge surge current will also flow through switch S_6; thus, switch S_6 may face the risk of second breakdown. The energy stored in the snubber capacitor must be discharged ahead of time. Thus, auxiliary switch S_b is turned on (ZCS is turned on as the current i_{Lr} cannot change suddenly due to the transformer inductance). As the transformer primary winding current i_{Lr} begins to increase, the current flowing through the freewheeling diode decays. The secondary winding current i_{Lrs} also begins to conduct through diode D_r to the DC link. Both of the terminal voltages of the primary and secondary windings are equal to the DC link voltage V_S. By neglecting the resistances of the windings and using the transformer equivalent circuit (referred to the primary side) [25], we get

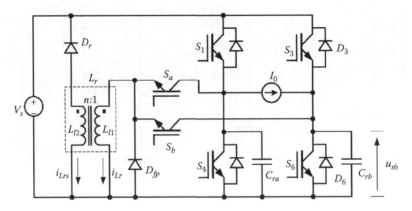

FIGURE 13.31 Equivalent circuit.

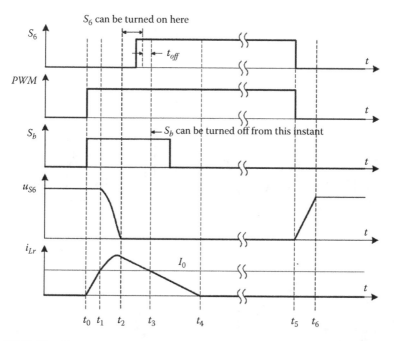

FIGURE 13.32 Key waveforms of the equivalent circuit.

$$V_S = L_{l1}\frac{di_{Lr}(t)}{dt} + a^2 L_{l2}\frac{d[i_{Lrs}(t)/a]}{dt} + aV_S \qquad (13.19)$$

where L_{l1} and L_{l2} are the primary and secondary winding leakage inductance, respectively, is the transformer turn ratio 1:n. The transformer has a high magnetizing inductance. We can assume that $i_{Lrs}=i_{Lr}/n$ and rewrite (13.19) as

$$\frac{di_{Lr}}{dt} = \frac{(n-1)V_S}{n(L_{l1}+(1/n^2)L_{l2})} = \frac{(n-1)V_S}{nL_r} \qquad (13.20)$$

where L_r is the equivalent inductance of the transformer $L_{l1} + L_{l2}/n^2$. The transformer primary winding current i_{Lr} increases linearly, and the mode is ended when $i_{Lr}=I_O$. The interval of this mode can be determined by

$$\Delta t_1 = t_1 - t_0 = \frac{nL_r I_O}{(n-1)V_S} \qquad (13.21)$$

Mode 2 (shown in Figure 13.33c) $t_1 < t < t_2$: At $t=t_1$, all load current flows through the transformer primary winding, and the freewheeling diode D_3 is turned off in the ZCS condition. The freewheeling diode reverse recovery problems are reduced greatly. The snubber capacitor C_{rb} resonates with the transformer, and the voltage

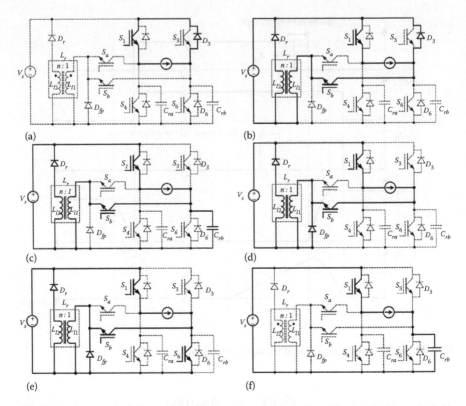

FIGURE 13.33 Operation modes of the resonant pole inverter: (a) mode 0, (b) mode 1, (c) mode 2, (d) mode 3, (e) mode 4, and (f) mode 6.

drop u_{S6} across switch S_6 decays. By redefining the initial time, the transformer current i_{Lr}, i_{Lrs} and capacitor voltage u_{S6} obey the equation

$$\begin{cases} u_{S6}(t) = L_{l1}\dfrac{di_{Lr}(t)}{dt} + a^2 L_{l2}\dfrac{d[i_{Lrs}(t)/a]}{dt} + aV_S \\[2mm] -C_r\dfrac{du_{S6}(t)}{dt} = i_{Lr}(t) - I_O \end{cases} \tag{13.22}$$

where C_r is the capacitance of snubber capacitor C_{rb}. The transformer current $i_{Lrs} = i_{Lr}/n$, as in mode 1, with initial conditions $u_{S6}(0) = V_S$, $i_{Lr}(0) = I_O$, then the solution of (13.22) is

$$\begin{cases} u_{S6}(t) = \dfrac{(n-1)V_S}{n}\cos(\omega_r t) + \dfrac{V_S}{n} \\[3mm] i_{Lrs}(t) = I_O + \dfrac{(n-1)V_S}{n}\sqrt{\dfrac{C_r}{L_r}}\,i_{Lr}\sin(\omega_r t) \end{cases} \tag{13.23}$$

where $\omega_r = \sqrt{1/(L_r C_r)}$. Let $u_{C_r}(t) = 0$, which gets the duration of the resonance

$$\Delta t_2 = t_2 - t_1 = \frac{1}{\omega_r} \arccos\left(-\frac{1}{n-1}\right) \tag{13.24}$$

The interval is independent from the load current. At $t = t_2$, the corresponding transformer primary current is

$$i_{Lr}(t_2) = I_O + V_S \sqrt{\frac{(n-2)C_r}{nL_r}} \tag{13.25}$$

The peak value of the transformer primary current can also be determined

$$i_{Lr-m} = I_O + \frac{n-1}{n} V_S \sqrt{\frac{C_r}{L_r}} \tag{13.26}$$

Mode 3 (shown in Figure 13.33d) $t_2 < t < t_3$: When the capacitor voltage u_{S6} reaches zero at $t = t_2$, the freewheeling diode D_{pf} begins to conduct. The current flowing through auxiliary switch S_b is the load current I_O. The sum current flowing through switch S_b and diode D_{pf} is the transformer primary winding current i_{Lr}. The transformer primary voltage is zero and the secondary voltage is V_S. By redefining the initial time, we obtain

$$0 = L_{l1} \frac{di_{Lr}(t)}{dt} + a^2 L_{l2} \frac{d[i_{Lrs}(t)/a]}{dt} + a V_S \tag{13.27}$$

Since the transformer current $i_{Lrs} = i_{Lr}/n$ as in mode 1, we deduce (13.27)

$$\frac{di_{Lr}}{dt} = -\frac{V_S}{nL_r} \tag{13.28}$$

The transformer primary current decays linearly, and the mode is ended while $i_{Lr} = I_0$. With the initial condition given by (13.25), the interval of this mode can be determined by

$$\Delta t_3 = t_3 - t_2 = \sqrt{n(n-2)L_r C_r} \tag{13.29}$$

The interval is also independent from the load current. During this mode, switch is turned on in ZVS condition.

   Mode 4 (shown in Figure 13.33e) $t_3 < t < t_4$: The transformer primary winding current i_{Lr} decays linearly from load current I_0 to zero. Partial load current flows through the main switch S_6. The sum current flowing through switch S_6 and S_b is equal to the load current I_0. The sum current flowing through switch S_b and diode D_{fp} is the transformer primary winding current i_{Lr}. By redefining the initial time, the transformer winding current agrees (13.28) with the initial condition $i_{Lr}(0)=I_O$. The interval of this mode is

$$\Delta t_4 = t_4 - t_3 = \frac{nL_r I_O}{V_S} \qquad (13.30)$$

The auxiliary switch S_b can be turned off in ZVS condition. In this case, after switch S_b is turned off, the transformer primary winding current flows through the freewheeling diode D_{fp}. The auxiliary switch S_b can be also turned off in ZVS and ZCS condition after i_{Lr} decays to zero.

Mode 5 $t_4 < t < t_5$: The transformer primary winding current decays to zero and the resonant circuit idles. This state is likely the same operational state as the conventional hard-switching inverter. The load current flows from DC link through the two switches S_1 and S_6, and the motor.

Mode 6 (shown in Figure 13.33f) $t_5 < t < t_6$: The main inverter switch S_6 is turned directly off and the resonant circuit does not work. The snubber capacitor C_{rb} can slow down the rise rate of u_{S6}, whereas the main switch S_6 operates in ZVS condition. The duration of the mode is

$$\Delta t_7 = t_7 - t_6 = \frac{C_r V_S}{I_O} \qquad (13.31)$$

The next period starts from mode 0 again, but the load current flows through freewheeling diode D_3. During phase current commutation, the switching state is changed from one lower switch to another (e.g., turn off S_6 and turn on S_2), S_6 can be turned off directly in ZVS condition (similar to mode 6), and by turning on auxiliary switch S_c to discharge the snubber capacitor C_{rc}, switch S_2 can get ZVS condition (similar to modes 1–4).

13.6.2.3 Design Considerations
It is assumed that the inductance of BDCM is much higher than the transformer leakage inductance. From the previous analysis, the design considerations can be summarized as follows:

1. Determine the value of snubber capacitor C_r and the parameter of transformer.
2. Select the main and auxiliary switches.
3. Design the control circuitry for the main and auxiliary switches.

The turn ratio (1:n) of the transformer can be determined ahead. From (13.24) must satisfy

$$n > 2 \qquad (13.32)$$

On the other hand, from (13.30) the transformer primary winding current i_{Lr} will take a long time to decay to zero if n is too big. So n must be a moderate number. The equivalent inductance of the transformer $L_r = L_{l1} + L_{l2}/n^2$ is inversely proportional to the rise rate of the switch current when turning on the auxiliary switches. It means that the equivalent inductance L_r should be big enough to limit the rising rate of the switch current to work in ZCS condition. The selection of L_r can be referenced from the rule depicted in Ref. [26]

$$L_r \approx 4t_{on}V_S/I_{O\,max} \tag{13.33}$$

where

t_{on} is the turn-on time of an IGBT
$I_{O\,max}$ is the maximum load current

The snubber capacitance C_r is inversely proportional to the rise rate of the switch voltage drop when turning off the lower main inverter switches. It means that the capacitance is as high as possible to limit the rising rate of the voltage to work in ZVS condition. The selection of the snubber capacitor can be determined as

$$C_r \approx 4t_{off}I_{O\,max}/V_S \tag{13.34}$$

where t_{off} is the turn-off time of an IGBT. However, as the capacitance increases, more energy is stored in it. This energy should be discharged when the lower main inverter switches are turned on. With high capacitance, the peak value of the transformer current will be also high. The peak value of i_{Lr} should be restricted to twice that of the maximum load current. From (13.26), we obtain

$$\sqrt{\frac{C_r}{L_r}} \leq \frac{nI_{O\,max}}{(n-1)V_S} \tag{13.35}$$

Three lower switches of the inverter (i.e., S_4, S_6, S_2) are turned on during mode 3 (i.e., lag rising edge of PWM at the time range $\Delta t_1 + \Delta t_2 \sim \Delta t_1 + \Delta t_2 + \Delta t_3$). In order to turn on these switches at a fixed time (say ΔT_1) lagging rising edge of PWM under various load currents for control convenience, the following condition should be satisfied:

$$\Delta t_1 + \Delta t_2 + \Delta t_3 \,|_{I_0=0} > (\Delta t_1 + \Delta t_2)\,|_{I_0=I_{0\,max}} + t_{off} \tag{13.36}$$

Substitute (13.21), (13.24), and (13.29) into (13.36)

$$\sqrt{n(n-2)L_rC_r} > \frac{nL_rI_{O\,max}}{(n-1)V_S} + t_{off} \tag{13.37}$$

The whole switching transition time is expressed as

$$T_w = \Delta t_1 + \Delta t_2 + \Delta t_3 + \Delta t_4 = \frac{nL_rI_0}{(n-1)V_S} + \sqrt{L_rC_r} \times \left[\arccos\left(-\frac{1}{n-1}\right) + \sqrt{n(n-2)}\right] \tag{13.38}$$

For high switching frequencies, T_w should be as short as possible. Select the equivalent inductance L_r and snubber capacitance C_r to satisfy (13.32) through (13.37), and L_r and C_r should be as small as possible.

As the transformer operates at high frequency (20 kHz), the magnetic core material can be ferrite. The design of the transformer needs the parameters of form factor, frequency, the input/output voltage, input/output maximum current, and ambient temperature. From Figure 13.31, the transformer current can be simplified as triangle waveforms, then the form factor can be determined as $2\sqrt{3}$. Ambient temperature is dependent on the application field. Other parameters can be obtained from the previous section. The transformer only carries current during the transition of turning on a switch in one cycle, so the winding can be a smaller-diameter one.

Main switches S_{1-6} work under ZVS condition; therefore, the voltage stress is equal to the DC link voltage V_S. The device current rate can be load current. Auxiliary switches S_{a-c} work under the ZCS or ZVS condition, and the voltage stress is also equal to the DC link voltage V_S. The peak current flowing through them is limited to double maximum load current. As the auxiliary switches S_{a-c} carry the peak current only during switch transitions, they can be rated with a lower continuous current rating. The additional cost will not be too much.

The gate signal generator circuit is shown in Figure 13.34. The rotor position signal decode module produces the typical gate signal of the main switches. The inputs of the module are rotor position signals, rotating direction of the motor, which "enable" the signal and PWM pulse train. The rotor position signals are three square-waves with a phase shift in 120°. The "enable" signal is used to disable all outputs in case of emergency (e.g., overcurrent, overvoltage, and overheat). The PWM signal

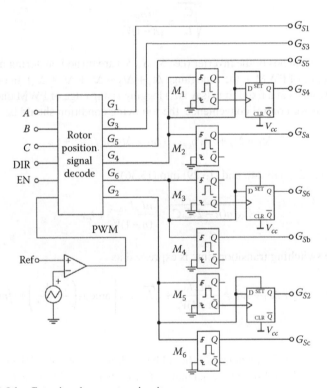

FIGURE 13.34 Gate signal generator circuit.

is the output of comparator, comparing the reference voltage signal with the triangular wave. The reference voltage signal is the output of the speed controller. The speed controller is a processor (single-chip computer or digital signal processor) and the PWM signal can be produced by software. The outputs $(G_1–G_6)$ of the module are the gate signals applied to the main inverter switches. The outputs $G_{1,3,5}$ are the required gate signals for three upper main inverter switches.

The gate signals of three lower main inverter switches and auxiliary switches can be deduced from the outputs $G_{4,6,2}$ as shown in Figure 13.35. The trailing edge of the gate signals for the three lower main inverter switches $G_{S4,6,2}$ is the same as that of $G_{4,6,2}$, and the leading edge of $G_{S4,6,2}$ lags $G_{4,6,2}$ for a short time ΔT_1. The gate signals for auxiliary switches $G_{S4,6,2}$ have a fixed pulse width (ΔT_2) with the leading edge, the same as that of $G_{4,6,2}$. In Figure 13.34, the gate signals $G_{Sa,b,c}$ are the outputs of monostable flip-flops $M_{4,6,2}$ with the inputs $G_{4,6,2}$. The three monostable flip-flops $M_{4,6,2}$ have the same pulse width ΔT_2. The gate signals $G_{S4,6,2}$ are combined by the negative outputs of monostable flip-flops $M_{1,3,5}$ and $G_{4,6,2}$. The combining logical controller can be implemented by a D flip-flop with "preset" and "clear" terminals. The three monostable flip-flops $M_{4,6,2}$ have the same pulse width ΔT_1. Determination of the pulse widths of ΔT_1 and ΔT_2 is referenced from theoretical analysis in Section 13.6.2.2. In order to get the ZVS condition of the main inverter switches under various load currents, the lag time should satisfy

$$(\Delta t_1 + \Delta t_2)\,|_{I_O=I_{O\max}} < \Delta T_1 < (\Delta t_1 + \Delta t_2 + \Delta t_3)\,|_{I_O=0} -t_{off} \qquad (13.39)$$

In order to get a soft-switching condition of the auxiliary switches, pulse width need only satisfy

$$\Delta T_2 > (\Delta t_1 + \Delta t_2 + \Delta t_3)\,|_{I_O=I_{O\max}} \qquad (13.40)$$

13.6.2.4 Simulation and Experimental Results

The proposed topology is verified by software simulation PSim. The DC link voltage is 300 V, and the maximum load current is 25 A. The parameters of the resonant circuit were determined from (13.32) through (13.38). The transformer turn ratio is 1:4,

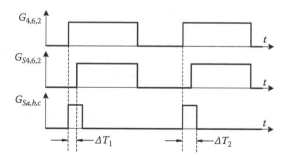

FIGURE 13.35 Gate signals $G_{S4,6,2}$ and $G_{Sa,b,c}$ from $G_{4,6,2}$.

and the leakage inductances of the primary secondary windings are 6 and 24 μH, respectively. Therefore, the equivalent transformer inductance L_r is 7.5 μH. The resonant capacitance C_r is 0.047 μF. Then, $\Delta t_1 + \Delta t_2$ and $\Delta t_1 + \Delta t_2 + \Delta t_3$ can be determined under various load currents I_O as shown in Figure 13.36, considering that the turn-off time of a switch lagging time ΔT_1 and pulse width ΔT_2 are set to 2.1 and 5 μs, respectively. The frequency of the PWM is 20 kHz. Waveforms of transformer primary winding current i_{Lr}, switch S_6 voltage drop u_{S6}, PWM, main switch S_6, auxiliary switch S_b, and the gate signal under low and high load current are shown in Figure 13.37. The figure shows that the inverter worked well under various load currents.

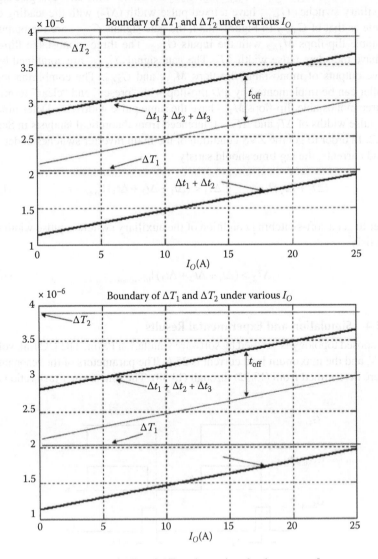

FIGURE 13.36 Boundary of ΔT_1 and ΔT_2 under various load currents, I_O.

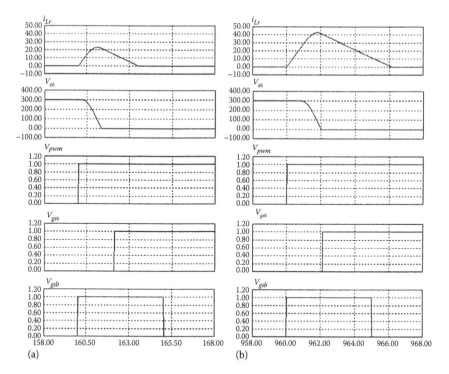

FIGURE 13.37 Simulation waveforms of i_{Lr}, u_{S6}, PWM, S_6 and S_b gate signals under various load currents: (a) under low load current ($I_O = 5$ A) and (b) under high load current ($I_O = 25$ A).

In order to verify the theoretical analysis and simulation results, the inverter was tested by experiment. The test conditions are

1. DC link voltage: 300 V
2. Power of the BDCM: 3.3 Hp
3. Rated phase current: 10.8 A
4. Switching frequency: 20 kHz

Select 50 A, 1200 V BSM 35 GB 120 DN2 dual IGBT module as main inverter switches and 30 A 600 V IMBH30D-060 IGBT as auxiliary switches. With datasheets of these switches and (13.32) through (13.38), the value of inductance and capacitance can be determined. Three polyester capacitors of 47 nF/630 V were adopted as snubber capacitor for three lower switches of the inverter. A high magnetizing inductance transformer with the turn ratio 1:4 was employed in the experiment. Fifty-two turns wires with size AWG 15 are selected as primary winding, and 208 turns wires with size AWG 20 are selected as secondary winding. The equivalent inductance is about 7 μH. The switching frequency is 20 kHz. The rotor position signal decode module is implemented by a 20 leads GAL IC GAL16V8. The monostable flip-flop is set up by IC 74LS123, variable resistor, and capacitor.

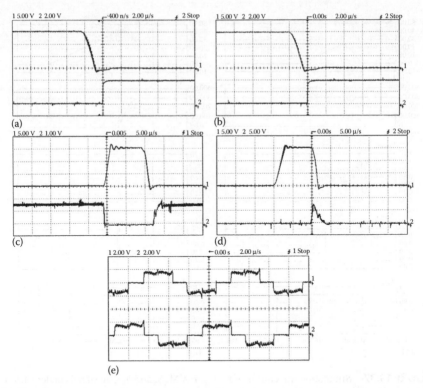

FIGURE 13.38 Experiment waveforms: (a) switch S_6 voltage u_{S6} (top) and its gate signal (bottom) under low load current (100 V/div), (b) switch S_6 voltage u_{S6} (top) and its gate signal (bottom) under high load current (100 V/div), (c) switch S_6 voltage u_{S6} (top) and its current i_{S6} (bottom) (100 V/div, 5 A/div), (d) switch S_6 voltage u_{S6} (top) and transformer current i_{S6} (bottom) (100 V/div, 25 A/div), and (e) waveforms of phase current (10 A/div).

With (13.21) and (13.22), lag time and pulse width are determined to be 2.5 and 5 μs, respectively.

The system is tested in light-load and full-load currents. The voltage waveforms across the main inverter switch u_{S6} and its gate signal in low- and high-load currents are shown in Figure 13.38a and b, respectively. All the voltage signals are measured by differential probes with a gain of 20, for voltage waveforms, 5.00 V/div and 100 V/div. The waveforms of u_{S6} and its current i_{S6} are shown in Figure 13.38c, and dv/dt and di/dt are reduced significantly. The waveforms of u_{S6} and transformer primary winding current i_{Lr} are shown in Figure 13.38d. The phase current is shown in Figure 13.38e. It can be seen that the resonant pole inverter works well under various load currents, and there is little overlap between the voltage and current waveforms during the switching under soft-switching condition; therefore, the switching power losses is low. The efficiency of hard switching and soft switching under rated speed and various load torques (p.u.) is shown in Figure 13.39. The efficiency improves with the soft-switching inverter. Therefore, the design of the system is successful.

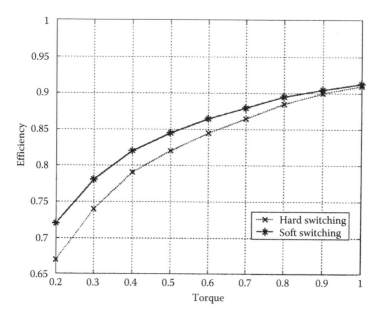

FIGURE 13.39 Efficiency of hard switching and soft switching under various load torques (p.u.).

13.6.3 TRANSFORMER-BASED RESONANT DC LINK INVERTER

In order to generate voltage notches of the DC link at controllable instants and reduce the power losses of the inductor, several quasi-parallel resonant schemes were proposed [9–11]. As a dwell time is generally required after every notch, severe interferences occur, mainly in multiphase inverters, appreciably worsening the modulation quality. A novel DC-rail parallel resonant ZVT voltage source inverter [27] is introduced; it overcomes many drawbacks mentioned earlier. However, it requires a stiff DC link capacitor bank that is center-taped to accomplish commutation. The center voltage of DC link is susceptible to drift, which may affect the operation of the resonant circuit. In addition, it requires two ZVT per PWM cycle, and it would worsen the output voltage and limit the switch frequency of the inverter.

On the other hand, the majority of soft-switching inverters proposed in the recent years have been considered for the induction motor drive applications. So it is necessary to study the novel topology of soft-switching inverter and special control circuit for BDCM drive systems. This chapter proposed a resonant DC link inverter based on transformer for BDCM drive system to solve the problems mentioned previously. The inverter possesses the advantages of low switching power loss, low inductor power loss, low DC link voltage ripple, small device voltage stress, and simple control scheme. The structure of the soft-switching inverter is shown in Figure 13.40 [28]. The system contains a diode bridge rectifier, a resonant circuit, a conventional three-phase inverter, and control circuit. The resonant circuit consists of three auxiliary switches (S_L, S_a, S_b) and corresponding built-in freewheeling diode (D_L, D_a, D_b),

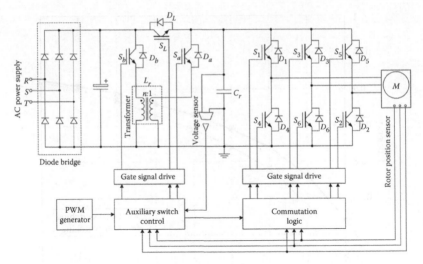

FIGURE 13.40 Structure of the resonant DC link inverter for BDCM drive system.

one transformer with turn ratio $1:n$, and one resonant capacitor. All auxiliary switches work under ZVS or ZCS condition. It generates voltage notches of the DC link to guarantee the main switches (S_1-S_6) of the inverter operating in ZVS condition.

13.6.3.1 Resonant Circuit

The resonant circuit consists of three auxiliary switches, one transformer, and one resonant capacitor. The auxiliary switches are controlled at certain instant to obtain the resonance between transformer and capacitor. Thus, the DC link voltage reaches zero temporarily (voltage notch) and the main switches of the inverter get ZVS condition for commutation. Since the resonant process is very short, the load current can be assumed constant. The equivalent circuit of the inverter is shown in Figure 13.41, where V_S is the DC power supply voltage and I_O is the load current. The corresponding waveforms of the auxiliary switches gate signal, PWM signal, resonant capacitor voltage u_{Cr} (i.e., DC link voltage), the transformer primary winding current i_{Lr}, and current i_{SL} of switch (S_L) are illustrated in Figure 13.42. The DC link voltage that reduces to zero and rising to supply voltage again is called one ZVT process or one DC link voltage notch. The operation of the ZVT process in one PWM cycle can be divided into eight modes.

Mode 0 (shown in Figure 13.43a) $0 < t < t_0$: Its operation is the same as conventional inverter. Current flows from DC power supply through S_L to the load. The voltage u_{Cr} across resonant capacitor C_r is equal to the supply voltage V_S. The auxiliary switches S_a and S_b are turned off.

Mode 1 (shown in Figure 13.43b) $t_0 < t < t_1$: When it is the instant for phase current commutation or PWM signal is flopped from high to low, auxiliary switch S_a is turned on with ZCS (as the i_{Lr} cannot suddenly change due to the transformer inductance) and switch S_L is turned off with ZVS (as they cannot change suddenly

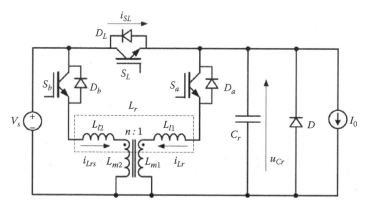

FIGURE 13.41 Equivalent circuit of the inverter.

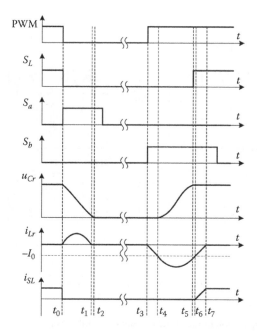

FIGURE 13.42 Key waveforms of the equivalent circuit.

due to resonant capacitor C_r) at the same time. The transformer primary winding current i_{Lr} begins to increase, and the secondary winding current i_{Lrs} also begins to build up through diode D_b to DC link. The terminal voltages of primary and secondary windings of the transformer are DC link voltage u_{Cr} and supply voltage V_S, respectively. Capacitor C_r resonates with transformer, and the DC link voltage u_{Cr} is decreased. Neglecting the resistances of windings, using the transformer equivalent

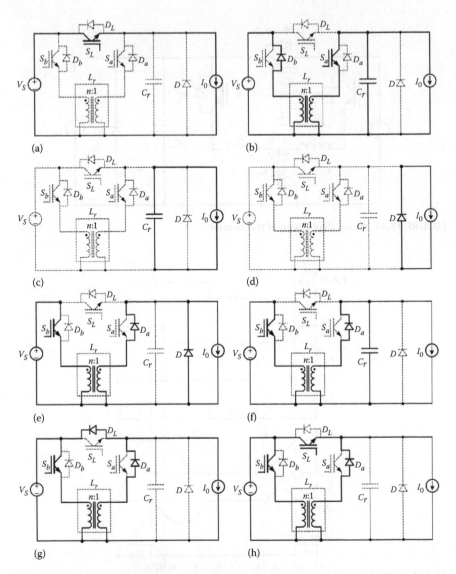

FIGURE 13.43 Operation mode of the resonant DC link inverter: (a) mode 0, (b) mode 1, (c) mode 2, (d) mode 3, (e) mode 4, (f) mode 5, (g) mode 6, and (h) mode 7.

circuit (referred to the primary side) [25], the transformer current i_{Lr}, i_{Lrs} and DC link voltage u_{Cr} obey the equation

$$
\begin{cases}
u_{Cr}(t) = L_{l1} \dfrac{di_{Lr}(t)}{dt} + a^2 L_{l2} \dfrac{d[i_{Lrs}(t)/a]}{dt} + a V_S \\[2mm]
i_{Lr}(t) + I_O + C_r \dfrac{du_{Cr}(t)}{dt} = 0
\end{cases}
\tag{13.41}
$$

where L_{l1} and L_{l2} are the primary and secondary winding leakage inductance, respectively; the transformer turn ratio is 1:n. The transformer has a high magnetizing inductance. Assuming $i_{Lrs}=i_{Lr}/n$, with initial condition $u_{Cr}(0)=V_S$, $i_{Lr}(0)=0$, to solve (13.41), we get

$$\begin{cases} u_{Cr}(t) = \dfrac{(n-1)V_S}{n}\cos(\omega_r t) - I_O\sqrt{\dfrac{L_r}{C_r}}\sin(\omega_r t) + \dfrac{V_S}{n} \\ i_{Lr}(t) = I_O\cos(\omega_r t) - I_O + \dfrac{(n-1)V_S}{n}\sqrt{\dfrac{L_r}{C_r}}\sin(\omega_r t) \end{cases} \tag{13.42}$$

where
$L_r = L_{l1} + L_{l2}/n^2$ is the equivalent inductance of the transformer
$w_r = \sqrt{(1/L_r C_r)}$ is the natural angular resonance frequency

Rewrite (13.42) to get

$$\begin{cases} u_{Cr}(t) = K\cos(\omega_r t + \alpha) + \dfrac{V_S}{n} \\ i_{Lr}(t) = K\sqrt{\dfrac{C_r}{L_r}}\sin(\omega_r t + \alpha) - I_O \end{cases} \tag{13.43}$$

where
$K = \sqrt{(n-1)^2 V_S^2/n^2 + (I_0^2 L_r/C_r)}$
$\alpha = \arctan[(nI_0\sqrt{L_r/C_r})/(n-1)V_S]$
n is a number that is slightly less than 2 (the selection of such a number will be explained later)
i_{Lr} will decay to zero faster than u_{Cr}

Let $i_{Lr}(t)=0$, the duration of the resonance can be determined:

$$\Delta t_1 = t_1 - t_0 = \frac{\pi - \alpha}{\omega_r} \tag{13.44}$$

When i_{Lr} is reduced to zero, auxiliary switch S_a can be turned off with ZCS condition. At $t=t_1$, the corresponding DC link voltage u_{Cr} is

$$u_{Cr}(t_1) = \frac{2-n}{n}V_S \tag{13.45}$$

Mode 2 (shown in Figure 13.43c) $t_1 < t < t_2$: When the transformer current is reduced to zero, the resonant capacitor is discharged through load from initial condition as (13.45). The interval of this mode can be determined by

$$\Delta t_2 = t_2 - t_1 = \frac{C_r V_S (2 - n)}{n I_0} \tag{13.46}$$

As it has already been mentioned that n is a number that is slightly less than 2, the interval is normally very short.

Mode 3 (shown in Figure 13.43d) $t_2 < t < t_3$: The DC link voltage u_{C_r} is zero. The main switches of the inverter can now be either turned on or turned off under ZVS condition during this mode. Load current flows through the freewheeling diode D.

Mode 4 (shown in Figure 13.43e) $t_3 < t < t_4$: As the main switches have turned on or turned off, auxiliary switch S_b is turned on with ZCS (as the i_{Lrs} cannot suddenly change due to the transformer inductance) and the transformer secondary current i_{Lrs} starts to build up linearly. The transformer primary current i_{Lr} also begins to conduct through diode D_a to the load. The current in the freewheeling diode D begins to fall linearly. The load current is slowly diverted from the freewheeling diodes to the resonant circuit. The DC link voltage u_{C_r} is still equal to zero before the transformer primary current becomes greater than load current. The terminal voltages of transformer primary and secondary windings are zero and DC power supply voltage V_S, respectively. Redefining the initial time, we obtain

$$0 = L_{l1} \frac{di_{Lr}(t)}{dt} + a^2 L_{l2} \frac{d[i_{Lrs}(t)/a]}{dt} + a V_S \tag{13.47}$$

Since the transformer current $i_{Lrs} = i_{Lr}/n$ as in mode 1, rewrite (13.47) as

$$\frac{di_{Lr}}{dt} = -\frac{V_S}{n L_r} \tag{13.48}$$

The transformer primary current is increased reverse linearly from zero, and the mode ends when $i_{Lr} = -I_O$; the interval of this mode can be determined as

$$\Delta t_4 = t_4 - t_3 = \frac{n L_r I_O}{V_S} \tag{13.49}$$

At t_4, i_{Lr} equals the negative load current $-I_O$, and the current through the diode D becomes zero. Thus, the freewheeling diode turns off under ZCS condition, and the diode reverse recovery problems are reduced.

Mode 5 (shown in Figure 13.43f) $t_4 < t < t_5$: Absolute value of i_{Lr} continuously increases from I_O, and u_{C_r} increases from zero when the freewheeling diode D is turned off. Redefining the initial time, we can get the same equation as (13.41).

The initial condition is $u_{Cr}(0)=0$, $i_{Lr}(0)=-I_O$; neglecting the inductor resistance and solving the equation, we get

$$
\begin{cases}
u_{Cr}(t) = -\dfrac{V_S}{n}\cos(\omega_r t)+\dfrac{V_S}{n} \\[3mm]
i_{Lr}(t) = -I_O - \dfrac{V_S}{n}\sqrt{\dfrac{C_r}{L_r}}\sin\left(\omega_r t\right)
\end{cases}
\tag{13.50}
$$

When

$$
\Delta t_5 = t_5 - t_4 = \frac{1}{\omega_r}\arccos(1-n)
\tag{13.51}
$$

$u_{Cr}=V_S$, auxiliary switch S_L is turned on with ZVS (due to C_r). The interval is independent from load current. At $t=t_5$, the corresponding transformer primary current i_{Lr} is

$$
i_{Lr}(t_5) = -I_O - V_S\sqrt{\frac{(2-n)C_r}{nL_r}}
\tag{13.52}
$$

The peak value of the transformer primary current can be also determined

$$
i_{Lr-m} = \left|-I_O - \frac{V_S}{n}\sqrt{\frac{C_r}{L_r}}\right| = I_O + \frac{V_S}{n}\sqrt{\frac{C_r}{L_r}}
\tag{13.53}
$$

Mode 6 (shown in Figure 13.43g) $t_5<t<t_6$: Both the terminal voltages of primary and secondary windings are equal to supply voltage V_S after auxiliary switch S_L is turned on. Redefining the initial time, we obtain

$$
V_S = L_{l1}\frac{di_{Lr}(t)}{dt}+a^2 L_{l2}\frac{d[i_{Lrs}(t)/a]}{dt}+aV_S
\tag{13.54}
$$

Since the transformer current $i_{Lrs}=i_{Lr}/n$ as in mode 1, rewrite (13.54) as

$$
\frac{di_{Lr}}{dt} = \frac{(n-1)V_S}{nL_r}
\tag{13.55}
$$

The transformer primary current i_{Lr} decays linearly, and the mode ends when $i_{Lr}=-I_O$ again. With initial condition (13.52), the interval of this mode can be determined

$$
\Delta t_6 = t_6 - t_5 = \frac{\sqrt{n(2-n)L_rC_r}}{n-1}
\tag{13.56}
$$

The interval is also independent from load current. As it has already been mentioned that n is a number that is slightly less than 2, the interval is also very short.

Mode 7 (shown in Figure 13.43h) $t_6 < t < t_7$: The transformer primary winding current i_{Lr} decays linearly from negative load current $-I_O$ to zero. Partial load current flows through the switch S_L. The sum current flowing through switch S_L and transformer is equal to the load current I_O. Redefining the initial time, the transformer winding current obeys (13.55) with the initial condition $i_{Lr}(0) = -I_O$. The interval of this mode is

$$\Delta t_7 = t_7 - t_6 = \frac{nL_r I_0}{(n-1)V_S} \qquad (13.57)$$

Then auxiliary switch S_b can be also turned off with ZCS condition after i_{Lr} decays to zero (at any time after t_7).

13.6.3.2 Design Consideration

It is assumed that the inductance of BDCM is much higher than transformer leakage inductance. From the analysis presented previously, the design considerations can be summarized as follows.

1. Determine the value of resonant capacitor C_r and the parameters of the transformer.
2. Select the main switches and auxiliary switches.
3. Design the gate signal for the auxiliary switches.

The turn ratio $1:n$ of the transformer can be determined ahead. From (13.51) n must satisfy

$$n < 2 \qquad (13.58)$$

On the other hand, from (13.45) and (13.46) it is expected that n is as close to 2 as possible so that the duration of mode 2 would be not very long and would be small enough at the end of mode 1.

Normally, n can be selected at the range of 1.7–1.9. The equivalent inductance of the transformer $L_r = L_{l1} + L_{l2}/n^2$ is inversely proportional to the rising rate of switch current when turning on the auxiliary switches. It means that the equivalent inductance L_r should be big enough to limit the rising rate of the switch current to work in ZCS condition. The selection of L_r can be referenced from the rule depicted in Ref. [29]:

$$L_r \geq \frac{4t_{on}V_S}{I_{O\max}} \qquad (13.59)$$

where
t_{on} is the turn-on time of switch S_a
$I_{O\max}$ is the maximum load current

The resonant capacitance C_r is inversely proportional to the rising rate of switch voltage drop when turning off the switch S_L. It means that the capacitance is as high as possible to limit the rising rate of the voltage to work in ZVS condition. The selection of the resonant capacitor can be determined as

$$C_r \geq \frac{4t_{off}I_{O\max}}{V_S} \qquad (13.60)$$

where t_{off} is the turn-off time of switch S_L. However, as the capacitance increases, more energy is stored in it, and the peak value of transformer current will be also high. The peak value of i_{Lr} should be limited to twice that of peak load current. From (13.53), we obtain

$$\sqrt{\frac{C_r}{L_r}} \leq \frac{nI_{O\max}}{V_S} \qquad (13.61)$$

The DC link voltage rising transition time is expressed as

$$T_w = \Delta t_4 + \Delta t_5 = \frac{nL_rI_{O\max}}{V_S} + \sqrt{L_rC_r} \, \arccos(1-n) \qquad (13.62)$$

For high switching frequency, T_w should be as short as possible. Select the equivalent inductance L_r and resonant capacitance C_r to satisfy the inequalities (13.58) through (13.61), and L_r and C_r should be as small as possible. L_r and C_r selection area is illustrated in Figure 13.44 to determine their values; the valid area is shadowed, where B_1–B_3 is boundary that is defined according inequalities (13.58) through (13.61)

$$B_1 : L_r = \frac{4t_{on}V_S}{I_{O\max}} \qquad (13.63)$$

$$B_2 : C_r = \frac{4t_{off}I_{O\max}}{V_S} \qquad (13.64)$$

$$B3 : \sqrt{\frac{C_r}{L_r}} = \frac{nI_{O\max}}{V_S} \qquad (13.65)$$

If boundary B_3 intersects B_1 first as shown in Figure 13.44a, the value of L_r and C_r in the intersection A_1 can be selected. Otherwise, the value of L_r and C_r in the intersection A_2 is selected as shown in Figure 13.44b.

Main switches S_1–S_6 work under ZVS condition, and the voltage stress is equal to the DC power supply voltage V_S. The device current rate can be load current. Auxiliary switch S_L works under ZVS condition; its voltage and current stress is the

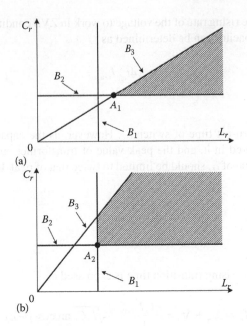

FIGURE 13.44 L and C selection area: (a) case 1: B intersects B first and (b) case 2: B intersects B first.

same as main switches. Auxiliary switches S_a and S_b work under ZCS or ZVS condition, and the voltage stress is also equal to the DC power supply voltage V_S. The peak current flowing through them is limited to double maximum load current. As the auxiliary switches S_a and S_b carry the peak current only during switch transitions, they can be rated lower continuous current rating.

The design of gate signal for auxiliary switches can be referenced from Figure 13.42. The trailing edge of the gate signal for auxiliary switch S_L is the same as that of PWM, and the leading edge is determined by the output of DC link voltage sensor. The gate signal for auxiliary switch S_a is a positive pulse with the leading edge being the same as PWM trailing edge; its width ΔT_a should be greater than Δt_1. From (13.44), Δt_1 is maximum when the load current is zero. So ΔT_a can be a fixed value determined by

$$\Delta T_a > \Delta t_1 \mid_{\max} = \frac{\pi}{\omega_r} = \pi\sqrt{L_r C_r} \qquad (13.66)$$

The gate signal for auxiliary switch S_b is also a pulse with the leading edge being the same as that of PWM; its width ΔT_b should be longer than t_7-t_3 (i.e., $\Delta t_4 + \Delta t_5 + \Delta t_6 + \Delta t_7$). ΔT_b can be determined from (13.49), (13.51), (13.56), and (13.57) that

$$\Delta T_b > \sum_{i=4}^{7} \Delta t_i \mid_{\max} = \frac{n^2 L_r I_{O\max}}{(n-1)V_S} + \sqrt{L_r C_r} \times \left[\arccos(1-n) + \frac{\sqrt{n(2-n)}}{n-1} \right] \qquad (13.67)$$

13.6.3.3 Control Scheme

When the duty of PWM is 100%, that is, full duty cycle, the main switches of the inverter work under the commutation frequency. When it is the instant to commutate the phase current of the BDCM, we control the auxiliary switches S_a, S_b, S_L, and resonance occurs between transformer inductor L_r and capacitor C_r. The DC link voltage reaches zero temporarily; thus, ZVS condition of the main switches is obtained. When the duty of PWM is less than 100%, the auxiliary switch S_L works as chopper. The main switches of the inverter do not switch within a PWM cycle when the phase current need not commutate. It has the benefit of reducing phase current drop when the PWM is off. The phase current is commutated when the DC link voltage becomes zero. There is only one DC link voltage notch per PWM cycle. It is very important especially for very low or very high duty of PWM. Otherwise the interval between two voltage notches is very short, even overlapped, which will limit the tuning range.

The commutation logical circuit of the system is shown in Figure 13.45. It is similar to conventional BDCM commutation logical circuit, except adding six D flip-flops to the output. Thus, the gate signal of the main switches is controlled by the synchronous pulse CK, which will be mentioned later, and the commutation can be synchronized with auxiliary switches control circuit (shown in Figure 13.46). The operation of the inverter can be divided into PWM operation and full duty cycle operation.

13.6.3.3.1 Full Duty Cycle Operation

When the duty of PWM is 100%, that is, full duty cycle, the whole ZVT process (modes 1–7) occurs when the phase current commutation is ongoing. The monostable flip-flop M_3 will generate one narrow negative pulse. The width of the pulse ΔT_3 is determined by $(\Delta t_1 + \Delta t_2 + T'_c)$; where T'_c is a constant, consider the turn-on and

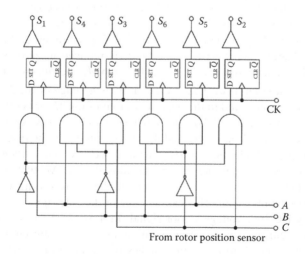

FIGURE 13.45 Commutation logical circuit for the main switches.

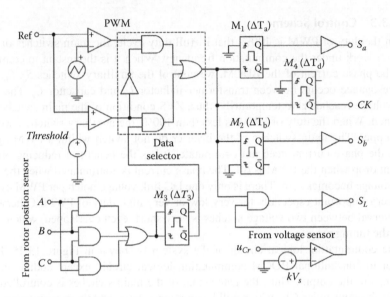

FIGURE 13.46 Control circuit for the auxiliary switches.

turn-off time of main switches. If n is close 2, Δt_2 would be very short or u_{Cr} would be small enough at the end of mode 1; thus, ΔT_3 can be determined by

$$\Delta T_3 = \Delta t_1 \mid_{max} + T_C = \pi\sqrt{L_r C_r} + T_C \qquad (13.68)$$

where T_c is another constant that is greater than T'_c. The data selector makes the output of monostable flip-flop M_3 active. The monostable flip-flop M_1 generates a positive pulse when the trailing edge of M_3 negative pulse is coming. The pulse is the gate signal for auxiliary switch S_a, and its width is ΔT_a, which is determined by inequality (13.66). The gate signal for switch S_L is flopped to low at the same time. Then mode 1 begins and the DC link voltage is reduced to zero. Synchronous pulse CK is also generated by a monostable flip-flop M_4; the pulsewidth ΔT_d should be greater than maximum Δt_1 (i.e., $p\sqrt{L_r C_r}$). If rising edge of the D flip-flops is active, then CK is connected to the negative output of the M_4 or is otherwise connected to the positive output. Thus, the active edge of pulse CK is within mode 3 when the voltage of DC link is zero and the main switches of the inverter get ZVS condition. The monostable flip-flop M_2 generates a positive pulse when the leading edge of negative pulse is coming. The pulsewidth of M_2 is ΔT_d that is determined by inequality (13.67). Then modes 4–7 occur and the DC link voltage is increased to match that of supply again. The leading edge of the gate signal for switch S_L is determined by DC link voltage sensor signal. In a word, in full-cycle operation when the phase current commutation is on going, the resonant circuit generates a DC link voltage notch to let main switches of the inverter switch under ZVS condition.

13.6.3.3.2   PWM Operation

In this operation, the data selector makes PWM signal active. The auxiliary switch S_L works as a chop, but the main switches of the inverter do not turn on or turn off within a single PWM cycle when the phase current need not commutate. The load current commutates when the DC link voltage becomes zero. (As the PWM cycle is very short, it does not affect the operation of the motor.)

1. When PWM signal is flopped down, mode 1 begins, pulse signal for switch S_a is generated by M_1, and gate signal for switch S_L drops to low. However, the voltage of DC link does not increase until PWM signal is flipped up. Pulse CK is also generated by M_4 to let active edge of CK locate in mode 3.
2. When PWM signal is flipped up, mode 4 begins, and pulse signal for switch S_b is generated at the moment. Then when the voltage of the DC link is increased to match the supply voltage V_S, the gate signal for switch S_L is flipped to high level.

Thus, only one ZVT occurs per PWM cycle: modes 1 and 2 for PWM turning off and modes 4–7 for PWM turning on. And the switching frequency would be not greater than PWM frequency.

13.6.3.4   Simulation and Experimental Results

The proposed system is verified by simulation software PSim. The DC power supply voltage V_S is 240 V, and the maximum load current is 12 A. The transformer turn ratio n is 1:1.8, and the leakage inductances of the primary secondary windings are selected as 4 and 12.96 μH, respectively. So the equivalent transformer inductance L_r is about 8 μH. The resonant capacitance C_r is 0.1 μF. Switch $S_{a,b}$ gate signal width ΔT_a and ΔT_b are set to be 3 and 6 μs, respectively. The narrow negative pulsewidth ΔT_3 in full duty cycle is set to be 4.5 μs, and the delay time for synchronous pulse CK is set to be 3.5 μs. The frequency of the PWM is 20 kHz. Waveforms of DC link voltage u_{Cr}, transformer primary winding current i_{Lr}, switch S_L and diode D_L current i_{SL}/i_{DL}, PWM, and auxiliary switch gate signal under low and high load current are shown in Figure 13.47. The figure shows that the inverter worked well under various load currents.

In order to verify the theoretical analysis and simulation results, the proposed soft-switching inverter was tested on an experimental prototype. The DC link voltage is 240 V, rated phase current is 10.8 A, and the switching frequency is 20 kHz. Select 50 A/1200 V BSM 35 GB 120 DN2 dual IGBT module as main inverter switches S_1–S_6 and auxiliary switch S_L; another switch in the same module of S_L can be used as auxiliary switch S_a, and 30 A/600 V IMBH30D-060 IGBT can be used as auxiliary switch S_b. With datasheets of these switches and (13.58) through (13.61), the value of capacitance and the parameter of transformer can be determined. A polyester capacitor of 0.1 μF, 1000 V, was adopted as DC link resonant capacitor C_r. A high-magnetizing-inductance transformer with turn ratio 1:1.8 was used in the experiment. The equivalent inductance is about 8 μH under short-circuit test [25]. The switching frequency is 20 kHz. The monostable flip-flop is set up by IC 74LS123, variable resistor and capacitor. The logical gate can be replaced by

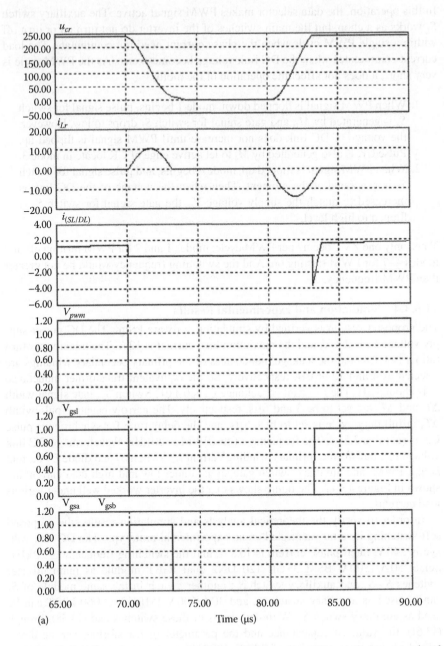

FIGURE 13.47 Waveforms of u_{Cr}, i_{Lr}, i_{SL}/i_{DL}, PWM, and auxiliary switches gate signal under various load currents: (a) under low load current ($I_O = 2$ A) and (b) under high load current ($I_O = 8$ A).

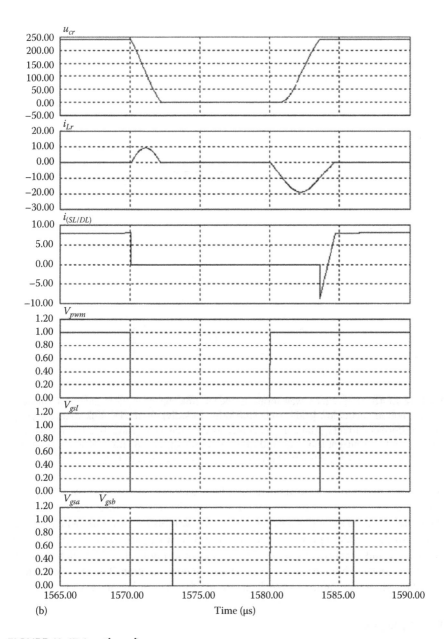

(b)

Time (μs)

FIGURE 13.47 (continued)

programmable logical device to reduce the number of IC. ΔT_a, ΔT_b, ΔT_3, and ΔT_d are set to be 3, 6, 4.5, and 3.5 μs, respectively.

The system is tested in light and heavy loads. The waveforms DC link voltage u_{Cr} and transformer primary winding current i_{Lr} in low- and high-load currents are shown in Figure 13.48a and b, respectively. The transformer-based resonant DC link inverter

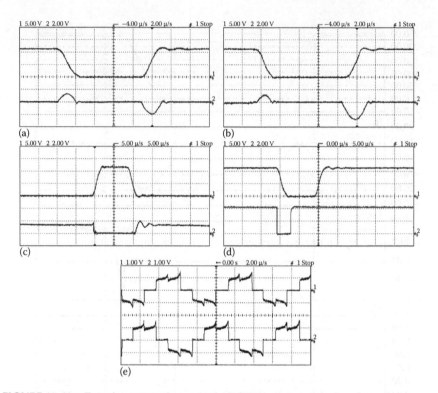

FIGURE 13.48 Experiment waveforms: (a) the DC link voltage u_{Cr} (top) and transformer current i_{Lr} (bottom) under low load current (100 V/div, 10 A/div), (b) the DC link voltage u_{Cr} (top) and transformer current i_{Lr} (bottom) under high load current (100 V/div, 10 A/div), (c) switch S_L voltage (top) and current (bottom) (100 V/div, 10 A/div), (d) the DC link voltage u_{Cr} (top) and synchronous signal CK (bottom) (100 V/div), and (e) phase current of BDCM (5 A/div).

works well under various load currents. The waveforms of auxiliary switch S_L voltage u_{SL} and its current i_{SL} are shown in Figure 13.48c. There is little overlap between the switch S_L voltage and its current during the switching under soft-switching condition, so the switching power losses are low. The waveforms of resonant DC link voltage u_{Cr} and synchronous signal CK are shown in Figure 13.48d, which the main switches can switch under ZVS condition during commutation. The phase current of BDCM is shown in Figure 13.48e. The design of the system is successful.

REFERENCES

1. Nabae, A., Takahashi, I., and Akagi, H. 1980. A neutral-point clamped PWM inverter. *Proceedings of IEEE APEC'80 Conference*, pp. 761–766.
2. Nabae, A., Takahashi, I., and Akagi, H. 1981. A neutral-point clamped PWM inverter. *IEEE Transactions on Industry Applications*, 17, 518–523.
3. Mohan, N., Undeland, T. M., and Robbins, W. P. 2003. *Power Electronics: Converters, Applications and Design*. New York: John Wiley & Sons, Inc.
4. Trzynadlowski, A. M. 1998. *Introduction to Modern Power Electronics*. New York: John Wiley & Sons, Inc.

5. Liu, Y. and Luo, F. L. 2006. Multilevel inverter with the ability of self voltage balancing. *IEE—Proceedings on Electric Power Applications*, 153, 105–115.
6. Pan, Z. Y. and Luo, F. L. 2004. Novel soft-switching inverter for brushless DC motor variable speed drive system. *IEEE—Transactions on Power Electronics*, 19, 280–288.
7. Divan, D. M. 1989. The resonant dc link converter—A new concept in static power conversion. *IEEE Transactions on Industry Applications*, 25, 317–325.
8. Divan, D. M. and Skibinski, G. 1989. Zero-switching-loss inverters for highpower applications. *IEEE Transactions on Industry Applications*, 25, 634–643.
9. Yi, W., Liu, H. L., Jung, Y. C., Cho, J. G., and Cho, G. H. 1992. Program-controlled soft switching PRDCL inverter with new space vector PWM algorithm. *Proceedings of IEEE PESC'92*, Toledo, Spain, pp. 313–319.
10. Malesani, L., Tenti, P., Tomasin, P., and Toigo, V. 1995. High efficiency quasiresonant dc link three-phase power inverter for full-range PWM. *IEEE Transactions on Industry Applications*, 31, 141–148.
11. Jung, Y. C., Liu, H. L., Cho, G. C., and Cho, G. H. 1995. Soft switching space vector PWM inverter using a new quasiparallel resonant dc link. *Proceedings of IEEE PESC'95*, Atlanta, GA, pp. 936–942.
12. Zhengfeng, M. and Yanru, Z. 2001. A novel dc-rail parallel resonant ZVT VSI for three-phases AC motor drive. *Proceedings of the International Conference on Electrical Machines and Systems (ICEMS'2001)*, Shenyang, China, pp. 492–495.
13. Murai, Y., Kawase, Y., Ohashi, K., Nagatake, K., and Okuyama, K. 1989. Torque ripple improvement for brushless dc miniature motors. *IEEE Transactions on Industry Applications*, 25, 441–450.
14. Chang-hee Won, C., Joong-ho Song, J., and Choy, I. 2002. Commutation torque ripple reduction in brushless dc motor drives using a single dc current sensor. *Proceedings of IEEE PESC*, Cairns, Australia, pp. 985–990.
15. Sebastian, T. and Gangla, V. 1996. Analysis of induced EMF waveforms and torque ripple in a brushless permanent magnet machine. *IEEE Transactions on Industry Applications*, 32, 195–200.
16. Pillay, P. P. and Krishnan, R. 1988. Modeling of permanent magnet motor drives. *IEEE Transactions on Industrial Electronics*, 35, 537–541.
17. Pan, Z. Y. and Luo, F. L. 2005. Novel resonant pole inverter for brushless DC motor drive system. *IEEE-Transactions on Power Electronics*, 20, 173–181.
18. De Doncker, R. W. and Lyons, J. P. 1990. The auxiliary resonant commutated pole converter. *Proceedings of the IEEE Industry Applications Society Annual Meeting*, Washington, DC, pp. 1228–1235.
19. McMurray, W. 1989. Resonant snubbers with auxiliary switches. *Proceedings of the IEEE Industry Applications Society Annual Meeting*, San Diego, CA, pp. 289–834.
20. Vlatkovic, V., Borojevic, D., Lee, F., Cuadros, C., and Gataric, S. 1993. A new zero-voltage transition, three-phase PWM rectifier/inverter circuit. *Proceedings of IEEE PESC*, Seattle, WA, pp. 868–873.
21. Cuadros, C., Borojevic, D., Gataric, S., and Vlatkovic, V. 1994. Space vector modulated, zero-voltage transition three-phase to DC bidirectional converter. *Proceedings of IEEE PESC*, Taipei, Taiwan, pp. 16–23.
22. Lai, J. S., Young Sr., R. W., Ott Jr., G. W., White, C. P., McKeever, J. W., and Chen, D. 1995. A novel resonant snubber based soft-switching inverter, *Proceedings of the Applied Power Electronics Conference*, Atlanta, GA, pp. 797–803.
23. Lai, J. S., Young Sr., R. W., Ott Jr., G. W., McKeever, J. W., and Peng, F. Z. 1996. A delta-configured auxiliary resonant snubber inverter. *IEEE Transactions on Industrial Applications*, 32, 518–525.
24. Miller, T. J. E. 1989. *Brushless Permanent-Magnet and Reluctance Motor Drives*. Oxford, U.K.: Clarendon.

25. Sen, P. C. 1997. *Principles of Electric Machines and Power Electronics*. New York: John Wiley.
26. Divan, D. M., Venkataramanan, G., and De Doncker, R. W. 1987. Design methodologies for soft switched inverters. *Proceedings of the IEEE Industry Applications Society Annual Meeting*, Atlanta, GA, pp. 626–639.
27. Ming, Z. Z. and Zhong, Y. R. 2001. A novel DC-rail parallel resonant ZVT VSI for three-phases ac motor drive. *Proceedings of the International Conference on Electronic Machines and Systems*, Incheon, Korea, pp. 492–495.
28. Pan, Z. Y. and Luo, F. L. 2005. Transformer based resonant DC link inverter for brushless DC motor drive system. *IEEE-Transactions on Power Electronics*, 20, 939–947.
29. Wang, K. R., Jiang, Y. M., Dubovsky, S., Hua, G. C., Boroyevich, D., and Lee, F. C. 1997. Novel DC-rail soft-switched three-phase voltage-source inverters. *IEEE Transactions on Industry Applications*, 33, 509–517.

14 Advanced Multilevel DC/AC Inverters Used in Solar Panel Energy Systems

Multilevel DC/AC Inverters have various structures. They have many advantages. Unfortunately, most existing inverters contain too many components (independent/floating batteries/sources, diodes, capacitors, and switches). We introduce a few types of inverters in this chapter that are new approaches of the development in this area. Their simple structure and clear operation are obviously different from those of the existing inverters. Their application in solar panel energy systems is successful. The simulation and experimental results strongly support our design. We believe that these inverters will draw much attention worldwide and will be applied in other renewable energy systems.

14.1 INTRODUCTION

In comparison with PWM DC/AC inverters, multilevel DC/AC inverters have the following advantages: (1) the switching flying voltage is low (from one level to next level), (2) the di/dt and dv/dt is low; (3) The switching frequency is low, (4) THD is better. Unfortunately, most existing inverters contain too many components (independent/floating batteries/sources, diodes, capacitors, and switches). For example, a diode-clamped inverter, also called the neutral-point clamped (NPC) inverter, has $n = (2b + 1)$ levels and needs the following components[1–3]:

- $4b$ switches
- $2b$ capacitors
- $(4b - 2)$ diodes

A linear H-bridged inverter has $n = (2b + 1)$ levels and needs the following components[4–8]:

- b floating batteries
- $4b$ switches
- $4b$ diodes

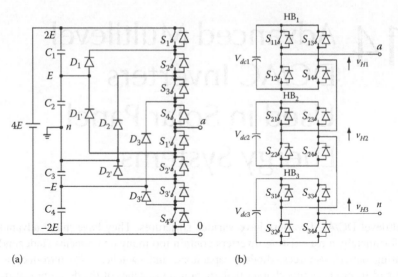

FIGURE 14.1 Example inverters. (a) Five-level NPC inverter and (b) three-bridge inverter.

where n is the level number that is always an odd number, and b is the stage number (from the neutral point to the top point) or bridge number. Figure 14.1 shows some examples: Five-level NPC inverter and three-bridge inverter.

From Figure 14.1, we can see that a five-level NPC inverter contains a 4E battery (or 22E batteries), 8 switches, 14 diodes, and 4 capacitors; a seven-level linear three-bridge inverter has three floating batteries ($V_{dc1} = V_{dc2} = V_{dc3} = E$), 12 diodes, and 12 switches.

We introduce a few new circuits of multilevel DC/AC inverters that are different from those of the existing multilevel DC/AC inverters. They are called laddered multilevel DC/AC inverters. Their structure is simple, and their operation is clear. It is a new approach to the development of DC/AC inverter.

14.2 PROGRESSIONS (SERIES)

In Mathematics, a progression is a series of numbers or quantities in which there is always the same relation between each quantity and the one succeeding it. We introduce several progressions in this section. We assume all progressions have the value of the general ith item as V_i, and the value of their first item V_1 is 1 for our laddered multilevel inverter design.

14.2.1 ARITHMETICAL PROGRESSIONS

Arithmetical progressions are general series. We usually see them as

- Unit progression
- Natural number progression
- Odd number progression

All arithmetical progressions have the same deference value between any two adjacent items. We define the value of the first item as V_1, the value of the general ith item as V_i, and the deference as d. Therefore, the value of the general item V_i is

$$V_i = V_1 + (i-1)d \tag{14.1}$$

The sum of the items from the first item to the ith item is S_i:

$$S_i = iV_1 + \frac{i(i-1)}{2}d \tag{14.2}$$

14.2.1.1 Unit Progression
The unit progression is listed in Table 14.1. We assume that the last item is b, and the value is V_b. The general item is the ith item, and the value is V_i. Since d is 0, the sum of the items S_i from 1 to the ith item is i.

14.2.1.2 Natural Number Progression
The natural number progression is listed in Table 14.2. Since the deference d is 1, the sum of the items S_i from 1 to the ith item is $S_i = i + (i(i-1)/2)$.

14.2.1.3 Odd Number Progression
The odd number progression is listed in Table 14.3. Since the difference d is 2, the sum of the items S_i from 1 to the ith item is $S_i = i^2$.

14.2.2 GEOMETRIC PROGRESSIONS

Geometric progressions are general series. We usually see them as

- Binary progression
- Trinary progression

TABLE 14.1
Unit Progression

Item no.	1	2	3	4	...	i	...	b
Value	1	1	1	1	...	1	...	1
S_i	1	2	3	4	...	i	...	b

TABLE 14.2
Natural Number Progression

Item no.	1	2	3	4	...	i	...	b
Value	1	2	3	4	...	i	...	b
S_i	1	3	6	10	...	$i+\frac{i(i-1)}{2}$	...	$b+\frac{b(b-1)}{2}$

TABLE 14.3
Odd Number Progression

Item no.	1	2	3	4	...	i	...	b
Value	1	3	5	7	...	$2i-1$	...	$2b-1$
S_i	1	4	9	16	...	i^2	...	b^2

All geometric progressions have the same proportion between any two adjacent items. We define the proportion as p. Therefore, the value of the general item V_i is

$$V_i = V_1 p^{i-1} \quad (14.3)$$

The sum of the items from the first item to the ith item is S_i:

$$S_i = \frac{p^i - 1}{p - 1} V_1 \quad (14.4)$$

14.2.2.1 Binary Progression
The binary progression is listed in Table 14.4. Since the proportion p is 2, the sum of the items S_i from 1 to the ith item is $S_i = 2^i - 1$.

14.2.2.2 Trinary Number Progression
The trinary progression is listed in Table 14.5. Since the proportion p is 3, the sum of the items S_i from 1 to the ith item is $S_i = (3^i - 1)/2$.

TABLE 14.4
Binary Progression

Item no.	1	2	3	4	...	i	...	B
Value, V_i	1	2	4	8	...	$2i-1$	...	$2b-1$
Sum, S_i	1	3	7	15	...	$2^i - 1$	...	$2^b - 1$

TABLE 14.5
Trinary Progression

Item no.	1	2	3	4	...	i	...	b
Value	1	3	9	27	...	i	...	b
S_i	1	4	13	40	...	$\frac{3^i-1}{2}$	...	$\frac{3^b-1}{2}$

14.2.3 SPECIAL PROGRESSIONS

Special progressions are designed for laddered multilevel inverters. We have

- Luo-Progression
- Ye-Progression

14.2.3.1 Luo-Progression

Luo-progression is a new progression that is deferent from any existing progression such as arithmetical progressions, geometric progressions, and so on. The value of each item is twice the sum of all previous items plus one from the third item:

$$V_i = \begin{cases} i & i \le 2 \\ 7 \times 3^{i-2} & i \ge 3 \end{cases} \tag{14.5}$$

The sum of the items from the first item to the ith item is S_i:

$$S_i = \sum_{k=1}^{i} V_k = \frac{7 \times 3^{i-2} - 1}{2} \quad i \ge 2 \tag{14.6}$$

Luo-progression is listed in Table 14.6. The sum of the items from 1 to the ith item S_i is $S_i = (7 \times 3^{i-2} - 1)/2$ with $i \ge 2$.

14.2.3.2 Ye-Progression

Ye-progression is also a new progression that is deferent from any existing ones. The value of each item is twice the sum of all previous items plus one from the fourth item:

$$V_i = \begin{cases} i^2 - u(i-2) & i \le 3 \\ 25 \times 3^{i-4} & i \ge 4 \end{cases} \tag{14.7}$$

where $u(i-2)$ is the unit-step function. It is

$$u(i-2) = \begin{cases} 0 & i = 1 \\ 1 & i \ge 2 \end{cases} \tag{14.8}$$

TABLE 14.6
Luo-Progression

Item no.	1	2	3	4	...	i $(i \ge 3)$	...	B
Value, V_i	1	2	7	21	...	$7 \times 3^{i-3}$	...	$7 \times 3^{b-3}$
Sum, S_i	1	3	10	31	...	$\dfrac{7 \times 3^{i-2} - 1}{2}$	...	$\dfrac{7 \times 3^{b-2} - 1}{2}$

TABLE 14.7

Ye-Progression

Item no.	1	2	3	4	5	...	$i\ (i \geq 4)$	...	b
Value, V_i	1	3	8	25	75	...	$25 \times 3^{i-4}$	...	$25 \times 3^{b-4}$
Sum, S_i	1	3	12	37	112	...	$\dfrac{25 \times 3^{i-3}-1}{2}$	...	$\dfrac{25 \times 3^{b-3}-1}{2}$

The sum of the items from the first item to the ith item is S_i:

$$S_i = \sum_{k=1}^{i} V_k = \frac{25 \times 3^{i-3}-1}{2} \quad i \geq 4 \tag{14.9}$$

Ye-progression is listed in Table 14.7. The sum of the items from 1 to the ith item S_i is $S_i = (25 \times 3^{i-3} - 1)/2$ with $i \geq 4$.

14.3 LADDERED MULTILEVEL DC/AC INVERTERS

14.3.1 Special Switches

14.3.1.1 Toggle Switch

We use Toggle switch as shown in Figure 14.2a; it is also called one-pole two-throw (1P2T) switch. The switch has one wiper pole (W) and two contact positions "a" and "b." We define the switch as the wiper pole linked to position "a"; otherwise, the switch is off, meaning the wiper pole linked to position "b."

The terminal voltage of V_{WN} is equal to V_{dc} during switch-on as shown in Figure 14.2c and 0 during switch-off as shown in Figure 14.2b.

14.3.1.2 Changeover Switch

A changeover switch is shown in Figure 14.3. It is also called the two-pole two-throw (2P2T) switch. It can reverse the output voltage from the input voltage. We define that the input voltage is V_{W1W2} and the output voltage is V_{N1N2}:

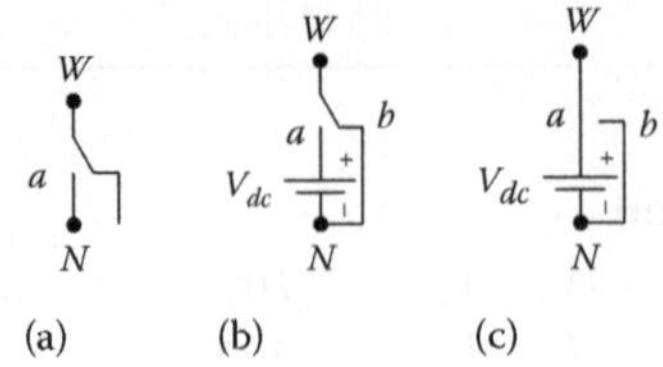

FIGURE 14.2 Toggle switch [or 1P2T switch] with a battery V_{dc}. (a) Toggle switch, (b) toggle switch off with a battery V_{dc}, and (c) switched on.

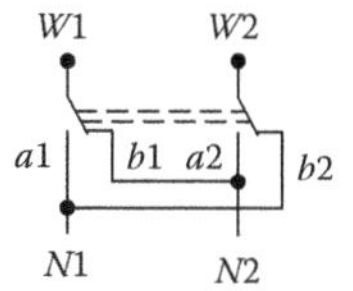

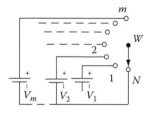

FIGURE 14.3   Changeover switch [or 2P2T switch].

$$V_{out} = V_{N1N2} = \begin{cases} V_{W1W2} & \text{switch is on} \\ -V_{W1W2} & \text{switch is off} \end{cases} \tag{14.10}$$

14.3.1.3   Band Switch

A band switch is shown in Figure 14.4. It is also called the one-pole multi-throw (1PMT) switch. It has one wiper pole (W) and multiple contact positions "0," "1," "2," ... "m."

We assume that there are m voltage sources linked to the bands as $V_1, V_2, \dots, V_m$, as shown in Figure 14.4. The terminal voltage of V_{WN} is equal to various source voltages as shown in Equation 14.11:

$$V_{WN} = \begin{cases} 0 & \text{the wiper is on position } N \\ V_1 & \text{the wiper is on position 1} \\ V_2 & \text{the wiper is on position 2} \\ \dots & \dots \\ V_m & \text{the wiper is on position } m \end{cases} \tag{14.11}$$

14.3.2   General Circuit of Laddered Inverters

The general circuit of laddered inverters is shown in Figure 14.5. It is a symmetrical circuit referring to the neutral point N, i.e., there are two wings: positive wing and negative wing. Each wing has b sets of the toggle switch with a switch Si and battery V_{dci} [where $i = -b, -(b-1), \dots -2, -1, 1, 2, \dots b-1$ and b]. "b" is the ladder stage number; "n" is the total level number. The positive wing contains the b sets; "i" is the ith set number. The negative wing contains the same sets (symmetrically). Therefore, we have

FIGURE 14.4   Band switch [or 1PMT switch] with m-throws.

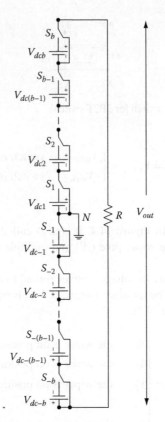

FIGURE 14.5   General circuit of ladder inverters.

$$|V_{dc-i}| = V_{dci} \quad i = 1, 2, \ldots, b-1, b \tag{14.12}$$

To simplify the analysis, we assume that the load is a purely resistive load R.

14.3.3 Linear Ladder Inverters

If we choose all DC voltages V_{dci} to be the same voltage E as in Figure 14.5, we obtain the linear ladder inverters (LLI). This ladder was constructed as a unit progression. The output voltage V_{out} has n levels, $n = 2b + 1$:

$$V_{dci} = E \quad i = -b, -(b-1), \ldots -2, -1, 1, 2, \ldots b-1, b \tag{14.13}$$

The operation status is shown here:

- $V_{out} = bE$: All positive wing switches are on (others are off).
- $V_{out} = (b-1)E$: Switches S_1—$Sb-1$ are on (others are off).
- …
- $V_{out} = 2E$: Switches S_1—S_2 are on (others are off).

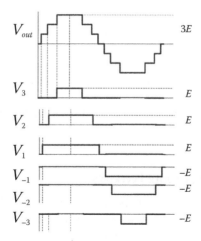

FIGURE 14.6 Seven-level output voltage waveform.

- $V_{out}=E$: Only switch S_1 is on (others are off).
- $V_{out}=0$: All switches are off.
- $V_{out}=-E$: Only Switch $S-1$ is on (others are off).
- $V_{out}=-2E$: Switches $S-1$—$S-2$ are on (others are off).
- ...
- $V_{out}=-(b-1)E$: Switches $S-1$—$S-(b-1)$ are on (others are off).
- $V_{out}=-bE$: All negative wing switches are on (others are off).

We obtain $n=2^{b+1}$ levels. For example, if $b=3$, we have the total level number $n=7$ levels. The output voltage waveform is shown in Figure 14.6.

14.3.4 NATURAL NUMBER LADDER INVERTERS

If we choose all DC voltages V_{dci} to be the voltage iE as in Figure 14.5, we obtain the natural number ladder inverters (NNLI). This ladder was constructed as a natural number progression. The output voltage V_{out} has n levels, $n=b^2+b+1$ (refer to Table 14.2):

$$V_{dci} =|i| E \quad i=-b,-(b-1),\ldots,-2,-1,1,2,\ldots b-1,b \qquad (14.14)$$

The operation status is shown as follows:

- $V_{out}=nE$: All positive wing switches are on (others are off).
- $V_{out}=(n-1)E$: Switches S_2—S_b are on (others are off).
- ...
- $V_{out}=2E$: Only switch S_2 is on (others are off).
- $V_{out}=E$: Only switch S_1 is on (others are off).
- $V_{out}=0$: All switches are off.

- $V_{out}=-E$: Only switch S_{-1} is on (others are off).
- $V_{out}=-2E$: Only switch S_{-2} is on (others are off).
- ...
- $V_{out}=-(n-1)E$: Switches S_{-2}—S_{-b} are on (others are off).
- $V_{out}=-nE$: All negative wing switches are on (others are off).

We obtain $n=b^2+b+1$ levels. For example, if $b=3$, we have the total level number $n=13$ levels.

14.3.5 Odd Number Ladder Inverters

If we choose all DC voltages V_{dci} to be the voltage $(2i-1)E$ (in positive wing) as in Figure 14.3, we obtain the odd number ladder inverters (ONLI). This ladder was constructed as an odd number progression. The output voltage V_{out} has n levels, $n=2b^2+1$ (refer to Table 14.3).

$$V_{dci} = \begin{cases} (2i-1)E & i\geq 1 \\ (2i+1)E & i\leq -1 \end{cases} \tag{14.15}$$

The operation status is shown as follows:

- $V_{out}=b^2E$: All positive wing switches are on (others are off).
- $V_{out}=(b^2-1)E$: Switches S_2—S_b are on (others are off).
- ...
- $V_{out}=3E$: Only switch S_2 is on (others are off).
- $V_{out}=2E$: Switch S_2 and S_{-1} are on (others are off).
- $V_{out}=E$: Only switch S_1 is on (others are off).
- $V_{out}=0$: All switches are off.
- $V_{out}=-E$: Only switch S_{-1} is on (others are off).
- $V_{out}=-2E$: Switch S_{-2} and S_1 are on (others are off).
- $V_{out}=-3E$: Only Switch S_{-2} is on (others are off).
- ...
- $V_{out}=-(b^2-1)E$: Switches S_{-2}—S_{-b} are on (others are off).
- $V_{out}=-b^2E$: All negative wing switches are on (others are off).

We obtain $n=2b^2+1$ levels. For example, if $b=3$, we have the total level number $n=19$ levels.

14.3.6 Binary Ladder Inverters

If we choose all DC voltages V_{dci} to be the voltage $(2^i-1)E$ as in Figure 14.5, we obtain the binary ladder inverters (BLI). This ladder was constructed as a binary progression. The output voltage V_{out} has n levels, $n=2^{b+1}-1$ (refer to Table 14.4).

The DC voltages V_{dci} to be binary voltage (in Figure 14.5),

$$V_{dci} = \begin{cases} 2^{i-1}E & i \geq 1 \\ 2^{-i-1}E & i \leq -1 \end{cases} \tag{14.16}$$

We obtain the BLI. The output voltage V_{out} has n levels, $n = 2^{b+1} - 1$. The operation status is shown as follows:

- $V_{out} = (2^b - 1)E$: All positive wing switches are on (others are off).
- ...
- $V_{out} = 3E$: Switches S_1—S_2 are on (others are off).
- $V_{out} = 2E$: Only switch S_2 is on (others are off).
- $V_{out} = E$: Only switch S_1 is on (others are off).
- $V_{out} = 0$: All switches are off.
- $V_{out} = -E$: Only switch S_{-1} is on (others are off).
- $V_{out} = -2E$: Only switch S_{-2} is on (others are off).
- $V_{out} = -3E$: Switches S_{-1}—S_{-2} are on (others are off).
- ...
- $V_{out} = -(2^b - 1)E$: All negative wing switches are on (others are off).

We obtain $n = 2^{b+1} - 1$ levels. For example, if $b = 3$, we have an $n = 15$ levels output voltage waveform as shown in Figure 14.7.

14.3.7   Modified Binary Ladder Inverters

The modified binary ladder inverter (MBLI) is shown in Figure 14.8. We used a change-over (2P2T) switch and saved half of the batteries and switches (in the negative wing).

We use a 2P2T switch exchange the output voltage polarity and can save half of the total 1P2T switches and batteries in the negative wing. The operation status and

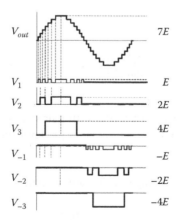

FIGURE 14.7   Fifteen-level output voltage waveform.

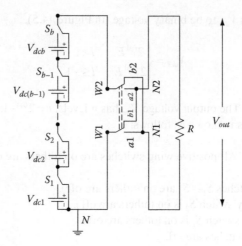

FIGURE 14.8 Modified binary ladder inverters.

output voltage waveform can be referred to in Section 14.3.6 (also Figure 14.7). The output voltage V_{out} has n levels, $n = 2^{b+1} - 1$.

14.3.8 LUO-PROGRESSION LADDER INVERTERS

In our research process, we invented and defined a new progression and called it Luo-Progression. It is different from all existing progressions such as the arithmetical progression, geometric progression, and so on. We still use the symbols as "b" is the progression stage number: "i" is the ith item; V_i is the ith item value; S_i is the sum value; "n" is the level number. We define

$$V_i = \begin{cases} i & i \leq 2 \\ 7 \times 3^{i-2} & i \geq 3 \end{cases} \tag{14.17}$$

The total levels, n, are

$$n = \left(2 \times \sum_{i=1}^{b} V_i \right) + 1 = 7 \times 3^{b-2} \quad b \geq 2 \tag{14.18}$$

From Table 14.8, if we construct a ladder with four stages ($b=4$) in both wings, we can obtain 63 levels ($n=63$). We still assume the level unit is "E." The operation status is shown as follows:

- $V_{out} = 31E$: All positive wing switches are on (others are off).
- ...
- $V_{out} = 4E$: Switches S_3, S_{-1}, S_{-2} are on (others are off).
- $V_{out} = 3E$: Switches S_1 and S_2 are on (others are off).

Luo-Progression (continued)

Stage number	1	2	3	4	5	6	...	ith $(i \geq 3)$	...	b
Item value, V_i	1	2	7	21	63	189	...	$7 \times 3^{i-3}$	...	$7 \times 3^{b-3}$
Sum, S_i	1	3	10	31	94	283	...	$\dfrac{7 \times 3^{i-2}-1}{2}$	...	$\dfrac{7 \times 3^{b-2}-1}{2}$
Total levels, n	3	7	21	63	189	567	...	$7 \times 3^{i-2}$	...	$7 \times 3^{b-2}$

- $V_{out}=2E$: Only switch S_2 is on (others are off).
- $V_{out}=E$: Only switch S_1 is on (others are off).
- $V_{out}=0$: All switches are off.
- $V_{out}=-E$: Only switch S_{-1} is on (others are off).
- $V_{out}=-2E$: Only switch S_{-2} is on (others are off).
- $V_{out}=-3E$: Switches S_{-1} and S_{-2} are on (others are off).
- $V_{out}=-4E$: Switches S_{-3}, S_1, and S_2 are on (others are off).
- ...
- $V_{out}=-31E$: All negative wing switches are on (others are off).

14.3.9 Ye-Progression Ladder Inverters

In our research process, we invented and defined another new progression and called it Ye-Progression. It is also different from all existing progressions. We still use the symbols as "b" is the progression stage number: "i" is the ith item; V_i is the ith item value; S_i is the sum value; "n" is the level number. We define

$$V_i = \begin{cases} i^2 - u(i-2) & i \leq 3 \\ 25 \times 3^{i-4} & i \geq 4 \end{cases} \qquad (14.19)$$

The total levels, n, are

$$n = \left(2 \times \sum_{i=1}^{b} V_i \right) + 1 = 25 \times 3^{b-4} \quad b \geq 4 \qquad (14.20)$$

From Table 14.9, if we construct a ladder with four stages ($b=4$) in both wings, we can obtain 75 levels ($n=75$). We still assume the level unit is "E." The operation status is shown as follows:

- $V_{out}=37E$: All positive wing switches are on (others are off).
- ...
- $V_{out}=4E$: Switches S_1 and S_3 are on (others are off).
- $V_{out}=3E$: Only switch S_2 is on (others are off).

TABLE 14.9

Ye-Progression (continued)

Stage number	1	2	3	4	5	6	...	ith ($i \geq 4$)	...	b
Item value, V_i	1	3	8	25	75	225	...	$25 \times 3^{i-4}$	...	$25 \times 3^{b-4}$
Sum, S_i	1	4	12	37	112	337	...	$\dfrac{25 \times 3^{i-3} - 1}{2}$	...	$\dfrac{25 \times 3^{b-3} - 1}{2}$
Total levels, n	3	9	25	75	225	675	...	$25 \times 3^{i-3}$	...	$25 \times 3^{b-3}$

- $V_{out} = 2E$: Switches S_2 and S_{-1} are on (others are off).
- $V_{out} = E$: Only switch S_1 is on (others are off).
- $V_{out} = 0$: All switches are off.
- $V_{out} = -E$: Only switch S_{-1} is on (others are off).
- $V_{out} = -2E$: Switches S_{-2} and S_1 are on (others are off).
- $V_{out} = -3E$: Only switch S_{-2} is on (others are off).
- $V_{out} = -4E$: Switches S_{-1} and S_{-2} are on (others are off).
- ...
- $V_{out} = -37E$: All negative wing switches are on (others are off).

14.3.10 TRINARY LADDER INVERTERS

If we choose the DC voltages V_{dci} to be the trinary voltage as in Figure 14.5,

$$V_{dci} = \begin{cases} 3^{i-1}E & i \geq 1 \\ -3^{-i-1}E & i \leq 1 \end{cases} \tag{14.21}$$

we obtain the trinary ladder inverters (TLI). The total level number $n = 3^b$. For example. If we choose $b = 4$, we obtain V_{out} with 81 levels.
The operation status is shown as follows:

- $V_{out} = 40E$: All positive wing switches are on (others are off).
- ...
- $V_{out} = 3E$: Only switch S_2 is on (others are off).
- $V_{out} = 2E$: Switches S_2 and S_{-1} are on (others are off).
- $V_{out} = E$: Only switch S_1 is on (others are off).
- $V_{out} = 0$: All switches are off.
- $V_{out} = -E$: Only switch S_{-1} is on (others are off).
- $V_{out} = -2E$: Switches S_1 and S_{-2} are on (others are off).
- $V_{out} = -3E$: Only switch S_{-2} is on (others are off).
- ...
- $V_{out} = -40E$: All negative wing switches are on (others are off).

14.4 COMPARISON OF ALL LADDERED INVERTERS

We introduced eight types of laddered inverters in Section 14.3. In order to understand ladder inverters and some other inverters, the following definitions are offered:

- LLI—Linear ladder inverters
- NNLI—Natural number ladder inverters
- ONLI—Odd number ladder inverters
- BLI—Binary ladder inverters
- MBLI—Modified binary ladder inverters
- LPLI—Luo-progression ladder inverters
- YPLI—Ye-Progression Ladder Inverters
- TLI—Trinary ladder inverters
- NPCI—Neutral-point-clamped inverters
- LHBI—Linear H-bridged inverters

We compare them in Table 14.10.

For example, if $b=3$, we obtain the level numbers for each inverter as shown in Table 14.11.

For example, if $b=5$, we obtain the level numbers for each inverter as shown in Table 14.12.

From Tables 14.10 through 14.12, it can obviously be seen that the TLI, YPLI,, LPLI use fewer components and yield higher number of levels.

14.5 SOLAR PANEL ENERGY SYSTEMS

The sun is a star in the universe; the earth is a planet surrounding the sun, as shown in Figure 14.9. The earth flies in an oval orbit surrounding the sun, and the sun is located as a focus of the oval orbit. The average distance between sun and earth is about 150 Mkm (150,000,000 km). The sun radiates a power of 3.8×10^{20} MW into the space. Our Earth has received a power of 174×10^9 MW from the sun.

The solar panel is constructed with the solar cell (or photovoltaic cell), which belongs to a wide multidisciplinary area. It converts solar energy into electricity by the photovoltaic effect. The solar cells can be divided into two groups: monocrystalline and multicrystalline silicon wafers. Figure 14.10 shows a solar cell made from a monocrystalline silicon wafer.

The overall approach is shown in Figure 14.11, which is assembled by a few solar cells. The theoretical solar panel system (a certain solar panel) power curve is shown in Figure 14.12. The current (from the solar panel) is nearly constant in the low-voltage (from the solar panel) regain. When the input voltage reaches 16.2 V, the input current sharply reduces to zero. The blue curve is the output power with its maximum power point (MPP) at about 132 W.

We set the voltage unit $E = 16.2$ V. It is easy to construct the batteries for a BLI:

- $V_{dc1} = |V_{dc-1}| = E = 16.2$ V
- $V_{dc2} = |V_{dc-2}| = 2E = 32.4$ V
- $V_{dc3} = |V_{dc-3}| = 4E = 64.8$ V

TABLE 14.10
Comparison with All Ladder Inverters and Some Other Inverters

Inverters	LLI	NNLI	ONLI	BLI	MBLI	LPLI	YPLI	TLI	NPCI	LHBI
b, Stage number	$2b$	$2b$	$2b$	$2b$	b	$2b$	$2b$	$2b$	$2b$	b
Battery number	$2b$	$2b$	$2b$	$2b$	b	$2b$	$2b$	$2b$	$1/2$	b
Switch number	$2b$	$2b$	$2b$	$2b$	$b+1$	$2b$	$2b$	$2b$	$4b$	$4b$
Capacitor number	0	0	0	0	0	0	0	0	$2b$	0
Diode number	0	0	0	0	0	0	0	0	$4b-2$	$4b$
n, Total levels	$2b+1$	b^2+b+1	$2b^2+1$	$2^{b+1}-1$	$2^{b+1}-1$	$7\times3^{b-2}$	$25\times3^{b-3}$	3^b	$2b+1$	$2b+1$

TABLE 14.11
Comparison with the Inverters (if $b = 3$)

Inverters	LLI	NNLI	ONLI	BLI	MBLI	LPLI	YPLI	TLI	NPCI	LHBI
Stage number	6	6	6	6	3	6	6	6	6	3
Battery number	6	6	6	6	3	6	6	6	1/2	3
Switch number	6	6	6	6	4	6	6	6	12	12
Capacitor number	0	0	0	0	0	0	0	0	6	0
Diode number	0	0	0	0	0	0	0	0	10	12
n, Total levels	7	13	19	15	15	21	25	27	7	7

TABLE 14.12
Comparison with the Inverters (if $b = 5$)

Inverters	LLI	NNLI	ONLI	BLI	MBLI	LPLI	YPLI	TLI	NPCI	LHBI
Stage number	10	10	10	10	5	10	10	10	10	5
Battery number	10	10	10	10	5	10	10	10	1/2	5
Switch number	10	10	10	10	6	10	10	10	20	20
Capacitor number	0	0	0	0	0	0	0	0	10	0
Diode number	0	0	0	0	0	0	0	0	18	20
n, Total levels	11	31	51	63	63	189	225	243	11	11

Another setting is the following to construct the batteries for an LPLI:

- $V_{dc1} = |V_{dc-1}| = E = 16.2\,\text{V}$
- $V_{dc2} = |V_{dc-2}| = 2E = 32.4\,\text{V}$
- $V_{dc3} = |V_{dc-3}| = 7E = 113.4\,\text{V}$

14.6 SIMULATION AND EXPERIMENTAL RESULTS

We use a BLI with $b = 3$ to conduct the simulation. The output voltage has 15 levels. The simulation result is shown in Figure 14.13. The corresponding experimental result is shown in Figure 14.14.

We use an LPLI with $b = 3$ to do the simulation again. The output voltage has 21 levels. The simulation result is shown in Figure 14.15. The corresponding experimental result is shown in Figure 14.16.

14.7 SWITCHED-CAPACITOR MULTILEVEL DC/AC INVERTERS

A switched capacitor (SC) is usually manufactured a switch and a capacitor together [9–12]. It is the device used in DC/DC converters for years. It can be integrated into power semiconductor IC chips. Hence, SC converters have limited size and work in high switching frequency. This technique opened the way to build converters with high power density. Therefore, they have drawn much attention from the researchers

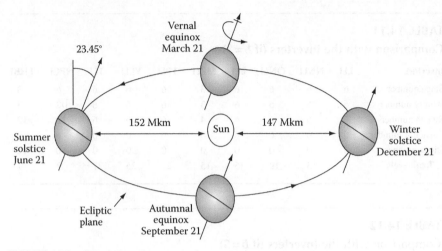

FIGURE 14.9 Sun and earth.

FIGURE 14.10 Monocrystalline solar cell.

and manufacturers. We are the first people to use a switched capacitor in DC/AC inverters.

For example, a switched-capacitor DC/DC converter is shown in Figure 14.17a. It has two SCs: C_1 and C_2; main switch S and two slave switches: S_1 and S_2; and three diodes. The main switch S and the slave switches are exclusively operated, i.e., the main switch is on, the slave switches are off; and vice versa, the main switch is off, the slave switches are on.

When the main switch S is on, the slave switches are off, and all diodes are conducted. The equivalent circuit is shown in Figure 14.17b. Both SCs are charged by the source voltage E in the steady state. When the main switch S is off, the slave switches are on, and all diodes are blocked. The equivalent circuit is shown in Figure 14.17c.

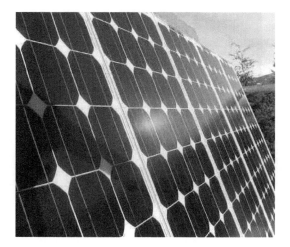

FIGURE 14.11   Overall approach of a solar panel.

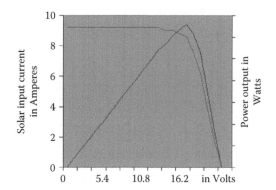

FIGURE 14.12   Theoretical system power curve.

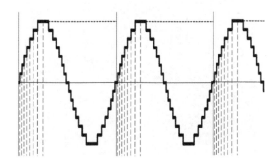

FIGURE 14.13   Simulation result of a 15-level BLI.

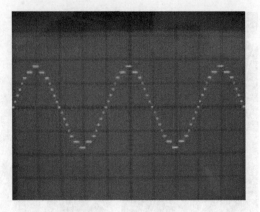

FIGURE 14.14 Experimental result of a 15-level BLI.

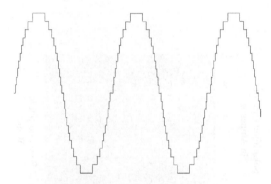

FIGURE 14.15 Simulation result of a 21-level LPLI.

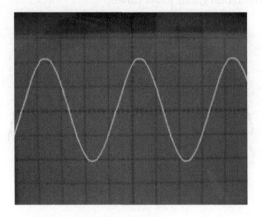

FIGURE 14.16 Experimental result of a 21-level BLI.

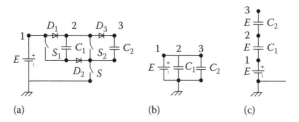

FIGURE 14.17 Switched capacitor DC/DC converter. (a) The circuit, (b) the equivalent circuit for switch S is on, and (c) switch S is off.

The voltage obtained at the point 1 is E, the voltage at the point 2 is $2E$, and the voltage at the point 3 is $3E$ in the steady state.

14.7.1 SWITCHED CAPACITOR USED IN MULTILEVEL DC/AC INVERTERS

We introduce several switched capacitors used in multilevel DC/AC inverters in this section. We can use the toggle switch [1P2T switch] or the one-pole multithrow (1PMT) switch [band switch]. We use the band switch in the following circuit.

14.7.1.1 Five-Level SC Inverter

A five-level switched-capacitor inverter is shown in Figure 14.18.

There is only one DC voltage source E, one three-position band switch, and one 2P2T switch in the circuit. Main switch is S, the slave switch S_1 is exclusively switching referring to the main switch S, i.e., when S is on S_1 is off, and alternatively S is off when S_1 is on. Capacitor C_1 is a switched capacitor. When S is on S_1 is off, and diode D_1 is conducted. Therefore, capacitor C_1 is charged by the voltage E in steady state. When S is off S_1 is on, and diode D_1 is blocked. The voltage at point 2 is $V_2 = 2 \times E$ (V_1 is always E).The operation status is shown as follows:

- $V_{out} = 2E$: The 2P2T is on, the band switch locates on position 2, and main switch S is off.
- $V_{out} = E$: The 2P2T is on, the band switch locates on position 1, main switch S is on.
- $V_{out} = 0$: The band switch locates on position 0 (i.e., N); all switches can be on.

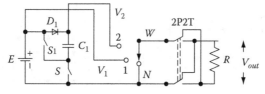

FIGURE 14.18 Five-level switched-capacitor inverter.

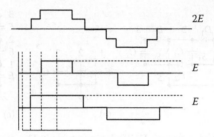

FIGURE 14.19   Five level waveform.

- $V_{out}=-E$: The 2P2T is off, the band switch locates on position 1, and main switch S is on.
- $V_{out}=-2E$: The 2P2T is off, the band switch locates on position 2, and main switch S is off.

We have obtained a five-level output AC voltage. The output voltage peak value is two times the input DC voltage E; the waveform is shown in Figure 14.19.

14.7.1.2   Nine-Level SC Inverter

A nine-level switched-capacitor inverter is shown in Figure 14.20.

There is only one DC voltage source E, one 5-position band switch, and one 2P2T switch in the circuit. Main switch is S, and the slave switches S_{1-3} are exclusively switching referring to the main switch S, i.e., when S is on all slave switches are off, and vice versa. Capacitors C_{1-3} are the switched capacitors. When S is on, all diodes are conducted. Therefore, all SCs are charged by the voltage E in steady state. When S is off, S_1 is on, and diode D_1 is blocked. The voltage at point 2 is $V_2=2\times E$; the voltage at point 2 is $V_3=3\times E$; the voltage at point 4 is $V_4=4\times E$; (V_1 is always E). Therefore, the operation status is as follows:

- $V_{out}=4E$: The 2P2T is on, the band switch locates on position 4, and main switch S is off.
- $V_{out}=3E$: The 2P2T is on, the band switch locates on position 3, and main switch S is off.
- $V_{out}=2E$: The 2P2T is on, the band switch locates on position 2, and main switch S is off.

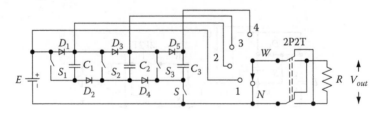

FIGURE 14.20   Nine-level switched-capacitor inverter.

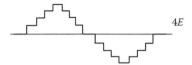

FIGURE 14.21   Nine-level waveform.

- $V_{out}=E$: The 2P2T is on, the band switch locates on position 1, and main switch S is on.
- $V_{out}=0$: The band switch locates on position 0 (i.e., N); all switches can be on.
- $V_{out}=-E$: The 2P2T is off, the band switch locates on position 1, and main switch S is on.
- $V_{out}=-2E$: The 2P2T is off, the band switch locates on position 2, and main switch S is off.
- $V_{out}=-3E$: The 2P2T is off, the band switch locates on position 3, and main switch S is off.
- $V_{out}=-4E$: The 2P2T is off, the band switch locates on position 4, and main switch S is off.

We have obtained a nine-level output AC voltage. The output voltage peak value is four times the input DC voltage E, and the waveform is shown in Figure 14.21.

14.7.1.3   Fifteen-Level SC Inverter

A 15-level switched-capacitor inverter is shown in Figure 14.22.

There are only one DC voltage source E, one 7-position band switch, and one 2P2T switch in the circuit. The main switch is S, and the slave switches S_{1-6} are exclusively switched referring to the main switch S, i.e., when S is on, all slave switches off, and vice versa. Capacitors C_{1-6} are SCs. When S is on and all slave switches are off, all diodes are conducted. Therefore, all SCs are charged by the voltage E in steady state. The voltage at point 2 is $V_2=2\times E$; the voltage at point 2 is $V_3=3\times E$; the voltage at point 4 is $V_4=4\times E$, and so on; (V_1 is always E). Therefore, the operation status is as follows:

- $V_{out}=7E$: The 2P2T is on, the band switch locates on position 7, and the main switch S is off.
- $V_{out}=6E$: The 2P2T is on, the band switch locates on position 6, and the main switch S is off.

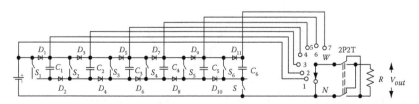

FIGURE 14.22   Fifteen-level switched-capacitor inverter.

- $V_{out}=5E$: The 2P2T is on, the band switch locates on position 5, and the main switch S is off.
- $V_{out}=4E$: The 2P2T is on, the band switch locates on position 4, and the main switch S is off.
- $V_{out}=3E$: The 2P2T is on, the band switch locates on position 3, and the main switch S is off.
- $V_{out}=2E$: The 2P2T is on, the band switch locates on position 2, and the main switch S is off.
- $V_{out}=E$: The 2P2T is on, the band switch locates on position 1, and the main switch S is on.
- $V_{out}=0$: The band switch locates on position 0 (i.e., N), and all switches are on.
- $V_{out}=-E$: The 2P2T is off, the band switch locates on position 1, and the main switch S is on.
- $V_{out}=-2E$: The 2P2T is off, the band switch locates on position 2, and the main switch S is off.
- $V_{out}=-3E$: The 2P2T is off, the band switch locates on position 3, and the main switch S is off.
- $V_{out}=-4E$: The 2P2T is off, the band switch locates on position 4, and the main switch S is off.
- $V_{out}=-5E$: The 2P2T is off, the band switch locates on position 5, and the main switch S is off.
- $V_{out}=-6E$: The 2P2T is off, the band switch locates on position 6, and the main switch S is off.
- $V_{out}=-7E$: The 2P2T is off, the band switch locates on position 7, and the main switch S is off.

We have obtained 15-level output AC voltage. The output voltage peak value is seven times the input DC voltage E, and the waveform is shown in Figure 14.23.

14.7.1.4 Higher-Level SC Inverter

Repeatedly add the components (S_1-C_1-D_1-D_2) as in Figure 14.18, and we can obtain the higher-level inverters. We believe the readers can catch the clue to construct the higher-level inverters. Therefore, we do not repeat the further circuits. We suggest that the readers try to construct the 21-level SC inverter.

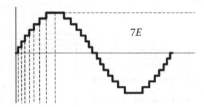

FIGURE 14.23   Fifteen-level waveform of a switched-capacitor inverter.

14.7.2   Simulation and Experimental Results

We use the circuit as shown in Figure 14.20 to conduct the simulation. The output voltage has 15 levels. The simulation result is shown in Figure 14.13. The corresponding experimental result is shown in Figure 14.14.

We use a circuit to produce a higher-level SC inverter (e.g., 21 level SC inverter) to do the simulation again. The output voltage has 21 levels. The simulation result is shown in Figure 14.15. The corresponding experimental result is shown in Figure 14.16.

14.8   SUPER-LIFT CONVERTER MULTILEVEL DC/AC INVERTERS

Super-Lift technique is the most outstanding contribution in DC/DC conversion technology [13–18]. Positive output super-lift Luo-converters can easily lift the DC input voltage into a higher-level output DC voltage [13,14]. In addition, they have many subseries and circuits. The voltage transfer gains of the circuits are increased at geometrical series (power series) stage by stage. Therefore, the super-lift technique has drawn much attention worldwide in recent decades. We introduce the "super-lift converter multilevel DC/AC inverters" in this section, which is new approach of the development in this area. Their simple structure and clear operation are obviously different from those of the existing multilevel inverters. Their application in solar panel energy systems is successful. The simulation and experimental results strongly support our design. We believe that these inverters will draw much attention over the world and be applied in other renewable energy systems.

14.8.1   Some P/O Super-Lift Luo-Converters

The elementary positive output super-lift Luo-converter is shown in Figure 14.24a. The equivalent circuits when switch S is on are shown in Figure 14.24b; and the equivalent circuits when switch S is off are shown in Figure 14.24c.

When the main switch S is on, the inductor L_1 and capacitor C_1 are charged by the source voltage V_{in}. In the steady state $V_{C1} = V_{in}$. When the main switch S is off, the output voltage is lifted by source voltage, inductor L_1, and capacitor C_1; hence, the output voltage V_O is highly lifted:

$$V_O = \frac{2-k}{1-k} V_{in}$$

(14.22a)

where k is the duty cycle. A re-lift positive output super-lift Luo-converter is shown in Figure 14.25.

Its output voltage V_O is

$$V_O = \left(\frac{2-k}{1-k}\right)^2 V_{in}$$

(14.22b)

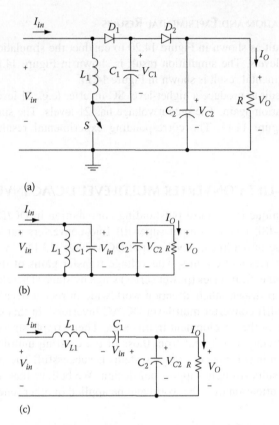

(a)

(b)

(c)

FIGURE 14.24 Elementary circuit of the positive output super-lift Luo-converter. (a) The circuit, (b) the equivalent circuit for switch S is on, and (c) switch S is off.

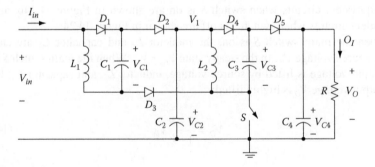

FIGURE 14.25 Re-lift circuit of the positive output super-lift Luo-converter.

14.8.2 SUPER-LIFT CONVERTERS USED IN MULTILEVEL DC/AC INVERTERS

We introduce several switched capacitors used in multilevel DC/AC inverters in this section. We can use the toggle switch [1P2T switch] or the one-pole multithrow (1PMT) switch [band switch]. We use the band switch in the following circuit.

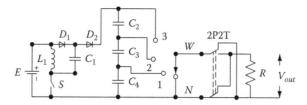

FIGURE 14.26   Seven-level super-lift inverter.

14.8.2.1   Seven-Level SL Inverter

A seven-level super-lift inverter is shown in Figure 14.26.

There is only one DC voltage source E, one three-position band switch, and one 2P2T switch in the circuit. We choose $k=0.5$, and the output voltage is $3E$. We use three capacitors, C_2, C_3, and C_4, to split the output voltage in three levels: E, $2E$, and $3E$. Therefore, the operation status is as follows:

- $V_{out}=3E$: The 2P2T is on, and the band switch locates on position 3.
- $V_{out}=2E$: The 2P2T is on, and the band switch locates on position 2.
- $V_{out}=E$: The 2P2T is on, and the band switch locates on position 1.
- $V_{out}=0$: The band switch locates on position 0 (i.e., N).
- $V_{out}=-E$: The 2P2T is off, and the band switch locates on position 1.
- $V_{out}=-2E$: The 2P2T is off, and the band switch locates on position 2.
- $V_{out}=-3E$: The 2P2T is off, and the band switch locates on position 3.

We have obtained seven-level output AC voltage. The output voltage peak value is three times the input DC voltage E, and the waveform is shown in Figure 14.27.

14.8.2.2   Fifteen-Level SL Inverter

A 15-level super-lift inverter is shown in Figure 14.28.

We choose $k=6/7$ (≈ 0.857), and the output voltage is $7E$. We use seven capacitors, C_2, C_3, C_4, C_5, C_6, C_7, and C_8, to split the output voltage in seven levels: E, $2E$, $3E$, $4E$, $5E$, $6E$, and $7E$. Therefore, the operation status is as follows:

- $V_{out}=7E$: The 2P2T is on, and the band switch locates on position 7.
- $V_{out}=6E$: The 2P2T is on, and the band switch locates on position 6.
- $V_{out}=5E$: The 2P2T is on, and the band switch locates on position 5.

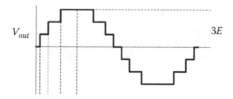

FIGURE 14.27   Seven-level waveform.

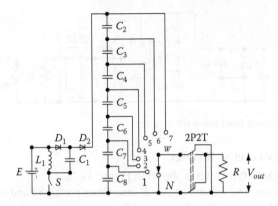

FIGURE 14.28   Fifteen-level super-lift inverter.

- $V_{out} = 4E$: The 2P2T is on, and the band switch locates on position 4.
- $V_{out} = 3E$: The 2P2T is on, and the band switch locates on position 3.
- $V_{out} = 2E$: The 2P2T is on, and the band switch locates on position 2.
- $V_{out} = E$: The 2P2T is on, and the band switch locates on position 1.
- $V_{out} = 0$: The band switch locates on position 0 (i.e., N).
- $V_{out} = -E$: The 2P2T is off, and the band switch locates on position 1.
- $V_{out} = -2E$: The 2P2T is off, and the band switch locates on position 2.
- $V_{out} = -3E$: The 2P2T is off, and the band switch locates on position 3.
- $V_{out} = -4E$: The 2P2T is off, and the band switch locates on position 4.
- $V_{out} = -5E$: The 2P2T is off, and the band switch locates on position 5.
- $V_{out} = -6E$: The 2P2T is off, and the band switch locates on position 6.
- $V_{out} = -7E$: The 2P2T is off, and the band switch locates on position 7.

We have obtained 15-level output AC voltage. The output voltage peak value is seven times the input DC voltage E, and the waveform is shown in Figure 14.29.

14.8.2.3   Twenty-One-Level SL Inverter

A positive output re-lift super-lift Luo-Converter is used to construct a 21-level inverter. Its circuit diagram is shown in Figure 14.30.

We choose $k = (\sqrt{10} - 2)/(\sqrt{10} - 1)$ ($k \approx 0.54$), and the output voltage is $10E$. We use 10 capacitors, C_4, C_5, C_6, C_7, C_8, C_9, C_{10}, C_{11}, C_{12}, and C_{13}, to split the output

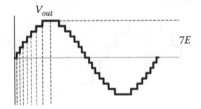

FIGURE 14.29   Fifteen-level waveform of a super-lift inverter.

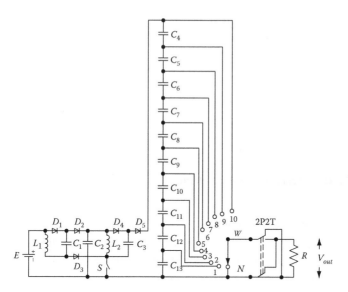

FIGURE 14.30   Twenty-one-level super-lift inverter.

voltage in 10 levels: E, $2E$, $3E$, $4E$, $5E$, $6E$, $7E$, $8E$, $9E$, and $10E$. Therefore, the operation status is as follows:

- $V_{out} = 10E$: The 2P2T is on, and the band switch locates on position 10.
- $V_{out} = 9E$: The 2P2T is on, and the band switch locates on position 9.
- $V_{out} = 8E$: The 2P2T is on, and the band switch locates on position 8.
- $V_{out} = 7E$: The 2P2T is on, and the band switch locates on position 7.
- $V_{out} = 6E$: The 2P2T is on, and the band switch locates on position 6.
- $V_{out} = 5E$: The 2P2T is on, and the band switch locates on position 5.
- $V_{out} = 4E$: The 2P2T is on, and the band switch locates on position 4.
- $V_{out} = 3E$: The 2P2T is on, and the band switch locates on position 3.
- $V_{out} = 2E$: The 2P2T is on, and the band switch locates on position 2.
- $V_{out} = E$: The 2P2T is on, and the band switch locates on position 1.
- $V_{out} = 0$: The band switch locates on position 0 (i.e., N).
- $V_{out} = -E$: The 2P2T is off, and the band switch locates on position 1.
- $V_{out} = -2E$: The 2P2T is off, and the band switch locates on position 2.
- $V_{out} = -3E$: The 2P2T is off, and the band switch locates on position 3.
- $V_{out} = -4E$: The 2P2T is off, and the band switch locates on position 4.
- $V_{out} = -5E$: The 2P2T is off, the band switch locates on position 5.
- $V_{out} = -6E$: The 2P2T is off, and the band switch locates on position 6.
- $V_{out} = -7E$: The 2P2T is off, and the band switch locates on position 7.
- $V_{out} = -8E$: The 2P2T is off, and the band switch locates on position 8.
- $V_{out} = -9E$: The 2P2T is off, and the band switch locates on position 9.
- $V_{out} = -10E$: The 2P2T is off, and the band switch locates on position 10.

We have obtained 21-level output AC voltage. The output voltage peak value is 10 times the input DC voltage E, and the waveform is shown in Figure 14.31.

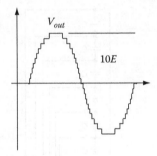

FIGURE 14.31   Twenty-one-level waveform.

14.8.2.4   Higher-Level SL Inverter

For use in higher stage super-lift converters, we can construct higher-level inverters. We believe the readers can catch the clue to construct the higher-level inverters. Therefore, we do not repeat the further circuits.

14.8.3   Simulation and Experimental Results

We use the circuit as shown in Figure 14.28 to perform the simulation. The output voltage has 15 levels. The simulation result is shown in Figure 14.13. The corresponding experimental result is shown in Figure 14.14.

We use a circuit to produce a higher-level SC inverter (e.g., 21 level SC inverter) to do the simulation again. The output voltage has 21 levels. The simulation result is shown in Figure 14.15. The corresponding experimental result is shown in Figure 14.16.

REFERENCES

1. Luo, F. L. and Ye, H. 2010. *Power Electronics: Advanced Conversion Technologies.* Boca Raton, FL: Taylor & Francis Group LLC, ISBN: 978-1-4200-9429-9.
2. Nabae, A., Takahashi, I., and Akagi, H. 1980. A neutral-point clamped PWM inverter. *Proceedings of IEEE APEC'80 Conference*, pp. 761–766.
3. Nabae, A., Takahashi, I., and Akagi, H. 1981. A neutral-point clamped PWM inverter. *IEEE Transactions on Industry Applications*, IA-17, 518–523.
4. Luo, F. L. and Ye, H. 2004. Advanced DC/DC converters. Boca Raton, FL; CRC Press LLC, ISBN: 0-8493-1956-0.
5. Peng, F. Z. 2001. A generalized multilevel inverter topology with self voltage balancing. *IEEE Transactions on Industry Applications*, 37(2), 611–618.
6. Liu, Y. and Luo, F. L. 2008. Trinary hybrid 81-level multilevel inverter for motor drive with zero common-mode voltage. *IEEE-Transactions on Industrial Electronics*, 55(3), 1014–1021.
7. Liu, Y. and Luo, F. L. 2006. Multilevel inverter with the ability of self voltage balancing. *IEE Proceedings on Electric Power Applications*, 153(1), 105–115.
8. Liu, Y. and Luo, F. L. 2005. Trinary hybrid multilevel inverter used in STATCOM with unbalanced voltages. *IEE Proceedings on Electric Power Applications*, 152(5), 1203–1222.

9. Luo, F. L. and Ye, H. 2004. Positive output multiple-lift push-pull switched-capacitor Luo-converters. *IEEE Transactions on Industrial Electronics*, 51(3), 594–602.
10. Gao, Y. and Luo, F. L. 2001. Theoretical analysis on performance of a 5V/12V push-pull switched capacitor DC/DC converter. *Proceedings of the IEE International Conference IPEC'2001*, Singapore, May 17–19, pp. 711–715.
11. Luo, F. L., Ye, H., and Rashid, M. H. 1999. Four-quadrant switched capacitor Luo-converter. *International Journal on Power Supply Technologies and Applications*, 2(3), 4–10, ISSN: 0219-2713.
12. Luo, F. L. and Ye, H. 1999. Two-quadrant switched capacitor converter. *Proceedings of the 13th Chinese Power Supply Society IAS Annual Meeting*, Shenzhen, China, November 15–18, pp. 164–168.
13. Luo, F. L. and Ye, H. 2002. Super-lift Luo-converters. *Proceedings of the IEEE International Conference PESC'2002*, Cairns, Queensland, Australia, June 23–27, pp. 425–430.
14. Luo, F. L. and Ye, H. 2003. Positive output super-lift converters. *IEEE Transactions on Power Electronics*, 18(1), 105–113.
15. Luo, F. L. and Ye, H. 2003. Negative output super-lift Luo-converters. *Proceedings of the IEEE International Conference PESC'03*, Acapulco, Mexico, June 15–19, pp. 1361–1366.
16. Luo, F. L. and Ye, H. 2003. Negative output super-lift converters. *IEEE Transactions on Power Electronics*, 18(5), 1113–1121.
17. Luo, F. L. and Ye, H. 2004. Positive output cascade boost converters. *IEE-EPA Proceedings*, 151(5), 590–606.
18. Luo, F. L. and Ye, H. 2005. Ultra-lift Luo-converter. *IEE-EPA Proceedings*, 152(1), 27–32.

15 Traditional AC/AC Converters

AC/AC conversion technology is an important subject in research and industrial applications. In recent decades, the AC/AC conversion technique has been greatly developed. We can sort these developments into two categories. The converters developed in the twentieth century can be called the traditional AC/AC converters; these are introduced in this chapter. The new technologies of AC/AC conversion technology are introduced in Chapter 16 [1–6].

15.1 INTRODUCTION

A power electronic AC/AC converter accepts electric power from one system and converts it for delivery to another AC system with different *amplitude, frequency,* and *phase*. The systems may be in single phase or three phase depending on their power ratings. The AC/AC converters used to vary the root-mean-square (rms) voltage across the load at constant frequency are known as *AC voltage controllers or AC regulators*. The voltage control is accomplished either by (1) *phase control* under natural commutation using pairs of *triacs*, silicon-controlled rectifiers (*SCRs*), or thyristors; or by (2) *on/off control* under forced commutation using fully controlled self-commutated switches such as gate turn-off thyristors (GTOs), power bipolar transistors (BTs), insulated gate bipolar transistors (IGBTs), MOS-controlled thyristors (MCTs), etc. [7,8].

The AC/AC power converters in which AC power at one frequency is directly converted to AC power at another frequency *without any intermediate DC conversion link* are known as *cycloconverters*, the majority of which use naturally commutated SCRs for their operation when the maximum output frequency is limited to a fraction of the input frequency. With the rapid advancement in fast-acting fully controlled switches, force-commutated cycloconverters (FCC) or recently developed *matrix converters* (MCs) with bidirectional on/off control switches provide independent control of the magnitude and frequency of the generated output voltage as well as sinusoidal modulation of output voltage and current.

Although typical applications of AC voltage controllers include lighting and heating control, online transformer tap changing, soft starting, and speed control of pump and fan drives, cycloconverters are used mainly for high-power low-speed large AC motor drives for application in cement kilns, rolling mills, and ship propellers. The power circuits, control methods, and the operation of the AC voltage controllers, cycloconverters, and MCs are introduced in this section. A brief review is also given regarding their applications.

The input voltage of a diode rectifier is an AC voltage, which can be single-phase or three-phase voltages. They are usually a pure sinusoidal wave. A single-phase input voltage can be expressed as follows:

$$v_s = \sqrt{2}V_{rms} \sin \omega t = V_m \sin \omega t$$

where
    v_s is the instantaneous input voltage
    V_m is its amplitude
    V_{rms} is its rms value

Traditional AC/AC converters are sorted in three groups:

1. Voltage-regulation converters
2. Cycloconverters
3. Matrix converters

Each group has single-phase and three-phase converters.

15.2 SINGLE-PHASE AC/AC VOLTAGE-REGULATION CONVERTERS

The basic power circuit of a single-phase AC/AC voltage converter, as shown in Figure 15.1a, is composed of a pair of SCRs connected back to back (also known as inverse parallel or antiparallel) between the AC supply and the load. This connection provides a *bidirectional full-wave symmetrical* control, and the SCR pair can be replaced by a triac in Figure 15.1b for low-power applications. Alternate arrangements are as shown in Figure 15.1c with two diodes and two SCRs to provide a common cathode connection for simplifying the gating circuit without needing isolation, and in Figure 15.1d with one SCR and four diodes to reduce the device cost but with increased device conduction loss. An SCR and diode combination, known as a *thyrode controller*, as shown in Figure 15.1e, provides a *unidirectional half-wave asymmetrical voltage* control with device economy but introduces a DC component and more harmonics and thus is not very practical to use except for a low-power heating load [1–5].

With *phase control*, the switches conduct the load current for a chosen period of each input cycle of voltage, and with *on/off* control the switches connect the load either for a few cycles of input voltage and disconnect it for the next few cycles (*integral cycle control*) or the switches are turned on and off several times within alternate half-cycles of input voltage (*AC chopper or PWM AC voltage controller*).

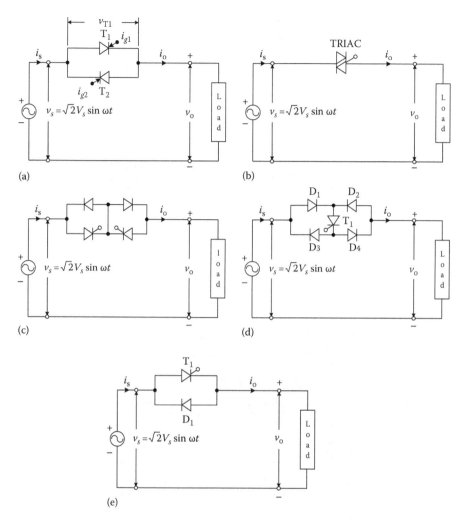

FIGURE 15.1 Single-phase AC voltage controllers: (a) full-wave, two SCRs in inverse parallel; (b) full wave with triac; (c) full wave with two SCRs and two diodes; (d) full wave with four diodes and one SCR; and (e) half wave with one SCR and one diode in antiparallel.

15.2.1 PHASE-CONTROLLED SINGLE-PHASE AC/AC VOLTAGE CONTROLLER

For a full-wave, symmetrical phase control, the SCRs T_1 and T_2 shown in Figure 15.1a are gated at α and $\pi + \alpha$, respectively, from the zero crossing of the input voltage, and by varying α, the power flow to the load is controlled through voltage control in alternate half-cycles. As long as one SCR is carrying current, the other SCR remains reverse biased by the voltage drop across the conducting SCR. The principle of operation in each half-cycle is similar to that of the controlled half-wave rectifier, and one can use the same approach for analysis of the circuit.

15.2.1.1 Operation with *R*-Load

Figure 15.2 shows the typical voltage and current waveforms for the single-phase bidirectional phase-controlled AC voltage controller of Figure 15.1a with resistive load. The output voltage and current waveforms have half-wave symmetry and thus no DC component.

If $v_s = \sqrt{2}V_s \sin \omega t$ is the source voltage, then the rms output voltage with T_1 triggered at α can be found from the half-wave symmetry as

$$V_o = \left[\frac{1}{\pi} \int_\alpha^\pi 2V_s^2 \sin^2 \omega t \, d(\omega t) \right]^{1/2} = V_s \left[1 - \frac{\alpha}{\pi} + \frac{\sin 2\alpha}{2\pi} \right]^{1/2} \qquad (15.1)$$

Note that V_o can be varied from V_s to 0 by varying α from 0 to π. The rms value of load current is

$$I_o = \frac{V_o}{R} \qquad (15.2)$$

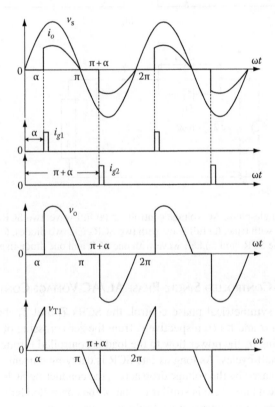

FIGURE 15.2 Waveforms for single-phase AC full-wave voltage controller with *R* load.

The input power factor is

$$\frac{P_o}{VA} = \frac{V_o}{V_s} = \left[1 - \frac{\alpha}{\pi} + \frac{\sin 2\alpha}{2\pi}\right]^{1/2} \tag{15.3}$$

The average SCR current is

$$I_{A,SCR} = \frac{1}{2\pi R}\int_{\alpha}^{\pi} \sqrt{2}V_s \sin \omega t\, d(\omega t) \tag{15.4}$$

As each SCR carries half the line current, the rms current in each SCR is

$$I_{o,SCR} = \frac{I_o}{\sqrt{2}} \tag{15.5}$$

Example 15.1 A single-phase full-wave AC/AC voltage controller shown in Figure 15.1a has input rms voltage $V_S = 220$ V/50 Hz, load $R = 100\,\Omega$, and the firing angle $\alpha = 60°$ for the thyristors T_1 and T_2. Determine the output rms voltage V_O and current I_O and the DPF.

Solution: From Equation 15.1, the output rms voltage is

$$V_o = V_s\left(1 - \frac{\alpha}{\pi} + \frac{\sin 2\alpha}{2\pi}\right)^{1/2} = 220\left(1 - \frac{1}{3} + \frac{\sqrt{3}}{4\pi}\right)^{1/2}$$

$$= 220(1 - 0.33333 + 0.13783)^{1/2} = 197.33\,\text{V}$$

The output rms current is

$$I_o = \frac{V_o}{R} = \frac{197.33}{100} = 1.9733\,\text{A}$$

The fundamental harmonic wave delayed to the supply voltage by the firing angle $\alpha = 60°$. Therefore, the DPF is DPF $= \cos\alpha = 0.5$.

From this example, we can recognize the fact that if the firing angle is greater than 90° it is possible to obtain leading power factor.

15.2.1.2 Operation with *RL* Load

Figure 15.3 shows the voltage and current waveforms for the controller in Figure 15.1a with *RL* load. Due to the inductance, the current carried by the SCR T_1 may

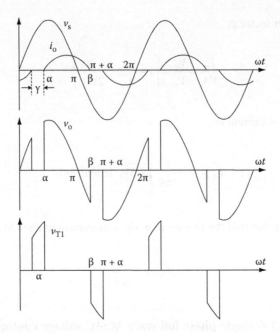

FIGURE 15.3 Typical waveforms of single-phase AC voltage controller with an *RL* load.

not fall to zero at $\omega t = \pi$ when the input voltage goes negative and may continue until $\omega t = \beta$, the extinction angle, as shown in Figure 15.3.

The conduction angle

$$\theta = \beta - \alpha \tag{15.6}$$

of the SCR depends on the firing delay angle α and the load impedance angle φ. The expression for the load current $I_o(\omega t)$ when conducting from α to β can be derived in the same way as that used for a phase-controlled rectifier in a discontinuous conduction mode (DCM) by solving the relevant Kirchhoff voltage equation:

$$i_o(\omega t) = \frac{\sqrt{2}V}{Z}[\sin(\omega t - \phi) - \sin(\alpha - \phi)e^{(\alpha - \omega t)/\tan\varphi}] \quad \alpha < \omega t < \beta \tag{15.7}$$

where

$Z(\text{load impedance}) = (R^2 + \omega^2 L^2)^{1/2}$

ϕ (load impedance angle) $= \tan^{-1}(\omega L/R)$

The angle β, when the current i_0 falls to zero, can be determined from the following transcendental equation obtained by putting $i_o(\omega t = \beta) = 0$ in Equation 15.7:

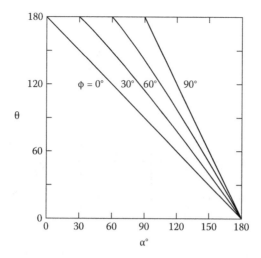

FIGURE 15.4 θ versus α curves for single-phase AC voltage controller with RL load.

$$\sin(\beta - \phi) = \sin(\alpha - \phi)e^{(\alpha - \beta)/\tan\phi} \tag{15.8}$$

From Equations 15.6 and 15.8, one can obtain a relationship between θ and α for a given value of ϕ as shown in Figure 15.4, which shows that as α is increased the conduction angle θ decreases and the rms value of the current decreases. The rms output voltage is

$$V_o = \left[\frac{1}{\pi}\int_\alpha^\beta 2V_s^2 \sin^2 \omega t \, d(\omega t)\right]^{1/2} = \frac{V_s}{\sqrt{\pi}}\left[\beta - \alpha + \frac{\sin 2\alpha}{2} + \frac{\sin 2\beta}{2}\right]^{1/2} \tag{15.9}$$

V_o can be evaluated for two possible extreme values of $\varphi = 0$ when $\beta = \pi$ and $\phi = \pi/2$ when $\beta = 2\pi - \alpha$, and the envelope of the voltage-control characteristics for this controller is shown in Figure 15.5.

The rms SCR current can be obtained from Equation 15.7 as

$$I_{o,\text{SCR}} = \left[\frac{1}{2\pi}\int_\alpha^\beta i_o^2 d(\omega t)\right] \tag{15.10}$$

The rms load current is

$$I_o = \sqrt{2}I_{o,\text{SCR}} \tag{15.11}$$

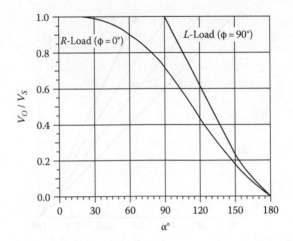

FIGURE 15.5 Envelope of control characteristics of a single-phase AC voltage controller with *RL* load.

The average value of SCR current is

$$I_{A,SCR} = \frac{1}{2\pi} \int_\alpha^\beta i_o d(\omega t) \qquad (15.12)$$

Example 15.2 A single-phase full-wave AC/AC voltage controller shown in Figure 15.1a has input rms voltage $V_s = 220\,V/50\,Hz$, load $R = 100\,\Omega$ and $L = 183.78\,mH$, and the firing angle $\alpha = 60°$ for the thyristors T_1 and T_2. Determine the extinction angle β, the output rms voltage V_O and current I_O, and the DPF.

Solution: Since the load is an *RL*-load, the output voltage is shown in Figure 15.3. The load impedance angle ϕ is

$$\phi = \tan^{-1}\frac{\omega L}{R} = \tan^{-1}\frac{100\pi \times 183.78m}{100} = \tan^{-1}0.57735 = 30°$$

The conduction angle θ is determined by Equation 15.6, or check the value from Figure 15.4. The conduction angle θ is about 150° (or $5\pi/6$). Therefore, the extinction angle β is

$$\beta = \theta + \alpha = \frac{5\pi}{6} + \frac{\pi}{3} = \frac{7}{6}\pi \text{ rad}$$

From Equation 15.1, the output rms voltage is

$$V_o = V_s\left(1 - \frac{\alpha}{\pi} + \frac{\sin 2\alpha}{2\pi}\right)^{1/2} = 220\left(1 - \frac{1}{3} + \frac{\sqrt{3}}{4\pi}\right)^{1/2}$$

$$= 220(1 - 0.33333 + 0.13783)^{1/2} = 197.33 \text{ V}$$

The output rms current is

$$I_o = \frac{V_o}{R} = \frac{197.33}{100} = 1.9733 \text{ A}$$

The fundamental harmonic wave delayed to the supply voltage by the firing angle $\alpha = 60°$. Therefore, the DPF is DPF $= \cos\alpha = 0.5$.

15.2.1.3 Gating Signal Requirements

For the inverse parallel SCRs as shown in Figure 15.1a, the gating signals of SCRs must be isolated from one another as there is no common cathode. For R load, each SCR stops conducting at the end of each half-cycle and under this condition, single short pulses may be used for gating as shown in Figure 15.2. With RL load, however, this single short pulse gating is not suitable as shown in Figure 15.6. When SCR T_2 is triggered at $\omega t = \pi + \alpha$, SCR T_1 is still conducting due to the load inductance. By the time the SCR T_1 stops conducting at β, the gate pulse for SCR T_2 has already ceased and T_2 will fail to turn on, causing the converter to operate as a single-phase rectifier with conduction of T_1 only. This necessitates application of a sustained gate pulse either in the form of a continuous signal for the half-cycle period, which increases the dissipation in SCR gate circuit and a large isolating pulse transformer or better a *train of pulses (carrier frequency gating)* to overcome these difficulties.

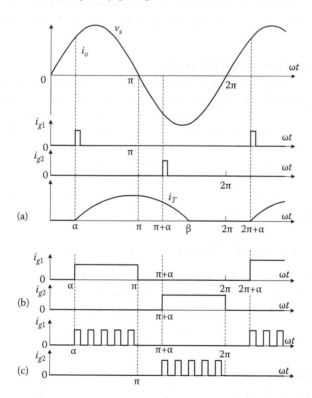

FIGURE 15.6 Single-phase full-wave controller with RL load: gate pulse requirements.

15.2.1.4 Operation with $\alpha < \phi$

If $\alpha = \phi$, then from Equation 15.8,

$$\sin(\beta - \phi) = \sin(\beta - \alpha) = 0 \tag{15.13}$$

and

$$\beta - \alpha = \theta = \pi \tag{15.14}$$

As the conduction angle θ cannot exceed π and the load current must pass through zero, the control range of the firing angle is $\phi \le \alpha \le \pi$. With narrow gating pulses and $\alpha < \phi$, only one SCR will conduct, resulting in a rectifier action as shown. Even with a train of pulses, if $\alpha < \phi$, the changes in the firing angle will not change the output voltage and current but both SCRs will conduct for the period π, with T_1 be on at $\omega t = \pi$ and T_2 be on at $\omega t + \pi + \alpha$. This *dead zone* ($\alpha = 0$ to ϕ), whose duration varies with the load impedance angle ϕ, is not a desirable feature in closed-loop control schemes. An alternative approach to the phase control with respect to the input voltage zero crossing has been reported in which the firing angle is defined with respect to the instant when it is the load current (not the input voltage) that reaches zero, this angle being called *the hold-off angle (γ) or the control angle* (as marked in Figure 15.3). This method requires sensing the load current—which may otherwise be required anyway in a closed-loop controller for monitoring or control purposes.

15.2.1.5 Power Factor and Harmonics

As in the case of phase-controlled rectifiers, the important limitations of the phase-controlled AC voltage controllers are the poor power factor and the introduction of harmonics in the source currents. As seen from Equation 15.3, the input power factor depends on α, and as α increases, the power factor decreases.

The harmonic distortion increases and the quality of the input current decreases with an increase in the firing angle. The variations in low-order harmonics with the firing angle as computed by Fourier analysis of the voltage waveform of Figure 15.2 (with R load) are shown in Figure 15.7. Only odd harmonics exist in the input current because of half-wave symmetry.

15.2.2 SINGLE-PHASE AC/AC VOLTAGE CONTROLLER WITH ON/OFF CONTROL

Figure 15.8 shows an on–off AC/AC voltage-regulation controller. In a period T, n cycles are on and m cycles are off. The conduction duty cycle is k:

$$k = \frac{n}{n + m} \tag{15.15}$$

15.2.2.1 Integral Cycle Control

As an alternative to the phase control, the method of integral cycle, control, or burst firing is used for heating loads. Here, the switch is turned on for a time t_n with n

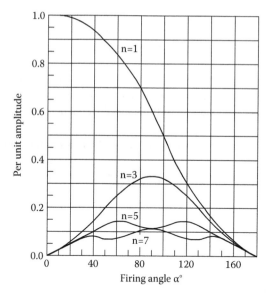

FIGURE 15.7 Harmonic content as a function of the firing angle for a single-phase voltage controller with RL load.

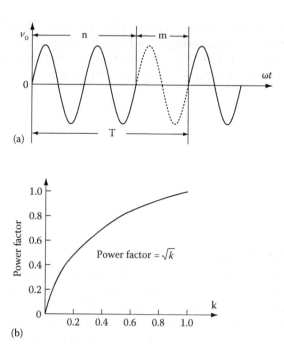

FIGURE 15.8 Integral cycle control: (a) typical load-voltage waveforms and (b) power factor with the duty cycle k.

integral cycles and turned off for a time t_m with m integral cycles (Figure 15.8). As the SCRs or triacs used here are turned on at the zero crossing of the input voltage and turn-off occurs at zero current, supply harmonics and radio frequency interference are very low.

However, subharmonic frequency components that are undesirable may be generated as they may set up subharmonic resonance in the power supply system, cause lamp flicker, and may interfere with the natural frequencies of motor loads causing shaft oscillations.

For sinusoidal input voltage $v = \sqrt{2}V_s \sin \omega t$, the rms output voltage is

$$V_o = V_s \sqrt{k} \qquad (15.16)$$

where
  $k = n/(n+m) = $ duty cycle
  $V_s = $ rms phase voltage

The power factor is

$$PF = \sqrt{k} \qquad (15.17)$$

which is poorer for lower values of the duty cycle k.

Example 15.3 A single-phase integral cycle controlled AC/AC controller has input rms voltage $V_S = 240$ V. It is turned on and off with a duty cycle $k = 0.4$ at a five cycles (refer to Figure 15.8). Determine the output rms voltage V_O and input-side PF.

Solution: Since the input rms voltage is 240 V and the duty cycle $k = 0.4$, the output rms voltage is

$$V_O = V_S \sqrt{k} = 240 \times \sqrt{0.4} = 151.79 \text{ V}$$

The power factor is

$$PF = \sqrt{k} = \sqrt{0.4} = 0.632$$

15.2.2.2 PWM AC Chopper

As in the case of controlled rectifier, the performance of AC voltage controllers can be improved in terms of harmonics, quality of output current, and input power factor by *pulse width modulation* (PWM) control in PWM AC choppers. The circuit configuration of one such single-phase unit is shown in Figure 15.9.

Here, fully controlled switches S_1 and S_2 connected in antiparallel are turned on and off many times during the positive and negative half-cycles of the input

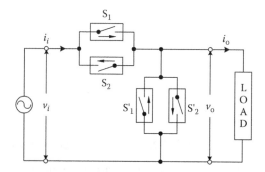

FIGURE 15.9   Single-phase PWM as chopper circuit.

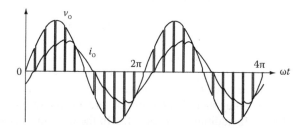

FIGURE 15.10   Typical output voltage and current waveforms of a single-phase PWM AC chopper.

voltage, respectively; S_1' and S_2' provide the freewheeling paths for the load current when S_1 and S_2 are off. An input capacitor filter may be provided to attenuate the high switching frequency current drawn from the supply and also to improve the input power factor. Figure 15.10 shows the typical output voltage and load-current waveform for a single-phase PWM AC chopper. It can be shown that the control characteristics of an AC chopper depend on the *modulation index k*, which theoretically varies from zero to unity. The relation between input and output voltages is expressed in Equation 15.16, and the power factor is calculated by Equation 15.17. By applying a low-pass filter in the output side of a PWM AC chopper, a good sine wave can be obtained.

Example 15.4 A single-phase PWM AC chopper has input rms voltage $V_S = 240\,V$. Its *modulation index k = 0.4* (refer to Figure 15.10). Determine the output rms voltage V_O and input-side PF.

Solution: Since the input rms voltage is 240 V and the *modulation index k = 0.4*, the output rms voltage is

$$V_O = V_S \sqrt{k} = 240 \times \sqrt{0.4} = 151.79\ V$$

The power factor is

$$PF = \sqrt{k} = \sqrt{0.4} = 0.632$$

Analogously, a three-phase PWM choppers consist of three single-phase choppers either delta connected or four-wire star connected.

15.3 THREE-PHASE AC/AC VOLTAGE-REGULATION CONVERTERS

Three-phase AC/AC voltage controllers have various circuits and configurations.

15.3.1 PHASE-CONTROLLED THREE-PHASE AC VOLTAGE CONTROLLERS

Several possible circuit configurations for three-phase phase-controlled AC regulators with star-or delta-connected loads are shown in Figure 15.11a through h. The configurations in Figure 15.11a and b can be realized by three single-phase AC regulators operating independently of each other, and they are easy to analyze. In Figure 15.11a, the SCRs are to be rated to carry line currents and withstand phase voltages, whereas in Figure 15.11b they should be capable of carrying phase currents and withstand the line voltages. In addition, in Figure 15.11b the line currents are free from triplen harmonics while these are present in the closed delta. The power factor in Figure 15.11b is slightly higher. The firing angle control range for both these circuits is 0°–180° for R load.

The circuits in Figure 15.11c and d are three-phase three-wire circuits and are difficult to analyze. In both these circuits, at least two SCRs (one in each phase) must be gated simultaneously to get the controller started by establishing a current path between the supply lines. This necessitates two firing pulses spaced 60° apart per cycle for firing each SCR. The operation modes are defined by the number of SCRs conducting in these modes. The firing control range is 0°–150°. The triplen harmonics are absent in both these configurations.

Another configuration is shown in Figure 15.11e when the controllers are delta connected and the load is connected between the supply and the converter. Here, current can flow between two lines even if one SCR is conducting, so each SCR requires one firing pulse per cycle. The voltage and current ratings of SCRs are nearly the same as those of the circuit in Figure 15.11b. It is also possible to reduce the number of devices to three SCRs in delta as shown in Figure 15.11f connecting one source terminal directly to one load circuit terminal. Each SCR is provided with gate pulses in each cycle spaced 120° apart. In both Figure 15.11e and f, each end of each phase must be accessible. The number of devices in Figure 15.11f is fewer but their current ratings must be higher.

As in the case of the single-phase phase-controlled voltage regulator, the total regulator cost can be reduced by replacing six SCRs by three SCRs and three diodes, resulting in three-phase *half-wave controlled* unidirectional AC regulators, as shown in Figure 15.11g and h for star- and delta-connected loads. The main drawback of these circuits is the large harmonic content in the output voltage,

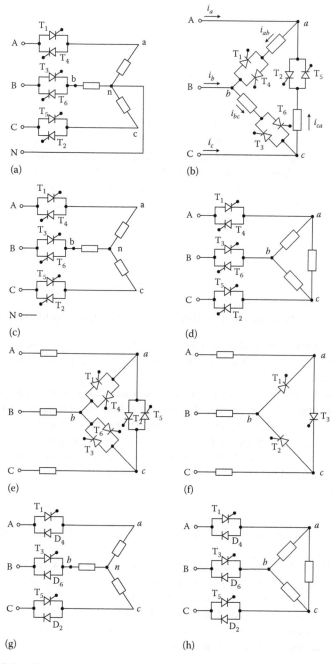

FIGURE 15.11 Three-phase AC voltage-controller circuit configurations: (a) fully-line-controlled Y-connection with neutral line, (b) fully-phase-controlled Δ-connection, (c) fully-line-controlled Y-connection without neutral line, (d) fully-line-controlled Δ-connection, (e) fully-phase-controlled Y-connection, (f) half-wave-controlled Y-connection, (g) half-line-controlled Y-connection, and (h) half-line controlled Δ-connection.

particularly the second harmonic because of the asymmetry. However, the DC components are absent in the line. The maximum firing angle in the half-wave controlled regulator is 210°.

15.3.2 FULLY CONTROLLED THREE-PHASE THREE-WIRE AC VOLTAGE CONTROLLER

15.3.2.1 Star-Connected Load with Isolated Neutral

The analysis of operation of the full-wave controller with isolated neutral as shown in Figure 15.11c is, as mentioned, quite complicated in comparison with that of a single-phase controller, particularly for an RL or motor load. As a simple example, the operation of this controller is considered here with a simple star-connected R load. The six SCRs are turned on in the sequence 1-2-3-4-5-6 at 60° intervals, and the gate signals are sustained throughout the possible conduction angle.

The output phase voltage waveforms for $\alpha=30°$, 75°, and 120° for a balanced three-phase R load are shown in Figure 15.12. At any interval, either three SCRs or two SCRs, or no SCRs may be on and the instantaneous output voltages to the load are either line-to-neutral voltages (three SCRs on) or one-half of the line-to-line voltage (two SCRs on) or zero (no SCR on).

Depending on the firing angle α, there may be *three* operating modes.

Mode I (also known as Mode 2/3): $0<\alpha<60°$. There are periods when *three* SCRs are conducting, one in each phase for either direction and periods when just *two* SCRs conduct.

For example, with $\alpha=30°$ in Figure 15.12a, assume that at $\omega t=0$, SCRs T_5 and T_6 are conducting, and the current through the R load in a-phase is zero, making $v_{an}=0$. At $\omega t=30°$, T_1 receives a gate pulse and starts conducting; T_5 and T_6 remain on and $v_{an}=v_{AN}$. The current in T_5 reaches zero at 60°, turning T_5 off. With T_1 and T_6 staying on, $v_{an}=1/2v_{AB}$. At 90°, T_2 is turned on, the three SCRs T_1, T_2, and T_6 are then conducting, and $v_{an}=v_{AN}$. At 120°, T_6 turns off, leaving T_1 and T_2 on, so $v_{an}=1/2$ v_{AC}. Thus with the progress of firing in sequence until $\alpha=60°$, the number of SCRs conducting at a particular instant alternates between two and three.

Mode II (also known as Mode 2/2): $60°<\alpha<90°$. *Two* SCRs, one in each phase, always conduct.

For $a=75°$ as shown in Figure 15.12b, just prior to $\alpha=75°$, SCRs T_5 and T_6 were conducting and $v_{an}=0$. At 75°, T_1 is turned on and T_6 continues to conduct while T_5 turns off as v_{CN} is negative; $v_{an}=1/2v_{AB}$. When T_2 is turned on at 135°, T_6 is turned off and $v_{an}=1/2v_{AC}$. The next SCR to turn on is T_3, which turns off T_1 and $v_{an}=0$. One SCR is always turned off when another is turned on in this range of α, and the output is either one-half line-to-line voltage or zero.

Mode III (also known as Mode 0/2): $90°<\alpha<150°$. When *none* or *two* SCRs conduct. For $\alpha=120°$ (Figure 15.12c), earlier no SCRs were on and $v_{an}=0$. At $\alpha=120°$, SCR T_1 is given a gate signal while T_6 has a gate signal already applied. As v_{AB} is positive, T_1 and T_6 are forward biased, and they begin to conduct and $v_{an}=1/2v_{AB}$. Both

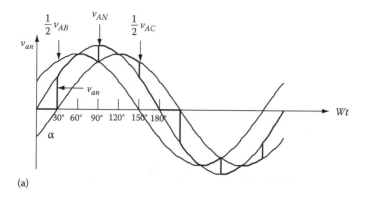

(a)

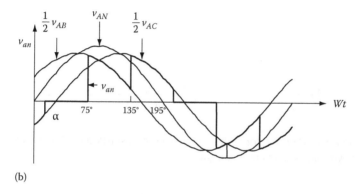

(b)

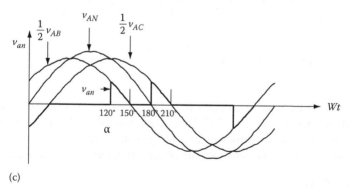

(c)

FIGURE 15.12 Output voltage waveforms for a three-phase AC voltage controller with star-connected R load: (a) v_{an} for $\alpha = 30°$; (b) v_{an} for $\alpha = 75°$; and (c) $\alpha = 120°$.

T_1 and T_6 turn off when v_{AB} becomes negative. When a gate signal is given to T_2, it turns on and T_1 turns on again.

For $\alpha > 150°$, there is no period when two SCRs are conducting and the output voltage is zero at $\alpha = 150°$. Thus, the range of the firing angle control is $0 \leq \alpha \leq 150°$.

For *star-connected R load*, assuming the instantaneous phase voltages as

$$v_{AN} = \sqrt{2}V_s \sin \omega t$$

$$v_{BN} = \sqrt{2}V_s \sin(\omega t - 120°) \qquad (15.18)$$

$$v_{CN} = \sqrt{2}V_s \sin(\omega t - 240°)$$

the expressions for the rms output phase voltage V_o can be derived for the three modes as

$$0 \le \alpha \le 60° \quad V_o = V_s \left[1 - \frac{3\alpha}{2\pi} + \frac{3}{4\pi} \sin 2\alpha \right]^{1/2} \qquad (15.19)$$

$$60° \le \alpha \le 90° \quad V_o = V_s \left[\frac{1}{2} + \frac{3}{4\pi} \sin 2\alpha + \sin(2\alpha + 60°) \right]^{1/2} \qquad (15.20)$$

$$90° \le \alpha \le 150° \quad V_o = V_s \left[\frac{5}{4} - \frac{3\alpha}{2\pi} + \frac{3}{4\pi} \sin(2\alpha + 60°) \right]^{1/2} \qquad (15.21)$$

For *star-connected pure L load*, the effective control starts at $\alpha > 90°$, and the expressions for two ranges of α are as follows:

$$90° \le \alpha \le 120° \quad V_o = V_s \left[\frac{5}{2} - \frac{3\alpha}{\pi} + \frac{3}{2\pi} \sin 2\alpha \right]^{1/2} \qquad (15.22)$$

$$120° \le \alpha \le 150° \quad V_o = V_s \left[\frac{5}{2} - \frac{3\alpha}{\pi} + \frac{3}{2\pi} \sin(2\alpha + 60°) \right]^{1/2} \qquad (15.23)$$

The control characteristics for these two limiting cases ($\phi = 0$ for R load and $\varphi = 90°$ for L load) are shown in Figure 15.13. Here, also, as in the single-phase case the dead zone may be avoided by controlling the voltage with respect to the control angle or hold-off angle (γ) from the zero crossing of current in place of the firing angle α.

15.3.2.2 *RL* Load

The analysis of the three-phase voltage controller with star-connected *RL* load with isolated neutral is quite complicated as the SCRs do not cease to conduct at voltage zero and the extinction angle β is to be known by solving the transcendental equation for the case. The Mode II operation, in this case, disappears and the operation shift from Mode I to Mode III depends on the so-called critical angle α_{crit}, which can be evaluated from a numerical solution of the relevant transcendental equations. Computer simulation either by PSPICE program or a switching-variable approach coupled with an iterative procedure is a practical means of obtaining the output

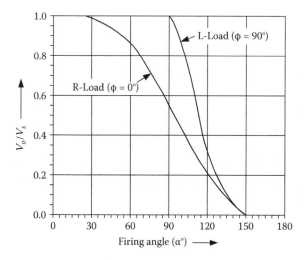

FIGURE 15.13 Envelope of control characteristics for a three-phase full-wave AC voltage controller.

voltage waveform in this case. Figure 15.14 shows typical simulation results, using the later approach for a three-phase voltage-controller-fed RL load for $\alpha = 60°$, $90°$, and $105°$, which agree with the corresponding practical oscillograms given.

15.3.2.3 Delta-Connected R-Load

The configuration is shown in Figure 15.11b. The voltage across an R load is the corresponding line-to-line voltage when one SCR in that phase is on. Figure 15.15 shows the line and phase currents for $\alpha = 120°$ and $90°$ with an R load. The firing angle α is measured from the zero crossing of the line-to-line voltage, and the SCRs are turned on in the sequence as they are numbered. As in the single-phase case, the range of firing angle is $0 \le \alpha \le 180°$. The line currents can be obtained from the phase currents as

$$i_a = i_{ab} - i_{ca}$$
$$i_b = i_{bc} - i_{ab} \qquad (15.24)$$
$$i_c = i_{ca} - i_{bc}$$

The line currents depend on the firing angle and may be discontinuous as shown. Due to the delta connection, the triplen harmonic currents flow around the closed delta and do not appear in the line. The rms value of the line current varies between the range

$$\sqrt{2}I_\Delta \le I_{L,rms} \le \sqrt{3}I_{\Delta,rms} \qquad (15.25)$$

as the conduction angle varies from very small (large α) to $180°$ ($\alpha = 0$).

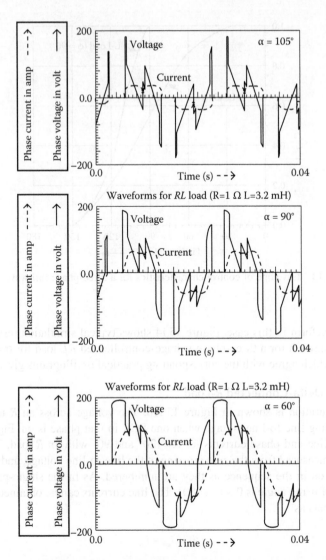

FIGURE 15.14 Typical simulation results for three-phase AC voltage-controller-fed *RL* load (*R* = 1 Ω, *L* = 3.2 mH) for α = 60°, 90°, and 105°.

15.4 CYCLOCONVERTERS

In contrast to the AC voltage controllers operating at constant frequency discussed so far, a cycloconverter operates as a direct AC/AC frequency changer with an inherent voltage control feature. The basic principle of this converter to construct an alternating voltage wave of lower frequency from successive segments of voltage waves of higher frequency AC supply by a switching arrangement was conceived and patented in the 1920s. Grid-controlled mercury-arc rectifiers were used

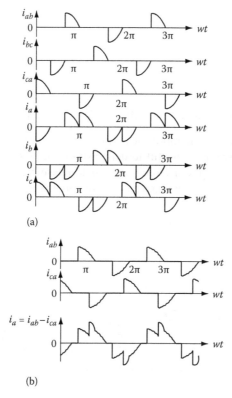

(a)

(b)

FIGURE 15.15 Waveforms of a three-phase AC voltage controller with a delta-connected R-load: (a) $\alpha = 120°$ and (b) $\alpha = 90°$.

in these converters installed in Germany in the 1930s to obtain $16\frac{2}{3}$ Hz single-phase supply for AC series traction motors from a three-phase 50 Hz system while at the same time a cycloconverter using 18 thyratrons supplying a 400 hp synchronous motor was in operation for some years as a power station auxiliary drive in the United States. However, the practical and commercial utilization of these schemes waited until the SCRs became available in the 1960s. With the development of large-power SCRs and microprocessor-based control, the cycloconverter today is a mature practical converter for application in large-power low-speed variable-voltage variable-frequency (VVVF) AC drives in cement and steel rolling mills as well as in variable-speed constant-frequency (VSCF) systems in aircraft and naval ships [9–11].

A cycloconverter is a naturally commuted converter with the inherent capability of bidirectional power flow, and there is no real limitation on its size unlike an SCR inverter with commutation elements. Here, the switching losses are considerably low, the regenerative operation at full power over complete speed range is inherent, and it delivers a nearly sinusoidal waveform, resulting in minimum torque pulsation and harmonic heating effects. It is capable of operating even with the blowing out of an individual SCR fuse (unlike the inverter), and the requirements regarding turn-off time, current rise time, and dv/dt sensitivity of SCRs are low. The main limitations of a naturally commutated cycloconverter (NCC) are (1) limited frequency range for subharmonic-free and efficient operation and (2) poor input displacement/power factor, particularly at low output voltages.

15.4.1 Single-Phase Input/Single-Phase Output Cycloconverter

Though rarely used, the operation of a single-phase input to single-phase output (SISO) cycloconverter is useful to demonstrate the basic principle involved. Figure 15.16a shows the power circuit of a single-phase bridge-type cycloconverter, which is the same arrangement as that of the dual converter.

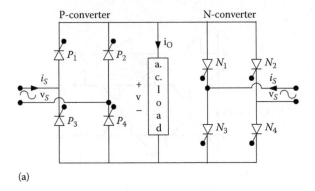

(a)

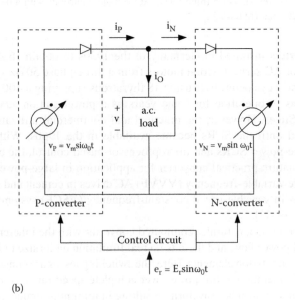

(b)

FIGURE 15.16 (a) Power circuit for a single-phase bridge cycloconverter and (b) simplified equivalent circuit of a cycloconverter.

The firing angles of the individual two-pulse two-quadrant bridge converters are continuously modulated here so that each ideally produces the same fundamental AC voltage at its output terminals as marked in the simplified equivalent circuit in Figure 15.16b. Because of the unidirectional current-carrying property of the individual converters, it is inherent that the positive half-cycle of the current is carried by the P-converter and the negative half-cycle of the current by the N-converter regardless of the phase of the current with respect to the voltage. This means that for a reactive load, each converter operates in both the rectifying and inverting region during the period of the associated half-cycle of the low-frequency output current.

15.4.1.1 Operation with R Load

Figure 15.17 shows the input and output voltage waveforms with a pure R load for a 50–16⅔ Hz cycloconverter. The P- and N-converters operate for all alternate $T_0/2$ periods. The output frequency $(1/T_0)$ can be varied by varying T_0 and the voltage magnitude by varying the firing angle α of the SCRs. As shown in the figure, three cycles of the AC input wave are combined to produce one cycle of the output frequency to reduce the supply frequency to one-third across the load.

For example, the waveforms of a SISO AC/AC cycloconverter with $T_o = 3\,T_s$ are shown in Figure 15.17. The firing angle α is listed in Tables 15.1 and 15.2 (the blank means no firing pulse applied).

Assuming the input voltage amplitude $\sqrt{2}V_S$ and output voltage amplitude $\sqrt{2}V_O$ keep the following relation for *full regulation*:

$$\frac{\sqrt{2}V_O}{\pi/3}\int_{\pi/3}^{2\pi/3}\sin\alpha\,d\alpha \leq \sqrt{2}V_S\,\frac{1}{\pi}\int_0^{\pi}\sin\alpha\,d\alpha \qquad (15.26)$$

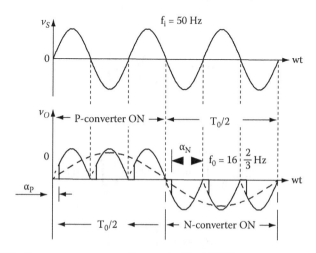

FIGURE 15.17 Input and output waveforms of a 50 − 16⅔ Hz cycloconverter with R load.

TABLE 15.1
The Firing Angle Set of the Positive Rectifier

Half-Cycle No. in f_O	1	2	3	4	5	6
SCR	P_1P_4	P_2P_3	P_1P_4	P_2P_3	P_1P_4	P_2P_3
α_p	α_1	α_2	α_1			

TABLE 15.2
The Firing Angle Set of the Negative Rectifier

Half-Cycle No. in f_O	1	2	3	4	5	6
SCR	N_1N_4	N_2N_3	N_1N_4	N_2N_3	N_1N_4	N_2N_3
α_n				α_1	α_2	α_1

i.e.,

$$3V_O \leq 2V_S \tag{15.27}$$

We then get the firing angles calculation formulae:

$$\sqrt{2}V_O \frac{3}{\pi} \int_0^{\pi/3} \sin\theta d\theta = \sqrt{2}V_S \frac{1}{\pi} \int_{\alpha1}^{\pi} \sin\theta d\theta \tag{15.28}$$

$$\alpha_1 = \cos^{-1}\left(\frac{3V_O}{2V_S} - 1\right) \tag{15.29}$$

and

$$\sqrt{2}V_O \frac{3}{\pi} \int_3^{2\pi/3} \sin\theta d\theta = \sqrt{2}V_S \frac{1}{\pi} \int_{\alpha2}^{\pi} \sin\theta d\theta \tag{15.30}$$

$$\alpha_2 = \cos^{-1}\left(\frac{3V_O}{V_S} - 1\right) \tag{15.31}$$

We also get

$$\alpha_3 = \alpha_1 = \cos^{-1}\left(\frac{3V_O}{2V_S} - 1\right) \tag{15.32}$$

The phase-angle shift (delay) in the frequency f_S is

$$\sigma = \frac{\alpha_1}{2} = \frac{1}{2}\cos^{-1}\left(\frac{3V_O}{2V_S} - 1\right) \tag{15.33}$$

and in the frequency f_O is

$$\sigma' = \frac{1}{3}\frac{\alpha_1}{2} = \frac{1}{6}\cos^{-1}\left(\frac{3V_O}{2V_S} - 1\right) \tag{15.34}$$

If the full regulation condition (15.27) is not satisfied, the modulation can still be done by another way; the limitation condition usually is

$$V_O \le 1.2V_S \tag{15.35}$$

If α_P is the firing angle of the P-converter, the firing angle of the N-converter α_N is $\pi - \alpha_P$ and the average voltage of the P-converter is equal and opposite to that of the N-converter. The inspection of the waveform with α remaining fixed in each half-cycle generates a square wave with a large low-order harmonic content. A near approximation to sine wave can be synthesized by a phase modulation of the firing angles as shown in Figure 15.18 for a 50–10 Hz cycloconverter. The harmonics in the load-voltage waveform are fewer compared to the earlier waveform. The supply current, however, contains a subharmonic at the output frequency for this case as shown.

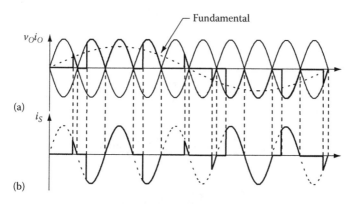

FIGURE 15.18 Waveforms of a single-phase/single-phase cycloconverter (50–10 Hz) with R load: (a) load voltage and load current and (b) input supply current.

Example 15.5 A full-wave single-phase input single-phase output (SISO) AC/AC cycloconverter. The input rms voltage $V_S = 140\,V/50\,Hz$ and the output voltage $V_o = 90\,V/16\frac{2}{3}\,Hz$, and the load is a resistance R with a low-pass filter. Assuming the filter is appropriately designed, only the fundamental component ($f_O = 16\frac{2}{3}\,Hz$) remains in the output voltage. Tabulate the firing angle (α in the period $T_s = 1/f_s = 20\,ms$) of both rectifiers' SCRs in a full period $T_o = 1/f_o = 60\,ms$, and calculate the phase-angle shift σ in the input voltage over the period $T_s = 1/f_s$.

Solution: The table is presented as follows (the blank means no firing pulse applied).

Positive rectifier

Half-Cycle No. in f_O	1	2	3	4	5	6
SCR	P_1P_4	P_2P_3	P_1P_4	P_2P_3	P_1P_4	P_2P_3
α_p	α_1	α_2	α_1			

Negative rectifier

Half-Cycle No. in f_O	1	2	3	4	5	6
SCR	N_1N_4	N_2N_3	N_1N_4	N_2N_3	N_1N_4	N_2N_3
α_n				α_1	α_2	α_1

The full regulation condition is

$$\frac{\sqrt{2}V_O}{\pi/3} \int_{\pi/3}^{2\pi/3} \sin\alpha\, d\alpha \leq \sqrt{2}V_S \frac{1}{\pi} \int_0^{\pi} \sin\alpha\, d\alpha$$

$$V_S \geq 3V_O \cos\frac{\pi}{3} = 1.5V_O$$

i.e.,

$$V_S = 140 \geq 3V_O \cos\frac{\pi}{3} = 1.5V_O = 135\ V$$

$$\sqrt{2}V_O \frac{3}{\pi} \int_0^{\pi/3} \sin\theta\, d\theta = \sqrt{2}V_S \frac{1}{\pi} \int_{\alpha 1}^{\pi} \sin\theta\, d\theta$$

$$3\left(1 - \cos\frac{\pi}{3}\right)V_O = (1 + \cos\alpha_1)V_S$$

$$\alpha_1 = \cos^{-1}\left(\frac{1.5V_O}{V_S} - 1\right) = \cos^{-1}(-0.0357) = 92.05°$$

$$\sqrt{2}V_O\frac{3}{\pi}\int_{\frac{\pi}{3}}^{2\pi/3} \sin\theta\, d\theta = \sqrt{2}V_S\frac{1}{\pi}\int_{\alpha 2}^{\pi} \sin\theta\, d\theta$$

$$3\left(\cos\frac{\pi}{3} - \cos\frac{2\pi}{3}\right)V_O = (1 + \cos\alpha_2)V_S$$

$$\alpha_2 = \cos^{-1}\left(\frac{3V_O}{V_S} - 1\right) = \cos^{-1}(0.9286) = 21.79°$$

The phase-angle shift σ in the input voltage over the period $T_S = 1/f_S$ is

$$\sigma = \frac{\alpha_1}{2} = \frac{1}{2} \times 92.05 = 46.02°$$

15.4.1.2 Operation with *RL* Load

The cycloconverter is capable of supplying loads of any power factor. Figure 15.19 shows the idealized output voltage and current waveforms for a lagging-power-factor load where both the converters are operating as rectifier and inverter at the intervals marked. The load current lags the output voltage and the load-current direction determines which converter is conducting. Each converter continues to conduct after its output voltage changes polarity, and during this period, the converter acts as an inverter and the power is returned to the AC source. The inverter operation continues until the other converter starts to conduct. By controlling the frequency of oscillation and the *depth of modulation* of the firing angles of the converters (as will be shown later), it is possible to control the frequency and the amplitude of the output voltage.

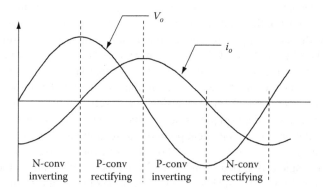

FIGURE 15.19 Load voltage and current waveform for a cycloconverter with *RL* load.

The load current with RL load may be continuous or discontinuous depending on the load phase angle φ. At light load inductance or for $\varphi \leq \alpha \leq \pi$, there may be discontinuous load current with short zero-voltage periods. The current wave may contain even harmonics as well as subharmonic components. Further, as in the case of a dual converter, though the mean output voltage of the two converters is equal and opposite, the instantaneous values may be unequal and a circulating current can flow within the converters. This circulating current can be limited by having a center-tapped reactor connected between the converters or can be completely eliminated by logical control similar to the dual converter case when the gate pulses to the converter remaining idle are suppressed when the other converter is active. A zero current interval of short duration is needed between the P- and N-converters to ensure that the supply lines of the two converters are not short-circuited.

For the circulating-current scheme, the converters are kept in virtually continuous conduction over the whole range and the control circuit is simple. To obtain a reasonably good sinusoidal voltage waveform using the line-commutated two-quadrant converters, and to eliminate the possibility of the short circuit of the supply voltages, the output frequency of the cycloconverter is limited to a much lower value of the supply frequency. The output voltage waveform and the output frequency range can be improved further by using converters of higher pulse numbers.

15.4.2 Three-Phase Cycloconverters

Three-phase cycloconverters have several circuits. For example, there are the 3-pulse cycloconverter, 6-pulse cycloconverter, and 12-pulse cycloconverter.

15.4.2.1 Three-Phase Three-Pulse Cycloconverter

Figure 15.20a shows a schematic diagram of a three-phase half-wave (three-pulse) cycloconverter feeding a single-phase load, and Figure 15.20b shows the configuration of a three-phase half-wave (three-pulse) cycloconverter feeding a three-phase load. The basic process of a three-phase cycloconversion is illustrated in Figure 15.20c at 15 Hz, 0.6 power factor lagging load from a 50 Hz supply. As the firing angle α is cycled from zero at "a" to 180° at "j," half a cycle of the output frequency is produced (the gating circuit is to be suitably designed to introduce this oscillation of the firing angle). For this load, it can be seen that although the mean output voltage reverses at X, the mean output current (assumed sinusoidal) remains positive until Y. During XY, the SCRs A, B, and C in the P-converter are "inverting." A similar period exists at the end of the negative half-cycle of the output voltage when D, E, and F SCRs in the N-converter are "inverting." Thus, the operation of the converter follows in the order of "rectification" and "inversion" in a cyclic manner, with the relative durations dependent on the load power factor. The output frequency is that of the firing angle oscillation about a quiescent point of 90° (condition when the mean output voltage, given by $V_o = V_{do} \cos \alpha$, is zero). For obtaining the positive half-cycle of the voltage, firing angle α is varied from 90° to 0° and then to 90°, and for the negative half-cycle, from 90° to 180° and back to 90°. Variation of α within the limits of 180° automatically provides for "natural" line commutation of the SCRs. It is shown that a complete cycle of low-frequency output voltage is fabricated from the segments

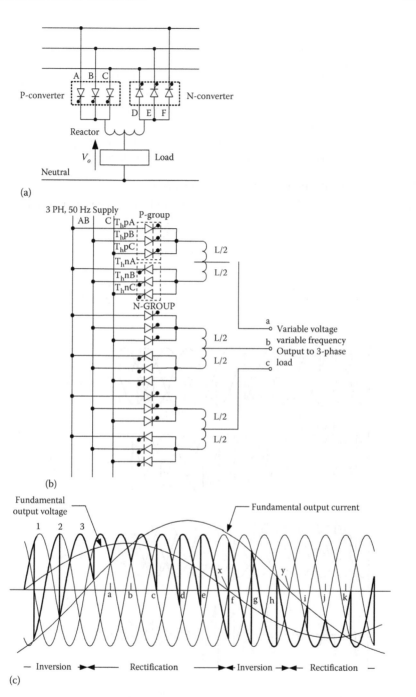

(a)

(b)

(c)

FIGURE 15.20 (a) Three-phase half-wave (three-pulse) cycloconverter supplying a single-phase load, (b) three-pulse cycloconverter supplying a three-phase load, and (c) output voltage waveform for one phase of a three-pulse cycloconverter operating at 15 Hz from a 50 Hz supply and 0.6 power factor lagging load.

of the three-phase input voltage by using the phase-controlled converters. The P- or N-converter SCRs receive firing pulses that are timed such that each converter delivers the same mean output voltage. This is achieved, as in the case of the single-phase cycloconverter or the dual converter, by maintaining the firing angle constraints of the two groups as $\alpha_P = (180° - \alpha_N)$. However, the instantaneous voltages of two converters are not identical, and a large circulating current may result unless limited by an intergroup reactor as shown (*circulating-current cycloconverter*) or completely suppressed by removing the gate pulses from the nonconducting converter by an intergroup blanking logic (*circulating-current-free cycloconverter*).

15.4.2.1.1 Circulating-Current Mode Operation

Figure 15.21 shows typical waveforms of a three-pulse cycloconverter operating with circulating current. Each converter conducts continuously with rectifying and inverting modes as shown, and the load is supplied with an average voltage of two converters reducing some of the ripple in the process, with the intergroup reactor behaving as a potential divider. The reactor limits the circulating current, with the value of its inductance to the flow of load current being one fourth of its value to the flow of circulating current, as the inductance is proportional to the square of the number of turns.

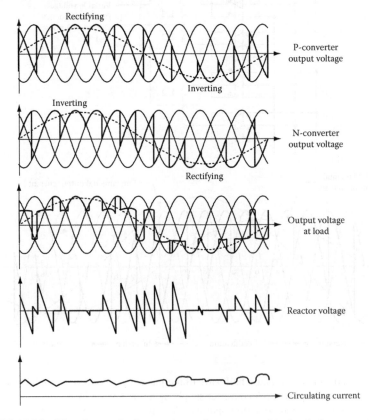

FIGURE 15.21 Waveforms of a three-pulse cycloconverter with circulating current.

The fundamental waves produced by both the converters are the same. The reactor voltage is the instantaneous difference between the converter voltages, and the time integral of this voltage divided by the inductance (assuming negligible circuit resistance) is the circulating current. For a three-pulse cycloconverter, it can be observed that this current reaches its peak value when $\alpha_p = 60°$ and $\alpha_N = 120°$.

15.4.2.1.2 Output Voltage Equation

A simple expression for the fundamental rms output voltage of the cycloconverter and the required variation of the firing angle α can be derived with the assumptions that (1) the firing angle α in successive half-cycles is varied slowly resulting in a low-frequency output; (2) the source impedance and the commutation overlap are neglected; (3) the SCRs are ideal switches; and (4) the current is continuous and ripple free. The average DC output voltage of a p-pulse dual converter with fixed α is

$$V_{do} = V_{do\,\max} \cos \alpha \tag{15.36}$$

where

$$V_{do\,\max} = \sqrt{2} V_{ph} \frac{p}{\pi} \sin \frac{p}{\pi}$$

For the p-pulse dual converter operating as a cycloconverter, the average phase voltage output at any point of the low frequency should vary according to the equation

$$V_{o,av} = V_{o1,\max} \sin \omega_0 t \tag{15.37}$$

where $V_{o1,\max}$ is the desired maximum value of the fundamental output voltage of the cycloconverter. Comparing Equation 15.36 with 15.37, the required variation of α to obtain a sinusoidal output is given by

$$\alpha = \cos^{-1}\left[\left(\frac{V_{o1,\max}}{V_{do\,\max}}\right)\sin \omega_0 t\right] = \cos^{-1}\left[r \sin \omega_0 t\right] \tag{15.38}$$

where r is the ratio ($V_{o1,\max}/V_{do\,\max}$), the *voltage magnitude control ratio*. Equation 15.38 shows α as a nonlinear function with r (≤ 1), as shown in Figure 15.22.

However, the firing angle α_P of the P-converter cannot be reduced to 0° as this corresponds to $\alpha_N = 180°$ for the N-converter, which, in practice, cannot be achieved because of allowance for commutation overlap and finite turn-off time of the SCRs. Thus, the firing angle α_P can be reduced to a certain finite value $\alpha_{\min}$, and the maximum output voltage is reduced by a factor $\cos \alpha_{\min}$.

The fundamental rms voltage per phase of either converter is

$$V_{or} = V_{oN} = V_{oP} = r V_{ph} \frac{p}{\pi} \sin \frac{\pi}{p} \tag{15.39}$$

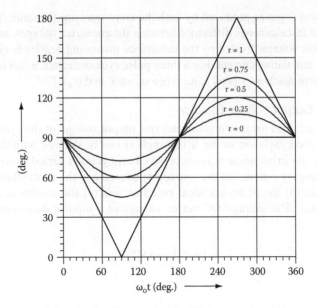

FIGURE 15.22 Variations of the firing angle (α) with *r* in a cycloconverter.

Although the rms values of the low-frequency output voltage of the P-converter and that of the N-converter are equal, the actual waveforms differ and the output voltage at the midpoint of the circulating-current limiting reactor (Figure 15.21), which is the same as the load voltage, is obtained as the mean of the instantaneous output voltages of the two converters.

15.4.2.1.3 Circulating-Current-Free Mode Operation
Figure 15.23 shows the typical waveforms for a three-pulse cycloconverter operating in this mode with *RL* load assuming continuous current operation. Depending on the load current direction, only one converter operates at a time and the load voltage is the same as the output voltage of the conducting converter. As explained earlier in the case of the single-phase cycloconverter, there is a possibility of a short circuit of the supply voltages at the crossover points of the converter unless care is taken in the control circuit. The waveforms drawn also neglect the effect of overlap due to the AC supply inductance. A reduction in the output voltage is possible by retarding the firing angle gradually at the points *a, b, c, d, e* in Figure 15.23 (this can easily be implemented by reducing the magnitude of the reference voltage in the control circuit). The circulating current is completely suppressed by blocking all the SCRs in the converter that is not delivering the load current. A current sensor is incorporated in each output phase of the cycloconverter that detects the direction of the output current and feeds an appropriate signal to the control circuit to inhibit or blank the gating pulses to the nonconducting converter in the same way as in the case of a dual converter for DC drives. The circulating-current-free operation improves the efficiency and the displacement factor of the cycloconverter and

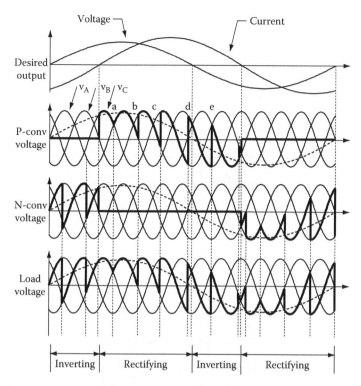

FIGURE 15.23 Waveforms for a three-pulse circulating-current-free cycloconverter with *RL* load.

also increases the maximum usable output frequency. The load voltage transfers smoothly from one converter to the other.

15.4.2.2 Three-Phase Six-Pulse and Twelve-Pulse Cycloconverter

A six-pulse cycloconverter circuit configuration is shown in Figure 15.24. Typical load-voltage waveforms for 6-pulse (with 36 SCRs) and 12-pulse (with 72 SCRs) cycloconverters are shown in Figure 15.25. The 12-pulse converter is obtained by connecting two 6-pulse configurations in series and appropriate transformer connections for the required phase shifted. It may be seen that the higher pulse numbers will generate waveforms closer to the desired sinusoidal form and thus permit higher frequency output. The phase loads may be isolated from each other as shown or interconnected with suitable secondary winding connections.

15.4.3 CYCLOCONVERTER CONTROL SCHEME

Various possible control schemes (analog as well as digital) for deriving trigger signals to control the basic cycloconverter have been developed over the years.

From the output of several possible signal combinations, it has been shown that a sinusoidal reference signal $(e_r = E_r \sin \omega_0 t)$ at desired output frequency f_0 and a

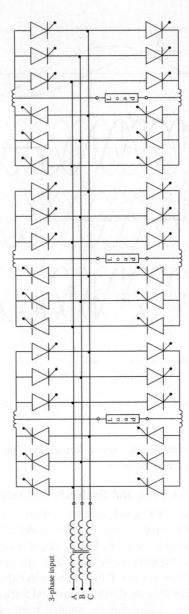

FIGURE 15.24 Three-phase six-pulse cycloconverter with isolated loads.

cosine modulating signal ($e_m = E_m \cos \omega_i t$) at input frequency f_i is the best combination possible for comparison to derive the trigger signals for the SCRs (Figure 15.26), which produces the output waveform with the lowest total harmonic distortion. The modulating voltages can be obtained as the phase-shifted voltages (B-phase for A-phase SCRs, C-phase voltage for B-phase SCRs, and so on) as explained in Figure 15.27, where at the intersection point "a."

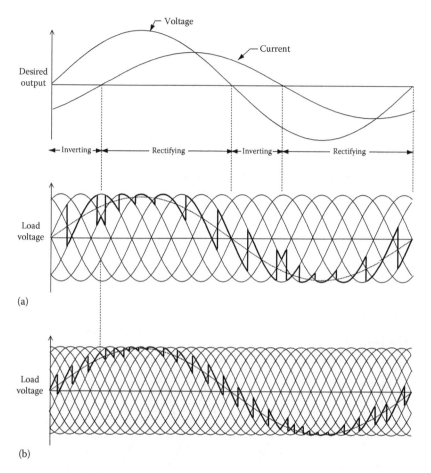

FIGURE 15.25 Cycloconverter load-voltage waveforms with lagging power factor load: (a) 6-pulse connection and (b) 12-pulse connection.

$$e_m = E_m \cos \omega_i t$$

or

$$E_m \sin(\omega_i t - 120°) = -E_r \sin(\omega_o t - \phi)$$

From Figure 15.27, the firing delay for A-phase SCR $\alpha = (\omega_i t - 30°)$. Thus,

$$\cos \alpha = \left(\frac{E_r}{E_m} \right) \sin(\omega_o t - \phi)$$

The cycloconverter output voltage for continuous current operation

$$V_o = V_{do} \cos \alpha = V_{do} \left(\frac{E_r}{E_m} \right) \sin(\omega_o t - \phi) \tag{15.40}$$

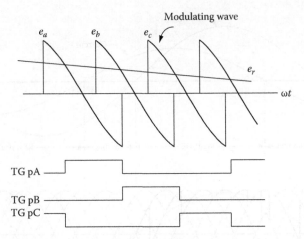

FIGURE 15.26 Deriving firing signals for one converter group of a three-pulse cycloconverter.

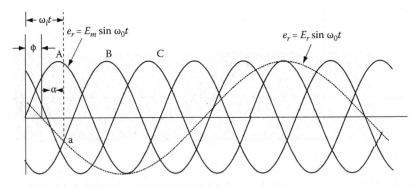

FIGURE 15.27 Derivation of the cosine modulating voltages.

which shows that the amplitude, frequency, and phase of the output voltage can be controlled by controlling correspondence parameters of the reference voltage, thus making the transfer characteristic of the cycloconverter linear. The derivation of the two complimentary voltage waveforms for the P-group or N-group converter "blanks" in this way is illustrated in Figure 15.28. The final cycloconverter output waveshape is composed of alternate half-cycle segments of the complementary P-converter and N-converter output voltage waveforms that coincide with the positive and negative current half-cycles, respectively.

15.4.3.1 Control Circuit Block Diagram

Figure 15.29 shows a simplified block diagram of the control circuit for a circulating-current-free cycloconverter. The same circuit is also applicable to a circulating-current cycloconverter with the omission of the *Converter Group Selection and Blanking Circuit*.

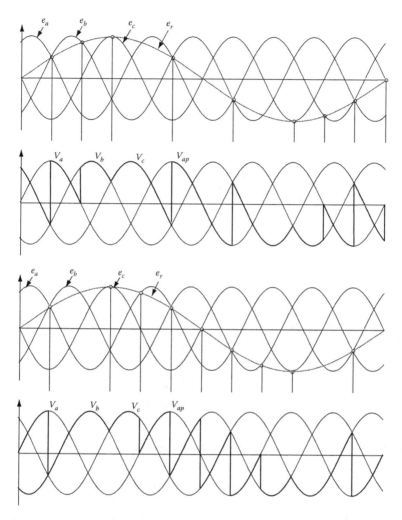

FIGURE 15.28 Derivation of P- and N-converter output voltages.

The *synchronizing circuit* produces the modulating voltages $(e_a = -Kv_b, e_b = -Kv_c, e_c = -Kv_a)$, synchronized with the mains through step-down transformers and proper filter circuits.

The *reference source* produces a VVVF reference signal (e_{ra}, e_{rb}, e_{rc}) (three-phase for a three-phase cycloconverter) for comparison with the modulation voltages. Various ways (analog or digital) have been developed to implement this reference source as in the case of the PWM inverter. In one of the early analog schemes (Figure 15.30) for a three-phase cycloconverter, a variable-frequency unijunction transistor (UJT) relaxation oscillator of the frequency $6f_d$ triggers a ring counter to produce a three-phase square-wave output of frequency (f_d), which is used to modulate a single-phase fixed frequency (f_c) variable amplitude sinusoidal voltage in a three-phase full-wave transistor chopper. The three-phase output contains $(f_c - f_d)$,

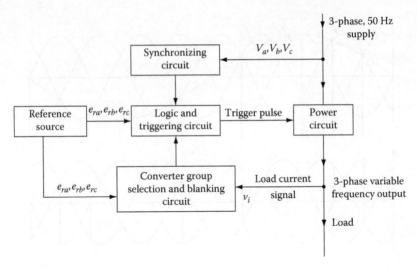

FIGURE 15.29 Block diagram for a circulating-current-free cycloconverter control circuit.

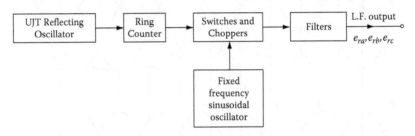

FIGURE 15.30 Block diagram of a VVVF three-phase reference source.

$(f_c + f_d)$, $(3f_d + f_c)$, and so forth, frequency components from where the "wanted" frequency component $(f_c - f_d)$ is filtered out for each phase using a low-pass filter. For example, with $f_c = 500\,\mathrm{Hz}$ and the frequency of the relaxation oscillator varying between 2820 and 3180 Hz, a three-phase 0–30 Hz reference output can be obtained with the facility for phase-sequence reversal.

The *logic* and *trigger* circuit for each phase involves comparators for comparison of the reference and modulating voltages and inverters acting as buffer stages. The outputs of the comparators are used to clock the flip-flops or latches whose outputs in turn feed the SCR gates through AND gates and pulse amplifying and isolation circuit. The second input to the AND gates is from the *converter group selection and blanking circuit*.

In the *converter group selection and blanking circuit*, the zero crossing of the current at the end of each half-cycle is detected and is used to regulate the control signals either to P-group or N-group converters depending on whether the current goes to zero from negative to positive or positive to negative, respectively. However, in practice, the current that is discontinuous passes through multiple zero crossings while changing direction, which may lead to undesirable switching of the converters.

Therefore in addition to the current signal, the reference voltage signal is also used for the group selection, and a *threshold* band is introduced in the current signal detection to avoid inadvertent switching of the converters. Further, a delay circuit provides a blanking period of appropriate duration between the converter group switching to avoid line-to-line short circuits. In some schemes, the delays are not introduced when a small circulating current is allowed during crossover instants limited by reactors of limited size, and this scheme operates in the so-called dual mode—circulating current as well as circulating-current-free mode for minor and major portions of the output cycle, respectively. A different approach to the converter group selection, based on the closed-loop control of the output voltage, where a bias voltage is introduced between the voltage transfer characteristics of the converters to reduce circulating current, is discussed.

15.4.3.2 Improved Control Schemes

With the development of microprocessors and PC-based systems, digital software control has taken over many tasks in modern cycloconverters, particularly in replacing the low-level reference waveform generation and analog signal comparison units. The reference waveforms can easily be generated in the computer, stored in the EPROMs, and accessed under the control of a stored program and microprocessor clock oscillator. The analog signal voltages can be converted to digital signals by using analog-to-digital converters (ADCs). The waveform comparison can then be made with the comparison features of the microprocessor system. The addition of time delays and intergroup blanking can also be achieved with digital techniques and computer software. A modification of the cosine firing control, using communication principles such as *regular sampling* in preference to the *natural sampling* of the reference waveform yielding a stepped sine wave before comparison with the cosine wave, has been shown to reduce the presence of *subharmonics* (to be discussed later) in the circulating-current cycloconverter and to facilitate microprocessor-based implementation, as in the case of PWM inverter.

15.4.4 Cycloconverter Harmonics and Input Current Waveform

The exact waveshape of the output voltage of the cycloconverter depends on (1) the pulse number of the converter, (2) the ratio of the output to input frequency (f_o/f_i), (3) the relative level of the output voltage, (4) load displacement angle, (5) circulating-current or circulating-current-free operation, and (6) the method of control of the firing instants. The harmonic spectrum of a cycloconverter output voltage is different and more complex than that of a phase-controlled converter. It has been revealed that because of the continuous "to-and-fro" phase modulation of the converter firing angles, the harmonic distortion components (known as *necessary distortion terms*) have frequencies that are sums and differences between multiples of output and input supply frequencies.

15.4.4.1 Circulating-Current-Free Operations

A derived general expression for the output voltage of a cycloconverter with circulating-current-free operation shows the following spectrum of harmonic frequencies for the 3-pulse, 6-pulse, and 12-pulse cycloconverters employing the cosine modulation technique:

$$\text{3-pulse:} \quad f_{oH} = \left|3(2k-1)f_i \pm 2nf_o\right| \quad \text{and} \quad \left|6kf_i \pm (2n+1)f_o\right|$$

$$\text{6-pulse:} \quad f_{oH} = \left|6kf_i \pm (2n+1)f_o\right| \tag{15.41}$$

$$\text{12-pulse:} \quad f_{oH} = \left|6kf_i \pm (2n+1)f_o\right|$$

where
  k is any integer from unity to infinity
  n is any integer from zero to infinity

It may be observed that for certain ratios of f_o/f_i, the order of harmonics may be less or equal to the desired output frequency. All such harmonics are known as *subharmonics* as they are not higher multiples of the input frequency. These subharmonics may have considerable amplitudes (e.g., with a 50 Hz input frequency and 35 Hz output frequency, a subharmonic of frequency $3 \times 50 - 4 \times 35 = 10$ Hz is produced whose magnitude is 12.5% of the 35 Hz component) and are difficult to filter and thus are objectionable. Their spectrum increases with the increase in the ratio f_o/f_i and thus limits its value at which a tolerable waveform can be generated.

15.4.4.2 Circulating-Current Operation

For circulating-current operation with continuous current, the harmonic spectrum in the output voltage is the same as that of the circulating-current-free operation except that each harmonic family now terminates at a definite term, rather than having an infinite number of components. They are

$$\text{3-pulse:} \quad f_{oH} = \begin{cases} \left|3(2k-1)f_i \pm 2nf_o\right| & n \le 3(2k-1)+1 \\ \left|6kf_i \pm (2n+1)f_o\right| & (2n+1) \le (6k+1) \end{cases}$$

$$\text{6-pulse:} \quad f_{oH} = \left|6kf_i \pm (2n+1)f_o\right|, \quad (2n+1) \le (6k+1) \tag{15.42}$$

$$\text{12-pulse:} \quad f_{oH} = \left|6kf_i \pm (2n+1)f_o\right|, \quad (2n+1) \le (12k+1)$$

The amplitude of each harmonic component is a function of the output voltage ratio for the circulating-current cycloconverter and the output voltage ratio as well as the load displacement angle for the circulating-current-free mode.

From the point of view of maximum useful attainable output-to-input frequency ratio (f_i/f_o) with the minimum amplitude of objectionable harmonic components, a guideline is available for it as 0.33, 0.5, and 0.75 for the 3-, 6-, and 12-pulse cycloconverters, respectively. However, with modification of the cosine wave modulation timings such as *regular sampling* in the case of circulating-current cycloconverters only and using a *subharmonic detection and feedback control concept* for both circulating- and circulating-current-free cases, the subharmonics can be suppressed and useful frequency range for the NCCs can be increased.

15.4.4.3 Other Harmonics Distortion Terms

Besides the harmonics as mentioned, other harmonic distortion terms consisting of frequencies of integral multiples of desired output frequency appear if the transfer characteristic between the output and reference voltages is not linear. These are called *unnecessary distortion terms*, which are absent when the output frequencies are much less than the input frequency. Further, some *practical distortion terms* may appear due to some practical nonlinearities and imperfections in the control circuits of the cycloconverter, particularly at relatively lower levels of output voltages.

15.4.4.4 Input Current Waveform

Although the load current, particularly for higher-pulse cycloconverters, can be assumed to be sinusoidal, the input current is more complex as it is made of pulses. Assuming the cycloconverter to be an ideal switching circuit without losses, it can be shown from the instantaneous power balance equation that in a cycloconverter supplying a single-phase load the input current has harmonic components of frequencies $(f_1 \pm 2f_o)$, called *characteristic harmonic frequencies* that are independent of pulse number, and they result in an oscillatory power transmittal to the AC supply system. In the case of a cycloconverter feeding a balanced three-phase load, the net instantaneous power is the sum of the three oscillating instantaneous powers when the resultant power is constant and the net harmonic component is greatly reduced compared with that of the single-phase load case. In general, the total rms value of the input current waveform consists of three components—in-phase, quadrature, and the harmonic. The in-phase component depends on the active power output, while the quadrature component depends on the net average of the oscillatory firing angle and is always lagging.

15.4.5 Cycloconverter Input Displacement/Power Factor

The input supply performance of a cycloconverter such as displacement factor or fundamental power factor, input power factor and the input current distortion factor are defined similarly to those of the phase-controlled converter. The harmonic factor for the case of a cycloconverter is relatively complex as the harmonic frequencies are not simple multiples of the input frequency but are sums and differences between multiples of output and input frequencies.

Irrespective of the nature of the load, leading, lagging, or unity power factor, the cycloconverter requires reactive power decided by the average firing angle. At low output voltage, the average phase displacement between the input current and the voltage is large and the cycloconverter has a low input displacement and power factor. Besides the load displacement factor and output voltage ratio, another component of the reactive current arises due to the modulation of the firing angle in the fabrication process of the output voltage. In a phase-controlled converter supplying DC load, the maximum displacement factor is unity for maximum DC output voltage. However, in the case of the cycloconverter, the maximum input displacement factor (IDF) is 0.843 with unity power factor load. The displacement factor decreases with reduction in the output voltage ratio. The distortion factor of the input current is given by (I_1/I), which is always less than unity and the resultant power factor (= distortion factor × displacement factor) is thus much lower (around 0.76 maximum) than the displacement factor, and this is a serious disadvantage of the NCC.

15.4.6 EFFECTS OF SOURCE IMPEDANCE

The source inductance introduces commutation overlap and affects the external characteristics of a cycloconverter similar to the case of a phase-controlled converter with DC output. It introduces delay in the transfer of current from one SCR to another and results in a voltage loss at the output and a modified harmonic distortion. At the input, the source impedance causes "rounding off" of the steep edges of the input current waveforms, resulting in reduction in the amplitudes of higher-order harmonic terms as well as a decrease in the IDF.

15.4.7 SIMULATION ANALYSIS OF CYCLOCONVERTER PERFORMANCE

The nonlinearity and discrete time nature of practical cycloconverter systems, particularly for discontinuous current conditions, make an exact analysis quite complex, and a valuable design and analytical tool is digital computer simulation of the system. Two general methods of computer simulation of the cycloconverter waveforms for *RL* and induction motor loads with circulating-current and circulating-current-free operation have been suggested. One of the methods, which is fast and convenient, is the *crossover point method*. This method gives the crossover points (intersections of the modulating and reference waveforms) and the conducting phase numbers for both P- and N-converters from which the output waveforms for a particular load can be digitally computed at any interval of time for a practical cycloconverter.

15.4.8 FORCED-COMMUTATED CYCLOCONVERTER

The NCC with SCRs as devices discussed so far is sometimes referred to as a *restricted frequency changer* as, in view of the allowance on the output voltage quality ratings, the maximum output voltage frequency is restricted ($f_o \ll f_i$) as mentioned earlier. With devices replaced by fully controlled switches such as forced-commutated SCRs, power transistors, IGBTs, GTOs, and so forth, an FCC can be built where the desired output frequency is given by $f_o = |f_s - f_i|$, when f_s = switching frequency, which may be larger or smaller than the f_i. In the case when $f_o \geq f_i$, the converter is called the *unrestricted frequency changer* (UFC) and when $f_o \leq f_i$, it is called a *slow switching frequency changer* (SSFC). The early FCC structures have been treated comprehensively. It has been shown that in contrast to the NCC, when the IDF is always lagging, in UFC it is leading when the load displacement factor is lagging and vice versa, and in SSFC, it is identical to that of the load. Further, with proper control in an FCC, the input displacement factor can be made unity (UDFFC) with concurrent composite voltage waveform or controllable (CDFFC) where P-converter and N-converter voltage segments can be shifted relative to the output current wave for the control of IDF continuously from lagging via unity to leading.

In addition to allowing bilateral power flow, UFCs offer an unlimited output frequency range, good input voltage utilization, do not generate input current and output voltage subharmonics, and require only nine bidirectional switches (Figure 15.31) for a three-phase to three-phase conversion. The main disadvantage of the structures treated is that they generate large unwanted low-order input current and output voltage harmonics that are difficult to filter out, particularly for low-output voltage conditions. This problem has largely been solved with the introduction of an imaginative PWM voltage-control scheme, which is the basis of a newly designated converter called the *matrix converter (also known as PWM cycloconverter)*, which operates as a *generalized solid-state transformer* with significant improvement in voltage and input current waveforms, resulting in sine-wave input and sine wave as discussed in the next session.

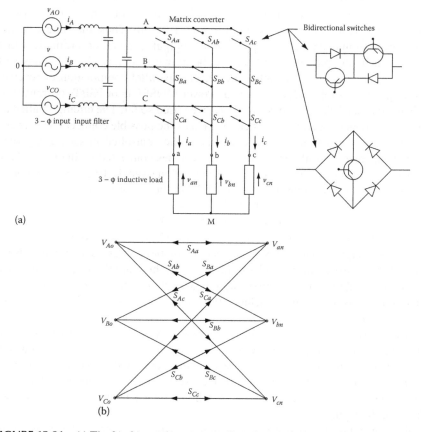

FIGURE 15.31   (a) The 3φ–3φ matrix converter (forced-commutated cycloconverter) circuit with input filter and (b) switching matrix symbol for converter.

15.5 MATRIX CONVERTERS

The MC is a development of the FCC based on bidirectional fully controlled switches, incorporating PWM voltage control, as mentioned earlier. This technique was developed by Venturine in 1980 [12]. With the initial progress reported, it has received considerable attention as it provides a good alterative to the double-sided PWM voltage-source rectifier-inverters, having the advantages of being a single-stage converter with only nine switches for three-phase to three-phase conversion and inherent bidirectional power flow, sinusoidal input/output waveforms with moderate switching frequency, the possibility of compact design due to the absence of DC link reactive components, and controllable input power factor independent of the output load current [12–21]. The main disadvantages of the MCs developed so far are the inherent restriction of the *voltage transfer ratio* (0.866), a more complex control and protection strategy, and above all the nonavailability of a fully controlled bidirectional high-frequency switch integrated in a silicon chip (triac, though bilateral, cannot be fully controlled).

The power circuit diagram of the most practical three-phase to three-phase (3ϕ–3ϕ) MC is shown in Figure 15.31a, which uses nine bidirectional switches so arranged that any of three input phases can be connected to any output phase as shown in the switching matrix symbol in Figure 15.31b. Thus, the voltage at any input terminal may be made to appear at any output terminal or terminals, while the current in any phase of the load may be drawn from any phase or phases of the input supply. For the switches, the inverse-parallel combination of reverse-blocking self-controlled devices such as Power MOSFETs or IGBTs or transistor-embedded diode bridge, as shown to have been used so far. The circuit is called an MC as it provides exactly one switch for each of the possible connections between the input and the output. The switches should be controlled in such a way that, at any time, one and only one of the three switches connected to an output phase must be closed to prevent "short circuiting" of the supply lines or interrupting the load-current flow in an inductive load. With these constraints, it can be visualized that from the possible 512 ($=2^9$) states of the converter, only 27 switch combinations are allowed as given in Table 15.3, which includes the resulting output line voltages and input phase currents. These combinations are divided into three groups. Group I consists of six combinations when each output phase is connected to a different input phase. In Group II, there are three subgroups, each having six combinations with two output phases short-circuited (connected to the same input phase). Group III includes three combinations with all output phases short-circuited.

With a given set of input three-phase voltages, any desired set of three-phase output voltages can be synthesized by adopting a suitable switching strategy. However, it has been shown that regardless of the switching strategy there are physical limits on the achievable output voltage with these converters as the maximum peak-to-peak output voltage cannot be greater than the minimum voltage difference between two phases of the input.

To have complete control of the synthesized output voltage, the envelope of the three-phase reference or target voltages must be fully contained within the

TABLE 15.3

Three-Phase/Three-Phase Matrix Converter Switching Combinations

Group	A	B	C	v_{ab}	v_{bc}	v_{ca}	i_A	i_B	i_C	S_{Aa}	S_{Ab}	S_{Ac}	S_{Ba}	S_{Bb}	S_{Bc}	S_{Ca}	S_{Cb}	S_{Cc}
I	A	B	C	v_{AB}	v_{BC}	v_{CA}	i_a	i_b	i_c	1	0	0	0	1	0	0	0	1
	A	C	B	$-v_{CA}$	$-v_{BC}$	$-v_{AB}$	i_a	i_c	i_b	1	0	0	0	0	1	0	1	0
	B	A	C	$-v_{AB}$	$-v_{CA}$	$-v_{BC}$	i_b	i_a	i_c	0	1	0	1	0	0	0	0	1
	B	C	A	v_{BC}	v_{CA}	v_{AB}	i_c	i_a	i_b	0	0	1	1	0	0	0	1	0
	C	A	B	v_{CA}	v_{AB}	v_{BC}	i_b	i_c	i_a	0	1	0	0	0	1	1	0	0
	C	B	A	$-v_{BC}$	$-v_{AB}$	$-v_{CA}$	i_c	i_b	i_a	0	0	1	0	1	0	1	0	0
II-A	A	B	B	v_{AB}	0	$-v_{AB}$	i_a	$-i_a$	0	1	0	0	0	1	1	0	0	0
	B	A	A	$-v_{AB}$	0	v_{AB}	$-i_a$	i_a	0	0	1	1	1	0	0	0	0	0
	B	C	C	v_{BC}	0	$-v_{BC}$	0	i_a	$-i_a$	0	0	0	1	0	0	0	1	1
	A	C	C	$-v_{CA}$	0	v_{CA}	i_a	0	$-i_a$	1	0	0	0	0	0	0	1	1
	C	A	A	v_{CA}	0	$-v_{CA}$	$-i_a$	0	i_a	0	1	1	0	0	0	1	0	0
	C	B	B	$-v_{BC}$	0	v_{BC}	0	$-i_a$	i_a	0	0	0	0	1	1	1	0	0
II-B	B	A	B	$-v_{AB}$	v_{AB}	0	i_b	$-i_b$	0	0	1	0	1	0	1	0	0	0
	A	B	A	v_{AB}	$-v_{AB}$	0	$-i_b$	i_b	0	1	0	1	0	1	0	0	0	0
	C	B	C	$-v_{BC}$	v_{BC}	0	0	i_b	$-i_b$	0	0	0	0	1	0	1	0	1
	C	A	C	v_{CA}	$-v_{CA}$	0	i_b	0	$-i_b$	0	1	0	0	0	0	1	0	1
	A	C	A	$-v_{CA}$	v_{CA}	0	$-i_b$	0	i_b	1	0	1	0	0	0	0	1	0
	B	C	B	v_{BC}	$-v_{BC}$	0	0	$-i_b$	i_b	0	0	0	1	0	1	0	1	0
II-C	B	B	A	0	$-v_{AB}$	v_{AB}	i_c	$-i_c$	0	0	0	1	1	1	0	0	0	0
	A	A	B	0	v_{AB}	$-v_{AB}$	$-i_c$	i_c	0	1	1	0	0	0	1	0	0	0
	C	C	B	0	$-v_{BC}$	v_{BC}	0	i_c	$-i_c$	0	0	0	0	0	1	1	1	0
	C	C	A	0	v_{CA}	$-v_{CA}$	i_c	0	$-i_c$	0	0	1	0	0	0	1	1	0
	A	A	C	0	$-v_{CA}$	v_{CA}	$-i_c$	0	i_c	1	1	0	0	0	0	0	0	1
	B	B	C	0	v_{BC}	$-v_{BC}$	0	$-i_c$	i_c	0	0	0	1	1	0	0	0	1
III	A	A	A	0	0	0	0	0	0	1	1	1	0	0	0	0	0	0
	B	B	B	0	0	0	0	0	0	0	0	0	1	1	1	0	0	0
	C	C	C	0	0	0	0	0	0	0	0	0	0	0	0	1	1	1

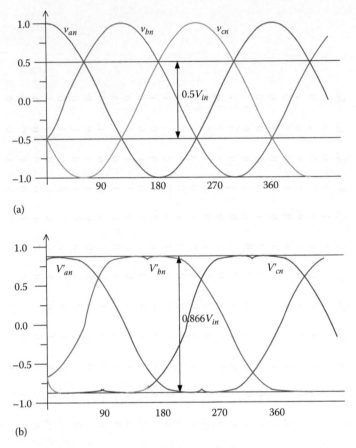

(a)

(b)

FIGURE 15.32 Output voltage limits for three-phase AC–AC matrix converter: (a) basic converter input voltages and (b) maximum attainable with inclusion of third harmonic voltages of input and output frequency to the target voltages.

continuous envelope of the three-phase input voltages. Initial strategy with the output frequency voltages as references reported the limit as 0.5 of the input, as shown in Figure 15.32a. This value can be increased to 0.866 by adding a third harmonic voltage of input frequency $(V_i/4)\cos3\omega_i t$, to all target output voltages, and subtracting from them a third harmonic voltage of output frequency $(V_o/6)\cos3\omega_o t$, as shown in Figure 15.32b. However, this process involves a considerable amount of additional computations in synthesizing the output voltages. The other alternative is to use the space vector modulation (SVM) strategy as used in PWM inverters without adding third harmonic components, but it also yields the maximum voltage transfer ratio as 0.866.

An AC input LC filter is used to eliminate the switching ripples generated in the converter, and the load is assumed to be sufficiently inductive to maintain continuity of the output currents.

15.5.1 OPERATION AND CONTROL METHODS OF THE MATRIX CONVERTER

The converter in Figure 15.31 connects any input phase (A, B, and C) to any output phase (a, b, and c) at any instant. When connected, the voltages v_{an}, v_{bn}, v_{cn} at the output terminals are related to the input voltages V_{Ao}, V_{Bo}, V_{Co} as

$$
\begin{bmatrix} v_{an} \\ v_{bn} \\ v_{cn} \end{bmatrix} = \begin{bmatrix} S_{Aa} & S_{Ba} & S_{Ca} \\ S_{Ab} & S_{Bb} & S_{Cb} \\ S_{Ac} & S_{Bc} & S_{Cc} \end{bmatrix} \begin{bmatrix} v_{Ao} \\ v_{Bo} \\ v_{Co} \end{bmatrix}
\tag{15.43}
$$

where S_{Aa} through S_{Cc} are the switching variables of the corresponding switches shown in Figure 15.31. For a balanced linear star-connected load at the output terminals, the input phase currents are related to the output phase current phase currents by

$$
\begin{bmatrix} i_A \\ i_B \\ i_C \end{bmatrix} = \begin{bmatrix} S_{Aa} & S_{Ab} & S_{Ac} \\ S_{Ba} & S_{Bb} & S_{Bc} \\ S_{Ca} & S_{Cb} & S_{Cc} \end{bmatrix} \begin{bmatrix} i_a \\ i_b \\ i_c \end{bmatrix}
\tag{15.44}
$$

Note that the matrix of the switching variables in Equation 15.44 is a transpose of the respective matrix in Equation 15.43. The MC should be controlled using a specific and appropriately timed sequence of the values of the switching variables, which will result in balanced output voltages having the desired frequency and amplitude, while the input currents are balanced and in phase (for unity IDF) or at an arbitrary angle (for controllable IDF) with respect to the input voltages. As the MC, in theory, can operate at any frequency, at the output or input, including zero, it can be employed as a three-phase AC/DC converter, DC/three-phase AC converter, or even a buck/boost DC chopper and thus as a *universal power converter*.

The control methods adopted so far for the MC are quite complex and are subjects of continuing research. On the methods proposed for independent control of the output voltages and input currents, two methods are of wide use and are reviewed briefly here: (1) the *Venturini* method based on a mathematical approach of transfer function analysis and (2) the SVM approach (as has been standardized now in the case of PWM control of the DC link inverter).

15.5.1.1 Venturini Method

Given a set of three-phase input voltages with constant amplitude V_i and frequency $f_i = \omega_i / 2\pi$, this method calculates a switching function involving the duty cycles of each of the nine bidirectional switches and generates the three-phase output voltages by sequential piecewise sampling of the input waveforms. These output voltages follow a predetermined set of reference or target voltage waveforms and with a three-phase load connected, a set of input currents I_i, and angular frequency ω_i, should be in phase for unity IDF or at a specific angle for controlled IDF.

A transfer function approach is employed to achieve the previously mentioned features by relating the input and output voltages and the output and input currents as

$$
\begin{bmatrix} V_{o1}(t) \\ V_{o2}(t) \\ V_{o3}(t) \end{bmatrix} = \begin{bmatrix} m_{11}(t) & m_{12}(t) & m_{13}(t) \\ m_{21}(t) & m_{22}(t) & m_{23}(t) \\ m_{31}(t) & m_{32}(t) & m_{33}(t) \end{bmatrix} \begin{bmatrix} V_{i1}(t) \\ V_{i2}(t) \\ V_{i3}(t) \end{bmatrix} \tag{15.45}
$$

$$
\begin{bmatrix} I_{i1}(t) \\ I_{i2}(t) \\ I_{i3}(t) \end{bmatrix} = \begin{bmatrix} m_{11}(t) & m_{21}(t) & m_{31}(t) \\ m_{12}(t) & m_{22}(t) & m_{32}(t) \\ m_{13}(t) & m_{23}(t) & m_{33}(t) \end{bmatrix} \begin{bmatrix} I_{o1}(t) \\ I_{o2}(t) \\ I_{o3}(t) \end{bmatrix} \tag{15.46}
$$

where the elements of the modulation matrix $m_{ij}(t)$ ($i, j = 1, 2, 3$) represent the duty cycles of a switch connecting output phase i to input phase j within a sample switching interval. The elements of $m_{ij}(t)$ are limited by the constraints

$$
0 \le m_{ij}(t) \le 1 \quad \text{and} \quad \sum_{j=1}^{3} m_{ij}(t) = 1 \quad (i = 1, 2, 3)
$$

The set of three-phase target or reference voltages to achieve the maximum voltage transfer ratio for unity IDF is

$$
\begin{bmatrix} V_{o1}(t) \\ V_{o2}(t) \\ V_{o3}(t) \end{bmatrix} = V_{om} \begin{bmatrix} \cos \omega_o t \\ \cos(\omega_o t - 120°) \\ \cos(\omega_o t - 240°) \end{bmatrix} + \frac{V_{im}}{4} \begin{bmatrix} \cos 3\omega_i t \\ \cos 3\omega_i t \\ \cos 3\omega_i t \end{bmatrix} - \frac{V_{om}}{6} \begin{bmatrix} \cos 3\omega_o t \\ \cos 3\omega_o t \\ \cos 3\omega_o t \end{bmatrix} \tag{15.47}
$$

where V_{om} and V_{im} are the magnitudes of output and input fundamental voltages of angular frequencies ω_o and ω_i, respectively. With $V_{om} \le 0.866 V_{im}$, a general formula for the duty cycles $m_{ij}(t)$ is derived. For unity IDF condition, a simplified formula is

$$
\begin{aligned}
m_{ij} = \frac{1}{3} \Big\{ & 1 + 2q \cos(\omega_i t - 2(j-1)60°) \Big[\cos(\omega_o t - 2(i-1)60°) \\
& + \frac{1}{2\sqrt{3}} \cos(3\omega_i t) - \frac{1}{6} \cos(3\omega_o t) \Big] \\
& - \frac{2q}{3\sqrt{3}} \big[\cos(4\omega_i t - 2(j-1)60°) - \cos(2\omega_i t - 2(1-j)60°) \big] \Big\}
\end{aligned} \tag{15.48}
$$

where
$i, j = 1, 2, 3$
$q = V_{om}/V_{im}$

The method developed is based on a *direct transfer function* (*DTF*) approach using a single modulation matrix for the MC and employing the switching combinations of all three groups in Table 15.3. Another approach called *indirect transfer function* (*ITF*) approach considers the MC as a combination of PWM voltage source rectifier–PWM voltage source inverter (VSR–VSI) and employs the already well-established VSR and VSI PWM techniques for MC control using the switching combinations of Group II and Group III only of Table 15.3. The drawback of this approach is that the IDF is limited to unity and the method also generates higher and fractional harmonic components in the input and the output waveforms.

15.5.1.2 SVM Method

The SVM is now a well-documented inverter PWM control technique that yields high voltage gain and less harmonic distortion compared to the other modulation techniques as discussed. Here, the three-phase input currents and output voltages are represented as space vectors, and SVM is applied simultaneously to the output voltage and input current space vectors. Applications of the SVM algorithm to the control of MCs have appeared in the literature and shown to have inherent capability to achieve full control of the instantaneous output voltage vector and the instantaneous current displacement angle even under supply voltage disturbances. The algorithm is based on the concept that the MC output line voltages for each switching combination can be represented as a voltage space vector denoted by

$$V_o = \frac{2}{3}\left[v_{ab} + v_{bc}\exp(j120°) + v_{ca}\exp(-j120°)\right] \tag{15.49}$$

Of the three groups in Table 15.3, only the switching combinations of Group II and Group III are employed for the SVM method. Group II consists of switching state voltage vectors with constant angular positions and are called *active* or *stationary* vectors. Each subgroup of Group II determines the position of the resulting output voltage space vector, and the six state space voltage vectors form a six-sextant hexagon used to synthesize the desired output voltage vector. Group III comprises the *zero* vectors positioned at the center of the output voltage hexagon, and these are suitably combined with the active vectors for the output voltage synthesis.

The modulation method involves selection of the vectors and their on-time computation. At each sampling period T_s, the algorithm selects four active vectors related to any possible combinations of output voltage and input current sectors in addition to the zero vector to construct a desired reference voltage. The amplitude and the phase angle of the reference voltage vector are calculated, and the desired phase angle of the input current vector is determined in advance. For computation of the on-time periods of the chosen vectors, these are combined into two sets leading to two new vectors adjacent to the reference voltage vector in the sextant and with the same direction as the reference voltage vector. Applying the standard SVM theory, the general formulas derived for the vector on-times, which satisfy, at the same time, the reference output voltage and input current displacement angle, are

$$t_1 = \frac{2qT_s}{\sqrt{3}\cos\varphi_i}\sin(60° - \theta_o)\sin(60° - \theta_i)$$

$$t_2 = \frac{2qT_s}{\sqrt{3}\cos\varphi_i}\sin(60° - \theta_o)\sin\theta_i$$

$$t_3 = \frac{2qT_s}{\sqrt{3}\cos\varphi_i}\sin\theta_o\sin(60° - \theta_i) \qquad (15.50)$$

$$t_4 = \frac{2qT_s}{\sqrt{3}\cos\varphi_i}\sin\theta_o\sin\theta_i$$

where

q is the voltage transfer ratio

φ_i is the input displacement angle chosen to achieve the desired input power factor (with $\varphi_i = 0$, a. Maximum value of $q = 0.866$ is obtained)

θ_o and θ_i are the phase displacement angles of the output voltage and input current vectors, respectively, whose values are limited within the 0°–60° range

The on time of the zero vector is

$$t_o = T_s - \sum_{i=1}^{4} t_i \qquad (15.51)$$

The integral value of the reference vector is calculated over one sample time interval as the sum of the products of the two adjacent vectors and their on-time ratios. The process is repeated at every sample instant.

15.5.1.3 Control Implementation and Comparison of the Two Methods

Both methods need a digital signal processor (DSP)-based system for their implementation. In one scheme for the Venturini method, the programmable timers, as available, are used to time out the PWM gating signals. The processor calculates the six switch duty cycles in each sampling interval, converts them to integer counts, and stores them in the memory for the next sampling period. In the SVM method, an EPROM is used to store the selected sets of active and zero vectors and the DSP calculates the on-times of the vectors. Then with an identical procedure as in the other method, the timers are loaded with the vector on-times to generate PWM waveforms through suitable output ports. The total computation time of the DSP for the SVM method has been found to be much less than that of the Venturini method. Comparison of the two schemes shows that while in the SVM method the switching losses are lower, the Venturini method shows better performance in terms of input current and output voltage harmonics

15.5.2 Commutation and Protection Issues in a Matrix Converter

As the MC has no DC link energy storage, any disturbance in the input supply voltage will affect the output voltage immediately and a proper protection mechanism has to be incorporated, particularly against overvoltage from the supply and overcurrent in the load side. As mentioned, two types of bidirectional switch configurations have hitherto been used—the transistor (now IGBT) embedded in a diode bridge and the two IGBTs in antiparallel with reverse voltage blocking diodes (shown in Figure 15.31). In the latter configuration, each diode and IGBT combination operates in two quadrants only, which eliminates the circulating currents otherwise built up in the diode-bridge configuration that can be limited by only bulky commutation inductors in the lines.

The MC does not contain freewheeling diodes that usually achieve safe commutation in the case of other converters. To maintain the continuity of the output current as each switch turns off, the next switch in sequence must be immediately turned on. In practice, with bidirectional switches, a momentary short circuit may develop between the input phases when the switches cross over, and one solution is to use a *semisoft current commutation* using a multistepped switching procedure to ensure safe commutation. This method requires independent control of each two-quadrant switches, sensing the direction of the load current and introducing a delay during the change of switching states.

A clamp capacitor connected through two three-phase full-bridge diode rectifiers involving an additional 12 diodes (a new configuration with the number of additional diodes reduced to six using the antiparallel switch diodes has been reported) at the input and output lines of the MC serves as a voltage clamp for possible voltage spikes under normal and fault conditions.

A three-phase single-stage LC filter consisting of three capacitors in star and three inductors in the line is used to adequately attenuate the higher-order harmonics and render sinusoidal input current. Typical values of L and C based on a 415 V converter with a maximum line current of 6.5 A and a switching frequency of 20 kHz are 3 mH and 1.5 µF only. The filter may cause a minor phase shift in the input displacement angle that needs correction. Figure 15.33 shows typical

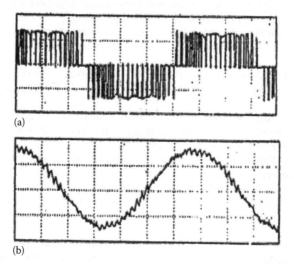

(a)

(b)

FIGURE 15.33 Experimental waveforms for an MC at 30 Hz frequency from 50 Hz input: (a) output line voltage and (b) output line current.

experimental waveforms of output line voltage and line current of an MC. The output line current is mostly sinusoidal except for a small ripple, when the switching frequency is around 1 kHz only.

REFERENCES

1. Luo, F. L., Ye, H., and Rashid, M. H. 2005. *Digital Power Electronics and Applications*. Boston, MA: Academic Press.
2. Rashid, M. H. 2001. *Power Electronics Handbook*, pp. 307–333. New York: Academic Press.
3. Agrawal, J. P. 2001. *Power Electronics Systems*, pp. 355–389. Eglewood Cliffs, NJ: Prentice Hall.
4. Rombaut, C., Seguier, G., and Bausiere, R. 1987. *Power Electronics Converters—AC/AC Converters*. New York: McGraw-Hill.
5. Lander, C. W. 1993. *Power Electronics*. London, U.K.: McGraw-Hill.
6. Dewan, S. B. and Straughen, A. 1975. *Power Semiconductor Circuits*. New York: John Wiley.
7. Hart, D. W. 1997. *Introduction to Power Electronics*. Eglewood Cliffs, NJ: Prentice-Hall.
8. Williams, B. W. 1987. *Power Electric Devices, Drivers and Applications*. London, U.K.: MacMillan.
9. Pelly, B. R. 1971. *Thyristor Phase-Controlled Converters and Cycloconverters*. New York: John Wiley.
10. McMurray, W. 1972. *The Theory and Design of Cycloconverters*. Cambridge, MA: MIT Press.
11. Syam, P., Nandi, P. K., and Chattopadhyay, A. K. 1998. An improvement feedback technique to suppress sub-harmonics in a naturally commutated cycloconverter. *IEEE Transactions on Industrial Electronics*, 45, 950–962.
12. Venturine, M. 1980. A new sine-wave in sine-wave out converter technique eliminated reactor elements. *Proceedings of Powercon 1980*, Munich, Germany, pp. E3-1–E3-15.
13. Alesina, A. and Venturine, M. 1980. The generalized transformer: A new bidirectional waveform frequency converter with continuously adjustable input power factor. *Proceedings of IEEE PESC 1980*, Atlanta, GA, pp. 242–252.
14. Alesina, A. and Venturine, M. 1989. Analysis and design of optimum amplitude nine-switch direct AC-AC converters. *IEEE Transactions on Power Electronics Letters*, 4, 101–112.
15. Ziogas, P. D., Khan, S. I., and Rashid, M. 1985. Some improved forced commutated cycloconverter structures. *IEEE Transactions on Industry Applications*, IA-21, 1242–1253.
16. Ziogas, P. D., Khan, S. I., and Rashid, M. 1986. Analysis and design of forced commutated cycloconverter structures and improved transfer characteristics. *IEEE Transactions on Industrial Electronics*, IE-33, 271–280.
17. Ishiguru, A., Furuhashi, T., and Okuma, S. 1991. A novel control method of forced commutated cycloconverter using instantaneous values of input line voltages. *IEEE Transactions on Industrial Electronics*, 38, 166–172.
18. Huber, L., Borojevic, D., and Burani, N. 1992. Analysis, design and implementation of the space-vector modulator for commutated cycloconverters. *IEE Proceedings Part B*, 139, 103–113.
19. Huber, L. and Borojevic, D. 1995. Space-vector modulated three-phase to three-phase matrix converter with input power factor correction. *IEEE Transactions on Industry Applications*, 31, 1234–1246.

20. Zhang, L., Watthanasarn, C., and Shepherd, W. 1998. Analysis and comparison of control strategies for AC-AC matrix converters. *IEE Proceedings on Electronic Power Applications*, 145, 284–294.
21. Das, S. P. and Chattopadhyay, A. K. 1997. Observer based stator flux oriented vector control of cycloconverter-fed synchronous motor drive. *IEEE Transactions on Industry Applications*, 22, 943–955.

16 Improved AC/AC Converters

Traditional methods of AC/AC converters are introduced in Chapter 15. All methods have general drawbacks:

1. Output voltage is lower than input voltage.
2. The input side THD is poor.
3. Output voltage frequency is lower than input voltage frequency when using voltage regulation method and cycloconverters.

Some new methods constructing AC/AC converters can overcome aforementioned issues. We introduce following converters in this chapter:

- DC-modulated AC/AC converters
- Sub-envelope modulation (SEM) method to reduce THD for matrix AC/AC converters

16.1 DC-MODULATED SINGLE-STAGE AC/AC CONVERTERS

Single-stage AC/AC converters are the most popular structure widely applied in various industrial applications [1–4]. These AC/AC converters are traditionally implemented by voltage regulation technique, cycloconverters, and matrix converters. However, they have high total harmonic distortion (THD), low power factor (PF), and poor power transfer efficiency (η). A typical single-stage AC/AC converter with voltage regulation technique and the corresponding waveforms are shown in Figure 16.1. The devices can be thyristors, IGBT, and MOSFET. For a clear example, MOSFETs are applied in the circuit with a pure resistive load R [5]. The input voltage is

$$v_S(t) = \sqrt{2}V_S \sin \omega t$$

where

V_S is the rms value

ω is the input voltage frequency $\omega = 2\pi f = 100\pi$

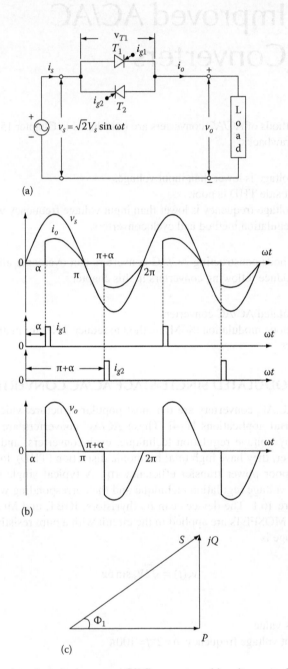

FIGURE 16.1 A typical single-stage AC/AC converter with voltage regulation technique. (a) Circuit diagram, (b) waveforms, and (c) power vectors.

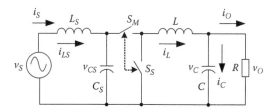

FIGURE 16.2   A DC-modulated single-stage buck-type AC/AC converter.

The PF is calculated by the formula [6]

$$PF = \frac{DPF}{\sqrt{1+THD^2}}$$

where
   DPF $= \cos \Phi_1$ is the displacement power factor
   THD is the total harmonic distortion

and the delay angle Φ_1 is the phase delay angle of the fundamental harmonic component.

For example, if the faring angle α is 30° (i.e., fundamental harmonic phase angle Φ_1 is 30°), the typical values are DPF $= \cos 30° = 0.866$ and THD is 0.15 (or 15%). Therefore, PF $= 0.856$. It is a low PF. The power vector diagram is shown in Figure 16.1c where P is the real power, jQ the reactive power, and S apparent power, $S = P + jQ$.

DC/DC conversion technology [7–10] can facilitate fast response and high efficiency. Our novel approach to PFC is *DC modulation power factor correction AC/AC conversion* [11–14]. This technique can help with reaching high PF. A DC-modulated single-stage buck-type AC/AC converter is shown in Figure 16.2.

We assume the input power supply rms voltage is 240 V with the frequency $f = 50$ Hz. The master switch S_M and slave switch S_S are bidirectional switches. They are working in the exclusive states. The DC modulation switching frequency f_m is usually high, say $f_m = 20$ kHz. Therefore, the input power supply voltage is quasi-static DC voltage (positive or negative value) in a DC modulation period $T_m = 1/f_m$ (50 μs). In a DC modulation period T_m, the input voltage is a quasi-constant DC value. Therefore, the converter works in a DC/DC conversion performance. We can use all conclusions of DC/DC conversion technology in this operation. The key device is the bidirectional exclusive switches S_M–S_S (even more multi-bidirectional switches) for the DC modulation operation. Since the buck converter input current is pulse train with the repeating frequency f_m, a low-pass input filter L_S–C_S is required.

16.1.1   Bidirectional Exclusive Switches S_M–S_S

The switching devices for bidirectional exclusive switches can be MOSFETs and IGBTs. MOSFETs are selected for our design. The bidirectional exclusive switches S_M–S_S for the DC modulation operation thus designed have the following technical features:

1. The master switch S_M is controlled by a PWM pulse train and conducts the input current to flow in forward direction in the positive input voltage. Vice versa, the S_M conducts the input current to flow in the reverse direction in the negative input voltage.
2. The slave switch S_S is conducted when the master switch S_M is switched off exclusively. It is the free-wheeling device that conducts the current to flow.

Figure 16.3 shows the circuit of the bidirectional exclusive switches S_M–S_S for the DC modulation operation. The switching control signal is a PWM pulse train that has adjustable frequency f_m and pulse width. The repeating period $T_m = 1/f_m$, and the conduction duty cycle $k = (\text{pulse-width})/T_m$.

If some converters require more than one bidirectional exclusive slave switches, the construction of further S_S just only need copy/repeat the existing one. If some converters require more than one bidirectional slave switches and one synchronously bidirectional slave switch, the construction of the synchronously bidirectional slave switch S_{S-S} just only need copy/repeat the master switch S_M. A group of master switches S_M with a synchronously bidirectional slave switch S_{S-S} plus two bidirectional exclusive slave switches S_{S1} and S_{S2} are shown in Figure 16.4.

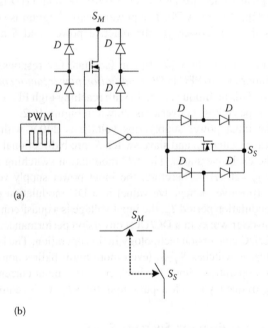

(a)

(b)

FIGURE 16.3 A bidirectional exclusive switches S_M–S_S for the DC modulation operation. (a) Circuit of bidirectional exclusive switches S_M–S_S, and (b) symbol of bidirectional exclusive switches S_M–S_S.

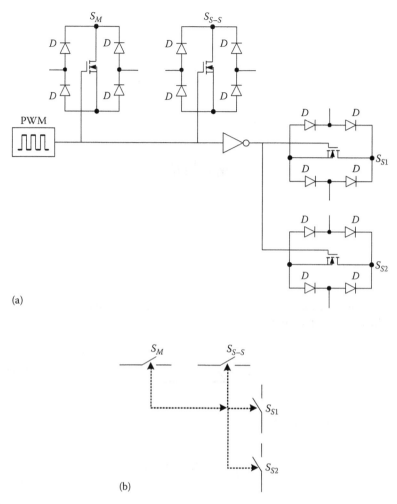

(a)

(b)

FIGURE 16.4 A bidirectional switches S_M–S_{S-S} plus exclusive switches S_{S1} and S_{S2}. (a) Circuit symbol of bidirectional switches S_M–S_{S-S} plus exclusive switches S_{S1} and S_{S2}, and (b) symbol of bidirectional switches S_M–S_{S-S} plus exclusive switches S_{S1} and S_{S2}.

16.1.2 MATHEMATICAL MODELING FOR DC/DC CONVERTERS

The mathematical modeling for DC/DC converters is a historic topic that accompanies the DC/DC converter development. This topic has been well discussed by Luo and Ye [8–10]. The main points are as follows:

1. The input pumping energy is PE:

$$PE = \int_0^{Tm} V_S i_S(t)\,dt = V_S \int_0^{Tm} i_S(t)\,dt = V_S I_S T_m \qquad (16.1)$$

where the average current I_S is

$$I_S = \frac{1}{T_m} \int_0^{Tm} i_S(t)\,dt \tag{16.2}$$

2. The stored energy in an inductor is

$$W_L = \frac{1}{2}LI_L^2 \tag{16.3}$$

The stored energy across a capacitor is

$$W_C = \frac{1}{2}CV_C^2 \tag{16.4}$$

Therefore, if there are n_L inductors and n_C capacitors the total *stored energy* (SE) in a DC/DC converter is

$$SE = \sum_{j=1}^{n_L} W_{Lj} + \sum_{j=1}^{n_C} W_{Cj} \tag{16.5}$$

3. The energy factor is EF:

$$EF = \frac{SE}{PE} = \frac{SE}{V_1 I_1 T_m} = \frac{\sum_{j=1}^{m} W_{Lj} + \sum_{j=1}^{n} W_{Cj}}{V_1 I_1 T_m} \tag{16.6}$$

4. The capacitor/inductor stored energy ratio (CIR) is defined as follows [7]:

$$CIR = \frac{\sum_{j=1}^{n_C} W_{Cj}}{\sum_{j=1}^{n_L} W_{Lj}} \tag{16.7}$$

5. The time constant τ is defined as follows:

$$\tau = \frac{2T_m \times EF}{1 + CIR}\left(1 + CIR\frac{1-\eta}{\eta}\right) \tag{16.8}$$

where η is the power transfer efficiency. If there are no power losses, $\eta = 1$.

$$\tau = \frac{2T_m \times \text{EF}}{1 + \text{CIR}} \qquad (16.9)$$

6. The damping time constant τ_d is defined as follows:

$$\tau_d = \frac{2T_m \times \text{EF}}{1 + \text{CIR}} \frac{\text{CIR}}{\eta + \text{CIR}(1 - \eta)} \qquad (16.10)$$

If there are no power losses,

$$\tau_d = \frac{2T_m \times \text{EF}}{1 + 1/\text{CIR}} \qquad (16.11)$$

7. The time constant ratio ξ is defined as follows:

$$\xi = \frac{\tau_d}{\tau} = \frac{\text{CIR}}{\eta\left(1 + \text{CIR}((1 - \eta)/\eta)\right)^2} \qquad (16.12)$$

If there are no power losses,

$$\xi = \frac{\tau_d}{\tau} = \text{CIR} \qquad (16.13)$$

8. A DC/DC converter has the transfer function

$$G(s) = \frac{M}{1 + s\tau + s^2\tau\tau_d} = \frac{M}{1 + s\tau + \xi s^2\tau^2} \qquad (16.14)$$

where M is the voltage transfer gain in a steady state, for example, $M = k$ for a buck converter.

Example 16.1 A buck converter in Figure 16.2 with $L = 1\,\text{mH}$, $C = 0.4\,\mu\text{F}$, and the load $R = 100\,\Omega$, shows input voltage and current, v_s and i_s, respectively; output voltage and current, v_O and i_O, respectively; no power losses, that is, $\eta = 1$; the switching frequency, f_m (the switching period $T_m = 1/f_m$); and the conduction duty cycle, k. Calculate the transfer function and its step response.

Solution: We obtain the following data:

$v_O = kv_S$	$i_S = ki_O$
$v_O = Ri_O$	$P_{in} = v_S i_S = v_O i_O = P_O$ with $\eta = 1$
$PE = \int_0^{Tm} V_S i_S(t)dt = V_S \int_0^{Tm} i_S(t)dt = V_S I_S T_m$	$W_L = \frac{1}{2}LI_L^2 = \frac{1}{2}Li_O^2$
$W_C = \frac{1}{2}CV_C^2 = \frac{1}{2}Cv_O^2$	$SE = \frac{1}{2}(Li_O^2 + Cv_O^2) = \frac{1}{2}(L + CR^2)i_O^2$
$EF = \dfrac{SE}{PE} = \dfrac{(L+CR^2)i_O^2}{2v_O i_O T_m} = \dfrac{L/R+CR}{2T_m}$	$CIR = \dfrac{0.5Li_O^2}{0.5Cv_O^2} = \dfrac{L}{CR^2} = \dfrac{1}{4}$
$\tau = \dfrac{2T_m \times EF}{1+CIR} = \dfrac{L/R+RC}{1+CIR} = 40\ \mu s$	$\tau_d = \dfrac{2T_m \times EF}{1+1/CIR} = \dfrac{L/R+RC}{1+1/CIR} = 10\ \mu s$

Therefore, the transfer function is

$$G(s) = \frac{k}{1+s\tau+0.25s^2\tau^2} = \frac{M}{(1+0.00002s)^2} \tag{16.15}$$

This transfer function is in the critical condition with two folded poles. The corresponding step response in the time domain has fast response without overshot and oscillation.

$$g(t) = k\left[1-\left(1+\frac{2t}{\tau}\right)e^{-\frac{2t}{\tau}}\right] = k\left[1-\left(1+\frac{2t}{0.00004}\right)e^{-2t/0.00004}\right] \tag{16.16}$$

The settling time from one steady state to another one is about $2.4\tau=0.000096\,s=0.096\,ms$. This time period is much smaller than the power supply period $T=1/f=20\,ms$. The corresponding radian distance is only $1.73°$. The statistic average delay angle Φ_1 is its $1/e$ time, that is, $\Phi_1 = 0.24\tau/e = 0.0353\,ms$ or $0.636°$, and $DPF = \cos\Phi_1 = 0.999938$. Consequently, we can assume the output voltage can follow the input voltage waveform.

16.1.3 DC-MODULATED SINGLE-STAGE BUCK-TYPE AC/AC CONVERTER

The DC-modulated single-stage buck-type AC/AC converter is shown in Figure 16.2. We have to investigate the operation during both positive and negative half-cycle of the input voltage.

16.1.3.1 Positive Input Voltage Half-Cycle

When the input voltage is positive, the buck converter operates as usual. The equivalent circuits during the switch-on and switch-off conditions are shown in Figure 16.5.

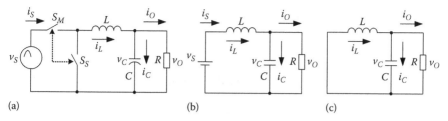

FIGURE 16.5 A DC-modulated buck-type AC/AC converter working in positive half-cycle. (a) Circuit diagram, (b) equivalent circuit during switch-on, and (c) equivalent circuit during switch-off.

The output voltage is calculated by

$$v_O = kv_S = k\sqrt{2}V_S \sin \omega t \quad 0 \le \omega t < \pi \tag{16.17}$$

where
   k is the conduction duty cycle of the buck converter
   V_S is the rms value of the input voltage
   ω is the power supply radian frequency

16.1.3.2 Negative Input Voltage Half-Cycle

When the input voltage is negative, the buck converter performs in the reverse. The equivalent circuits during the switch-on and switch-off conditions are shown in Figure 16.6.
   The output voltage is calculated by

$$v_O = -k \mid v_S \mid = k\sqrt{2}V_S \sin \omega t \quad \pi \le \omega t < 2\pi \tag{16.18}$$

where
   k is the conduction duty cycle
   V_S is the rms value
   ω is the power supply radian frequency

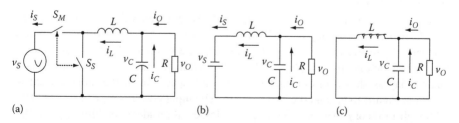

FIGURE 16.6 A DC-modulated buck-type AC/AC converter working in negative half-cycle. (a) The circuit diagram, (b) equivalent circuit during switch-on, and (c) equivalent circuit during switch-off.

16.1.3.3 Whole-Cycle Operation

Combining the previous two operation states, we can summarize the whole-cycle operation. The output voltage is calculated by

$$v_O = kv_S = k\sqrt{2}V_S \sin \omega t \qquad (16.19)$$

where
   k is the conduction duty cycle
   V_S is the rms value
   ω is the power supply radian frequency

The whole-cycle input and output voltage waveforms are shown in Figure 16.7a with the duty cycle $k=0.75$. The voltage transfer gain $M=V_O/V_S$ versus the DC/DC converter conduction duty cycle k is shown in Figure 16.7b. It is easy to obtain variable output voltage with very high PF and high efficiency.

The whole-cycle input voltage and current waveforms are shown in Figure 16.8a. The spectrum of the input current is shown in Figure 16.8b. The spectrum is very clean and there is little distortion from the harmonic component I_M at 20 kHz that is far away from the fundamental frequency component I_S at 50 Hz. Its value is only 0.5%, that is, $I_M/I_S=0.005$. Therefore, THD = 1.0000125. Considering the DPF=0.9998, we obtain the final PF=0.99979.

16.1.3.4 Simulation and Experimental Results

The simulation and experimental results are shown in order to verify the design.

16.1.3.4.1 Simulation Results

The DC-modulated buck-type AC/AC converter has the following components: $L_S=1$ mH, $C_S=10$ μF, $L=10$ mH, and $C=3$ μF. The conduction duty cycle is selected as $k=0.75$. The simulation results are shown in Figure 16.9. The output voltage $V_O=0.75 \times V_S=150$ Vrms (peak value is approximately 212 V) with the frequency $f=50$ Hz. The waveforms of the input and output voltages v_S (t) and v_O (t) are shown as Channel 1 and Channel 2 in Figure 16.9a. It can be seen that there is not any phase delay, although there may be about 3.374° phase angle delay as shown in our analysis. The output current I_O should be 1 Arms, and the output power $P_O=V_O^2/R=150$ W.

The input current is measured as $I_S=0.95$ Arms (peak value is approximately 1.34 A) with the frequency $f=50$ Hz. The waveforms of the input voltage $v_S(t)$ and current $i_S(t)$ are shown as Channel 1 and Channel 2 in Figure 16.9b. It can be seen that there is not nearly any phase delay, although there may be about 3.374° phase angle delay as shown in our analysis. The FFT spectrum of the input current is shown in Figure 16.9c, and the THD=0.015. The input power is $P_{in}=V_S \times I_S=190$ W. Although in theory there can be no power losses in an ideal condition ($\eta=1$), the particular test shows there can indeed be power losses, which are mainly caused by the switches power losses. From the test results we obtain the final PF=0.9979 and the power transfer efficiency $\eta=P_o/P_{in}$ 190/200=0.95 or 95%.

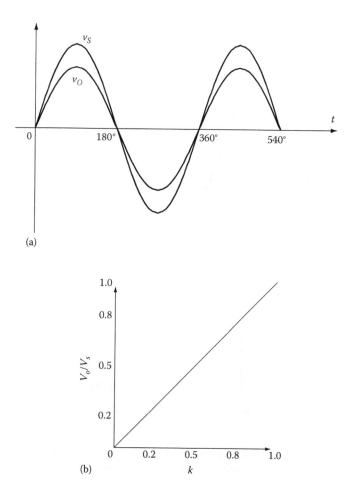

FIGURE 16.7 The input/output voltage waveforms of the DC-modulated buck-type AC/AC converter. (a) The input/output voltage waveforms, and (b) the voltage transfer gain versus conduction duty cycle k.

16.1.3.4.2 Experimental Results

The experimental results of the DC-modulated buck-type AC/AC converter are shown in Figure 16.10.

Example 16.2 A buck-type DC-modulated AC/AC converter in Figure 16.2 has input rms voltage $v_s = 240\,V$ and a dimmer load with $R = 100\,\Omega$. In order to adjust the light, the output rms voltage v_O varies in the range of 100–200 V. Calculate the range of the duty cycle k, the output current, and power.

Solution: Since the output voltage is calculated as

$$v_O = kv_s = 240k$$

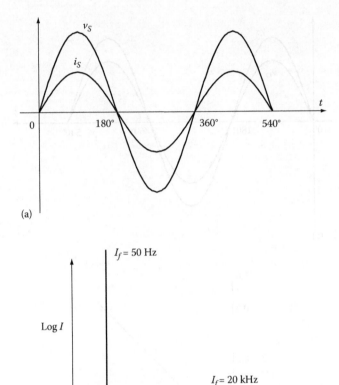

FIGURE 16.8 The input voltage/current waveforms of the DC-modulated buck-type AC/AC converter. (a) The input voltage and current waveforms, and (b) the spectrum of input current.

the duty cycle is calculated as

$$k = \frac{v_O}{v_S} = \begin{cases} \dfrac{100}{240} = 0.42 \\[2mm] \dfrac{200}{240} = 0.83 \end{cases}$$

The range of the duty cycle k is 0.42–0.83.
The output rms current is 1–2 A, and the output power is 100–400 W.

16.1.4 DC-MODULATED SINGLE-STAGE BOOST-TYPE AC/AC CONVERTER

The DC-modulated single-stage buck-type AC/AC converter can only convert an input voltage to a lower output voltage. For certain applications, the output voltage

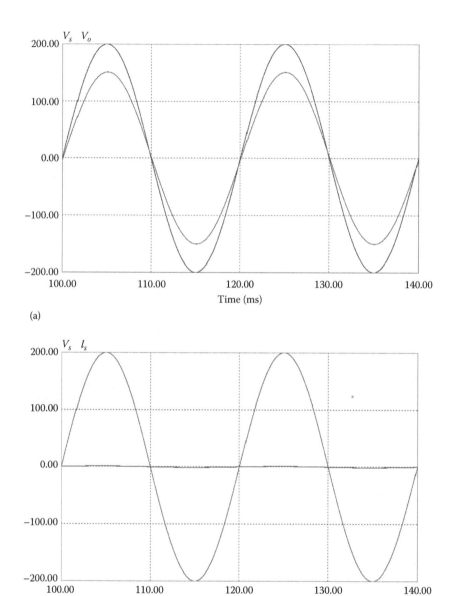

FIGURE 16.9 Test results of the DC-modulated buck-type AC/AC converter. (a) Input/output voltage waveforms of the DC-modulated buck-type AC/AC converter, (b) input voltage and current waveforms of the DC-modulated buck-type AC/AC converter, and (c) spectrum of input current of the DC-modulated buck-type AC/AC converter.

(continued)

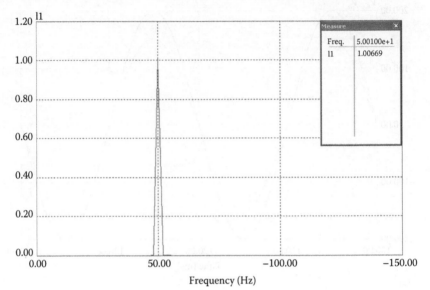

Test result spectrum first harmonic measurement of the
DC-modulated buck-type AC / AC converter
$f = 50$ kHz, $k = 0.75$ (zoom 1)

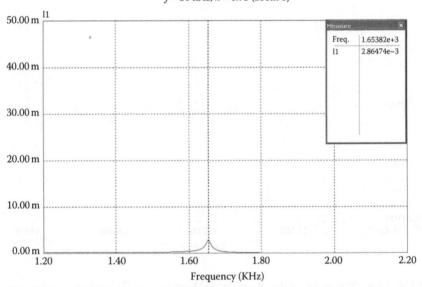

Test result spectrum second harmonic measurement of the DC-modulated
buck-type AC / AC converter $f = 50$ kHz, $k = 0.75$ (zoom 2)

FIGURE 16.9 (continued)

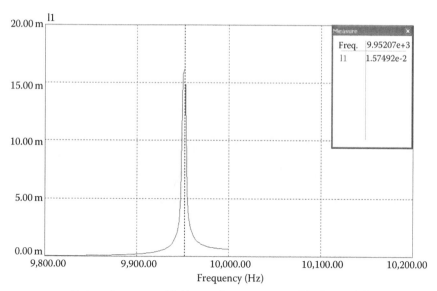

Measure	⊠
Freq.	9.95207e+3
I1	1.57492e-2

Test result spectrum third harmonic measurement of the DC-modulated buck-type AC/AC converter f = 50 kHz, k = 0.75 (zoom 3)

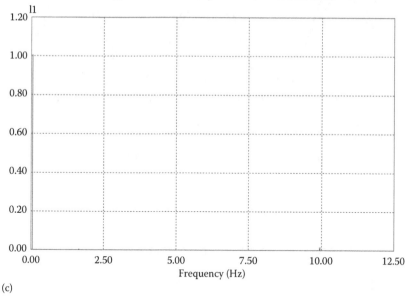

(c)

FIGURE 16.9 (continued)

needs to be higher than the input voltage. For this purpose, the DC-modulated single-stage boost-Type AC/AC converter has been designed and is shown in Figure 16.11. Since the input current is continuous current, there is no need to set a low-pass filter. We have to investigate the operation during both positive and negative half-cycle of the input voltage.

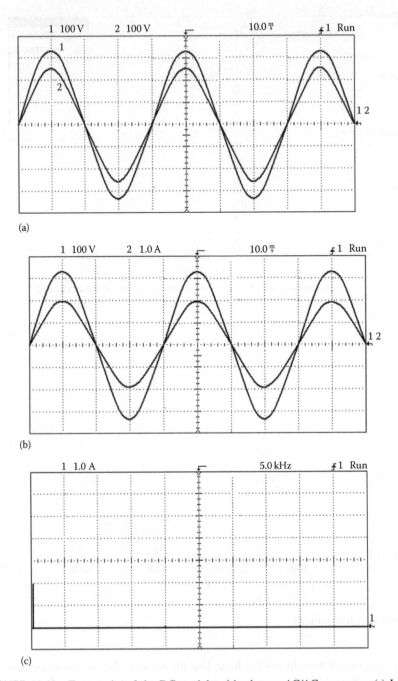

(a)

(b)

(c)

FIGURE 16.10 Test results of the DC-modulated buck-type AC/AC converter. (a) Input/output voltage waveforms, (b) input voltage and current waveforms, and (c) spectrum of the input current.

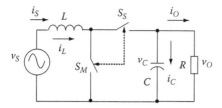

FIGURE 16.11 A DC-modulated single-stage boost-type AC/AC converter.

16.1.4.1 Positive Input Voltage Half-Cycle

During the input voltage is positive, the boost converter operates as usual. The equivalent circuits during the switch-on and switch-off conditions are shown in Figure 16.12.

The output voltage is calculated by

$$v_O = \frac{v_S}{1-k} = \frac{\sqrt{2}}{1-k} V_S \sin \omega t \quad 0 \le \omega t < \pi \tag{16.20}$$

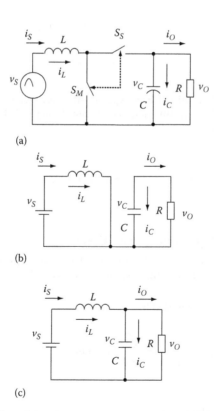

(a)

(b)

(c)

FIGURE 16.12 A DC-modulated boost-type AC/AC converter working in positive half-cycle. (a) Circuit diagram, (b) equivalent circuit during switch-on, and (c) equivalent circuit during switch-off.

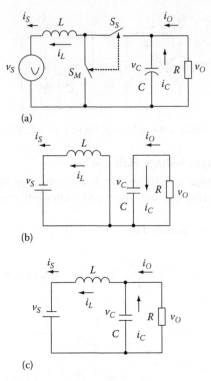

FIGURE 16.13 A DC-modulated buck-type AC/AC converter working in negative half-cycle. (a) The circuit diagram, (b) equivalent circuit during switch-on, and (c) equivalent circuit during switch-off.

where
    k is the conduction duty cycle of the buck converter
    V_S is the rms value of the input voltage
    ω is the power supply radian frequency

16.1.4.2 Negative Input Voltage Half-Cycle
When the input voltage is negative, the boost converter performs in the reverse. The equivalent circuits during the switch-on and switch-off conditions are shown in Figure 16.13.

The output voltage is calculated by

$$v_O = -\frac{|v_S|}{1-k} = \frac{\sqrt{2}}{1-k} V_S \sin \omega t \quad \pi \le \omega t < 2\pi \tag{16.21}$$

where
    k is the conduction duty cycle
    V_S is the rms value
    ω is the power supply radian frequency

16.1.4.3 Whole-Cycle Operation

Combining aforementioned two operation states, we can summarize the whole-cycle operation. The output voltage is calculated by

$$v_O = \frac{v_S}{1-k} = \frac{\sqrt{2}}{1-k} V_S \sin \omega t \qquad (16.22)$$

where

k is the conduction duty cycle
V_S is the rms value
ω is the power supply radian frequency

The whole-cycle input and output voltage waveforms are shown in Figure 16.14a with the duty cycle $k = 0.25$. The voltage transfer gain $M = V_O/V_S$ versus the DC/DC converter conduction duty cycle k is shown in Figure 16.14b. It is easy to obtain higher variable output voltage than the input voltage with very high PF and high efficiency.

The whole-cycle input voltage and current waveforms are shown in Figure 16.15a. The spectrum of the input current is shown in Figure 16.15b. The spectrum is very clean and there is little distortion from the harmonic component I_M at 20 kHz that is far away from the fundamental frequency component I_S at 50 Hz. Its value is only 0.5%, that is, $I_M/I_S = 0.005$. Therefore, THD = 1.0000125. Considering the DPF = 0.9998, we obtain the final PF = 0.99979.

16.1.4.4 Simulation and Experimental Results

The simulation and experimental results are shown in order to verify the design.

16.1.4.4.1 Simulation Results

The DC-modulated boost-type AC/AC converter shown in Figure 16.11 has the following components: $L = 10$ mH and $C = 3$ μF. The conduction duty cycle is selected as $k = 0.25$. The experimental results are shown in Figure 16.16. The output voltage $V_o = V_S/(1-k) = 267$ Vrms (peak value is approximately 377 V) with the frequency $f = 50$ Hz. The waveforms of the input and output voltages $v_S(t)$ and $v_o(t)$ are shown as Channel 1 and Channel 2 in Figure 16.16a. From it, we can see that there is not any phase delay, although there may be about 3.374° phase angle delay as shown in our analysis. The output current I_o should be 1.8 Arms, and the output power $P_o = V_o^2/R = 475$ W.

The input current is measured as $I_S = 2.4$ Arms (peak value is approximately 3.39 A) with the frequency $f = 50$ Hz. The waveforms of the input voltage $v_S(t)$ and current $i_S(t)$ are shown as Channel 1 and Channel 2, as shown in Figure 16.16b. From the figure it can be seen that there is not any phase delay, although there may be about 3.374° phase angle delay as shown in our analysis in Section 16.3. The FFT spectrum of the input current is shown in Figure 16.16c, and the THD = 0.015. The input power $P_{in} = V_S \times I_S = 480$ W. Although in theory there are no power losses in an ideal condition ($\eta = 1$), the test shows there are power losses. From the results, we obtain the final PF = 0.9979 and the power transfer efficiency $\eta = P_o/P_{in} = 475/480 = 0.989$ or 98.9%.

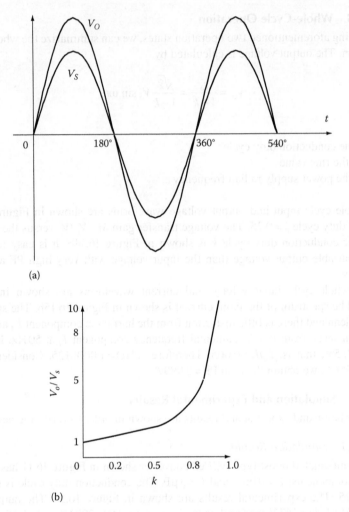

(a)

(b)

FIGURE 16.14 The input/output voltage waveforms of the DC-modulated boost-type AC/AC converter. (a) The input/output voltage waveforms, and (b) the voltage transfer gain versus conduction duty cycle k.

16.1.4.4.2 Experimental Results

The experimental results are shown in Figure 16.17.

16.1.5 DC-Modulated Single-Stage Buck-Boost-Type AC/AC Converter

For certain applications, the output voltage is required to be either lower or higher than input voltage. For this purpose, the DC-modulated single-stage buck-boost-type AC/AC converter has been designed and is shown in Figure 16.18. Since the input current is pulse train with the repeating frequency f_m, a low-pass filter L_S–C_S is required. We have to investigate the operation during both positive and negative half-cycle of the input voltage.

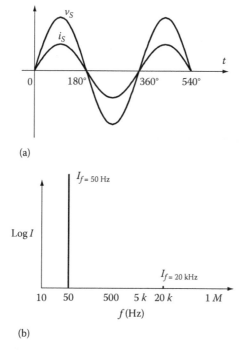

(a)

(b)

FIGURE 16.15   The input voltage/current waveforms of the DC-modulated boost-type AC/ AC converter. (a) The input voltage and current waveforms, and (b) the spectrum of input current.

16.1.5.1   Positive Input Voltage Half-Cycle

When the input voltage is positive, the buck-boost converter operates as usual. The equivalent circuits during the switch-on and switch-off conditions are shown in Figure 16.19.

The output voltage is calculated by

$$v_O = \frac{kv_S}{1-k} = \frac{k\sqrt{2}}{1-k} V_S \sin \omega t \quad 0 \le \omega t < \pi \tag{16.23}$$

where
   k is the conduction duty cycle of the buck converter
   V_S is the rms value of the input voltage
   ω is the power supply radian frequency

16.1.5.2   Negative Input Voltage Half-Cycle

When the input voltage is negative, the buck-boost converter operates in the reverse. The equivalent circuits during the switch-on and switch-off conditions are shown in Figure 16.20.

The output voltage is calculated by

$$v_O = -\frac{k}{1-k}\,|\,v_S\,| = \frac{k\sqrt{2}}{1-k}\,V_S\sin\omega t \quad \pi \le \omega t < 2\pi \qquad (16.24)$$

where
   k is the conduction duty cycle
   V_S is the rms value
   ω is the power supply radian frequency

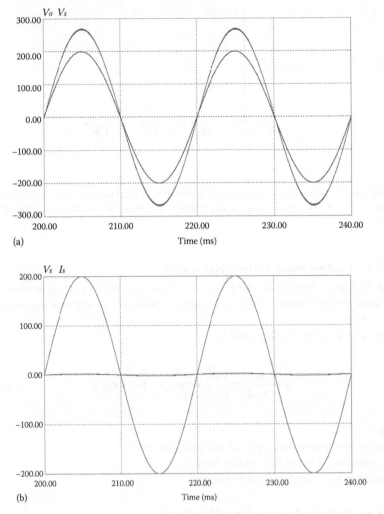

(a)

(b)

FIGURE 16.16 Test results of the DC-modulated boost-type AC/AC converter. (a) Input/output voltage waveforms of the DC-modulated boost-type AC/AC converter, (b) input voltage and current waveforms of the DC-modulated boost-type AC/AC converter, and (c) spectrum of input current of the DC-modulated boost-type AC/AC converter.

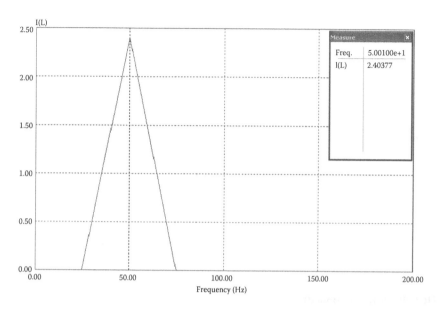

Test result spectrum first harmonic measurement of the DC-modulated boost-type
AC / AC converter f = 50 kHz, k = 0.25 (zoom 1)

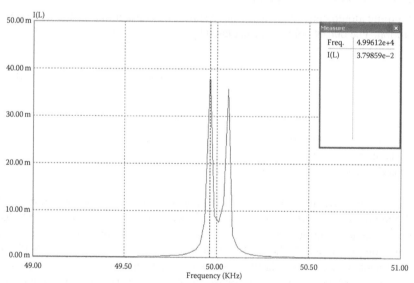

Test result spectrum second harmonic measurement of the DC-modulated
boost-type AC / AC converter f = 50 kHz, k = 0.25 (zoom 2)

FIGURE 16.16 (continued)

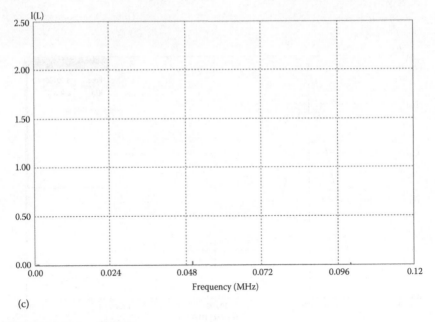

(c)

FIGURE 16.16 (continued)

16.1.5.3 Whole-Cycle Operation

Combining aforementioned two operation states, we can summarize the whole-cycle operation. The output voltage is calculated by

$$v_O = kv_S = \frac{k\sqrt{2}}{1-k} V_S \sin \omega t \qquad (16.25)$$

where
    k is the conduction duty cycle
    V_S is the rms value
    ω is the power supply radian frequency

The whole-cycle input and output voltage waveforms are shown in Figure 16.21a with the duty cycle $k=0.43$. The voltage transfer gain $M=V_O/V_S$ versus the DC/DC converter conduction duty cycle k is shown in Figure 16.21b. It is easy to obtain variable output voltage higher or lower than input voltage, with very high PF and high efficiency. There is polarity reversal between input and output voltages, or phase angle shift 180°. Usually, this phase angle shift does not affect most industrial applications.

The whole-cycle input voltage and current waveforms are shown in Figure 16.22a. The spectrum of the input current is shown in Figure 16.22b. The spectrum is very clean, and there is little distortion from the harmonic component I_M at 20 kHz

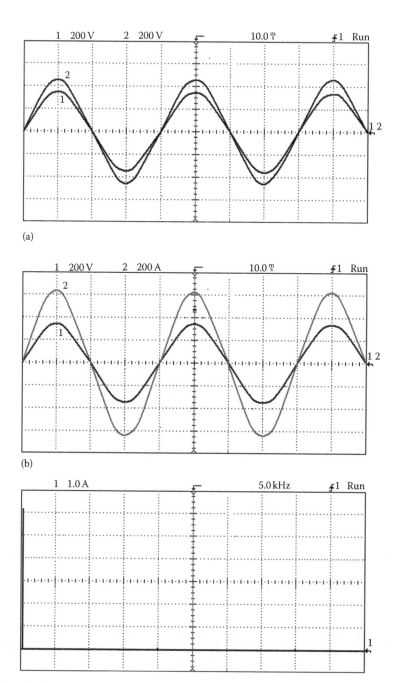

FIGURE 16.17 Test results of the DC-modulated boost-type AC/AC converter. (a) Input/output voltage waveforms, and (b) input voltage and current waveforms.

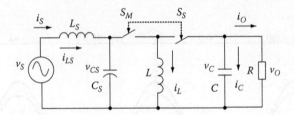

FIGURE 16.18 A DC-modulated single-stage buck-boost-type AC/AC converter.

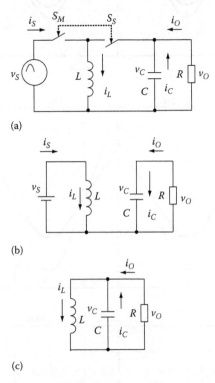

(a)

(b)

(c)

FIGURE 16.19 A DC-modulated buck-boost-type AC/AC converter working in positive half-cycle. (a) Circuit diagram, (b) equivalent circuit during switch-on, and (c) equivalent circuit during switch-off.

that is far away from the fundamental frequency component I_S at 50 Hz. Its value is only 0.5%, that is, $I_M/I_S=0.005$. Therefore, THD$=1.0000125$. Considering the DPF$=0.9998$, we obtain the final PF$=0.99979$.

16.1.5.4 Simulation and Experimental Results

The simulation and experimental results are shown in order to verify the design.

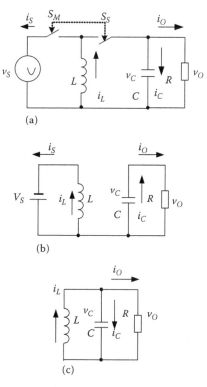

FIGURE 16.20 A DC-modulated buck-boost-type AC/AC converter working in negative half-cycle. (a) The circuit diagram, (b) equivalent circuit during switch-on, and (c) equivalent circuit during switch-off.

16.1.5.4.1 Simulation Results

The DC-modulated buck-boost-type AC/AC converter shown in Figure 16.18 has the following components: $L_S = 1$ mH, $C_S = 10\,\mu$F, $L = 10$ mH, and $C = 3000$ nF. The conduction duty cycle is selected as $k = 0.45$. The simulation results are shown in Figure 16.23. The output voltage $V_o = k/(1 - k) \times V_S = 0.818 \times V_S = 163.6$ Vrms (peak value is approximately 231.4 V) with the frequency $f = 50$ Hz. The waveforms of the input and output voltages $v_S(t)$ and $v_o(t)$ are shown as Channel 1 and Channel 2 in Figure 16.23a. It can be seen that there is not any phase delay, although there may be about 3.374° phase angle delay as shown in our analysis in Section 16.3. The output current I_o should be 1.1 Arms, and the output power $P_o = V_o^2/R = 178.5$ W.

The input current is measured as $I_S = 0.9$ Arms (peak value is approximately 1.27 A) with the frequency $f = 50$ Hz. The waveforms of the input voltage $v_S(t)$ and current $i_S(t)$ are shown as Channel 1 and Channel 2 in Figure 16.23b. It can be seen that there is not nearly any phase delay, although there may be about 3.374° phase angle delay as shown in our analysis. The FFT spectrum of the input current is shown in Figure 16.23c, and the THD = 0.1375. The input power $P_{in} = V_S \times I_S = 180$ W.

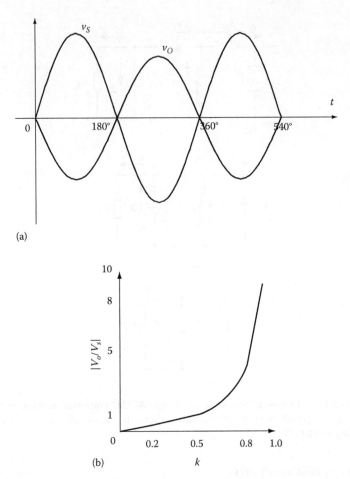

FIGURE 16.21 The input/output voltage waveforms of the DC-modulated boost-type AC/ AC converter. (a) The input/output voltage waveforms, and (b) the voltage transfer gain versus conduction duty cycle k.

Although in theory there are no power losses in an ideal condition ($\eta = 1$), the test shows there are power losses, which are mainly caused by the switches power losses. From the test results, we obtain the final PF = 0.9939 and the power transfer efficiency $\eta = P_o/P_{in} = 178.5/180 = 0.9917$ or 99.17%.

16.1.5.4.2 Experimental Results
The experimental results are shown in Figure 16.24.

16.2 OTHER TYPES OF DC-MODULATED AC/AC CONVERTERS

By using the knowledge gained with designing and constructing a DC-modulated single-stage AC/AC converter, we can easily design and construct a two-stage AC/AC

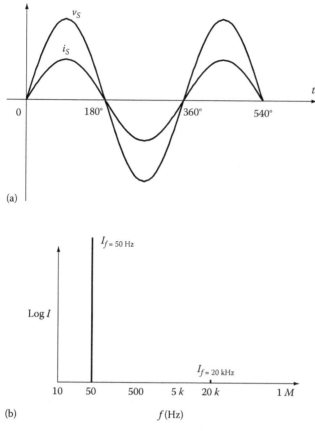

FIGURE 16.22 The input voltage/current waveforms of the DC-modulated boost-type AC/AC converter. (a) Input voltage and current waveforms of the DC-modulated buck-boost-type AC/AC converter, and (b) the spectrum of input current.

converter. Some converters have more complex structure such as Luo converters, super-lift Luo converters, and multistage cascaded boost converters [8,15–19]. In order to offer more information to readers, a DC-modulated positive output Luo converter and a two-stage boost-type AC/AC converter have been designed.

16.2.1 DC-Modulated Positive Output Luo-Converter-Type AC/AC Converter

The DC-modulated positive output (P/O) Luo-converter-type AC/AC converter is shown in Figure 16.25. Its output voltage has the same polarity as the input voltage.

The P/O Luo converter has more complex circuit than buck-boost converter. An input low-pass filter is required. All components values are shown in the circuit diagram in Figure 16.25. The simulation circuit is shown in Figure 16.26.

The simulation waveforms with $k=0.7$ and $f=80\,\text{kHz}$ are shown in Figure 16.27.

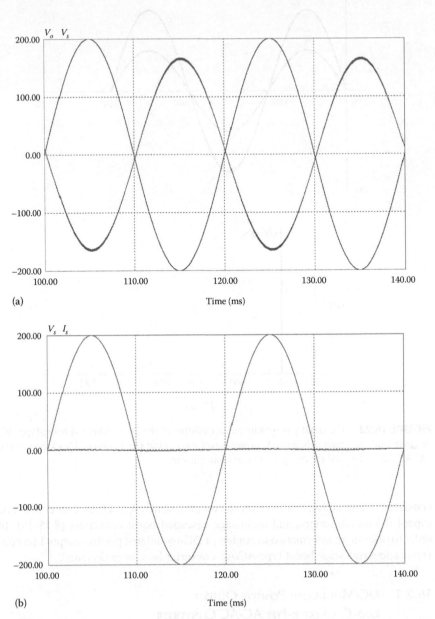

(a)   Time (ms)

(b)   Time (ms)

FIGURE 16.23 Test results of the DC-modulated buck-boost-type AC/AC converter. (a) Input/output voltage waveforms of the DC-modulated buck-boost-type AC/AC converter, (b) input voltage and current waveforms of the DC-modulated buck-boost-type AC/AC converter, and (c) spectrum of input current of the DC-modulated buck-boost-type AC/AC converter.

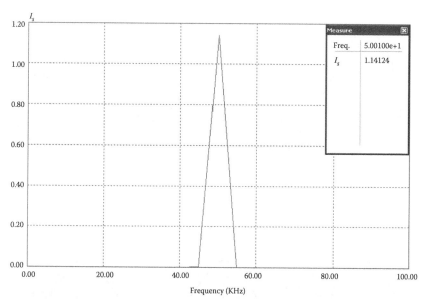

Test result spectrum first harmonic measurement of the DC-modulated
buck-boost-type AC/AC converter f = 50 kHz, k = 0.45 (zoom 1)

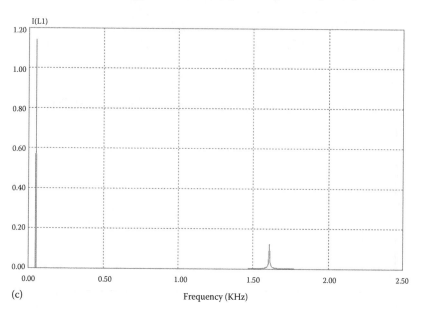

(c)

FIGURE 16.23 (continued)

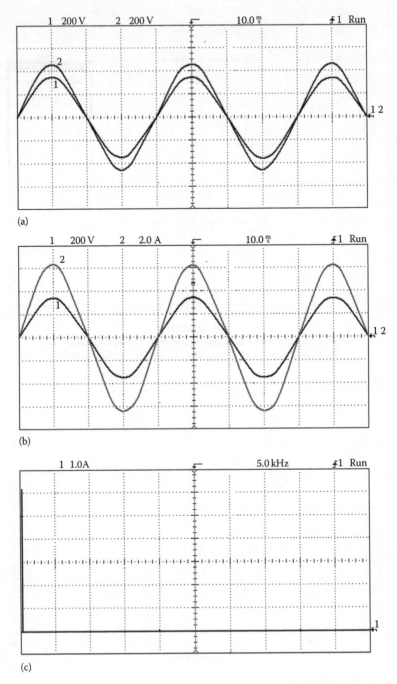

FIGURE 16.24 Test results of the DC-modulated boost-type AC/AC converter. (a) Input/output voltage waveforms, (b) input voltage and current waveforms, and (c) spectrum of the input current.

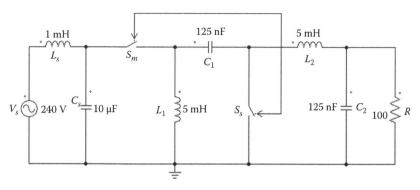

FIGURE 16.25 DC-modulated positive output Luo-converter-type AC/AC converter.

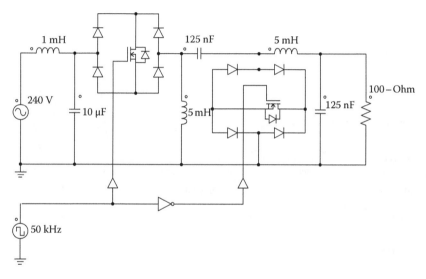

FIGURE 16.26 Simulation circuit of DC-modulated P/O Luo-converter-type AC/AC converter.

Example 16.3 A DC-modulated positive output Luo-converter-type AC/AC converter in Figure 16.25 has input rms voltage $v_S = 240\,V$ and a dimmer load with $R = 100\,\Omega$. In order to obtain the output rms voltage v_O, which varies in 100–400 V (can be higher or lower than the input voltage), calculate the range of the duty-cycle k, the output current, and power.

Solution: Since the output voltage is calculated as

$$v_O = \frac{k}{1-k}v_S = \frac{k}{1-k}240$$

the duty cycle is calculated as

$$k = \frac{v_O}{v_O + v_S} = \begin{cases} \dfrac{100}{240 + 100} = 0.294 \\ \dfrac{400}{240 + 400} = 0.625 \end{cases}$$

The range of the duty cycle k is 0.294–0.625.
The output rms current is 1–4 A, and the output power is 100–1600 W.

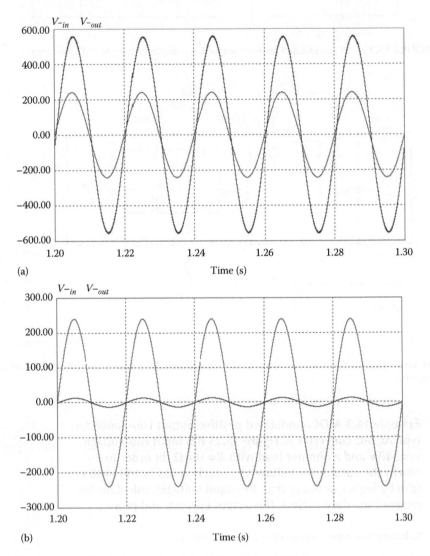

(a)

(b)

FIGURE 16.27 Simulation results of the DC-modulated P/O Luo-converter-type AC/AC converter. (a) Input/output voltage waveforms, (b) input voltage and current waveforms, and (c) spectrum of the input current.

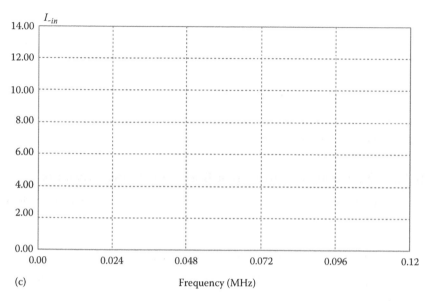

(c)   Frequency (MHz)

FIGURE 16.27 (continued)

16.2.2 DC-Modulated Two-Stage Boost-Type AC/AC Converter

A DC-modulated two-stage boost-type AC/AC converter is shown in Figure 16.28.

The four bidirectional switches (S_M–S_{S-S} plus S_{S1}–S_{S2}) in Figure 16.4 are applied. The output voltage v_O is

$$v_O(t) = \left(\frac{1}{1-k}\right)^2 v_S = \left(\frac{1}{1-k}\right)^2 \sqrt{2}V_S \sin \omega t \qquad (16.26)$$

The voltage transfer gain M is

$$M = \frac{v_O(t)}{v_S(t)} = \left(\frac{1}{1-k}\right)^2 \qquad (16.27)$$

From this calculation formula, the output voltage can be easily increased in high voltage. For example, $k = 0.7$ results in the voltage transfer gain $M = 16.11$.

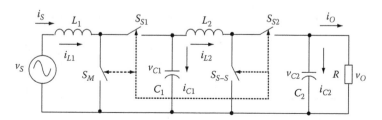

FIGURE 16.28   Circuit diagram of a DC-modulated two-stage boost-type AC/AC converter.

16.3 DC-MODULATED MULTIPHASE AC/AC CONVERTERS

Using the same technique, we can construct DC-modulated multiphase AC/AC converters.

16.3.1 DC-MODULATED THREE-PHASE BUCK-TYPE AC/AC CONVERTER

A DC-modulated three-phase buck-type AC/AC converter is shown in Figure 16.29. The simulation results are shown in Figure 16.30.

16.3.2 DC-MODULATED THREE-PHASE BOOST-TYPE AC/AC CONVERTER

A DC-modulated three-phase boost-type AC/AC converter is shown in Figure 16.31. The simulation results are shown in Figure 16.32.

16.3.3 DC-MODULATED THREE-PHASE BUCK-BOOST-TYPE AC/AC CONVERTER

A DC-modulated three-phase buck-boost-type AC/AC converter is shown in Figure 16.33.
The simulation results are shown in Figure 16.34.

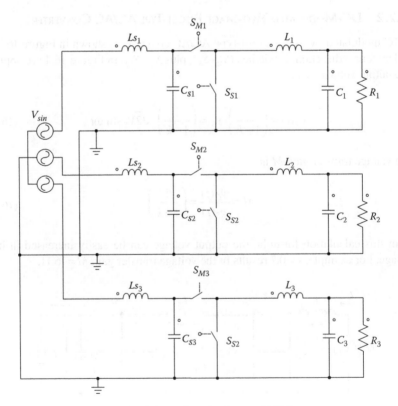

FIGURE 16.29   A DC-modulated three-phase buck-type AC/AC converter.

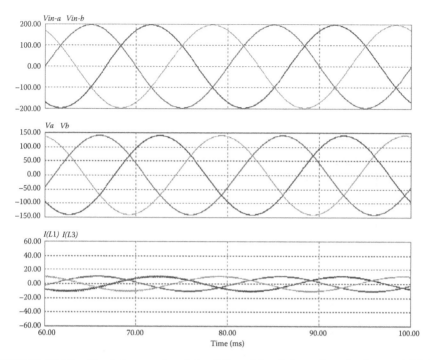

FIGURE 16.30 Simulation results of the DC-modulated three-phase buck-type AC/AC converter.

16.4   SUB-ENVELOPE MODULATION METHOD TO REDUCE THD OF AC/AC MATRIX CONVERTERS

AC/AC matrix converter is an array of power semiconductor switches that connects directly a three-phase AC source to another three-phase load. It can convert an AC power source with certain voltage and frequency to another AC load with variable voltage and variable frequency directly without DC link and bulk energy storage component. Classical modulation methods such as Venturini method and space vector modulation (SVM) method using AC network maximum-envelop modulation can execute matrix conversion successfully. However, they cause very high THD. This paper presents a novel approach: SEM method to reduce THD of matrix converters effectively [20,21]. The approach is extended to an improved version of matrix converters, and the THD can be reduced further. The algorithm of the SEM method is described in detail. Simulation and experiment results are also presented to verify the feasibility of the SEM approach. The results will be very helpful for industry applications.

AC/AC matrix converter [22–24] is an array of power semiconductor switches that connect directly a three-phase AC source to another three-phase load. This converter has several attractive features that have been investigated in recent decades. It can convert an AC power source with certain voltage and frequency to another AC load with variable voltage and variable frequency directly without DC link and

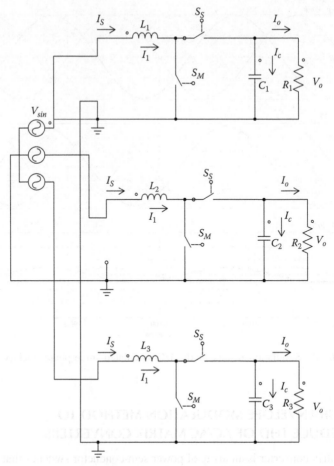

FIGURE 16.31 DC-modulated three-phase boost-type AC/AC converter.

bulk energy storage component. It eliminates large energy storage components, such as bulk inductor or electrolytic capacitor. The structure of the classical 3×3 matrix converter is shown in Figure 16.35. The semiconductor switches are marked with S_{Jk}, which means the switch is connected between the input phase J and the output phase k, where $J = \{A, B, C\}$, $k = \{a, b, c\}$.

All the switches S_{Jk} in matrix converters require a bidirectional-switch capability of blocking voltage and conducting current in both directions. Unfortunately, currently there are no such devices available; therefore, discrete devices need to be used to construct suitable switch cells. One option is diode bridge bidirectional switch cell arrangement, which consists of an insulated gate bipolar transistor (IGBT) (or other full control power semiconductor switches) at the center of a single-phase diode bridge [25]. The main advantage is that both current directions are carried by the same switching device; therefore, only one gate driver is required per switch cell. Device power losses are relatively high since there are three devices in each conduction path. The current direction through the switch cell cannot be controlled. This

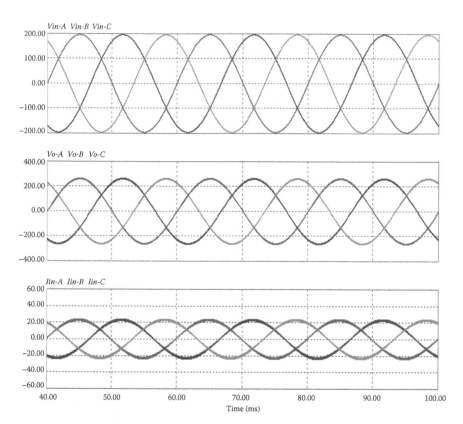

FIGURE 16.32   Simulation results of the DC-modulated three-phase boost-type AC/AC converter.

is a disadvantage since many advanced commutation methods require the current direction of the switch cell to be controllable.

The common emitter bidirectional switch cell arrangement consisting of two IGBTs and two diodes is another option. The diodes provide the reverse blocking capability. There are several advantages using this arrangement when compared to the diode bridge bidirectional switch cell. First, it is possible to independently control the direction of the current. Second, conduction power losses are also reduced since only two devices carry the current. Third, each bidirectional switch cell requires an isolated power supply for the gate drive. The switch cell can be connected in common collector. The conduction power losses are the same as that of the common emitter configuration. An often-quoted advantage of this method is that only six isolated power supplies are needed to supply the gate driver [26]. Therefore, the common emitter configuration is generally preferred for creating the matrix converter bidirectional switch cells.

Normally, the matrix converter is fed by a three-phase voltage source and, for this reason, the input terminals should not be in short circuit (Rule 1). On the other hand, the load has typically an inductive nature and, for this reason, an output phase must

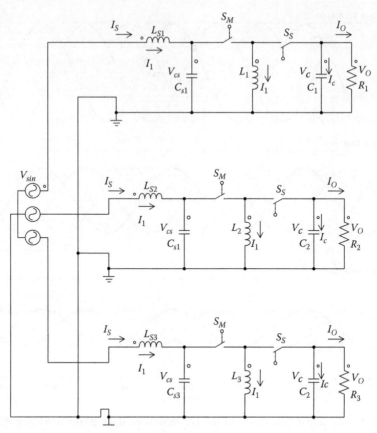

FIGURE 16.33   A DC-modulated three-phase buck-boost-type AC/AC converter.

never be in open circuit (Rule 2). Reliable current commutation between switches
in matrix converters is more difficult to achieve than in conventional voltage source
inverter (VSI) since there are no natural freewheeling paths [22]. The commutation
has to be actively controlled at all times with respect to the two basic rules. These
rules can be visualized by considering just two switch cells and one output phase of
a matrix converter. It is important that no two bidirectional switches are switched on
at any time, as shown in Figure 16.36a. This would result in line-to-line short circuiting and the destruction of the converter due to rush current. Also, the bidirectional
switches for each output phase should not be turned off at any instant, as shown
in Figure 16.36b. This would result in the absence of a path for the inductive load
current, causing rush voltage. These two considerations cause a conflict since semiconductor devices cannot be switched instantaneously due to propagation delays
and finite switching times. There are some successful approaches to avoid these
two cases: basic current commutation [22], current direction–based commutation
[26–28], relative voltage magnitude–based commutation [29–31], and soft-switching
techniques [32,33].

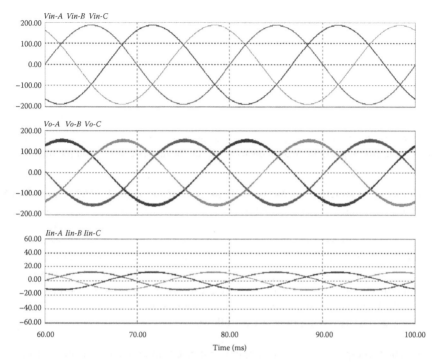

FIGURE 16.34 Simulation results of DC-modulated three-phase buck-boost-type AC/AC converter.

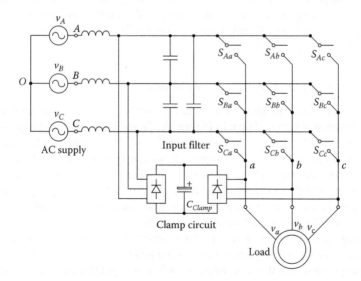

FIGURE 16.35 The structure of conventional matrix converter.

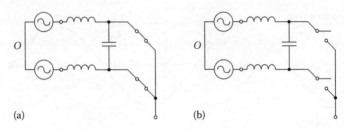

(a)   (b)

FIGURE 16.36 Two cases that the matrix converter should avoid. (a) Short circuits on the matrix converter input lines, and (b) open circuits on the matrix converter output lines.

Classical modulation methods such as Venturini method [23,24] and SVM method [30] using AC-network maximum-envelop modulation execute matrix conversion successfully. However, they cause very high THD. This paper presents a novel approach: SEM method to reduce THD for matrix converters effectively. The approach is extended to an improved version of matrix converters, and the THD can be reduced further. The algorithm of the SEM method is described in detail. Simulation and experiment results are also presented to verify the feasibility of SEM approach. The results will be very helpful to industry applications.

Some of our are assumptions are as follows:

- The three-phase AC input supply is balanced; input phase voltages are v_A, v_B, and v_C.
- The input AC supply frequency is f_i, and its corresponding angular speed is $\omega_i = 2\pi f_i$.
- The output phase voltages are v_a, v_b, and v_c.
- The output frequency is f_o, and its corresponding angular speed is $\omega_o = 2\pi f_o$.
- The switching frequency is f, and period is T. Usually, $f \gg f_i$ and f_o.
- V_{dc} is the *imaginary* DC link voltage, corresponding to the maximum-envelope rectifying average voltage.

16.4.1 SUB-ENVELOPE MODULATION METHOD

One commonly used modulation method for matrix converters is maximum-envelope modulation, which is shown in Figure 16.37a; the output phase voltage is pulse-width-modulated between maximum input phase and minimum input phase. The disadvantages are obvious: The magnitude of the output pulse is the difference between maximum input phase and minimum input phase, so there is output pulse with high magnitude and narrow width. Therefore, there are many high-frequency components in the output voltage, and they will cause very high THD. Moreover, there is high dv/dt, which will induce severe EMI.

Actually, in matrix converters, the three output phases can be connected to any input phases. So the output phase can be modulated between any two input phases. If the output phase is modulated between two *adjacent* input phases as shown in Figure 16.37b, the pulse magnitude of the output voltage can be low. Correspondingly, the

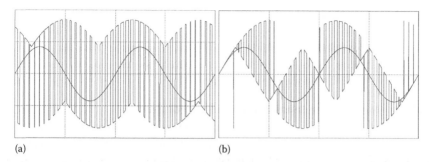

(a)   (b)

FIGURE 16.37 Modulation method for conventional matrix converter. (a) Maximum-envelope modulation method and (b) SEM method.

high-frequency components of the output voltage can be reduced. Thus, THD and dv/dt are also reduced. The input line current pulses are smaller and wider, and the THD of the input line current are also reduced. The approach is called sub-envelope modulation method.

The structure of the matrix converter implementing SEM method is shown in Figure 16.38. The matrix converter comprises 18 power semiconductor switches (9 switch cells) so that all the output phases can connect to any input phases with bidirectional current capability. The switches are marked with S_{Kjr} or S_{Kjf}, where $K = \{A, B, C\}$ is the input phase, $j = \{a, b, c\}$ is the output phase, r means the switch carries current from output to the input (reverse), and f means the switch carries the current from input to the output (forward). It has been mentioned that the common

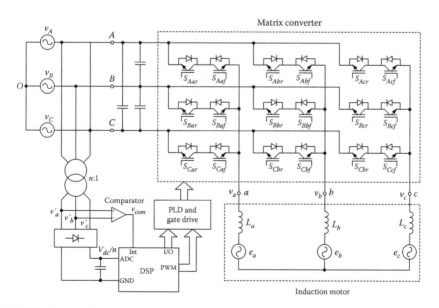

FIGURE 16.38 The structure of matrix converter drive system.

collector switch cell has the advantage of requiring fewer isolated DC power supplies for the gate drives. The 9 switch cells (18 switches) are a common collector configuration. The 18 switches can be divided into 6 switch groups S_{Ajr}, S_{Bjr}, S_{Cjr}, S_{Kaf}, S_{Kbf} and S_{Kcf}. Each group comprises three common emitter switches and uses a common gate drive output DC floating power supply.

16.4.1.1 Measure the Input Instantaneous Voltage

It is required to know the instantaneous phase voltage of AC supply. One approach is to measure the input voltage with three voltage sensors. If the AC supply is balanced pure sinusoidal supply, one simple approach to get the instantaneous phase voltage is to calculate the input voltage in real time. If the magnitude and time base of three-phase supply are known, the instantaneous-phase voltage can be determined. Thus, a three-phase transformer and a rectifier are adopted. The turn ratio of the transformer is $n{:}1$. The function of the transformer is to insulate control circuit from power stage. The scaled DC link voltage V_{dc}/n can be obtained by a small rate rectifier and an electrolytic capacitor. In order to get the time base, a comparator is introduced. The input of the comparator can be via two input phases (such as Phase A and Phase B). The output waveform v_{com} of the comparator is shown in Figure 16.39.

At the falling edge of v_{com} (such as t_1), the instantaneous phase voltage of AC supply can be obtained:

$$\begin{cases} v_A(t_1) = V_m \sin(5\pi/6) \\ v_B(t_1) = V_m \sin(\pi/6) \\ v_C(t_1) = V_m \sin(-\pi/2) \end{cases} \qquad (16.28)$$

where

$$V_m = \frac{\pi V_{dc}}{3\sqrt{3}}$$

where V_{dc} is imaginary DC link voltage.

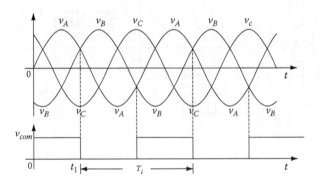

FIGURE 16.39 The output of the comparator.

The frequency of AC power supply is known as f_i, and the angular frequency $\omega_i = 2\pi f_i$. By redefining the initial time, the instantaneous input voltage during one cycle ($1/f_i = T_i$) can be expressed as follows:

$$\begin{cases} v_A(t) = \dfrac{\pi V_{dc}}{3\sqrt{3}} \sin\left(\omega_i t + \dfrac{5\pi}{6}\right) \\[4mm] v_B(t) = \dfrac{\pi V_{dc}}{3\sqrt{3}} \sin\left(\omega_i t + \dfrac{\pi}{6}\right) \\[4mm] v_C(t) = \dfrac{\pi V_{dc}}{3\sqrt{3}} \sin\left(\omega_i t - \dfrac{\pi}{2}\right) \end{cases} \tag{16.29}$$

In a discrete system with sampling frequency f (sampling time T), the voltage sequence of the AC power supply can be obtained as shown in Equation 16.29:

$$\begin{cases} v_A(kT) = \dfrac{\pi V_{dc}}{3\sqrt{3}} \sin\left(\omega_i kT + \dfrac{5\pi}{6}\right) \\[4mm] v_B(kT) = \dfrac{\pi V_{dc}}{3\sqrt{3}} \sin\left(\omega_i kT + \dfrac{\pi}{6}\right) \\[4mm] v_C(kT) = \dfrac{\pi V_{dc}}{3\sqrt{3}} \sin\left(\omega_i kT - \dfrac{\pi}{2}\right) \end{cases} \tag{16.30}$$

With the help of signal v_{com}, the voltage sequence of the AC power supply can be calculated rigorously without error accumulation.

16.4.1.2 Modulation Algorithm

An SEM example (only v_a is illustrated) is shown in Figure 16.40. The output is modulated between two closest input phases.

The modulation rule of the example is shown in Table 16.1; v_{high} is the smallest one, which is greater than v_{ra} (v_{ra} is reference voltage of output phase a), and v_{low} is the biggest one, which is less than v_{ra}; that is, output phase a is connected to two closest phases alternately, and thus, the duty cycle of PWM δ can be determined as follows:

$$\delta = \frac{v_a - v_{low}}{v_{high} - v_{low}} \tag{16.31}$$

Assume the output frequency is f_o, the magnitude of reference output phase voltage is V_o, the angular frequency $\omega_o = 2\pi f_o$, and the initial phase angle is ϕ_0. The PWM switching frequency is f, and period is T. The modulation algorithm of the system can be accomplished by

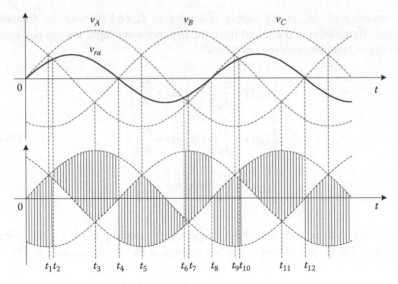

FIGURE 16.40   One SEM example.

TABLE 16.1

SEM Rule of the Example in Figure 16.40

Time		$-t_1$	t_1-t_2	t_2-t_3	t_3-t_4	t_4-t_5	t_5-t_6	t_6-t_7
Modulation	v_{high}	v_A	v_C	v_A	v_A	v_B	v_A	v_B
phase	v_{low}	v_B	v_B	v_C	v_B	v_C	v_C	v_A
Time		t_7-t_8	t_8-t_9	t_9-t_{10}	$t_{10}-t_{11}$	$t_{11}-t_{12}$	$t_{12}-t_{13}$	$t_{13}-$
Modulation	v_{high}	v_B	v_C	v_B	v_C	v_C	v_A	v_C
phase	v_{low}	v_C	v_A	v_A	v_B	v_A	v_B	v_B

Algorithm 16.1

Define the variable θ_i, θ_o, and their initial value are set to zero, that is,

$$\theta_i = 0, \theta_o = 0.$$

Define array $v_i[3]$ to store input phase voltage.

Define array *order*[3] to map v_i to the input phase, and the initial array with {1, 2, 3} represents {A, B, C}.

Do nothing until the first falling edge of the signal v_{com} comes. When the first falling edge comes, for every PWM cycle T, do the following loop:

a. Calculate AC power supply voltage $v_i[1] = v_A(kT)$, $v_i[2] = v_B(kT)$, and $v_i[3] = v_C(kT)$ according to Equation 16.30

$$
\begin{cases}
v_i[1] = v_A(kT) = V_m \sin\left(\theta_i + 5\pi/6\right) \\
v_i[2] = v_B(kT) = V_m \sin\left(\theta_i + \pi/6\right) \\
v_i[3] = v_C(kT) = V_m \sin\left(\theta_i - \pi/2\right)
\end{cases}
\tag{16.32}
$$

where $V_m = \pi V_{dc}/3\sqrt{3}$ is the magnitude of input phase voltage that is measured from the transformer. If the AC power supply is measured by three voltage sensors, this step can be ignored.

b. Sort the voltage $v_i[1]$, $v_i[2]$, and $v_i[3]$ in the descending order:

   i.   If $v_i[1] < v_i[2]$, then exchange $v_i[1]$ and $v_i[2]$, and also exchange $order[1]$ and $order[2]$; else do nothing
   ii. If $v_i[1] < v_i[3]$, then exchange $v_i[1]$ and $v_i[3]$, and also exchange $order[1]$ and $order[3]$; else do nothing
   iii. If $v_i[2] < v_i[3]$, then exchange $v_i[2]$ and $v_i[3]$, and also exchange $order[2]$ and $order[3]$; else do nothing

   Thus, condition $v_i[1] \geq v_i[2] \geq v_i[3]$ is satisfied, and then the variable $order$ will also map the input phases.

c. Calculate three output voltages v_a, v_b, v_c with the following equation:

$$
\begin{cases}
v_a = V_o \sin\left(\theta_o + \varphi_0\right) \\
v_b = V_o \sin\left(\theta_o - 2\pi/3 + \varphi_0\right) \\
v_c = V_o \sin\left(\theta_o - 4\pi/3 + \varphi_0\right)
\end{cases}
\tag{16.33}
$$

where
  V_o is the magnitude of reference output phase voltage
  φ_0 is the initial phase angle

d. For the value of v_a do the following:

   i. If $v_a \geq v_i[2]$, it means that v_a is between $v_i[1]$ and $v_i[2]$, then output phase a is modulated between input phase $order[1]$ and $order[2]$, and the PWM duty cycle δ is

$$
\delta = \frac{v_a - v_i[2]}{v_i[1] - v_i[2]}
\tag{16.34}
$$

   ii. Or else it means that v_a is between $v_i[2]$ and $v_i[3]$, then output phase a is modulated between input phase $order[2]$ and $order[3]$, and PWM duty cycle δ is

$$
\delta = \frac{v_a - v_i[3]}{v_i[2] - v_i[3]}
\tag{16.35}
$$

e. Follow a procedure similar to (d) for v_b and v_c.
f. Increase θ_i by $\omega_i T$.

g. Add θ_o with $\omega_o T$; if θ_o is greater than 2π, then subtract 2π from θ_o.
h. Wait for the next PWM cycle, carry out (a)–(g) again.

In the interim, if falling edge of the signal v_{com} comes, set variable θ_i with zero.

The algorithm can be easily implemented by a microprocessor.

16.4.1.3 Improve Voltage Ratio

Assume input AC supply–phase voltage is

$$
\begin{cases}
v_A(t) = V_m \sin(\omega_i t) \\
v_B(t) = V_m \sin(\omega_i t - 2\pi/3) \\
v_C(t) = V_m \sin(\omega_i t - 4\pi/3)
\end{cases}
\tag{16.36}
$$

and output phase voltage is

$$
\begin{cases}
v_a(t) = qV_m \sin(\omega_o t) \\
v_b(t) = qV_m \sin(\omega_o t - 2\pi/3) \\
v_c(t) = qV_m \sin(\omega_o t - 4\pi/3)
\end{cases}
\tag{16.37}
$$

where q is the voltage ratio of the output voltage (voltage transfer gain, usually $q<1$).
The direct phase voltage modulation with Equation 16.37 has a maximum voltage
ratio of 50% as illustrated in Figure 16.41.

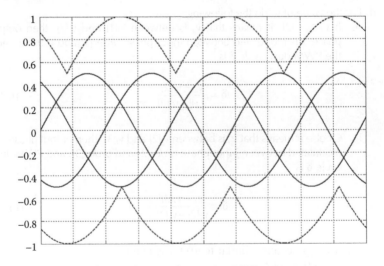

FIGURE 16.41 Illustrating maximum voltage ratio of 50%.

An improvement in voltage ratio to $\sqrt{3}/2$ (or 87%) is possible [34,35] by adding common-mode voltages to the target outputs as follows:

$$\begin{cases} v_a(t) = qV_m \left[\sin(\omega_o t) + \dfrac{\sin(3\omega_o t)}{6} - \dfrac{\sin(3\omega_i t)}{2\sqrt{3}} \right] \\[3mm] v_b(t) = qV_m \left[\sin(\omega_o t - 2\pi/3) + \dfrac{\sin(3\omega_o t)}{6} - \dfrac{\sin(3\omega_i t)}{2\sqrt{3}} \right] \\[3mm] v_c(t) = qV_m \left[\sin(\omega_o t - 4\pi/3) + \dfrac{\sin(3\omega_o t)}{6} - \dfrac{\sin(3\omega_i t)}{2\sqrt{3}} \right] \end{cases} \qquad (16.38)$$

The common-mode voltages have no effect on the output line-to-line voltages but allow the target outputs to fit within the input voltage envelope with a value of up to 87% as illustrated in Figure 16.42.

The improvement in voltage ratio is achieved by redistributing the null output states of the converter (all output lines connected to the same input line) and is analogous to similar well-established techniques used in conventional DC link PWM converters. It should be noted that a voltage rate of 87% is the intrinsic maximum voltage to be gained in any modulation method. Venturini provides a rigorous proof of this fact in Refs. [34,35].

To increase the voltage ratio, the Equation 16.33 in Algorithm 16.1 should be changed to

$$\begin{cases} v_{co} = \sin(3\theta_o)/6 - \sin(3\theta_i)/2\sqrt{3} \\[2mm] v_a = V_o \left[\sin(\theta_o + \phi_0) + v_{co} \right] \\[2mm] v_b = V_o \left[\sin(\theta_o - 2\pi/3 + \phi_0) + v_{co} \right] \\[2mm] v_c = V_o \left[\sin(\theta_o - 4\pi/3 + \phi_0) + v_{co} \right] \end{cases} \qquad (16.39)$$

16.4.2 TWENTY-FOUR-SWITCHES MATRIX CONVERTER

From Figure 16.40 it can be seen that if the reference voltage is greater than zero, the output is modulated between two positive phases or neutral point and one positive phase, vice versa, and then the THD can be further reduced. One example of the modulation is shown in Figure 16.43.

The modulation can be accomplished by 12-switch-cells (24-switches) matrix converter as shown in Figure 16.44. The structure is similar to Figure 16.38, except for the phase lag added from all the output to the neutral point. The added switches are S_{Oif} and S_{Ojr}, where $j = \{a, b, c\}$ is the output phase. An example of the modulation rule is shown in Table 16.2.

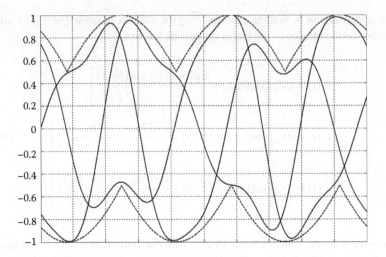

FIGURE 16.42 Illustrating maximum voltage ratio of 87%.

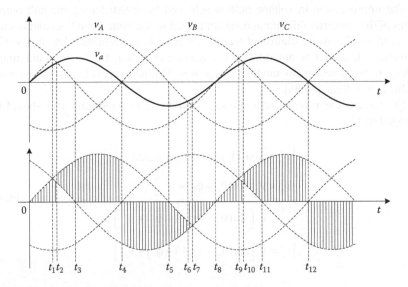

FIGURE 16.43 One SEM example for 12-switch-cells matrix converter.

The modulation algorithm is similar to Algorithm 16.1, except for procedure (d).

Algorithm 16.2

Define the variable θ_i, θ_o, and their initial value are set to zero, that is,

$$\theta_i = 0, \theta_o = 0.$$

Define array $v_i[3]$ to store input phase voltage.

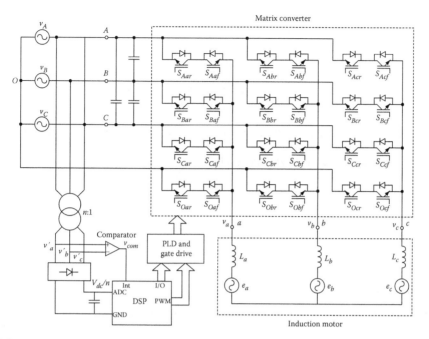

FIGURE 16.44 The structure of 24-switches matrix converter.

TABLE 16.2

Modulation Rule of 24-Switches Matrix Converter

Time		$-t_1$	t_1-t_2	t_2-t_3	t_3-t_4	t_4-t_5	t_5-t_6	t_6-t_7
Modulation	v_{high}	v_A	v_C	v_A	v_A	0	v_A	0
phase	v_{low}	0	0	v_C	0	v_C	v_C	v_A
Time		t_7-t_8	t_8-t_9	t_9-t_{10}	$t_{10}-t_{11}$	$t_{11}-t_{12}$	$t_{12}-$	
Modulation	v_{high}	0	v_C	v_B	v_C	v_C	0	
phase	v_{low}	v_C	0	0	v_B	0	v_B	

Define array *order*[3] to map v_i to the input phase, and the initial array with {1, 2, 3}, which represent {A, B, C}.

Define variables v_{high} and v_{low} to store the voltages that have been modulated.

Do nothing until the first falling edge of the signal v_{com} comes. When the first falling edge comes, for every PWM cycle T, do the following loop:

(a), (b), and (c) are the same as that of Algorithm 16.1.

d. For the value of v_a do the following:

   i. If $v_a \geq v_i[2]$, it means that v_a is between $v_i[1]$ and $v_i[2]$, then store the $v_i[1]$ to v_{high} and store $v_i[2]$ to v_{low}.

   ii. Or else it means that v_a is between $v_i[2]$ and $v_i[3]$, then store the $v_i[2]$ to v_{high} and store $v_i[3]$ to v_{low}.

iii. If $v_a > 0$ and $v_{low} < 0$, then store the v_{low} with zero. It means that v_a is modulated between neutral pint and lower positive phase, which is higher than v_a.

iv. Else do nothing. It means that v_a is modulated between two positive phases.

v. If $v_a < 0$ and $v_{high} > 0$, then store the v_{high} with zero. It means that v_a is modulated between neutral point at one higher negative phase, which is lower than v_a.

vi. Else do nothing. It means that v_a is modulated between two negative phases.

vii. Calculate the PWM duty cycle δ.

$$\delta = \frac{v_a - v_{low}}{v_{high} - v_{low}} \qquad (16.40)$$

(e), (f), (g), and (h) are also the same as that of Algorithm 16.1.

16.4.3 CURRENT COMMUTATION

We need investigate current commutation between input phases.

16.4.3.1 Current Commutation between Two Input Phases

The current commutation must obey two rules: avoid two input phases in short circuit and avoid any output phase in open circuit. Relative voltage magnitude–based commutation [30–32] will be introduced to this system. For one output phase, it is always modulated between two input phases (neutral point is also considered as one phase, so in total there are four phases). When PWM signal is high, it connected to the smallest input phase (v_{high}), which is higher than the reference voltage. When PWM signal is low, it is connected to the largest input phase (v_{low}), which is lower than the reference voltage. For convenience, assume the output phase is a, v_{high} is Phase A, and v_{low} is Phase B, then the gate signal for one PWM cycle is shown in Figure 16.45a. All switches related to the commutation are also shown in Figure 16.45b. Switching state transition is also illustrated in Figure 16.45c.

Commutation details are given as follows:

- When PWM signal is high, switches S_{Aaf}, S_{Aar}, and S_{Baf} are all turned on. As $v_A > v_B$ and switch S_{Bar} is turned off, there is no short circuit for two input phases. Output phase current i_a always flows through Phase A: When $i_a > 0$, i_a flows from Phase A because of $v_A > v_B$; when $i_a < 0$, i_a flows to Phase A because switch S_{Bar} is turned off.

- When PWM signal is flopped down, switch S_{Aaf} is turned off instantly, and the output current i_a transfers from Phase A to Phase B; there is no open circuit of Phase A. After a short delay t_d, switch S_{Bar} is turned on; there is no short circuit of the input phases. The delay t_d between each switching event is determined by the device characteristics, such as turned off time of a device.

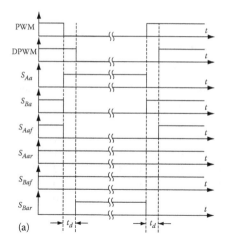

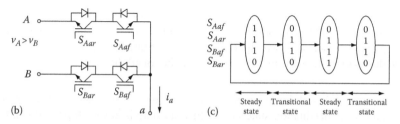

FIGURE 16.45 Gate signal of current commutation. (a) Gate signal, (b) switches related to the commutation, and (c) switching state transition.

- When PWM signal is low, switches S_{Aar}, S_{Baf} and S_{Bar} are all turned on. As $v_A > v_B$ and switch S_{Aaf} is turned off, there is no short circuit for two input phases. Similarly, output current i_a always flows through Phase B.
- When PWM signal is flipped up, switch S_{Bar} is turned off instantly, and output phase current i_a transfers from Phase B to Phase A; there is no open circuit of Phase A. After a short delay t_d, switch S_{Aaf} is turned on; there is no short circuit of the input phases.

16.4.3.2 Current-Commutation-Related Three Input Phases

Within a PWM cycle, that is, when PWM signal is flipped high, current commutation takes place only between two adjacent input phases. However, when PWM signal is flipped down, current commutation may take place among the other two of the three input phases and neutral O. For example, before t_4 in Figure 16.43, modulation takes place between Phases A and O; after t_4 modulation takes place between Phases O and C. We assumed that during latest PWM cycle, the modulation takes place between M and N (voltage of M > voltage of N). In next PWM cycle, Phase P will be related to the modulation. Then all the possible switching states among M, N, and P are shown in Table 16.3, where State 0 is the initial state of the switches.

TABLE 16.3
All Possible Switching State Transition

Condition	M > P > N				P > M > N				M > N > P				M > P > N			
Mode	Mode 1				Mode 2				Mode 3				Mode 4			
State	0	1	2	3	0	1	2	3	0	1	2	3	0	1	2	3
S_{Mjf}	1	0	0		1	1			1	0	0	0	1	0	0	0
S_{Mjr}	1	1	1		1	1			1	1	1	0	1	1	1	0
S_{Njf}	1	1	0		1	0			1	1	0	0	1	1	1	1
S_{Njr}	0	0	0		0	0			0	0	1	1	0	0	0	1
S_{Pjf}	0	1	1		0	0			0	1	1	1	0	0	0	0
S_{Pjr}	0	0	1		0	1			0	0	0	1	0	0	1	1
Phase	$MN \rightarrow MP$				$MN \rightarrow PM$				$MN \rightarrow NP$				$MN \rightarrow PN$			

Rewrite the switching state transition in Table 16.4; the functions of mode 1 to mode 4 are the same as those in Table 16.3, and in mode 5 the current commutation takes place only between two input phases, where State 0 is the initial state. State 4 is the destination state. From Table 16.4, it can be found that

- For switch S_{Mjf}, if the destination state is "1," then its state is kept unchanged, and if the destination state is "0," then its state is always flipped down to "0" immediately
- For switch S_{Mjr}, if the destination state is "1," then its state is kept unchanged, and if the destination state is "0," then its state is always flipped down to "0" in step 4
- For switch S_{Njf}, if the destination state is "1," then its state is kept unchanged, and if the destination state is "0," then its state is always flipped down to "0" in step 2
- For switch S_{Njr}, if the destination state is "0," then its state is kept unchanged, and if the destination state is "1," then its state is always flipped down to "1" in step 3
- For switch S_{Pjf}, if the destination state is "0," then its state is kept unchanged, and if the destination state is "1," then its state is always flipped down to "1" immediately
- For switch S_{Pjr}, if the destination state is "0," then its state is kept unchanged, and if the destination state is "1," then its state is always flipped down to "1" in step 3

For all switches S_{Kjf} and S_{Kjr}, the general switching state transition true table can be summarized as Table 16.4. Where Q_{fn} is the initial state of the switches S_{Kjf}, Q_{rn} is the initial state of the switches S_{Kjr}, Q_{fn+1} is the destination state of the switches S_{Kjf}, and Q_{rn+1} is the destination state of the switches S_{Kjr}.

TABLE 16.4
Modified Switching State Transition

Condition	M>P>N					P>M>N					M>N>P					M>P>N					M>N				
Mode	Mode 1					Mode 2					Mode 3					Mode 4					Mode 5				
State	0	1	2	3	4	0	1	2	3	4	0	1	2	3	4	0	1	2	3	4	0	1	2	3	4
S_{Mjf}	1	0	0	0	0	1	1	1	0	0	1	0	0	0	0	1	0	0	0	0	1	0	0	0	0
S_{Mjr}	1	1	1	0	1	1	1	1	0	1	1	0	1	0	0	1	0	1	1	0	1	0	1	1	1
S_{Njf}	1	1	0	0	0	1	1	1	0	0	1	1	1	0	0	1	1	1	1	0	1	1	1	1	1
S_{Njr}	0	0	0	1	1	0	0	0	0	0	0	0	0	1	1	0	0	0	0	1	0	0	0	0	0
S_{Pjf}	0	1	1	1	0	0	0	0	0	0	0	1	1	1	0	0	0	0	0	0	0	0	0	0	0
S_{Pjr}	0	0	0	1	1	0	0	0	0	1	0	0	0	0	1	0	0	0	1	1	0	0	0	0	0
Phase	$MN \rightarrow MP$					$MN \rightarrow PM$					$MN \rightarrow NP$					$MN \rightarrow PN$					$MN \rightarrow MN$				

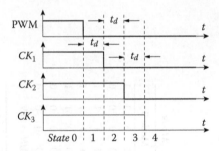

FIGURE 16.46 Signal of delayed PWM.

The state transition can be implemented by combinational logic circuit by adding three delayed PWM signals CK_1–CK_3 as shown in Figure 16.46. Then the logical equation of the switches S_{Kjf} and S_{Kjr} can be obtained:

$$S_{Kjf} = Q_{fn} \cdot Q_{fn+1} + Q_{fn} \cdot Q_{rn} \cdot \bar{Q}_{fn+1} \cdot \text{PWM}$$
$$+ Q_{fn} \cdot \bar{Q}_{rn} \cdot \bar{Q}_{fn+1} \cdot CK_1 + \bar{Q}_{fn} \cdot \bar{Q}_{rn} \cdot Q_{fn+1} \cdot \text{PWM} \qquad (16.41)$$

$$S_{Kjr} = Q_{fn} \cdot Q_{rn} \cdot Q_{rn+1} + Q_{fn} \cdot Q_{rn} \cdot \bar{Q}_{rn+1} \cdot CK_3 + \bar{Q}_{rn} \cdot Q_{rn+1} \cdot CK_2 \qquad (16.42)$$

16.4.4 SIMULATION AND EXPERIMENTAL RESULTS

The direct phase voltage modulation has a maximum voltage proportion of 50%. The simulation and experimental results are based on three-phase voltage modulation methods:

- Modulation I: maximum-envelop modulation for conventional nine-switch-cells matrix converter
- Modulation II: SEM method for conventional nine-switch-cells matrix converter
- Modulation III: SEM method for 12-switch-cells matrix converter

16.4.4.1 Simulation Results

The simulation results of output phase voltage (only phase a is illustrated), output line-to-line voltage, input line current, and their FFT under three-phase modulation method are shown in Figures 16.47 through 16.49, respectively. The input phase voltage is 240 V (rms value, the same for the following), and frequency is 50 Hz; output phase voltage is 120 V, and frequency is 100 Hz; and switching frequency is 2 kHz. From the FFT figure, we can see that harmonics with 24-switches converter are reduced significantly.

With improved ratio modulation, the voltage ratio can reach about 87%. The simulation results of the phase voltage, line-to-line voltage, and FFT of line-to-line voltage with modulation I and modulation III are shown in Figure 16.50e and f. The corresponding THD of these two waveforms are 63.4% and 20.8%, respectively.

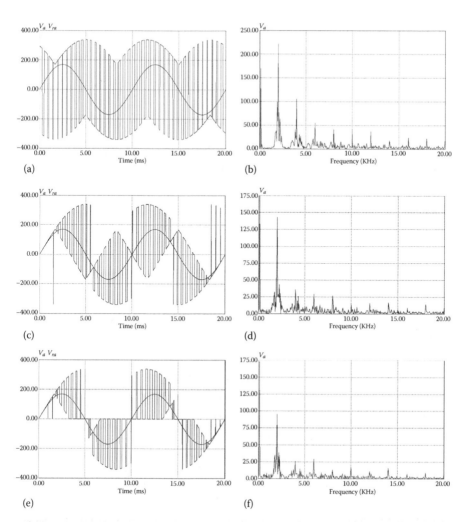

FIGURE 16.47 Simulation of output phase voltage and their FFT. (a) Output phase voltage with modulation I, (b) FFT of figure (a), (c) output phase voltage with modulation II, (d) FFT of figure (c), (e) output phase voltage with modulation II, and (f) FFT of figure (e).

16.4.4.2 Experimental Results

In order to verify the feasibility of the proposed scheme, a 12-switch-cells matrix converter is built up. The modulation algorithm is implemented by a DSP TMS320F2407, which is specially designed for power electronics and electric drive. The digital signal processor (DSP) comprises dual 10-bits 16-channel analog-to-digital converter (ADC), PWM generator, digital I/O, and other modules. So the DSP can measure input voltages, generate three required PWM signal for three output phases, and indicate peripheral circuit whose phases are to be modulated. The peripheral circuit is to generate gate signal for matrix converter (include current commutation), which is built up by GAL PLD, logic gate, monostable flip-flops, and so on. 15 A/1200 V

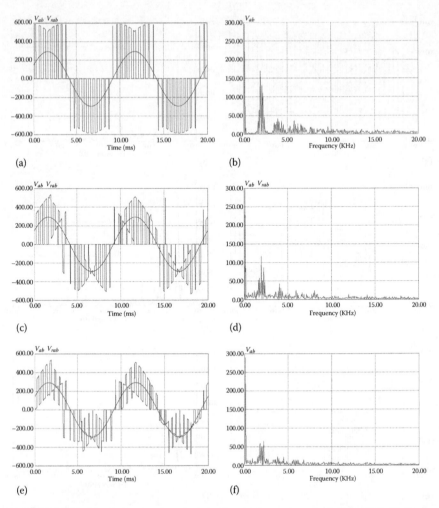

FIGURE 16.48 Simulation results of output line-to-line voltage and their FFT. (a) Output line-to-line voltage with modulation I, (b) FFT of figure (a), (c) output line-to-line voltage with modulation II, (d) FFT of figure (c), (e) output line-to-line voltage with modulation III, and (f) FFT of figure (e).

1MBH15-120 IGBT is adopted as main switches. A photo-coupled gate driver, TLP250, is used to implement the gate drive circuit. This gate driver provides a peak output current of 1.5 A. It also isolates the input signal from the output; thus, common-mode noise is reduced. An IGBT needs +15 V to +20 V voltage to turn on and −5 V to −10 V to turn off. To get the required voltage with single DC power supply, a +7 V stabilivolt tube (Zener diode) is used to get the required negative turn-off voltage. Thus, the driver circuit can provide +17 V for turn-on and −7 V for turn-off with single +24 V output DC power supply. Gate driver circuit for one IGBT is shown in Figure 16.51. The switches with emitter connected together can use the same gate

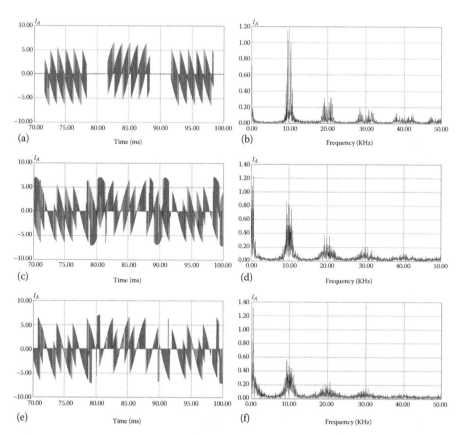

FIGURE 16.49 Simulation results of input line current and their FFT. (a) Input-line current with modulation I, (b) FFT of figure (a), (c) input-line current with modulation II, (d) FFT of figure (c), (e) input-line current with modulation III, and (f) FFT of figure (e).

driver output DC power supply. For 12-switch-cells matrix converter, 7 insulated DC power supplies are required.

The phase voltage of AC power supply is 240 V, line-to-line voltage is about 415 V, and frequency f_i is 50 Hz. Switching frequency f is 10 kHz. Modulation methods I to III can be implemented by the same hardware by changing only the software of the DSP. A 2.2 kW three-phase induction motor is connected to the output of the matrix converter as load. Experimental waveforms of output line-to-line voltage, phase current of the induction motor, and FFT of line-to-line voltage under modulation methods I–III are shown in Figure 16.52. All the voltage signals are measured by a differential probe with a gain of 20, the voltage scale is 400 V/div, current scale is 8 A/div, the magnitude scale of FFT waveform is 20 dB/div, and frequency scale is 5 kHz/div. Figure 16.52a shows the modulation I, output frequency f_O is 130 Hz, the voltage ratio δ is $\sqrt{3}/2$, and the magnitude of output line-to-line voltage is 360 V with THD = 62.7%. The experiment results implementing modulation II and III with same

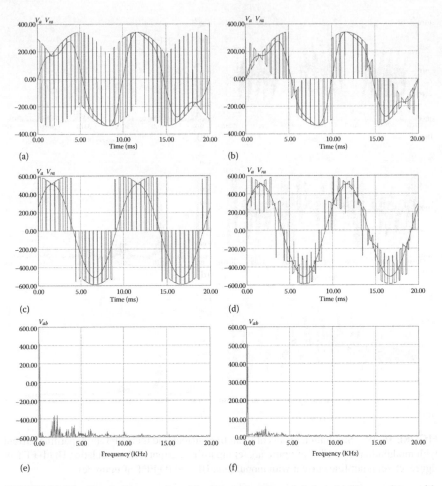

FIGURE 16.50 Simulations results with enhanced ratio modulation. (a) Phase voltage with modulation I, (b) phase voltage with modulation III, (c) line-to-line voltage of with modulation I, (d) line-to-line voltage with modulation III, (e) FFT figure (c) (THD: 63.4%), and (f) FFT figure (d) (THD: 20.8%).

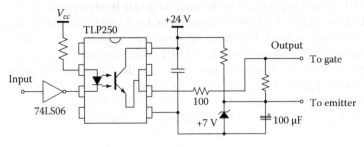

FIGURE 16.51 Gate driver circuit for one IGBT.

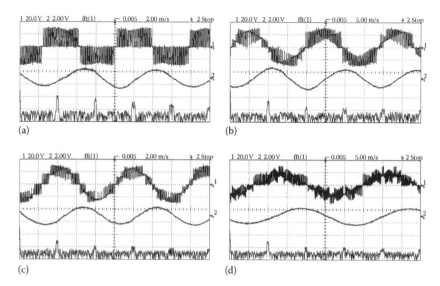

FIGURE 16.52 Experimental waveforms of output line-to-line voltage (top, 400 V/div), phase current (middle, 8 A/div), and FFT of line-to-line voltage (bottom, 20 dB/div, 5 kHz/div) under various modulation methods. (a) Modulation method I (output frequency: 130 Hz, line-to-line voltage 360 V with THD = 62.7%), (b) modulation method II (output frequency: 130 Hz, line-to-line voltage 360 V with THD = 28.3%), (c) modulation method III (output frequency: 130 Hz, line-to-line voltage 360 V with THD = 17.2%), and (d) modulation method III (output frequency: 35 Hz, line-to-line voltage 252 V with THD = 15.3%).

frequency and magnitude are shown in Figures 16.52b and c, respectively. From the figures we can see that the output under SEM method has lower THD (28.3% and 17.2%) than that of maximum-envelop modulation. If the method is applied to the 12-switch-cells matrix converter, the THD can be further reduced. Figure 16.52d is also with modulation III, output frequency is 35 Hz, the magnitude of line-to-line voltage is 252 V with THD = 15.3%.

REFERENCES

1. Luo, F. L. 2009. Switched-capacitorized DC/DC converters. *Proceedings of the IEEE-ICIEA'2009*, Xi'an, China, pp. 377–382.
2. Shen, M. and Qian, Z. 2002. A novel high-efficiency single-stage PFC converter with reduced voltage stress. *IEEE Transactions on Industry Applications*, 38(2), 507–513.
3. Qiu, M., Moschopoulos, G., Pinheiro, H., and Jain, P. 1999. Analysis and design of a single stage power factor corrected full-bridge converter. *Proceedings of the IEEE APEC*, Dallas, TX, pp. 119–125.
4. Qiao, C. and Smedley, K. M. 2001. A topology survey of single-stage power factor correction with a boost type input-current-shaper. *IEEE-Transactions on Power Electronics*, 16(3), 360–368.
5. Rashid, M. H. 2001. *Power Electronics Handbook*. San Diego, CA: Academic Press.
6. Mohan, N., Undeland, T. M., and Robbins, W. P. 2003. *Power Electronics* (3rd edn.). John Wiley & Sons.

7. Cheng, K. W. E. 2003. Storage energy for classical switched mode power converters. *Proceedings of IEE–EPA*, 150(4), 439–446.

8. Luo, F. L. and Ye, H. 2004. *Advanced DC/DC Converters*. Boca Raton, FL: CRC Press LLC.

9. Luo, F. L. and Ye, H. 2005. Energy factor and mathematical modeling for power DC/DC converters. *IEE-Proceedings on EPA*, 152(2), 191–198.

10. Luo, F. L., Ye, H., and Rashid, M. H. 2005. *Digital Power Electronics and Applications*. Boston, MA: Academic Press, Elsevier.

11. Luo, F. L. and Ye, H. 2007. Research on DC-modulated power factor correction AC/AC converters. *Proceedings of IEEE IECON'2007*, Taipei, Taiwan, pp. 1478–1484.

12. Luo, F. L. and Ye, H. 2007. DC-modulated single-stage power factor correction AC/AC converters. *Proceedings of IEEE ICIEA'2007*, Harbin, China, pp. 1477–1483.

13. Luo, F. L. and Ye, H. 2006. DC-modulated power factor correction on AC/AC Luo-converter. *Proceedings of ICARCV 2006*, Singapore, pp. 1791–1796.

14. Luo, F. L. and Ye, H. 2006. DC-modulated single-stage power factor correction AC/AC converters (key notes). *Proceedings of the 10th CPESAM*, Xi'an, China, pp. 21–32.

15. Luo, F. L. and Ye, H. 2002. Positive output super-lift Luo-converters. *Proceedings of the IEEE International Conference PESC'2002*, Cairns, Queensland, Australia, pp. 425–430.

16. Luo, F. L. and Ye, H. 2003. Positive output super-lift converters. *IEEE-Transactions on Power Electronics*, 18(1), 105–113.

17. Luo, F. L. and Ye, H. 2003. Negative output super-lift Luo-converters. *Proceedings of the IEEE International Conference PESC 2003*, Acapulco, Mexico, pp. 1361–1366.

18. Luo, F. L. and Ye, H. 2003. Negative output super-lift converters. *IEEE-Transactions on Power Electronics*, 18(5), 1113–1121.

19. Luo, F. L. and Ye, H. 2004. Positive output cascade boost converters. *IEE-EPA Proceedings*, 151(5), 590–606.

20. Luo, F. L. and Pan, Z. Y. 2006. Sub-envelope modulation method to reduce total harmonic distortion of AC/AC matrix converters. *IEE-Proceedings on Electric Power Applications*, 153(6), 856–863.

21. Luo, F. L. and Pan, Z. Y. 2006. Sub-envelope modulation method to reduce total harmonic distortion of AC/AC matrix converters. *Proceedings of the IEEE Conference PESC'2006*, Jeju, Korea, pp. 2260–2265.

22. Wheeler, P. W., Rodríguez, J., Clare, J. C., Empringham, L., and Weinstein, A. 2002. Matrix converters: A technology review. *IEEE Transactions on Industrial Electronics*, 49(2), 276–288.

23. Venturini, M. 1980. A new sine wave in sine wave out, conversion technique which eliminates reactive elements. *Proceedings of POWERCON 7*, San Diego, CA, pp. E3-1–E3-15.

24. Venturini, M. and Alesina, A. 1980. The generalized transformer: A new bidirectional sinusoidal waveform frequency converter with continuously adjustable input power factor. *Proceedings of IEEE PESC 1980*, Atlanta, GA, pp. 242–252.

25. Neft, C. L. and Schauder, C. D. 1992. Theory and design of a 30-HP matrix converter. *IEEE Transactions on Industry Applications*, 28, 546–551.

26. Wheeler, P. and Grant, D. 1997. Optimized input filter design and low loss switching techniques for a practical matrix converter. *Proceedings of the Institute of Electrical Engineering, Part B*, 144, 53–60.

27. Svensson, T. and Alakula, M. 1991. The modulation and control of a matrix converter synchronous machine drive. *Proceedings of the EPE 1991*, Florence, Italy, pp. 469–476.

28. Empringham, L., Wheeler, P., and Clare, J. 1998. Intelligent commutation of matrix converter bi-directional switch cells using novel gate drive techniques. *Proceedings of IEEE PESC 1998*, Fukuoka, Japan, pp. 707–713.

29. Ziegler, M. and Hofmann, W. 1998. Semi natural two steps commutation strategy for matrix converters. *Proceedings of IEEE PESC 1998*, Fukuoka, Japan, pp. 727–731.
30. Casadei, D., Serra, G., Tani, A., and Zarri, L. 2002. Matrix converter modulation strategies: A new general approach based on space-vector representation of the switch state. *IEEE Transactions on Industrial Electronics*, 49(2), 370–381.
31. Kwon, B. H., Min, B. H., and Kim, J. H. 1998. Novel commutation technique of AC–AC converters. *Proceedings of the Institute of Electrical Engineering, Part B*, 145, 295–300.
32. Pan, C. T., Chen, T. C., and Shieh, J. J. 1993. A zero switching loss matrix converter. *Proceedings of IEEE PESC 1993*, Seattle, WA, pp. 545–550.
33. Villaça, M. V. M. and Perin, A. J. 1995. A soft switched direct frequency changer. *Conference on Record IEEE-IAS Annual Meeting*, Orlando, FL, pp. 2321–2326.
34. Alesina, A. and Venturini, M. 1988. Intrinsic amplitude limits and optimum design of 9-switches direct PWM AC–AC converters. *Proceedings of IEEE PESC 1988*, Kyoto, Japan, pp. 1284–1291.
35. Alesina, A. and Venturini, M. G. B. 1989. Analysis and design of optimum amplitude nine-switch direct AC–AC converters. *IEEE Transactions on Power Electronics*, 4(1), 101–112.

17 AC/DC/AC and DC/ AC/DC Converters

AC/DC/AC and DC/AC/DC conversion technologies are a special subject area in research and industrial applications. AC/DC/AC converters are usually used in synchronous and asynchronous AC motor adjustable speed drives (ASD). In recent years, they have been widely used in renewable energy systems. DC/AC/DC converters are usually used in high-voltage equipment to isolate the source side and load side. They are used in medium and large power systems as well.

17.1   INTRODUCTION

Renewable energy sources have become popular in recent years. Most renewable energy sources involve distributed generations (DG). These sources, such as fuel cells, solar panels, photovoltaic cells, wind turbines, are not standard general sources with stable output voltage and frequency.

Some renewable energy sources are AC voltage sources, for example, wind turbines. The output AC voltage (single phase or three phases) of a wind turbine depends on wind speed and other factors. Definitely, its output AC voltage amplitude is unstable, and the output frequency and phase are unstable as well. Consequently, directly using this output AC voltage is inconvenient.

The necessities for using AC/DC/AC converter are listed as follows:

- The AC source voltage is unstable.
- The AC source frequency is unstable.
- The AC source phase number does not match on the load requirement.

Some renewable energy sources are DC voltage sources, for example, fuel cells and solar panels. The output DC voltage of a solar panel depends on weather, temperature, and sunlight. Definitely, its output DC voltage and power are unstable. Consequently, directly using this output DC voltage is inconvenient. Normalized DC/DC converters are restricted by power limitations. DC/AC/DC converters can transfer larger power than normal DC/DC converters.

The necessities for using DC/AC/DC converter are listed as follows:

- The DC source voltage is unstable.
- The DC source power is unstable.
- The DC source impedance is not much different from the requirement.

17.2 AC/DC/AC CONVERTERS USED IN WIND TURBINE SYSTEMS

Wind turbines are one of the most promising energy sources, which have gained attraction over the past few decades and are more widely used in utility systems compared to other renewable sources [1–3]. Unfortunately, the output voltage and frequency of the wind turbines are unstable, since the wind speed varies. These turbines are installed onshore or offshore, or sometimes as a wind farm where a large number of turbines are installed and connected together. A single wind-tower structure is shown in Figure 17.1.

German scientist Albert Betz proved that wind turbine is most efficient when the wind slows down to $2v/3$ of its speed just before the rotor and decreases to $v/3$ after the rotor, before regaining its original speed v due to surrounding winds. Therefore, Betz's Law states that the maximum power that can be extracted from the wind is 59% of the total power available in the wind, ignoring mechanical and aerodynamic losses. Wind turbine would transfer the linear moving wind energy into rotational energy by the function

$$P = 0.5\rho\pi R^2 v^3 Cp \tag{17.1}$$

where
P is the power
ρ is the air density
R is the turbine radius
v is the wind speed
Cp is the turbine power coefficient

On the other hand, Cp is a function only of the tip speed ratio λ where the variation of Cp with λ is given. The tip speed needs to be maintained at optimal value in order to extract maximum power.

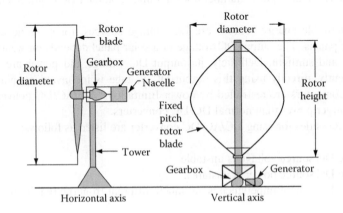

FIGURE 17.1 Wind turbine configuration.

Typically, wind turbines consist of aerodynamically designed three blades, which are positioned in horizontal axis and whole system is mounted on a tower. The rotational mechanical energy is converted to electrical energy with the use of a generator. In some cases, energy is transmitted through a gear box to change the speed. Basically, wind turbines are controlled mechanically, either by pitch controlling or stall controlling. Pitch controlling is more complex where the wind speed is continuously measured and blades are adjusted accordingly, in order to capture energy efficiently. Also, it would protect the turbine from high wind speeds. This control method is more efficient compared to stall control. The stall-controlled blades are fixed at a constant pitch angle, and it is not changed during the operation. Stall is a simple aerodynamic effect that separates the airflow from aerofoil when the turbine runs at a constant speed and when the wind speed increases. This would change the angle of attack and limit the wind power captured, thereby protecting the turbine from high wind speed. However, due to the randomness of the wind availability and, again, when these wind turbines are operated to capture maximum power, the operating voltage and frequency tend to vary, making the output unsuitable for grid connection demanding power conditioning before consumption.

There are many generating topologies commonly used in wind turbines such as induction generators, synchronous generators, and permanent magnet synchronous generators. Some generators are connected directly to the grid while others use power electronic interfaces. Power electronic interfaces have to be selected depending on the generator used and controlling method adopted. In general, induction generators are used with fixed speed wind turbines and power is limited mechanically with pitch or stall controlling. The other type is the variable speed wind turbine, which controls the pitch and use power electronic interface at the output of the generator, which can be a synchronous generator, permanent magnet synchronous generator, or doubly fed induction generator. There are different power electronic converters used in interfacing these wind generators to overcome problems of variation in frequency and voltage. In the case of synchronous generators, full-rated power electronic converters are used. Usually, they can be AC-to-DC converters followed by inverters or simple rectifiers followed by DC-to-DC converters and then inverters. For induction generators, there are two possibilities; it can have an AC-to-DC converter followed by a DC-to-AC inverter in both stator and rotor or only AC to DC followed by a DC-to-AC inverter connected in the rotor of induction generators. To sum up, all those topologies use a combination of two or more power electronic converters, making the overall process inefficient and difficult to control as identified in the first section of this chapter. This sets the stage for developing single-stage topologies in integrating wind power generating systems.

The output AC voltage of wind turbine can be single phase or three phase or multiphase. Its output voltage and frequency are usually not sable. Some industrial applications require AC/AC converters to transfer unstable AC energy source to a stable AC load. Most of AC/AC converters are not suitable for these applications. We need use AC/DC converters to implement the work.

17.2.1 Review of Traditional AC/AC Converters

Traditional AC/AC inverters are introduced in Chapter 15. There are three methods to implement AC/AC conversion:

1. Voltage regulation converters
2. Cycloconverters
3. Matrix converters

Voltage regulation converters are usually used in the applications with a stable input AC source, unchanged output frequency (fundamental harmonic frequency), and adjustable output voltage. These converters have the following advantages: simple structure, lower cost, and easy control. The drawbacks are poor power factor, heavy distorted waveform (poor THD), lower power transfer efficiency, and low voltage transfer gain (lower than unity).

Cycloconverters are usually used in the applications with a stable input AC source, adjustable output voltage, and output frequency (lower than input frequency). These converters have the following advantages: good power factor, waveform with little distortion (good THD), and adjustable output voltage and frequency. The drawbacks are complex structure, higher cost, and complex control circuitry. The output voltage and frequency are lower than input voltage and frequency.

Matrix converters are usually used in the applications with a stable input AC source, adjustable output voltage, and frequency. These converters have the following advantages: adjustable output voltage and frequency, simple structure, and lower cost. The drawbacks are poor power factor, waveform with heavy distortion (bad THD), need for bidirectional switches, heavy network pollution, and complex control circuitry. The main characteristics are output voltage is lower than the input voltage and the output frequency can be higher and lower than the input frequency.

17.2.2 New AC/DC/AC Converters

AC/DC/AC converters can absorb the energy from a random input AC voltage source with unstable voltage and frequency to a fixed DC link voltage (AC/DC conversion) and then convert the energy to a required AC output voltage with adjustable frequency and voltage (DC/AC conversion). The uncertainty of the input voltage from a random input AC voltage source has been dispelled by a controlled AC/DC converter, whose output DC voltage is DC link voltage. The DC link voltage is stable and very good for use with DC/AC inverter. The typical application is ASD. In this section we introduce three circuits:

1. AC/DC/AC boost-type converter
2. Three-level AC/DC/AC converter
3. Three-level AC/DC/AC ZSI

17.2.2.1 AC/DC/AC Boost-Type Converter

An AC/DC/AC boost-type converter is shown in Figure 17.2. The input AC source may be a single-phase or three-phase energy source with unstable voltage and

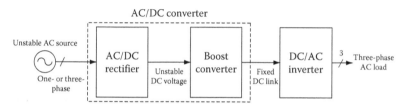

FIGURE 17.2 AC/DC/AC boost-type converter.

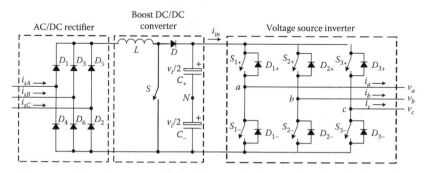

FIGURE 17.3 Circuit of an AC/DC/AC boost-type converter.

frequency. If it is a wind turbine, its voltage can vary in ±25%, and its frequency can change ±15%. The AC/DC converter has two blocks. The first one is an AC/DC rectifier. It can be a diode rectifier, and the output DC voltage is unstable with high efficiency, but independent from the frequency. It may be a controllable rectifier, and the output DC voltage is little stable with lower efficiency, but independent of the frequency as well. The second block is a boost DC/DC converter. It can convert an unstable DC voltage to a fixed DC link voltage, for example, 660 VDC.

The real-end block is a DC/AC inverter. Usually it is a VSI with three-phase output AC voltage with 400 V/60 Hz.

A particular circuit diagram is shown in Figure 17.3.

Example 17.1 A wind turbine has three-phase output voltage 230 V ± 25% and frequency 60 Hz ± 15%; the power rate is 5 kW. The end-user is a three-phase load with voltage 400 V. Design an AC/DC/AC boost-type converter for this application.

Solution: Use a diode rectifier to rectify the input AC voltage to an unstable output DC voltage; the efficiency η can be 92%–97%. Thus, the wind turbine three-phase output voltage is 230 V ± 25% independent from the frequency 60 Hz ± 15%. The rectified output DC voltage can be 311 V ± 25%, that is, 233–389 VDC.

Use a boost DC/DC converter to convert the unstable 233–389 VDC to a fixed 660 VDC; that is, V_{in} is 233–389 V and V_O is 660 VDC. V_O is the fixed DC link voltage. The corresponding duty cycle k can be set as follows:

$$k = \frac{V_O - V_{in}}{V_O} = \begin{cases} \dfrac{660 - 233}{660} = 0.647 \\[2mm] \dfrac{660 - 389}{660} = 0.410 \end{cases} \tag{17.2}$$

A voltage source inverter (VSI) is selected for DC/AC inverter. In linear operation region the output maximum line-to-line peak voltage is 0.866 × link voltage. Therefore, the output maximum line-to-line rms voltage is as follows:

$$V_{AC} = \frac{0.866 V_{link}}{\sqrt{2}} = \frac{0.866 \times 660}{\sqrt{2}} = 404 \text{ V} \tag{17.3}$$

This output three-phase voltage is satisfactory.

17.2.2.2 Three-Level Diode-Clamped AC/DC/AC Converter

A three-level AC/DC/AC converter is shown in Figure 17.4 [4]. The AC source can be a single-phase wind turbine generator. A single-phase half-wave AC/DC diode rectifier is used to obtain the DC link voltage v_d. Two balanced capacitors $C+$ and $C-$ ($C+=C-=C$) are charged in the voltage $v_d/2$. A three-phase three-level diode-clamped voltage source DC/AC inverter converts the DC link voltage to the load.

Usually, if the single-phase wind turbine output voltage has smaller voltage variation, for example, ±5%–10%, and the applications are not so serious, we can directly link the AC/DC rectifier to DC/AC diode-clamped inverter. Therefore, this is the simplest form of AC/DC/AC converter, but it works well and is easily controlled.

17.2.2.2.1 AC/DC Half-Wave Rectifier

The diode AC/DC rectifier has a source inductor L and two identical half-wave diode rectifiers. It converts the wind turbine voltage

$$v_w(t) = V_m \sin(\omega t) \tag{17.4}$$

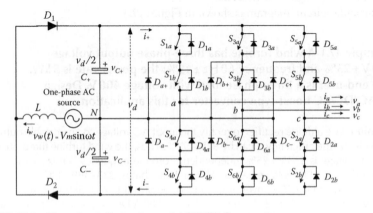

FIGURE 17.4 Three-level diode-clamped AC/DC/AC converter.

to the DC link voltage v_d. Since there are source inductor and two capacitors, the DC link voltage is

$$v_d \approx 0.9V_m \qquad (17.5)$$

The average voltage across both capacitors $C+$ and $C-$ is half of the DC link voltage $v_d/2$. The differential coefficient of the source current and capacitor voltages are as follows:

$$\frac{di_w}{dt} = \frac{v_w - sv_{C+} + (1-s)v_{C-}}{L}$$

$$\frac{dv_{C+}}{dt} = \frac{si_w - i_+}{C} \qquad (17.6)$$

$$\frac{dv_{C-}}{dt} = \frac{-(1-s)i_w - i_-}{C}$$

where $s=1$ for positive half-cycle of wind turbine voltage and $s=0$ for negative half-cycle of wind turbine voltage.

17.2.2.2.2 Three-Level Diode-Clamped DC/AC Inverter

The three-level diode clamped DC/AC inverter is shown in the right-hand part of Figure 17.5. There are two fast-recovery diodes, four power switch devices, and four freewheeling diodes in each leg of the three-level inverter. There are eight (2^3) switching states in the traditional two-level inverter. However, there are 27 (3^3) switching states in the three-level inverter. The switching states of each phase of the three-level inverter are expressed as follows:

$$v_{xN} = \begin{cases} v_d/2 & v_{ref,x} > v_{tri,1} \\ 0 & v_{tri,1} > v_{ref,x} > v_{tri,2} \\ -v_d/2 & v_{tri,2} > v_{ref,x} \end{cases} \qquad (17.7)$$

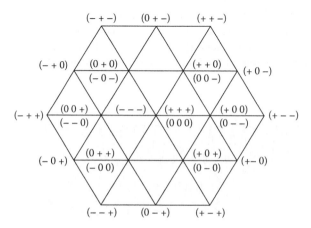

FIGURE 17.5 Space vector representation of the three-phase three-level inverter.

where
   $x = a$, b, or c
   $v_{ref,x}$ is the phase x reference
   $v_{tri,1}$ is the upper triangle pulse
   $v_{tri,2}$ is the lower triangle pulse

A space voltage vector V can be used to represent the output voltages of the three-phase inverter

$$\overline{V} = \sqrt{\frac{2}{3}}(v_{aN} + v_{bN}e^{j120°} + v_{cN}e^{j240°}) \tag{17.8}$$

The space vector representation of output voltages of the inverter in the two-axis coordinate system is shown in Figure 17.5. According to the magnitude of the voltage vectors, the possible switching states can be classified into four groups: large voltage vector $\left(|\overline{V}| = \sqrt{2/3}v_d\right)$ such as (+ − −) (+ + −) (− + −) (− + +) (− − +) (+ − +), middle voltage vector $\left(|\overline{V}| = v_d/\sqrt{2}\right)$ such as (+ 0 −) (0 + −) (− + 0) (− 0 +) (0 − +) (+ − 0), small voltage vector $\left(|\overline{V}| = v_d/\sqrt{6}\right)$ such as (+ 0 0) (+ + 0) (0 + 0) (0 0 +) (0 + +) (+ 0 +), and zero voltage vector $\left(|\overline{V}| = 0\right)$ (+ + +)/(− − −)/(0 0 0). The DC link capacitor voltages are usually regulated and maintained in balanced condition in the three-level inverter. To reduce the voltage imbalance in the capacitors, the redundant switching states can be used to provide some degree of freedom.

17.2.2.2.3 Waveforms
The three-level pulse-width-modulated waveforms are generated by comparing three reference control signals with two triangular carrier waves shown in Figure 17.6.

17.2.3 WIND TURBINE SYSTEM LINKING TO UTILITY NETWORK

Distributed generation sources usually have no standard and stable output voltage and frequency. Wind turbine is a typical example, although it is an AC voltage source. In order to link its output power to utility network, an AC/DC/AC converter is necessarily used for implementing the synchronization. The synchronization conditions for an AC generator to link to utility network are as follows:

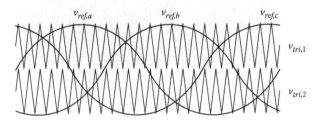

FIGURE 17.6   Waveforms.

- The output voltage amplitude of the AC generator is the same as the voltage amplitude of the utility network
- The frequency of the AC generator is the same as the frequency of the utility network
- The voltage phase of the AC generator is the same as the voltage phase of the utility network

For a wind turbine system to be linked to utility network, the AC/DC/AC inverter has to adjust its output voltage, frequency, and phase angle. By carefully controlling the DC/AC inverter, the synchronization condition can be achieved without much difficulty.

17.3   DC/AC/DC CONVERTERS

There are more than 500 topologies of DC/DC converters existing for DC voltage conversion [5]. Usually, they are enough for research and industrial applications. DC/AC/DC converter is used in some special applications such as high power transformation.

17.3.1   Review of Traditional DC/DC Converters

There are three traditional converters: buck converter, boost converter, and buck-boost converter. They have simple structures and clear operational procedures. One inductor plays the role of pumping circuit. The maximum power transferred from the source to the load is restricted by the pumping energy.

For example, a buck converter as shown in Figure 17.7a converts the energy from source V_1 to load R (voltage is V_2). The inductor current increases when the switch S is on and decreases when the switch S is off. In steady state, the inductor current changes from I_{min} to I_{max} when the switch S is on and from I_{max} to I_{min} when the switch S is off. In a switching cycle T in steady state, the inductor L absorbs the energy from the source, which can be determined as follows:

$$PE = \frac{1}{2}L(I_{max}^2 - I_{min}^2) \tag{17.9}$$

The total power transferred from source to load is defined as follows:

$$P = f \times PE = \frac{1}{2}fL(I_{max}^2 - I_{min}^2) \tag{17.10}$$

The maximum power corresponds $I_{min}=0$. It means the converter works in a discontinuous conduction (DCM) mode.

$$P_{max} = \frac{1}{2}fLI_{max}^2 \tag{17.11}$$

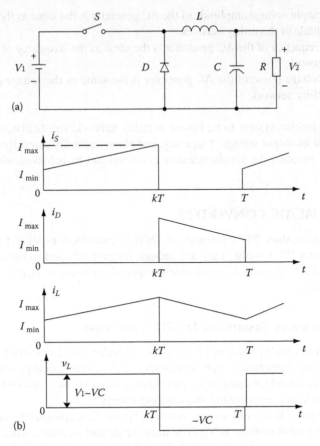

FIGURE 17.7   Buck converter. (a) Circuit and (b) waveforms.

Example 17.2 A buck converter has $V_1 = 40$ V, $L = 10$ mH, $C = 20\,\mu$F, $R = 10\,\Omega$, switching frequency $f = 20$ kHz, and duty cycle $k = 0.5$. Calculate the power transferred to load.

Solution: The output voltage V_2 is

$$V_2 = kV_1 = 0.5 \times 40 = 20 \text{ V}$$

Therefore, the power is

$$P = \frac{V_2^2}{R} = \frac{20^2}{10} = 40 \text{ W}$$

From the known data $T = 1/f = 50\,\mu$s, and using the formulae (5.13) and (5.14), we get

$$I_{max} = kV_1\left(\frac{1}{R} + \frac{1-k}{2L}T\right) = 20\frac{41}{400} = 2.05 \text{ A}$$

$$I_{min} = kV_1\left(\frac{1}{R} - \frac{1-k}{2L}T\right) = 20\frac{39}{400} = 1.95 \text{ A}$$

Substituting the values into (17.10), we obtain the output power as follows:

$$P = \frac{1}{2}fL(I_{max}^2 - I_{min}^2) = \frac{20k \times 10m}{2}(2.05^2 - 1.95^2) = 40 \text{ W}$$

It is verified.

The same operation is available for boost and buck-boost converters. From this example we know that the power delivered from source to load is restricted by the pumping circuit.

17.3.2 CHOPPER-TYPE DC/AC/DC CONVERTERS

In order to increase the power delivered from source to load, we have to avoid using inductor-pumping circuit. A good way is to apply the choppers to chop the DC source voltage to AC pulse train and then rectify the AC waveform back to DC voltage. The rectifiers can be diode rectifiers or transformer plus diode rectifiers. Figure 17.8 shows a DC/AC/DC converter with a dual-polarity chopper plus a diode rectifier circuit [6]. The copper has two pairs of switches (S_{1+}, S_{2-}) and (S_{2+}, S_{1-}). Each pair's switches switch on and switch off simultaneously. The output AC voltage v_{ac} is an AC voltage with positive and negative peak values of $+v_i$ and $-v_i$. A diode rectifier $(D_1 - D_4)$ is applied to rectify the AC voltage v_{ac} to the DC output voltage v_O.

Using this DC/AC/DC converter the power delivered from source v_i to load is not restricted since there is no pumping circuit. The output voltage v_O is lower than the input voltage v_i. The switching duty cycle of the pair (S_{1+}, S_{2-}) is k_1 and the pair's switching duty cycle of the pair (S_{2+}, S_{1-}) is k_2. Usually, $k_1 + k_2 \leq 1$. The output voltage v_O is

$$v_O = (k_1 + k_2)v_i \tag{17.12}$$

We can add a transformer in the circuit and then obtain random output voltage depending on the transformer turns ratio. Figure 17.9 shows the DC/AC/DC converter with a dual-polarity chopper plus a transformer and diode rectifier circuit. Using this DC/AC/DC converter the power delivered from source v_i to load is not restricted since there is no pumping circuit. The output voltage v_O can be higher or lower than the input voltage v_i.

The switching duty cycle of the pair S_{1+}, S_{2-} is k_1, and the pair's switching duty cycle of the pair S_{2+}, S_{1-} is k_2. Usually, $k_1 + k_2 \leq 1$. The transformer winding turn's ratio is n; n can be greater or smaller than unity. If the turn's ratio n is greater than unity, it is very easy to obtain the output voltage v_O that is higher than the input voltage v_i. The output voltage v_O is

$$v_O = n(k_1 + k_2)v_i \tag{17.13}$$

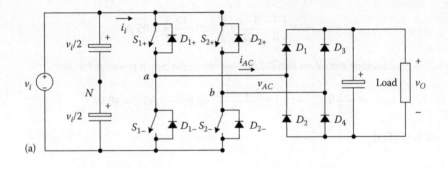

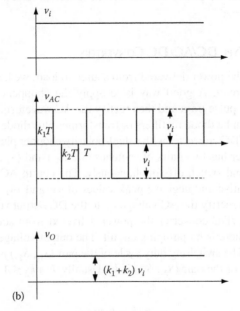

FIGURE 17.8 DC/AC/DC converter with a dual-polarity chopper plus a diode rectifier. (a) Circuit and (b) waveforms.

17.3.3 SWITCHED-CAPACITOR DC/AC/DC CONVERTERS

Switched capacitor can be used to build DC/AC/DC converter [7]. Since switched capacitor can be integrated into a power IC chip, consequently, these converters have small size and high power density. In this section we introduce several switched-capacitor DC/AC/DC converters:

- Single-stage switched-capacitor DC/AC/DC converter
- Three-stage switched-capacitor DC/AC/DC converter
- Four-stage switched-capacitor DC/AC/DC converter

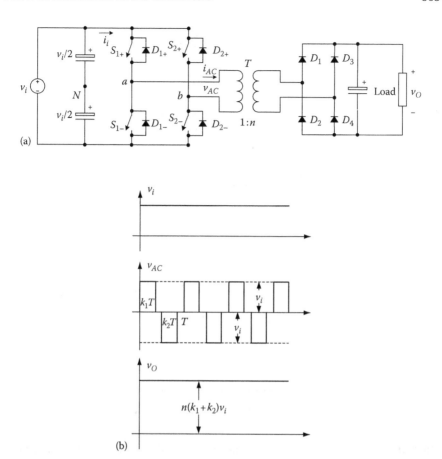

FIGURE 17.9 DC/AC/DC converter with a dual-polarity chopper plus a transformer and diode rectifier circuit. (a) Circuit and (b) waveforms.

17.3.3.1 Single-Stage Switched-Capacitor DC/AC/DC Converter

A single-stage switched-capacitor DC/AC/DC converter is shown in Figure 17.10.

This single-stage switched-capacitor DC/AC/DC converter has the input voltage v_i, middle AC voltage v_{ac}, and output DC voltage v_O. The switching frequency is f, and its corresponding period is $T = 1/f$. The conduction duty cycle of main switches S_1 and S_2 is k_1. Therefore, the main switches S_1 and S_2 switch on during the period k_1T. In the meantime, the auxiliary switches S_3 and S_4 switch on, which changes the switched capacitor voltage v_C to the source voltage v_i. From $t = k_1T$ to T, the main switches S_1 and S_2 switch off. The auxiliary switches S_5 and S_6 switch on during the period k_2T. In the meantime, the auxiliary switches S_3 and S_4 must switch off. We can arrange the main switches S_1 and S_2 to switch on at $t = 0$ to k_1T and the auxiliary switches S_5 and S_6 to switch on at $t = 0.5\,T$ to $(0.5 + k_2)T$. The auxiliary switches S_3 and S_4 can switch on from $t = 0$ to $0.5\,T$. In other words, the auxiliary switches S_3 and S_4

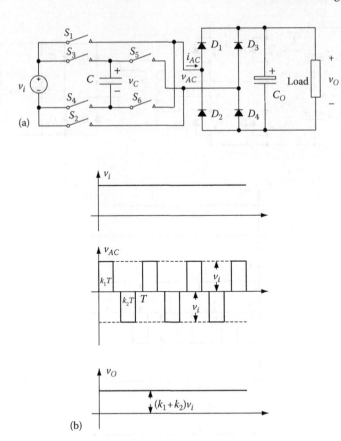

FIGURE 17.10 Single-stage switched-capacitor DC/AC/DC converter. (a) Circuit and (b) waveforms.

can switch on simultaneously with the main switches S_1 and S_2, but not necessarily switch off simultaneously with the main switches S_1 and S_2. The auxiliary switches S_3 and S_4 can switch off at any moment from $t>0$ to 0.5 T. Usually, $(k_1+k_2) \le 1$. The output voltage v_O is

$$v_O = (k_1 + k_2)v_i \tag{17.14}$$

The corresponding waveforms are shown in Figure 17.10b.

We can add a transformer in the circuit and then obtain random output voltage depending on the transformer turn ratio. Figure 17.11 shows the DC/AC/DC converter with a switched capacitor plus a transformer and diode rectifier circuit. Using this DC/AC/DC converter, the power delivered from source v_i to load is not restricted since there is no inductor pumping circuit. The output voltage v_O can be higher or lower than the input voltage v_i. The transformer winding turn's ratio is n; n can be greater or smaller than unity. If the turn's ratio n is greater than unity, it is very easy to obtain the output voltage v_O that is higher than the input voltage v_i. The output voltage v_O is

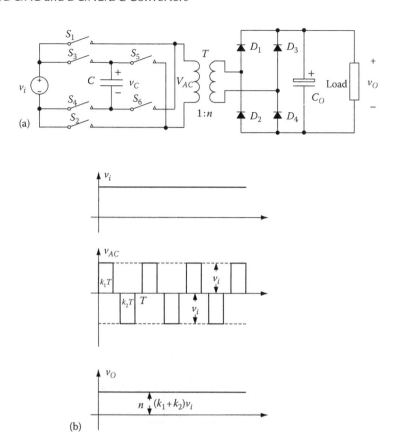

FIGURE 17.11 DC/AC/DC converter with a switched capacitor plus a transformer and diode rectifier circuit. (a) Circuit and (b) waveforms.

$$v_O = n(k_1 + k_2)v_i \qquad (17.15)$$

The corresponding waveforms are shown in Figure 17.11b.

17.3.3.2 Three-Stage Switched-Capacitor DC/AC/DC Converter

In order to use more switched capacitors, we can design other types of switched-capacitor DC/AC/DC converters. A three-stage switched-capacitor DC/AC/DC converter is shown in Figure 17.12.

This three-stage switched-capacitor DC/AC/DC converter has three switched capacitors C_1, C_2, and C_3, and the input voltage v_i, middle AC voltage v_{ac}, and output DC voltage v_O. The switching frequency is f, and its corresponding period is $T = 1/f$. The conduction duty cycle of the main switches S_1, S_2, and S_3 is k_1. Therefore, the main switches S_1, S_2, and S_3 switch on during the period $k_1 T$ to provide AC voltage $v_{ac} = 2v_i$. In the meantime, the auxiliary switches S_4–S_7 switch on, which changes the switched capacitors C_1 and C_2 (voltages v_{C1} and v_{C2}) to the source voltage v_i. From $t = k_1 T$ to T, the main switches S_1–S_3 switch off. The auxiliary switches S_{10}–S_{12} switch on from $t = 0.5\,T$ (in the period $k_2 T$) to provide AC voltage $v_{ac} = -2v_i$. In the meantime,

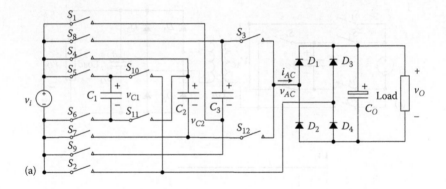

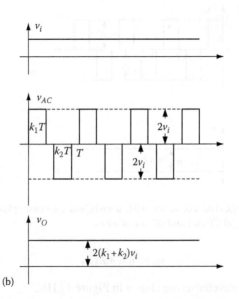

(b)

FIGURE 17.12 Three-stage switched-capacitor DC/AC/DC converter. (a) Circuit and (b) waveforms.

the auxiliary switches S_8 and S_9 switch on to charge the switched capacitor C_3 to the source voltage v_i. The auxiliary switches S_4–S_7 must switch off in this period. We can arrange the main switches S_1–S_3 to switch on from $t=0$ to k_1T, and the auxiliary switches S_{10}–S_{12} to switch on from $t=0.5\ T$ to $(0.5+k_2)T$. The auxiliary switches S_4–S_7 can switch on from $t=0$ to $0.5\ T$. In other words, the auxiliary switches S_4–S_7 can switch on simultaneously with the main switches S_1–S_3, but not necessarily switch off simultaneously with the main switches S_1–S_3. The auxiliary switches S_4–S_7 can switch off at any moment from $t>0$ to $0.5\ T$. Similarly, the auxiliary switches S_8 and S_9 can switch off at any moment from $t>0.5\ T$ to T. Usually, $(k_1+k_2) \leq 1$. The output voltage v_O is

$$v_O = 2(k_1 + k_2)v_i \qquad (17.16)$$

The corresponding waveforms are shown in Figure 17.12b.

We can add a transformer in the circuit and then obtain random output voltage depending on the transformer turn ratio. Figure 17.13 shows the DC/AC/DC converter with a three-stage switched capacitor plus a transformer and diode rectifier circuit. Using this DC/AC/DC converter the power delivered from source v_i to load is not restricted since there is no inductor pumping circuit. The output voltage v_O can be higher or lower than the input voltage v_i. The transformer winding turn's ratio is n; n can be greater or smaller than unity. If the turn's ratio n is greater than unity, it is very easy to obtain the output voltage v_O that is higher than the input voltage v_i. The output voltage v_O is

$$v_O = n(k_1 + k_2)v_i \qquad (17.17)$$

The corresponding waveforms are shown in Figure 17.13b.

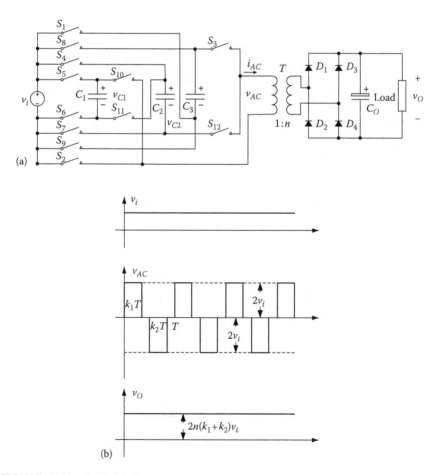

FIGURE 17.13 DC/AC/DC converter with a three-stage switched capacitor plus a transformer and diode rectifier circuit. (a) Circuit and (b) waveforms.

It is possible to design other odd-number-stage (n is the odd number, $m>3$) switched-capacitor DC/AC/DC converters; $(m-1)/2$ switched capacitors plus the source voltage supply the positive half-cycle of the intermediate AC voltage; other $(m+1)/2$ switched capacitors supply the negative half-cycle of the intermediate AC voltage.

17.3.3.3 Four-Stage Switched-Capacitor DC/AC/DC Converter

It is possible to design the even-number-stage (n is the even number, $m \geq 2$) switched-capacitor DC/AC/DC converters. Considering the symmetry of the intermediate AC voltage, half of the switched capacitors ($m/2$ switched capacitors) supply the positive half-cycle of the intermediate AC voltage; other half of switched capacitors supply the negative half-cycle of the intermediate AC voltage. The source voltage can only be used to change the two groups of the switched capacitors alternatively. Figure 17.14 shows the four-stage DC/AC/DC switched-capacitor converter.

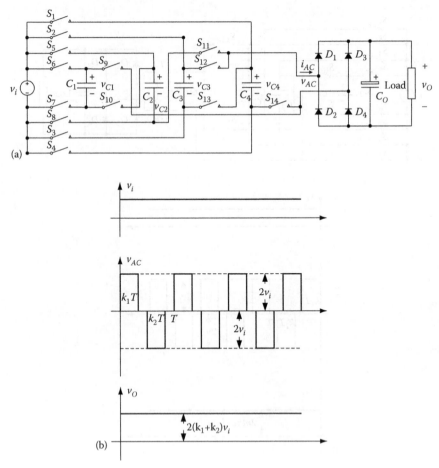

FIGURE 17.14 Four-stage DC/AC/DC switched-capacitor converter. (a) Circuit and (b) waveforms.

When $t=0$ the switches S_5–S_8 switch on to charge the capacitor C_1 and C_2 to the source voltage v_i. When $t=k_1 T$, the switches S_{12}–S_{14} switch on and the capacitors C_3 and C_4 supply $+2v_i$ to v_{ac}. When $t=0.5\ T$, the switches S_1–S_4 switch on to charge the capacitor C_3 and C_4 to the source voltage v_i. When $t=k_2 T$, the switches S_9–S_{11} switch on and the capacitor C_1 and C_2 supply $-2v_i$ to v_{ac}. The waveforms are shown in Figure 17.14b. After diode rectifier, we obtain the output voltage v_O is

$$v_O = (k_1 + k_2)v_i \qquad (17.18)$$

We still can add a transformer in the circuit to enlarge the output voltage. A four-stage switched-capacitor with transformer DC/AC/DC converter is shown in Figure 17.15. The waveforms are shown in Figure 17.15b.

The transformer turn's ratio is n. The output voltage is

$$v_O = n(k_1 + k_2)v_i \qquad (17.19)$$

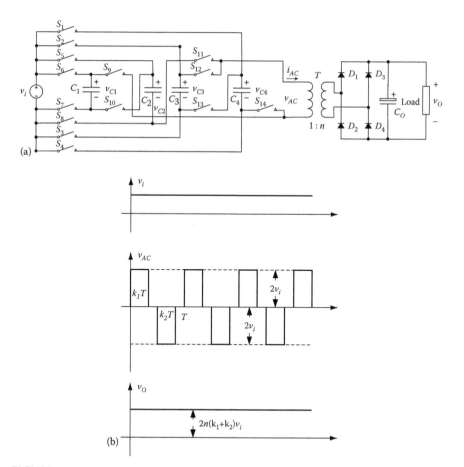

FIGURE 17.15 Four-stage switched capacitor with transformer DC/AC/DC converter. (a) Circuit and (b) waveforms.

REFERENCES

1. Masters, G. M. 2005. *Renewable and Efficient Electric Power Systems*. New York: John Wiley.
2. Ackermann, T. 2005. *Wind Power in Power Systems*. New York: John Wiley.
3. Johnson, G. L. 1985. *Wind Energy Systems*. Englewood Cliffs, NJ: Prentice-Hall.
4. Lin, B. R., Lu, H. H., and Chen, Y. M. 1998. Implementation of three-level AC/DC/AC converter with power factor correction and harmonic reduction. *Proceedings of IEEE PEDES*, Perth, Australia, pp. 768–773.
5. Luo, F. L. and Ye, H. 2004. *Advanced DC/DC Converters*. Boca Raton, FL: CRC Press.
6. Luo, F. L. and Ye, H. 2009. Chopper-type DC/AC/DC converters. *Technical Talk of ICIEA'2009*, Xi'an, China, pp. 1356–1368.
7. Luo, F. L. and Ye, H. 2009. Switched-capacitor DC/AC/DC converters. *Technical Talk of ICIEA'2009*, Xi'an, China, pp. 163–168.

18 Designs of Solar Panel and Wind Turbine Energy Systems

Solar panel and wind turbine energy are clean and renewable energy sources. In recent years their applications have drawn much attention all over the world. Therefore, the topic of design of solar panel and wind turbine energy systems is a very popular research field.

18.1 INTRODUCTION

First, we introduce sum units to measure large power. They are grouped by orders of magnitude as follows. The energy unit Joule ($1\,J = 1\,W \times 1\,s$):

- kW—kilowatt ($10^3\,W$)
- MW—megawatt ($10^6\,W$)
- GW—gigawatt ($10^9\,W$)
- TW—terawatt ($10^{12}\,W$)
- PW—petawatt ($10^{15}\,W$)
- EW—exawatt ($10^{18}\,W$)
- ZW—zettawatt ($10^{21}\,W$)
- YW—yottawatt ($10^{24}\,W$)

The Sun radiates the power of $3.8 \times 10^{20}\,MW$ into space. The Earth receives $174\,PW$ of incoming solar radiation (insolation) at the upper atmosphere. Approximately, 30% is reflected back to space and the rest is absorbed by clouds, oceans, and land masses. The spectrum of solar light at the Earth's surface is mostly spread across the visible and near-infrared ranges with a small part in the near-ultraviolet range.

Earth's land surface, oceans, and atmosphere absorb solar radiation, and this raises their temperature (Figure 18.1). Warm air containing evaporated water from the oceans rises, causing atmospheric circulation or convection. When the air reaches a high altitude, where the temperature is low, water vapor condenses into clouds, which rain onto the Earth's surface, completing the water cycle. The latent heat of water condensation amplifies convection, producing atmospheric phenomena such as wind, cyclones, and anticyclones. Sunlight absorbed by the oceans and land masses keeps the surface at an average temperature of 14°C. By photosynthesis, green plants convert solar energy into chemical energy, which produces food, wood, and the biomass from which fossil fuels are derived.

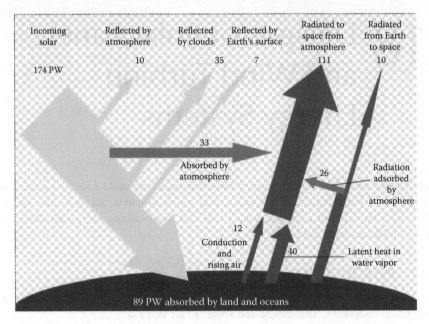

FIGURE 18.1   About half of the incoming solar energy reaches the Earth's surface.

TABLE 18.1
Yearly Solar Fluxes and Human
Energy Consumption (in 2005)

Solar	3,850,000 EJ
Wind	2,250 EJ
Biomass	3,000 EJ
Primary energy use	487 EJ
Electricity	56.7 EJ

The total solar energy absorbed by Earth's atmosphere, oceans, and land masses is approximately 3,850,000 exajoules (EJ) per year, which is shown in Table 18.1. In 2002, this was more energy in 1 h than the world used in 1 year; that is, the total world energy consumption is less than 1/10,000 of the energy that the Earth received from the Sun. Photosynthesis captures approximately 3000 EJ per year in biomass. The amount of solar energy reaching the surface of the planet is so vast that in 1 year it is about twice as much as will ever be obtained from all of the Earth's nonrenewable resources of coal, oil, natural gas, and mined uranium combined. Solar energy can be harnessed in different levels around the world. Depending on a geographical location, the closer the location to the equator, the more solar energy received.

From Table 18.1 it can be seen that the whole world electricity energy (56.7 EJ) is much less than the 1/10,000 of the solar energy (385 EJ). There is another information: The wind energy is more than the "primary energy use." It indicates that if we can effectively use the solar energy and the wind energy, the energy requirement is no problem.

18.2   WIND TURBINE ENERGY SYSTEMS

Wind turbines are one of the most promising energy sources, which have gained much attention in a recent few decades and penetrated utility systems deeply compared to other renewable sources [1–3]. Unfortunately, the output voltage and frequency of the wind turbines are unstable since the wind speed is variable. These turbines are installed onshore or offshore, or sometimes as a wind farm where large number of turbines are installed and connected together.

The air flowing is called the wind. Because of the Sun, the wind always exists. The wind energy is from the Sun; it is a renewable energy resource. Figure 18.2 shows the wind production [1]. The whole Earth air is like boiler. The air becomes light in the equator and becomes heavy in the two poles. The wind flows day and night.

The wind speed is uncertain; its probability is called Weibull probability density function, which is shown in Figure 18.3.

The use of wind energy is traced back to thousands of years. In the ancient ages, the windmill has been used for water pumping and grain grinding for more than 3000 years. The first proof of the use of wind energy has been found in Persia in around AD 500–900, which is shown in Figure 18.4.

Wind turbine is the modern equipment to convert the wind dynamic energy into the electrical energy. Two types of wind turbines are used: vertical axis and horizontal axis, which are shown in Figure 18.5. The horizontal-axis three-blade wind turbine is popular, which is shown in Figure 18.6 [1–3].

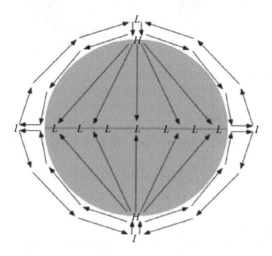

FIGURE 18.2   The wind production.

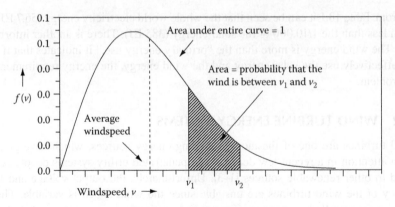

FIGURE 18.3 The Weibull probability density function.

FIGURE 18.4 Persia in around AD 500–900.

18.2.1 Technical Features

The wind speed changes from time to time. The probability distribution function is

$$f(v) = \frac{2v}{c^2} e^{-(v/c)^2} \tag{18.1}$$

where
  v is the wind speed
  c is the scale parameter, which can be 4, 6, 8, and so on, depending in the particular position, for example, Singapore $c \approx 6$ (see Figure 18.7)

FIGURE 18.5   The modern wind turbines.

FIGURE 18.6   The wind turbine structure.

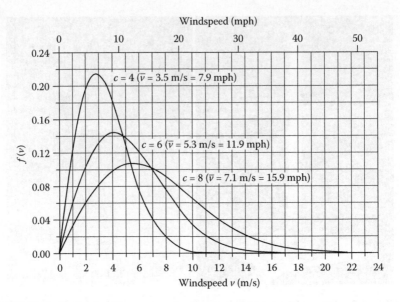

FIGURE 18.7   Rayleigh pdf.

Equation 18.1 is called the Weibull probability distribution function (pdf) or Rayleigh pdf. The mass of air that flows through a wind turbine is

$$m = A_1 v = A v_b = A_2 v_d \tag{18.2}$$

As shown in Figure 18.8.
   The power extracted by the turbine is

$$P = \frac{1}{2} m (v^2 - v_d^2) \tag{18.3}$$

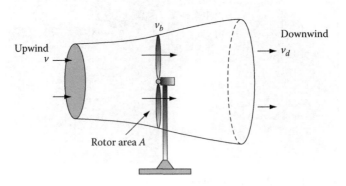

FIGURE 18.8   The mass of air flows through a wind turbine.

Since

$$m = \rho A v_b \quad \text{and} \quad v_b \approx \frac{1}{2}(v + v_d) \tag{18.4}$$

$$P = \frac{1}{2}m(v^2 - v_d^2) \tag{18.5}$$

or

$$P = 0.5\rho A v^3 \left[\frac{1}{2}(1 + \lambda)(1 - \lambda^2) \right] \tag{18.6}$$

where
  ρ is the air density
  A is the section area
  v_b is the wind speed through the turbine
  v is the blow-in wind speed
  v_d is the blow-out wind speed

To describe the power produced by the wind turbine we use the Betz's law that is

$$P = 0.5\rho \pi R^2 v^3 C_p \tag{18.7}$$

where
  R is the radius of the windmill (or the length of the blade)
  C_p is the power coefficient

$$C_p(\lambda) = 0.5(1 + \lambda)(1 - \rho \lambda^2) \tag{18.8}$$

With $\lambda = v_d/v$ is the wind speed deduction ratio.
  The effect of temperature and pressure of air density is described by gas state equation:

$$PV = nRT \quad \Rightarrow \quad \frac{n}{V} = \frac{P}{RT} \tag{18.9}$$

where
  n is the mass of air in mols
  V is the volume of air in m^3
  P is the pressure in atm
  R is the ideal gas constant = 8.2056×10^{-5} in m$^3 \cdot$atm/K/mol
  T is the absolute temperature in K

The air density is given by

$$\rho = \left(\frac{n}{V}\right)M$$

where M is molecular weight of air in kg/mol$=0.02897$.

$$\rho = 353\frac{P}{T} \tag{18.10}$$

In kg/m^3.

The atmospheric pressure depends on altitude; it is 1 atm at the sites on mean sea level. At sites above sea level, pressure is less, and it can be shown that

$$\frac{dP}{dh} = -\rho \cdot g\left[\frac{(N/m^2)}{m}\right] \tag{18.11}$$

where

h is the height above mean sea level
g is the gravitational acceleration constant (9.806 m/s^2)

1 atm$=1.01325 \times 105$ N/m^2. Therefore,

$$P = e^{-(0.0341/T)h} \tag{18.12}$$

The wind speed at a location varies by height h. Thus, the relation is

$$\frac{v}{v_0} = \left(\frac{h}{h_0}\right)^\alpha \tag{18.13}$$

where

v is the wind speed at height h
v_0 is the wind speed at height h_0 (usually $h_0=10$ m)
α is the friction coefficient (see Table 18.2):

Usually, the wind turbine works in certain ranges of wind speed as shown in Figure 18.9 [3]. The wind speed change produces the output voltage and variation in frequency.

18.2.2 DESIGN EXAMPLE

A wind turbine feeds power to a three-phase, 11 kV, 50 Hz grid through a wound-rotor induction generator operating with slip-power control. The data of the system are given in the following [3–6]:

TABLE 18.2
Friction Coefficient α for Various Terrains

Terrain Characteristics	Friction Coefficient α
Smooth, hard ground, calm water	0.10
Tall grass on level ground	0.15
High crops, hedges, and shrubs	0.20
Wooded countryside, many trees	0.25
Small town with trees and shrubs	0.30
Large city with tall buildings	0.40

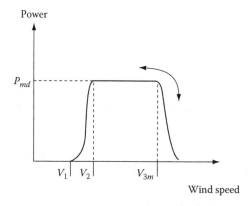

FIGURE 18.9   The wind speed range (power vs. wind speed).

1. Induction generator
   a. Three-phase, 11 kV, 50 Hz, four-pole, delta-connected
   b. Per-phase magnetizing inductance referred to the stator = 7 H
   c. Per-phase rotor winding resistance referred to the stator = 30 Ω; stator to rotor turns ratio = 3:1
2. Gear box data: power efficiency = 85%; speed ratio = 65
   a. Site
      i. Altitude, 500 m above mean sea level
      ii. Average temperature, 30°C ≈ 303 K
      iii. Friction coefficient of a terrain is 0.15
      iv. Wind speed 7.79 m/s at a height of 10 m above ground
   b. Wind turbine
      i. Horizontal-axis, two-blade
      ii. Diameter 50 m ($R = 25$ m)
      iii. Tower height 70 m
      iv. Efficiency = 45% at tip-speed ratio (TSR) of 5.5

Determine the following:

1. Slip of the generator
2. Mechanical power converted to electrical form
3. The magnitude, phase, and the frequency of the phase voltage injected into the rotor by taking stator terminal voltage as the reference phasor
4. Real and reactive power supplied by the rotor-side converter
5. Real power supplied to the grid by assuming no losses in the converters and in the stator winding

Solution:

1. Slip of the generator

$$s = \frac{\omega_s - \omega_m}{\omega_s}; \quad \omega_s = \frac{4\pi f}{P} = \frac{4\pi \times 50}{4} = 157.1 \text{ rad/s}$$

Shaft speed ω_m is directly determined by the wind speed. The wind speed at height of 10 m above ground is given. Average wind speed is calculated as measured at the turbine midpoint, that is, at the top of the tower

$$\frac{v}{v_0} = \left(\frac{h}{h_0}\right)^\alpha \quad \Rightarrow \quad \frac{v}{7.79} = \left(\frac{70}{10}\right)^{0.15} \quad \Rightarrow \quad v = 10.43 \text{ m/s}$$

TSR = 5.5 = tip speed/wind speed

$$\omega_t \cdot \frac{D}{2} = v \cdot \text{TSR} \quad \Rightarrow \quad \omega_t = 10.43 \times 5.5 \times \frac{2}{50} = 2.295 \text{ rad/s}$$

Speed conversion through gear box:

$$\omega_m = 65\omega_t = 65 \times 2.295 = 149.175 \text{ rad/s}$$

Therefore, the slip of the generator

$$s = \frac{\omega_s - \omega_m}{\omega_s} = \frac{157.1 - 149.175}{157.1} = 0.05$$

2. Mechanical power converted to electrical form

$$P_m = \eta_g P_t = \eta_g \eta_t P_w = \eta_g \eta_t \left(\frac{1}{2}\rho A v^3\right)$$

Hence,

$$p_m = \frac{1}{2}\eta_g\eta_t \left(\frac{353}{T} e^{-0.0341h/T}\right)\frac{\pi D^2}{4} v^3$$

where

T= 303 K
h= 500+70= 570 m
v is the wind speed at turbine midpoint= 1.043 m/s
D= 50 m
η_g = efficiency of gearbox = 0.85
η_t = efficiency of turbine = 0.45

Substituting the data into the formula,

$$p_m = 465,540 \text{ W} = 465.54 \text{ kW}$$

3. Rotor-injected voltage
   Delta-connected stator winding: $V_{phase} = V_{line} = 11,000\,\text{V}$
   We choose it as reference phasor: $V_1 = 11,000\,\angle 0°\,\text{V}$
   Since there are no stator losses,

$$P_1 = P_g = \frac{P_m}{1-s} = \frac{-465,540}{1-0.05} = -490,042 \text{ W}$$

and

$$P_1 = 3V_1 I_1 \cos\Phi_1 \quad \Rightarrow \quad I_1 = \frac{-49,0042}{3\times 11,000 \times 1.0} = -14.85 \text{ A}$$

$$I_m = V_1 / jX_m = \frac{11,000}{2\pi \times 50 \times 7} = -j5 \text{ A}$$

$$I_2 = I_1 - I_m = -14.85 + j5 \text{ A}$$

$$V_1 = \frac{V_2}{s} + I_2 \frac{R_2}{s} \quad \Rightarrow \quad \frac{V_2}{0.05} = V - (-14.85 + j5)\frac{30}{0.05} = 19,910 - j3,000$$

$$V_2 = 0.05(19,910 - j3,000) = 995.5 - j150 = 1,006.7\angle -8.57° \text{ V}$$

Rotor-injected voltage is calculated using the turn's ratio as

$$V_{1'} = \frac{1}{3}V_2 = 335.6 \angle -8.57° \text{ V}$$

Rotor-injected voltage should have the frequency given by

$$f_2 = sf_1 = 0.05 \times 50 = 2.5 \text{ Hz}$$

4. Complex power absorbed by the rotor converter

$$S_r = 3V_r I_r^* = 3V_2 I_2^* = 3 \times 1006.7 \angle -8.57° \times (-14.85 - j5)$$

$$S_r = 3 \times 1006.7 \angle -8.57° \times -15.67\angle18.6° = -47325\angle -10.04°$$

$$S_r = P_r + jQ_r = -46,600 + j8,250$$

$$P_r = -46.6 \text{ kW} \quad \text{and} \quad Q_r = 8.25 \text{ kVAr}$$

Real power and reactive power injected into the rotor winding by the rotor-side converter are 46.6 kW and −8.25 kVAr.

5. Real power drawn from the grid

$$P_{grid} = P_1 - P_r = -49,0042 + 46,600 = -443,442 \text{ W}$$

This is also equal to the sum of the power converted to mechanical form (P_m) and the rotor C_u loss.

18.2.3 Converters' Design

Since the wind speed is unstable, it changes in a range. The output voltage and frequency of the double-feed induction generator (DFIG) change about ±20%. In order to transfer the unstable electrical energy generated from DFIG to the grid, we design our converter system as described in the following [3–6].

We assume the output voltage (e.g., line-to-line rms 11,000 V) and frequency (e.g., 50 Hz) of the DFIG change about ±20%, and the grid voltage (e.g., line-to-line rms 11,000 V) and frequency (e.g., 50 Hz) are very stable with ±1% variation [7]. The converters system design includes three parts: AC/DC rectifier, DC/DC converter, and DC/AC inverter. The block diagram is shown in Figure 18.10. The AC/DC

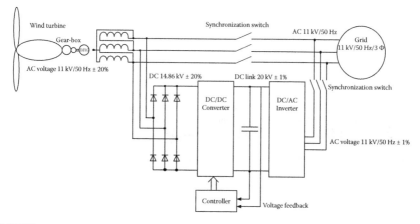

FIGURE 18.10 Block diagram of the wind turbine power system.

rectifier is an uncontrolled diode full-bridge rectifier. Its output is an unstable DC voltage about $14.86\,\text{kV}\pm 20\%$. It avoids the input frequency. The DC/DC converter is a boost converter with closed-loop PI control. Its output voltage is $20\,\text{kV}\pm 1\%$. It is very stable. The DC/AC inverter is a VSI. Its output is as follows: three phase, $50\,\text{Hz}$, $11\,\text{kV}$ (line-to-line rms).

18.2.4 SIMULATION RESULTS

The simulation diagram is shown in Figure 18.11. The simulation results are shown in Figure 18.12.

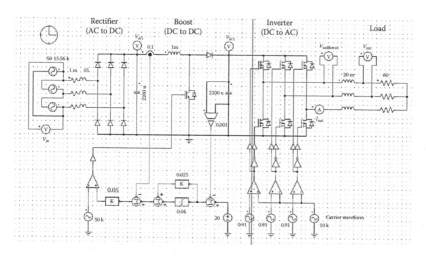

FIGURE 18.11 Simulation diagram of the wind turbine power system.

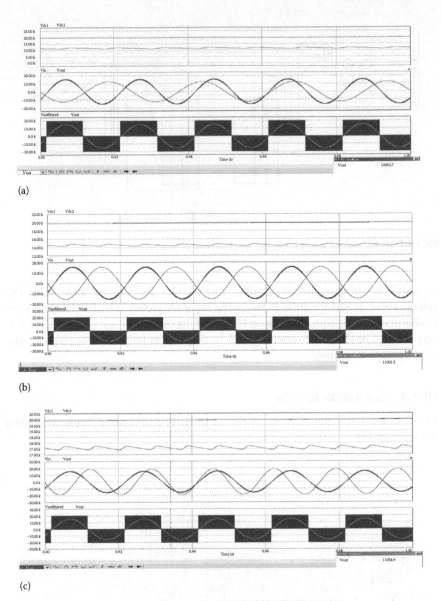

FIGURE 18.12 Simulation results of the wind turbine power system. (a) V_{in}=8.8kV (line-to-line, rms)/40Hz. (b) V_{in}=11kV (line-to-line, rms)/50Hz. (c) V_{in}=13.2kV (line-to-line, rms)/60Hz.

18.3 SOLAR PANEL ENERGY SYSTEMS

The Sun is the source of light and also chemical effects. The tremendous energy offered by Sun is thousand times higher than the total energy consumption used by the world in the present time.

18.3.1 Technical Features

The sunlight changes from time to time. For example, we buy a solar panel. It works from 8 a.m. to 8 p.m. The rated voltage from the panel is 186 V (plus the current is about 13 A), but it varies from 186 − 20% to 186 + 20%, that is, from 148.8 to 223.2 V. In order to convert this energy into a grid, 400 V/50 Hz/ 3 Φ, we have to design our power electronic circuits to deal with it.

The objectives to design the circuit include the following:

1. To convert an unstable DC voltage into a stable DC voltage
2. To invert a stable DC voltage into 3 Φ AC voltage
3. To link the solar panel system to main grid

Considering the aforementioned objectives, we arrange the technical features as follows:

1. To match the grid data, we need an inverter to provide its output of 400 V/50 Hz/3 Φ with 1% variation.
2. To provide the inverter with output of 400 V/50 Hz/ 3 Φ, we need to offer a DC link voltage 700 V with 1% variation.
3. Since the input voltage is 186 V with ±20% variation, we need a high-voltage-transfer-gain DC/DC converter. The positive-output super-lift Luo converter is selected.
4. To keep the link voltage 700 V with 1% variation, we need a closed-loop control for the DC/DC converter.

18.3.2 P/O Super-Lift Luo Converter

The positive-output super-lift Luo converter [4–6] is shown in Figure 18.13. It consists of a switch S, an inductor L, two capacitors C_1 and C_2, two diodes D_1 and D_2, a resistive load R, and input voltage V_{in} and output voltage V_O. The switch frequency is f and the period $T = 1/f$, and the switch-on duty cycle is k (to avoid the parasitic effect, k should be 0.1–0.9).

When the switch S is on, the source voltage V_{in} charges the capacitor C_1 to V_{in} and provides the current flow through the inductor L. Then the inductor current increases:

$$\Delta I_L = \frac{V_{in}}{L} kT \qquad (18.14)$$

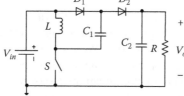

FIGURE 18.13   Positive-output super-lift Luo converter.

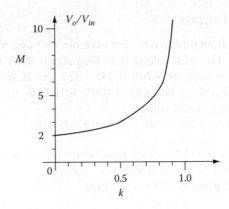

FIGURE 18.14 The voltage transfer gain M versus the duty cycle k.

When the switch S is off, the inductor current decreases with the applied voltage $(V_O - 2V_{in})$. Therefore, the inductor current decrement is

$$\Delta I_L = \frac{V_O - 2V_{in}}{L}(1-k)T \qquad (18.15)$$

In the steady state, the inductor current increment must be equal to its decrement. Therefore, we obtain the voltage transfer gain M as follows:

$$M = \frac{V_O}{V_{in}} = \frac{2-k}{1-k} \qquad (18.16)$$

This voltage transfer gain is much higher than that of the boost converter and positive-output Luo converter. When k is very small, the voltage transfer gain $M \approx 2$. When $k=0.5$, the output voltage V_O is equal to $3 \times V_{in}$. The voltage transfer gain M versus the duty cycle k is shown in Figure 18.14.

It is very good for high voltage transformation and is used in solar panel energy system. For example, in our system, $V_{in}=186\,V$ and $V_O=700\,V$. The voltage transfer gain is $M=3.76$, and the duty cycle is $k=0.638$.

Since the input voltage varies from 148.8 to 223.2 V, the voltage transfer gain is changing $M=3.136-4.704$, and the duty cycle $k=0.532-0.73$. The values are very good inside the range of variations.

18.3.3 CLOSED-LOOP CONTROL

The input voltage from the solar panel varies in the range of 148.8–223.2 V. In order to obtain a stable output voltage, we have to design a closed-loop control for the positive-output super-lift Luo converter. We use a proportional plus integral (PI) controller for outer voltage loop control and a proportional (P) controller for inner current loop control. The control block diagram is shown in Figure 18.15.

The output PWM signal is used to apply to control the duty cycle k for the positive-output super-lift Luo converter. The switching frequency is usually chosen in

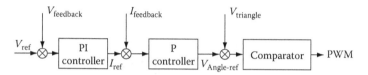

FIGURE 18.15   Double closed-loop controller.

the range of 50–500 kHz. Since this is automatically controlled, we should not preset any value for k.

18.3.4 PWM INVERTER

The pulse-width modulation technique is a popular method to implement DC/AC inversion technology [7,8]. A pulse-width-modulated (PWM) voltage source inverter (VSI) is used for this design, which is shown in Figure 18.16.

The triangular and modulating signals are shown in Figure 18.17.

There are two important modulation ratios for the PWM technique. We define the amplitude modulation ratio m_a,

$$m_a = \frac{V_{in\text{-}m}}{V_{tri\text{-}m}} \qquad (18.17)$$

where
   $V_{in\text{-}m}$ is the amplitude of the control signer (sine) waveform
   $V_{tri\text{-}m}$ is the amplitude of the triangle waveform

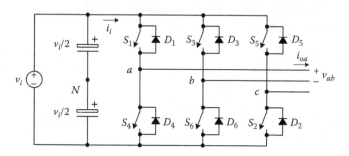

FIGURE 18.16   Three-phase full-bridge VSI.

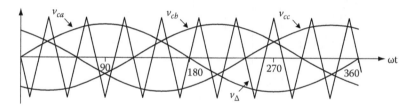

FIGURE 18.17   The triangular and modulating signals.

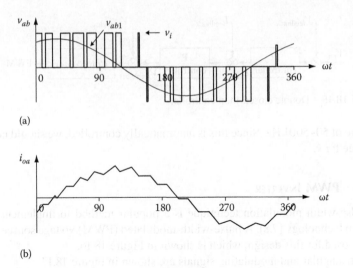

(a)

(b)

FIGURE 18.18 The AC output (a) voltage and (b) current (each phase).

Usually, for nondistorted inversion the amplitude modulation ratio m_a is selected to be smaller than 1.0.

We also define the frequency modulation ratio m_f,

$$m_f = \frac{f_{tri\text{-}m}}{f_{in\text{-}m}} \tag{18.18}$$

where
$f_{in\text{-}m}$ is the frequency of the control signer (sine) waveform
$f_{tri\text{-}m}$ is the frequency of the triangle waveform

Usually, for nondistorted inversion the frequency modulation ratio m_f is selected to be a large figure that is greater than 21. The AC output voltage and current (each phase) are shown in Figure 18.18.

In order to produce the three-phase AC voltage to synchronize to the main grid voltage, we take the grid signals as the control signer. Therefore, there is no trouble with the synchronization.

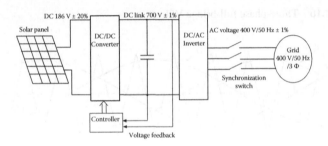

FIGURE 18.19 Block diagram of the solar panel power system.

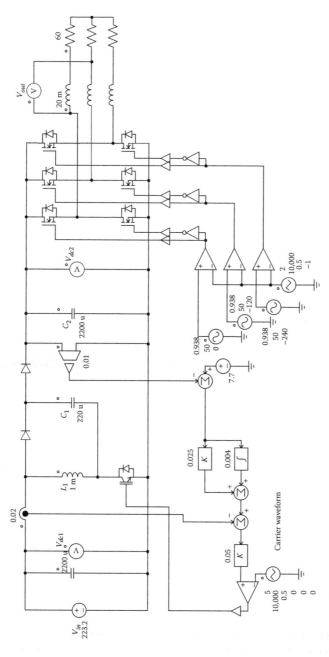

FIGURE 18.20 Simulation diagram of the solar panel power system.

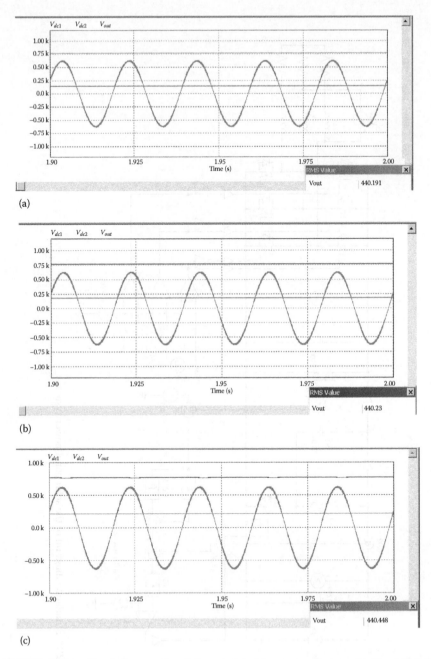

FIGURE 18.21 Simulation results of the wind turbine power system. (a) $V_{in} = 148.8$ V, $V_{dc2} = 770$ V, and $V_o = 440.191$ and V/50 Hz/3 Φ. (b) $V_{in} = 186$ V, $V_{dc2} = 770$ V, and $V_o = 440.23$ and V/50 Hz/3 Φ. (c) $V_{in} = 223.2$ V, $Vd_{c2} = 770$ V, and $V_o = 440.448$ and V/50 Hz/3 Φ.

18.3.5   SYSTEM DESIGN

After all blocks are prepared, we can install our system design. The block diagram is shown in Figure 18.19. The solar panel yields the input voltage of $186\,V \pm 20\%$. The DC/DC converter is the positive-output super-lift Luo converter with double closed-loop control. Its output voltage is the DC link voltage with $700\,V \pm 1\%$. Since the DC link voltage is quite stable, there is no need for any closed-loop control for the DC/AC VSI. Considering the synchronization, we use the grid voltage as the control signal of the VSI. Its output is as follows: three phase, $50\,Hz$, $400\,kV$ (line-to-line rms).

18.3.6   SIMULATION RESULTS

The simulation diagram is shown in Figure 18.20. The simulation results are shown in Figure 18.21. Figure 18.21a shows that the input voltage V_{in} is $V_{in} = 186 - 20\% = 148.8\,V$. After the double closed-loop control, the output voltage of the P/O super-lift Luo converter $V_{dc2} = 770\,V$. We then obtain $V_O = 440.191\,V/50\,Hz/\,3\,\Phi$. Figure 18.21b shows the input voltages $V_{in} = 186\,V$, $V_{dc2} = 770\,V$, and $V_O = 440.23\,V/50\,Hz/\,3\,\Phi$. Figure 18.21c shows the input voltages $V_{in} = 186\,V + 20\% = 223.2\,V$, $V_{dc2} = 770\,V$, and $Vo = 440.448\,V/50\,Hz/\,3\,\Phi$. All simulation results satisfy our requirement.

REFERENCES

1. Masters, G. M. 2005. *Renewable and Efficient Electric Power Systems*. New York: John Wiley.
2. Ackermann, T. 2005. *Wind Power in Power Systems*. New York: John Wiley.
3. Johnson, G. L. 1985. *Wind Energy Systems*. Englewood Cliffs, NJ: Prentice-Hall.
4. Lin, B. R., Lu, H. H., and Chen, Y. M. 1998. Implementation of three-level AC/DC/AC converter with power factor correction and harmonic reduction. *Proceedings of IEEE PEDES*, Perth, Australia, pp. 768–773.
5. Luo, F. L. and Ye, H. 2004. *Advanced DC/DC Converters*. Boca Raton, FL: CRC Press.
6. Luo, F. L. and Ye, H. 2009. Chopper-type DC/AC/DC converters. *Technical Talk of ICIEA'2009*, Xi'an, China, pp. 1356–1368.
7. Luo, F. L. and Ye, H. 2009. Switched-capacitor DC/AC/DC converters. *Technical Talk of ICIEA'2009*, Xi'an, China, pp. 163–168.

Index